Lecture Notes in Earth System Sciences

For further volumes:
http://www.springer.com/series/10529

Eulogio Pardo-Igúzquiza
Carolina Guardiola-Albert
Javier Heredia · Luis Moreno-Merino
Juan José Durán · Jose Antonio Vargas-Guzmán
Editors

Mathematics of Planet Earth

Proceedings of the 15th Annual Conference of the International Association for Mathematical Geosciences

Springer

Editors
Eulogio Pardo-Igúzquiza
Carolina Guardiola-Albert
Javier Heredia
Luis Moreno-Merino
Juan José Durán
Departamento de Investigación y Prospectiva Geocientífica
Instituto Geológico y Minero de España
Madrid
Spain

Jose Antonio Vargas-Guzmán
Saudi Aramco
Dhahran
Saudi Arabia

ISSN 2193-8571 ISSN 2193-858X (electronic)
ISBN 978-3-642-32407-9 ISBN 978-3-642-32408-6 (eBook)
DOI 10.1007/978-3-642-32408-6
Springer Heidelberg New York Dordrecht London

Library of Congress Control Number: 2013943567

Printed on acid-free paper

Springer is part of Springer Science+Business Media (www.springer.com)

Preface

This book is a comprehensive summary of the 15th annual conference of the International Association for Mathematical Geosciences (IAMG Madrid 2013) held in Madrid (Spain). The IAMG Madrid 2013 has been an international forum of scientific debate on the research progress made worldwide on theoretical developments and practical applications of geomathematics. The theme of the conference, "Frontiers of Mathematical Geosciences: New Approaches to Understand the Natural World," makes emphasis on the need for new paradigms, methodologies, and detailed earth models at multiple scales in order to solve important technological problems that the humankind is facing in relation to sustainable water, energy, minerals, and multiple environmental resources considering climate and natural hazards concerns.

Planet Earth is a complex system where the lithosphere, atmosphere, hydrosphere, cryosphere, and biosphere interact among them to build products from complex processes. Furthermore, the increase of human population is associated to new issues in relation to fast lessening of earth resources and habitats; e.g., depletion of nonrenewable resources, contamination of water, soils and air pollution, environmental impact, deforestation, endangered species, climate change, and geological hazards (e.g., earthquakes, tsunamis, volcanic eruptions, flooding, potential loss of glaciers and ice caps, etc.). Some of these problems may be also associated to the increase of urban settlements around the world. In consequence, governmental agencies, the industry, and society in general are demanding the scientific community to increase the quantitative understanding of how the complex earth system works. Computers, satellite imagery, and new geophysical techniques have been deployed to gather and process large amounts of new data, expecting to facilitate the planetary monitoring and modeling. Nevertheless, the scientific basis for interpreting such data is well established in the young discipline of geomathematics, some of the challenges in relation to earth processes are very complex, and their solution will require the best of human ability and creativity, through numerical models and associated methods. Therefore, geomathematics has been raised as the essential science to model processes and resources in geosciences for solving fundamental resource and planetary sustainability problems at multiple scales. Moreover, the application and use of mathematics in geological research and technology is not only focused on dealing with specific issues of our planet. Study findings on mathematical geosciences are also extremely important for the planetary geology research.

The main duty of IAMG 2013 has been to discuss and disseminate the latest trends in research about leading problems. This proceeding book contains 184 contributions grouped adequately in 27 sessions.

Acknowledgments

This book was made possible due to the dedicated and combined effort of the Conference conveners, authors, and reviewers. The editors of this book offer to all of them their sincere gratitude. Especially we would like to thank the 150 "silent" reviewers who spent much time commenting the manuscripts.

The 15th Annual Conference of the International Association for Mathematical Geosciences (IAMG) has been organized by the Spanish Geological Survey (Instituto Geológico y Minero de España, IGME). We acknowledge the IAMG for its support. In particular, we are grateful to Dr. Vera Pawlowsky-Glahn who encouraged us to organize this Conference and Dr. Ricardo Olea for his continuous guidance.

We would like to thank the Madrid Convention Bureau for its assistance in preparing the proposal to organize the Conference and help with the meeting logistics.

Appreciation is also extended to the following sponsors, without their support it would not have been possible to publish this book:

- Instituto Geológico y Minero de España
- Saudi Aramco
- Gas Natural Fenosa
- Gamma Geofísica
- KARSTINV

Organizing Committee

The 15th Annual Conference of the International Association for Mathematical Geosciences (IAMG) has been organized by researches from the Spanish Geological Survey (Instituto Geológico y Minero de España, IGME).

Eulogio Pardo Igúzquiza
Carolina Guardiola Albert
Javier Heredia Díaz
Luís Moreno Merino
Juan José Durán Valsero

Scientific Committee

Kanevski, M. (University of Lausanne, Switzerland)
Karimi, Hassan A. (University of Pittsburgh, USA)
Kaufman, Gordon M. (Sloan School of Management MIT, USA)
Mardia, Kanti V. (University of Leeds, UK)
Martín, Miguel Angel (Universidad Politécnica de Madrid, Spain)
Martín Alfageme, Santiago (Instituto Geológico y Minero de España, Spain)
Martínez Frías, Jesús (Centro de Astrobiología, Spain)
Mateu, Jorge (Universidad Jaume I, Spain)
Moreno, José F. (Universidad de Valencia, Spain)
Mosegaard, Klaus (Technical University of Denmark, Denmark)
Nashed, Zuhair (University of Central Florida, USA)
Neto de Carvalho, Carlos (UNESCO Geopark Naturtejo Meseta Meridional, Portugal)
Olea, Ricardo (U.S. Geological Survey, USA)
Pachepsky, Yakov (United States Department of Agriculture, USA)
Paredes, Carlos (Universidad Politécnica de Madrid, Spain)
Pawlowsky-Glahn, Vera (Universidad de Girona, Spain)
Pérez Cerdan, Fernando (Instituto Geológico y Minero de España, Spain)
Poblet Esplugas, Josep (Oviedo University, Spain)
Ramón Francés García, Felix (Universidad Politécnica de Valencia, Spain)
Renard, Philip (University of Neuchâtel, Switzerland)
Rivas, Raul (Instituto de Hidrología de Llanuras, Argentina)
Rodríguez, Leticia (Centro de Estudio Hidro-Ambientales, Universidad Nacional del Litoral, Argentina)
Rodríguez-Tovar, Francisco Javier (Universidad de Granada, Spain)
Romero Pérez, M^{a} del Pilar (Universidad Complutense de Madrid, Spain)
Samper, Javier (Universidad de La Coruña, Spain)
San Jose Martínez, Fernando (Universidad Politécnica de Madrid, Spain)
Schaeben, Helmut (Technische Universität Bergakademie, Germany)
Schuenemeyer, Jack (Southwest Statistical Consulting, LLC, USA)
Soares, Amilcar (Instituo Superior Tecnico Lisboa, Portugal)
Soto, Juan I. (Granada University, Spain)
Vargas Guzmán, J. A. (Aramco, Saudi Arabia)
Victorov, Alexey S. (Institute of Environmental Geoscience of Russian Academy of Science, Russia)
Wackernagel, Hans (Centre Geosciences Ecole de Mines de Paris, France)
Yongjiu, Dai (Beijing Normal University, China)

Sponsored by

Aramco: Saudi Arabian Oil Company that produces, manufactures, markets and ships crude oil, natural gas and petroleum products to meet the global demand.

Gamma Geofísica: Gamma Geofísica is company committed to the development of projects and geophysical consultancy.

Gas Natural Fenosa: Gas Natural Fenosa is a leading group in the energy sector, pioneering in gas and electricity integration.

KARSTINV: KARSTINV is a Research Project (CGL2010-15498) supported by the "Ministerio de Economía y Competitividad" of Spain.

Madrid Convention Bureau: The Madrid Convention Bureau provides a wide range of services for any event hosted in Madrid.

Contents

Part I Advances in Classical Statistics Relevant to the Geosciences

Performance Evaluation of Swanson's Rule for the Case of Log-Normal Populations 1
Maryam Moghadasi and Jerry L. Jensen

Modelling Ore Bodies of High-Nugget Gold Using Conditional Probability 5
Evelyn June Hill, Nicholas H. S. Oliver, James Cleverley and Michael Nugus

The Extraction of Long-Term Distribution Trends of Ore-Sourced Geochemical Elements from an Drill Core in the Loess Covered Area 9
Deyi Xu, Hongwei Pei, Ningqiang Liu, Zhaoxian Yuan, Shuyun Xie and Qiuming Cheng

Evaluating Predictive Performance 15
Michael Scheuerer and Tilmann Gneiting

At the Interface Between Mathematical Geoscience and Classical Statistics 19
Frits Agterberg

Estimating the Number and Locations of Euler Poles 23
Florian Bachmann, Peter Jupp and Helmut Schaeben

Testing for Microhomogeneity in Reference Materials for Microanalytical Methods 27
R. Tolosana-Delgado, A. D. Renno, P. P. Michalak and K. G. van den Boogaart

Recent Univariate and Multivariate Statistical Techniques Applied to Gold Exploration in the Amapari Area, Amazon Region, Brazil 33
Luis Paulo Braga, Francisco José da Silva and Claudio Gerheim Porto

Maximum Likelihood Inference of Spatial Covariance Parameters of Large Data Sets in Geosciences 37
Kanti V. Mardia and Eulogio Pardo-Igúzquiza

A Cross-Polar Modeling Approach to Hindcast Paleo-Arctic Mega Icebergs: A Storyboard 41
Reinhard Furrer, Nina Kirchner and Martin Jakobsson

Towards a Statistical Treatment of Images Acquired by Automated Mineralogy 45
Sandra Birtel, Raimon Tolosana Delgado, Stephan Matos Camacho, Jens Gutzmer and K. Gerald van den Boogaart

Distributional Assumptions and Parametric Uncertainties in the Aggregation of Geologic Resources 49
John H. Schuenemeyer and Ricardo A. Olea

Modeling Extremal Dependence Using Copulas. Application to Rainfall Data 53
M. I. Ortego, J. J. Egozcue and R. Tolosana-Delgado

Part II Frontier Geostatistics

Revisiting "Estimating and Choosing" 57
Jean Serra

Pragmatic Bayesian Kriging for Non-Stationary and Moderately Non-Gaussian Data 61
Konstantin Krivoruchko and Alexander Gribov

Geostatistical History Matching Conditioned to Seismic Data 65
Amílcar Soares, Leonardo Azevedo, Sara Focaccia and João Carneiro

Handling Soft Probabilities in Multiple Point Statistics Simulation . . . 69
Pierre Biver, Gregoire Mariethoz, Julien Straubhaar, Tatiana Chugunova and Philippe Renard

An Application of Equal-Area-Growing Window for Calculating Local Singularity for Mapping Granites in Inner Mongolia 73
Daojun Zhang, Qiuming Cheng and Frits Agterberg

Semi-Variogram Model Inference Using a Median Bootstrap Statistics 79
Eulogio Pardo-Igúzquiza, Ricardo A. Olea and Peter A. Dowd

Part III Compositional Data Analysis Applied to Geochemistry

Geometric and Statistical Modeling of Fractures in the 3D Disturbed Zone of a Claystone Around a Cylindrical Gallery (Meuse-Haute Marne Underground Research Laboratory, France) 83
Rachid Ababou, Israel Cañamón and Adrien Poutrel

Robust Regression with Compositional Response: Application to Geosciences. 87
Karel Hron, Peter Filzmoser, Matthias Templ, Karl Gerald van den Boogaart and Raimon Tolosana-Delgado

Compositional Analysis in the Study of Mineralization Based on Stream Sediment Data 91
Renguang Zuo

Modelling World Energy Applying Simplicial Linear Ordinary Differential Equations 95
Eusebi Jarauta-Bragulat, Carme Hervada-Sala and Juan José Egozcue

Structural Analysis of the National Geochemical Survey of Australia Data. 99
Ute Mueller, Johnny Lo, Patrice de Caritat and Eric Grunsky

Discriminant Analysis of Palaeogene Basalt Lavas, Northern Ireland, Using Soil Geochemistry 103
Jennifer M. McKinley, Sam Roberson, Mark Cooper and Raimon Tolosana-Delgado

Chemical Equilibria in Compositional Data 107
K. Gerald van den Boogaart, Raimon Tolosana Delgado and Silke Konsulke

Analysis of Total Abundances of Phytoplankton Compositions in a River 111
Vera Pawlowsky-Glahn, Juan J. Egozcue and David Lovell

Part IV Data Assimilation in Geosciences

Recursive Upward Sweeping and Updating Method on Ensemble Based Multiscale Algorithm in Data Assimilation 115
Chen Li, Shihua Chen, Chunlin Huang and Wei Gong

Corrected Kriging Update Formulae for Batch-Sequential Data Assimilation 119
Clément Chevalier, David Ginsbourger and Xavier Emery

Part V Machine Learning Geoscience Applications

Analysis of the Parametrization Needs of Different Land Cover Classifiers: The Case Study of Granda Province (Spain) 123
Victor F. Rodriguez-Galiano and Mario Chica-Olmo

Automatic Raman Spectra Processing for Exomars 127
Isaac Hermosilla Rodriguez, Guillermo Lopez-Reyes, D. R. Llanos and Fernando Rull Perez

Fuzzy Parameterization of a Filtration Model for a Non-homogeneous Sedimentary Rock 131
Elena Savelyeva and Aleksander Rastorguyev

Application of Multivariate Analysis Techniques for the Identification of Sulfates From Raman Spectra 135
Guillermo Lopez-Reyes, Pablo Sobron, Catherine Lefebvre and Fernando Rull

Comparison of Micro X-ray Computer Tomography Image Segmentation Methods: Artificial Neural Networks Versus Least Square Support Vector Machine 141
Swarup Chauhan, Wolfram Rühaak, Frieder Enzmann, Faisal Khan, Philipp Mielke, Michael Kersten and Ingo Sass

Learning Uncertainty from Training Images for Reservoir Predictions 147
Temitocles Rojas, Varily Demyanov, Mike Christie and Darn Arnold

A Particle Swarm Optimization for Parameter Estimation of a Rainfall-Runoff Model 153
Frédéric Bardolle, Frédérick Delay, Francis Bichot, Gilles Porel and Nathalie Dörfliger

Sequential Weights of Evidence as a Machine Learning Model for Mineral Deposits Prediction 157
Qiuming Cheng

Identification of Spatial Models of δ^{18} O in Precipitation of the Wet Season Over Spain Using Genetic Algorithms 163
Javier Heredia, Eulogio Pardo-Igúzquiza, Javier Rodríguez-Arévalo, Silvino Castaño, Maria F. Díaz-Teijeiro, Jose E. Capilla, Aantonio Prado and Lara Bardasano

Part VI Spatiotemporal Analysis: Structural Complexity and Extreme Behaviour

Space-Time Prediction of Extreme Events by Using Mathematical Logic Modeling of Cause-Effect Relations 167
Susanna Sirotinskaya

Local Clustering in Spatio-Temporal Point Patterns 171
Jorge Mateu and Francisco J. Rodríguez-Cortés

The Use of Remote Sensing Data in a Colombian Andean Basin for Risk Analysis 175
Olga Lucía Ocampo López and Jorge Julián Vélez Upegui

Spatio-Temporal Analysis of Earthquake Occurrences Using a Multiresolution Approach 179
Orietta Nicolis

Statistical Complexity Analysis of Spatiotemporal Dynamics 185
José M. Angulo and Francisco J. Esquivel

Part VII Parameterization of Soil Systems at Different Scales

Frequency Distributions and Scaling of Soil Texture and Hydraulic Properties in a Stratified Deep Vadose Zone Near Maricopa, Arizona 189
Alberto Guadagnini, Shlomo P. Neuman, Marcel G. Schaap and Monica Riva

Cascade of Proppant-Sandwiched Silt Blocks as a Double-Continuum: From Discovery to Mathematical Modeling 193
Anvar Kacimov, Said Al-Ismaily, Ali Al-Maktoumi, Hamed Al-Busaidi and Said Al-Saqri

Parameterization of Soil Thermal Diffusivity Versus Moisture Content Dependencies and Modeling Spatial Heterogeneity of Soil Temperature 197
Tatiana Arkhangelskaya

A Packing Computational Method Relating Fractal Particle Size Distribution and Void Fraction in Granular Media 201
Carlos García-Gutiérrez, Miguel A. Martín, Francisco Muñoz, Miguel Reyes and Francisco J. Taguas

Parallel Sets and Morphological Measurements of CT Images of Soil Pore Structure in a Vineyard 205
Fernando San José Martínez, Francisco J. Muñoz, Francisco J. Caniego and Fernando Peregrina

Part VIII Fractal, Chaos, and Complexity in the Earth System

Evaluation of Fractal Dimension in Karst Aquifers 211
Pedro A. Robledo Ardila, Juan José Durán Valsero and Eulogio Pardo Igúzquiza

Fractal Modelling of Karst Conduits 217
Eulogio Pardo-Igúzquiza, Juan José Durán, Pedro Robledo, Carolina Guardiola, Juan Antonio Luque and y Sergio Martos

Frequency–Area Distribution of Historical Landslides in the Sannio Apennine (Southern Italy) 221
Angelo Donnarumma, Paola Revellino and Francesco Maria Guadagno

Part IX Remote Sensing a Changing World

A Spatiotemporal Remotely-Sensed Assessment of Peat Covered Areas Using Airborne Radiometrics 225
Jennifer M. McKinley, Antoinette Keaney and Alastair Ruffell

Monitoring of Urban-Damaging Landslides with Satellite Radar Missions: Arcos de la Frontera (Spain) 229
Guadalupe Bru, José Fernández, Pablo J. González and Kristy F. Tiampo

Multivariate Variogram and Madogram: Tools for Quantifying Diversity/Dissimilarity in Spatiotemporal Data 235
Phaedon Kyriakidis, Dimitra Kitsiou and Dimitris Kavroudakis

Spatiotemporal Interactions for Daily Mapping of PM_{10} with MODIS and Meteorological Data 239
Piero Campalani, Simone Mantovani and Peter Baumann

Part X Radar Remote Sensing for the Detection, Monitoring and Modelling of Ground Instabilities

Non Linear PS Time Series: Analysis and Post-Processing for Landslides Studies 245
Davide Notti, Claudia Meisina, Francesco Zucca and Alessio Colombo

Observation of the Mining-Induced Surface Deformations Using C and L SAR Bands: The Upper Silesian Coal Basin (Poland) Case Study 249
Marek Graniczny, Zbigniew Kowalski, Anna Piątkowska and Maria Przyłucka

Multi-Temporal Evaluation of Landslide-Induced Movements and Damage Assessment in San Fratello (Italy) by Means of C- and X-Band PSI Data 257
Silvia Bianchini, Deodato Tapete, Andrea Ciampalini, Federico Di Traglia, Chiara Del Ventisette, Sandro Moretti and Nicola Casagli

Integration of Earth Observation and Ground-Based HR Data in the Civil Protection Emergency Cycle: The Case of the DORIS Project 263
Paola Pagliara, Giuseppe Basile, Pierluigi Cara, Angelo Corazza, Andrea Duro, Bruno Manfrè, Roberta Onori, Chiara Proietti and Vincenzo Sansone

Geostatistical Analysis of PSI Radar Data: A Methodology to Assess Serviceable Limit State of Buildings 267
Margarita P. Sanabria, Carolina Guardiola-Albert, Roberto S. Tomás, Geraint Cooksley and Gerardo Herrera

Using Modern Sensor Data and Advanced Numerical Modelling for Slow Landslides Motion Forecasting. 271
Jose A. Fernández-Merodo, Juan C. García-Davalillo and Gerardo Herrera

Characterization of Underground Cellars in the Duero Basin by GNSS, LIDAR and GPR Techniques. 277
Miguel Angel Conejo-Martín, Tomás Ramón Herrero-Tejedor, Enrique Pérez-Martín, Javier Lapazaran-Izargain, Jaime Otero-García, Juan Francisco Prieto-Morín and Jesús Velasco-Gómez

Study of Subsidence on Aquifers Having Undergone Extraction and Inactive Cycles 281
Rubén Martínez-Marín, Pablo Ezquerro-Martín, Juan Carlos Ibáñez-Carranza, Juan Gregorio Rejas-Ayuga and Maguel Marchamalo-Sacristán

Advanced InSAR Techniques to Support Landslide Monitoring 287
Fernando Bellotti, Marco Bianchi, Davide Colombo, Alessandro Ferretti and Andrea Tamburini

Part XI Geographic Information Systems/Geoinformatics

Estimation of Information Loss When Masking Conditional Dependence and Categorizing Continuous Data: Further Experiments on a Database for Spatial Prediction Modelling in Northern Italy. 291
Andrea G. Fabbri, Simone Poli, Antonio Patera, Angelo Cavallin and Chang-Jo Chung

3D-GIS Analysis for Mineral Resources Exploration in Luanchuan, China. 295
Gongwen Wang, Shouting Zhang, Changhai Yan, Yaowu Song, Jianan Qu, Yanyan Zhu and Dong Li

Animation of Groundwater Flow with STRING 299
Isabel Ostermann and Torsten Seidel

SHEE Program, a Tool for the Display, Analysis and Interpretation of Hydrological Processes in Watersheds 303
Jesús Mateo Lázaro, José Ángel Sánchez Navarro, Alejandro García Gil and Vanesa Edo Romero

Mathematical Methods of the Data Analysis in a Prospecting Database for Geological Mapping 309
Oleg Mironov

Geoprocessing Tool Regenerat: Characterization of Mineral Resource Quality of Renewable Sediment Deposits 315
Heinz Reitner, Sebastian Pfleiderer, Maria Heinrich, Irena Lipiarska, Piotr Lipiarski, Julia Rabeder, Thomas Untersweg and Ingeborg Wimmer-Frey

Spatiotemporal Data Model for Multi-factor Geological Process Analysis with Case Study 319
Gang Liu, Xiang Que, Xiaonan Hu, Shanjun Tian and Jiacheng Zhu

Use of Variational Methods in Geological Mapping 325
Andrey N. Sidorov, Andrey G. Plavnik, Andrei A. Sidorov and Michail S. Shutov

Comparison of Methods for Depth to Groundwater Calculation in Hard Rock Areas 329
Ronny Laehne, Dorothee Altenstein and Wolfgang Gossel

Part XII Quantitavie Methods in Geomorphology and Land Surface Processes

Line-Geometry-Based Inverse Distance Weighted Interpolation (L-IDW): Geoscientific Case Studies 333
Wolfgang Gossel and Michael Falkenhagen

Two-Dimensional Hydraulic Modeling and Analysis of Morphological Changes in the Palancia River (Spain) During a Severe Flood Event on October 2000 339
Beatriz Nácher-Rodríguez, Ignacio Andrés-Doménech, Carles Sanchis-Ibor, Francisca Segura-Beltrán, Francisco J. Vallés-Morán and Eduardo Albentosa Hernández

Stream Length-Gradient Index Mapping as a Tool for Landslides Identification . . . 343
Jorge Pedro Galve, Daniela Piacentini, Francesco Troiani and Marta Della Seta

Quantitative Method on Historical Reconstruction of Coastal Geomorphological Change on Wave-Dominated Coast: A Case Study of the Pomeranian Bay, Southern Baltic Sea . . . 347
Junjie Deng, Jan Harff and Joanna Dudzinska-Nowak

Linfo: A Visual Basic Program for Analysis of Spatial Properties of Lineaments . . . 351
A. C. Dinesh, Vipin Joseph Markose and K. S. Jayappa

Numerical Karst: Spatio-temporal Modelling of Karst Aquifer Systems . . . 355
Eulogio Pardo-Igúzquiza, Sergio Martos, Juan Antonio Luque, Juan José Durán, Carolina Guardiola-Alberto and Pedro Robledo

Surface Insights of Structural Relief Distribution Within the Madrid Cenozoic Basin from Fluvial and Terrain Morphometric Index . . . 359
Julio Garrote and Guillermina Garzón

Part XIII Deformation Modeling, Geodynamics and Natural Hazards

Reinterpretation of Teide 2004–2005 Gravity Changes by 3D Line Segments Approximation . . . 363
Peter Vajda, Ilya Prutkin and Jo Gottsmann

An Update GPS Velocity and Strain Rate Fields for the Iberian Region . . . 369
Mimmo Palano, Pablo J. González and José Fernández

The 2012 Kilauea Volcano, Hawai'i, Slow-Slip Event Captured by cGPS and Satellite Radar Interferometry . . . 373
Pablo J. González, Sergey V. Samsonov and Mimmo Palano

Spatiotemporal Analysis of Ground Deformation at Campi Flegrei and Mt Vesuvius, Italy, Observed by Envisat and Radarsat-2 InSAR During 2003–2013 . . . 377
Sergey V. Samsonov, Pablo J. González, Kristy F. Tiampo, Antonio G. Camacho and José Fernández

Anthropogenic and Natural Ground Deformation Near Bologna, Italy Observed by Radarsat-2 InSAR During 2008–2013 383
Sergey V. Samsonov, Pablo J. González and Kristy F. Tiampo

Surface Displacements, Deformations and Gravity Changes Due to Underground Heat Source 387
Ladislav Brimich and Igor Kohut

Polarization Persistent Scatterer InSAR Analysis on the Hayward Fault, CA, 2008–2011 393
Kristy F. Tiampo, Pablo J. González and Sergey S. Samsonov

Earth Models at the Low Andarax River Valley (SE Spain) by Means of Cross-Correlation of Ambient Noise 399
Abigail Jiménez, Antonio García-Jerez, Enrique Carmona, Francisco Sánchez-Martos and Francisco Luzón

Some Insights About Volcano Deformation Interpretation 403
María Charco and Pedro Galán del Sastre

Part XIV Hydrogeology: From Process Understanding to Improved Predictions

Ensemble Kalman Filter Assimilation of Transient Groundwater Flow Data: Stochastic Moment Solution Versus Traditional Monte Carlo Approach 407
Marco Panzeri, Monica Riva, Alberto Guadagnini and Shlomo P. Neuman

New Analytical Solutions for Phreatic Darcian Flows Over Non-Planar Bedrocks 411
Anvar Kacimov, Yurii Obnosov and Osman Abdalla

When Steady-State is Not Enough 415
J. Jaime Gómez-Hernández, Teng Xu, Haiyan Zhou and Liangping Li

Hydrogeological and Thermal Modelling of an Underground Mining Reservoir 419
Clara Andrés Arias, Almudena Ordóñez Alonso and Rodrigo Álvarez García

Effect of Entrapped Gas Below the Phreatic Surface on Pressure Propagation and Soil Deformation 425
Héctor Montenegro, Oliver Stelzer and Bernhard Odenwald

Time Series Analysis of Groundwater Hydrographs: Case Study From a Hardrock Area 431
Wolfgang Gossel and Ronny Laehne

Coupled Hydrogeophysical Simulation of a Pumping Test in an Unconfined Aquifer and its Associated Gravimetric Anomaly 435
Andrés González-Quirós and José Paulino Fernández-Álvarez

Towards a More Efficient Simulation of Surface–Groundwater Interaction in Conjunctive Use Systems: Selective Compression and Modal Masking in the Eigenvalue Method 439
Oscar David Álvarez-Villa, Eduardo Cassiraga and Andrés Sahuquillo

Rational Lanczos Reduction of Groundwater Flow Models to Perform Efficient Simulations of Surface-Ground Water Interaction in Conjunctive Use Systems 443
Oscar David Álvarez-Villa, Eduardo Cassiraga and Andrés Sahuquillo

Efficient Generation of Effective Modes and Spectral Masks to Build Reduced Groundwater Flow Models Using the Eigenvalue Method with Selective Compression and Modal Masking 447
Oscar David Álvarez-Villa, Eduardo Cassiraga and Andrés Sahuquillo

Application of Eigenvalue Method with Selective Compression and Modal Masking and Rational Lanczos Reduction Method to the Efficient Simulation of Surface–Ground Water Interactions in Heterogeneous Aquifers 451
Oscar David Álvarez-Villa, Eduardo Cassiraga and Andrés Sahuquillo

A Multipoint Flux Domain Decomposition Method for Transient Flow Modeling in Complex Porous Media 457
Andrés Arrarás, Laura Portero and Ivan Yotov

Optimal Reconstruction of 3D Fracture Networks (FEBEX Field Test, Grimsel Site, Swiss Alps) 463
Israel Cañamón, Rachid Ababou and F. Javier Elorza

Part XV Quantitative Hydrology: Working Across Scientific Disciplines and Time-space Scales

Numerical Analysis of Stream–Groundwater Exchange in a Floodplain Following a Dike Opening 467
Héctor Montenegro

Models of Parallel Linear Reservoirs (PLR) with Watershed Traversal Algorithm (WTA) in Behaviour Research of Hydrological Processes in Catchments 471
Jesús Mateo Lázaro, José Ángel Sánchez Navarro, Vanesa Edo Romero and Alejandro García Gil

An Analytical Solution of Tide-Induced Head Fluctuations in an Inhomogeneous Coastal Aquifer 475
Luis Guarracino and Leonardo Monachesi

Mapping of Flood-Plain by Processing of Elevation Data from Remote Sensing 481
Viviana Aguilar Muñoz and Márcio de Morisson Valeriano

Effect of Spatial Heterogeneity on Rate of Sedimentary O_2 Consumption Reaction 485
Tanushree Dutta and Simonetta Rubol

Optimal Development of Regional Rain Network Using Entropy and Geostatistics 491
Hadi Mahmoudi-Meimand, Sara Nazif and Hasan-Ali Faraji-Sabokbar

Part XVI Quantitative Environmental Geology

Application of Drastic Methodology for Evaluation of Guarani Aquifer Vulnerability: Study Case in Ribeirão Bonito, SP, Brazil 497
Rafael Gonçalves Santos, Mara Lúcia Marques, Gabriela Trigo Ferreira and José Ricardo Sturaro

A Case Study of Geometric Modelling via 3-D Point Interpolation for the Bathymetry of the Rabasa Lakes (Alicante, Spain) 503
África de la Hera, Enrique López-Pamo, Esther Santofimia, Guillermo Gallego, Raquel Morales, Juan José Durán-Valsero and José Manuel Murillo-Díaz

Analysis of Groundwater Monitoring Data Sets with Non-Detect Observations: Application to the Plana de Sagunto (Valencia, Spain) Groundwater Body 507
Juan Grima, Juan Antonio Luque-Espinar, Juan Ángel Mejía-Gómez and Ramiro Rodríguez

The Mesh Optimization of the Environmental Investigation Applied to the Diagnosis of the Quality of the Basement (Algerian Experience) 513
Abderrahim Gheris

Data Archives: Development and Application in Environmental Management 517
Konstantin Alexeev

Ecological Remediation Volume (ERV) in Coastal Aquifers Affected by Seawater Intrusion. Methodology and Application in the Oropesa-Torreblanca Plain (Spain) 521
Arianna Renau-Pruñonosa, Ignacio Morell, David Pulido and Jorge Mateu

A Radon Risk Map of Germany Based on the Geogenic Radon Potential 527
Peter Bossew

Quantitative Risk Management of Groundwater Contamination by Nitrates Using Indicator Geostatistics 533
Mario Chica-Olmo, Eulogio Pardo-Igúzquiza, Juan Antonio Luque-Espinar, Víctor Rodríguez-Galiano and Lucía Chica-Rivas

Geologically Constrained Groundwater Monitoring Network Optimization in Halle (Germany): A Case Study with Saltwater Intrusion Along a Fault System 537
Thomas Horschig, Ronny Lähne, Michael Falkenhagen and Wolfgang Gossel

Informative Trace-Element Features of Cassiterite from Tin Mineralized Zones and Breccias in the Russian Far East: Application of Logic-Informational Analysis 541
Nina Gorelikova, Irina Chizhova and Filipp Balashov

A Spatial Statistical Approach for Sedimentary Gold Exploration: A Portuguese Case Study 545
Pierre Goovaerts, Teresa Albuquerque and Margarida Antunes

Part XVII Modeling of Energy Resources

Coalbed Methane Production Analysis and Filter Simulation for Quantifying Gas Drainage from Coal Seams 549
C. Özgen Karacan and Ricardo A. Olea

The Relationship Between Skin and Apparent Wellbore Radius in Fractal Drainage Areas 553
Tom Aage Jelmert

Parameterization of Channelized Training Images: A Novel Approach for Multiple-Point Simulations of Fluvial Reservoirs 557
Mohamed M. Fadlelmula, Serhat Akin and Sebnem Duzgun

Mathematical and Physical Models for the Estimation of Wind-Wave Power Potential in the Eastern Mediterranean Sea 561
George Galanis, George Emmanouil, Christina Kalogeri and George Kallos

Geostatistical AVO Direct Facies Inversion 565
Leonardo Azevedo, Pedro Correia, Rúben Nunes and Amílcar Soares

Optimized History Matching with Stochastic Image Transforming of a Deltaic Reservoir 571
Maria Helena Caeiro, Amilcar Soares, Vasily Demyanov and Mike Christie

Conditioning 3D Object Based Models to a Large Number of Wells: A Channel Example 575
Jeff B. Boisvert and Michael J. Pyrcz

Modeling Channel Forms Using a Boundary Representation Based on Non-uniform Rational B-Splines 581
Jeremy Ruiu, Guillaume Caumon, Sophie Viseur and Christophe Antoine

Moving Away from Distance Classifications as Measures of Resource Uncertainty 585
Ricardo A. Olea, James A. Luppens and Susan J. Tewalt

Part XVIII New Developments in Oil and Gas Discovery Modeling

Reliability Analysis of Least Squares Estimation and Prediction of a Non-parametric Discovery Process Model 589
Mi Shi-yun and Zhang Qian

Application and Comparison of Discovery Model and Others in Petroleum Resource Assessment 595
Guo Qiu-Lin, Yan Wei and Chen Zhuo-Heng

Handling Seismic Anomalies in Multiple Segment Prospects with Graphical Models 601
Gabriele Martinelli, Charles Stabell and Espen Langlie

The North American Shale Resource: Characterization of Spatial and Temporal Variation in Productivity 607
Qudsia Ejaz and Francis O'Sullivan

A Pyramid Scheme: Integrating Petroleum Systems Analysis into Probabilistic Petroleum Resource Assessments 613
Kirk G. Osadetz and Zhuoheng Chen

Application of a Least Square Non-parametric Discovery Process Model to Colorado Group Mixed Conventional and Unconventional Oil Plays, Western Canada Sedimentary Basin 617
Zhuoheng Chen, Kirk G. Osadetz and Gemai Chen

A Further Investigation of Local Nonparametric Estimation Techniques in Shale Gas Resource Assessment 621
Emil D. Attanasi, Timothy C. Coburn and Philip A. Freeman

Part XIX Open Session on Mathematics of Oil Recovery (OSMOR)

Fractal Analysis of AVO Seismic Attributes for Oil/Water Contact Identification 625
Sid-Ali Ouadfeul and Leila Aliouane

Upscaling on Anelastic Vertically Heterogeneous Reservoirs 629
Alexey Stovas

Lithofacies Prediction from Well-Logs Data Using Hybrid Neural Network Model: A Case Study from Algerian Sahara 633
Sid-Ali Ouadfeul and Leila Aliouane

Stochastic Simulation of the Morphology of Fluvial Sand Channel Reservoirs 639
Alexandra Kuznetsova, José A. Almeida and Paulo Legoinha

Sorting Reservoir Models According to Flow Criteria: A Methodology, Using Fast Marching Methods and Multi-Dimensional Scaling 643
Gaétan Bardy and Pierre Biver

Integrate Facies Clustering Feature Information in Reservoir Modeling 647
Yupeng Li and Lihui Geng

Permeability Prediction Using Artificial Neural Networks. A Comparative Study Between Back Propagation and Levenberg–Marquardt Learning Algorithms 653
Leila Aliouane, Sid-Ali Ouadfeul, Noureddine Djarfour and Amar Boudella

A Method for Multi-Level Probabilistic History Matching and Production Forecasting: Application in a Major Middle East Carbonate Reservoir 659
Marko Maučec, Ajay Singh, Gustavo Carvajal, Seyed Mirzadeh, Steven Knabe, Richard Chambers, Genbao Shi, Ahmad Al-Jasmi, Harish Kumar Goel and Hossam El-Din

Causal Analysis and Data Mining of Well Stimulation Data Using Classification and Regression Tree with Enhancements 665
Srimoyee Bhattacharya, Marko Maučec, Jeffrey Yarus, Dwight Fulton, Jon Orth and Ajay Singh

Seismic Data Interpretation Improvement by the Multiscale Analysis of Gravity Data Using the Wavelet Transform: Application to Algerian Sahara 669
Sid-Ali Ouadfeul and Leila Aliouane

Part XX Geostatistical Priors in Inversion of Geophysical and Engineering Data

History Matching of Channelized Reservoir Using Ensemble Smoother with Clustered Covariance 675
Kyungbook Lee and Jonggeun Choe

Assessing the Probability of Training Image-Based Geological Scenarios Using Geophysical Data 679
Thomas Hermans, Jef Caers and Frédéric Nguyen

Reservoir Modeling Combining Geostatistics with Markov Chain Monte Carlo Inversion 683
Andrea Zunino, Katrine Lange, Yulia Melnikova, Thomas Mejer Hansen and Klaus Mosegaard

Geostatistical Inversion of 3D Post-stack Seismic and Well Data for the Characterization of Acoustic Impedance in Oil Fields 689
Fernando Alves, José A. Almeida and António Ferreira

The Effect of the Noise and the Regularization in Inverse Problems: Geophysical Implications 695
José Luis GarcÍa Pallero, Juan Luis Fernández-Martínez, Zulima Fernández-Muñiz and Luis Mariano Pedruelo-González

A Generalized Local Gradual Deformation Method for History Matching 699
Benjamin Marteau, Didier Yu Ding and Laurent Dumas

History Matching with Geostatistical Prior: A Smooth Formulation 703
Yulia Melnikova, Katrine Lange, Andrea Zunino, Knud Skou Cordua and Klaus Mosegaard

Part XXI Mineral and Energy Resources for Planet Earth: Evaluations, Extraction and Optimal Management

Factor Analysis for Metal Grade Exploration at Pallancata Vein in Peru 709
Jorge E. Gamarra-Urrunaga, Ricardo Castroviejo and Jesús Domínguez

Compositional Block Cokriging 713
Raimon Tolosana-Delgado, Ute Mueller, K. Gerald van den Boogaart and Clint Ward

Milling Result Prediction 717
Stephen Matos Camacho, Thomas Leißner, Petya Atanasova, Andre Kamptner, Martin Rudolph, Urs Alaxander Peuker and K. Gerald van den Boogaart

A Geostatistical Study of Tertiary Coal Fields in Turkey 723
Fırat Atalay, A. Erhan Tercan, Bahtiyar Ünver, Mehmet Ali Hindistan and Güneş Ertunç

Placer Deposit: From Modeling to Evaluation 727
Nikolay Laverov, Irina Chizhova and Elena Matveeva

Uncertainty Assessment of the Orebodies Geometry by Using Block Indicator Simulation 731
Julia Carvalho, Pedro Correia, Sofia Menezes, Cláudia Peixoto and Amilcar Soares

Part XXII Recent Advances in Quantitative Methods Applied to Stratigraphy and Paleontology

"Horse-Shoe" Cu-Au Porphyry Orebody Modeling Based on Blasthole Data Using Unfolding Technique 735
Mohamad Nur Heriawan, Loya Jirga and Anton Perdana

A New Mathematical Approach to Model Trophic Dynamics of Mammalian Palaeocommunities. The Case of Atapuerca-TD6 739
Guillermo Rodríguez-Gómez, Jesús A. Martín-González, Idoia Goikoetxea, Ana Mateos and Jesús Rodríguez

Multifractals and Capacity Dimension as Measures of Disturbance Patch Dynamics in *Daedalus* Ichnofabrics 747
Carlos Neto de Carvalho and Andrea Baucon

Part XXIII Geo-mathematical Models of Folds and Folding

Non-Linear Thermo-Mechanics of Folding in Geomaterials 753
Martin K. Paesold, Ali Karrech, Tim Dodwell, Klaus Regenauer-Lieb, Andrew P. Bassom, Alison Ord and Bruce E. Hobbs

Analytical and Numerical Investigation of 3D Multilayer Detachment Folding 757
Naiara Fernández and Boris Kaus

Mechanics of Fold Development in Pure- and Simple Shear 763
Maria-Gema Llorens, Paul D. Bons, Albert Griera and Enrique Gomez-Rivas

Stochastic Modelling of the 3D Geometry of a Faulted and Folded Deep Carbonate Aquifer: Loma de Úbeda (Southern Spain) 767
Javier Heredia, Eulogio Pardo-Igúzquiza and Antonio González-Ramón

3D Geometry of an Active Shale-Cored Anticline in the Western South Caspian Basin 771
Idaira Santos-Betancor, Juan Ignacio Soto, Lidia Lonergan, Ismael Sánchez-Borrego and Carlos Macellari

Gravity-Instabilities Processes and Mass-Transport Complexes During Folding: The Case of a Shale-Cored Anticline in the Western South Caspian Basin 775
Idaira Santos-Betancor, Juan Ignacio Soto, Lidia Lonergan and Carlos Macellari

Part XXIV Mathematical Geosciences and Planetary Geology

Domaining Bi-modal Data Sets Geostatistically Using a Directional Neighborhood Search 779
Steffen Brammer

FRISER-IRMIX Database: A Web-based Support System with Implications in Planetary Mineralogical Studies, Ground Temperature Measurements and Astrobiology 783
Jesús Martínez-Frias, María Serrano Rubio, F. Javier Martín-Torres, María Paz Zorzano, José A. Rodríguez-Manfredi, Javier Gómez-Elvira and REMS Team

Spectral Analysis for Anomaly Detection in the Central Volcanic Range, Costa Rica. Implications for Planetary Geology 787
Juan Gregario Rejas, Ruben Martínez, Miguel Marchamalo, Javier Bonatti and Jesús Martínez-Frías

A Mathematical Algorithm to Simulate the Growth and Transformation of Framboidal Pyrite: Characterization of the Biogenic Influence in Their Size Distributions 793
Raúl Merinero Palomares, Rosario Lunar Hernández, Francisco-Javier González-Sanz, Luis Somoza Losada and Jesús Martínez-Frías

Computational Models and Simulations of Meteor Impacts as Tools for Analysing and Evaluating Management of Crisis Scenarios 797
Jesús Martínez-Frías and Alain Leppinette Malvite

In-situ Raman Analysis of the Precipitation Sequence of Sulphate Minerals Using Small Droplets: Application to Rio Tinto (Spain) 801
Fernando Rull, Francisco Sobrón, Julia Guerrero, Jesús Medina, Gloria Venegas, Fernando Gázquez and Jesús Martínez-Frías

Part XXV Mathematics of Planet Earth

Signal Analysis by Means of Multi-Scale Methods 807
Christian Blick

Modeling and Simulation of Forest Fire Spreading 811
Sarah Eberle

The Finite Pointset Method (FPM) and an Application in Soil Mechanics 815
Jörg Kuhnert and Isabel Ostermann

Geostatistical Estimation of the Crustal Thickness Inferred from the Geometry of Monogenetic Volcanoes (Central Mexico Case) 819
Raúl Pérez-López, Carolina Guardiola-Albert and José Luis Macias

Mathematical Modelling of Geochemical Processes Applied to Cenozoic Iberian Volcanics: A Review 823
Raúl Benito and Jesús Martínez-Frías

Part XXVI Geoscience Data Models for Practical Interoperability

Management of Geological Field Data: The Open Solution MO2GEO FieldModule 827
Lars Schimpf and Wolfgang Gossel

Swiss Data Models for Geology: from 2D Data Towards Geological 3D Models 831
Cristina Salomè Michael, Nils Oesterling, Stefan Strasky and Roland Baumberger

Virtual Collaboration of Geoscientists of Russian Academy of Sciences in the Far East of Russia 835
Vera V. Naumova, Mike I. Patuk and Valentin V. Nichepkov

Geological 3D Modeling (Processes) and Future Needs for 3D Data and Model Storage at Geological Survey of Finland 839
Eevaliisa Laine

Part XXVII Advances on Stochastic Non Linear Methods and Inverse Problems for Dynamic Models

Estimation of Parameters in Random Dynamical Systems 843
Silke Konsulke, K. Gerard van den Boogaart, Fellix Ballani, Markus Franke and Martin Sauke

Nonlinear Estimation with Gaussian Kriging and Riemann Sums 847
K. Daniel Khan

Modeling Trace Element Concentration in Vertical Regolith Profile Over Mineral Deposits 853
Qiuming Cheng

Unified Principles for Nonlinear Nonstationary Random Fields in Stochastic Geosciences 857
José A. Vargas-Guzmán

Performance Evaluation of Swanson's Rule for the Case of Log-Normal Populations

Maryam Moghadasi and Jerry L. Jensen

1 Introduction

The arithmetic average (AA) and Swanson's Rule (SR) are commonly used in the petroleum industry to estimate the mean value of reservoir properties. The AA estimates the mean based on assigning equal weight to all n samples, $X_1, X_2, \ldots, X_n$:

$$x_A = \frac{1}{n}\sum_{i=1}^{n} X_i. \tag{1}$$

Numerous studies, for example [1], prove the AA is unbiased with standard error $[var(X)/n]^{1/2}$, where $var(X)$ is the variance of X.

SR is a discretization method and is based on unequally weighted sample percentiles to estimate the mean value by

$$x_{SR} = 0.3\,x_{10} + 0.4\,x_{50} + 0.3\,x_{90}, \tag{2}$$

where x_i is the ith percentile. Swanson first introduced SR as an alternative mean estimator in 1972 [2]. Subsequently, SR has often been used in the oil and gas industry, perhaps because the 10th, 50th, and 90th percentiles have associations with possible, probable, and proved reserves, respectively.

For the case $\ln(X) \sim N(\mu, \sigma^2)$, Megill used the ratio x_{90}/x_{50} $(= e^{1.28\sigma})$ varying from one to 15 ($\sigma = 0$ to 2.1) as a measure of the sample variability and showed that SR underestimates the mean value [2]. He concluded that, for modestly skewed distributions, SR estimates the mean with acceptable error but the bias becomes large

M. Moghadasi (✉) · J. L. Jensen
Chemical and Petroleum Engineering, University of Calgary, Calgary, AB T2N 1N4, Canada
e-mail: mmoghada@ucalgary.ca

J. L. Jensen
e-mail: jjensen@ucalgary.ca

E. Pardo-Igúzquiza et al. (eds.), *Mathematics of Planet Earth*,
Lecture Notes in Earth System Sciences, DOI: 10.1007/978-3-642-32408-6_1,

as sample variability increases. Nevertheless, [3–5] advocate SR instead of the AA without mentioning the potential for large bias. For example, according to Megill's results, SR underestimates the population means of Delfiner's [3] and Hurst et al. [4] data sets by 16 % and 10 %, respectively.

Besides Megill [2], others have reported on SR bias [6, 7]. Bickel et al. [7] found the bias may be so large as to render the SR unacceptable for "professional practice". To our knowledge, however, other SR properties such as efficiency and uncertainty have not been assessed. This study attempts a more balanced SR evaluation, comparing it to the properties of the AA when the population is log-normal. We show that SR has a smaller standard error than the AA when σ becomes large and SR becomes more efficient when σ exceeds a value which depends on n.

2 Properties of the Arithmetic Average and Swanson's Rule

In this study, $X_1, X_2, \ldots, X_n$ are assumed independent and identically log-normally distributed (LND) with $E[\ln(X)] = \mu$ and $var[\ln(X)] = \sigma^2$. Many reservoir parameters, e.g., drainage area, gross and net pay, reserves, recoveries, and permeability, have been found to be LND [2, 8–11]. The uth sample percentile of X is assumed to be distributed as

$$\hat{x}_u \sim N\left(e^{\mu+\mathrm{w}_u\sigma}, \frac{2\pi\sigma^2}{n}u(1-u)e^{2\mu+2\mathrm{w}_u\sigma+\mathrm{w}_u^2}\right), \tag{3}$$

where $w_{\mathrm{u}} = \Phi^{-1}(u/100)$, and Φ is the standard normal cum. distribution function [1].

A good estimator should have simultaneously small bias, good efficiency, and small standard error (SE). Efficiency is evaluated based on comparing the root mean square error (RMSE, S) of estimators. S incorporates the bias and SE of the estimator and an estimator with small S is more efficient.

Analytical expressions of S for SR and the AA are [12]:

$$S_{SR} = e^{\mu}\left\{\left(0.3e^{\sigma\mathrm{w}_{50}} + 0.4e^{\sigma\mathrm{w}_{50}} + 0.3e^{\sigma\mathrm{w}_{95}} - e^{\sigma^2/2}\right)^2 + (2\pi\sigma^2/n)\mathrm{f}_{SR}\right\}^{1/2}, \tag{4}$$

where $\mathrm{f}_{\mathrm{SR}} = 0.0081(e^{2\sigma\mathrm{w}_{10}{}^2} + e^{2\sigma\mathrm{w}_{90}+\mathrm{w}_{90}{}^2}) + 0.04 + 0.012(e^{\sigma\mathrm{w}_{10}{}^2/2} + e^{\sigma\mathrm{w}_{90}} + \mathrm{w}_{90}{}^2/2) + 0.0018e^{(\mathrm{w}_{10}{}^2+\mathrm{w}_{90}{}^2)/2}$, and

$$S_A = e^{\mu}\left\{e^{\sigma^2/2}\sqrt{e^{\sigma^2}-1}/\sqrt{n}\right\}. \tag{5}$$

Neither the AA nor SR emerges as the best for all conditions of variability and sample size (Fig. 1). SR is biased, even for small σ whereas the AA is unbiased.

However, SR has smaller SE than the AA when $\sigma > 0.94$ for any n. Setting $S_{SR} = S_{AA}$ gives the line in Fig. 1 which has the approximate form

$$\sigma = -3182.5/n^2 + 1319.8/n^{3/2} - 97.05/n - 18.66/\sqrt{n} + 3.131. \tag{6}$$

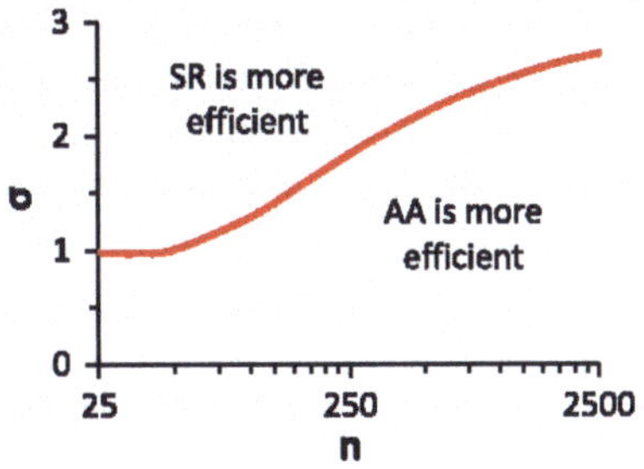

Fig. 1 σ versus n showing region in which SR is more efficient ($S_{SR} < S_A$) than the AA

Hence, SR becomes statistically more appropriate than the AA for some ranges of variability and sample size. Moghadasi [12] proposes a bias-corrected version which improves SR performance based on the RMSE.

3 Applications

We illustrate our findings using two data sets. One concerns North Sea prospect sizes [4] and the other is a set of Cleveland Formation permeability measurements [10].

The oil prospect data set has $n = 21$ points, which are LND with sample $\mu = 3.7$ and sample $\sigma = 1.1$. $x_A = 76$ and $x_{SR} = 67$ million barrels (mmbbl). The analyticallyderived bias for SR is 10 %, in close agreement with $x_{SR} = 0.91x_A$. The AA and SR SEs are similar (29 vs. 25 mmbbl, respectively) and a hypothesis test shows x_A and x_{SR} are not significantly different statistically. In economic terms however, $x_A - x_{SR}$ reflects a US$ 0.9 billion change in prospect value (at $100/bbl). Thus, despite being somewhat less efficient, using the unbiased estimator AA may be preferable.

The permeability dataset ($n = 319$) is LND with the sample $\mu = -3.6$ and sample $\sigma = 1.7$. This data set gives $x_{SR} = 0.09$ md and $x_A = 0.18$ md; there is a clear preference for x_{SR} as the SR bias (28 %) is compensated by the 55 % smaller SE. For production prediction, the two-fold difference is relatively modest [3, 10], so SR is preferable. The difference is important, however, for tax and regulatory purposes as it changes the cleveland from a tight (<0.1 md) to a conventional classification.

4 Conclusions

A good estimator should simultaneously have small bias and small SE. Zero bias is a desirable property; however, other mean estimator properties such as efficiency and uncertainty also require evaluation to assess more completely estimator merits. Some reports have advocated the use of the SR in place of the AA but bias and efficiency were apparently overlooked. We find that, although the AA is unbiased, it has larger standard error and smaller efficiency than SR for large-variability cases. Hence, there are statistical benefits to replacing the AA with SR under some conditions but the SR must be used with care as its bias can diminish its other advantages. A modified SR with bias reduction could be a better option to either the SR or AA in petroleum reservoir characterization.

Acknowledgments MM was funded by the Natural Sciences and Engineering Research Council of Canada. JLJ holds the Schulich Chair in Geostatistics.

References

1. Stuart, A., & Ord, J. K. (1987). *Kendall's advanced theory of statistics* (5th ed). New York: Oxford University Press.
2. Megill, R. E. (1984). *An introduction to risk analysis* (2nd ed.). Tulsa: PennWell Publishing Company.
3. Delfiner, P. (2007). Three statistical pitfalls of Phi-K transforms. *SPE Reservoir Evaluation & Engineering, 10*, 609–617.
4. Hurst, A., Brown, G. C., & Swanson, R. I. (2000). Swanson's 30–40-30 rule. *AAPG Bulletin, 84*, 1883–1891.
5. Rose, P. R. (2001). *Risk Analysis and Management of Petroleum Exploration Ventures, Methods in Exploration Series*. Tulsa: American Association of Petroleum Geologists.
6. Bickel, J. E., Lake, L. W., & Lehman, J. (2011). Discretization, simulation, and Swanson's (inaccurate) mean. *SPE Economics and Management, 3*, 128–140.
7. Keefer, D. L., & Bodily, S. E. (1983). Three-point approximations for continuous random variables. *Management Science, 29*, 595–609.
8. 8 Kaufman, G. M. (1965). *Statistical analysis of the size distribution of oil and gas fields*: SPE 1096-MS.
9. 9 Bennion, D. W.(1965). *A stochastic model for predicting variations in reservoir rock properties*. Ph. D. dissertation, USA: Pennsylvania State University, University Park.
10. 10 Rollins, J. B., Holditch, S. A., Lee, W. J. (1992). Characterizing average permeability in oil and gas formations. *SPE Formation Evaluation*, 99–105
11. 11 Seidle, J. P., & O'Connor, L.S. (2003). *Production based probabilistic economics for unconventional gas*: SPE paper 82024-MS.
12. Moghadasi, M. *Evaluating averages for reservoir characterization*. Ph.D. Dissertation, The University of Calgary, Calgary (in preparation).

Modelling Ore Bodies of High-Nugget Gold Using Conditional Probability

Evelyn June Hill, Nicholas H. S. Oliver, James Cleverley and Michael Nugus

1 Introduction

In vein-hosted gold deposits, gold distribution typically shows very low spatial continuity, i.e., a very high-nugget effect, which makes it unreliable to model the ore-body using traditional geostatistical techniques [1, 2]. In such situations geologists may resort to using a proxy for gold, such as presence of alteration or deformation, to outline the ore-body extents. The requirement is that the proxy is significantly more spatially continuous than the distribution of high grade gold [3, 4]. We present a method for evaluating and combining several geological features or geochemical elements to provide a proxy for mineralisation using conditional probability.

The study is based on drill hole data (geological logging and geochemical analyses) from Sunrise Dam Gold Mine. The mine lies within an Archaean greenstone belt in the Yilgarn Craton of Western Australia. The host rocks of the gold mineralisation are structurally complex and deposits include vein stockwork lodes and shear hosted lodes [2, 5]. Gold grades tend to be elevated in regions of strong sericitic alteration and are closely associated with high arsenic values. This geological information is used to decide which geological and geochemical features may be useful proxies for predicting the extent of the gold ore-bodies.

E. J. Hill (✉) · J. Cleverley
CSIRO, Kensington, WA, Australia
e-mail: June.Hill@csiro.au

J. Cleverley
e-mail: James.Cleverley@csiro.au

N. H. S. Oliver
Holcombe Coughlin Oliver, Hemit Park, QLDPO Box 3533, Australia
e-mail: nick@hco.net.au

M. Nugus
AngloGold Ashanti, 44 St Georges Tce, Perth, WA, Australia
e-mail: MNugus@AngloGoldAshanti.com.au

E. Pardo-Igúzquiza et al. (eds.), *Mathematics of Planet Earth*,
Lecture Notes in Earth System Sciences, DOI: 10.1007/978-3-642-32408-6_2,

2 Calculating Conditional Probability

Conditional probability provides a data-driven method for determining which rock types and spatial regions are most favourable for mineralisation. Conditional probability is defined as the probability of an event occurring given that another event has already occurred. In this case we calculate the probability of a sample returning a gold assay value (Au) that exceeds a pre-determined cut-off value (v) given certain features observed in the sample. This can be expressed in the form:

$$p(\mathrm{Au} \geq \mathrm{v}|\mathrm{F}_1, \mathrm{F}_2, \ldots, \mathrm{F}_\mathrm{n}) \tag{1}$$

where $\mathrm{F}_1, \mathrm{F}_2, \ldots, \mathrm{F}_\mathrm{n}$ represents a set of geological features.

These features may be data logged by geologists in a categorical form, such as alteration intensity or vein density. For categorical data the conditional probability is calculated by creating an n-dimensional matrix of counts of samples; where n is the number of features used in the calculation. Each cell of the matrix represents a class. Conditional probability is calculated by dividing the number of samples in a class with gold assay values exceeding the cut-off value ($\mathrm{Au} \geq \mathrm{v}$) by the total number of samples in that class [1, 2].

Features may also be in the form of continuous numerical data such as geochemical data. For continuous numerical values the conditional probability is calculated using the density of data points in feature space defined by $\mathrm{F}_1, \mathrm{F}_2, \ldots, \mathrm{F}_\mathrm{n}$ (FS):

$$\mathrm{p}(\mathrm{Au} \geq \mathrm{v}|\mathrm{F}_1, \mathrm{F}_2, \ldots, \mathrm{F}_\mathrm{n}) = \text{prior probability} \times \frac{\text{density in FS (points Au} \geq \text{v)}}{\text{density in FS (all points)}} \tag{2}$$

The density of the data is calculated using a kernel density estimator [6]. The prior probability is the ratio of the number of samples with $\mathrm{Au} \geq \mathrm{v}$ to the total number of samples.

3 Results

In our first example we use conditional probability to predict favourable rock types for gold mineralisation from categorical data: foliation intensity (F), alteration intensity (A) and vein density (V) logged in drill holes. The prior probability value is used as the value which divides regions of favourable rock type from unfavourable rock type. Higher probabilities will delineate increasingly favourable regions. Figure 1 shows that the favourable regions in the drill holes are substantially more spatially continuous than the original gold assays at a cut off of 1 ppm Au. An isotropic interpolation has been performed on the conditional probability values, illustrated in Fig. 1 as traces of isosurfaces, to provide a first-pass orebody outline. Alternatively,

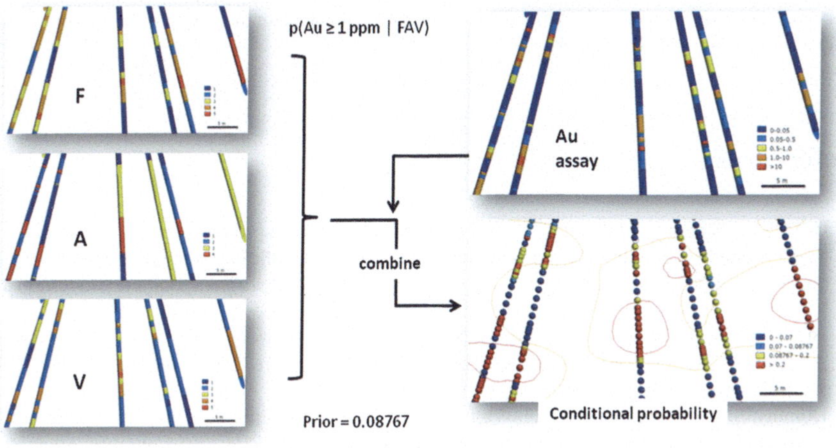

Fig. 1 F, A and V categories (*left*) are combined with gold assay data (*top right*) to generate conditional probability (*bottom right*). F, A, V: warmer colours represent stronger intensity. Au assay: *orange* >1 ppm, *red* >10 ppm. Conditional probability (of Au ≥ 1ppm): *yellow* >0.08767 (value of prior), *red* >0.2

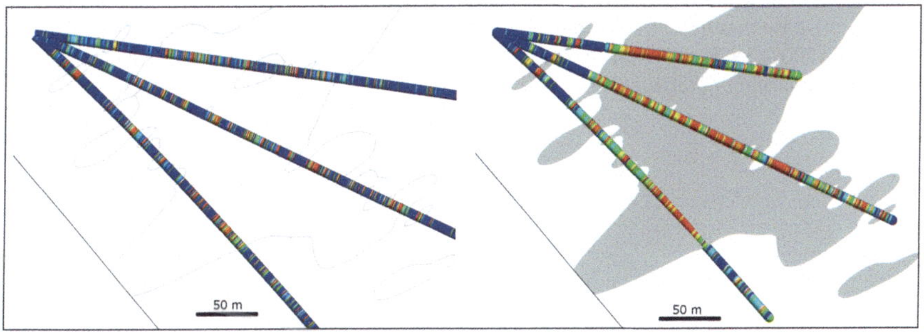

Fig. 2 Underground drill fan showing (*left*) distribution of gold assays and (*right*) conditional probability values (*blue = low* values, *red = high*). Grey domain in right image is the interpolated region where p > 0.308 (value of prior)

if the local structural trend was taken into account (left-to right horizontally across the figure) then 2 distinct bands of favourable rock could be interpreted.

In the second example we present results using conditional probability to predict favourable rock types from multi-element geochemical data; arsenic (As ppm), rubidium (Rb ppm) and potassium (K_2O %) reflect mineralisation-related alteration and are more spatially continuous than gold. The favourable rock types defined by this method appear to be more spatially continuous than the original gold assay values and should make the definition of the ore-body outline substantially easier (Fig. 2). Traces of isosurfaces are shown for an interpolation using a strong anisotropy parallel to the dominant structural trend in the region.

Acknowledgments AngloGold Ashanti Sunrise Dam are thanked for their support for this research and permission to publish this work.

References

1. Hill, E. J., Cleverley, J. Oliver, N .H. S., Nugus, M. (2012). Using structural and geological features encoded as a conditional probability to predict 3D gold in the high-nugget vein-dominated Sunrise Dam GQ lodes. *Structural Geology and Resources Symposium Extended Abstracts* (p. 56). Kalgoorlie: AIG Bulletin.
2. Hill, E. J., Oliver, N. H. S., Cleverley, J. S., Nugus, M. J., Carswell, J., Clark, F. (2012). Characterisation and 3D modelling of a nuggety, vein-hosted gold ore body, Sunrise Dam, Western Australia. *Journal of Structural Geology* (in review).
3. Dominy, S. C., Johansen, G. F. (2004). Reducing grade uncertainty in high-nugget effect gold veins-application of geological and geochemical proxies. In *PACRIM 2004 Congress*, Adelaide, 19–22 Sep 2004.
4. Dominy, S. C., Platten, I. M., Raine, M. D. (2003). Grade and geological continuity in high-nugget effect gold-quartz reefs: Implications for resource estimation and reporting. *Applied Earth Science Transactions of the Institution of Mining and Metallurgy B*, *112*, 239–259.
5. Blenkinsop, T., Baker, T., McLellan, J., Cleverley, J., & Nugus, M. (2007). *Final Report Sunrise Dam Gold Mine Geological Study Project (Project G15)*. Predictive Mineral Discovery Cooperative Research Centre: Technical report.
6. Duon, T. (2007). ks: Kernel Density Estimation and kernel discriminants analysis for multivariate data in R. *Journal of Statistical Software*, *21*, 1–16.

The Extraction of Long-Term Distribution Trends of Ore-Sourced Geochemical Elements from an Drill Core in the Loess Covered Area

Deyi Xu, Hongwei Pei, Ningqiang Liu, Zhaoxian Yuan, Shuyun Xie and Qiuming Cheng

1 Introduction

The vertical distributions of ore-forming related elements may statistically reveal the transportation mechanisms of these elements. Apparently, to explore the mechanisms may be helpful for us to detect deposits through surface geochemical anomalies. That's one of the main reasons that more and more scientists devote their efforts to this study area [1–6]. In this paper, we tried to detach the long-term trends of ore-sourced elements from a vertical drill-core of a loess cover to study the transportation modes of deposit-related elements (Fig. 1).

2 Sample Collecting and Data Processing

Loess covers a very big area in the northwest of China, as well as other areas around the world [7]. The resources under the loess cover are abundant. The thickness of the cover varies from meters to kilometers. For the study of the transportation modes of deposit-related elements, 106 samples were sequentially collected from a 106.5 m vertical drill core of the loess cover on the Diyanqinamu Mo-Ag deposit in the

D. Xu (✉) · Z. Yuan · S. Xie · Q. Cheng
China University of Geosciences, 388 Lumo Rd., 430074 Wuhan, China
e-mail: xudeyi.ca@gmail.com

Q. Cheng
e-mail: qiuming@cug.edu.cn

H. Pei
Xiamen University, 422 Siming South Rd., 361005 Xiamen, China
e-mail: xdy@cug.edu.cn

N. Liu · Q. Cheng
China University of Geosciences, 28 Xueyuan Rd., 100083 Beijing, China

E. Pardo-Igúzquiza et al. (eds.), *Mathematics of Planet Earth*,
Lecture Notes in Earth System Sciences, DOI: 10.1007/978-3-642-32408-6_3,

Fig. 1 Loess cover *Right*: cover profile *Up-left*: ground surface *Down-left*: raw sample

grassland area in the Inner Mongolia of China. They were pounded to pieces and filtered into 60-mesh powder. The contents of 33 elements were measured by a Niton XL3t X-ray fluorescence analyzer (Fig. 2).

The vertical distributions of elements are determined mainly by two processes, one is stochastic, and another is deterministic. We think that if the vertical distribution of an element is dominated by a deterministic process, the contents of this element should probably be correlated with the depth. The nonlinear Kendall τ correlation coefficient method was used to test the correlations and we found that 18 elements Mo, S, W, Te, Fe, Cu, Ni, Mn, Sb, Rb, Cs, Sn, Zn, Ti, V, Pb, Th and As are correlated with the depth.

3 Long-Term Trends Detaching

A HP filter [8] is applied to detach the long-term trends of the 18 elements. The HP filter is used to separate cycles and trends in a time series. We regard the vertical distributions of the elements as spatial series and apply the HP filter to them as applied in a time series to get the long-term trends. The trends of 18 elements are shown in Fig. 3. The trends of Mo, S, W are approximately of exponential decreasing

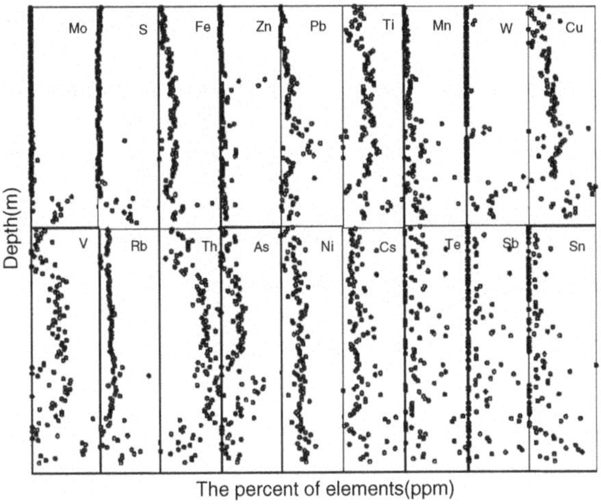

Fig. 2 Vertical distributions of 18 elements

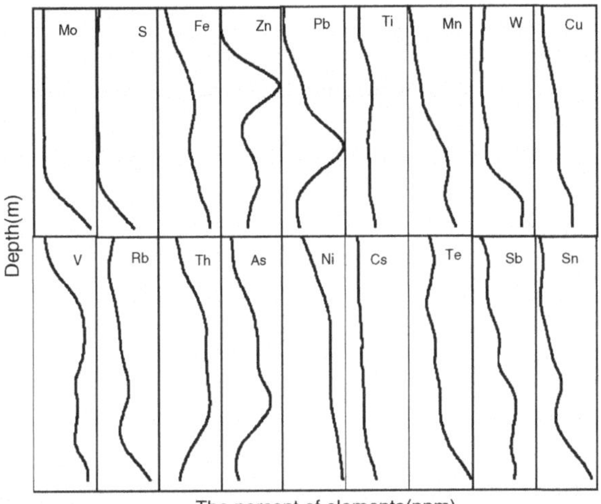

Fig. 3 Long-term distribution trends

with distance from ore body. The contents of these elements decrease from ore layer to the surface, which in some extent indicate that the detached trends can reflect the deterministic process for forming the distributions. The trends can be classified into three clusters, $C_1 = \{Mo, S, W, Te, Rb, Cs, Sn\}$, $C_2 = \{Fe, Cu, Ni, Mn, Sb\}$ and $C_3 = \{V, Pb, Zn, Ti, As, Th\}$. Statistically we verified that C_1 is of exponential decreasing mode, while C_2 is of linearly decreasing mode. C_3 is thought to be over-affected by other stochastic processes, which need other methods to study.

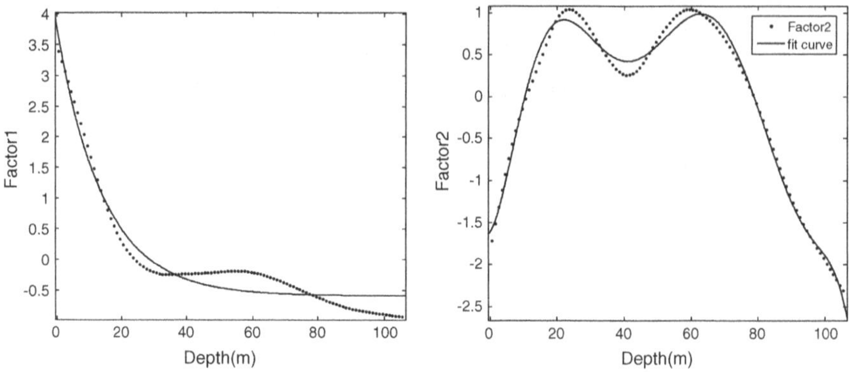

Fig. 4 Factor scores and fitting curves *dash lines* represent factor scores, *solid lines* represent fitted curves

4 The Transportation Modes of the Ore-Forming Factors

Factor analysis was preformed to get two factors which were named as Mo-W factor and Pb-Zn factor. Factor scores were calculated and the trends of these two factors were detached as did in Sect. 3. Factor 1 is fitted with an exponential function very well (Eq. 1, Fig. 4), the goodness of fit is $R^2 = 0.9569$. Factor 2 can be fitted by a polynomial model.

$$C_{Factor1} = 4.387e^{-0.06989h} - 0.5917 \quad (1)$$

5 Discussions

The ore-sourced or ore-related elements diffuse from the ore layer to the ground surface could be regarded as one of the main transportation modes. However, random factors, such as the heterogeneity of the cover medium, surface water infiltration, other effects on the surface, may affect or conceal the main process (as seen in Fig. 2). For the data collected from a long vertical drill-core of loess cover, to test the contents of what elements may correlate with the distance from the ore, to detach the main trends of the contents by a suitable filter and to fit the main trend with curves might be a good line of thinking to help us to find the transportation modes of a deposit-related elements. These modes are apparently useful for deposit detecting. This study is only an experiment and detailed work should be done in the future.

Acknowledgments This study was financially supported by NSFC (No. 40972205, 41272362), China Geological Survey (No. 1212010633910, 1212011121101) and the “863” project (2009AA06Z110).

References

1. Xu, D., Pei, H., et al. (2012). Extraction and applications of the long-term vertical distribution trends of geochemical elements in the loess cover. *Earth Science–Journal of China University of Geosccienes*, *37*(6), 1135–1139 (in Chinese).
2. Cheng, Q. (2012). Singularity theory and methods for mapping geochemical anomalies caused by buried sources and for predicting undiscovered mineral deposits in covered areas. *Journal of Geochemical Exploration*, *122*, 55–70.
3. Cheng, Q., & Xia, Q. (2011). Thoughts and practice for the integrated assessment of mineral resources in covered areas. *Mineral Journal*, (suppl.) 755–756 (in Chinese).
4. Anand, R. R., & Paine, M. (2012). Regolith geology of the Yilgarn Craton, Western Australia: implications for exploration. *Australian Journal of Earth Sciences*, *49*, 3–162.
5. Wang, X. (2005). Conceptual model of deep-penetrating geochemical migration. *Geological Bulletin of China*, *24*(10–11):892–896 (in Chinese)
6. Tyler, G. (2004). Vertical distribution of major, minor, and rare elements in a Haplic Podzol. *Environmental Pollution*, *119*, 277–290.
7. Liu, D. (1965). *Chinese loess*. Beijing: Science press.
8. Ahumada, H., & Garegnani, M. L. (1999). Hodrick-prescott filter in practice. *Jornadas de Economia Monetaria e International*, FCE-UNLP.

Evaluating Predictive Performance

Michael Scheuerer and Tilmann Gneiting

1 Probabilistic Forecasting

A probabilistic forecast is a prediction that takes the form of a probability distribution for a future quantity. What makes up a good probabilistic forecast? Gneiting et al. [4] contend that predictive distributions should have *maximal sharpness subject to calibration*. Sharpness refers to the concentration of the predictive distributions, with very sharp distributions implying little uncertainty. It can be assessed via numerical and graphical summaries of the width of prediction intervals.

While sharp predictions are desired, they are only useful if statistically consistent with the values that materialize. The notion of calibration formalizes this property. Denote by $(F_i)_{i=1,2,\ldots}$ the sequence of probabilistic forecasts and by $(y_i)_{i=1,2,\ldots}$ the associated sequence of observations. If the true data-generating process creating the observations had the same distribution as the forecasts for all i, those would be considered ideal. In practice, the data-generating distribution remains hypothetical, and the predictive distribution F_i is an expert opinion, e.g. obtained from a statistical prediction algorithm. Hence, the predictive distributions need to be assessed on the basis of the forecast-observation pairs (F_i, y_i) only, regardless of their origins.

2 Assessing Calibration: Diagnostic Methods

Given such a sequence $(F_i, y_i)_{i=1,2,\ldots}$, the notion of probabilistic calibration boils down to the study of the probability integral transform (PIT) values, $p_i = F_i(y_i)$. If the forecasts are ideal and F_i is continuous, then p_i has a uniform distribution.

M. Scheuerer (✉) · T. Gneiting (✉)
Institute of Applied Mathematics, University of Heidelberg, Im Neuenheimer Feld 294, Heidelberg, Germany
e-mail: michael.scheuerer@uni-heidelberg.de

T. Gneiting
e-mail: t.gneiting@uni-heidelberg.de

E. Pardo-Igúzquiza et al. (eds.), *Mathematics of Planet Earth*,
Lecture Notes in Earth System Sciences, DOI: 10.1007/978-3-642-32408-6_4,

Uniformity is usually assessed either by plotting the empirical CDF of the PIT values and comparing it with the CDF of the uniform distribution (e.g. [6]), or by studying histograms of the PIT values. A flat histogram is necessary but not sufficient for the forecaster to be ideal. Nevertheless, PIT histograms are very useful to detect biases, dispersion errors and other distributional misfits, and thus form a corner-stone of forecast evaluation.

The notion of marginal calibration addresses a different facet of predictive performance; it concerns the equality of forecast climate and actual climate. Gneiting et al. [4] propose a comparison of the average predictive CDF $\bar{F}_n$ with the empirical CDF $\hat{G}_n$ of the observations:

$$\bar{F}_n(y) = \frac{1}{n}\sum_{i=1}^{n} F_i(y), \qquad \hat{G}_n(y) = \frac{1}{n}\sum_{i=1}^{n} \mathbb{1}_{\{y_i \le y\}}, \qquad y \in \mathbb{R}. \tag{1}$$

3 Proper Scoring Rules

If a more quantitative measure of predictive performance is sought, e.g. for comparing different forecasting methods, proper scoring rules are an attractive tool. A negatively oriented score $s(F, y)$ is assigned to each forecast-observation pair (F, y) and interpreted as a penalty the forecaster wishes to minimize. A scoring rule is proper if the expected value of $s(F, y)$ for an observation y drawn from G is minimized if $F = G$ and strictly proper if the minimum is unique. Proper scoring rules encourage honest and sharp forecasts; in contrast, improper scores may result in misguided scientific inferences as illustrated in [5]. The most popular and most important examples of proper scoring rules are arguably the logarithmic and the continuous ranked probability score

$$\mathrm{LogS}(f, y) = -\log f(y), \qquad \mathrm{CRPS}(F, y) = \int_{-\infty}^{\infty} \left(F(t) - \mathbb{1}_{\{t \ge y\}}\right)^2 dt, \tag{2}$$

where the former requires that a density f of F exists, and the latter can be viewed as an integral of the proper quadratic or Brier score [1] for probability forecasts of the binary event that y exceeds the respective threshold value t. For further discussion of these and other proper scoring rules see [5] and references therein.

4 Making and Evaluating Point Forecasts

In various applications, interest is still in single-valued point forecasts. These are usually assessed by means of an error measure or scoring function, such as the mean absolute error (MAE) or the mean squared error (MSE). Gneiting [3] argues that the scoring function ought to be specified a priori in order to permit the forecaster to issue an optimal point forecast by the Bayes rule, minimizing the expected loss under the forecaster's predictive distribution. Or, conversely, if the forecaster receives a directive in the form of a functional, her forecasts should be judged on the basis of a scoring function that is consistent for it. For example, if the median of the predictive distribution is used as a point forecast, it is consequent to use the MAE for forecast evaluation, while the mean links up naturally with the MSE. Similar relations also exist for quantiles and other functionals. The examples in [3] illustrate how the use of inconsistent—though commonly used—scoring functions can lead to grossly misguided inferences about predictive abilities.

5 Data Example

We illustrate the evaluating techniques reviewed above with a data example on one day ahead predictions of ozone concentrations in downtown San Francisco [2]. We first fit an $ARMA(2, 2)$ time series model to daily maximum eight-hour ozone concentration measurements between January 1, 1980 and December 31, 1999 in ppm, employing several covariates: the temperature on the forecast day, a linear function of time, a sine and a cosine function of time to capture a seasonal cycle, and a separate offset for each weekday. Keeping the fitted parameters fixed and assuming a Gaussian distribution, we then calculate one day ahead predictive distributions for daily ozone concentrations between January 1, 2000 and December 31, 2012. Now, since ozone concentration is a positive quantity, would it be more appropriate to model its logarithm instead, thus obtaining a log-normal predictive distribution? We implement and evaluate both options and draw conclusions.

Figure 1 shows PIT histograms for both normal and log-normal predictive distributions. Neither of the two is ideally suited for ozone concentrations, which seem to call for a distribution with positive excess kurtosis. The positive skew of the log-normal distribution seems inappropriate, and in spite of violating the non-negativity constraint, the normal predictions are better calibrated. This is confirmed by the consistently better (lower) scores in Table 1, where in the case of the MAE the median of the predictive distribution was used as point forecast, and in the case of the RMSE the mean. For lognormal distributions, median and mean are distinct, with the mean leading to an MAE of 0.00492 and the median to an RMSE of 0.00648. While the

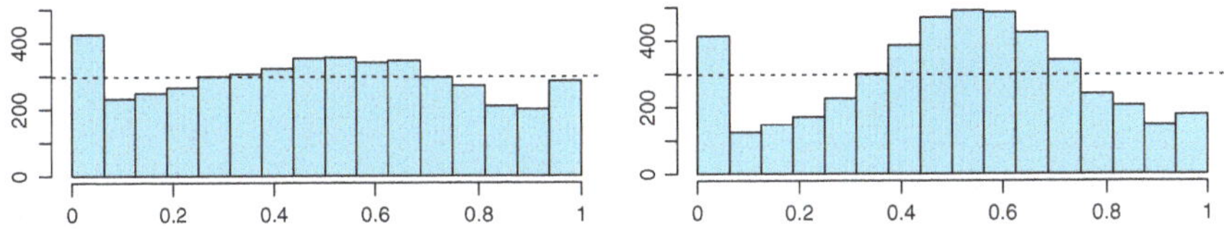

Fig. 1 PIT histograms for normal (*left*) and log-normal (*right*) predictive distribution

Table 1 Average logarithmic and continuous ranked probability score, mean absolute error, root mean squared error, average width and empirical coverage for 50 and 90 % prediction intervals

	LogS	CRPS	MAE	RMSE	50 % pred. int.		90 % pred. int.	
					Width	Coverage (%)	Width	Coverage (%)
Normal	−3.53	0.00329	0.00467	0.00622	0.0081	55.0	0.0198	86.9
Lognormal	−3.28	0.00371	0.00485	0.00650	0.0116	65.7	0.0295	88.7

increase in the MAE could be expected, the slight decrease in the RMSE appears to contradict the above guidelines for making optimal point forecasts. However, the guidelines assume calibrated predictive distributions. In the present case the unwarranted skew of the lognormal distribution might divert the mean from the optimal point forecast. Moreover, it leads to noticeably wider prediction intervals.

References

1. Brier, G. W. (1950). Verification of forecasts expressed in terms of probability. *Monthly Weather Review, 78*, 1–3.
2. Data were obtained from http://www.epa.gov/airdata/ad_data_daily.html and http://www.ncdc.noaa.gov/cdo-web/
3. Gneiting, T. (2011). Making and evaluating point forecasts. *Journal of American Statistical Association, 106*, 746–762.
4. Gneiting, T., Balabdaoui, F., & Raftery, A. E. (2007). Probabilistic forecasts, calibration and sharpness. *Journal of the Royal Statistical Society: Series B, 69*, 243–268.
5. Gneiting, T., & Raftery, A. E. (2007). Strictly proper scoring rules, prediction and estimation. *Journal of American Statistical Association, 102*, 359–378.
6. Moyeed, R. A., & Papritz, A. (2002). An empirical comparison of Kriging methods for nonlinear spatial point prediction. *Mathematical Geology, 34*, 365–386.

At the Interface Between Mathematical Geoscience and Classical Statistics

Frits Agterberg

1 Introduction

The geosciences continually benefit from the use of mathematics. An important objective of geology is to construct three-dimensional maps of the upper part of the Earth's crust with hypothetical delineation of various rock units including ore and hydrocarbon deposits. Uncertainties in predictive geology are very large but need to be quantified nevertheless. Use of geological concepts is essential in this endeavor involving a good understanding of the underlying physical and chemical processes with detailed knowledge of the ages of both rock units and processes. Classical statistics is especially important because it helps to quantify the large uncertainties geoscientists normally have to cope with. Geologists have had a long history of interaction with mathematical statisticians. During his long and distinguished career, William Krumbein regularly consulted with John Tukey. Krumbein also wrote one of the first geomathematical textbooks together with Franklin Graybill [1]. John Griffiths based much of his work on advanced statistical sampling techniques, especially as they had been developed by Ronald Fisher. Andrew Vistelius worked closely with Andrey Kolmogorov, already when he was preparing his PhD thesis in the 1940s. Felix Chayes developed modal analysis and took Karl Pearson's admonishment regarding spurious correlations that could result from closed-number systems to heart. This later led John Aitchison to develop compositional data analysis, which continues to be an important research topic within IAMG.

In turn, geoscientists have inspired statisticians to pursue new research directions. Georges Matheron introduced the idea of regionalized random variables and this has helped to found the important field of spatial statistics. Ronald Fisher became interested in statistics of directional features when a geophysics student at Cambridge University asked him for help in dealing with greatly dispersed paleomag-

F. Agterberg (✉)
Geological Survey of Canada, 601 Booth Street, OttawaK1A 0E8, Canada
e-mail: agterber@nrcan.gc.ca

E. Pardo-Igúzquiza et al. (eds.), *Mathematics of Planet Earth*,
Lecture Notes in Earth System Sciences, DOI: 10.1007/978-3-642-32408-6_5,

netic measurements. It led to the cone of confidence for unit vectors. Geoffrey Watson was instrumental in developing statistical significance tests for directional features. Geoscientists often work with very large data sets to be subjected to exploratory data analysis with use of jackknife and bootstrap techniques for uncertainty estimation. John Tukey pioneered these approaches and advised mathematical geologists to use them on many occasions. GIS and its 3-D extensions, for which credit is owed to engineers and computer scientists, are widely used in geological map modeling and image analysis. Promising new developments include projection pursuit, boosting and applications of radial basis function theory.

2 History of Statistics from a Geoscientist's Perspective

Ian Hacking [2] has pointed out that the concept of "probability", which is a cornerstone of classical statistics, emerged in the middle of the middle of the seventeenth century, gradually assuming its dual objective and subjective meanings. In its objective sense, probability is related to stable experimental frequencies. Subjective frequencies aim to quantify degrees of belief. Advocates of these two types of probabilities have often disagreed in the past. For example, Ronald Fisher strongly opposed Bayesians who used Thomas Bayes' rule to update initially subjective probabilities in an objective manner. His colleague, the geophysicist Harald Jeffreys had introduced inductive logic that was later refined by others such as Bruno de Finetti. To-day, primarily deductive and subjective reasoning both continue to be practiced. Possibility theory as developed by Didier Dubois and colleagues [3] takes a new type of axiomatic approach in uncertainty theory. In many Bayesian approaches, the starting point continues to be based on the concept of equipossibility as originally used by Pierre, Marquis de Laplace.

A simple example of equipossibility followed by deductive reasoning based on traditional axioms of probability theory is Weights-of-Evidence (WofE) modeling to estimate probabilities of occurrences of discrete events such as mineral deposits in a study area. The initial hypothesis of equipossibility is that the probability that a mineral deposit occurs within a small unit area is the same everywhere within the study area. It gives the prior probability that only depends on size of unit area. Using Bayes' rule, this prior probability is updated repeatedly by using as evidence features of the unit area that differ from place to place in the study area. The final WofE product is a map of posterior probabilities for occurrences of mineral deposits.

The geologist Georgius Agricola (in 1556) developed methods of reading signs on the surface of the Earth such as occurrences of faults or anticlines as indicators for the occurrence of mineralization. Nevertheless, as Hacking [2] points out, Agricola had no idea that there could have been a process of mineralization that took place millions of years ago. Such concepts had not yet been developed. In WofE theories of process-modeling are taken into account when map layers are selected for improving the posterior probabilities.

It is well known that Blaise Pascal around 1650 was adept in solving problems related to the rolling of one or more dice. In [2], Christiaan Huygens is credited with introducing statistical inference in the first probability textbook published in 1657. Statistical reasoning, however, became only slowly accepted by scientists and later by the public. This is evident from the history of the arithmetic mean. Early calculations that resemble the process of estimating the mean are reviewed in [4]. The first average on record was taken by William Borough in 1581 for a set of compass readings. The procedure of averaging numbers was regarded with suspicion for a long period of time. Thomas Simpson in 1755 advocated the approach in a paper entitled: "On the advantage of taking the mean of a number of observations in practical astronomy", stating: "It is well-known that the method practiced by astronomers to diminish the errors arising from the imperfections of instrument and of the organs of sense by taking the mean of several observations has not so generally received but that some persons of note have publicly maintained that one single observation, taken with due care, was as much to be relied on, as the mean of a great number." The invention of the normal distribution, originally derived from the binomial distribution by Abraham de Moivre in 1718 and becoming more widely known after its use by Friedrich Gauss (in 1809) and derivation of the central-limit theorem, helped to popularize the idea that many different random errors combine to produce errors that are normally distributed. The normal distribution became another corner stone of mathematical statistics with the development of Student's t-test, analysis of variance and the chi-square test for goodness of fit. Many methods of mathematical statistics were developed for statistical populations of independent (uncorrelated) identically distributed objects from which random samples can be drawn to estimate parameters such as the mean, variance and covariance. Generalization to multivariate analysis followed naturally. In [1], the "general linear model" was introduced as a basic tool of mathematical geology.

One of the early applications of the general linear model is trend surface analysis [1]. In the late 1960s, this technique was competing with universal kriging originally developed by Charles Huijbregts and Georges Matheron. To-day, both techniques remain in use for describing spatial trends or "drifts" in variables with a mean that changes systematically in two- or three-dimensional space. Simple moving averaging as practiced by Daniel Krige or inverse distance weighting methods can be equally effective when there are many observations.

Traditionally, the upper part of the Earth's crust has been viewed as a complex three-dimensional mosaic of numerous rock units with different compositions and ages. However, as emphasized in this paper, many geological features display random characteristics that could be modeled by adapting methods of mathematical statistics.

3 Nonlinear Modeling

A question to which new answers are being sought is: Where does the randomness in Nature come from? Nonlinear process modeling is providing new clues to answers. Benoit Mandelbrot discovered about 50 years ago that many objects on Earth can be modeled as fractals with non-Euclidean dimensions. Other scientists including Donald Turcotte have pointed out that, as in chaos theory, otherwise deterministic Earth process models can contain terms that generate purely random responses [5]. Well-known examples are the logistic equation and the van der Pol equation with solutions that contain unstable fixed points or bifurcations. Multifractals, which are spatially intertwined fractals and were anticipated by Henri de Wijs (in 1948), provide a novel way of approach to problem-solving in situations where the attributes display strongly positively skewed frequency distributions. The standard geostatistical model used in ordinary kriging assumes a semivariogram with both range and nugget effect. The range extends to distances at which results from other deterministic or random processes begin to overshadow local variability. The nugget effect often is due to relatively wide sampling between points that hides short-distance variability. The multifractal semivariogram shows sharp decrease toward zero near the origin. Local singularity mapping uses this short-distance variability to delineate places with relatively strong enrichment or depletion of element concentration on geochemical maps and in other applications [6]. This method provides a new approach for mineral exploration and regional environmental assessment. These nonlinear developments are closely related to the statistical theory of extreme events.

References

1. Krumbein, W. C., & Graybill, F. A. (1965). *An introduction to statistical models in geology*. New York: McGraw-Hill.
2. Hacking, I. (2006). *The emergence of probability* (2nd ed.). Cambridge: Cambridge University Press.
3. Dubois, D., & Prade, H. (2000). Possibility theory, probability theory and multiple-valued logics: A clarification. *Annals of Mathematics and Artificial Intelligence*, *32*, 35–66.
4. Plackett, R. L. (1970). The principle of the arithmetic mean. In E. S. Pearson & M. G. Kendall (Eds.), *Studies in the history of statistics and probability* (pp. 121–126). Darien: Hafner.
5. Turcotte, D. L. (1997). *Fractals and chaos in geology and geophysics*. (2nd ed.). Cambridge: Cambridge University Press.
6. Cheng, Q., & Agterberg, F. P. (2009). Singularity analysis of ore-mineral and toxic trace elements in stream sediments. *Computers and Geosciences*, *35*(2), 234–244.

Estimating the Number and Locations of Euler Poles

Florian Bachmann, Peter Jupp and Helmut Schaeben

1 A Model for a Single Euler Pole

In plate tectonics, the geometric evolution of plates is modelled in terms of rotations about their Euler poles, i.e. axes (unsigned unit vectors). If a point $\mathbf{r}$ on the sphere, S^2, is rotated about an axis $\pm\mathbf{p}$ then the image of $\mathbf{r}$ traces out an arc of a small circle which is tangent to $\mathbf{p} \times \mathbf{r}$ at $\mathbf{r}$, and so the axis tangent to this small circle at $\mathbf{r}$ is $\pm\mathbf{u}(\mathbf{p} \times \mathbf{r})$, where $\mathbf{u}(\mathbf{p} \times \mathbf{r}) = \|\mathbf{p} \times \mathbf{r}\|^{-1}\mathbf{p} \times \mathbf{r}$ is the unit vector in the direction of $\mathbf{p} \times \mathbf{r}$. Data on Euler poles usually take the form $(\mathbf{r}_1, \pm\mathbf{n}_1), \ldots, (\mathbf{r}_n, \pm\mathbf{n}_n)$, where $\mathbf{r}_1, \ldots, \mathbf{r}_n$ are positions on S^2, and (for $i = 1, \ldots, n$) $\pm\mathbf{n}_i$ is an axis tangent to S^2 at $\mathbf{r}_i$. The data can be viewed as regression data, in which $\mathbf{r}_1, \ldots, \mathbf{r}_n$ are values of a predictor on S^2, while $\pm\mathbf{n}_1, \ldots, \pm\mathbf{n}_n$ are responses.

It is reasonable to assume that the observed tangential axes $\pm\mathbf{n}_1, \ldots, \pm\mathbf{n}_n$ are observations on independent random axes. We propose the statistical model in which $\pm\mathbf{n}_i$ has the bipolar Watson distribution (see Sect. 9.4.2 of Mardia & Jupp [3], with $p = 2$) with mean direction $\mathbf{u}(\mathbf{p} \times \mathbf{r}_i)$ and concentration $\gamma\|\mathbf{p} \times \mathbf{r}_i\|^{2\beta}$ for some positive constants γ and β. Then the density of $\pm\mathbf{n}_i$ is

$$g(\pm\mathbf{n}_i; \pm\mathbf{p}, \beta, \gamma|\mathbf{r}_i) = \left\{e^{\kappa_i/2} I_0(\kappa_i/2)\right\}^{-1} \exp\left\{\kappa_i[\mathbf{n}_i \cdot \mathbf{u}(\mathbf{p} \times \mathbf{r}_i)]^2\right\}, \quad (1)$$

where

$$\kappa_i = \gamma\|\mathbf{p} \times \mathbf{r}_i\|^{2\beta} \quad (2)$$

F. Bachmann (✉) · H. Schaeben
Geophysics and Geoscience Informatics, TU Bergakademie Freiberg, Germany
e-mail: fl.bachmann@googlemail.com

H. Schaeben
e-mail: schaeben@mailserver.tu-freiberg.de

P. Jupp
School of Mathematics and Statistics University of St. Andrews, Glasgow, Scotland
e-mail: pej@st-andrews.ac.uk

E. Pardo-Igúzquiza et al. (eds.), *Mathematics of Planet Earth*,
Lecture Notes in Earth System Sciences, DOI: 10.1007/978-3-642-32408-6_6,

and $I_0(\cdot)$ denotes a Bessel function. Equation (1) formalises the intuitive idea that, when $\mathbf{r}_i$ is nearer $\pm\mathbf{p}$, the small circles are smaller, and so $\pm\mathbf{n}_i$ is more poorly defined.

The unknown parameters $\pm\mathbf{p}$, β, γ can be estimated by maximum likelihood. Approximate $100(1-\alpha)\%$ confidence regions for $\pm\mathbf{p}$ are

$$\left\{\pm\mathbf{p} : \mathbf{p}'\hat{I}^{+}_{\mathbf{pp}\cdot\gamma\beta}\mathbf{p} \le \chi^2_{2;\alpha}\right\} \quad \text{and} \quad \left\{\pm\mathbf{p} : \mathbf{p}'\hat{J}^{+}_{\mathbf{pp}\cdot\gamma\beta}\mathbf{p} \le \chi^2_{2;\alpha}\right\}, \tag{3}$$

where $\hat{I}^{+}_{\mathbf{pp}\cdot\gamma\beta}$ denotes the generalised inverse of the $\mathbf{p}$-part $I_{\mathbf{pp}\cdot\gamma\beta}$ of the Fisher information evaluated at the maximum likelihood estimate, $\hat{J}^{+}_{\mathbf{pp}\cdot\gamma\beta}$ denotes the generalised inverse of the $\mathbf{p}$-part $J_{\mathbf{pp}\cdot\gamma\beta}$ of the negative of the Hessian of the log-likelihood evaluated at the maximum likelihood estimate, and $\chi^2_{2;\alpha}$ denotes the upper α quantile of the χ^2_2 distribution.

The goodness of fit of the model to the data can be assessed using the residuals $\delta_1, \ldots, \delta_n$, defined by

$$\delta_i = 2 \times \{\text{anti-clockwise angle from } \pm\mathbf{u}(\hat{\mathbf{p}} \times \mathbf{r}_i) \text{to } \pm\mathbf{n}_i\},$$

where $\pm\mathbf{u}(\hat{\mathbf{p}} \times \mathbf{r}_1), \ldots, \pm\mathbf{u}(\hat{\mathbf{p}} \times \mathbf{r}_n)$ are the fitted tangential axes corresponding to observations $\pm\mathbf{n}_1, \ldots, \pm\mathbf{n}_n$. Inspection of the residuals can indicate how well the model fits the data and how it can be modified to improve the fit. Assuming that most of the estimated concentrations are large, goodness of fit can be measured using various standard statistics. Thus, at the cost of an additional assumption, corresponding tests provide the quantitative assurance of p-values.

2 A Model for Several Euler Poles

For many data sets, model (2) fits badly and there is no reason to assume that there is only one Euler pole. It may then be helpful to fit a mixture model, in which the underlying population of potential observations is a mixture of c populations, with population j (for $j = 1, \ldots, c$) making up proportion π_j of the total population and having Euler pole $\pm\mathbf{p}_j$. Then the density of $\pm\mathbf{n}_i$ given $\mathbf{r}_i$ is

$$f(\pm\mathbf{n}_i; \pm\mathbf{p}_1, \ldots, \pm\mathbf{p}_c, \pi_1, \ldots, \pi_c, \beta, \gamma|\mathbf{r}_i) = \sum_{j=1}^{c} \pi_j g(\pm\mathbf{n}_i; \pm\mathbf{p}_j, \beta, \gamma|\mathbf{r}_i),$$

where $g(\cdot)$ is given by (1) and (2).

The maximum likelihood estimates of $\pm\mathbf{p}_1, \ldots, \pm\mathbf{p}_c, \pi_1, \ldots, \pi_c, \beta, \gamma$ can be found using the EM algorithm. (See, e.g., Sect. 1.4.1 of Everitt & Hand [2], or Sect. 4.3.2 of Titterington et al. [4].) Generalisations of (3) give approximate confidence regions for the Euler poles $\pm\mathbf{p}_1, \ldots, \pm\mathbf{p}_c$ based on the Fisher information or the

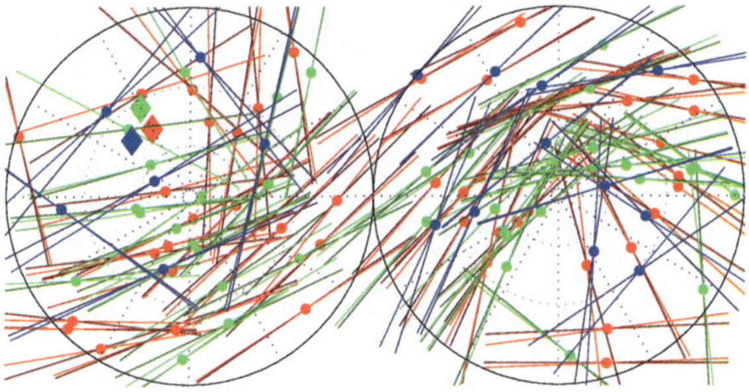

Fig. 1 Fabricated dataset with three given "true" Euler poles $\mathbf{p}_1 = (45, 120)$, $\mathbf{p}_2 = (57, 120)$, $\mathbf{p}_3 = (47, 137)$ (*diamonds*), uniformly distributed locations $\mathbf{r}_i$ (*dots*) of observations $\mathbf{n}_i$ tangential to S^2 at $\mathbf{r}_i$, and subsamples of proportions $\pi_1 = 0.47$, $\pi_2 = 0.37$, $\pi_3 = 0.17$ referring to Euler poles, respectively

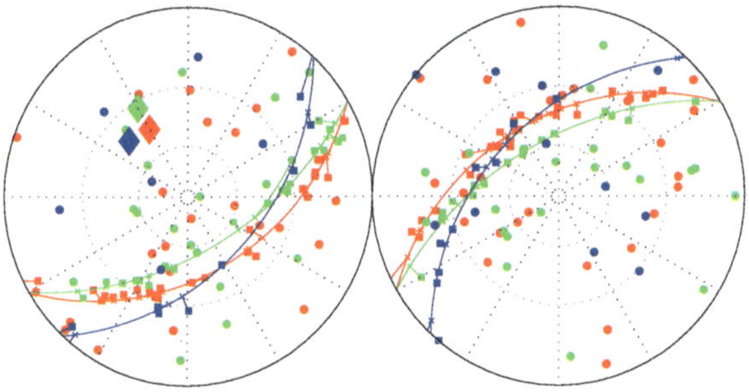

Fig. 2 Fabricated dataset with three given "true" Euler poles $\mathbf{p}_1 = (45, 120)$, $\mathbf{p}_2 = (57, 120)$, $\mathbf{p}_3 = (47, 137)$ (*diamonds*), uniformly distributed locations $\mathbf{r}_i$ (*dots*) of observations $\mathbf{n}_i$ (*squares*), and subsamples of proportions $\pi_1 = 0.47$, $\pi_2 = 0.37$, $\pi_3 = 0.17$ referring to Euler poles, respectively

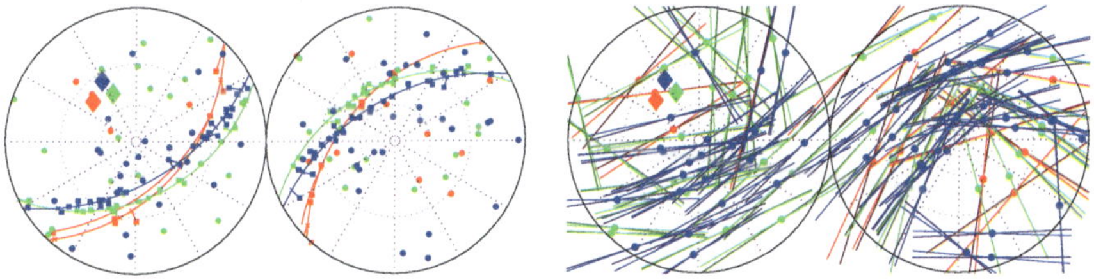

Fig. 3 Estimated Euler poles with $\hat{\pi}_1 = 0.22$, $\hat{\pi}_2 = 0.34$, $\hat{\pi}_3 = 0.45$, locations $\mathbf{r}_i$ (*dots*) of observations $\mathbf{n}_i$ and fitted normals $\widehat{\mathbf{n}}_i$ tangential to S^2 at $\mathbf{r}_i$

Hessian of the log-likelihood. Observations can be allocated to the estimated Euler poles using probabilities provided by the EM algorithm.

Testing the null hypothesis that there is a single Euler pole (i.e. $c = 1$) versus a non-trivial mixture of Euler poles (i.e. $c > 1$) can be carried out using the modified likelihood ratio test of chen et al. [1]. If the hypothesis is rejected then c can be estimated using a penalised likelihood.

3 Examples

To check our approach we apply it to samples simulated with systematically varying parameters, guided by geological experience, i.e. with varying (Figs. 1, 2, 3)

- the total number c of Euler poles $\mathbf{p}_i$, $i = 1, 2, 3$, and their relative location (close to or apart of each other),
- their corresponding proportions of the sample π_i, $i = 1, 2, 3$, for moderate sample sizes $n < 100$,
- the locations $\mathbf{r}_1, \ldots, \mathbf{r}_n$ on S^2 of the observations with respect to the Euler poles (uniform or clustered around Euler poles),
- the shape parameter of a bipolar Watson distribution to draw $\mathbf{n}_{ij} = \pm\mathbf{u}(\mathbf{p}_j \times \mathbf{r}_i)$ and generate $(\mathbf{r}_1, \pm\mathbf{n}_1), \ldots, (\mathbf{r}_n, \pm\mathbf{n}_n)$,

and compile the results. A typical example and its preliminary analysis is presented as follows.

Comprehensive results will be presented in the oral presentation and published elsewhere later.

References

1. Chen, H., Chen, J., & Kalbfleisch, J. D. (2001). A modified likelihood ratio test for homogeneity in finite mixture models. *Journal of the Royal Statistical Society: Series B*, *63*, 19–29.
2. Everitt, B. S., & Hand, D. J. (1981). *Finite mixture distributions*. London: Chapman & Hall.
3. Mardia, K. V., & Jupp, P. E. (2000). *Directional statistics*. Chichester: Wiley.
4. Titterington, D. M., Makov, U. E., & Smith, A. F. (1985). *Statistical analysis of finite mixture distributions*. Chichester: Wiley.

Testing for Microhomogeneity in Reference Materials for Microanalytical Methods

Raimon Tolosana-Delgado, Aaxel D. Renno, Przemlaw P. Michalak and K. Gerald van den Boogaart

1 Introduction

Homogeneity is a relative property of a sample in relation to the measurement (analytical method), the analyte, and its intended purpose (e.g. the usage as a reference material). In a measuring batch, some reference material samples are intercalated between samples of unknown concentration, and the readings of the analytical method for the reference are used to build a bottom line over which the unknown concentrations of the samples are determined. Of course, the reference concentration must be known and homogeneous up to a precision level given by the analytical method. Thus, the verification of homogeneity (at that precision) becomes essential to prove that a reference material is fit-for-purpose. For microanalytical methods, this becomes particularly tricky, because the reference material concentration must be determined at spots where it cannot be measured, as most measuring methods at such small scale are destructive.

This contribution proposes a solution to this problem. Consider the concentration at any spot on the reference material as a random value. Then, one can determine the whole distribution of the random concentration by taking a random sample of spots and (destructively) measuring it. Finally, making use of the standard error of the mean one can determine how many (unmeasured) spots of the reference material

R. Tolosana-Delgado (✉) · A. D. Renno (✉) · K. G. van den Boogaart
Helmholtz-Institute Freiberg for Resource Technology, Halsbrücker Strae 34, 09599 Freiberg, Germany
e-mail: tolosana@hzdr.de

A. D. Renno · P. P. Michalak
Institute for Mineralogy, Technical University Bergakademie Freiberg, Brennhausgasse 14, 09599 Freiberg, Germany
e-mail: a.renno@hzdr.de

K. G. van den Boogaart
Institute for Stochastics, Technical University Bergakademie Freiberg, Prüferstr. 9, 09599 Freiberg, Germany

E. Pardo-Igúzquiza et al. (eds.), *Mathematics of Planet Earth*, Lecture Notes in Earth System Sciences, DOI: 10.1007/978-3-642-32408-6_7,

must be visited (and burnt) in a row during a measuring batch to derive a sufficiently precise reading of the reference.

These calculations can be done in a first approximation considering that the concentration (and the sampling sequence) on the reference material is completely random, but they can be greatly improved by considering realistic spatial patterns of variation of the concentration. To show the fitness of any of these methods for providing reliable reference values one can statistically check values against further measurements, as it is needed to prove the fitness of material through its values in metrology with homogeneous materials.

2 Types of Microheterogeneity of Reference Materials

We consider four types of microheterogeneity [1–3], depending on which is the source of spatial dependence exhibited (Fig. 1):

Random: The concentration varies as a random field around an average value.

Systematic: The material presents a consistent trend of enrichment in the element.

Periodic: The material presents banding or zonation of concentration with enrichment zones surrounded by depletion zones and vice-versa; these zones may be well-structured bands of different width, or else represent growth periods of the crystal (in which case, the zones are quasi-concentric).

Nugget: A relatively homogeneous material is spotted with small highly enriched (or depleted) particles, internally homogeneous and of a totally different nature.

3 The Complete Random Case

Consider first a reference material, which average concentration on the relevant element has been adequately characterized. Its microscale variability has been characterized by a destructive method at a series of grid spots. From the obtained measurements a mean z and a standard deviation σ_z are calculated. For this contribution we consider these measurement error-free, and this mean and variance as fixed (not uncertain). Then, the property of the standard erorr of the mean ensures that the average obtained from a future sample $z_1, z_2, \ldots, z_N$ will estimate z with a precision

$$\sigma_\varepsilon = \frac{\sigma_z}{\sqrt{N}}.$$

Thus, knowing the natural material variability σ_z and the desired minimal precision level σ_ε, the number of spots that one must visit in the future to determine one single standard reaidng in a measuring batch is bounded by

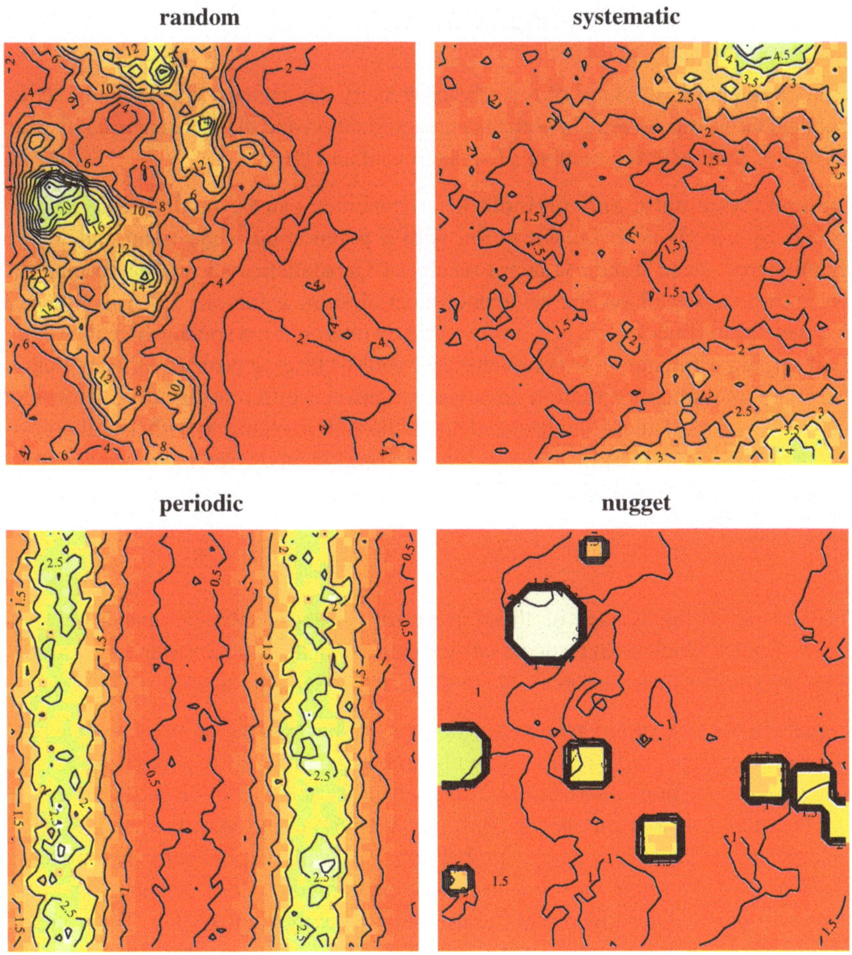

Fig. 1 Typical examples of microheterogeneity

$$N \geq \left(\frac{\sigma_z}{\sigma_\varepsilon}\right)^2. \tag{1}$$

Note that this procedure is always valid, whenever the heterogeneity value distribution of the reference material has a variance, but at the price that N will be huge if it is large. It may be thus convenient to reduce that number.

4 Improving the Sampling Strategy

In this framework, the kind of microheterogeneity determines which should be the adequate sampling strategy in both the characterization and the measuring batch step, in order to minimize the number of spots to visit and burn.

Random, systematic and periodic heterogeneities can be modelled as random fields, respectively with known mean, with a polynomial mean trend and with a trigonometric mean trend. Standard results of Geostatistics allow thus to map the concentration overall on the reference material from a sample $z_1, z_2, \ldots, z_M$ of observations at spots $x_1, x_2, \ldots, x_M$ to a grid on interpolation spots, where the vector of all kriging estimates $\hat{\mathbf{z}}_0$ is complemented with its kriging error covariance matrix $\mathbf{S}_0$ [4], i.e. including estimation covariances between different locations. Consider a selection of N of these interpolated spots through a vector $\mathbf{w}$ with 1 at the position of a selected spot and 0 at a non-selected one. Then the reference value and its uncertainty are obtained with

$$\hat{z} = \mathbf{w}^t \cdot \hat{\mathbf{z}}_0 / N, \quad \sigma_z^2 = \mathbf{w}^t \cdot \mathbf{S}_0 \cdot \mathbf{w} / N^2.$$

This variance is the average of all kriging error variances and covariances between selected spots, which must be necessarily smaller than the global variance (as covariances are smaller than variances due to the Cauchy-Schwarz theorem).

Nugget: In the case of nugget heterogeneities, one has two populations, with a quite different size: a spot falls in a nugget with quite a low probability p. To work around this problem, one must also estimate this probability at the calibration step, together with the true mean and variance. Robust methods [5] are the best choice in this case. Once mean, variance and nugget probability are available, the reference material can be used in the same way as with the complete random case, just that mean and variance must be determined with robust methods with a cutoff of p. This roughly corresponds to increasing the sample size N obtained from Eq. (1) by a factor $1/p$. In all cases, to reduce the effects of further drifts on the batch measuring methods, spots should always be randomly visited, not orderly rastered.

References

1. Danzer, K. (1995). Sampling of inhomogeneous materials statistical models and their practical relevance. *Chemia Analityczna*, *40*, 429–453.
2. Kempenaers, L., et al. (2002). A Monte Carlo model for studying the microheterogeneity of trace elements in reference materials by means of synchrotron microscopic X-ray fluorescence. *Anal. Chem.*, *79*, 5017–5026.
3. Inczédy, J. (1999). Application of the wave function model for the detection of inhomogeneity of solids in the case of periodical concentration fluctuations. *Chemia Analityczna. 44(3B)*, 591–596.

4. Chilès, J. P., & Delfiner, P. (1999). *Geostatistics, modelling spatial uncertainty*. New York: Wiley.
5. Maronna, R., Martin, R. D., & Yohai, V. J. (2006). *Robust statistics: Theory and methods*. New York: Wiley.

Recent Univariate and Multivariate Statistical Techniques Applied to Gold Exploration in the Amapari Area, Amazon Region, Brazil

Luis Paulo Braga, Francisco José da Silva and Claudio Gerheim Porto

1 Introduction

Geochemical exploration in deeply weathered terrains is often faced with the problem of strong modification of the geochemical signal at surface, making it difficult to identify primary mineralized source rocks from metal concentrations of soil samples. This problem is very pertinent to the Amapari gold prospect where 14,263 soil samples were collected in a large gridded area covering nearly 500 km^2. These samples were analysed by ICP-MS after an Aqua Regia digest producing results for 51 elements. The effect of closure, in this particular case, may have little effect on the outcome of most statistical procedures.

Subsequently, the samples were classified into five groups according to the geological domain: (i) the anomalous area (TARGET), centered around a gold deposit, with 999 samples and (ii) the non-anomalous areas GRAN1 (115 samples), GRAN2 (47 samples) and GREEN (678 samples). The NONE code was attributed to the remaining samples.

L. P. Braga (✉) · C. G. Porto
Universidade Federal do Rio de Janeiro, Rio de Janeiro, Brazil
e-mail: lpbraga@geologia.ufrj.br

C. G. Porto
e-mail: porto@geologia.ufrj.br

F. J. da Silva
Universidade Federal Rural do Rio de Janeiro, Rio de Janeiro, Brazil
e-mail: fjosilva@ufrrj.br

E. Pardo-Igúzquiza et al. (eds.), *Mathematics of Planet Earth*,
Lecture Notes in Earth System Sciences, DOI: 10.1007/978-3-642-32408-6_8,

2 Selecting Variables for the Multivariate Analysis

The EDA phase revealed that 12 out of 51 assayed metals show more than 20 % of the results below the analytical detection limit where gold can reach as high as 74 %. Within this percentage of non-detects or censored data a lot of information is hidden. Efficient procedures for dealing with censored data combine the values above the detection limit with those below it. The non-detects can be incorporated into boxplots which maintain the percentiles data structure. The percentiles are estimated using either the Kaplan-Meier or the more robust Regression on Order Statistics (ROS) method [1].

Figures 1a and b illustrate the Au and Be censored boxplots for the five selected areas. The horizontal line is drawn to represent the detection limit. Although most data are under the detection limit in Fig. 1a, it was possible to estimate the quartiles and where the lowest median is related to the GRAN1 area. In Fig. 1b most data is above the detection limit and the medians of the GREEN, NONE and TARGET areas are all very similar.

A censored boxplot was estimated for every variable and the medians of each area were compared in order to pre-select the variables for the canonical discriminant analysis. Variables which showed the highest or the lowest median value and are clearly contrasting with the medians of each other areas were selected. Following this criteria, Au was retained (Fig. 1a) and Be excluded (Fig. 1b). Based on trial and error runs of the canonical analysis (see below) elements with more than 50 % of the observations below the detection limit were also discarded.

3 Discriminant Canonical Analysis and HE Plots

Canonical discriminant analysis is a linear transformation of the response variable into a canonical space in which (a) each successive canonical variate produces maximum separation among the groups and (b) are mutually uncorrelated. The response variable is the non-metric areas of GRAN1, GRAN2, GREEN and TARGET and the covariates the concentration of the selected 29 elements, the values below the detection limits being computed as one half of these values.

The HE plot function draws a representation of the covariance ellipse for the hypothesized model terms and linear hypotheses (H). It also shows the corresponding error (E) for a response variable in a multivariate linear model. Yeo-Johnson power transformations [2] were applied to the data in order to improve normality.

The first two canonical components explain approximately 95 % of the variance and the null hypothesis were accepted. Figure 2a is the spatial representation of the canonical variates including the centroids. The HE plot diagram in Fig. 2b shows the centroids of the anomalous TARGET area as well as the non-anomalous ones. The drawn ellipse encompasses about 65 % of the observations and, in support of the model, the error circle does not contain any of the centroids.

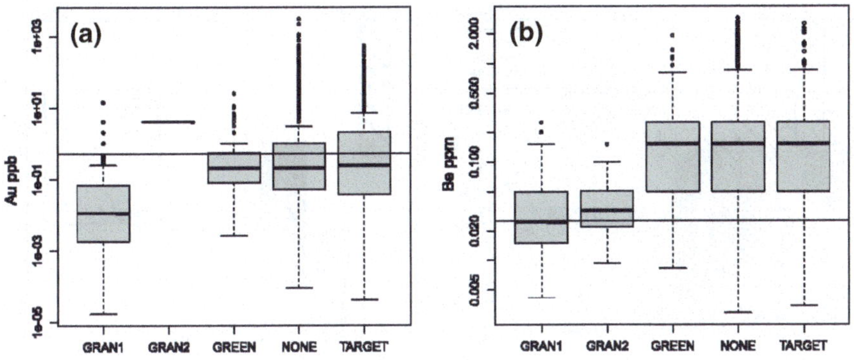

Fig. 1 **a** Censored boxplot of Au. **b** Censored boxplot of Be

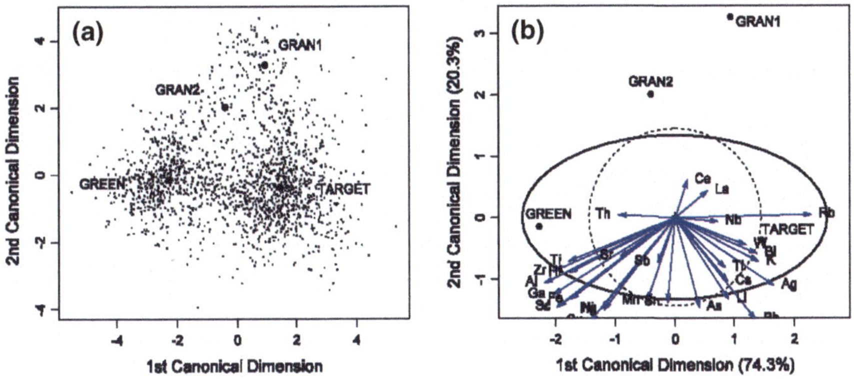

Fig. 2 **a** Samples plotted in the canonical space and corresponding centroids of the areas. **b** The ellipsoid represents the size and dimensionality of the variation in the group means regarding a spherical error ellipsoid

4 Validating the Discriminant Predictive Model

In order to validate the model, the observations close to the centroids in the canonical coordinates should be compared to their corresponding areas in the real world.

Figure 3a presents all 14,263 soil samples from where 1,839 were coded and pertain to the GRAN1, GRAN2, GREEN and TARGET areas accordingly. The TARGET area alone contains 999 samples.

Based on the distance to the centroid of the TARGET area in the canonical bi-dimension (search parameter = 3), the model was able to predict most TARGET points (~84 %) but with 179 of those falling into the other areas (Fig. 3b). The identification of non-TARGET points was about 79 % while the overall matching index of the model was approximately 82 %.

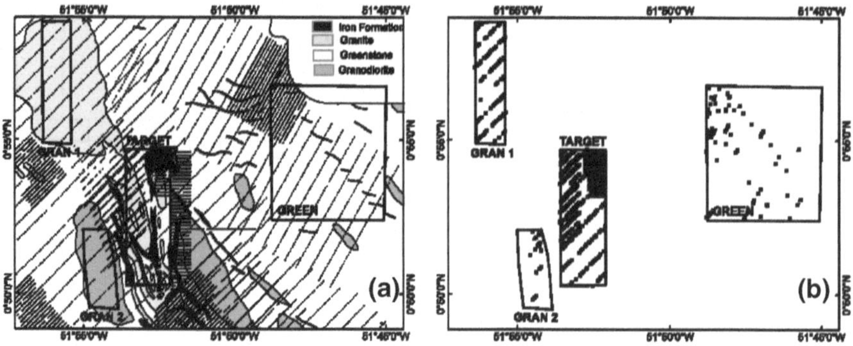

Fig. 3 **a** Geological map showing areas GRAN1, GRAN2, GREEN and TARGET plus the location of soil sampling traverses. **b** Predictive model points coded TARGET for each area

5 Conclusions

The applied method was able to provide a discriminant function to select between TARGET and non-TARGET areas. An applicable canonical search parameter obtained by trial and error, and specific to the Amapari study area, opens the possibility for exploration of new areas based on the presented methodology.

The canonical vectors (Fig. 2b) are divided into two main orthogonal groups. The southwest group (Ti, Zr, Hf, Al, Ga, Fe, Sc and V) probably indicates a soil derived from mafic rocks. Although not so evident, the southeast group (W, Bi Ag, K) is probably associated to the TARGET mineralization. The independent Ce and La vectors are related to the GRAN1 and GRAN2 rock types.

References

1. Helsel, D. R. (2005). *Nondetects and data analysis: Statistics for censored environmental data*. Hoboken: Wiley-Interscience.
2. Yeo, I. K., & Johnson, R. A. (2000). A new family of power transformations to improve normality or symmetry. *Biometrika*, *7*, 954–959.

Maximum Likelihood Inference of Spatial Covariance Parameters of Large Data Sets in Geosciences

Kanti V. Mardia and Eulogio Pardo-Igúzquiza

1 Introduction

The main advantage of maximum likelihood estimation (MLE) over other methods of estimation such as method of moments and least squares is that the MLE gives directly the estimates of the covariance parameters for a given model. This method is well suited in practice since it also provides the uncertainty of the estimated parameters. However, the main problem with MLE for large data sets is the computational burden because a large matrix must be inverted many times which makes this approach unfeasible in practice. Different methods have been proposed to approximate the exact likelihood which then provides an "approximate" MLE (AMLE). A historical background and review of AMLE as composite likelihood methodology can be found in [1] and [2]. In spatial statistics, Vecchia [3] has proposed an AMLE which will be extended below for large data sets (called Resampling AMLE method) though [4] has given an algorithm for Vecchia's AMLE which is applicable for relatively moderately large data. [5] has extended the method of Vecchia to restricted AMLE where as [6] has given the spectral AMLE to consider irregularly spaced spatial data with the concept of gridding the data. A hybrid method using large and small blocks is given in [7].

K. V. Mardia
Department of Statistics, School of Mathematics, University of Leeds, Leeds LS2 9JT, Leeds, United Kingdom
e-mail: K.V.Mardia@leeds.ac.uk

E. Pardo-Igúzquiza(✉)
Instituto Geológico y Minero de España (IGME), Geological Survey of Spain, Ríos Rosas 23, 28003 Madrid, Spain
e-mail: e.pardo@igme.es

E. Pardo-Igúzquiza et al. (eds.), *Mathematics of Planet Earth*, Lecture Notes in Earth System Sciences, DOI: 10.1007/978-3-642-32408-6_9,

2 Methodology

The AMLE of Vecchia (1988) relies on the multiplication theorem: the multivariate probability density function for the data vector $\mathbf{y} = (y_1, y_2, \ldots, y_n)$ can be written as:

$$p(\mathbf{y}) = p(y_1)\prod_{i=2}^{n} p(y_i\,|y_1, \ldots, y_{i-1}) \tag{1}$$

Using the argument that some information provided by the data will be somewhat redundant, Vecchia (1988) has suggested using the following approximation for the above conditional probability:

$$p(y_i\,|y_1, \ldots, y_{i-1}) \cong p(y_i\,|y_1, \ldots, y_m) \tag{2}$$

with $m < (i-1)$. In fact, this expression can be generalized as:

$$p(y_i\,|\mathbf{y}_{i-1}) \cong p(y_i\,|\mathbf{y}_m) \tag{3}$$

where $\mathbf{y}_{i-1} = (y_1, y_2, \ldots, y_{i-1})$ but $\mathbf{y}_m$ has now any m elements, without repetition from the set of values $(y_1, y_2, \ldots, y_{i-1})$. Thus for a fixed m, the procedure of Vecchia can be repeated many times as a type of resampling or bootstrap procedure where for each conditional term different m elements are taken without repetition from the set of values $(y_1, y_2, \ldots, y_{i-1})$. To sum up, our approach for AMLE is to take a random permutation of $\mathbf{y} = (y_1, y_2, \ldots, y_n)$ and treat this as a new time series and work unilaterally. This procedure is repeated several times (the resampling step) and then the empirical joint distribution is obtained of the estimates.

3 Results

We give a realization of a second order stationary Gaussian random field with an isotropic exponential covariance in Fig. 1 (left) while the same realization plus a second order drift $m(x, y) = 10 + 0.1x + 0.1y + 0.01x^2 + 0.01y^2 + 0.01xy$ is shown in Fig. 1 (right). The realizations are on a 200×200 grid so using the exact MLE with 40000 data points is impractical.

Figure 2 shows the results on applying our proposed resampling AMLE method for a simulated random field with no drift (Fig. 2 left) and with a second order drift (Fig. 2 right). The number of samples used here in resampling was 100. When there is no drift, the estimates are unbiased and consistent (the variance decreases with increasing m). The results are acceptable even for $m = 1$ so we can use m = 1 for virtually any data size since the procedure requires only $O(n)$ operations. On the other hand, when a drift is present, the procedure gives good estimates of drift and

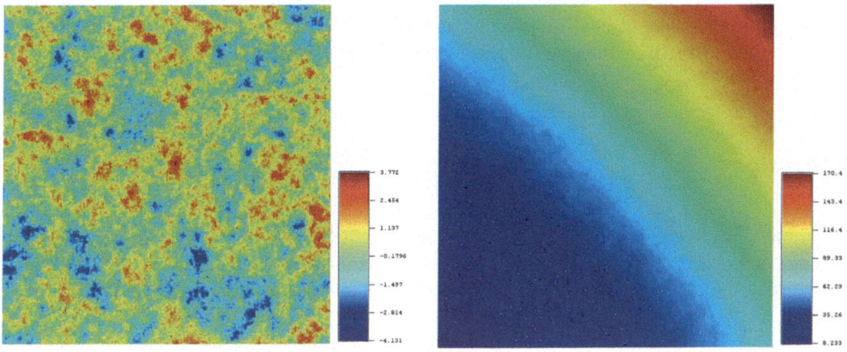

Fig. 1 *Left*: Realization of a second order stationary Gaussian random field with isotropic exponential semivariogram. *Right*: Realization of the left plus a second order drift

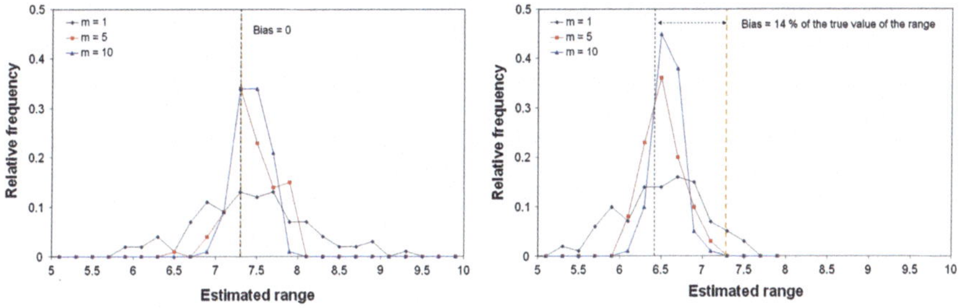

Fig. 2 Sampling distributions of the estimated range by our AMLE method for different values of *m* with (1) no drift (*left*) and (2) the second order drift (*right*)

variance but the estimate of the range is biased (with a negative bias of a 14 % of the true value of the parameter in the present example). However, the bias can be easily corrected on following a computational procedure described in [8].

4 Conclusion

Different approaches of AMLE have been proposed in the scientific literature. We have extended the approach of Vecchia through resampling; our approach can be used for any large data sets (lattice data or irregularly spaced data). Further, our approach provides some measures of uncertainty. In the drift case, the estimates are biased (as it happens with exact MLE), and so some correction procedure needs to be used, e.g., use the method of [8]. Although only one experiment has been described here, some further experiments have given similar results. In future, we intend to investigate analytical properties of our resampling AMLE method, and carry out a comparison with alternative methods.

Acknowledgments The work of the second author has been supported by the research project CGL2010-15498 from the Ministerio de Economía y Competitividad of Spain.

References

1. Mardia, V. M., Kent, J. T., Hughes, G., & Taylor, C. C. (2009). Maximum likelihood estimation using composite likelihoods for closed exponential families. *Biometrika*, *96*(4), 975–982.
2. Varin, C., Reid, N., & Firth, D. (2011). An overview of composite likelihood methods. *Statistica Sinica*, *21*, 5–42.
3. Vecchia, A. V. (1988). Estimation and model identification for continuous spatial processes. *Journal Royal Statistical Society*, *50*(2), 297–312.
4. Pardo-Igúzquiza, E., & Dowd, P. (1997). AMLE3D: A computer program for the inference of spatial covariance parameters by approximate maximum likelihood estimation. *Computers and Geosciences*, *23*, 793–805.
5. Stein, M. L., Chi, Z., & Welty, L. J. (2004). Approximating likelihoods for large spatial data sets. *Journal Royal Statistical Society*, *66*(2), 275–296.
6. Fuentes, M. (2004). Approximate likelihood for large irregularly spaced spatial data. *Journal of the American Statistical Association*, *102*(477), 321–331.
7. Caragea, P. C., Smith, R. L. (2002). Approximate likelihoods for spatial processes. *Joint Statistical Meetings—Section on Statistics and the Environment*, 385–289.
8. Dowd, P., & Pardo-Igúzquiza, E. (2012). Extensions of the parametric inference of spatial covariances by maximum likelihood. Geostatistics Oslo. (2012). *Quantitative Geology and Geostatistics 17* (pp. 129–141). New York: Springer.

A Cross-Polar Modeling Approach to Hindcast Paleo-Arctic Mega Icebergs: A Storyboard

Reinhard Furrer, Nina Kirchner and Martin Jakobsson

Plow marks, observed at present water depths shallower than $\approx$1000 m (Fig. 1a) and caused by icebergs originating possibly from a large ice shelf complex confined to the Amerasian sector of the Arctic Ocean roughly 135,000 years ago (Fig. 1b), provide clear evidence of glacial erosion at the Arctic Ocean seafloor.

The statistical approach to assess the likelihood of such paleo-arctic mega icebergs is based on estimating characteristic properties from 12 present day Antarctic ice shelves (Fig. 2), for which estimates are pooled (Fig. 3). Assuming that similar relations held for the Paleo-Arctic ice shelves, we predict the maximal draft (depth below sea level) along the calving front of the paleo ice shelf complex (Fig. 4).

We do not claim present-day Antarctic conditions to be a modern analogue for paleo-Arctic glacial conditions, yet our modeling approach provides a range of possible Arctic Ocean paleo ice shelf drafts, based on considering present dimensions of Antarctic ice shelves as well as firm-based statistical techniques. To model extreme draft values we use a peak over threshold modeling, illustrated here with the Amery Ice Shelf (69°45′S 71°0′E, Fig. 2a). The draft values are cleaned from spurious outliers and only values larger than some threshold are considered (blue line, Fig. 2b). To address the spatial dependencies in the values, we decluster the data (run length 2) and retain the cluster maxima (red dots, Fig. 2b) to which we fit a generalized Pareto distribution (GPD; red line Fig. 2c). The choice of the GPD is mathematically driven by limiting arguments of peak over threshold models. The fit is then assessed

R. Furrer (✉)
Institute of Mathematics, University of Zurich, 8057 Zurich, Switzerland
e-mail: reinhard.furrer@math.uzh.ch

N. Kirchner
Department of Physical Geography and Quaternary Geology, Stockholm University, 106 91 Stockholm, Sweden
e-mail: nina.kirchner@natgeo.su.se

M. Jakobsson
Department of Geological Sciences, Stockholm University, 106 91Stockholm, Sweden
e-mail: martin.jakobsson@geo.su.se

E. Pardo-Igúzquiza et al. (eds.), *Mathematics of Planet Earth*,
Lecture Notes in Earth System Sciences, DOI: 10.1007/978-3-642-32408-6_10,

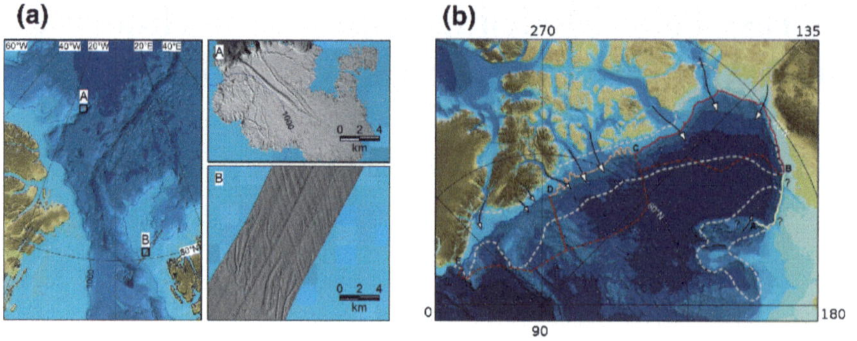

Fig. 1 **a** Iceberg scour marks on the Arctic Ocean seafloor (Moris Jessup Rise north of Greenland A and Yermak Plateau north of Svalbard B), and **b** hypothesized ice shelf complexes in the Amerasian sector of the Arctic Ocean, ca, 135,000 years ago

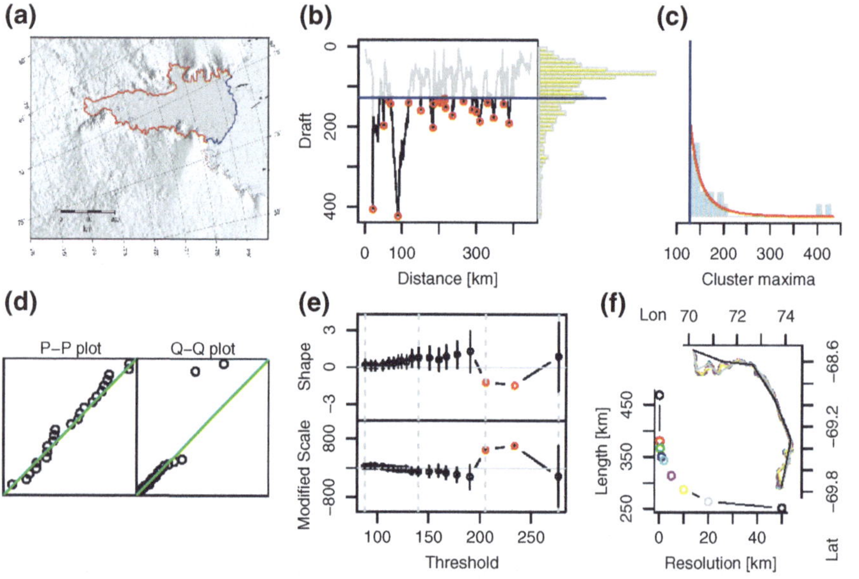

Fig. 2 Modeling extreme drafts of a contemporary Antarctic ice shelf (Amery)

and validated (P-P and Q-Q plots, Fig. 2d and parameter estimates against threshold, Fig. 2e) and the threshold may be adjusted (iteratively). Measured calving front lengths are, due to their wiggliness, roughly 1.8 times longer (black dots, Fig. 2f) compared to the sketched smooth fronts of the hypothetical paleo-arctic ice shelf complex (Fig. 1b).

We have made a similar analysis for the following Antarctic ice shelves: Brunt, Drygalski, Ekström, Filchner, Fimbul, Mertz Glacier, Riiser-Larsen, Ronne, Ross, Shackleton and West. Estimates of the scale and shape parameters (including bivariate

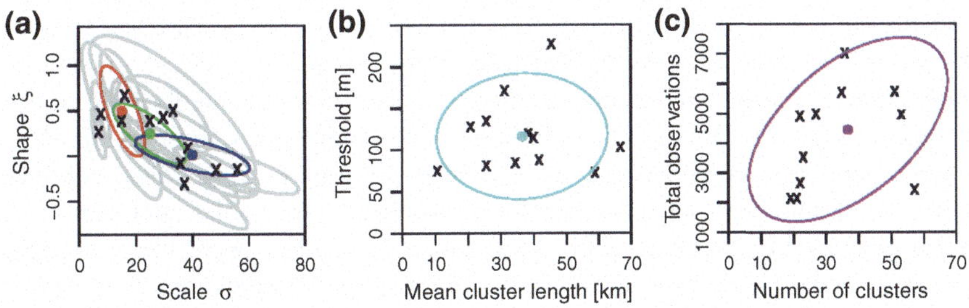

Fig. 3 Borrowing strengths: pooling information across Antarctic ice shelves

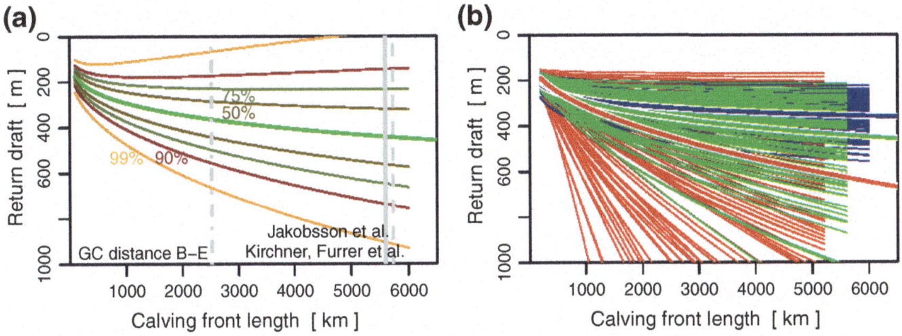

Fig. 4 Paleo-Arctic mega iceberg return drafts as a function of calving front length

uncertainty), mean cluster size (measured by the inverse of the extremal index), threshold, number of clusters and relative calving front length are given in Fig. 3.

We assume that the draft can be modeled by a stationary series for which the $D(u)$ condition [2] is satisfied. Draft prediction is based on the concept of return levels [2], where the return period is associated with length of calving front instead of time and here termed return draft. Thus, the m-kilometer return draft is expected to be exceeded on average once every m kilometers along the calving front. The return draft quantifies thus the deepest reaching icebergs to have calved off a paleo-arctic ice shelf front.

The return drafts strongly depend on the shape and scale parameters of the GPD. Therefore, the 12 estimates are classified into three groups and each is represented by an appropriate value for scale/shape and its uncertainty (colored dots and 75 % confidence ellipses based on Gaussian approximation Fig. 3a). The remaining estimates have a weaker influence and we use the mean values (dots Fig. 3b, c).

Figure 4 gives various return drafts that can be compared to calving front lengths of hypothetical ice shelves. Figure 4a gives for an "average" parameter set (green, cyan, magenta cases in Fig. 3) the return draft and pointwise symmetric confidence interval approximations which are—by nature of extrapolation—quite large. The vertical lines in Fig. 4a are front lengths from [1, 3].

Figure 4b shows return drafts of 100 parametric bootstrap samples of three different parameter settings (means and uncertainties given in Fig. 3a; color matches). Fixing shape and scale but varying threshold, mean cluster size, number of clusters, etc. (according to Fig. 3b, c) leads to tighter bounds compared to those in Fig. 4b.

Assuming large ice shelf complexes (in terms of long calving fronts) confined to the Amerasian sector of the Arctic Ocean, and the cross-polar approach to be a valid analogy [4], ice bergs with drafts larger than 800 m were likely to exist. Obviously, some parameters in the statistical (frequentist) model are heuristically determined. However, preliminary analysis suggests that reasonable priors assigned, the conclusions would be very similar in a fully Bayesian model.

Acknowledgments We greatly acknowledge the contributions of John W. Robbins and H. Jay Zwally, NASA Goddard Space Flight Center, Cryospheric Sciences Branch, Greenbelt, MD, USA. RF acknowledges funding from SNSF 129782; 143282; URPP Systems Biology and URPP GCB.

References

1. Kirchner, N., Furrer, R., Jakobsson, M., Zwally, H. J., & Robbins, J. W. (2013). Statistical modeling of a former Arctic Ocean ice shelf complex using Antarctic analogies. *Journal of Geophysical Research., Earth Surface*. doi:10.1002/jgrf.20077.
2. Coles, S. (2001). *An introduction to statistical modeling of extreme values*. New York: Springer.
3. Jakobsson, M., Nilsson, J., O'Regan, M., Backman, J., Löwemark, L., Dowdeswell, J. A., et al. (2010). An Arctic Ocean ice shelf during MIS 6 constrained by new geophysical and geological data. *Quaternary Science Reviews*, *29*, 3505–3517.
4. Mercer, J. H. (1970). A former ice sheet in the Arctic Ocean? *Palaeogeography, Palaeoclimatology, Palaeoecology*, *8*, 19–27.

Towards a Statistical Treatment of Images Acquired by Automated Mineralogy

Sandra Birtel, Raimon Tolosana Delgado, Stephan Matos Camacho, Jens Gutzmer and K. Gerald van den Boogaart

1 Introduction and Definitions

The aim of this contribution is to introduce a kind of statistical data to the mathematical geosciences community: image databases acquired by automated mineralogy. Similar to compositional data, a deeper understanding of this kind of data is necessary to develop an adequate treatment, as straight forward applications of standard tools of classical statistic are misleading. The Mineral Liberation Analyzer (MLA) provided by the company FEI is a scanning electron microscope that functions largely automated, collecting and processing data based on backscattered electron images (BSE). An electron beam scans the samples at high resolution to define regions of homogeneous grey tones (as a result of electron interaction with atoms of the sample). Each of these regions is then classified as a possible mineral, and additional energy dispersive X-ray spectroscopic (EDS) sensor allows to qualitatively identify the elemental content at the beam location. This combination allows a fast distinction of different minerals on a polished 2D surface. For further information of MLA analyses see [1, 2].

S. Birtel (✉) · R. T. Delgado · S. M. Camacho · J. Gutzmer and K. G. van den Boogaart
Helmholtz Institute Freiberg for Resources Technology, Halsbrücker Str. 34,
09599 Freiberg, Germany
e-mail: s.birtel@hzdr.de

R.T. Delgado
e-mail: r.tolosana@hzdr.de

S.M. Camacho
e-mail: s.matos@hzdr.de

J. Gutzmer
e-mail: j.gutzmer@hzdr.de

K. Gerald van den Boogaart
e-mail: boogaart@hzdr.de

E. Pardo-Igúzquiza et al. (eds.), *Mathematics of Planet Earth*,
Lecture Notes in Earth System Sciences, DOI: 10.1007/978-3-642-32408-6_11,

Two different types of samples are typically analyzed: The MLA was originally designed for liberation analysis of processing products, e.g. granular material. Here, particles are embedded into an epoxy matrix. This is called a *particle sample*. A particle is said to be *liberated* if it only consists of value minerals or waste minerals. That is relevant because only liberated particles can be correctly sorted in mechanical separation during ore processing, without wasting or diluting the value mineral. In this usage, the MLA is often used to quantify the degree of liberation, and to investigate, the reasons for the incomplete separation in the mechanical separation steps of ore processing, hence the name *liberation*. Each single observation (in a statistical sense) consists thus of an image of all particles and their identified minerals. However, measurements can be performed on samples before and after a certain sorting or processing step, or at intermediate products along a process chain. Moreover, for reasons mentioned later on, it is also often recommended to measure a single size fraction, called a *sized sample*. Thus, multiple measurements can have complex relationships among them, with several sized samples belonging together to a larger meta-observation, or several observations with external control variables.

The automated mineralogy can also be applied to undisturbed samples, to characterize the microstructure of the original ore, i.e. at the start of the processing chain, leading to a 2D image of the undisturbed ore's microstructure. These can be called *in-situ samples*. Microstructure refers to the appearance of mineral grains and relation of grains to each other, such as grain shape, grain boundaries, grain size, but not on their crystallographic orientation (i.e. microtexture). The same dataset can thus include several MLA-measurements of different data structure, different interpretations and different observation-observation relations (e.g. different size fractions of the same sample, or same size fraction along a process). Automated mineralogy generates a hierarchy of different objects: A sample consists of *particles*, i.e. physically bounded objects, which might consist of multiple minerals. A connected area of the same mineral is a *mineral grain*. An *in-situ* sample consists exclusively of non-liberated mineral grains, i.e. one particle. Mineral grains may consist of monomineralic crystals, called *(crystal) grains*. Crystal grains can consist of grains with high angle or low angle grain boundaries, the latter called *'subgrains'*. Moreover mineral grains and crystal grains of the same phase can show intra-grain and inter-grain chemical variations, through solid solutions (e.g. feldspar, sphalerite). These variations might be identified relative to each other, however the mineral chemistry cannot be quantitatively characterized, neither can single crystals within crystal grains currently be identified by MLA.

2 Stereological 2D Analysis of a Single MLA Image

Particles, mineral grains, and crystal grains are 3 dimensional objects. A mineral particle is liberated if it only contains one type of mineral (value or waste). However these 3D objects are only observed in a 2D section, considered random. A liberated particle will always appear liberated in the 2D cross section, but a non-liberated one might appear as liberated. Inferring the proportion of actual liberation (3D) from the proportion of apparent liberation (2D) is another important task of automated mineralogy.

The stereological theorem [3] allows to map several 2D portions to corresponding 3D portions, on average. This theorem directly implies that volume percentage of each mineral present in the 3D material (the *modal mineralogy*) can be estimated from their expected area percentage. Under the additional assumption of isotropy the theorem yields that the surface boundary proportions between minerals can be estimated from the length proportions of boundaries between mineral grains observed in the 2D sections. Finally, the 0-D version of the theorem provides a similar result linking the apparent surface density of triple points in 2D to the volumetric density of edges connecting three different minerals. Moreover, any kind of planar multiple point statistics can be directly estimated from its apparent counterpart.

Some of these proportions and densities must be corrected to account for the average projection area of a unit square,

$$\int_0^{\frac{pi}{2}} 2\pi \sin t \cos t \, dt = 2\pi \left[\frac{1}{2} \cos^2 t \right]_0^{\frac{\pi}{2}} = \pi.$$

Thus, to get the corresponding boundary surface density, the apparent contact length density has to be multiplied by a factor of $\frac{1}{\pi}$ in the in-situ samples; and reduced even more in particle samples by multiplying it by the particle volume portion of the grain mount, estimated by the area portion of the particles in the image. Other 2D averages typically do not correspond to interpretable 3D averages: the average portion of liberated grains, the portions of neighboring grains of given types, the average particle size, the proportions of particle sizes or grain sizes, correlations of particle sizes and type, etc. The current data analysis software available for automated mineralogy only covers some of these statistics, but none of the following issues.

3 Stochastic Inversion of the Stereological Problem

The properties relevant for processing like, e.g. the portion of particles liberated in 3 dimensions, the portion of mineral B getting into a concentrate of mineral A, the relative chance that mineral A behaves more brittle than mineral B during a specific milling process, etc. are not directly accessible by MLA or any other analytical procedure. All this information is, however, in principle qualitatively available. The challenge is to get quantitative information on such properties, which could e.g. be used to optimize mineral processing chains, to prove an improvement through a statistical test, or to predict material streams. Matos Camacho et al. [4] proposes a method to predict the actual liberation from apparent liberation of a comminuted (i.e. crushed and milled) rock.

4 Subsample Recombination

A further issue is the combination of several subsamples of different typical particle size fractions into a single observation. This can only be achieved with information on the relative masses of each fraction, i.e. with the particle size mass distribution (akin to the grain size distribution in sedimentology). However, even with that information, the recovery of 3D global properties from 2D subsample properties might not be a simple convex combination of the 3D properties inferred from each subsample, because many of them are ratios and may therefore be subject to compositional analysis rules [5].

5 Particle Tracking and Process Prediction

Finally, several samples may be related through a process, a chain of processes or a batch of alternative processes. In other words, we might have a set of MLA runs where each of them has also some control variables recorded. It is thus natural to devise explanatory models for the MLA output as a function of the explanatory variables. In this way, regression can be used with respect to a milling time or milling energy to predict the optimal milling. Or the output of several flotation treatments can be compared within a typical ANOVA scheme, to determine significant differences between treatments or perhaps interactions between controls.

References

1. Gu, Y. (2003). Automated scanning electron microscope based mineral liberation analysis. An introduction to JKMRC/FEI mineral liberation analyser. *Journal of Mineral and Materials Characterization*, 2(1), 33–41.
2. Fandrich, R., Gu, Y., Burrows, D., & Moeller, K. (2007). Modern SEM-based mineral liberation analysis. *International Journal of Mineral Processing*, *84*, 310–320.
3. John, E., & Hilliard, L. R. (2003). *Lawson, stereology and stochastic geometry (Computational Imaging and Vision)*. The Netherlands: Kluver Academic Publishers, 488 pp.
4. Matos Camacho, S., Leißner, T., Atanasova, P., Kamptner, A., Rudolph, M., Peuker, U. A., et al. (2013). *Milling result prediction*. Heidelberg: Springer.
5. Aitchison, J. (1986). *The statistical analysis of compositional data*. London: Chapman.

Distributional Assumptions and Parametric Uncertainties in the Aggregation of Geologic Resources

John H. Schuenemeyer and Ricardo A. Olea

1 Introduction

Concern over climate change has triggered modeling of the ability of porous saline formations as well as exhausted oil and gas fields to store CO2 from industrial processes in what are referred to as storage assessment units (SAUs) [1]. Most often policy considerations call for the aggregation of these resources to a basin, regional or national level. The first level of difficulty in a stochastic aggregation results from a recognition that adjacent units are rarely independent or totally dependent. A second level of difficulty results from a lack of hard data, which necessities that dependency be estimated by expert judgment. Another complexity is the fact that the probability distributions of resources made independently for each AU or SAU tend to be skewed.

2 Elicitation and Specification of Dependency

Expert judgment is used in the earth sciences in multiple ways. Among those is specifying the form of the distribution of the size of frequency of an energy or mineral resource within a play or basin. For example, the expert may know that the same level of porosity permeates adjacent SAUs. Thus if the estimate of one is biased, the other will also be biased, although not necessarily by the same amount unless the two SAUs are totally dependent with respect to porosity. An approach we used in the aggregation part of the CO2 assessment was to ask assessors independently, if say,

J. H. Schuenemeyer (✉)
Southwest Statistical Consulting, LLC, 960 Sligo St., Cortez , CO 81321, USA
e-mail: jackswsc@q.com

R. A. Olea
U. S. Geological Survey, 12201 Sunrise Valley Drive, MS 956, Reston, VA 20192, USA
e-mail: olea@usgs.gov

E. Pardo-Igúzquiza et al. (eds.), *Mathematics of Planet Earth*,
Lecture Notes in Earth System Sciences, DOI: 10.1007/978-3-642-32408-6_12,

the porosity in one SAU was misspecified, how would it affect their estimate in an adjacent SAU [2]. If the answer is not at all, then it can be assumed that the two SAUs are independent. If the answer is totally, then it can be assumed that the two SAUs are totally dependent (again with respect to porosity). Unfortunately, the answer is often somewhat. The next task it to try to have the assessor quantify "somewhat" by the use of ordinals, low, medium, and high. Subsequently correlations will be assigned to these ordinals.

It is our experience that when scientists think of correlation, it is the Pearsonian correlation. This is appropriate when the correlation pattern is generated by a bivariate elliptical distribution of which the bivariate normal is a special case. An alternative is the Spearman correlation, which is based upon rank order, and sometimes is a more appropriate measure of association for skewed distributions.

For the study of CO2 storage capacity recently completed [2], the approach to specifying dependency was via pairwise correlation coefficients initially expressed as ordinals. Of course, choosing pairwise correlation coefficients does not necessarily yield a proper correlation matrix. An iterative projection procedure [3], which minimizes a Frobenius norm using projections, made the needed minor adjustments when the specified matrix was improper. There are several alternative procedures to adjust an improper correlation matrix [4, 5].

3 Correlation and Conditional Probability

The perception of correlation for a skewed bivariate distribution is different and less meaningful than for an elliptical distribution. The correlations for bivariate normal and lognormal distributions are shown in Fig. 1a and b respectively. The correlation from Gumbel copulas for normal and lognormal marginals is shown in Fig. 1c and d respectively. A person viewing the above diagrams will recognize the association between the variables in Fig. 1a as being moderate to high, while it is much more difficult to make the same judgment for the other (Fig. 1b, c, and d) scatter plots. *Non-elliptically shaped bivariate distributions do not easily lend themselves to visual interpretation of correlations.*

There are also other issues associated with the relationship between marginals and their joint distributions. Embrechts et al. [6], show that random variates from two distributions with identical marginals can have different joint distributions and patterns of dependency with the same correlation. Notice that the dependency in the bivariate lognormal distribution (Fig. 1b) and Gumbel copula with lognormal marginals (Fig. 1d) is mostly in the left tails (small values). The dependency in the bivariate Gumbel with normal marginals (Fig. 1c) is stronger in the right tails. Thus, if the dependence is correctly specified and the bivariate distribution is Gumbel, then a large value in say variable V1G (Fig. 1c) would be accompanied by a large value of V2G. For the lognormal distributions (Fig. 1b and d), the dependency occurs for small values, and little if any inference could be made about the association between large values from V1L and V2L or V1GLN and V2GLN. In some sense then the

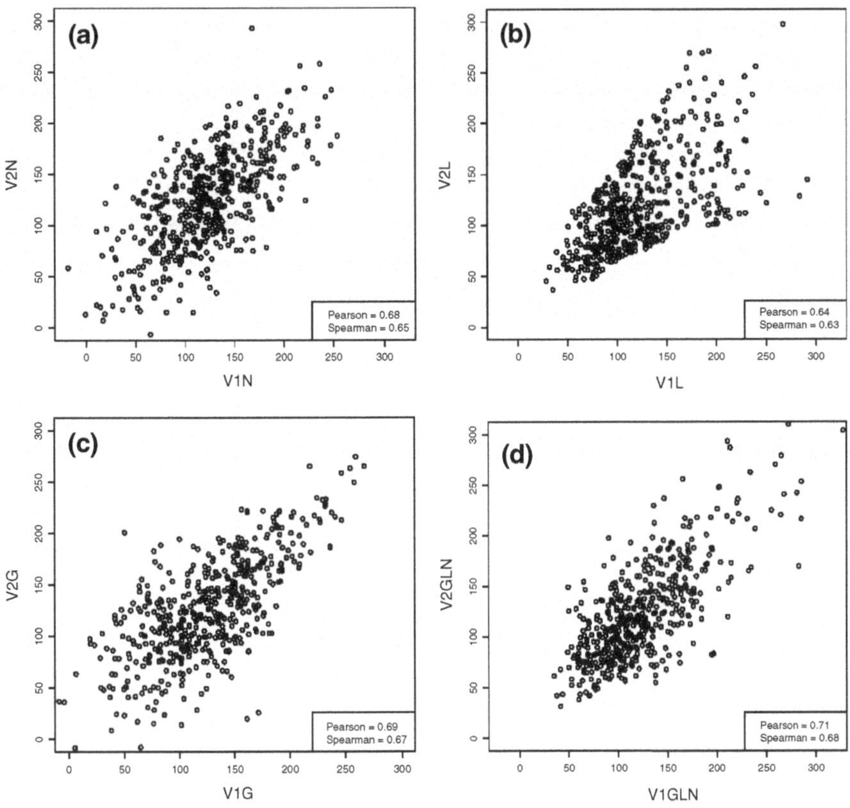

Fig. 1 Correlations: **a** Bivariate normal; **b** bivariate lognormal; **c** bivariate Gumbel copula with normal marginal; **d** bivariate Gumbel copula with lognormal marginals, all with $\rho = 0.7$ and $n = 500$. The variables identify the associated marginal distributions

bivariate normal distribution is a compromise between dependencies in the right and left tails. (See Chen et al. [7]) for a discussion of copulas in petroleum resource assessment.

4 Correlation for Lognormal Versus Normal Distributions

Consider *g*(*X*,*Y*) a bivariate normal distribution and arbitrary transformations T, U: $R \to R$, then Kendall et al. [8] showed that $\left|\rho(T(X), \tilde{T}(Y))\right| \leq \rho(X, Y)$. Thomopoulos and Johnson [9] discuss this relationship as it applies to the bivariate lognormal distribution. The specific relationships between correlations for the bivariate lognormal and bivariate normal vary as a function of the bivariate normal standard deviations, however, the inequality shown above is true for all. As an example, if the two standard deviations of the bivariate normal were both one and the corresponding correlation was 0.8, the bivariate lognormal correlation would be 0.71. So that the correlation supplied by an assessor with respect to a bivariate normal distribution would serve as an upper bound on other distributions.

An obvious conclusion is that correlations can be misleading when dealing with non-normal distributions. An important point made by Kat [10] is that for non elliptical distribution, the range may not be between −1 and +1 but some much narrower interval.

Acknowledgments This research was partly supported by the U.S. Geological Survey. Suggestions of Madalyn Blondes and Emil Attanasi are greatly appreciated.

References

1. U.S. Geological Survey Geologic Carbon Dioxide Storage Resources Assessment Team (2013). National assessment of the geologic carbon dioxide storage resources—Results. U.S. Geological Survey Circular 1386.
2. Blondes, M. S., Schuenemeyer, J. H., Olea, R. A., & Drew, L. J. (2013). Aggregation of carbon dioxide sequestrations storage assessment units, stochastic rnvironmental research and risk assessment. DOI 10.1007/s00477-013-0718 (2013).
3. Higham, N. J. (2002). Computing the nearest correlation matrix–a problem from finance. *IMA Journal of Numerical Analysis*, *22*(3), 329–343.
4. Frigessi, A., Løland, A., Pievatolo, A., & Ruggeri, F. (2010). Statistical rehabilitation of improper correlation matrices. *Quantitative Finance*, *11*(7), 1081–1090.
5. Kaufman, G. M., & Schuenemeyer, J. H. (2012). Petroleum assessment via hierarchical modeling. AAPG Search and Discovery article 40558. Accessed December 7, 2010, from http://www.searchanddiscovery.com/documents/2010/40558kaufman/ndx_kaufman.pdf
6. Embrechts, P., McNeil, A., & Straumann, D. (2002). Correlation and dependence in risk management: Properties and pitfalls. In M. A. H. Dempster (ed.), *Risk management: Value at risk and beyond* (pp. 176–223) Cambridge: Cambridge University Press.
7. Chen, Z., Osadetz, K. G., Dixon, J., & Dietrich, J. (2012). Using copulas for implementation of variable dependencies in petroleum resource assessment: Example from beaufort-mackenzie basin, Canada. *AAPG Bulletin*, *96*(3): 439–457.
8. Kendall, M. G., Stuart, A., & Ord, J. K. (1979). The Advanced Theory of Statistics (748 p.). New York: Griffin.
9. Thomopoulos, N. T., & Johnson, A. C. (2004). *Some measures on the standard bivariate lognormal distribution*. Boston: Decision Sciences Institute Proceedings.
10. Kat, H. M. (2002). *The dangers of using correlation to measure dependence*. (pp. 2002–2023). Finance: ISMA Center Discussion Papers in.

Modeling Extremal Dependence Using Copulas. Application to Rainfall Data

María I. Ortego, Juan J. Egozcue and Raimon Tolosana-Delgado

1 Introduction

Describing dependence between two or more variables is a key problem in applications. In this work, the dependence between daily rainfall in two nearby locations (Vergel de Recons and Simat de Valldigna, Spain) is analysed. The dataset consists of 30 years daily rainfall record [2, 7]. For each location, log-precipitation over a high enough threshold (Fig. 1) has been modeled using a Generalized Pareto Distribution (GPD) [1]. The dependence between these rainfall variables can be described through copulas. A new family of copulas (CrEnC copulas) is introduced and applied to the dataset.

The GPD provides a suitable model for excesses of X over a high enough threshold x_0, $Y = X - x_0$ given $X > x_0$ [6]. The GPD cumulative probability function is

$$F_Y(y|\beta,\xi) = 1 - \left(1 + \frac{\xi}{\beta} y\right)^{-\frac{1}{\xi}}, \quad 0 \leq y < y_{sup} \tag{1}$$

M. I. Ortego (✉)· J. J. Egozcue
Departament de Matemàtica Aplicada III, Universitat Politècnica de Catalunya, C/ Jordi Girona, 1–3. Campus Nord UPC. Edifici C2, Barcelona, Spain
e-mail: maria.isabel.ortego@upc.edu

J. J. Egozcue
e-mail: juan.jose.egozcue@upc.edu

R. Tolosana-Delgado
Departament of Modelling and Evaluation, Helmholtz-Institut Freiberg fr Ressourcentechnologie, Halsbrcker Strae 34, Freiberg, Germany
e-mail: r.tolosana@hzdr.de

R. Tolosana-Delagado
LIM-UPC, Barcelona, Spain

E. Pardo-Igúzquiza et al. (eds.), *Mathematics of Planet Earth*,
Lecture Notes in Earth System Sciences, DOI: 10.1007/978-3-642-32408-6_13,

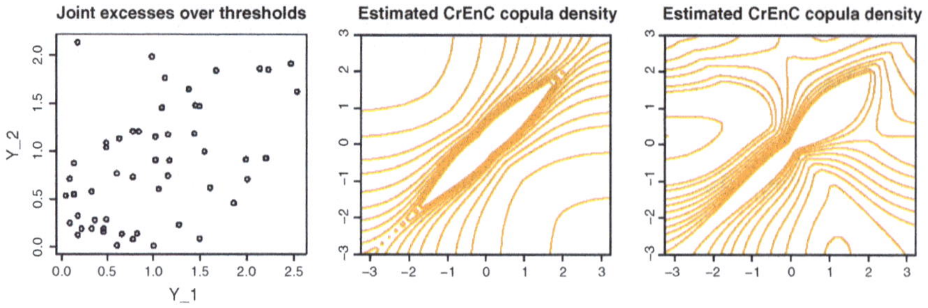

Fig. 1 Joint excesses of log-rainfall over the selected threshold ($\mathbb{R}^{+2}$ scale) (*left figure*). Contours of the estimated CrEnC copula density for a value of the sample of the posteriori distribution of parameters ($\mathbb{R}^2$ scale) (*center figure*). Contours of the geometric average of the densities of the estimated CrEnC copula density for the sample of the posteriori distribution of parameters ($\mathbb{R}^2$ scale) (*right figure*)

For $\xi = 0$, the cdf (1) is an exponential model. For $\xi < 0$ the support has the upper limit $y_{sup} = -\beta/\xi$ (Weibull domain of attraction).

Copula distributions are a useful tool to describe dependence between two or more random variables [4]. Copula functions join univariate cumulative distribution functions to obtain their multivariate CDF. The use of copula functions allows to treat separately the marginal distributions of the variables and the dependence structure among them. Sklar's theorem [10], ensures that under continuity hypothesis, the bivariate distribution between two variables can be expressed in a unique manner as a combination of the marginal distributions and their copula. The most popular version of Sklar's Theorem appears in [9]. This version of the theorem is expressed in terms of random variables.

One of the most useful properties of copulas is their invariance for strictly monotonous transformations of the random variables: the copula captures the form in which Z_1, Z_2 are related, independently of the scale in which each random variable is measured. Copulas capture nonparametric aspects of the relationship between variables, and therefore, association measures are properties of the copula.

2 CrEnC Copulas

Let Z_1, Z_2 be two random variables with joint density function $f_{12}(z_1, z_2)$. Let $\vartheta = \{\vartheta_1, \dots, \vartheta_J\}$ be a set of joint moments of these variables, with the conditions of [8] and invariant for monotonous transformations of the variables [9]. If ϑ_{Z_1,Z_2} is one of these moments and α and β are strictly monotonous functions (a.s.) over $\text{Sup}(Z_1)$ and $\text{Sup}(Z_2)$, respectively, then $\vartheta_{\alpha(Z_1)\beta(Z_2)} = \vartheta_{Z_1,Z_2}$.

The mutual information of two r.v. Z_1, Z_2 is based on the Kullback-Leibler divergence [3]. It measures de distance between both variables in the sense of information. This measure is independent from the marginal CDFs.

Given two absolutely continous r.v. Z_1,Z_2, with marginal densities $\varphi_1(z_1)$, $\varphi_2(z_2)$, the joint density of Z_1,Z_2 of minimum cross-entropy, given marginal densities $\varphi_1(z_1)$, $\varphi_2(z_2)$ and a given set of invariant for monotonous transformations of the variables moments ϑ has the form [8][Th. 3]:

$$f_{12}(z_1, z_2) = \widetilde{a}_\alpha(z_1)\widetilde{b}_\alpha(z_2) \exp\left[\sum_{k=1}^{J} \alpha_k T_k(z_1, z_2)\right], \quad (z_1, z_2) \in \mathbb{R}^2, \qquad (2)$$

where $\{T_k(z_1, z_2), k = 1, \dots, J\}$ are the corresponding statistics for these joint moments, i.e., $T_k(z_1, z_2)$ is the function that

$$\int\int_{R^2} T(z_1, z_2) f_{12}(z_1, z_2|\theta) dz_1 dz_2 = \vartheta_k.$$

Definition 1 (CrEnC Copula) The copula of minimum cross-entropy, given uniform marginal densities and a given set of invariant for monotonous transformations of the variables moments ϑ has the form:

$$C_{12}(c_1, c_2) = \widetilde{a}_\alpha(u_1)\widetilde{b}_\alpha(u_2) \exp\left[\sum_{k=1}^{J} \alpha_k T_k(u_1, u_2)\right], \quad (u_1, u_2) \in [0, 1]^2. \qquad (3)$$

3 Results

The main goal is to model dependence between log-rainfall in two locations (X_1, X_2) through a CrEnC copula. Excesses of log-rainfall over a suitable high enough threshold (Y_1, Y_2) are modelled using GPD variables, with marginal densities $\text{gpd}_1(y_1)$, $\text{gpd}_2(y_2)$. The joint density of minimum cross-entropy of (Y_1, Y_2) given a set of invariant for monotonous transformations of the variables moments ϑ and the marginal densities has the form of Eq. (2), and their CrEnC copula has the form of Eq. (3). As variables with support in [0, 1] often have a relative scale [5], both variables are transformed into $\mathbb{R}^2$ through a probit transformation: $Z_i = \Phi^{-1}(Y_i)$, $i = 1, 2$ where $\Phi(\cdot)$ is the CDF of the $N(0, 1)$ distribution. In this way, the suitable scale for data is taken into account, and due to the invariance property of copulas and the selected moments, the expression of the joint density of (Z_1, Z_2) of minimum cross-entropy, given the set of moments ϑ and the normal marginals has the form of Eq. (2).

The parameters of the marginal GPD distributions (ξ, β) and parameters α_k, $k = 1, \dots, J$ of CrEnC copulas have been estimated using a Bayesian Gibbs sampler. The estimated density for a set of simulated values of the posteriori distribution of parameters are shown in Fig. 1. Figure 1 shows a geometric average of the densities of the posteriori sample. CrEnC copulas are flexible and can describe different types of dependence.

Acknowledgments This research has received funding from the Spanish Government, projects COVARIANCE (CTM2010-19709), CODA-RSS (MTM2009-13272) and Metrics (MTM2012-33236).

References

1. Egozcue, J. J., Pawlowsky-Glahn, V., Ortego, M. I., & Tolosana Delgado, R. (2006). The effect of scale in daily precipitation hazard assessment. *Natural Hazards and Earth Systems Sciences*, *6*, 459–470.
2. Egozcue, J. J., & Ramis, C. (2001). Bayesian hazard analysis of heavy precipitation in eastern Spain. *International Journal of Climatology*, *21*, 1263–1279.
3. Kullback, S., & Leibler, R. A. (1951). On information and sufficiency. *The Annals of Mathematical Statistics*, *22*(1), 79–86.
4. Nelsen, R. B. (1999). *An introduction to copulas*. New York: Springer.
5. Pawlowsky-Glahn, V., & Egozcue, J. J. (2001). Geometric approach to statistical analysis on the simplex. *Stochastic Environmental Research and Risk Assessment*, *15*(5), 384–398.
6. Pickands, J., III. (1975). Statistical inference using extreme order statistics. *Annals of Statistics*, *3*(1), 119–131.
7. Romero, R., Guijarro J. A., Ramis, C., & Alonso, S. (1998). A 30-years (1964–93) daily rainfall data base for the Spanish Mediterranean regions: First exploratory study. *International Journal of Climatology*, *18*,541–560.
8. Rumsey, H, Jr, & Posner, E. C. (1965). Joint distributions with prescribed moments. *The Annals of Mathematical Statistics*, *1*(36), 286–298.
9. Schweizer, B., & Wolff, E. F. (1981). On nonparametric measures of dependence for random variables. *Annals of Statistics*, *9*(4), 870–885.
10. Sklar, A. (1959). Fonctions de répartition à n dimensions et leurs marges. *Publications of the Institute of Statistics of the University of Paris*, *8*(1), 229–231.

Revisiting "Estimating and Choosing"

Jean Serra

1 Introduction

For most of the people who immerge themselves in Geostatistics, the work of Georges Matheron gives the impression of a huge mathematical cathedral. Its large dome rests on dense mathematical formulae, rarely followed by experiments, but often associated with nice approximations which make them tractable.

Then the reader puts Matheron's book back, and thanks the geostatistical softwares which take all that staff in hands, and deliver well kriged maps or sound estimations.

However, when, 25 years after its publication, we reopen "Estimating and Choosing" [1], it is a quite different personality which emerges from the reading. It is a physicist who reveals himself, and neither a mathematician nor a geologist. It is a man who strongly takes in each hand one extremity of a long chain, and tries to discover what lies in the middle. The phenomena he studies are so erratic that it becomes imperative to have recourse to probability, but on the other hand each geological phenomenon, each ore body is unique and will never be reproduced in the future. How to conciliate both demands?

A physical concept is a formal relation between observable variables, called regionals in [1]. It ranges from the simplest linear regression to the most sophisticated minimizations. The three flashes presented below aim to illustrate G. Matheron's physical approach.

2 The Range

Intuitively, the range a of the covariance $C(h)$ of a stationary R. F. is the distance after which correlations become negligible. Theoretically, the fluctuation variance of the local mean $\overline{z}$ inside the domain S is:

J. Serra (✉)
Laboratoire d'Informatique Gaspard Monge, A3SI, ESIEE, Université Paris-Est, Créteil, France
e-mail: j.serra@esiee.fr

E. Pardo-Igúzquiza et al. (eds.), *Mathematics of Planet Earth*,
Lecture Notes in Earth System Sciences, DOI: 10.1007/978-3-642-32408-6_14,

$$Var(\overline{z} - m) = \frac{1}{S^2} \int C(h)K(h)dh = \frac{1}{S^2} \int C(h)[K(0) + hK'(0) + \varepsilon]dh$$

where $K(h)$ stands for the geometric covariogram of the domain S. To say that the covariance becomes negligible comes back to replacing the covariogram by its first term $Var(\overline{z} - m) \simeq \frac{1}{S} \int C(h)dh$. By introducing then the range

$$A = \int C(h)dh, \tag{1}$$

we obtain the very simple form

$$Var(\overline{z} - m) \simeq \frac{1}{S} \int C(h)dh = \frac{A}{S} \tag{2}$$

where A appears as the intrinsic scale of the regionalized variable.

A problem arises here. By using an integral over the whole space in Relation (1), we are appealing to characteristics of the model which do not correspond to any objective properties of the real phenomenon. Different covariance may be practically identical inside S, but lead to completely different values of the range A, or even to an infinite value.

Matheron overcomes the objection by carrying out an operational reconstruction of the concept. Divide mentally the domain S into a certain number of elements s of approximately equal size. It is then easy to define numerically the variance $v(s/S)$ of the finite population of mean values of the REV z over each of the elements s whose union is equal to S. Associated with the regional $v(s/S)$ there is as usual, in the model, a random variable $V(s/S)$ whose expectation $E[V(s/S)] = \sigma^2(s/S)$ may be called the theoretical variance of s in S. The latter quantity is a parameter of our stationary model, which can be easily expressed in terms of the covariance alone, from Eq. (2):

$$\sigma^2(s/S) = A\left(\frac{1}{s} - \frac{1}{S}\right)$$

Therefore, if the model is good, we should expect to observe a relationship of the form $v(s/S) = A(\frac{1}{s} - \frac{1}{S})$. This relationship is a physical law since $v(s/S)$ is a regional magnitude. Since we can experimentally check (after the fact) whether beyond some value of s this law is actually valid, we have succeeded in re-constructing the concept of integral range $\sigma^2 A$ in operational terms. It is now the physical law $v(s/S)$ that constitutes itself the operational definition of the concept.

3 Covariance or Variogram?

Certain physical phenomena can exhibit a quasi infinite range of fluctuations, and present a variance which increases without limit with the size of the domain of experiments (ore deposits, rains...). Then the estimated variance $[\sigma^2_{0/Z}]^*$ of a point inside the domain Z increases indefinitely with Z, at any scale of analysis. This means that the stationary model is just inadequate: $\sigma^2_{0/\infty} = C(0)$ being infinite, the covariance no longer exists. However, the increments of f may still exist and have a meaning. By assuming them stationary, we can study their variances, also called *variogram* $\gamma(h)$

$$\gamma(h) = E[f(x) - f(x+h)]^2$$

Clearly, we always have $\gamma(0) = 0$, and

$$\sigma^2 < \infty \Leftrightarrow \gamma(h) = 2[C(0) - C(h)]$$

When $\sigma^2 = \infty$, although the covariance vanishes, the variogram remains defined and new behaviors at infinity appear. Of course, one could argue that numerical data are always finite! We will answer the objection by means of a very suggestive model that lends itself to easy calculations, namely the Poisson steps. Start from a Poisson point process of variance λ in $\mathbb{R}^1$, and place a jump of amplitude δ at each Poisson point. The random variable δ follows a law of mean 0 and of variance σ^2. In such a model the increments only are defined, their average is equal to zero and their variogram to

$$\gamma(h) = \lambda\sigma^2 h.$$

When the data are known along a finite segment of length L, the variogram is estimated from experimental data by the relationship

$$\gamma^*(h) = 1/2(L-h)\int_0^{L-h} [f(x+h) - f(x)]^2 dx$$

where the right member admits for expectation $E[\gamma^*(h)] = \gamma(h) = \lambda\sigma^2 h$. But whereas the mean of the phenomenon does not exists, it is always possible to put $f_L = 1/L \int_0^L f(x)dx$, and to introduce the pseudo-covariance

$$Cov^*(h) = 1/(L-h)\int_0^{L-h} (f(x+h) - f_L)(f(x) - f_L)dx.$$

Such a pseudo-covariance admits for expectation

$$E[Cov^*(h)] = \lambda\sigma^2(L/3 - 4/3.h + 2/3.h^2/L) \qquad (0 \le h \le L)$$

where the actual phenomenon (behaviors at the origin and at infinity, parabolic increasingness instead of a linear one) is completely distorted. The physicist, here, will begin by checking whether the law of Eq. (2) is satisfied, and when not, he will adopt the variogram approach.

4 The Stationarity

In geostatistics, all local models involve stationarity, for the variable itself, or for its increments, or for the residuals of order k. But what can mean stationarity. in modelling a phenomenon which is not only unique, but moreover always bounded? Here again, G. Matheron builds an operational reconstruction of physicist.

Consider for example, the linear estimate $z^*(x)$ of the REV z at point x of the type

$$z^*(x) = \sum_i \lambda^i z(x + h_i) \tag{3}$$

We have chosen a translation-invariant algorithm: the weights that appear in Eq. (3) may depend on the vectors h_i of a moving neighbourhood V but not on the point x to be estimated. We will interpret the algorithm by choosing a moving neighbourhood V, a domain S_0 of the REV z and introducing the RF

$$Z(h) = z(\underline{x} + h)$$

where x is the random point obtained by equipping S_0 with the uniform probability law. If the REV is is identically zero outside S_0, then the RF Z is stationary over V, with the covariance

$$C(h, h') = \frac{1}{S_0} \int z(x) z(x + h - h') dx$$

By so doing, we make stationarity. a (restrictive) characteristic, chosen by us, of the class of estimators to be used rather than considering it as a property of physical reality, the REV, or an a priori characteristic of a RF model. The decision to only use estimators that have a local character and are associated with translation-invariant algorithms does not in any way presuppose either the stationarity. or the homogeneity of the phenomenon in space. The decision appear rather judicious if the phenomenon is actually homogeneous in space, but it will not by any means appear absurd if the phenomenon is heterogeneous.

Reference

1. Matheron, G. (1989). *Estimating and choosing: An essay on probability in practice*. Berlin: Springer.

Pragmatic Bayesian Kriging for Non-Stationary and Moderately Non-Gaussian Data

Konstantin Krivoruchko and Alexander Gribov

1 Introduction

Development of reliable real time automatic statistical interpolation model is a common issue in the GIS community. In the Geostatistical Analyst, we regularly update and improve the semivariogram model fitting algorithm and the most recent version of the software performs much better than the first one released in 2001 and described in [1]. However, there is little hope that the default semivariogram model will be close to optimal because in practice the observations may include erroneous values, measurement errors can be large and vary across space, the data are often contaminated by trend, and the data are rarely Gaussian.

In practice, interactive data exploration and modeling allows the researcher to find a reliable geostatistical model when the number of samples is relatively small, usually less than a thousand. However, a large number of GIS users prefer non-interactive interpolation for both small and very large datasets.

In the case of nearly perfect data (several hundred samples of nearly stationary Gaussian data without errors), a known solution for the automatic data interpolation problem is fully Bayesian kriging with non-informative or objective prior distributions for the parameters, although there are still some problems, including large computational time and understanding of the meaning of the prior distributions.

Since there are many Geostatistical Analyst users who asked for automatic interpolation, our goal became the development of a series of semi-automatic empirical Bayesian kriging models that work reasonably fast; do not require specification of the prior distributions for the model parameters; produce reliable outputs with default parameters; allow moderate local and large global data non-stationarity; allow for

K. Krivoruchko (✉) · A. Gribov
Environmental Systems Research Institute, 380 New York St, 92373 Redlands, CA, USA
e-mail: kkrivoruchko@esri.com

A. Gribov
e-mail: agribov@esri.com

E. Pardo-Igúzquiza et al. (eds.), *Mathematics of Planet Earth*,
Lecture Notes in Earth System Sciences, DOI: 10.1007/978-3-642-32408-6_15,

varying measurement error; locally transform data to Gaussian distribution, if needed; use explanatory variables; work with counts and proportions collected in points or polygons; simulate (transformed) Gaussian, binomial, and negative binomial fields; and potentially can be used on supercomputers or clusters.

A non-technical description of Geostatistical Analyst's empirical Bayesian kriging models can be found in [2] and an article with mathematical details is under review. In this paper we briefly describe our basic models and illustrate their usage using 1.35 billion samples collected using LiDAR technology.

2 Empirical Bayesian Intrinsic Random Function Kriging and Kriging with Local Data Transformation

There are several model candidates for semi-automatic interpolation, including objective Bayesian kriging [3], low-rank kriging [4], approximations by Gaussian Markov random fields [5], and Bayesian bootstrap [6]. Although all these models are valuable, we found that they cannot be safely used in general purpose commercial software for reasons specific to each model.

We decided to work on the Bayesian bootstrap model enhancements and generalization. The authors of [6] use informative priors simulated from the data by simple kriging with estimated covariance model. It was assumed that the data are normally distributed and the mean value is constant and independent from the covariance model parameters. The main differences between our implementation and the model discussed in [6] are the following:

- Our calculations are done locally on spatially-contiguous subsets, possibly overlapping, and the simulations from each subset are mixed together to capture the local spatial structure. The authors of [6] do calculations globally (on the entire dataset).
- All model parameters are estimated simultaneously.
- After each simulation, we do a Bayesian model update while the authors of [6] keeps simulating from the same model without updating. The difference between prior and posterior semivariograms is usually large as shown in Fig. 1.
- We support intrinsic random function kriging (IRFK) of order zero and one and also simple kriging with or without first order trend removal. The authors of [6] considers simple kriging without detrending. We use the following generalized covariance models: linear, power, thin plate spline (with IRFK) and the K-Bessel covariance model (with simple kriging.)
- The authors of [6] supports Box-Cox transformation. We support a much more flexible data transformation described in [7].
- Our software is interactive (the user can see a large number of graphs at the specified locations, including those that are shown in Figs. 1 and 2) and it works very fast due to large effort spent on algorithms optimization.

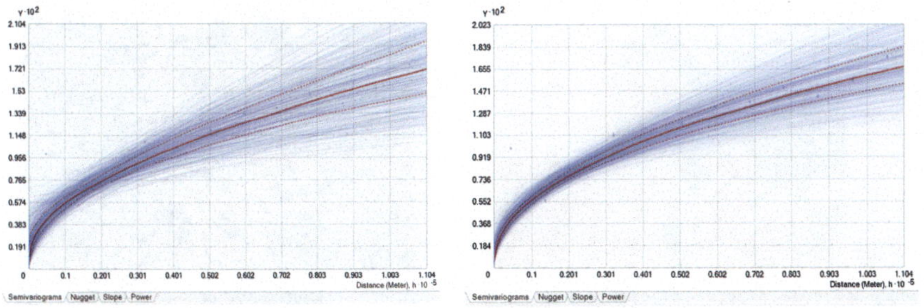

Fig. 1 An example of prior (*left*) and *posterior* (*right*) power semivariogram models. *Red lines* show quartiles of the distributions. *Prior* models are weighted equally. The weight of each *posterior* model is represented by line darkness

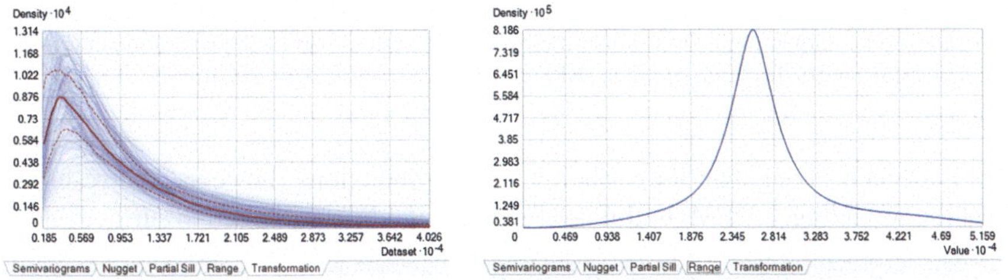

Fig. 2 Distributions of the data transformations (*left*) and the range parameter (*right*)

Prediction with variable model parameters is very useful when the number of samples is insufficient for accurate model fitting. It can be even more useful in the case of large datasets when the model is changing across space and the changes can be non-smooth, as in the case study in [8].

When the dataset is large, the software creates subsets with specified number of samples M (M can vary from 30 to 500) recursively by splitting the data locations on two or three subsets. The division on two subsets is a straight line which divides the data into two equal parts in such a way that the sum of the squares of the deviation of each point from the corresponding center of mass is the lowest. In the case of three subsets, three beams from the center of the area divide the plane. For each subset with the number of samples not equals to M, the samples are added or removed to get exactly M samples, if possible. When the prediction searching neighborhood includes observations from several data subsets, the datum contribution to the prediction is divided between the subsets.

Figure 3 shows an example of predictions and prediction standard errors produced using more than a billion samples collected using LiDAR technology in part of Ohio, USA [9]. We used a model with the least number of parameters, IRFK, with linear semivariogram (this model describes Brownian motion, which can be relevant to elevation formation). It is interesting that the variation of the prediction standard

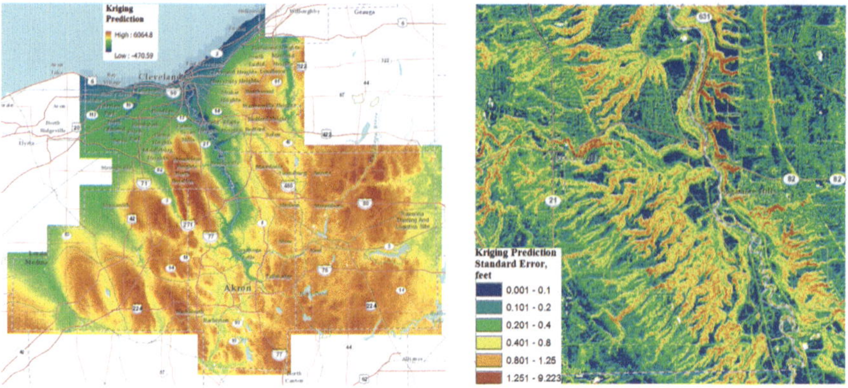

Fig. 3 Predictions using 1.35 billion samples collected using LiDAR technology in part of Ohio, USA (*left*) and enlarged prediction standard errors in the central part of the data domain (*right*). The data are from [9]. The units are feet. IRFK with linear semivariogram was used

errors is much larger than the variation of the predictions (however, the result of elevation modeling, a DEM, is almost always available as a prediction surface only.)

Extensive testing using a large variety of data showed that our model is a reliable automatic interpolator, which produces accurate predictions with non-stationary and non-Gaussian data even when the data vary non-smoothly across space.

References

1. Gribov, A., Krivoruchko, K., & Ver Hoef, J. M. (2006). Modeling the semivariogram: New approach, methods comparison, and simulation study. In T. C. Coburn, J. M. Yarus, & R. L. Chambers (Eds.), *Stochastic modeling and geostatistics: Principles, methods, and Case Studies, Volume II: AAPG Computer Applications in Geology 5* (pp. 45–57).
2. Krivoruchko, K. (2012). Empirical Bayesian kriging. ArcUser, Fall 2012. http://www.esri.com/news/arcuser/1012/empirical-byesian-kriging.html
3. Kazianka, H., & Pilz, J. (2012). Objective Bayesian analysis of spatial data with uncertain nugget and range parameters. *Canadian Journal of Statistics*, *40*(2), 304–327.
4. Katzfuss M. (2013). Bayesian nonstationary spatial modeling for very large datasets. (wileyonlinelibrary.com) doi:10.1002/env.2200.
5. Lindgren, F., Rue, H., & Lindström, J. (2011). An explicit link between Gaussian fields and Gaussian Markov random fields: The stochastic partial differential equation approach. *Journal of the Royal Statistical Society, Series B*, *73*(4), 423–498.
6. Pilz, J., & Spöck, G. (2008). Why do we need and how should we implement Bayesian Kriging methods. *Stochastic Environmental Research and Risk Assessment*, *22*(5), 621–632.
7. Gribov, A., & Krivoruchko, K. (2012). New flexible non-parametric data transformation for trans-Gaussian kriging. *Geostatistics Oslo 2012, quantitative geology and geostatistics* (Vol. 17, Part 1, pp.. 51–65). Netherlands: Springer.
8. Krivoruchko, K. (2012). Modeling contamination using empirical Bayesian kriging. ArcUser, Fall 2012, http://www.esri.com/news/arcuser/1012/modeling-contamination-using-empirical-bayesian-kriging.html
9. Ohio Geographically Referenced Information Program (OGRIP) (2011). Ohio statewide imagery program II. http://gis3.oit.ohio.gov/geodata

Geostatistical History Matching Conditioned to Seismic Data

Amílcar Soares, Leonardo Azevedo, Sara Focaccia and João Carneiro

1 Introduction

In both stages of hydrocarbon reservoir characterization and development it is essential to get reliable and detailed reservoir models in terms of the spatial distribution of its internal petrophysical properties (e.g. porosity, permeability). To decrease the uncertainty related with these models, and therefore better assess the associated risk, they should match all the available information from the subsurface: the observed production data and the seismic reflection data. In both cases, to retrieve reliable subsurface models we are facing highly non-linear inverse problems with non-unique solutions. These inverse problems are commonly designated by history matching and seismic inversion respectively.

History matching is based on two main features: the highly non-linear relationship between the local petrophysical properties, the models parameters that one aims to optimize and the dynamic response of the reservoir, the observed production data; and the scale problem between petrophysical properties, simulated for the entire reservoir grid, and the dynamic responses that are obtained locally at the wells location.

Seismic inversion methods are based on the physical relations between earth models (acoustic and/or elastic impedance models), which are intrinsic of the subsurface

A. Soares, L. Azevedo(✉), S. Focaccia and J. Carneiro
Center for Modeling Petroleum Reservoirs, CERENA/DECivil, Instituto Superior Técnico, Universidade Técnica de Lisboa, Lisbon, Portugal
e-mail: asoares@ist.utl.pt

L. Azevedo
CESAM & Departamento de Geociências, Universidade de Aveiro Campus Universitário de Santiago, 3810-193 Aveiro, Portugal
e-mail: leonardo.azevedo@ua.pt

Sara Focaccia
e-mail: sara.focaccia@ist.utl.pt

João Carneiro
e-mail: joao.carneiro@ist.utl.pt

E. Pardo-Igúzquiza et al. (eds.), *Mathematics of Planet Earth*, Lecture Notes in Earth System Sciences, DOI: 10.1007/978-3-642-32408-6_16,

geology, and the seismic amplitudes (the seismic reflection data itself). The latter are obtained through convolution of the reflectivity coefficients, easily derived from the impedance models, with a known wavelet, estimated from each seismic dataset. Seismic inversion problems may be posed in a geostatistical framework. Within this basis one aims to simulate stochastic acoustic impedance models (and/or elastic models), derive the corresponding synthetic seismic which is then compared with the recorded seismic data. The objective is to iteratively minimize the differences between the synthetic and the real seismic data until a threshold is reached.

However, if the physical nature of these inverse problems is very distinct, both have the same parameter solution space: the reservoir grid itself.

We propose herein a new geostatistical methodology for history matching conditioned to seismic data. With this approach the stochastic Earth models simultaneously match the production data for a given reservoir and produce a synthetic seismic reflection data that matches the observed one.

2 Methodology

We propose a new methodology based on a genetic algorithm where the perturbation of the models is done recurring to the simulation and co-simulation of the reservoir properties using Direct Sequential Simulation (DSS) and co-Simulation (co-DSS) [1]. Our iterative inversion process can be summarized by the following steps:

1. Sampling of the parameters space by simulation of N_s facies images using the direct seismic inversion of facies algorithm [2]. Briefly, this stochastic inversion algorithm starts by defining a certain number of lithofacies that may be identified in a petrophysical domain such as Vp/Vs versus Density. Then, from available well-log data and using DSS, we simulate *Ns* density images that are then used as secondary images in the co-simulation of Vp/Vs models. For each duplet of images we classify it into a new lithofacies volume (Fig. 1, stage 1). On a second stage, Vp and Vs images are simulated conditioned to the previously computed facies cube (Fig. 1, step2). For the simulated elastic models (*Ns* triplets of density, Vp and Vs) we compute synthetic seismograms, following Shuey's approximation [3], and compare the synthetic seismic response with the observed one (Fig. 1, step 3). At the end of the inversion process we are able to retrieve the best elastic models that better match the recorded data as well as the corresponding petrophysical models (derived from the facies model by an appropriate Rock Physics model [4]).
2. For each petrophysical model created in (1) compute the respective flow simulations and compare the obtained dynamic responses with the observed ones from the production data;
3. From previous steps we are able to calculate the deviations between inverted seismic data and recorded one, and between simulated dynamic responses and real ones;

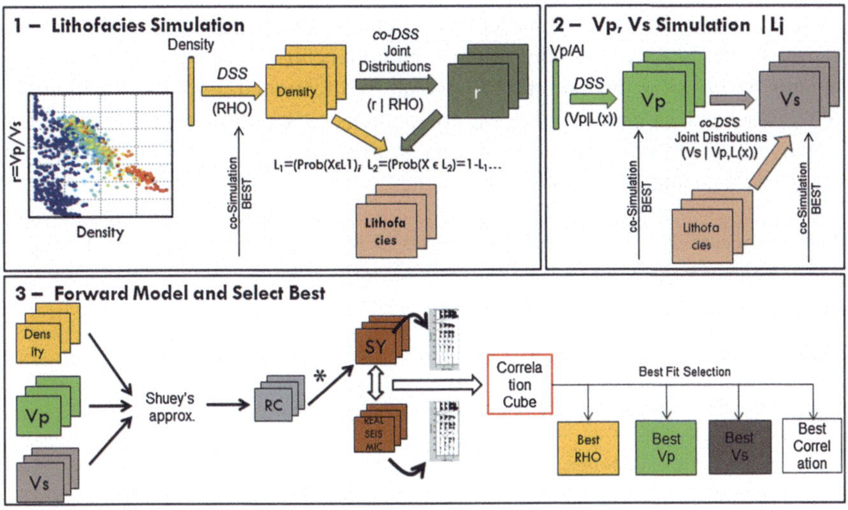

Fig. 1 Direct seismic inversion of facies workflow [2]

4. Compose new "best" elastic and petrophysical models, from the *Ns* set of simulated images in (1), that simultaneously better match the production data and the observed seismic data. This "best" model is a composite model created by selecting patches from the set of simulated models that locally ensure the lowest misfit between observed and synthetic data. For the areas surrounding the wells location the preference will go to the models best fitting the production data, while for the other areas we choose the models best fitting the real seismic data. The selection of these patches is done by normalizing both the deviations for all the simulated models in a multidimensional scaling referential;
5. Using a genetic algorithm concept, use the "best" models created in (4) as secondary variables in the parameters perturbation process by using a co-simulation methodology [5] and return to (1).

This optimization iterative inversion process stops when the global match, of both inverse problems, reaches a pre-defined target. Notice that being a stochastic approach the final reservoir models not only match the seismic and the production data but in addition the uncertainty is part of the solution itself.

3 Final Remarks

From the preliminary results, this new simultaneous inverse methodology is able to retrieve subsurface Earth models with less uncertainty when compared with the individual solutions retrieved by the traditional geostatistical approaches for both history matching and seismic inversion. Also of notice is that most of the times the elastic models inferred by the seismic inversion solution disagree with the ones obtained by the history matching solution. Simultaneously solving both the inverse problems we ensure a joint solution for the above mentioned problems. In addition, using a stochastic approach allows a comprehensive sampling of the uncertainty space, as revealed by the multidimensional scaling approach, when compared with single linearized and optimization approaches.

Acknowledgments The authors would like to thank CERENA/CMRP for supporting this work.

References

1. Soares, A. (2001). Direct sequential simulation and cosimulation. *Mathematical Geology, 33*(8), 911–926.
2. Azevedo, L., Correia, P., Nunes, R., & Soares, A. (2013). *Geostatistical AVO Direct Facies Inversion*. In *15th Annual Conference of the International Association of Mathematical Geosciences*. Madrid: Spain.
3. Shuey, R. T. (1985). A simplification of the Zoeppritz equations. *Geophysics, 50*(4), 609–614.
4. Avseth, P., Mukerji, T., & Mavko, G. (2005). Quantitative seismic interpretation. Cambridge: Cambrige University Press.
5. Mata-Lima, H. (2008). Reservoir characterization with iterative direct sequential co-simulation: Integrating fluid dynamic data into stochastic model. *Journal of Petroleum Science and Engineering, 62*(3–4), 59–72.

Handling Soft Probabilities in Multiple Point Statistics Simulation

Pierre Biver, Gregoire Mariethoz, Julien Straubhaar, Tatiana Chugunova and Philippe Renard

1 Introduction

Mutiple Point Statistics (MPS) simulation is a geostatistical technique to simulate complex lithology architectures using Training Images (TI's). The first generation of efficient MPS algorithms (Strebelle [4]) is based on the storage of Multiple Point Statistics in a search tree; these statistics are subsequently used in a sequential simulation process. A second generation of efficient algorithms (Mariethoz et al. [3]) is based on the direct sampling of the training image during the simulation process conditional to the pattern of the neighboring nodes of a point to simulate. The core of the method is that once an appropriate configuration is found, the central point is pasted on the simulation grid.

Whatever the technique used, the reproduction of prior proportions P_{target} is always a source of debates and approximations. Combinations between prior proportions and probabilities according to neighbors are often envisioned (Allard et al. [1]).

Other authors are suggesting to use non stationary training images (Chugunova and Hu [2]). In this paper we present a new coherent approach allowing to impose

P. Biver (✉) · T. Chugunova
Total SA CSTJF, av. Larribau, 64000 Pau, France
e-mail: Pierre.Biver@Total.com

T. Chugunova
e-mail: Tatiana.Chugunova@Total.com

J. Straubhaar · P. Renard
CHYN University of Neuchatel, 11 rue Emile Argand, 2000 Neuchâtel, Switzerland
e-mail: julien.straubhaar@unine.ch

P. Renard
e-mail: philippe.renard@unine.ch

G. Mariethoz
University of New South Wales, Sydney,UNSW 2052 , Australia
e-mail: gregoire.mariethoz@unsw.edu.au

E. Pardo-Igúzquiza et al. (eds.), *Mathematics of Planet Earth*,
Lecture Notes in Earth System Sciences, DOI: 10.1007/978-3-642-32408-6_17,

local or global proportions for the Direct Sampling simulation method. The efficiency of the approach is illustrated in complex configurations and is proved to be operational without the introduction of arbitrary parameters.

2 Distance Perturbation Methodology

2.1 Recall of Direct Sampling MPS Algorithm

Direct Sampling (DS) algorithm works as follows: for each simulated node (in the simulation grid) a local data event is extracted from the nearest previously simulated nodes. The training image is then scanned randomly with this data event. As soon as the distance between the data event searched for and the one currently scanned are below a given threshold, the corresponding training image value at the central pixel is then pasted on the simulation grid location. A simplified pseudo-code of the algorithm is given below. If no distance under the threshold is found after visiting a maximal fraction of the TI, the scan is interrupted and the best candidate so far is used. A simplified pseudo-code of the algorithm is given in Fig. 1. The key principle is that as soon as the condition mentioned at point 4 is not honored, the corresponding loop is interrupted, drastically accelerating the sampling procedure compared to an exhaustive scan.

2.2 Adaptation to Honor Target Proportions

In the DS algorithm, the distance between data events $d\{Nx, Ny\}$, used in point 4, is crucial. We propose to modify this distance criterion in order to honor prior proportions P_{target}. It is assumed that P_{target} characterizes a support size corresponding to

```
Input: Simulation grid SG, training image TI, distance function d(.), distance
threshold t (distance under which two neighborhoods are considered identical).
1. while # non-informed locations in SG > 0 do
2.             Choose a non-informed location x in SG and identify the
               neighborhood N_x centered on x.
3.             Initialize d{N_x, N_y}=∞
4.             while d{N_x, N_y} ≥ t do *
5.                     Sample a location y in TI.
6.                     Define N_y as the neighborhood centered on y
                       having identical lag vectors as N_x.
7.                     Compute the distance d{N_x, N_y}.
8.             end while
9.             Assign Z(x) = Z(y)
10.        end while
Output: Completed simulation SG
```

Fig. 1 Simplified pseudo-code of the direct sampling algorithm a radius R. For any new node to assign in the sequential simulation process, all the previously informed nodes that are inside a neighborhood of radius R are considered.

They are used to calculate an estimated current local proportions vector $P_{current}$ for the central node to assign. The difference between $P_{current}$ and P_{target} vectors is calculated using the following formula (some others could be envisioned):

$$diff\ P_{=} \sum_{k=1}^{K} P_{k\ target} \cdot \log \frac{P_{k\ target}}{P_{k\ current}} \tag{1}$$

The data event is accepted only if the calculated difference is below a fixed threshold. Hence two criteria are simultaneously considered to accept a TI node:

- the distance between simulation grid data event and TI data event,
- the distance between proportions vectors *diff P*.

If the scan of the TI does not yield any data event with both distances below prescribed thresholds, the best event regarding the sum of the two distances is kept. In addition, we define a minimum radius *Rmin*. If all the data event nodes are inside a radius smaller than *Rmin*, the distance between proportions is not considered because the estimated proportions vector $P_{current}$ is not representative of support size R.

2.3 Application Example

The modified Direct Sampling algorithm has been tested on several synthetic but realistic cases. One example is shown in this extended abstract. The Training Image is a stationary picture and represents a channelized system with 4 facies (channel axis, channel margin, crevasses and matrix). The target proportion maps are issued from seismic attribute and are complex maps with fine details.

Some results of the test are presented on Fig. 2. The parameters for this test are the following ones: number of neighbors is set to 32, the radius for target proportions is set to 12 cells, and the threshold for proportions vector difference is set to 0.1. Two realizations are shown where the extension of facies bodies is limited by the target proportions fluctuations, maintaining however good pattern reproduction. The posterior proportion maps are calculated on the basis of an ensemble of 100 realizations, they are compared to the prior proportion maps by the histograms of differences. The method provides a good accuracy and allows reproducing detailed features of proportion maps.

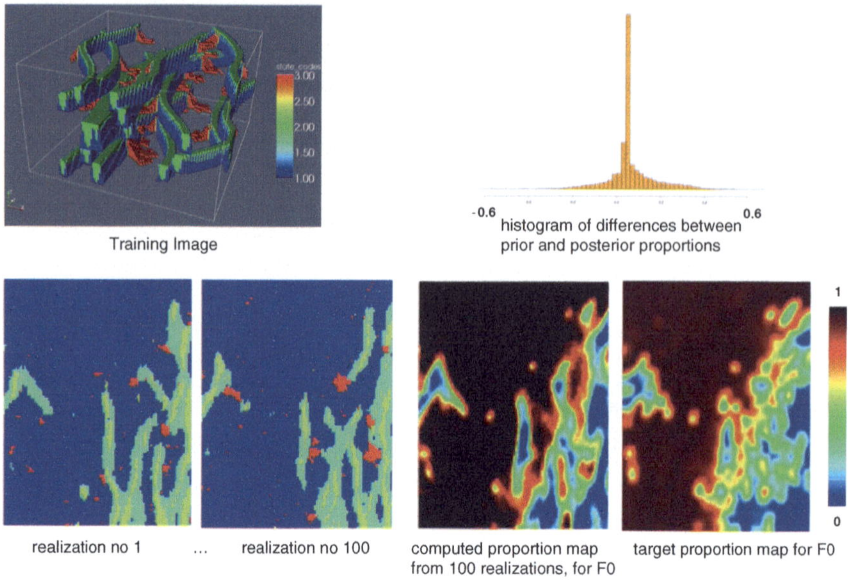

Fig. 2 Simulation example of modified MPDS algorithm to fit to target proportion map

3 Conclusion and Future Work

We have presented a method to generalize the DS algorithm in order to reproduce target proportions. It enables more flexibility to integrate, in MPS models, non stationary prior proportions (based on auxiliary data or conceptual geology), whatever the Taining Image is (stationary or not). Importantly, since our method relies on perturbing the distance used to compare data events, it can be applied to all algorithms based on distances. This includes a number of patch-based approaches such as FilterSim, SimPat, and CCSim.

References

1. Allard, D., Comunian, A., & Renard, P. (2012). Probability aggregation methods in geoscience. *Mathematical Geosciences*, *44*(5), 545–581.
2. Chugunova, T., & Hu, L. (2008). Multiple-point simulations constrained by continuous auxiliary data. *Mathematical Biosciences*, *40*(2), 133–146.
3. Mariethoz, G., Renard, P., & Straubhaar, J. (2010). The direct sampling method to perform multiple-point geostatistical simulations. ltextitWater Resource Research, 46, (11).
4. Strebelle, S. (2002). Conditional simulation of complex geological structures using multiple-point statistics. *Mathematical Geosciences*, *34*(1), 1–22.

An Application of Equal-Area-Growing Window for Calculating Local Singularity for Mapping Granites in Inner Mongolia

Daojun Zhang, Qiuming Cheng and Frits Agterberg

1 Introduction

Local singularity analysis (LSA) was firstly proposed by Cheng (1997) based on fractal/multifractal theory [1, 2], which can be seen as the application of the concentration-area (C-A) model within a small window for local anomaly information extraction [3]. LSA is essentially a spatial neighborhood-window statistical method, considering not only the original value of each spatial location, but also the variation trends of these values within a local window, which supplements classical geological statistics. LSA has been confirmed as a powerful tool for information extraction of geochemical data [4, 5], remote sensing data [6], and tectonics data [7].

The core of LSA is to calculate the local singularity index (LSI) for each location (grid) on a map. LSI is obtained through a nonlinear metallogenic based theory that complex geological processes can lead to anomalous amounts of energy release or material accumulation, further resulting in a fractal/multifractal spatial distribution of these anomalous amounts. A power-law model is used for the calculation of LSI within a local window

$$\rho(\varepsilon) = c\varepsilon^{\alpha-2} \tag{1}$$

where c is a constant, ε represents window size, ρ represents an average local density within a local window with side of size ε, and α is the singularity index that

D. Zhang (✉) · Q. Cheng
Faculty of Earth Resources, China University of Geosciences, Lumo Road, 388, 430074 Wuhan, China
e-mail: cugzdj@gmail.com

Q. Cheng
e-mail: qiuming@yorku.ca

F. Agterberg
Geological Survey of Canada, Booth Street, 601, K1A0E8, Ottawa, ON, Canada
e-mail: Frits.Agterberg@NRCan-RNCan.gc.ca

E. Pardo-Igúzquiza et al. (eds.), *Mathematics of Planet Earth*,
Lecture Notes in Earth System Sciences, DOI: 10.1007/978-3-642-32408-6_18,

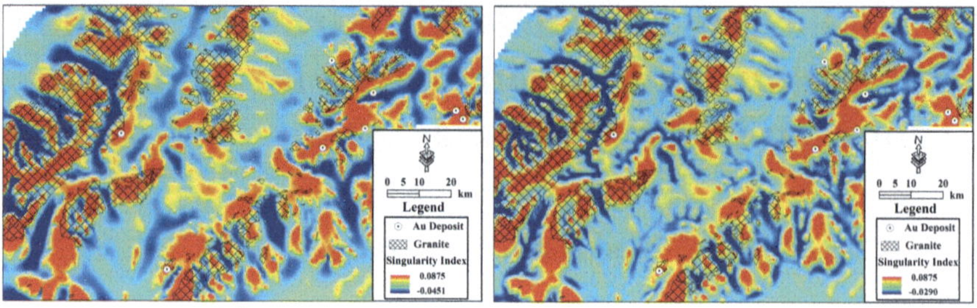

(a) (b)

Fig. 1 Multi-scale obtained using rules of (**a**) absolute maximum LSI and (**b**) maximum LSI

can be estimated by least square method as the slope of a straight line fitted to the relation between log ρ and log ε. See [1, 2] for more details on the meaning of α.

2 A Modified Model for LSA

Although the initial concept of singularity index was developed on the basis of an elliptical moving window determined by three parameters [2], it is more common and easier to implement to use an isotropical window in GIS environment. In GeoDAS [8], one should decide the maximum window size based on a global scale, with the minimum window specified as 1 grid $\times$ 1 grid, and the step increase of window side as 2 grids. There are some disadvantages with this approach, such as the number of grids entered in a group is increasing exponentially with the increase of window size, so that the weights of grids which are far from the currently processed grid are reduced greatly; on the other hand, new entered grids may be of very different distances from currently processed one, which becomes more serious with increased window size. In this study, LSI is calculated based on equal-area-growing concentric circular window, keeping grids entered simultaneously are at the same distance from currently processed grid, so as to avoid these two problems.

The spatial U-statistic method was developed by Cheng and Agterberg (1996) for the separation of different spatial populations [9, 10], and has been used to obtain optimal anisotropic local window in LSA [2, 3]. Inspired by the spatial U-statistic for LSA, a simple solution is provided to avoid the problem that a globally optimal LSI may not accommodate each location. That is to calculate a series of LSI maps based on different maximum size windows, and then integrate these maps through an appropriate algorithm. The algorithm can be designed according to various criteria, such as spatial U-statistic [2, 3], goodness of fit (R^2) and F test in statistics. In this study, LSI of each grid is determined by taking the maximum LSI (Fig. 1a) and absolute maximum LSI (Fig. 1b) from a series of LSIs based on different local window scales (Fig. 2).

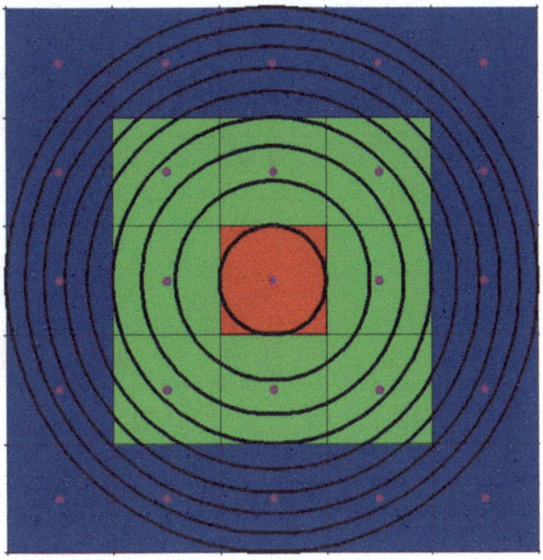

Fig. 2 The present (equal-distance, different *colors* indicate different groups of window) and improved (equal-area, using the center point to represent the whole grid) method for window division

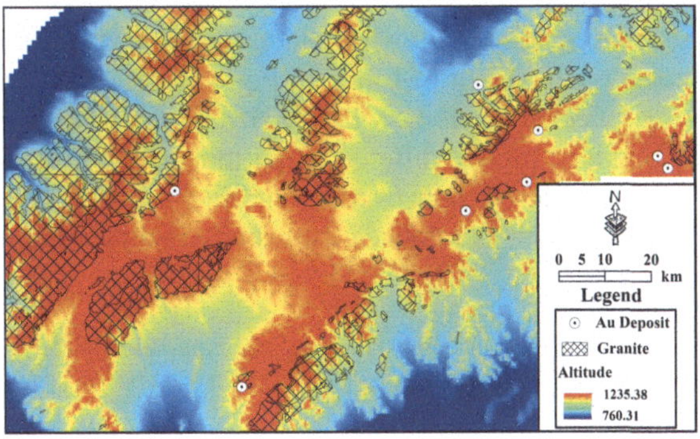

Fig. 3 Spatial relationship between altitude and granite in study area in Inner Mongolia

2.1 Case Study

In Fig. 3, it can be seen that granite masses are generally located areas with altitude values higher than in their neighborhood because of their physical and geochemical characteristics; therefore, LSA can be used to delineate granite masses.

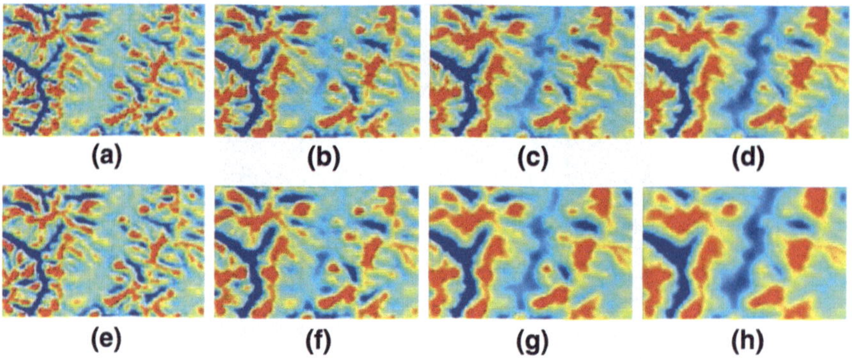

Fig. 4 LSI maps, **a–d** are print screens calculated by equal-distance window with maximum lengths of side of 4, 6, 8 and 10 km; and **e–h** are calculated by equal-area window with maximum diameters of 4, 6, 8 and 10 km

3 Discussions and Conclusions

In Fig. 4 we can see that equal-area-growing window method results in smoother LSI maps with less noise points. Each LSI map in Fig. 4 is obtained at a single scale: more details can be reflected when the window size is small. In Fig. 1, multi-scale LSI maps can be obtained by two different algorithms, which provide singularity information at different scales, so both overall trend and local details can be captured in Fig. 1. Based on Fig. 1 one can delineate covered granites more accurately and efficiently.

Acknowledgments This research has benefited from financial support of the Program of Integrated Prediction of Mineral Resources in Covered Areas (Nos. 1212011085468) in China, and Laboratory Open Fund of China University of Geosciences.

References

1. Cheng, Q. (1997). Fractal/multifractal modeling and spatial analysis. *In: Keynote Lecture in Proceedings of the International Mathematical Geology Association Conference*, *1* (pp. 57–72).
2. Cheng, Q. (1999). Spatial and scaling modelling for geochemical anomaly separation. *Journal of Geochemical Exploration*, *65*, 175–194.
3. Chen, Z. (2007). *Multifractal Theory Based Local Singularity Analysis Method and Its Application in Spatial Information Extraction for Mineral Exploration*. Doctoral Dissertation, China University of Geosciences, Wuhan (in Chinese with English abstract).
4. Cheng, Q. (2007). Mapping singularities with stream sediment geochemical data for prediction of undiscovered mineral deposits in Gejiu, Yunnan province. *China Ore Geology Reviews*, *32*, 314–324.
5. Cheng, Q., & Agterberg, F. P. (2009). Singularity analysis of ore-mineral and toxic trace elements in stream sediments. *Computers and Geosciences*, *35*, 234–244.

6. Cheng, Q. (1999). Multifractality and spatial statistics. *Computers and Geosciences*, *25*, 949–961.
7. Wang, W., Zhao, J., Cheng, Q., & Liu, J. (2012). Tectonic-geochemical exploration modeling for characterizing geo-anomalies in southeastern Yunnan district. *China Journal of Geochemical Exploration*, *122*, 71–80.
8. Cheng, Q. (2006). GIS-based multifractal anomaly analysis for prediction of mineralization and mineral deposits. In J. Harris, (Ed.), *GIS Applications in Earth Sciences* (pp. 289–300). Geological Association of Canada Special Paper.
9. Cheng, Q., Bonham-Carter, G. F., Hall, G. E. M., & Bajc, A. (1997). Statistical study of trace elements in the soluble organic and amorphous Fe- Mn phases of surficial sediments, Sudbury Basin. 1. Multivariate and spatial analysis. *Journal of Geochemical Exploration*, *59*, 27–46.
10. Cheng, Q., Agterberg, F. P., & Bonham-Carter, G. F. (1996). A spatial analysis method for geochemical anomaly separation. *Journal of Geochemical Exploration*, *56*, 183–195.

Semi-Variogram Model Inference Using a Median Bootstrap Statistics

Eulogio Pardo-Igúzquiza, Ricardo A. Olea and Peter A. Dowd

1 Introduction

The increasing availability of powerful and affordable computers has opened up new areas of research for processing-intensive procedures such as genetic algorithms, neural networks, simulated annealing and the bootstrap. In particular, the bootstrap is a well-known method to quantify the uncertainty of complex models [1]. Another of its uses is as a method for obtaining estimates of statistical parameters. For example, the median of the bootstrap distribution has been proposed as a robust estimate of location [2]. The median of the predictive distribution is the optimal point forecast when the loss function is linear [3, 4], which is sometimes a better estimate than the mean. The bootstrap has been extensively used in classical statistics [5] and introduced in geostatistics by Solow [6]. Olea and Pardo-Igúzquiza generalized the bootstrap to attach confidence levels to the experimental semi-variogram [7, 8].

The objective is to formulate a method that: (a) offers performance advantages over the classical approach in two stages of preparing an empirical semi-variogram of mean values for various distances and then fit a model; (b) demands minimum information from the user for finding the parameters of the semi-variogram model; (c) provides a measure of performance besides empirical and analytical semi-variogram fitting; and (d) works for any type of distribution followed by the attribute.

E. Pardo-Igúzquiza (✉)
Instituto Geológico y Minero de España (IGME), Ríos Rosas 23, 28003 Madrid, Spain
e-mail: e.pardo@igme.es

R. A. Olea
U.S. Geological Survey, 12201 Sunrise Valley Drive, MS 956,
Reston VA 20192, USA
e-mail: rolea@usgs.go

P. A. Dowd
Faculty of Engineering, Computer and Mathematical Sciences, University of Adelaide,
Adelaide SA 5000, Australia
e-mail: peter.dowd@adelaide.edu.au

E. Pardo-Igúzquiza et al. (eds.), *Mathematics of Planet Earth*,
Lecture Notes in Earth System Sciences, DOI: 10.1007/978-3-642-32408-6_19,

2 Methodology

For each lag, the conventional empirical semi-variogram uses the mean of the square differences for finding the model parameters. Here, we use the generalized bootstrap to numerically model the distribution for such mean, then selecting the median as the estimate for the true mean square difference. Implementation comprises eight steps in two embedded loops for generating bootstrap resamples and modeling two semi-variograms by successive approximations, one for the attribute and another one for its normal scores. The steps are: (a) obtain the normal scores of the n data values; (b) use the normal scores semi-variogram model from the previous iteration to calculate the covariance matrix for all pairs of data locations. For the initial iteration, it is assumed that the normal scores follow a pure nugget effect model; (c) using the Cholesky method, spatially de-correlate the normal scores to provide a set of n independent and identically distributed normal values; (d) take a bootstrap resample; (e) restore the spatial correlation given by the semi-variogram by inverting the Cholesky method; (f) calculate an empirical semivariogram for these spatially correlated normal scores; (g) backtransform the resample values to the original data space; (h) calculate the empirical semi-variogram for the attribute using the back-transformed resample. Resampling steps (d) through (h) are repeated at least thousand times yielding as a result an equal number of empirical semi-variograms. At the conclusion of this resampling loop, for each lag, prepare two empirical semi-variograms given by the median of the set of values, one for the normal scores and another one for the attribute. Complete the resampling loop by fitting semi-variogram models of the types specified by the user. If the discrepancies with the previous models are larger than a specified tolerance, go back to Step (b) to start another iteration of the model loop, or stop after an arbitrarily specified number of maximum model iterations.

In practice, the procedure converges after a few iterations, but mathematical proof of convergence is pending. The de-correlation in Step (c) works properly only if the covariance matrix is the right one, which can be tested [9], offering modeling quality control in the absence of an exhaustive sample.

3 Testing

A realization of a Gaussian random field with an isotropic exponential model with range 10 units of distance (effective range 30 units), nugget variance 0.3 and partial sill 0.7 was used as an exhaustive sample (Fig. 1a). The realization was generated on a 128 by 128 grid with unit spacing. From this realization four random samples of sizes 50, 100, 150 and 200 were selected, where each sample includes the previous sample plus 50 additional points. The semi-variogram model fitting was done using program VARFIT [10]. The resampling loop was repeated two thousand times. The model loop converged at most after 39 iterations. In the consistency testing at Step

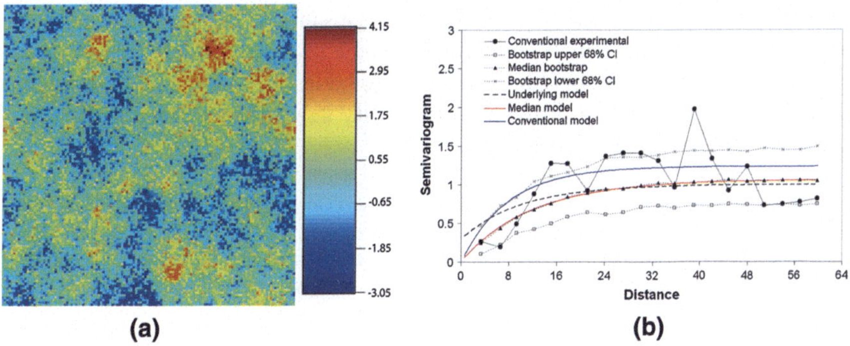

Fig. 1 **a** Exhaustive sample with isotropic exponential semi-variogram and nugget variance. **b** Semi-variograms for the sample of size 50

Table 1 Estimated semi-variogram parameters using the median semi-variogram and the empirical semi-variogram (values in brackets)

Sample size	Nugget (Target = 0.3)	Variance (Target = 1.00)	Range (Target = 10.0)	Discrepancy (Target = 0)
50	0.010 [0.000]	1.059 [1.237]	11.88 [7.81]	1.39 [10.36]
100	0.428 [0.538]	0.913 [0.997]	13.76 [12.61]	1.67 [0.35]
150	0.399 [0.416]	0.999 [0.966]	7.409 [7.482]	0.40 [0.41]
200	0.215 [0.232]	0.928 [1.082]	7.85 [10.95]	0.74 [0.96]

The last column shows the discrepancy between target and fitted models according to Eq. (1)

(c), in all four cases, 19 out of the 20 lags passed a test for randomness at a 95 % confidence level.

Figure 1b shows several types of semi-variograms and some confidence intervals for the smallest sample. Table 1 shows the models fitted to each experimental sample and the discrepancy between the target model and a fitted model according to the integrated distance between models calculated according to the equation:

$$D(\gamma(h), \gamma^*(h)) = \int_0^{\infty} \left[\gamma(h) - \gamma^*(h)\right]^2 dh \tag{1}$$

4 Summary and Conclusions

Extending previous results [7, 8], we apply generalized bootstrapping to automatically estimate the parameters of a given type of permissible semi-variogram model. The modeler needs only to specify the analytical type of model for the attribute and another one for its normal scores. The method is independent of the data distribution and provides a measure of modeling consistency based on the Cholesky decomposition of the correlation matrix. Performance is comparable to that of the conventional approach for generously sampled attributes (large sample size) and more accurate for low density samples (small sample size). Future research includes mathematical proof of convergence of approach used to automatically find the semi-variogram model parameters, removal of the need to specify the analytical form of the semi-variogram and additional case studies.

Acknowledgments The work of the first author was supported by research project CGL2010-15498 from the Ministerio de Economía y Competitividad of Spain. The work of the third author was supported by Australian Research Council Discovery Grant DP110014766. Comments by Özgen Karacan, Goeffrey Phelps and by an anonymous reviewer contributed to improve earlier versions of the manuscript.

References

1. Efron, B. (1979). Bootstrap methods: Another look at the jackknife. *Annals of Statistics*, *7*, 1–26.
2. Cox, M., & Pardo-Igúzquiza, E. (2001). The total median and its uncertainty. In: P. Ciarlini, M. G. Cox, E. Filipe, F. Pavese, & D. Ritchter (Eds.), *Advanced mathematical and computational tools in metrology V* (pp. 106–117). Singapore: World Scientific Publishing Company.
3. Yu, C. W., & Clarke, B. (2010). Asymptotics of bayesian median loss estimation. *Journal of Multivariate Analysis*, *101*, 1950–1958.
4. Gneiting, T. (2011). Quantiles as optimal point forecasts. *International Journal of Forecasting*, *27*, 197–207.
5. Davison, A. C., & Hinkley, D. V. (1997). *Bootstrap methods and their application* (p. 582). New York: Cambridge University Press.
6. Solow, A. R. (1985). Bootstrapping correlated data. *Mathematical Geology*, *17*(7), 769–775.
7. Olea, R. A., & Pardo-Igúzquiza, E. (2011). Generalized bootstrap method for assessment of uncertainty in semivariogram inference. *Mathematical Geosciences*, *43*, 203–228.
8. Pardo-Igúzquiza, E., & Olea, R. A. (2012). VARBOOT: A spatial bootstrap program for semi-variogram uncertainty assessment. *Computers and Geosciences*, *41*, 188–198.
9. Hoeksema, R. J., & Kitanidis, P. K. (1985). Analysis of the spatial structure of properties of selected aquifers. *Water Resources Research*, *21*(4), 563–572.
10. Pardo-Igúzquiza, E. (1999). VARFIT: A fortran-77 program for fitting variogram models by weighted least squares. *Computers and Geosciences*, *25*(3), 251–261.

Geometric and Statistical Modeling of Fractures in the 3D Disturbed Zone of a Claystone Around a Cylindrical Gallery (Meuse-Haute Marne Underground Research Laboratory, France)

Rachid Ababou, Israel Cañamón and Adrien Poutrel

1 Statistical Network of Small Fractures (Random Planar Discs)

The micro-fissures and small fractures around the drift were modeled as a statistical set of planar discs, with randomly isotropic orientations, and with radially decreasing density, diameters and apertures away from the drift wall. The resulting statistical set is visualized in Fig. 1. Apertures are not depicted, but it can be clearly seen that density and size decrease away from the drift wall.

The main hypotheses for generating this statistical set of fractures were as follows:

- The fissures are plane discs with statistically isotropic orientations in 3D space.
- Their planar size is a random variable (random radius R or random diameter D),
- Their thickness or 'aperture' is a random variable (random aperture 'a'),
- Their euclidian positions (X, Y, Z) are random: they follow a modified Poisson process such that the volumetric density of fissures (ρ_{03}) is radially inhomogeneous (decreasing).

The volumetric density $\rho_{03}(r)$ expresses the number of 'objects' per m^3 of euclidian space. For more details, the reader is referred to Appendix A1 of Ref. [1]. The

R. Ababou (✉)
Institut de Mécanique des Fluides de Toulouse, University of Toulouse,
1 Allée Camille Soula, 31400 Toulouse, France
e-mail: ababou@imft.fr

I. Cañamón (✉)
Departamento de Matemática Aplicada y Médodos Informáticos, Universidad Politécnica de Madrid (UPM), E.T.S.I. Minas, C./ Ríos Rosas 21, 28003 Madrid, Spain
e-mail: israel.canamon@upm.es

A. Poutrel
Agence Nationale pour la Gestion des Déchets Radioactifs (ANDRA),
1/7 rue Jean Monnet, Parc de la Croix Blanche, 92298 Châtenay-Malabry cedex, France
e-mail: adrien.poutrel@andra.fr

E. Pardo-Igúzquiza et al. (eds.), *Mathematics of Planet Earth*,
Lecture Notes in Earth System Sciences, DOI: 10.1007/978-3-642-32408-6_20,

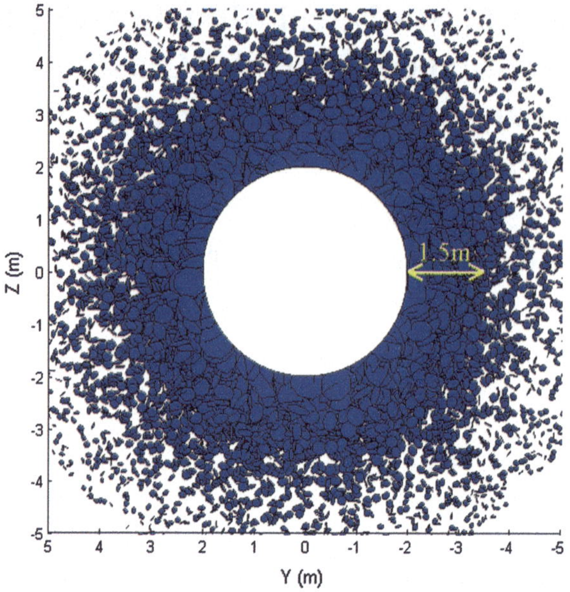

Fig. 1 Front view of the statistical set of 10,000 planar disc fissures, shown on the transverse vertical plane located at the drift entrance (3D view with hidden parts). The density and diameters decrease radially from the drift wall (so do the apertures, although not visible in this view)

latter provides a "geometric probability" formulation of the radial inhomogeneity of fissures around the drift, and particularly, of the relation between the volumetric density ρ_{03} (r) and the Probability Density Function (PDF) of the random positions (X, Y, Z) of fissure centers.

2 Deterministic Set of Large Curved 'Chevron' Fractures (Parametric Surfaces)

In situ observations at the Meuse/Haute Marne Underground Research Laboratory (URL), indicate the existence of a quasi-periodic set of large curved fractures organized as a 3D herringbone pattern (or chevron pattern) around each gallery and ahead of the excavation front (they are caused by excavation). It was decided, in this work, that each "chevron" fracture should be represented by a curved parametric surface, resembling as much as possible the complex 3D shapes observed *in situ*.

The parametric surface model retained here is a "generalized elliptic conoïd", developed by us as an extension of other known conoïdal surfaces, such as the elliptic conoïd of Isawi (it.wikipedia.org/wiki/Conoide_ellittico). The latter is itself derived from the classical conoïdal surface of Wallis, which has the shape of the closed end of a toothpaste tube. Wallis' conoïd has been known for centuries under various names:

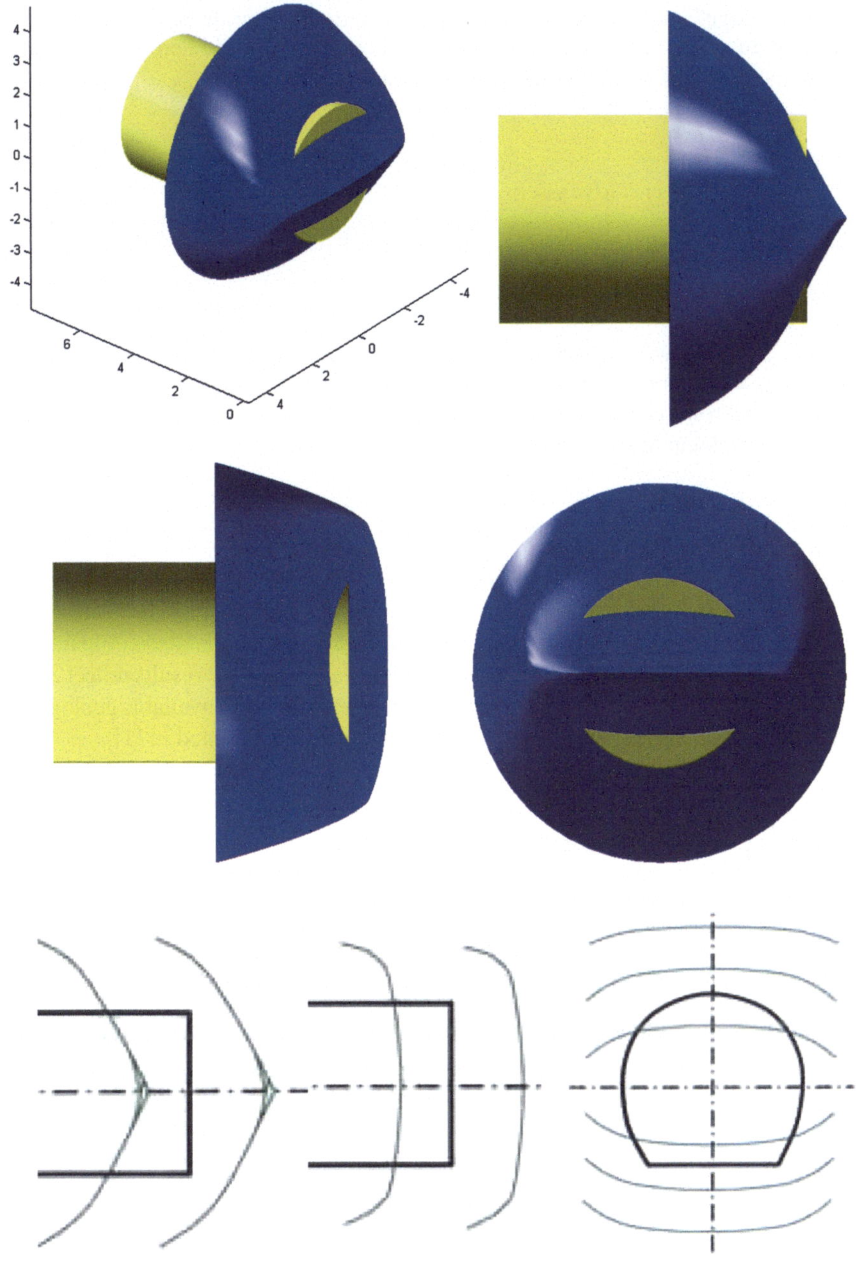

Fig. 2 Morphology of the *curved* surfaces of large shear fractures ('chevrons'). *Above*. Various 3D perspective views of the "generalized elliptic conoïd", the surface model retained for describing the *curved* 'chevrons' fractures at the claystone site (Meuse/Haute Marne URL). *Below*. Schematic traces of chevron fractures, as observed at the in the GMR gallery (which is aligned with the minor horizontal principal stress σh). The schematics show the traces projected on the axial vertical plane, axial horizontal plane, and transverse plane, respectively

"conical wedge", "conocuneus", and also, "arrière-voussure de Saint-Antoine" in French architecture. The "generalized elliptic conoïd" proposed here for the shape of chevron fractures has the parametric equation:

$$\vec{S} = \begin{pmatrix} x \\ y \\ z \end{pmatrix} = \begin{pmatrix} (a \cdot (1-u) + c \cdot u) \cdot \cos(\nu) \\ \left(d \cdot u^{(d1 \cdot (1-EX) + d2 \cdot EX)}\right) \cdot \sin(\nu) \\ b \cdot \left(1-u^2\right) \cdot |\cos(\nu)|^g + e \cdot u^{(e1 \cdot (1-EX)+e2 \cdot EX)} + f \cdot u^{(f1 \cdot (1-EX)+f2 \cdot EX)} \end{pmatrix}$$

$$EX(\nu, n) = \left[1 - \left|(\cos(\nu))^{|\tan(n \cdot \pi + \pi/2)+1|}\right|\right] \cdot (1-n) + \left|(\sin(\nu))^{|\tan(n \cdot \pi + \pi/2)+1|}\right| \cdot n$$

$$\nu \in [0,\ 2\pi]; \quad u \in [0,\ 1]$$

where 'n' is a real number between 0 and 1 (we selected here the value $n = 0.1$). The following parameters were finally selected after trial and error comparisons with the observed chevron traces to represent typical chevron fractures for the GMR gallery at the Meuse/Haute Marne URL:

a=2.75, b=0.4, c=4, d=4, e=1.5, f=2, g=3,
d1=0.5, d2=1.75, e1=2, e2=8, f1=1.25, f2=1.5, n=0.1.

The resulting fracture surface is depicted in Fig. 2, where it is also compared to schematic representations of observed fracture traces. Note: these results reflect 2009 site-specific observations in the GMR gallery; the complete set of available geological and geo-mechanical references used for these comparisons is listed in [1].

Reference

1. Ababou R., Cañamón Valera, I., & Poutrel, A. (2011). Macro-permeability distribution and anisotropy in a 3D fissured and fractured clay rock: 'Excavation damaged zone' around a cylindrical drift in callovo-oxfordian argilite (Bure). *Special Issue on "Clays in Natural & Engineered Barriers for Radioactive Waste Confinement" (4th Internat. Meeting CLAYS 2010, Nantes, 29 March 1st April 2010) : Journal of Physics and Chemistry of the Earth*. ISSN 1474–7065, DOI: 10.1016/j.pce.2011.07.032.

Robust Regression with Compositional Response: Application to Geosciences

Karel Hron, Peter Filzmoser, Matthias Templ, Karl Gerald van den Boogaart and Raimon Tolosana-Delgado

1 Compositional Data and Their Geometrical Properties

Data sets in geosciences are frequently characterized by variables that quantitatively describe parts of a whole, carrying exclusively relative information. We call them compositional data [2, 7]. From this point of view, any representation of the variables, for example in mg/kg, proportions, or percentages, should not alter the results of the corresponding statistical analysis, even if the scaling is performed for a subcomposition of the original composition. This is due to the inherent property of compositional data, where all the relevant information is contained in ratios between compositional variables (parts). In other words, compositional data induce their own geometry, nowadays called the Aitchison geometry. Unfortunately, most statistical methods are designed for the standard Euclidean geometry in real space, thus compositional data need to be expressed in orthonormal coordinates prior to their statistical analysis. The orthonormal coordinates (one less than the number of parts in the original composition) are obtained via the isometric logratio (ilr) transformation [4].

K. Hron (✉)
Department of Mathematical Analysis and Applications of Mathematics, Palacký University, Faculty of Science, 17. listopadu 12, Olomouc, Czech Republic
e-mail: hronk@seznam.cz

K. Hron
Department of Geoinformatics, Palacký University, tř, Svobody 26, Olomouc, Czech Republic

P. Filzmoser (✉) · M. Templ (✉)
Department of Statistics and Probability Theory, Vienna University of Technology, Wiedner Hauptstrasse 8-10, Vienna, Austria
e-mail: P.Filzmoser@tuwien.ac.at

K. G. van den Boogaart (✉) · R. Tolosana-Delgado (✉)
Department of Modelling and Evaluation, Helmholtz-Institut Freiberg for Resources Technology, Halsbrücker-Strasse 34, Freiberg, Germany
e-mail: k.van-den-boogaart@hzdr.de; r.tolosana@hzdr.de

E. Pardo-Igúzquiza et al. (eds.), *Mathematics of Planet Earth*, Lecture Notes in Earth System Sciences, DOI: 10.1007/978-3-642-32408-6_21,

The concrete choice of the coordinates for a D-part composition $\mathbf{x} = (x_1, \ldots, x_D)'$ usually depends on good interpretability of results of a statistical analysis. For the purpose of this study, that is to test the influence of an explanatory variable on a chosen compositional part as a response of a regression model, we introduce the following coordinates,

$$x_i^* = \sqrt{\frac{D-i}{D-i+1}} \ln \frac{x_i}{\sqrt[D-i]{\prod_{j=i+1}^{D} x_j}}, \; i = 1, \ldots, D-1. \tag{1}$$

In such a configuration, the first ilr variable x_1^* explains all the relative information (log-ratios) about the original compositional part x_1 *with respect to the other parts*, and the coordinates $x_2^*, \ldots, x_{D-1}^*$ explain the remaining log-ratios in the composition (after a permutation of parts in the composition, also any other part can play the role of x_1). Of course, these ilr coordinates can be inverted to recover a composition. Expressions for this back-transformation are given in [4].

2 Robust Regression with Compositional Response

In geosciences, frequently situations occur where a compositional response depends on one or more (non-compositional) explanatory variables. Similar to other standard statistical methods, also linear regression cannot be applied directly to compositional data. Although the regression model can be even expressed directly for the original compositional data [3, 8], for the purpose of statistical inference it is necessary to express the response variables in coordinates. Thus, we arrive at a multivariate linear regression model with p explanatory and q response variables,

$$\mathbf{y}_i = \mathbf{b}_0 + \sum_{j=1}^{p} (x_{ij} \cdot \mathbf{b}_j) + \mathbf{e}_i \; , \; i = 1, 2, \ldots, n \tag{2}$$

(the errors $\mathbf{e}_i$ are independent and identically distributed) whose interpretation depends on the particular regression type. Accordingly, $\mathbf{y}_i$, $i = 1, \ldots, n$, collected as rows in an $n \times q$ matrix $\mathbf{Y}$, stand for the coordinate representation (like (1) from Sect. 1) of the corresponding compositional variables; $\mathbf{x}_i = (x_{i1}, \ldots, x_{ip})'$ form the rows of the $n \times (p+1)$ design matrix $\mathbf{X}$ (whose first column corresponds to the absolute term parameters and is formed by vector of ones), and the parameters $\mathbf{b}_j = (x_{j1}, \ldots, x_{jq})'$ are collected as rows in the $(p+1) \times q$ matrix $\mathbf{B}$.

Usually, the regresion parameters $\mathbf{B}$ are estimated by the standard least-squares method [5]; the result is denoted by $\widehat{\mathbf{B}}$. The above stated hypothesis that the coordinate x_1^* from (1), explaining all the relative information on x_1, does not depend on the k-th explanatory variable, can be tested with the test statistic

$$F_k = \frac{\mathbf{h}'\widehat{\mathbf{B}}\mathbf{m}_k(\mathbf{m}'_k\mathbf{Y}'\mathbf{M}_X\mathbf{Y}\mathbf{m}_k)^{-1}\mathbf{m}'_k\widehat{\mathbf{B}}'\mathbf{h}}{\mathbf{h}'(\mathbf{X}'\mathbf{X})^{-1}\mathbf{h}}\frac{n-p-1}{1} \sim F_{1,n-p-1}, \quad (3)$$

where the q-part vector $\mathbf{m}_k$ has zero entries except a 1 at position k, and $\mathbf{M}_X = \mathbf{I}_n - \mathbf{X}(\mathbf{X}'\mathbf{X})^{-1}\mathbf{X}'$ $\mathbf{I}_n$ stands for the identity matrix of order n).

It is well known that outliers—deviating points from the main linear trend—in either the response or the explanatory variables, or in both, can lead to biased least-square estimates. For this reason, it is necessary to use a proper robust counterpart to the classical regression estimator that tolerates a certain amount of outlying observations. For several reasons there exists no straightforward solution for the robust regression problem with compositional response. Similarly as in the classical case, the step from the multivariate to the multiple model is not possible if the response ilr coordinates are not independent [3]. Even more, in the robust case, to regress the response variables separately would result in ignoring the multivariate outliers. An additional challange is the proper choice of the ilr transformation that is crucial for an appropriate interpretation of the results. Finally, a simplified approach to implement robust methods to ilr transformed data may produce transformation-dependent results, an undesirable characteristic.

A solution is provided by the multivariate least trimmed squares (MLTS) method that fulfills all required concepts of robustness for regression with compositional data, in particular equivariance properties that are necessary for a reasonable regression model behaviour with any coordinate representation of the compositional response [1, 3]. With the MLTS estimates of **B** even the test statistic (3) can be used, nevertheless, its F-distribution is achieved only approximatively.

To illustrate the above theoretical developments, we employ the Kola data set which resulted from a large geochemical mapping project, carried out from 1992 to 1998 by the Geological Surveys of Finland and Norway, and the Central Kola Expedition, Russia. An area covering 1,88,000 km^2 in the Kola peninsula of Northern Europe was sampled (approximately 600 samples were taken) in four different layers. The data are available in the package `StatDA` of the statistical software R. For our study, 27 chemical elements (Ag, Al, As, B, Ba, Bi, Ca, Cd, Co, Cu, Fe, K, Mg, Mn, Na, Ni, P, Pb, Rb, S, Sb, Sr, Th, Tl, V, Y, Zn) from the O-horizon are taken [3]. Their concentrations were measured in mg/kg; however, not the absolute values but the ratios among them are of interest. The concentrations of the elements are primarily influenced by geochemical processes (water/rock interactions, pollution, etc.), however, elements like phosphor (P) are likely to depend on elevation, because phosphor forms a nutrient base for plants. With higher elevation the climate changes (for example temperature and presence of water) so that the presence of plants decreases. The behaviour of P is also found for K, another fundamental nutrient for plants. However, from the raw element concentrations it is not easy to detect whether elevation is indeed a dominant effect here. By forming a regression model, where the first response coordinate x_1^* explains the relative information (ratios) concerning phosphor in the whole composition, we can use the statistic (3) to test for the significance of elevation on the concentrations on this element. Even when using the classical least-squares estimator, the elevation is highly significant, with a p-value of 1.60×10^{-6}. However, due to many outlying observations in the data set, robust regression using the MLTS estimator is required. The variable elevation is then even more significant (p-value 4.84×10^{-9}), the regression model was thus able to confirm our expectations that phosphor as plant nutrient indeed depends on the chosen explanatory variable.

Acknowledgments The authors gratefully acknowledge the support by the Operational Program Education for Competitiveness—European Social Fund (project CZ.1.07/2.3.00/20.0170 of the Ministry of Education, Youth and Sports of the Czech Republic).

References

1. Agulló, J., Croux, C., & Van Aelst, S. (2008). The multivariate least-trimmed squares estimator. *Journal of Multivariate Analysis*, *99*, 311–338.
2. Aitchison, J. (1986). *The statistical analysis of compositional data*. London: Chapman and Hall.
3. Egozcue, J. J., Daunis-i-Estadella, J., Pawlowsky-Glahn, V., Hron, K., & Filzmoser, P. (2011). Simplicial regression. The normal model. *Journal of Applied Probability and Statistics*, *6*, 87–108.
4. Egozcue, J. J., Pawlowsky-Glahn, V., Mateu-Figueras, G., & Barceló-Vidal, C. (2003). Isometric logratio transformations for compositional data analysis. *Mathematical Geology*, *35*, 279–300.
5. Johnson, R., & Wichern, D. (2007). *Applied multivariate statistical analysis* (6th ed.). London: Prentice-Hall.
6. Maronna, R., Martin, R. D., & Yohai, V. J. (2006). *Robust statistics: Theory and methods*. New York: Wiley.
7. Pawlowsky-Glahn, V., & Buccianti, A. (2011). *Compositional data analysis: Theory and applications*. Chichester: Wiley.
8. Tolosana-Delgado, R., & Van den Boogaart, K. G. (2011). Linear models with compositions in R. In V. Pawlowsky-Glahn & A. Buccianti (Eds.), *Compositional data analysis: Theory and applications* (pp. 356–372). Chichester: Wiley.

Compositional Analysis in the Study of Mineralization Based on Stream Sediment Data

Renguang Zuo

1 Introduction

Compositional data with components measured as proportions or percentages of some whole only carry relative and limited information [2]. The main problem in compositional data analysis, referred as data closure problem or negative bias problem, provides difficulty in the interpretation of correlation coefficients among variables. There are three methods for reducing the effects of compositional data closure problem, viz.: additive logratio (***alr***) transformation [1], centered logratio (***clr***) transformation [2], and isometric log ratio (***ilr***) transformation [3].

Stream sediment geochemical data, often used in mapping of anomalies for mineral exploration, are typical compositional data that should be opened prior to any analysis and interpretation even when only one variable is involved. However, few studies focused on the effects of data closure problem on the geochemical exploration. In this study, a stream sediment dataset from southwestern Fujian depression belt (SFDB) was used to investigate this problem and to document results.

2 Data

The datasets used in this study consist of a geological map (1:200,000) and stream sediment multi-element concentration data. The stream sediment samples were collected at a density of one sample per 4 km^2 and were selected, according to the mineral assemblages of skarn iron deposit in this study region, from the China National Geochemical Mapping Project database that comprises 39 major, minor and trace

R. Zuo (✉)
State Key Laboratory of Geological Processes and Mineral Resources,
China University of Geosciences, Wuhan 430074, China
e-mail: zrguang@cug.edu.cn

E. Pardo-Igúzquiza et al. (eds.), *Mathematics of Planet Earth*,
Lecture Notes in Earth System Sciences, DOI: 10.1007/978-3-642-32408-6_22,

elements. The vector points of the selected point stream sediment sample data are almost regularly distributed and the distance between any two points is roughly 2 km. The main ore-forming elements selected from the data are Au, Ag, Cu, Mo, Pb, Zn, W, Sn, Sb, and Fe_2O_3.

3 Effects of Data Closure Problem

3.1 Univariate Analysis

Classical estimators like mean and standard deviation (SD) are sensitive to outliers, resulting in unreliable estimates of the center and spread of data distribution due to outliers. If the selected data is approximately symmetrical, various estimators of original data and their transformations (log-transformed or ***ilr*** transformed) give similar results [4]. Although the mean, median, SD, and median absolute deviation (MAD) can be applied for the original and log-transformed data, they do not account for the data closure problem. Therefore, the median and MAD of ***ilr*** transformed data are recommended to characterize the center and spread of selected univariate data.

3.2 Bivariate Analysis

In bivariate analysis, the concern is about the strength of relationship or correlation between two geochemical variables. However, correlation coefficients estimated by classical methods are unreliable because they do not account for the data closure problem [5]. The variability and stability of the ***ilr*** transformed variables show the largest variance for Cu and Sn and the smallest variance for Zn and Fe_2O_3. The largest correlation coefficient for log-transformed Pb and Zn does not result in the smallest ***ilr*** variance for this pair of element data (Table 1). The classical linear or nonlinear relationships between two variables are meaningless for compositional data because the data points only present relative information. Therefore, it is imperative to measure the variability and stability of ***ilr*** transformed data in bivariate analysis.

3.3 Multivariate Analysis

The robust principal component analysis (RPCA) based on robust statistical analysis can obtain principal components that are not much influenced by outliers through replacement of the classical covariance matrix with the minimum covariance determinant estimator. Here, the RPCA based on the ***ilr*** transformed data is used to combine

Table 1 Variability and stability of ***ilr*** variables. The first number denotes the variability, and the inside parentheses number denotes the stability

	Cu	Pb	Zn	Sn	Fe_2O_3
Cu	0 (1)				
Pb	1.47 (0.23)	0 (1)			
Zn	0.95 (0.39)	0.44 (0.65)	0 (1)		
Sn	1.76 (0.17)	0.91 (0.40)	0.79 (0.45)	0 (1)	
Fe_2O_3	0.82 (0.44)	0.838 (0.432)	0.37 (0.69)	0.92 (0.40)	0 (1)

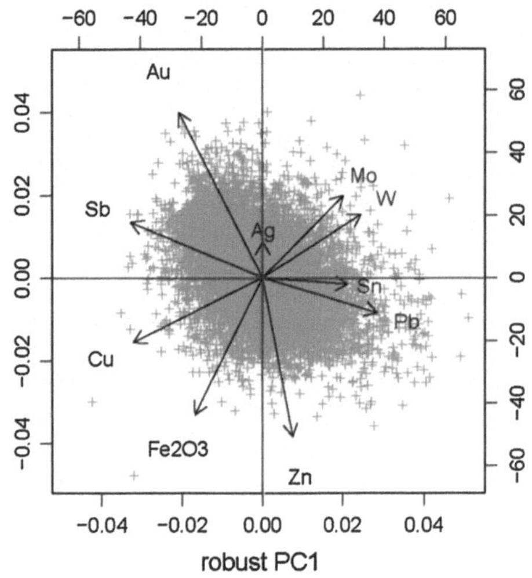

Fig. 1 Biplot of results of RPCA

multiple variables. The PC1 and PC2 obtained through the RPCA account for 27.67 % of the total variances of ***ilr*** transformed data. The eigenvalues of either PC1 or PC2 are > 1, suggesting they are significant components. The PC2 of the ***ilr*** transformed data show two geochemical assemblages. The first assemblage consists of Au, Ag, Mo, Sb, and W with positive loadings on PC2, likely representing Au-related mineralization. The second assemblage consists of Cu, Pb, Zn, Sn, and Fe_2O_3 with negative loadings on PC2 possibly representing Fe-related mineralization (Fig. 1).

4 Conclusions

Geochemical exploration plays a crucial role in discovering new mineral deposits. However, the data closure problem of geochemical data is seldom addressed. In this study, univariate, bivariate and multivariate data analysis of geochemical data were investigated. The following conclusions were obtained from the results of the study: (1) geochemical data, like any other compositional data, should be opened prior to analysis; (2) for univariate analysis, the robust estimators like median and median absolute deviation should be used for measuring the center and spread of the selected ***ilr*** transformed data; (3) for bivariate analysis, the ***ilr*** information should be used to measure the variability and stability between any two variables; and (4) for multivariate analysis, the ***ilr*** transformation and robust multivariate analytical tools, such as RPCA, should be applied.

Acknowledgments This research benefited from the Fundamental Research Funds for the Central Universities, China University of Geosciences (Wuhan) (Nos.CUG120501 and CUG120116).

References

1. Aitchison, J. (1982). The statistical analysis of compositional data (with discussion). *Journal of the Royal Statistical Society, Series B44*, 139–177.
2. Aitchison, J. (1986). *The statistical analysis of compositional data* (p. 416). London: Chapman & Hall.
3. Egozcue, J. J., Pawlowsky-Glahn, V., Mateu-Figueras, G., & Barcelo-Vidal, C. (2003). Isometric logratio transformations for compositional data analysis. *Mathematical Geology*, *35*, 279–300.
4. Filzmoser, P., Hron, K., & Reimann, C. (2009). Univariate statistical analysis of environmental (compositional) data: Problems and possibilities. *Science of the Total Environment*, *407*, 6100–6108.
5. Filzmoser, P., Hron, K., & Reimann, C. (2010). The bivariate statistical analysis of environmental (compositional) data. *Science of the Total Environment*, *408*, 4230–4238.

Modelling World Energy Applying Simplicial Linear Ordinary Differential Equations

Eusebi Jarauta-Bragulat, Carme Hervada-Sala and Juan J. Egozcue

1 Introduction

The evolution of a dynamic system, whose status evolves in time, can be approached from different perspectives. The most usual one is to identify the elements or parts of interest and consider relevant data for each element, with an appropriate scale to the study of the process. The information (data) related to the process, in general is a matrix of data, which may become assimilated to a role of vector function of one real variable (time) and as much components as elements or interesting parts to study. If we wish to apply a model which enables the characterization of the process studied, usually the matrix of data is fitted to a certain model through the estimation of some parameters and is analyzed by means of some criterion of goodness of that adjustment. The model applied can be used not only to describe the process, but also to predict the evolution of the studied dynamic system.

2 Methodologies

We consider a data set consisting on several rows (time values) and columns corresponding to studied variables. We compute the corresponding proportions and then we apply to these proportions the methodology of analysis of compositional data. Proportions are in the simplex of n+1 parts $\vec{x}(t) = (\in x_1(t), \ldots, x_{n+1}(t)) \quad \mathcal{S}^{n+1}$ and they can be modeled as evolutionary compositions with a simplex-valued

E. Jarauta-Bragulat (✉) · J. José Egozcue
Department of Applied Mathematics III, School of Civil Engineering,
Technical University of Catalonia (UPC), Terrassa, Spain
e-mail: eusebi.jarauta@upc.edu

C. Hervada-Sala
Department of Physics and Nuclear Engineering, School of Engineering,
Technical University of Catalonia (UPC), Terrassa, Spain

E. Pardo-Igúzquiza et al. (eds.), *Mathematics of Planet Earth*,
Lecture Notes in Earth System Sciences, DOI: 10.1007/978-3-642-32408-6_23,

function of one real variable, $\vec{x}: [0, T] \subset \mathbb{R}^+ \to \mathcal{S}^{n+1}$ following [1]. The isometric log-ratio transformation (ilr-coordinates or balances) are calculated from the proportions, accordingly with [2]. The simplicial linear ordinary differential equation model in the simplex is:

$$D^{\oplus}\vec{x}(t) = \left[M \boxdot \vec{x}(t)\right] \oplus \vec{c}. \tag{1}$$

Applying ilr transformation, this equation is transformed to a linear first-order ordinary differential equation in a Euclidean space. The equation is:

$$\frac{d\vec{u}(t)}{dt} = A\vec{u}(t) + \vec{b}. \tag{2}$$

In this equation, $\vec{u}(t) = \text{ilr}(\vec{x}(t)) = V^{\top}\log(\vec{x}(t))$, being V the chosen contrast matrix ($\top$ indicates transposition), $A = \left(a_{kj}\right)_{1 \le k, j \le n}$, satisfying $M = VAV^{\top}$ and $\vec{b} = \text{ilr}(\vec{c})$. Developing this equation we obtain the linear system:

$$\frac{du_k(t)}{dt} = a_{k1}u_1(t) + \cdots + a_{kn}u_n(t) + b_k, \; k = 1, 2, \ldots, n. \tag{3}$$

The estimation of the coefficients of the matrix and the independent applying (3), presents some problems of numerical stability. To avoid this, we work with the integral form of (3):

$$\begin{aligned} u_k(t) - u_k(0) &= a_{k1}U_1(t) + \cdots + a_{kn}U_n(t) + b_k t, \; U_k(t) \\ &= \int_0^t u_k(z)\,dz, \; k = 1, 2, \ldots, n. \end{aligned} \tag{4}$$

We use linear regression to estimate coefficients, following the methodology described in [3]. Three cases can be found: $a_{kj} = 0, b_j \neq 0$ (Model 0); $a_{kj} \neq 0, b_j = 0$ (Model 1); $a_{kj} \neq 0, b_j \neq 0$ (Model 2). The simplest appropriate model is adopted, following the parsimony principle.

3 The Data Sets

Data sets are taken from BP workbook [4]. For the first group, it consists of values of total world energy production (Mtoe, mega tones oil equivalent) by five different sources: hydraulic (HYD), nuclear (NUC), coal (COAL), oil (OIL) and gas (GAS), from years 1965 to 2011. For the second group, it consists of values of total world energy production (TWh, Terawatt-hours) by three different sources: solar (SOL), wind (WIN) and others (OTH) from years 1990 to 2011.

Table 1 Multiple regression coefficients R^2 obtained for the three models and the corresponding balances for each group

	First group					Second group		
	*b*1	*b*2	*b*3	*b*4	*Param.*	*b*1	*b*2	*Param.*
Model 0	0.145	0.921	0.713	0.633	4	0.142	0.951	2
Model 1	0.653	0.985	0.995	0.973	16	0.773	0.993	4
Model 2	0.742	0.991	0.995	0.973	20	0.965	0.999	6

The number of parameters fitted in each regression is shown under *Param*

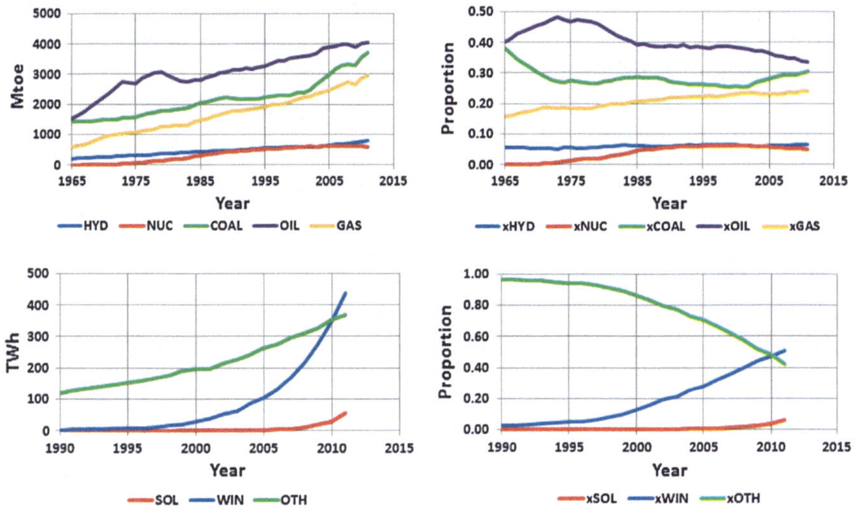

Fig. 1 Time evolution of energy production values (*left*) and their proportions (*right*), corresponding to each group

4 Results and Discussion

Data have been processed according to exposed methodology. For both groups, model 1 seems to be the appropriate one, according to values of multiple regression coefficients R^2 shown in Table 1. Predicted values are 2012–2030 for the first group and 2012–2025 for the second group (renewables). In both series, the ratio number of years predicted / number of data is 0.40 (Fig. 1).

Figure 2 shows the prediction of balances and proportions for each group. For the first one, an increasing proportion of oil can be detected and certain constant values for coal, gas and hydraulic; a decreasing proportion for nuclear is obtained. In the second group (renewables), a very appreciable increase in solar energy is obtained, a certain stabilization of wind power after reaching a maximum about 2017 and a significant decrease in the proportion of use of other renewable energy sources is observed (Table 2).

Table 2 Evolution in time of proportions of energy sources in each group

	First group						Second group		
	HYD	NUC	COAL	OIL	GAS		SOL	WIN	OTH
1965	0.056	0.002	0.381	0.403	0.158	1990	0.003	0.029	0.968
2011	0.065	0.049	0.307	0.335	0.244	2011	0.065	0.508	0.427
2030	0.049	0.028	0.266	0.416	0.241	2020	0.185	0.614	0.200

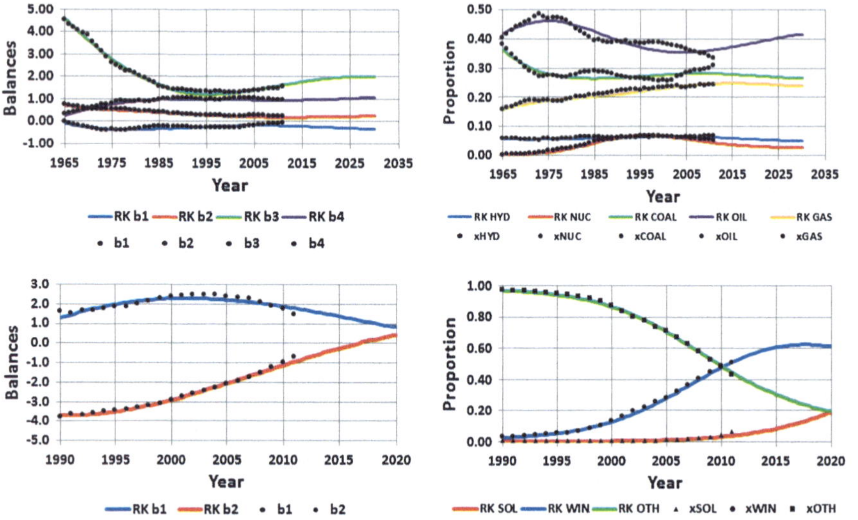

Fig. 2 Time evolution of energy production balances (*left*) and their proportions (*right*), corresponding to each group. Dots are data balances (*left*) and data values (*right*)

References

1. Egozcue, J. J., Jarauta-Bragulat, E., & Díaz-Barrero, J. L. (2011). Compositional data analysis, theory and applications. In *Calculus of simplex-valued functions* (pp. 158–175). New Jersey: John Wiley & Sons.
2. Egozcue, J. J. et al. (2003). Isometric logratio transformation for compositional data analysis. *Mathematical Geology*, *35*(3), 279–300.
3. Egozcue, J. J., & Jarauta-Bragulat, E. (2013). Differential models for evolutionary compositions. *Mathematical Geosciences*, Paper submitted in January 2013.
4. BP Statistical Review of World Energy (June 2012), http://www.bp.com/statisticalreview.

Structural Analysis of the National Geochemical Survey of Australia Data

Ute Mueller, Johnny Lo, Patrice de Caritat and Eric Grunsky

1 Introduction

A low-density geochemical survey, the National Geochemical Survey of Australia (NGSA), covering > 6 million km^2 has been completed for the first time in Australia [1]. The NGSA was conducted between 2007 and 2009 by Geoscience Australia and all States/NT (http://www.ga.gov.au/ngsa). Catchment outlet sediment samples were collected at 1186 sites (ignoring duplicate sites) from both top (0 – 10cm) and bottom (∼ 60 – 80cm) soil. These were sieved to two grain-size fractions (<2 mm, and <75 μm), resulting in four data sets. Here data for 51 elements (Al, As, Au, Ba, Be, Bi, Ca, Ce, Co, Cr, Cs, Cu, Dy, Er, Eu, F, FeT, Ga, Gd, Ge, Hf, Ho, K, La, Lu, Mg, Mn, Na, Nb, Nd, Ni, P, Pb, Pr, Rb, Sc, Se, Si, Sm, Sn, Sr, Tb, Th, Ti, U, V, Y, Yb, Zn and Zr) determined by total analysis, plus Loss on Ignition (LOI), were selected.

As geochemical data are compositional, the survey data form regionalised compositions and prior to any analysis a transform is applied to account for the constant sum constraint and avoid spurious correlations. Following [2] the centered logratio (clr) transform [3] was used. Caritat and Grunsky [2] applied Principal Component Analysis (PCA) with subsequent ordinary kriging to the clr transformed data. They found that the first four factors account for 59 % of the total variability of the data

U. Mueller (✉) · J. Lo
School of Engineering, Edith Cowan University, Joondalup, WA 6027, Australia
e-mail: u.mueller@ecu.edu.au

J. Lo
e-mail: j.lo@ecu.edu.au

P. de Caritate
Geoscience Australia, GPO Box 378, Canberra ACT 2601, Australia
e-mail: Patrice.deCaritat@ga.gov.au

E. Grunsky
Geological Survey of Canada, Natural Resources Canada, 601 Booth St., Ottawa K1A 0E8, Ontario, Canada
e-mail: Eric.Grunsky@NRCan-RNCan.gc.ca

E. Pardo-Igúzquiza et al. (eds.), *Mathematics of Planet Earth*,
Lecture Notes in Earth System Sciences, DOI: 10.1007/978-3-642-32408-6_24,

and that the smoothed maps obtained through ordinary kriging are suitable for the identification of multivariate associations and patterns that are broadly consistent with the knowledge of geological processes.

2 Objectives

In [2] detailed structural analysis was performed on the PCA factors. Whilst PCA factors are uncorrelated at separation distance 0, there is still remnant spatial correlation between distinct factors, which require a treatment via multivariate geostatistical techniques. The application of the method of minimum/maximum autocorrelation factors (MAF) results in spatially decorrelated factors which will form the basis of the structural analysis performed here. At the same time an assessment will be made of the suitability of MAF factors for delineating geochemical regions.

3 MAF-Transform

Let $\mathbf{Z}(\mathbf{u}) = [Z_1(\mathbf{u}), Z_2(\mathbf{u}), \ldots, Z_K(\mathbf{u})]^T$ be the multivariate random function with correlation matrix M where K is the number of attributes, and $\mathbf{z}(\mathbf{u}) = [z_1(\mathbf{u}_\alpha), z_2(\mathbf{u}_\alpha), \ldots, z_K(\mathbf{u}_\alpha)]^T, \alpha = 1, \ldots, n$, be the corresponding sample vectors.

For MAF it is assumed that the semivariogram function, $\boldsymbol{\Gamma}(\mathbf{h})$, of the multivariate random function $\mathbf{Z}(\mathbf{u})$ can be modelled by a two structure linear model of coregionalisation $\boldsymbol{\Gamma}(\mathbf{h}) = M_1 g_1(\mathbf{h}) + (M - M_1) g_2(\mathbf{h})$ where the functions $g_1(\mathbf{h})$ and $g_2(\mathbf{h})$ are permissible semivariogram models and the matrix M_1 contains the sills of first model structure [4]. The matrix M_1 is approximated by the experimental semivariogram matrix $\hat{\Gamma}(\mathbf{h}_1)$ at a suitably chosen lag $\mathbf{h}_1$, usually a vector whose length reflects the sampling distance. To jointly diagonalise M and $\hat{\boldsymbol{\Gamma}}(\mathbf{h}_1)$, a PCA is performed on M followed by a spectral decomposition of $V_1 = \Lambda^{-1/2} Q^T \hat{\boldsymbol{\Gamma}}(\mathbf{h}_1) Q \Lambda^{-1/2}$. The matrix Q is the PCA factor matrix and Λ is the corresponding diagonal matrix factor variances. MAF factors derived in this way are ordered so that the strength of autocorrelation decreases with increasing index with the last few factors often exhibiting no spatial structure at all.

In the case of clr-transformed data the correlation matrix M is singular since the sum of the clr-transformed variables is 0 by construction. One method that may be used to avoid the singularity is to remove one of the clr-transformed variables from the data set prior to the calculation of the MAF and this approach has been used here.

4 Results and Discussion

To ensure nonsingularity of the correlation matrix, the clr-transformed LOI was excluded from the calculation of the MAF factors. This is an arbitrary choice and any other element could have been used instead. However, given that LOI was computed as the difference between 100 % and the sum of major oxides and major trace elements [2], its choice seemed the most appropriate. It is to be noted that LOI was included in the calculation of the geometric mean (and so in the clr transformation), and only discarded for the calculation of the MAF factors. As the sampling distance was roughly 75 km, the lag distance at which $\hat{\boldsymbol{\Gamma}}(\mathbf{h}_1)$ was calculated was set to 100 km with a tolerance of 50 km. The angular tolerance was set to 90°, so that effectively $\hat{\boldsymbol{\Gamma}}(\mathbf{h}_1)$ was calculated as omnidirectional.

An examination of the semivariogram maps and experimental semivariograms of the MAF factors shows that the first 12 factors exhibit zonal anisotropy of decreasing strength, followed by isotropic behaviour up to factor 40 and then finally pure nugget. The relative contribution of the nugget component exceeds 80 % from factor 21 onwards for the coarse grain size, and from factor 25 for the fine grain size. For MAF factors from 30 onwards, ranges are typically around 100 km.

Directions of greatest continuity for the first six factors are shown in Table 1. These are reasonably consistent when considered across soil type and grain size. Exceptions are factors two and three where Bc shows different continuity compared to the others. Except for these two factors the direction of greatest continuity is towards NW to NNW.

Linear kriging was used to generate spatial maps of the MAF factors. In the first factor the western Yilgarn Craton stands out in warm colours. This area is characterised by granites (and rare mafic dykes). The clarity of the appearance is dependent on both soil layer and grain size. The fabric in the Yilgarn is compatible with N155 shown on the variograms; that is the strike of the mafic/ultramafic dykes, which are especially abundant in the eastern Yilgarn. The cool colours outline the shape of the Eromanga (Great Artesian) Basin in Queensland and surrounding or underlying basins (Bowen/Surat Basins).

In the second MAF factor positive values highlight the New England Fold Belt (north of Sydney) and its extension into southeastern Queensland (north of Brisbane).

Table 1 Directions of greatest continuity for the first six MAF factors by soil layer and grain size (B = bottom soil, T = top soil, f = fine and c = coarse)

Factor	Tc	Tf	Bc	Bf
1	N150	N120, N150	N120, N160	N115, N160
2	N10	N15	N90	N10
3	N95	N95	N5	N120
4	N145	N100, N130	N145	N110
5	N140	N145	N130	N145
6	N140	N140	N135	N125

Lithologies there are mostly Palaeozoic to Mesozoic sediments and granites. Other areas with positive estimates are in southern and southeastern Victoria, where the Otway Basin is located (Mesozoic to Cainozoic sediments).

The higher factors can similarly be used for feature identification; however the feature identification is sensitive to the soil horizon and grain size.

5 Conclusion

A comparison with the maps in [2] shows that the application of a MAF transform results in more clearly delineated regions. It should be noted that in [2] a single PCA for all soil layers was implemented, while here the layers and fractions were treated separately. While the MAF transform allows for the generation of factors with decreasing continuity, the factor loadings cannot be interpreted in the same manner as those of a PCA, as was already observed in [4]. Notwithstanding this shortcoming the results are promising, but further work on choice of separation distance and potential combined layers is required.

References

1. Caritat, P., de Cooper, M. (2013). National geochemical survey of Australia: The geochemical atlas of Australia. Geoscience Australia Record, 2011/20, https://www.ga.gov.au/products/servlet/controller?event=GEOCAT_DETAILS&catno=71973 (last accessed 15 January 2013).
2. Grunsky, E. C., & de Caritat, P. (2013). Defining element associations and inferring geological processes from total element concentrations in Australian catchment outlet sediments: multivariate analysis of continental-scale geochemical data. *Applied Geochemistry*, *33*, 104–126.
3. Egozcue, J.J., Pawlowsky-Glahn, V. (2011). Basic concepts and procedures. In V. Pawlowsky-Glahn and A. Buccianti (Eds.), Compositional data analysis, theory and applications. (pp. 12–28) New York: Wiley & Sons.
4. Desbarats, J. A., & Dimitrakopoulos, R. (2000). Geostatistical simulation of regionalized pore-size distributions using Min/Max autocorrelation factors. *Mathematical Geology*, *32*(8), 919–942.

Discriminant Analysis of Palaeogene Basalt Lavas, Northern Ireland, Using Soil Geochemistry

Jennifer M. McKinley, Sam Roberson, Mark Cooper and Raimon Tolosana-Delgado

1 Introduction

The Palaeogene, comprising basalt lavas of the Antrim Lava Group, formed between *circa* 55 and 62 million years ago [1]. The Antrim Lava Group extends over an area of 4,009 km^2 in the north eastern corner of Northern Ireland [2] and is composed of olivine tholeiite Lower and Upper Basalt formations (LBF and UBF) that are separated by the Interbasaltic Formation. The latter represents a period of relative volcanic quiescence, and includes the quartz tholeiitic Causeway Tholeiitic Member (CTM) in north County Antrim. As a result there are mineralogical and geochemical differences observed between the olivine tholeiites of the LBF and UBF and the quartz olivine tholeiitic CTM.

1.1 Soil Geochemistry

A number of soil geochemical surveys have been completed across Northern Ireland, most recently this includes the Tellus Project which was completed between 2004

J. M. McKinley (✉)
School of Geography, Archaeology and Palaeoecology, Queen's University Belfast, Belfast BT7 1NN, UK
e-mail: j.mckinley@qub.ac.uk

S. Roberson · M. Cooper
Geological Survey of Northern Ireland, Belfast BT9 5BF, UK
e-mail: sam.roberson@detini.gov.uk

M. Cooper
e-mail: mark.cooper@detini.gov.uk

R. Tolosana-Delgado
Helmholtz Institute Freiberg for Resources Technology, Halsbrueckerstr. 34, D-09599 Freiberg, Germany
e-mail: r.tolosana@hzdr.de

E. Pardo-Igúzquiza et al. (eds.), *Mathematics of Planet Earth*,
Lecture Notes in Earth System Sciences, DOI: 10.1007/978-3-642-32408-6_25,

and 2006. This comprehensive soil geochemical dataset, comprising a total of 6,862 soil samples analysed by XRF for 52 elements, provides a means to characterise the basalts and identify potential differences between the formations. Analysis of trace metal concentrations in the shallow soils Tellus dataset revealed elevated Ni and Cr concentrations across the Antrim Lava Group [3] that exceed current soil guidelines values (SGVs), which represent generic assessment criteria with regard to identified risks posed to human health by chronic exposure to contaminated soil. The findings from bioaccessibility testing [3] indicated that trace element bioaccessibility is specific to individual geologic formations. However, the contaminants of Ni and Cr were not found to be readily bioaccessible, reducing the risk to human health. Specific soil properties including parent rock material and the co-occurrence of major element oxides can influence the mobility and bioavailability of PTEs. Therefore greater understanding of the factors controlling the geochemical composition of soils would elucidate the risks posed by elevated geogenic PTEs in soils overlying the Antrim Lava Group. With this aim, we apply a linear discriminant analysis (LDA) to the Tellus geochemical database. This is based on a logratio representation of the components, to comply with the obvious compositional nature of the data set. LDA provides a set of directions on the space of compositions which optimally separate each pair of formations. These directions do not depend on which logratio transformation was used to do the computations.

1.2 Spatial Variation in Total Concentrations and Bioaccessibility

Spatial prediction using kriging reveals that the presence of the basaltic lavas of the Antrim Lava Group has a strong control over the spatial distribution of several of the PTE concentrations, though a detailed differentiation of UBF and LBF was not possible. The olivine rich basalts (mostly the UBF and LBF) contain high concentrations of Ni (Fig. 1). Samples overlying the LBF contained the greatest amount of bioaccessible Ni (Fig. 1a), while the UBF showed greater amounts of bioaccessible Cr (Fig. 1b). This suggests that distinct bioacessibility fractions are associated with different lithological controls [3].

2 Statistical Methods

2.1 Linear Discriminant Scores

Discriminant analysis (Fig. 2) indicates that there is no difference between the UBF and LBF groups, as a reclassification table yields a misclassification rate comparable to a random classification.

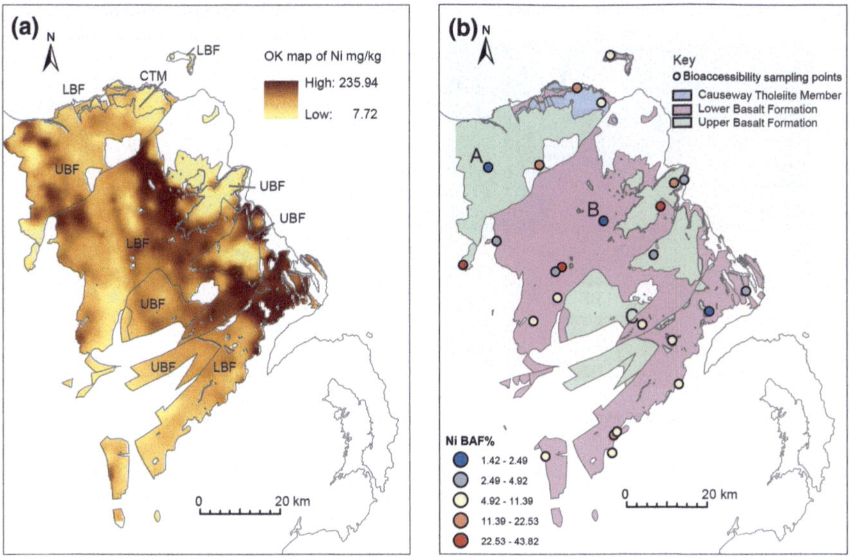

Fig. 1 **a** Kriged output map (*Ni mg/kg*), **b** Bioacessibility (Ni BAF%) for the Basalt formations [3]

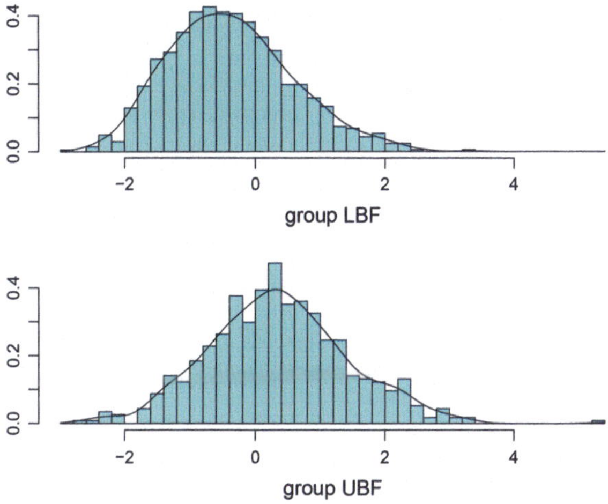

Fig. 2 Histograms of the linear discriminant scores for the Basalt formations (UBF and LBF)

In Fig. 3a the areas represent the conditional probabilities that samples belong to the two groups, the Basalt Formations, fixed a value of the logratio K/Mn (the most informative pairwise ratio). It can be seen that the data are mixed and no separation is possible. A comparison of the centered-log ratio-average composition of the soils over each Basalt Formation (Fig. 3b) shows no difference between them.

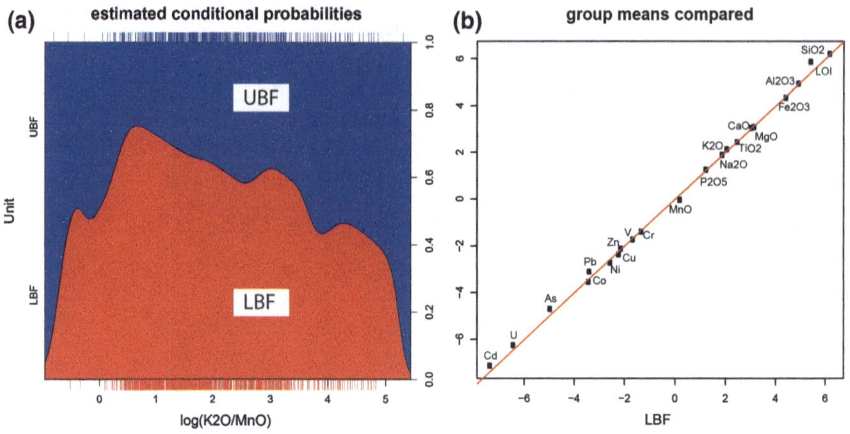

Fig. 3 **a** Non-parametric estimation of the probability that samples belong to either Basalt Formation given a certain value of the most informative pairwise ratio of the soil composition (in this case, K to Mn), **b** A comparison of the centered-log ratio-average composition of the soils over each Basalt Formation

Results show that any posterior statistical treatment, including mapping and subsequent analysis of bioaccessibility, can be safely done with the whole data base. The two main formations are geochemically equivalent.

Acknowledgments The project was carried out by the Geological Survey of Northern Ireland (GSNI) and was funded by The Department for Enterprise, Trade and Investment (DETINI) and The Rural Programme through the Northern Ireland Programme for Building Sustainable Prosperity.

References

1. Cooper, M. R. (2004). Palaeogene extrusive igneous rocks (Chap. 14). In W. I. Mitchell (Ed.), *The geology of Northern Ireland—Our natural foundation* (2nd ed., pp. 167–178). Belfast: Geological Survey of Northern Ireland.
2. Geological Survey of Northern Ireland. (1997). *Northern Ireland Solid Geology, 1:250 000 Keyworth* (2nd ed.). Nottingham: British Geological Survey.
3. Barsby, A., McKinley, J. M., Ofterdinger, U., Young, M., Cave, M. R., & Wragg, J. (2012). Bioaccessibility of trace elements in soils in Northern Ireland. *STOTEN*, *433*, 398–417.

Chemical Equilibria in Compositional Data

K. Gerald van den Boogaart, Raimon Tolosana Delgado and Silke Konsulke

1 Introduction: The Chemical Master Equation

This contribution is concerned with the chemical composition of aqueous solutions, which are typically observed compositionally, i.e. the total is not properly observed and not all chemical species are quantified. For D molecular species $S_1, \ldots, S_D$ equilibrium reactions might govern the interdependences between the various components:

$$-\nu_1 S_1 + \cdots + -\nu_r S_r \leftrightarrow \nu_{r+1} S_{r+1} + \cdots + \nu_D S_D$$

Here $|\nu_i| \in \mathbb{N}$ denotes the stochiometric multiplicity in which a chemical species S_i is used in the equation. The left hand side constants (those of reactants) are defined negative to keep notation simple hereafter.

Various approaches can be used to describe the equilibrium. The law of mass action predicts the speeds of reaction based on collision theory and the molar fractions n_i by: $s_{\rightarrow} = c_{\rightarrow} \prod_{i=1}^{r} n_i^{-\nu_i}$, and $s_{\leftarrow} = c_{\leftarrow} \prod_{i=r+1}^{D} n_i^{\nu_i}$, where $c_{\rightarrow}$ and $c_{\leftarrow}$ are two temperature-dependent constants. Consequently we reach equilibrium when $s_{\leftarrow} = s_{\rightarrow}$, i.e. when the chemical master equation holds (see e.g. [1]):

K. G. van den Boogaart (✉) · R. T. Delgado (✉) · S. Konsulke (✉)
Helmholtz Institute Freiberg for Resource Technology, Halsbrücker Street. 34, 09599 Freiberg, Germany
e-mail: boogaart@hzdr.de; boogaart@math.tu-freiberg.de

R. T. Delgado
e-mail: r.tolosana@hzdr.de

S. Konsulke
e-mail: s.konsulke@hzdr.de

K. G. van den Boogaart
Institute for Stochastics, TU Bergakademie Freiberg, PrüferstraŠe 9, 09599 Freiberg, Germany

E. Pardo-Igúzquiza et al. (eds.), *Mathematics of Planet Earth*,
Lecture Notes in Earth System Sciences, DOI: 10.1007/978-3-642-32408-6_26,

$$K_{eq} := \frac{c_{\rightarrow}}{c_{\leftarrow}} = \frac{\prod_{i=r+1}^{D} n_i^{\nu_i}}{\prod_{i=1}^{r} n_i^{-\nu_i}} = \prod_{i=1}^{D} n_i^{\nu_i} \tag{1}$$

A thermodynamic analysis shows the same result. A system is in equilibrium when it shows the minimal Gibbs energy for its configuration. The derivative of the Gibbs energy G with respect to each of these state variables is thus zero

$$dG = V \cdot dp - S \cdot dT + \sum_{i=1}^{D} \mu_i \cdot dN_i = V \cdot dp - S \cdot dT + \sum_{i=1}^{D} \mu_i \cdot \nu_i d\nu_0 = 0, \tag{2}$$

where V, S, and $N_1, N_2, \ldots, N_D$ are the extensive variables volume, entropy and number of mols of each of the D species; p, T and $\mu_1, \mu_2, \ldots, \mu_D$ are respectively pressure, temperature and the chemical potentials, and $\nu_0 = f(c_{\rightarrow}, c_{\leftarrow})$ is the one-dimensional parameter describing the *extent of reaction*, i.e. its degree of completeness towards the right hand side (products). The chemical potential of each species is linked to its activity a_i through $\mu_i = \mu_i^0 + RT \ln a_i$, where μ_i^0 is a reference potential, and R the universal constant of gases. For this contribution we assume a closed aqueous solution system at a constant temperature and pressure (since they are controlled by the local geological environment, not by the chemical reaction itself). In this case, $dp = dT = 0$ and the activities are often considered proportional to a unit of concentration, e.g. molar fraction $a_i = \gamma_i n_i$. If activity coefficients γ_i can be taken as (approximately) constant, then potentials can be rewritten $\mu_i = \mu_i^{\varnothing} + RT \ln n_i$, with different reference levels $\mu_i^{\varnothing}$. Feeding them into (2), this gives $0 = dG = V \cdot 0 + S \cdot 0 + d\nu_0 \cdot \sum_{i=1}^{D} \left[\nu_i \mu_i^{\varnothing} + RT \nu_i \ln n_i\right]$ which implies

$$-\frac{\Delta G^{\varnothing}}{RT} = \sum_{i=1}^{D} \nu_i \ln n_i, \text{ with } \Delta G^{\varnothing} := \sum_{i=1}^{D} \nu_i \mu_i^{\varnothing} \tag{3}$$

thus again implying Eq. (1) with $K_{eq} := \exp\left(-\Delta G^{\varnothing}/RT\right)$. Note that the actual value of $\Delta G^{\varnothing}$ depends on which concentration units have been used to replace activities, and equivalent expressions can be derived for molar volume N_i/V.

2 A Compositional View to the Chemical Master Equation

Equation (3) can be written in vectorial form, with $\nu = (\nu_1, \ldots, \nu_D) \in \mathbb{N}^D \subset \mathbb{R}^D$, $\ln \mathbf{n} := (\ln n_1, \ldots, \ln n_D)$ and $(\cdot, \cdot)$ denoting a scalar product, as $\ln K_{eq} = (\nu, \ln \mathbf{n})$, i.e. the equilibrium ondition is a known hyperplane in the vector space of log-transformed molar fractions. When the concentrations are not given by molar concentrations, but some other measure like mass density, the n_i are rescaled and replaced by $n_i = k_i y_i$ and thus:

$$\ln K_{eq} = \sum_{i=1}^{D} \nu_i \ln(y_i k_i) = \sum_{i=1}^{D} \nu_i \ln y_i + \sum_{i=1}^{D} \nu_i \ln k_i$$

i.e. with logs applied component-wise to the vectors $(\nu, \ln \mathbf{y}) = \ln K_{eq} - (\nu, \ln \mathbf{k}) =: \ln K$. Thus the same linear projection has again a fixed but different value.

However, in many natural systems, concentrations are observed compositionally, i.e. only up to joint scaling constant, e.g. as equivalent to $z_i = y_i / \sum_j y_j$. Due to the scaling invariance requirement, the quantities $\ln z_i$ are meaningless in general, and should be replaced by the centered log ratio transform (see e.g. [2] or [3] for details),

$$\text{clr}(\mathbf{z}) = \ln \mathbf{z} - \frac{(\mathbf{1}_D, \ln \mathbf{z})}{D}\mathbf{1}_D = \ln \mathbf{y} - \frac{(\mathbf{1}_D, \ln \mathbf{y})}{D}\mathbf{1}_D = \text{clr}(\mathbf{y}),$$

where $\mathbf{1}_D$ is a vector of D ones, i.e. the clr and the log-transformed concentrations differ only by a constant value $\ln g\,(\mathbf{y}) = (\mathbf{1}_D, \ln \mathbf{y})/D$, which corresponds to the log-geometric mean of the observed components, and is a form of total concentration. Actually, the real concentrations $\mathbf{y}$ and their compositional representation $\mathbf{z}$ differ only on that constant, so that their log-transformed vectors are different, but their clr-transformed vectors are the same. We can then distinguish two kinds of reactions, in the way they behave with respect to $\mathbf{y}$ and $\mathbf{z}$.

3 Reactions Preserving the Amount of Matter

In the case of $\sum_i \nu_i = (\mathbf{1}_D, \nu) = 0$, the reaction preserves the amount of matter and we get

$$(\nu, \text{clr}\,(\mathbf{z})) = (\nu, \text{clr}\,(\mathbf{y})) = (\nu, \ln \mathbf{y}) - \ln g\,(\mathbf{y}) \cdot \overbrace{(\nu, \mathbf{1}_D)}^{=0} = (\nu, \ln \mathbf{y}) = \ln K$$

i.e. the scalar product of clr $(\mathbf{z})$ with the stochiometric vector gives the same result as for unclosed, unobservable concentration $\mathbf{y}$. Moreover, the equilibrium condition defines a hyperplane of the Aitchison geometry. The equilibrium constant can thus be computed through $(\nu, \text{clr}\,(\mathbf{z})) = \ln K_{eq} - (\nu, \ln \mathbf{k})$, independently of further unobserved components, other reactions, etc. That, for instance, allows to estimate the temperature of a system using Eq. (3): $\hat{T}\,(\mathbf{z}) = \Delta G^{\varnothing} / \left(R\,((\nu, \text{clr}\,(\mathbf{z})) + (\nu, \ln \mathbf{k}))\right)$. Of course, this may offer ways to derive thermodynamically based geothermometers, though further research is needed to assess its practical relevance.

4 Reactions Not Preserving the Amount of Matter

In the other case $\sum_i \nu_i = (\mathbf{1}_D, \nu) =: \kappa \neq 0$, the projection of clr transformed concentrations and compositions provide:

$$(\nu, \operatorname{clr}(\mathbf{z})) = (\nu, \operatorname{clr}(\mathbf{y})) = (\nu, \ln \mathbf{y}) - \ln g(\mathbf{y}) \cdot (\nu, \mathbf{1}_D) = \ln K_{eq} - (\nu, \ln \mathbf{k}) - \kappa \cdot \ln g(\mathbf{y})$$

i.e. for systems in equilibrium at known temperature

$$\hat{\tau}(\mathbf{z}) := \ln g(\mathbf{y}) = \frac{1}{\kappa}\left(\ln K_{eq} - (\nu, \ln \mathbf{k}) - (\nu, \operatorname{clr}(\mathbf{z}))\right)$$

where the right hand side only contains information we typically have. It is thus possible to reconstruct the total from the compositional data if the equilibrium constant K_{eq} is known. Knowing that total, one can then recover the individual values of concentration from the proportions through their equivalence in clr's: $\hat{y}_i = \exp\left(\operatorname{clr}(\mathbf{z})_i + \hat{\tau}(\mathbf{z})\right)$. Note that this may allow to estimate the total matter of species S_i per unit of volume if the concentration y_i is molar volume.

5 Conclusions

In simple aqueous systems, an iso-molar equilibrium represents a hyperplane in the Aitchison geometry for compositional data. Its leading vector is the vector of stochiometric coefficients of the reaction, and its position depends on the temperature, standard Gibbs energy of the reaction and the units of concentration used. This may allow to estimate temperature from the proportions, thus build a geothermometer. In equilibrium reactions not preserving the total number of mols, the unobservable amounts of matter of the elements in play can be inferred from their proportions, if the equilibrium constant is known. In both cases, the leading vectors of the reactions in play may define an appealing basis or coordinate reference system for the composition.

References

1. Gillespie, D. T. (1992). A rigorous derivation of the chemical master equation. *Physica A, 188*, 404–425.
2. van den Boogaart, K. G., & Tolosana-Delgado, R. (2013).Analyzing compositional data with R (Use R!). (p. 272). The Netherlands: Springer.
3. Pawlowsky-Glahn, V.,& Buccianti, A.(2011). Compositional data analysis: Theory and Applications. Chichester: John Wiley, DOI: 10.1002/9781119976462.

Analysis of Total Abundances of Phytoplankton Compositions in a River

Vera Pawlowsky-Glahn, Juan J. Egozcue and David Lovell

1 Introduction: Theory and Methods

We deal with data $\mathbf{x} = [x_1, x_2, \ldots, x_D]$ in the real positive orthant ($\mathscr{R}_+^D$). Taking logarithms on each component, $\mathscr{R}_+^D$ has a D-dimensional real Euclidean vector space structure [4]. The closure of the original data leads to a composition in the simplex $\mathscr{S}^D$, where we assume the Aitchison geometry [3] is appropriate. The total is described by a product or a sum and, in general, by a positive function of the original components with a relative scale in $\mathscr{R}_+$. Jointly, the total and the composition are elements of the product space $\mathscr{R}_+ \times \mathscr{S}^D = \mathscr{T}$. A D-dimensional real Euclidean vector space structure is obtained in $\mathscr{T}$ considering the Aitchison geometry in $\mathscr{S}^D$, and logarithms in $\mathscr{R}_+$ [4]. To apply standard multivariate analyses, this product space $\mathscr{T}$ must have Euclidean structure [2]. But not all total functions are compatible with the assumed vector space structure in $\mathscr{R}_+^D$ and $\mathscr{S}^D$. Inconsistencies typically manifest when comparing the mean value calculated in $\mathscr{T}$ and in $\mathscr{R}_+^D$. For example, inconsistencies arise with a total function that is the sum of all original components, but not with a total function $t_p = \left(\prod_{i=1}^{D} x_i\right)^{1/\sqrt{D}}$ which ensures $\mathscr{T}$ and $\mathscr{R}_+^D$ are isometric. The following analysis is performed using (a) $\mathscr{R}_+^D$; (b) $\mathscr{T}_p = \left(\mathscr{R}_+ \times \mathscr{S}^D\right)$ (with the total t_p); and (c) $\mathscr{T}_s = \left(\mathscr{R}_+ \times \mathscr{S}^D\right)$ (with the total $t_s = \sum_{i=1}^{D} x_i$). The number of parts is $D = 8$, and equivalences and differences of the three cases are shown.

V. Pawlowsky-Glahn(✉)
Universitat de Girona, Girona, Spain
e-mail: vera.pawlowsky@udg.edu

J. J. Egozcue
Universitat Politècnica de Catalunya, Barcelona, Spain
e-mail: juan.jose.egozcue@upc.edu

D. Lovell
CSIRO-CMIS, Canberra, Australia
e-mail: David.Lovell@csiro.au

E. Pardo-Igúzquiza et al. (eds.), *Mathematics of Planet Earth*,
Lecture Notes in Earth System Sciences, DOI: 10.1007/978-3-642-32408-6_27,

2 Phytoplankton Abundances

Phytoplankton abundances were measured over a period of 14 years in an Australian river. Interest was in the evolution of the total abundance in time, and its possible relation to taxa present. After removal of incomplete compositions, data consisted of 173 samples of the following taxa: *Anabaena* (ana), *Aphanizomenon* (aph) and *Other Cyanophyceae* (ocy) of the group of *Cyanobacteria*, and *Actinastrum* (act), *Cryptophyceae* (cry), *Ankistrodemus* (ank), *Aulacoseira distans* (aud) and *Aulacoseira granulata* (aug) of the group of *Algae*. The exploratory analysis suggested two different periods, before 1998 (labelled B, 58 samples), and after 1998 (labelled A, 115 samples). Our aim is to check whether there is a statistical difference in the centre values [3] of samples B and A, as there is a suspicious increment of the total abundance.

A first step is computing the centres for samples B and A, and considering all the samples belonging to the same sample, labelled BA (Table 1). The upper part of the table is the centre, Cen_+, computed in $\mathscr{R}_+^D$, and given in abundances per liter. Under the labels t_s and t_p, the totals corresponding to Cen_+ are specified. The lower part of the table shows the centre of the composition, which is common to the three approaches $\mathscr{R}_+^D$, $\mathscr{T}_p$, $\mathscr{T}_s$, with the centre of t_s and t_p. The inspection of Table 1 reveals that the total sum of the centre in $\mathscr{R}_+^D$ is not equal to the centre total in $\mathscr{T}_s$. However, t_p of the centre in $\mathscr{R}_+^D$ is equal to the centre t_p, as predicted by the isometry between $\mathscr{R}_+^D$ and $\mathscr{T}_p$. Moreover, the differences between the centres appear to be substantial, thus announcing significant differences in the centres for all approaches.

Concerning metric variances in Table 2, $\mathscr{R}_+^D$ and $\mathscr{T}_p$ have the same metric variances as expected which, in turn, differ from the metric variance in $\mathscr{T}_s$.

The biplot representation in the three approaches Fig. 1 provides an insight about the features of the samples. The comparison shows that the data points are exactly the same in $\mathscr{R}_+^D$ and $\mathscr{T}_p$, as predicted by the isometry. The samples B and A appear quite well separated. In $\mathscr{T}_s$ the obtained projection is different, and it represents a lower percentage of metric variance. In $\mathscr{T}_p$ the first principal component is dominated by the total $\prod x_i^{1/\sqrt{D}}$, whereas the total sum is relatively less important in the first principal

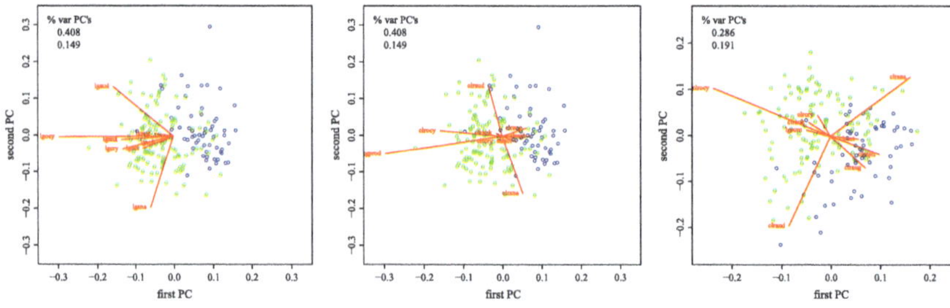

Fig. 1 Phytoplankton data biplots. Sample B: *blue*; A: *green*. *Left*: $\mathscr{R}_+^D$. *Middle*: $\mathscr{T}_p$. *Right*: $\mathscr{T}_s$

Table 1 Centres for samples B, A, and joint sample (BA)

Centre in $\mathscr{R}_+^D$										
	t_s	t_p	ana	aph	ocy	aug	aud	act	ank	cry
B	4244	7573866	627	491	94	2373	67	193	287	111
A	7530	69151799	957	671	856	3597	108	320	715	305
BA	6079	32945104	830	604	408	3129	92	270	526	218
Centres in $\mathscr{T}$										
	t_s	t_p	ana	aph	ocy	aug	aud	act	ank	cry
B	5456	7573866	0.148	0.116	0.022	0.559	0.016	0.046	0.068	0.026
A	9654	69151799	0.127	0.089	0.114	0.478	0.014	0.043	0.095	0.041
BA	7973	32945104	0.137	0.099	0.067	0.515	0.015	0.045	0.087	0.036

Upper table: abundances (counts per liter) in $\mathscr{R}_+^D$ and totals of the centre with the sum t_s and product t_p. Lower table: compositional centre and centre of the totals, t_s and t_p. The composition of the centre in $\mathscr{R}_+^D$ equals to the compositional part of centres in the product space

Table 2 Metric variances in $\mathscr{R}_+^D$, $\mathscr{T}_p$, and $\mathscr{T}_s$, for the three samples B, A, BA

Sample	$\mathscr{R}_+^D$	$\mathscr{T}_p$	$\mathscr{T}_s$	Sample	$\mathscr{R}_+^D$	$\mathscr{T}_p$	$\mathscr{T}_s$	Sample	$\mathscr{R}_+^D$	$\mathscr{T}_p$	$\mathscr{T}_s$
B	7.268	7.268	6.039	A	6.867	6.867	5.311	BA	8.705	8.705	6.241

component in $\mathscr{T}_s$. This comparison is reinforced by the fact that the compositional part coincides in $\mathscr{T}_p$ and $\mathscr{T}_s$. In the biplots the totals appear almost orthogonal to the separation of samples B and A, i.e. they play an important role in the discrimination of the two samples, thus confirming that differences between B and A are not only compositional but also in total abundances.

MANOVA techniques can be used for testing equal centres in samples B and A. The vectors of abundances are represented in coordinates and then treated as multivariate real variables. Certainly, in $\mathscr{R}_+^D$, $\mathscr{T}_p$ and $\mathscr{T}_s$, the difference of the centres is clearly significant, as expected after the representation in the biplots. Also biplots suggest coordinates with significant and non-significant different means. For instance, the balance of ana vs. aud has means in B and A which are not significative in an ANOVA F-test, while balances between variables pointing at opposite sides of the axis ana-aud have significant different means. The log-totals also have significant different means for the two samples. As a conclusion, the hypothesis of equal centres for the samples B and A should be rejected with a very low significance, much less than 10^{-4}.

In the studied case, the relevant total was assumed to be the sum of the abundances, and the approach in $\mathscr{T}_s$ provides appropriate information for the study of the Phytoplankton data set. The approach in $\mathscr{R}_+^D$ also provides similar information, but the product total dominates the analysis, as the first PC is essentially the product total.

3 Conclusions

Data carrying compositional information plus a total are frequently analysed as vectors in $\mathscr{R}_+^D$, taking logarithms and then using standard multivariate methods designed for real space $\mathscr{R}^D$. This practice is not compatible with the also common idea that the sum of all components, in $\mathscr{R}_+$, is a relevant total. Assuming the log-geometry in $\mathscr{R}_+^D$, and, simultaneously, that closed data are compositions, implicitly implies that the total is the product of all components to the power $1/\sqrt{D}$.

As a conclusion, if a data set in $\mathscr{R}_+^D$ is assumed compositional and the relevant total is the sum, to perform the analysis in the product space $\mathscr{T}_s = \mathscr{R}_+ \times \mathscr{S}^D$ is advisable. Sometimes, an analysis in $\mathscr{R}_+^D$ can give similar results to those obtained in $\mathscr{T}_s$, but circumstances in which they are similar are not clear.

One of the central issues in compositional data analysis is subcompositional coherence [1]. Therefore, the question arises on how subcompositional coherence is reflected in $\mathscr{R}_+^D$ and $\mathscr{T}_s$ or $\mathscr{T}_p$. Also, the basic approach to sampling zeros will be analogous to the usual approach in $\mathscr{S}^D$, although a detailed study is needed concerning the impact of substitution techniques on the total function chosen for analysis.

Acknowledgments This research has been supported by the Spanish Ministry of Education, Culture and Sports under a *Salvador de Madariaga* grant (Ref. PR2011-0290); by the Spanish Ministry of Economy and Competitiveness under the project METRICS Ref. MTM2012-33236.; and by the *Agència de Gestió d'Ajuts Universitaris i de Recerca* of the *Generalitat de Catalunya* under project Ref: 2009SGR424.

References

1. Aitchison, J. (2003). *The statistical analysis of compositional data*. (p. 416). USA: The Blackburn Press.
2. Mateu-Figueras, G., Pawlowsky-Glahn, V & Egozcue, J. J. (2011). The principle of working on coordinates. In V. Pawlowsky-Glahn & A. Buccianti (Eds.),*Compositional data analysis: Theory and applications*. (pp. 31–42). Chichester: Wiley (p. 378).
3. Pawlowsky-Glahn, V., & Egozcue, J. J. (2001). Geometric approach to statistical analysis on the simplex. *SERRA*, *15*(5), 384–398.
4. Pawlowsky-Glahn, V., Egozcue, J. J & Lovell, D. (2013). The product space $\mathscr{T}$ (tools for compositional data with a total). CoDaWork'2011, Vorau (A), June 2013.

Recursive Upward Sweeping and Updating Method on Ensemble Based Multiscale Algorithm in Data Assimilation

Chen Li, Shihua Chen, Chunlin Huang and Wei Gong

1 Introduction

Multiple measurements at different scales may contain implicitly redundant information which subdues the impact of model predictions or first-guess can lead to over-calibration. The ensemble multiscale Kalman filter(EnMSF), an ensemble approach based on MAR presented in [1] provided a multiscale update in data assimilation had been used for hydrologic data assimilation in [2]. However, the update at different scales can counteract interactively due to random error in observation information to result in under-estimate. Another ensemble multiscale filter using a recursive upward sweeping and updating method (Rec-EnMSF), is proposed to utilize the observational information at different scales asynchronously to get precise estimation. This improvement in upward sweeping steps is designed to utilize observational information hierarchically to avoid under-estimate. In this paper, EnMSF and Rec-EnMSF are introduced and compared, and numerical experiments with two-dimensional turbulent flow model with virtual measurements at different scales are used to illustrate efficiency and effectiveness of the algorithms.

C. Li (✉) · S. Chen
School of Mathematics and Statistics, Wuhan University, Wuhan 430072, China
e-mail: peterli@whu.edu.cn

S. Chen
e-mail: shcheng@whu.edu.cn

C. Li · C. Huang
Cold and Arid Regions Environmental and Engineering Research Institute,
Chinese Academy of Sciences, Lanzhou 730000, China
e-mail: huangcl@lzb.ac.cn

C. Li · W. Gong
State Key Laboratory of Information Engineering in Surveying, Mapping & Remote Sensing, Wuhan University, Wuhan 430079, China
e-mail: weigong@lmars.whu.edu.cn

E. Pardo-Igúzquiza et al. (eds.), *Mathematics of Planet Earth*,
Lecture Notes in Earth System Sciences, DOI: 10.1007/978-3-642-32408-6_28,

2 Principles of the EnMSF and Rec-EnMSF Algorithm

A typical two dimensional multiscale tree can be delineated in Fig. 1a, more details and properties can be found in [3, 4]. In traditional EnMSF in [1], model states variables at the finest scale are decomposed to scales firstly, and updated with observations at the corresponding scale if available. Then coarser nodes are used to smooth nodes at finer scales, namely downward sweep, see in Fig. 1a.

Then we propose an improved Upward Sweep steps, i.e. when nodes at scale m(sa) are decomposed into coarse scale m(s), if observations are available, then update the status of the nodes at scale m(s) and decompose them into coarse scale m(sγ); else if observations are not available, nodes at scale m(s) are decomposed into scale m(sγ). The cycle terminates until decomposition and update if available at the coarsest scale is finished and traditional downward sweep steps are also used (to smooth fine scale) This process is shown in Fig. 1c.

3 Results and Discussion

In this paper, the two-dimensional incompressible turbulent flow model (Re = 4000) is simulated in a square domain range of [0,1]×[0,1] with a 64×64 grid, i.e.Δx to 1/64 , and the number of total time steps is 100.

In the cases to compare EnKF and EnMSFs, multiscale tree is established as shown in Fig. 1a. The measurements are generated from a synthetic data set when the model runs forward sequentially without model error and assimilated into the model integrated forward with model error later.

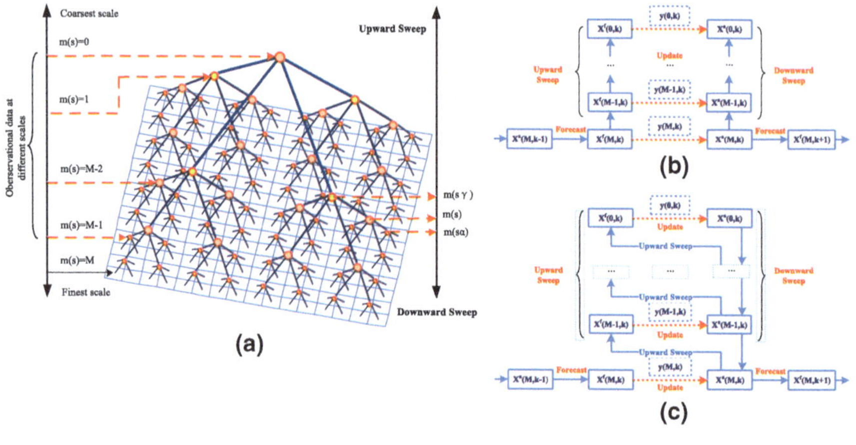

Fig. 1 **a** A typical multiscale tree; **b** Flow diagram for origin EnMSF; **c** Flow diagram for Recursive EnMSF

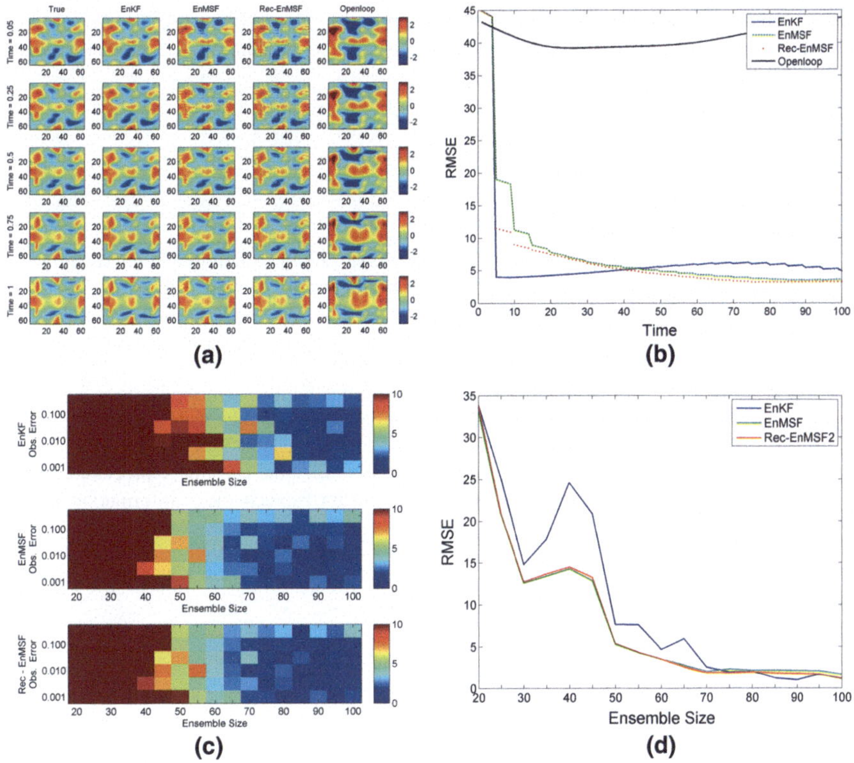

Fig. 2 **a** Evolution of model states at specific times for EnKF, EnMSFs, reference solution (true states) and open-loop, with ensemble size equals to 60 and variance of observation error is 0.1; **b** RMSE from initial time to the final are used to test the performance of these three assimilation algorithms; **c** Mean RMSE for EnMSFs and EnKF with different ensemble size varies from 20 to 100 by 5 and different coefficient σ_R^2 . **d** Mean RMSE for EnMSFs and EnKF with coefficient $\sigma_R^2 = 0.1$ and different ensemble size varies from 20 to 100 by 5

Experiments results in Fig. 2a indicate that both EnKF and EnMSFs can assimilate the observational information effectively, i.e. estimated analytical states with three methods can converge to the reference state gradually when models evolve forward in time. In Fig. 2b, it is obvious that trends of both the RMSE and model evolution are similar, and the EnMSFs has a better performance in this case with lower dispersion of ensemble samples, faster convergence and less deviation at the final. In Fig. 2c, d, mean RMSE declines with increasing ensemble size and the EnMSF also presented a better result than the EnKF here with consistently less deviations when ensemble size changes, and the EnMSFs also presented a better result than the EnKF here with consistently less deviations when ensemble size changes. When the coefficient is relatively large i.e. measurements are relatively inaccurate, the Rec-EnMSF is better than origin EnMSF with less RMSE.

Acknowledgments This work was supported by 973 Programs (2009CB723905, 2011CB707106), the NSFC (61273215, 10978003), One Hundred Person Project of the Chinese Academy of Sciences (29Y127D01).

References

1. Zhou, Y., McLaughlin, D., Entekhabi, D., & Ng, G.-H. (2008). An ensemble multiscale filter for large nonlinear data assimilation problems. *Monthly Weather Review*, *136*, 678–698.
2. Pan, M., Wood, E. F., McLaughlin, D. B., Entekhabi, D., & Luo, L. (2009). A multiscale ensemble filtering system for hydrologic data assimilation. Part I: Implementation and synthetic ex-periment. *Journal of Hydrometeorology*, *10*, 794–806.
3. Frakt, A. B., & Willsky, A. S. (2001). Computationally efficient stochastic realization for internal multiscale autoregressive models. *Multidimensional Systems and Signal Processing*, *12*, 109–142.
4. Willsky, A. S. (2002). Multiresolution Markov models for signal and image processing, *Proceedings of the IEEE*, *90*, 1396–1458. National Center for Biotechnology Information, http://www.ncbi.nlm.nih.gov.

Corrected Kriging Update Formulae for Batch-Sequential Data Assimilation

Clément Chevalier, David Ginsbourger and Xavier Emery

1 Kriging Update Formulae for Batch-Sequential Data Assimilation

Let us consider a real-valued second-order random field Z indexed by $D \subset \mathbb{R}^d$. The term kriging is often used when one aims at calculating a linear predictor $\hat{Z}_n(\mathbf{x})$ and the associated prediction variance $\sigma_n^2(\mathbf{x})$ (often called kriging mean and variance), of the field Z at a point $\mathbf{x} \in D$, from a set of n observations, at locations $\mathbf{x}_1, \ldots, \mathbf{x}_n \in D$. Also, denote by $\sigma_n : (\mathbf{x}, \mathbf{y}) \in D^2 \to \sigma_n(\mathbf{x}, \mathbf{y}) := \mathbb{E}[(Z(\mathbf{x}) - \hat{Z}_n(\mathbf{x}))(Z(\mathbf{y}) - \hat{Z}_n(\mathbf{y}))]$ the *kriging covariance* function, giving covariances between kriging errors. Kriging means, variances and covariances can be computed using the so-called kriging equations, given in, e.g., [2]. Recently, a lot of effort has been put in reducing the cost for computing kriging means, variances and covariances when the observations are assimilated sequentially. In particular, when n observations are available, one may take advantage of previous computations to reduce the calculation cost of the kriging predictors when $k > 1$ additional observations are available, at locations $\mathbf{x}_{n+1}, \ldots, \mathbf{x}_{n+k}$. In that setup, [3] recently proposed the following kriging update formulae:

$$\hat{Z}_{n+k}(\mathbf{x}) = \hat{Z}_n(\mathbf{x}) + \sum_{i=1}^{k} \lambda_{n+i|n+k}(\mathbf{x}) \Big(Z(\mathbf{x}_{n+i}) - \hat{Z}_n(\mathbf{x}_{n+i}) \Big), \quad (1)$$

C. Chevalier (✉) · D. Ginsbourger
IMSV, University of Bern, Alpeneggstrasse 22, 3012 Bern, Switzerland
e-mail: clement.chevalier@stat.unibe.ch

D. Ginsbourger
e-mail: david.ginsbourger@stat.unibe.ch

X. Emery
Department of Mining Engineering and Advanced Mining Technology Center, University of Chile, Avenida Tupper, 2069 Santiago, Chile
e-mail: xemery@ing.uchile.cl

E. Pardo-Igúzquiza et al. (eds.), *Mathematics of Planet Earth*,
Lecture Notes in Earth System Sciences, DOI: 10.1007/978-3-642-32408-6_29,

$$\sigma^2_{n+k}(\mathbf{x}) = \sigma^2_n(\mathbf{x}) - \sum_{i=1}^{k} \lambda^2_{n+i|n+k}(\mathbf{x})\sigma^2_n(\mathbf{x}_{n+i}), \tag{2}$$

$$\sigma_{n+k}(\mathbf{x},\mathbf{y}) = \sigma_n(\mathbf{x},\mathbf{y}) - \sum_{i=1}^{k} \lambda_{n+i|n+k}(\mathbf{x})\lambda_{n+i|n+k}(\mathbf{y})\sigma^2_n(\mathbf{x}_{n+i}) \tag{3}$$

where $\lambda_{n+i|n+k}(\mathbf{x})$ denotes the kriging weight of $Z(\mathbf{x}_{n+i})$ when predicting $Z(\mathbf{x})$ relying on $Z(\mathbf{x}_1), \ldots, Z(\mathbf{x}_{n+k})$. In [3], Eqs. (2, 3) are proven only for $k = 1$. In fact, for $k > 1$, a counterexample for Eq. (2) can be obtained with the Brownian motion, as shown in a draft version of the present paper [1]. The next sections provide corrected formulae and an application in Gaussian field simulation.

2 Corrected Kriging Update Formulae

We now propose corrected expressions that replace Eqs. (2, 3). To improve the readability, we adopt the following simplified notations:

- $\mathbf{X}_{\text{old}} := \{\mathbf{x}_1, \ldots, \mathbf{x}_n\}$, and $\mathbf{X}_{\text{new}} := \{\mathbf{x}_{n+1}, \ldots, \mathbf{x}_{n+k}\}$,
- $\mathbf{Z}_{\text{old}} := (Z(\mathbf{x}_1), \ldots, Z(\mathbf{x}_n))$, and $\mathbf{Z}_{\text{new}} := (Z(\mathbf{x}_{n+1}), \ldots, Z(\mathbf{x}_{n+k}))$,
- $\boldsymbol{\lambda}_{\text{new,old}}(\mathbf{x}) := (\lambda_{1|n+k}(\mathbf{x}), \ldots, \lambda_{n|n+k}(\mathbf{x}))^\top$,
- $\boldsymbol{\lambda}_{\text{new,new}}(\mathbf{x}) := (\lambda_{n+1|n+k}(\mathbf{x}), \ldots, \lambda_{n+k|n+k}(\mathbf{x}))^\top$,
- $\sigma^2_{\text{old}}(\mathbf{x}) := \sigma^2_n(\mathbf{x})$, $\sigma^2_{\text{new}}(\mathbf{x}) := \sigma^2_{n+k}(\mathbf{x})$, and similarly for the covariances.

For conciseness and coherence, $\hat{Z}_n(\mathbf{x})$ and $\hat{Z}_{n+k}(\mathbf{x})$ are also denoted by $\hat{Z}_{\text{old}}(\mathbf{x})$ and $\hat{Z}_{\text{new}}(\mathbf{x})$, respectively. The corrected update formulae are given below:

Proposition 1 *(Corrected kriging update equations for the batch-sequential case)*

$$\hat{Z}_{new}(\mathbf{x}) = \hat{Z}_{old}(\mathbf{x}) + \boldsymbol{\lambda}_{new,new}(\mathbf{x})^\top(\mathbf{Z}_{new} - \hat{Z}_{old}(\mathbf{X}_{new})) \tag{4}$$

$$\sigma_{new}(\mathbf{x},\mathbf{y}) = \sigma_{old}(\mathbf{x},\mathbf{y}) - \boldsymbol{\lambda}_{new,new}(\mathbf{x})^\top \Sigma_{new}\boldsymbol{\lambda}_{new,new}(\mathbf{y}) \tag{5}$$

where $\Sigma_{new} := \text{Cov}[\mathbf{Z}_{new} - \hat{Z}_{old}(\mathbf{X}_{new})]$ *is the covariance matrix of kriging errors.*

Note that Eq. 4 is exactly the same result as Eq. 1 (which original proof is correct).

Proof Subtracting $Z(\mathbf{x})$ from both sides of Eq. 4, simple manipulations give

$$\hat{Z}_{\text{old}}(\mathbf{x}) - Z(\mathbf{x}) = (\hat{Z}_{\text{new}}(\mathbf{x}) - Z(\mathbf{x})) - \boldsymbol{\lambda}_{\text{new,new}}(\mathbf{x})^\top(\mathbf{Z}_{\text{new}} - \hat{Z}_{\text{old}}(\mathbf{X}_{\text{new}})).$$

Using the uncorrelatedness between kriging errors and observations, we then obtain

$$Var[\hat{Z}_{\text{old}}(\mathbf{x}) - Z(\mathbf{x})] = Var[\hat{Z}_{\text{new}}(\mathbf{x}) - Z(\mathbf{x})] + Var[\boldsymbol{\lambda}_{\text{new,new}}(\mathbf{x})^\top(\mathbf{Z}_{\text{new}} - \hat{Z}_{\text{old}}(\mathbf{X}_{\text{new}}))]$$

$$\sigma^2_{\text{old}}(\mathbf{x}) = \sigma^2_{\text{new}}(\mathbf{x}) + \boldsymbol{\lambda}_{\text{new,new}}(\mathbf{x})^\top \Sigma_{\text{new}}\boldsymbol{\lambda}_{\text{new,new}}(\mathbf{x})$$

which proves Eq. (5) for $\mathbf{x} = \mathbf{y}$. A proof with $\mathbf{x} \neq \mathbf{y}$ can be obtained similarly. □

Proposition 2 *(Kriging update equations in terms of kriging covariance)*

$$\Sigma_{new}\boldsymbol{\lambda}_{new,new}(\mathbf{x}) = \sigma_{old}(\mathbf{X}_{new}, \mathbf{x}) \tag{6}$$

$$\hat{Z}_{new}(\mathbf{x}) = \hat{Z}_{old}(\mathbf{x}) + \sigma_{old}(\mathbf{X}_{new}, \mathbf{x})^T \Sigma_{new}^{-1}(\mathbf{Z}_{new} - \hat{Z}_{old}(\mathbf{X}_{new})) \tag{7}$$

$$\sigma_{new}(\mathbf{x}, \mathbf{y}) = \sigma_{old}(\mathbf{x}, \mathbf{y}) - \sigma_{old}(\mathbf{X}_{new}, \mathbf{x})^T \Sigma_{new}^{-1}\sigma_{old}(\mathbf{X}_{new}, \mathbf{y}) \tag{8}$$

Proof We prove Eq. (6) using a Gaussian assumption on the field Z. The formula remains valid in non-Gaussian cases as the best linear prediction and the conditional expectation coincide in the Gaussian case. Using the orthogonal projection interpretation of the conditional expectation,

$$\begin{aligned} Z(\mathbf{x}) &= \mathbb{E}(Z(\mathbf{x})|\mathbf{Z}_{\text{old}}, \mathbf{Z}_{\text{new}}) + \overbrace{Z(\mathbf{x}) - \mathbb{E}(Z(\mathbf{x})|\mathbf{Z}_{\text{old}}, \mathbf{Z}_{\text{new}})}^{=:\varepsilon} \\ &= \boldsymbol{\lambda}_{\text{new,old}}(\mathbf{x})^\top \mathbf{Z}_{\text{old}} + \boldsymbol{\lambda}_{\text{new,new}}(\mathbf{x})^\top \mathbf{Z}_{\text{new}} + \varepsilon, \end{aligned}$$

with ε centered, and independent of $\mathbf{Z}_{\text{old}}$ and $\mathbf{Z}_{\text{new}}$. Let us now calculate the conditional covariance between $Z(\mathbf{x})$ and $\mathbf{Z}_{\text{new}}$ knowing the observations $\mathbf{Z}_{\text{old}}$:

$$\begin{aligned} \sigma_{\text{old}}(\mathbf{X}_{\text{new}}, \mathbf{x}) &:= Cov(\mathbf{Z}_{\text{new}}, Z(\mathbf{x})|\mathbf{Z}_{\text{old}}) \\ &= 0 + Cov\left(\mathbf{Z}_{\text{new}}, \boldsymbol{\lambda}_{\text{new,new}}(\mathbf{x})^\top \mathbf{Z}_{\text{new}} \mid \mathbf{Z}_{\text{old}}\right) + Cov(\mathbf{Z}_{\text{new}}, \varepsilon|\mathbf{Z}_{\text{old}}) \\ &= \Sigma_{\text{new}}\boldsymbol{\lambda}_{\text{new,new}}(\mathbf{x}) + Cov(\mathbf{Z}_{\text{new}}, \varepsilon|\mathbf{Z}_{\text{old}}) \end{aligned}$$

Noting that $Cov(\mathbf{Z}_{\text{new}}, \varepsilon|\mathbf{Z}_{\text{old}}) = \mathbf{0}$, the latter equation proves Eq. (6). Equations (7, 8) follow by plugging in Eq. (6) into Eqs. (4, 5). □

3 GP Simulation, with Batch-Sequential Data Assimilation

A well known algorithm for simulating M Gaussian process (GP) realizations in p points conditionally on n observations consists in adding to the kriging mean obtained with the n real observations M kriging residual functions artificially obtained based on non-conditional realizations [2, 4]. The kriging update formulae can be used in this algorithm to reduce computation costs in the case where one aims at smoothly "converting" GP realizations conditioned on n observations to realizations conditioned on $n+k$ observations (see, Fig. 1, with $n = 6$ and $k = 3$). Computing kriging means knowing n observations and k real or simulated new observations (denoted by $\mathbf{Z}_{\text{new}}$ and $\mathbf{Z}_{\text{sim}}$ respectively) requires to use Eq. (4). It appears that the difference between these two updated kriging mean functions (i.e. with k observations equal to $\mathbf{Z}_{\text{new}}$ and $\mathbf{Z}_{\text{sim}}$) only depends on $\mathbf{Z}_{\text{new}} - \mathbf{Z}_{\text{sim}}$ and on $\boldsymbol{\lambda}_{\text{new,new}}$, which can be obtained from Eq. 6. Then, the calculation of $\boldsymbol{\lambda}_{\text{new,new}}^\top(\mathbf{Z}_{\text{new}} - \mathbf{Z}_{\text{sim}})$ for p points has $O(pk)$

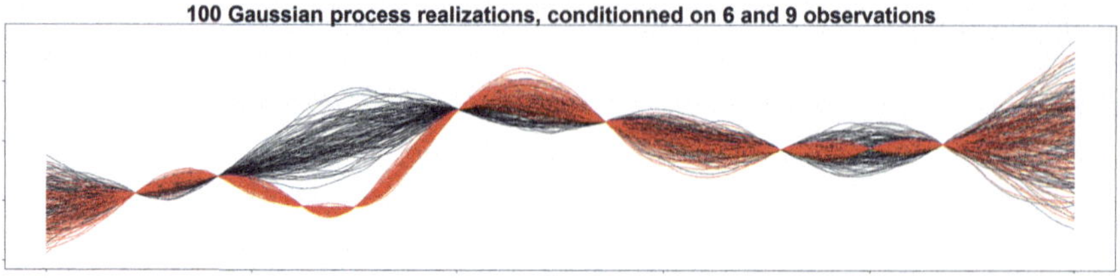

Fig. 1 100 GP realizations conditioned on 6 data (*black lines*) and 9 data (*red lines*)

complexity and $O(Mpk)$ for M simulations. This is faster than standard algorithms based on a decomposition (e.g., LU or Cholesky) of the $p \times p$ covariance matrix which require M matrix-vector product for a cost of $O(Mp^2)$. The gain of $O(p/k)$ can be substantial: in an application set up, with $M = 1, 00, 000$, $p = 200$, $n = 6$, $k = 3$, the computation time is divided by more than 10.

Acknowledgments Part of this work has been conducted within the frame of the ReDice Consortium, gathering industrial (CEA, EDF, IFPEN, IRSN, Renault) and academic (Ecole des Mines de Saint-Etienne, INRIA, and the University of Bern) partners around advanced methods for Computer Experiments. Clément Chevalier also gratefully acknowledges support from the French Nuclear Safety Institute (IRSN) and warmly thanks Prof. Julien Bect for fruitful discussions on Gaussian process simulations. David acknowledges support from the IMSV.

References

1. Chevalier, C., & Ginsbourger, D. (2012): Corrected kriging update formulae for batch-sequential data assimilation. http://arxiv.org/abs/1203.6452
2. Chilès, J. P., & Delfiner, P. (2012). *Geostatistics: Modeling spatial uncertainty* (2nd edn.). New York: Wiley.
3. Emery, X. (2009). The kriging update equations and their application to the selection of neighboring data. *Computational Geosciences*, *13*(1), 211–219.
4. Hoshiya, M. (1995). Kriging and conditional simulation of gaussian field. *Journal of Engineering Mechanics*, *121*(2), 181–186.

Analysis of the Parametrization Needs of Different Land Cover Classifiers: the Case Study of Granda Province (Spain)

Víctor F. Rodriguez-Galiano and MarioChica-Olmo

1 Introduction

The application of classification methods for the mapping and monitoring of land covers is one of the most relevant remote sensing applications. The accuracy of the generated mapping can be assessed quantitatively using the kappa index (K) or the overall accuracy (OA) [1]. This accuracy is not only dependent on the classification scene or the data themselves, but it is also strongly bound to the applied classification method. The accuracy of models induced by means of ML classifiers strongly depends of the combination of parameters used. Therefore, a detailed comparison cannot be carried out without previously establishing the optimal parameterization of each model.

Most studies focus on land cover mapping accuracy only, avoiding details about the parameter setting used in the training of classifiers. Additionally, it is also frequent to apply the default settings recommended by the commercial software used, or only a very limited number of experiments are carried out to determine the optimal parameters [5]. However, few are the studies based on remote sensing data which analyse the effect of parameter selection in ML algorithms [3]. Hence there is a need to study the impact of the parameterization of these algorithms for the classification of land covers and land uses in depth. This chapter discusses the crucial issues related to the parametrization of different up-to-date ML classifiers: classification trees (CT), artificial neural networks (ANN), support vector machines (SVM) and Random Forest (RF) [2].

V. F. Rodriguez-Galiano (✉) · M. Chica-Olmo
Dpto. de Geodinámica, Universidad de Granada, 18071Granada, Spain
e-mail: vrgaliano@ugr.es

M. Chica-Olmo
e-mail: mchica@ugr.es

E. Pardo-Igúzquiza et al. (eds.), *Mathematics of Planet Earth*,
Lecture Notes in Earth System Sciences, DOI: 10.1007/978-3-642-32408-6_30,

2 Data and Methods

Two Landsat Thematic Mapper-5 scenes of the same area in southeast Spain were captured (spring and summer). Both, spectral and auxiliary variables were used. On the one hand, spectral variables consisted of the Kauth Thomas multi-seasonal components of the summer and spring images. On the other hand, the auxiliary variables included in the analysis were elevation, slope and aspect, derived from the digital terrain model of spatial resolution equal to 20 m. These auxiliary variables were rescaled to the spatial resolution of the spectral variables (30 m). An exhaustive database for training and testing was obtained by visual interpretation of digital true-colour orthophotos (1:10000). The number of the training sites per class was kept equal (100 training sites and 50 testing sites per land cover category).

3 Results and Conclusions

The parametrization of classification algorithms has a great influence on their robustness and generalization capacity. Figure 1 shows significant differences in the accuracy obtained by the different ML methods according to the parameter setting used. CT classifiers were less accurate than the rest of methods, reaching moderate accuracy levels ($K < 0.85$). However, RF was very robust, with K values over 0.89 for most parameter combinations studied. RF, apart from being an operative method in terms of the simplicity of its parameters, also presented a greater stability against variations in its internal configuration. SVM and ANN obtained values equivalent to RF accuracy levels only for very specific parameter combinations. In the case of ANN this effect was much more noticeable: of the 15,580 different classifiers which were trained, only one reached a value for K equal to 0.89. Below mapping accuracy is quantitatively analysed with relation to the different parameters used in the building of each type of classifier [2].

The CT classification with a greater K index (0.85) was created by using the Gini index as a measure of heterogeneity, six variables at least in each node, and a depth of the tree of seven levels. The classification error increased, once a minimum of about six levels in depth was obtained, until reaching a maximum around 10, from which the error converged. Hence, it is preferable to limit the depth of trees so that these do not overfit data and, hence, the model does not lose generality in turn.

Regarding RF, it incorporates an additional parameter which is not considered in traditional decision trees: the m parameter. This m value remains constant while the tree grows, and the selection of variables is random. From about 100 trees the K value converged up to 0.91 and 0.89 for m equal to 1 and 9, respectively. The addition of more trees neither increased nor decreased the generalization error. However, an important increase in computation time was observed when a high number of trees was considered. Ensembles made up of few classification trees produced poor results, while greater ensembles produced more accurate classifications. The highest value

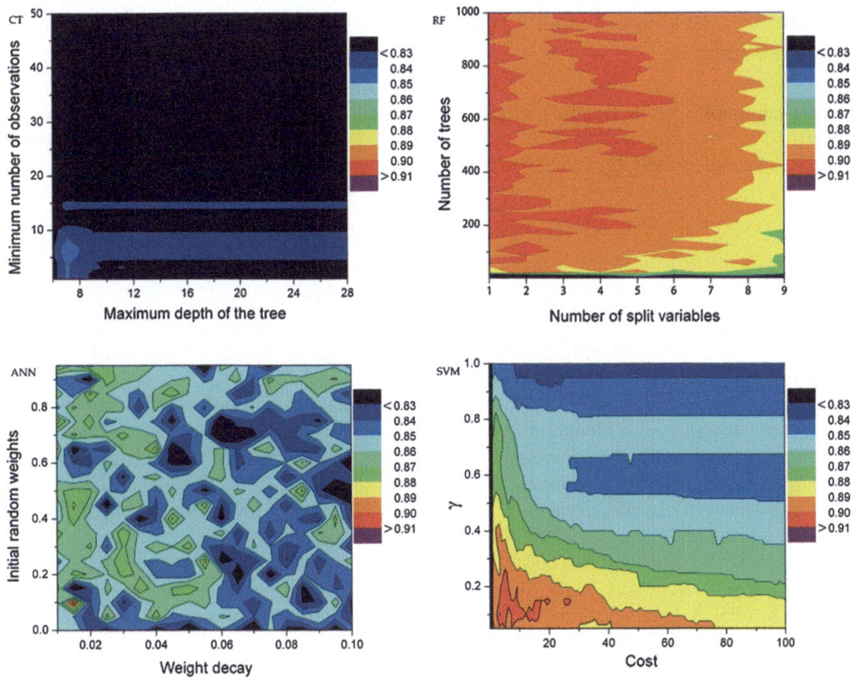

Fig. 1 Mapping accuracy (K) of all classifiers considering different parameter combinations. RF is the most stable classifier, reaching high accuracy values for most generated models. On the contrary, CT and ANN were extremely sensitive to the chosen combination of parameters and overfitting. SVMs show an average sensitivity to the chosen combination of parameters

for K (0.92) was obtained when 479 decision trees and only one random variable were used.

The maximum value for K was obtained for neural networks with a number of units in the hidden layer equal or over five. From this value the network error converged and the addition of more units did not have a significant impact on the accuracy of classifications. The initial value of weights had a significant impact on the K index. For negative weight values a very low K was obtained, while for positive values over 0.5 the highest K was obtained (Fig. 1). These results differ from those recommendations by [4] who suggest setting the range of weight values between −1 and 1. ANNs reached a maximum K for very low weight decay values (0.01).

K maximum values were obtained for SVMs with costs below 10, which indicates that the training and test data used in the calibration of the classifiers had a very low number of outliers. SVMs underwent a decreasing tendency of the K index with relation to cost. The best K values were obtained for very small gamma values of between 0.05 and 0.1, from which a more important decrease than that of the cost parameter took place. This parameter, gamma, is traditionally fixed to the value of the inverse of the number of input variables into the classifier [5]. However, in view of our results, we believe the joint adjustment of both parameters, cost and gamma, to be more suitable.

Acknowledgments We are grateful for the financial support given by the Spanish MICINN (Project CGL2010-17629), Junta de Andalucía (Group RNM122) and the Portuguese Fundação para a Ciência e a Tecnologia (SFRH/BPD/89082/2012).

References

1. Congalton, R. G., & Green, K. (2009). *Assessing the accuracy of remotely sensed data: Principles and practices* (2nd ed.). Boca Raton, Florida: CRC Press.
2. Rodriguez-Galiano, V. F., & Chica-Rivas, M. (2012). Evaluation of different ML methods for land cover mapping of a Mediterranean area using multi-seasonal Landsat images and digital terrain models. *International Journal of Digital Earth*, doi:10.1080/17538947.2012.748848.
3. Rodriguez-Galiano, V. F., Ghimire, B., Rogan, J., Chica-Olmo, M., & Rigol-Sánchez, J. P. (2012). An assessment of the effectiveness of a random forest classifier for land-cover classification. *ISPRS-Journal of Photogramm Remote Sensing*, *67*, 93–104.
4. Venables, W. N., & Ripley, B. D. (2002). *Modern applied statistics with s. statistics and computing* (4th ed.). New York, USA: Springer.
5. Yang, X. (2011). Parameterizing support vector machines for land cover classification. *Photogrammetric Engineering and Remote Sensing*, *77*(1), 27–37.

Automatic Raman Spectra Processing for Exomars

Isaac Hermosilla Rodriguez, Guillermo Lopez-Reyes, D. R. Llanos and Fernando Rull Perez

1 Introduction

Raman spectroscopy is a spectroscopic technique used to study vibrational, rotational, and other low-frequency modes in a system [1]. It is widely used to provide information on chemical structures and physical forms, to identify substances from the characteristic spectral patterns (fingerprinting). So, it is possible to identify these materials through the study of their Raman spectrum.

The automated identification of Raman spectra is not an easy task due to the variability between different spectral acquisitions of the same sample, influenced by the different acquisition conditions (for example, spectra acquired with different equipment or laser wavelength). The identification problem is even harder when the spectrum corresponds to mixtures of more than one mineral. This work is developed for its use with the RLS instrument of the ExoMars mission, in which the apparatus function will be known and the necessary spectral corrections can be performed. However, the algorithmic here presented is intended at identifying the mineral phases of Raman spectra without the need to perform these corrections, providing a fast and robust identification tool from Raman spectra.

I. H. Rodriguez (✉) · G. Lopez-Reyes · D. R. Llanos · F. R. Perez
Unidad Asociada UVA—CSIC, Valladolid, Spain
e-mail: ihrog85@gmail.com

G. Lopez-Reyes (✉)
e-mail: guillermo.lopez.reyes@cab.inta-csic.es

D. R. Llanos
Department of Computer Science,Universidad de Valladolid (UVa), Valladolid, Spain

E. Pardo-Igúzquiza et al. (eds.), *Mathematics of Planet Earth*,
Lecture Notes in Earth System Sciences, DOI: 10.1007/978-3-642-32408-6_31,

2 Pre-Processing of Raman Spectra

The pre-processing of Raman spectra is intended at extracting the spectral information of interest from the spectra for its subsequent analysis and identification. Mainly, the information that has to be extracted from the spectrum is the peaks Raman shift (cm^{-1}) and intensity. This task is divided into three steps: noise reduction, baseline removal and peak detection, and it is performed automatically or with very little interaction with the user.

Noise reduction main goal is to improve the signal-to-noise ratio. Depending on the characteristics of the spectrum, different well-known smoothing filters as Savitzky-Golay or local regression (also known as loess) can be used [2].

Raman spectra are characterized by the presence of a background that provides the spectra a sample-dependent characteristic curvature which does not correspond to Raman emissions. The background removal is necessary because it can hinder the presentation, visualization and processing of the Raman spectral information. Several methods of automatic baseline determination can be found in the literature [3] or [4]. However, we have implemented a method to calculate the baseline based on an algorithm with an adjustable window of delta points that provides accurate results for spectra with different spectral characteristics. This algorithm calculates the baseline at each point as a function of the closest delta surrounding points. This delta value is adjusted depending on the spectral region to address for the different peak characteristics in the different regions. Thanks to this adjustment, the algorithm can provide very accurate baselines, as shown in Fig. 1. Its main advantages are a low number of iterations and computation that make it suitable to be included in the RLS instrument of the ExoMars mission [5]. The algorithm optimal configuration was established by minimizing the error in several dozens of cases, and its robustness was checked with Leave-one-out Cross-Validation.

The last step consists on the automatic location of the Raman peaks. This identification is currently performed by finding relative maxima over a threshold within the baseline corrected spectrum. This threshold is defined as a function of the spectrum

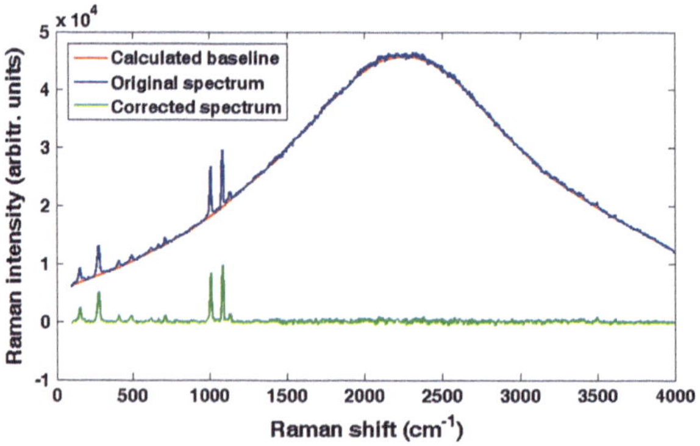

Fig. 1 Automatic baseline removal of a calcite + gypsum mixture

noise, allowing the algorithm to automatically adjust its behavior as a function of the spectral characteristics. The final product of the pre-processing routine consists on a list of peak positions and intensities that will be used to uniquely identify the materials the sample is comprised of.

3 Raman Spectra Identification

Once the spectral information of interest (peak positions and intensities) have been correctly detected, the recognition process kicks-off. The aim is to identify a spectrum of unknown materials, referred to as the problem spectrum. Our algorithm is based on band comparisons as proposed in [6], but our proposal also includes the intensity information (in addition to the Raman shift) to improve the identification accuracy. The algorithm aim is to create a tree of solutions capable of identifying both pure materials and mixtures in two steps.

The first step consists on retrieving the spectra that have a peak that matches with the most intense band in the problem spectrum. If all the bands of the problem spectrum match, it is therefore considered as a pure material, and the algorithm ends. On the contrary, if there are still unmatched bands, the problem spectrum will probably belong to a mixture of two or more materials, and the algorithm advances to the next step.

The second step consists on a recursive search which retrieves all the spectra that have coincidences with any of the unmatching bands from the previous step. Before each new iteration, the spectrum problem peak set is modified taking into account the results from previous iterations: if a band from the problem spectrum matches, the corresponding intensity of the reference spectrum is subtracted to address for mixtures in which there are overlapping bands. Once this has been made for all the matching peaks, the spectrum is renormalized. The recursive search will finish when all the problem spectrum bands are identified, or when there are not more candidates in the database.

For testing purposes, the algorithm has been run with 100 randomly selected spectra from the set of Crystal Sleuth software which is part of the RRUFF project [7], with pretty relaxed parameters for the band matching conditions to validate in a more realistic scenario. The results showed that 100 % of the spectra were included among the possible spectra reference sets provided by the software. All results lists included the original spectrum as a suggestion that matched with the Raman bands of the problem spectrum. In 55 % of the cases only one suggestion was returned, so, in these cases, the problem spectrum was unequivocally identified. However, in 21 % of the cases the number of possible results was too large (ranging from 28 to 628 suggestions), but this only happened when the problem spectrum had only one or two Raman bands. In the remaining cases, the number of suggestions was lower than 15. In contrast, when intensity was not taken into account (as in [6]), the results were far worse: no unequivocal identifications, and 45 % (instead of 21 %) of results with too many possible coincident spectra sets.

4 Conclusion

The algorithms presented here are capable of processing and identifying Raman spectra. Some of them will be used on-board or in the ground segment of the ExoMars mission for the analysis of the Raman spectra provided by the RLS instrument.

Besides the algorithms presented here, a software platform has been created to test them and to provide users with a tool for processing Raman spectra.

The development of these algorithms is still in progress, but it has been shown that this kind of analytical tools can provide reliable pseudo-automated means for the analysis of Raman spectra.

References

1. Long, D. A. (2002). *The Raman effect: A unified treatment of the theory of Raman scattering by molecules* (Vol. 1, 1st ed.). New York: Wiley.
2. Quintero, A. L., Hunt, S., & Diem, M. (2007). Denoising of Raman spectroscopy signals, Research Thrust R2 Publications, Paper 2007.
3. Weakley, A. T., Griffiths, P. R., & Aston, D. E. (2012). Automatic baseline subtraction of vibrational spectra using minima identification and discrimination via adaptive, least-squares thresholding. *Applied Spectroscopy*. *66*, 519–529.
4. Jirasek, A., Schulze, G., Yu, M. M. L., Blades, M. W., & Turner, R. F. B. (2004). Accuracy and precision of manual baseline determination. *Applied Spectroscopy*, *58*, 1488–1499.
5. Rull Perez, F., & Martinez-Frias, J. (2006). Raman spectroscopy goes to mars. *Spectroscopy Europe*, *18*, 18–21.
6. Vandenabeele, P., Hardy, A., Edwards, H. G. M., & Moens, L. (2011). Evaluation of a spectral searching algorithm for the comparison of Raman band positions. *Spectrochimica Acta Part A*, *80*, 8–13.
7. Database of Raman spectra, X-ray diffraction and chemistry data for minerals. Retrieved January 27, 2013, from http://rruff.info.

Fuzzy Parameterization of a Filtration Model for a Non-homogeneous Sedimentary Rock

Elena Savelyeva and Aleksander Rastorguyev

1 Introduction

Usual procedure to fit parameters of a filtration model deals with inverse modeling [1] under the assumption of homogeneity of the predefined sedimentary layers. It means that each sedimentary layer is characterized by single values or tensors of so called effective parameters. Such assumption can be crucial in the case of strongly non-homogeneous sedimentary layers, as for example, an aquifer with rather large crops of low conducting clay.

As an alternative to sedimentary layers one can consider lithological types (lithofaces) forming the geological structure. The information on their vertical distribution together with samples can be obtained during borehole drilling. Laboratory experiments allow estimation of the filtration parameters. Thus the parameterization problem seems to transform into construction of a 3D model of lithofaces, or in other words a 3D classification problem. But the real situation is more complicated due to uncertainties: experimental (core permutations during drilling and extraction, non-accuracy of the laboratory devices, etc) and modeling (sparse distribution of boreholes) causes. So, parameters still require fitting before modeling. So, this fitting should also take into the account classification errors.

This work presents an approach uniting classification and fitting of the class related parameters. The approach was tested and applied to real data.

E. Savelyeva (✉) · A. Rastorguyev
Nuclear Safety Institute Russian Academy of Sciences (IBRAE RAS),
B.Tulskaya, 52, 113191 Moscow, Russia
e-mail: esav@ibrae.ac.ru

A. Rastorguyev
e-mail: alvr9@mail.ru

E. Pardo-Igúzquiza et al. (eds.), *Mathematics of Planet Earth*,
Lecture Notes in Earth System Sciences, DOI: 10.1007/978-3-642-32408-6_32,

2 Description of the Approach

The parameterization is an iterative approach. It starts by identification of an initial state defined by: (a) spatial distribution of lithofaces (classification); (b) intervals of values for the filtration parameters.

The classification problem is solved by probabilistic neural network (PNN) [2]. PNN estimates a posteriori probabilities for all classes (3). Which are the basis of the reclassification procedure—the substitution of lithological class-winner (class with the maximum probability (2)) by a class second to the winner. Bootstrap approach allows to identify the average number and positioning of reclassifications improving the classification result. For example, for data from presented below case study, a bootstrap approach with 1,000 repeats showed the best result was achieved after changing ≈31 % of classes with probability ≤0.6 also considering the continuity of classes.

Lithoface dependent filtration parameters are obtained by laboratory experiments or from reliable references. They are given in the form of intervals of admissible values accompanied by a likelihood function in a triangular form which in the mathematical theory of fuzzy sets is referred to as a grade of membership ($\mu_{K1}(y)$). According to the value of $\mu_{K1}(y)$ a value is attributed to one of fuzzy categories: nearly impossible, low possible, possible and highly possible. As the problem is a multiparametric one all intervals of values of parameters form a multidimensional cube. The grade of membership for a multidimensional case is defined by algebraic multiplication of the one-dimensional functions:

$$\mu_{K1 \cdot K2 \cdot \dots \cdot Kn}(y) = \mu_{K1}(y) \circ \cdots \circ \mu_{Kn}(y) \tag{1}$$

The first step of iteration procedure is to search a set of filtration parameters minimizing a misfit function (5) in the multidimensional cube for the given geological structure. For minimization of the misfit a rude regular grid is stretched over the multidimensional cube of parameters' values. Misfit function is estimated for all nodes and those nodes where misfit's decreasing is observed are noted. Around them a finer grid is stretched, local minimums are discovered, and so forth till predefined definition of the grid is obtained. A minimum (or several minimums) found out for the finest grid is the result of minimization.

The second step deals with establishing the correspondence between the positions of the misfit's minimum within the multidimensional cube and a multidimensional fuzzy categories due to the grade of membership (1). If minimum has possible or highly possible category the fitting procedure finishes, otherwise the lithological structure changes (reclassification) and process continues with the first step.

2.1 Probability Neural Network

PNN is a supervised neural network constructing a classifier using Bayesian optimal or maximum a posteriori decision rule:

$$C(X) = \underset{i, i=1,\ldots,K}{argmax}\ P(C_i)p(X|C_i) \tag{2}$$

where $P(C_i)$ is a prior probability of class C_i, $p(X|C_i)$ is density of class conditional distribution for all locations X and K is a number of classes. The density $p(X|C_i)$ is estimated by non-parametric kernel method [3]. In this work 3-dimensional anisotropic Gaussian kernel ($W(X)$) is used:

$$p(X|C_i) = \frac{1}{N}\sum_{n=1}^{N} W(||X - X_n||) = \frac{1}{(2\pi)^{3/2} N}\sum_{n=1}^{N}\prod_{j=1}^{3}\left(\frac{1}{\sigma_j} e^{\frac{-\left\|x_j - x_j^{(n)}\right\|^2}{2\sigma_j^2}}\right) \tag{3}$$

where N is a number of samples in the training part of data and $\sigma_j > 0$, $j = 1, 2, 3$ are the scaling parameters (kernel bandwidths) incorporating anisotropic features. Training of PNN is a procedure to adjust the optimal $\sigma_j (j = 1, 2, 3)$ values for current data, so as to minimize the classification error. It is performed by a cross-validation approach.

2.2 Minimization of the Misfit Function

Filtration parameters of lithofaces are fitted basing on experimental data. Fitting procedure includes numerical solution of the spatial filtration equation in terms of decreasing of hydraulic heads using different values of parameters (filtration coefficients and coefficients of the resilient capacity) from the multidimensional cube:

$$\frac{\partial}{\partial x} k_h \frac{\partial S}{\partial x} + \frac{\partial}{\partial y} k_h \frac{\partial S}{\partial y} + \frac{\partial}{\partial z} k_v \frac{\partial S}{\partial z} + Q = \eta \frac{\partial S}{\partial t} \tag{4}$$

where x, y, z and t are spatial and temporal coordinates; S is a decrease of a hydraulic head while pumping out; Q is a discharge of the pumping. Within a lithoface filtration coefficients are considered as isotropic ($k_h = k_v$). The numerical solution of the filtration problem (4) is performed by ModFlow program [4].

The solution of Eq. (4) allows estimating the misfit function—a subject of minimization:

$$\varphi(k, \eta) = \frac{1}{N}\sum_{i=1}^{N} \frac{(S_i - S_i')^2}{|S_i|} \tag{5}$$

Table 1 Dynamics of misfit function during reclassification iterations

Iteration	Case 1	Case 2
Starting geology	0.46	0.25
1st-iteration	0.34	0.31
2nd-iteration	0.27	0.12

where N is the total number of control measurements (in all wells for all temporal interval), S is for measured head's decrease, S' is for estimated head's decrease.

3 The Case Study

The proposed approach was applied to two real examples from the same place but different depths. The both cases in general deal with three hydrogeologic layers (an aquifer surrounded by aquitards), but the lithology is presented by six lithofaces: two types of sands, a clay, coal, a product of the hard rock corrosion and a hard rock. Classification was based on data from 118 wells, total number of samples—11386.

Filtration parameters are fitted on pumping-down experiments, which were performed regularly during 20 days. Measurements of water level dynamics were made in five adjacent wells five times per day. Initial filtration parameters were given by minimal, maximal and the most probable values for corresponding lithofaces due to laboratory experiments.

The both cases required two reclassification iterations for obtaining parameters from probable fuzzy category. The dynamics of misfit function over these iterations is presented in Table 1. Obtained filtration parameters differ from the most probable. The resultant geology presents rather long high permeable riders within traditional aquifers.

References

1. Jacob, D. J. (2007). Lectures on inverse modeling. http://www-as.harvard.edu/education/jacob_lectures_inverse_modeling.pdf
2. Specht, D. (1990). Probabilistic neural networks. *Neural Networks*, *3*, 109–118.
3. Parzen, E. (1962). On estimation of a probability density function and model. *Annals of Mathematical Statistics*, *33*, 1065–1076.
4. McDonald, M. C., & Harbaugh A. W. (1988). *MODFLOW, a modular three-dimensional finite difference ground-water flow model*. Open-file report 83–875, U. S. Geological Survey.

Application of Multivariate Analysis Techniques for the Identification of Sulfates From Raman Spectra

Guillermo Lopez-Reyes, Pablo Sobron, Catherine Lefebvre and Fernando Rull

1 Introduction

The Raman instrument (RLS) onboard the 2018 ExoMars rover will determine the structural and compositional features of Mars' surface and subsurface samples [1]. In its current configuration, samples will be collected using a drill, then crushed and finally delivered to the analytical laboratory [1]. While such concept of operation will enable adequate management of collected samples, geological and morphological context will be lost. Thus, innovative spectral treatment methods must be developed that will enable unambiguous qualitative and quantitative identification.

Multivariate analysis (MVA) techniques for the quantitative analysis of the Raman spectra of several minerals and rocks show, for instance, that principal component analysis (PCA) is capable of differentiating mineral species such as carbonates, sulfates, oxides and silicates in geological samples [2]. Partial least squares regression (PLS) has been used to determine the quality of biodiesel fuels [3]. Artificial neural networks (ANN) have been designed for the identification and quantification of inorganic salts in water solutions [4]. Also, Combinations of these techniques for chemometrical analyses from Raman spectra have been attempted [5].

The aim of this study is to evaluate the aforementioned MVA techniques for the analysis of mineral samples in the framework of the operation of the RLS instrument. Sulfates are used because they are one of the two major types of secondary minerals found on Mars that may provide potentially habitable environments. Due to the relation of sulfate salts with ancient aqueous environments in which life might have thrived, it is expected that sulfates will be priority targets of the ExoMars mission.

G. Lopez-Reyes (✉) · F. Rull
Unidad Asociada Universidad de Valladolid – CSIC – Centro de Astrobiología, Valladolid, Spain
e-mail: guillermo.lopez.reyes@cab.inta-csic.es

P. Sobron · C. Lefebvre
Space science and technology, Canadian Space Agency, Montreal, Canada

E. Pardo-Igúzquiza et al. (eds.), *Mathematics of Planet Earth*,
Lecture Notes in Earth System Sciences, DOI: 10.1007/978-3-642-32408-6_33,

A set of 17 spectra of sulfates were used as input for training all the three techniques. The set was divided in four subsets, each grouping sulfates with different hydration state: $FeSO_4$ (1, 4 and 7w), $MgSO_4$ (1–7w, and 12w), $CaSO_4$ (0, 1/2 and 2w) and $Na_2\,SO_4$ (0 and 10w). A set of mixed spectra was generated by computing linear combinations of the original ones, parameterized with the expected proportion of the mixture and the cross-section of the mixed materials. Random noise was added to the synthetic spectra to guarantee differentiation. This set of mixtures was used to validate the models, as well as spectra from a mixture of Anhydrite (dehydrated $CaSO_4$) and Thenardite (dehydrated $Na_2\,SO_4$) powders in different proportions.

2 Principal Component Analysis (PCA)

PCA analysis of the samples showed that the first two components, PC1 and PC2, represent ~80 % of the variance in the data set: 61 and 17 %, respectively. The scores are plotted in Fig. 1. Dehydrated or poorly hydrated salts are somewhat differentiated from the highly hydrated salts, for both the training and test samples.

3 Partial Least Squares (PLS)

The PLS responses were chosen to indicate the hydration level and the relative abundance of the various cations. The criterion to optimize the model behavior was to evaluate the Mean Square Error (MSE) of the difference between the expected regression responses and those fitted by the model.

Model 1: When using all the points of the spectrum as input variables, the optimum number of components in the PLSR model was found to be 7. In this case, a 93 % mean correlation coefficient was achieved for the pure samples spectra, and 89 % for the 1:1 mixtures.

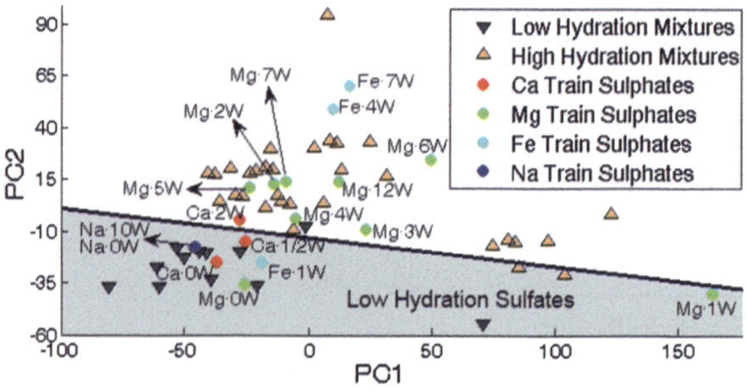

Fig. 1 PCA scores for pure samples (*color*) and mixtures (*triangles*)

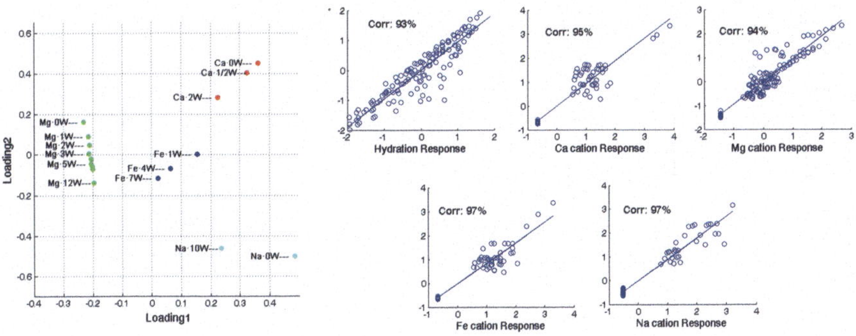

Fig. 2 PLS scores for pure samples (*left*). Predicted versus real responses for mixtures (*right*)

Model 2: A new model was created based only on selected peaks from the spectra. In this case, the optimum number of components was 4, and also provided good prediction: linear unitary slope and 99 % mean correlation coefficients for all the responses of the pure samples, and 95 % mean correlation for the 1:1 proportion mixtures spectra (Fig. 2). However, the tail values for these mixtures are mostly responsible for this figure (see Ca and Fe). This value is a worst case for binary mixtures: the most unbalanced the proportion of the mixtures, the better correlation is obtained (as the spectrum is more similar to the pure spectra used for training).

4 Artificial Neural Networks (ANN)

We have designed a three-layer Feed-Forward Back-Propagation network with 33 neurons in the input layer, 33 neurons in the hidden layer and 17 outputs corresponding to each of the samples, with log sigmoid neuron transfer functions. The input data for the ANN consisted on 33 Raman intensities at selected wavenumbers of non-overlapping peaks (the same as in the second PLS model previously described) to improve performance [6]. The training and validation sets consisted of the pure samples spectra, as well as of computed mixed spectra in proportions 25:75, 50:50 and 75:25. The test sets consisted on computed binary mixtures in proportions ranging from 5:95 to 95:5, as well as spectra from a binary mixture of Anhydrite and Thenardite with varying proportions.

The results showed that, establishing a detection threshold of 0.05, the identification was of 100 % for pure samples, and, for binary mixtures, a fail ratio of 3 % for proportions between 10:90 and 90:10. In addition, the outputs of the network were somewhat proportional to the actual concentration of the sample, providing some kind of quantification of the abundance in binary mixtures. The results were outstanding with computed mixtures, and provided a fair quantification when using spectra obtained from the RLS instrument (Fig. 3), especially when taking into account that the training spectra were obtained with a different spectrometer.

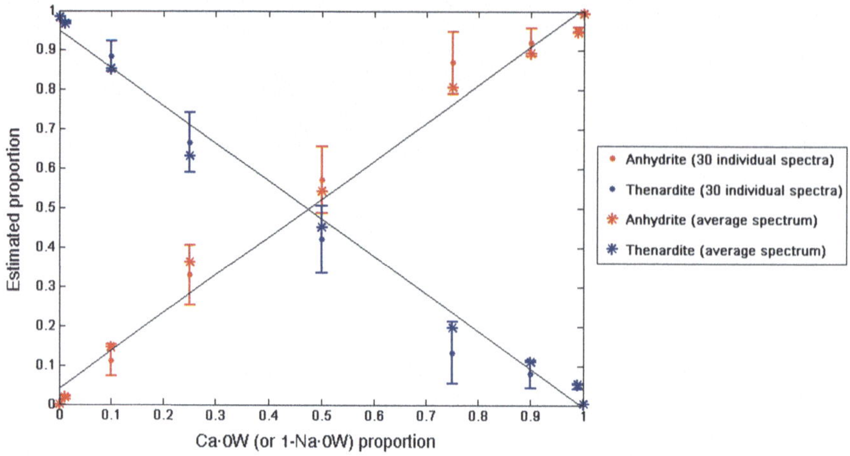

Fig. 3 ANN output for spectra from real mixtures of Anhydrite and Thenardite with RLS

5 Conclusions and Acknowledgements

The application of unsupervised multivariate techniques for the processing of RLS products is a must in order to provide science support to the ExoMars mission. These techniques will provide fast identification of materials and quantification of mineral species during the tactical operations of the rover-based mission. We have compared the performance of three MVA mineral ID techniques. The results presented here are encouraging, and demonstrate that these techniques can provide critical information for the identification and quantification of mineral phases in geological samples relevant for ExoMars, from Raman spectra. PCA differentiated hydrated sulfates from dehydrated ones. PLS proved to be a good chemomentric tool for the detection and quantification of binary mixtures, with better results when applied to unbalanced mixtures or pure samples. ANN has proven to be a powerful tool for the identification of sulfates: the model proved to be able to detect and somewhat quantify the abundance of both elements in binary mixtures with promising figures and quite unbalanced abundances in the mixtures.

The authors acknowledge the support for the project from R. Leveille, A. Wang and A. Koujelev. GLR acknowledges the University of Valladolid (Spain) for providing funding for the project. PS and CL acknowledge support from NSERC.

References

1. Rull, F. et al. (2010). 42nd Lunar and Planetary Science Conference, Abstract #2400.
2. Lafuente, B. et al. (2012). Georaman Xth, pp. 149–150.
3. Ghesti et al. (2007). *Energy & Fuels*, *21*(5), 2475–2480.

4. Burikov et al. (2010). *Optical Memory & Neural Networks*, *19*, 140–148
5. Dorfer, T., et al. (2010). *Journal of Raman Spectroscopy*, *41*(6), 684–689.
6. Koujelev, A., et al. (2010). *Planet. Space Sci.*, *58*(4), 682–690.

Comparison of Micro X-ray Computer Tomography Image Segmentation Methods: Artificial Neural Networks Versus Least Square Support Vector Machine

Swarup Chauhan, Wolfram Rühaak, Frieder Enzmann, Faisal Khan, Philipp Mielke, Michael Kersten and Ingo Sass

1 Introduction

Determination of petrophysical rock properties at meso-, macro- and microscale is an important process in geothermal reservoir evaluation. The analysis of the micro pore network can effectively contribute to the forward modeling and history matching, thereby enhancing the prognosis on reservoir exploitation. At this scale micro X-ray computer tomography (XCT) serves as a powerful non-destructive method for obtaining information about rock structures and mineralogy. However, obtaining rock properties such as porosity and permeability from XCT digital images is still a challenging task [9]. In this work artificial neural networks (ANN) and least square support vector machine (LS-SVM) for image segmentation of 2D XCT images has been exploited. The results will give insight in the correctness of pore-space classification by this two approaches and the uncertainty within which porosity can be estimated. ANN and LS-SVM despite different operating principles are characterized by a great deal of modularity and flexibility and are well suited for demanding large-scale and pattern recognition and pixel based classification applications [7, 8].

S. Chauhan(✉) · W. Rühaak · P. Mielke · I. Sass
Department of Geothermal Science and Technology, Institute of Applied Geosciences, Technische Universität Darmstadt, Darmstadt, Germany
e-mail: chauhan@geo.tu-darmstadt.de

F. Enzmann · F. Khan · M. Kersten
Department of Environmental Mineralogy, Institute for Geosciences, Johannes Gutenberg-Universität Mainz, Mainz, Germany

E. Pardo-Igúzquiza et al. (eds.), *Mathematics of Planet Earth*,
Lecture Notes in Earth System Sciences, DOI: 10.1007/978-3-642-32408-6_34,

2 Methodology

2.1 XCT Image

For this study a tuff rock sample was imaged using a high resolution XCT scanner (www.hrxct.geo.uni-mainz.de), applying X-ray energy of 110 keV and pre-filter of 0.3 copper is used. During the reconstruction of the projections no noise filter was used. The projections were Radon-transformed in sinograms, thereafter converted through back-projection into tomograms. These stacked tomograms resulted in a 16-bit 3D imagery, with a resulting voxel resolution of 43.9 μ m. Later beam hardening correction was performed using regression analysis using least squares 2D-parabolid fitting. Finally, the data was stored in raw format and AVIZO software was used for visualization.

2.2 Image Processing and Classification Techniques

The image pre-processing and segmentation of XCT images was done according to [7]. The K-Means Clustering (KM) [4], Fuzzy C-Means (FCM) [1, 6] and Self Organized Map Algorithms (SOM) [6] were used for deriving feature vectors and image clustering. The initial condition for KM, FCM and SOM clustering was followed as described in [7].

ANN was used to classify rock images into pore space and solid space. The ANN was tested for each segmentation method mentioned above. MATLAB® software nftool box was used to setup ANN and perform classification.

SVM are well known binary classifiers and often known as "Kernel-based classifiers" [8, 10], In our case, a least square support vector machine (LS-SVM), approach is used for classification [8]. It uses a non-linear radial basis function (RBF) for improving accuracy [10].

2.3 Image Segmentation

The 2D XCT images used for pore space classification have an image resolution of 43.9 μm per pixel with an image size of 905 pixels $\times$ 905 pixels. Twenty images were used to develop a feature vector (FV) comprising of a sum of 16,380,500 pixels. The FV was clustered and labeled using KM and FCM and SOM algorithm and segmented into clusters of 5, 7 and 9. Figure 1 shows the preliminary results obtained from KM and FCM clustering for seven classes. Further, ANN was used for image classification.

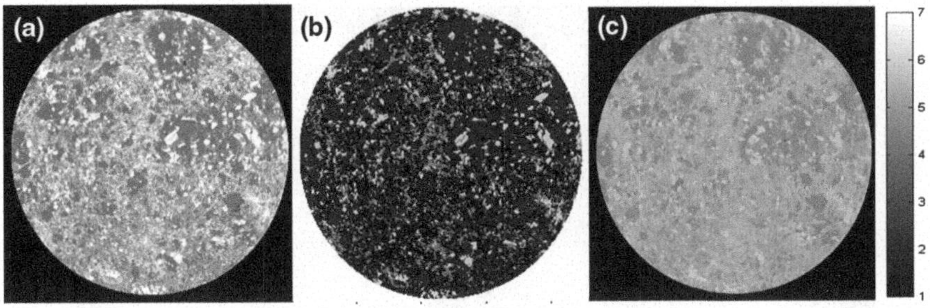

Fig. 1 **a** Segmentation image with KM, **b** segemented image with FCM, **c** classfied image obtained from ANN. Pores are represented by *grey color* and solid matrix as *white*

The ANN was trained using ten images (8,190,250 pixels), validated using one image (81,9025 pixels) and a testing on ten images (8,190,250 pixels). The best classification rate was obtained using the FV of KM algorithm.

In case of LS-SVM, 20 images were utilized to perform pixel-by-pixel classification. Based on this classification, segmentation of two classes: solid phase and pore phase was obtained. Among twenty images the best representative image was selected. From this image, certain regions were manually chosen for training, shown in Fig. 2a and Table 1. Care was taken, such that, the chosen regions corresponding to each class did not contain overlapping boundaries between solid and pore phase, so as to restrict misclassifications. Furthermore, LS-SVM was trained using the pixel values of the chosen classes. The training data contained 2,366 pixels (less than 1 % of the total image pixels) and are validated with the remaining 59,4457 pixels of the same 2D image. The LS-SVM was trained until it reached optimal performance thresholds. Thereafter, it was tested on the remaining 19 images.

3 Results

Artificial Neural Network: Figure 1c shows a reconstructed image obtained after ANN classification. In the reconstructed image, the grey region corresponds to pore space and the white region corresponds to solid minerals. Comparison with KM, FCM and LS-SVM shows that ANN overestimates pore-space.

Least Square Support Vector Machine: The result of binary classification by LS-SVM is illustrated in Fig. 2b. The performance of LS-SVM is verified by determining the pore space which is 0.30 cm^{-3} (30 %).

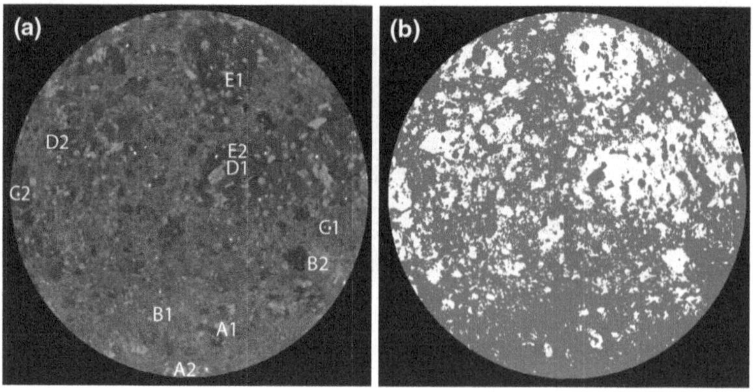

Fig. 2 **a** Trained data with regions marked by letters. **b** segmented data; with *white color* represent pores (porosity = 30 %), *grey color* is solid matrix

Table 1 Selection of regions and pixels for training set data

	Solid (minerals)						Pores			
Training region's	A1	A2	B1	B2	C1	C2	D1	D2	E1	E2
Pixels trained	195	171	600	400	40	24	400	400	72	64
Total trained pixels	1430						936			

4 Conclusions

In this work we have used two classifiers ANN and LS-SVM for pore space detection and classification in the CT images using data clustering and unsupervised machine learning. In our case the LS-SVM classification was accurate but was computationally slow. Whereas, ANN was used for classification but not for image segmentation and labeling, and was computationally fast.

Within the scope of the study, the observed results are interesting. The example presented here, shows that the result of LS-SVM is more accurate than that of the ANN. In future work, ANN will be configured with a radial basis function (RBF) function, this will help, to quantitatively compare between ANN and LS-SVM. ANN will be configured to also perform image segmentation of 2D and 3D XCT images.

Additionally, the use of an ensemble learning algorithm [3] might give further insight if such approaches improve performance and accuracy in comparison to single classifiers (LS-SVM and ANN). The data mining software WEKA [5] is a potential tool box for use of ensemble of classifiers.

References

1. Bezdek, J. C., Ehrlich, R., & Full, W. (1984). FCM: The fuzzy c-means clustering algorithm. *Computers and Geosciences*, *10*(2–3), 191–203.
2. Cortina-Januchs, M. G., Quintanilla-Dominguez, J., Vega-Corona, A., Tarquis, A. M., & Andina, D. (2011). Detection of pore space in CT soil images using artificial neural networks. *Biogeosciences*, *8*(2), 279–288.
3. Dietterich, T. G. (2000). Ensemble methods in machine learning. Multiple classifier systems (pp. 1–15). Berlin: Springer.
4. Dunn, J. C. (1973). A fuzzy relative of the ISODATA process and its use in detecting compact well-separated clusters. *Journal of Cybernetics*, *3*, 32–57.
5. Hall, M., Frank, E., Holmes, G., Pfahringer, B., Reutemann, P., Witten, I. H. (2009). The WEKA data mining software: An update. *SIGKDD Explorations*, *11*(1), 10–18.
6. Kohonen, T. (1990). The self-organizing map. *In: Proceedings of the IEEE*, *78*(9), 1464–1480.
7. MacQueen, J. B. (Ed.) (1967). Some methods for classification and analysis of multivariate observations (Vol. 1). Berkeley, CA: University of California Press.
8. Selvaraj, H., Selvi, S., Selvathi, D., Gewali, L. (2007). Brain MRI slices classification using least squares support vector machine. *International Journal of Intelligent Computing in Medical Sciences & Image Processing*, 1(1), 21–33.
9. Soille, P. (2nd ed.) (2003). Morphological image analysis: Principles and applications. Secaucus, NJ: Springer.
10. Suykens, J. A. K., & Vandewal, J. (1999). Least squares support vector machine classifiers. *Neural Processing Letters*, *9*(3), 293300.

Learning Uncertainty from Training Images for Reservoir Predictions

Temitocles Rojas, Vasily Demyanov, Mike Christie and Darn Arnold

1 Introduction

One of the largest uncertainties in reservoir modelling is the interpretation of the sedimentary environment and the assumptions about its description. Uncertainty in interpretation of a depositional environment affects the description of facies and petrophysical distribution in a reservoir model. These factors become important in assessing uncertainty of reservoir prediction and may impact reservoir development decisions.

One of the contemporary challenges is to find a way to integrate multiple reservoir description concepts or multiple sedimentological interpretations of a model using multipoint statistics or another algorithm for reservoir description. The pioneering work by [6] proposed a distance metric approach using multi-dimensional scaling (MDS) to navigate in the space of reservoir realizations. The approach was developed further in [3], where multidimensional scaling (MDS) was used to distinguish between the production responses of models generated with multiples training images and the actual field production data. Later, [3] used MDS to compare different production responses obtained from different simulations (discretized by the training image that was used to generate the models) with the actual production history data.

In the present work we propose a way to quantify uncertainty of reservoir prediction by differentiating between facies scenarios represented by multiple training images. This is done by solving a classification problem in the metric space of the reservoir model realisations using a multi-class Support Vector Machine classifier [1]. SVM classifier provides a decision boundary in the metric space of stochastic model realisations based on different training images and transformed by MDS.

T. Rojas
Xodus Group, The Hague, The Netherland

V. Demyanov (✉) · M. Christie · D. Arnold
Heriot-Watt University, Edinburgh EH14 4AS, UK
e-mail: vasily.demyanov@pet.hw.ac.uk

E. Pardo-Igúzquiza et al. (eds.), *Mathematics of Planet Earth*,
Lecture Notes in Earth System Sciences, DOI: 10.1007/978-3-642-32408-6_35,

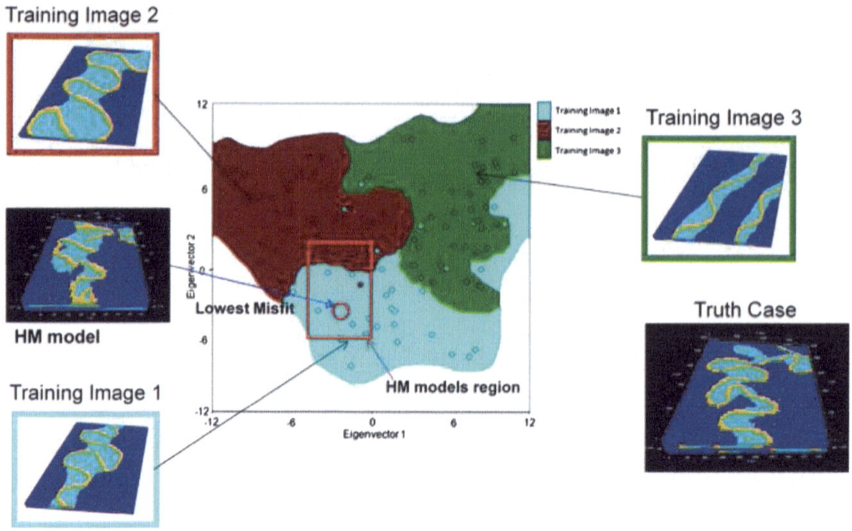

Fig. 1 MPS realizations with the corresponding training images mapped into the MDS space (○ depict locations of the initial realization (in a 2d projection) used for SVM training). SVM classifier provides regions corresponding to each of the three training images used. Location of the best HM model (○) and the region, where HM homed in, is identified in *red*

Multiple history matched models are then generated based on the likelihood in the classified metric space. The approach was applied to a synthetic fluvial case study also taking into account the variability of other geomodel parameters like channel geometry and proportions.

2 Methodology of Classifying Realization in the Model Space

Uncertainty in the interpretation of the depositional environment properties in reservoir modelling can be addressed within a multipoint statistics (MPS) approach by using different training images and their modifications through rotation and scaling (affinity) to vary the body size in geomodelling. Accounting for uncertainty in training image modifications with rotation and affinity was proposed in [5], where a machine learning technique (artificial neural network) was applied to predict the range for geomorphic parameters of fluvial channels based on the MPS training image transformation parameters—rotation and scaling.

In the present work MDS was applied to the MPS reservoir model realisations based on different training images in order to map them into a unique model space and then classify the model space according to the chosen training image concepts. Classification in the model space is performed using Support Vector Machine (SVM)

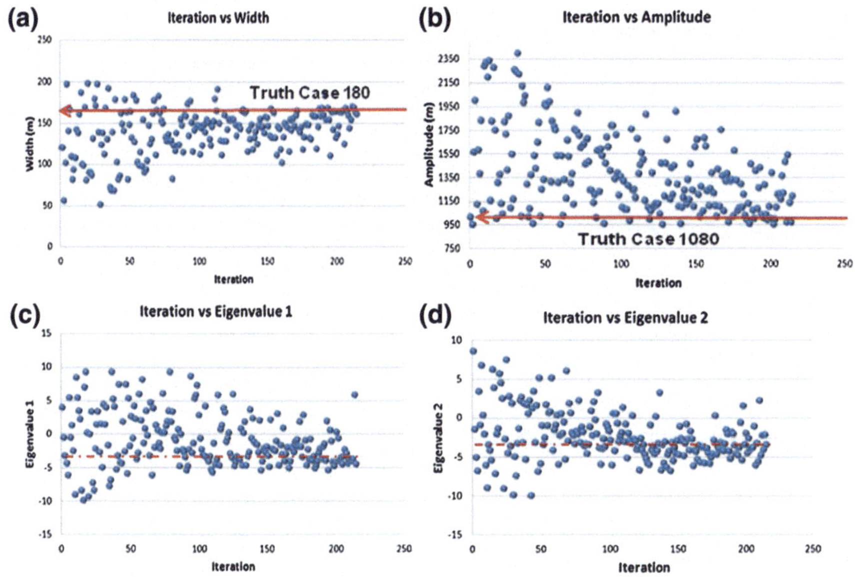

Fig. 2 Evolution of the reservoir model parameters in history matching: channel width (**a**), channel amplitude (**b**), eigenvalues corresponding to the coordinates of the realizations in the MDS space (**c**, **d**). Solid line depicts the parameter values for the truth case

trained on the initial set of realizations. SVM classifier contoured the regions where the model realizations are based on one of the training images used. Then, it is possible to search the classified space in order to obtain the model realization that would match best the historical production data. This history matching (HM) task is performed by using adaptive stochastic algorithms, such as Particle Swarm Optimisation [2].

MPS facies reservoir models were built based on three training image representing different variations fluvial channels depositional environment: sinuous parallel and two types of meanders (see Fig. 1). Three sets of stochastic reservoir model realizations were produced with the MPS algorithm SNESIM [7] and mapped into the MDS Eigen vector space (see a 2D projection in Fig. 1). The dimensionality of the Eigen vector space depends on the number of components being considered. The choice of the detail (number of the Eigen vectors taken into account) is guided by the stress which increases with the dimensionality of the MDS space [8]. SVM applied for classification in the MDS space is good at tackling the high dimensionality problem by controlling the complexity of the class boundaries.

3 History Matching Results

A history matching study was set up to illustrate how the approach helps to quantify uncertainty between different training images. A simplified (3 facie only model) 3D

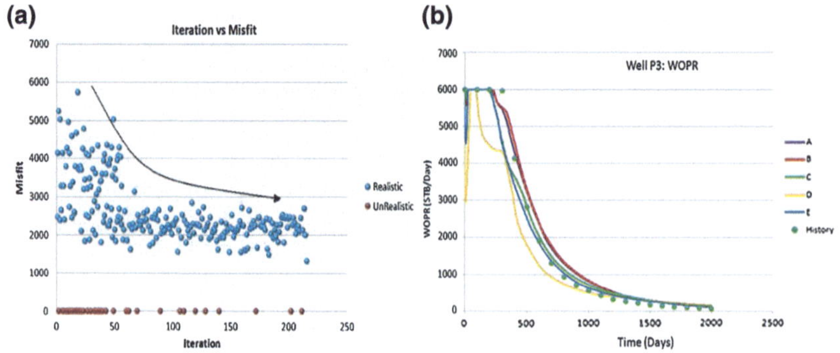

Fig. 3 Evolution of the misfit function in history matching with adaptive sampling iterations: **a** match quality improves with the decrease of the misfit as the sampling evolves in the model space; **b** oil production from one of the well for 5 best HM models vs the history data

section from a synthetic Stanford VI reservoir (see Fig. 1) was selected as the truth case to generate the history production profiles.

History matching was performed by varying the Eigen vectors to distinguish between the three training images to choose from in the classified MDS space and the channel geometry parameters (wavelength, amplitude, width and thickness). Channel geometry parameters are then used to estimate the MPS algorithm input (affinity) to build realistic reservoir geomodels, which are consistent with the prior information from modern river analogues [5]. Figure 2 shows the evolution of the uncertain parameters while they are being optimized though the history matching. The points, which correspond to each of the newly generated model, are homing in around the better solutions close to the truth case parameter values. The decrease of the misfit function (least square norm of the production history) is shown in Fig. 3 along with production profiles for multiple good HM models. The region of the low misfit models includes the models based on the two out of three training images (see in Fig. 1).

4 Conclusions

The paper presents a way to account for different geological concepts in uncertainty quantification study with multiple training images. Uncertainty quantification based on multiple training images demonstrated that adaptive stochastic sampling is capable of matching closer the observed production and finding more realistic range possible model realizations.

Automated history matching with geologically realistic priors for the MPS parameterisation was able to identify a region of good fitting models, which corresponds to the two out of the three training images used. This methodology can be seen as a way to select from different sedimentological interpretations by various geoscientists, which is a common issue in reservoir characterization.

Acknowledgments The funding for this work was provided by the industrial sponsors of the Heriot-Watt Uncertainty Project. We would like to thank J. Caers and Stanford University for sharing the MDS use of SGems software [4] for MPS simulations and for providing Stanford VI case study. We would like to thank M. Kanevski and University of Lausanne for using MLOffice for neural network and SVM modelling [1]. We appreciate Epistemy for providing Raven history-matching and uncertainty quantification software. We appreciate Schlumberger for providing of Eclipse reservoir simulator.

References

1. Kanevski, M., Pozdnoukhov, A., & Timonin, V. (2009). *Machine learning for spatial environmental data. Theory, applications and software* (p. 377). Lausanne Switzerland: EPFL Press.
2. Mohamed, L., Christie, M., & Demyanov, V. (2010). Comparison of stochastic sampling algorithms for uncertainty quantification. *SPE Journal*, *15*(1), 31–38.
3. Park, H., Scheidt, C., Fenwick, D., Boucher, A., & Caers, J. (2013). *History matching and uncertainty quantification of facies models with multiple geological interpretations*. Submitted to Computational Geosciences, February (online).
4. Remy, N., Boucher, A., & Jianbing, W. (2009). Applied geostatistics with SGEMS (p. 264). Cambridge: Cambridge University Press.
5. Rojas T., Demyanov V., Christie M., & Arnold D. (2011). Use of geological prior information in reservoir facies modelling. *Proceedings of IAMG, Salzburg*. http://www.cogeo.at/publications/iamg2011/IAMG2011_Salzburg_093.pdf
6. Scheidt, C., & Caers, J. (2008). A new method for uncertainty quantification using distances and kernel methods: Application to a deepwater turbidite reservoir. *SPE Journal*, *14*(4), 680–692.
7. Strebelle, S. (2002). Conditional simulation of complex geological structures using multiple-points statistics. *Mathematical Geology*, *34*(1), 1–21.
8. Wickelmaier, F. (2003). *An Introduction to MDS* (p. 26). Sound Quality Research Unit, Aalborg University, Denmark.

A Particle Swarm Optimization for Parameter Estimation of a Rainfall-Runoff Model

Frédéric Bardolle, Frédérick Delay, Francis Bichot, Gilles Porel and Nathalie Dörfliger

1 Model Presentation

The hydrological model presented in this paper is based on a systemic approach establishing relationships between hydrological signals by means of transfer functions. The first innovation is the versatile architecture which can associate three elementary transfer functions in series along four parallel branches (Fig. 1).

Consequently, an output $O(t)$ (e.g., river discharge) is the sum of up to four convolution products:

$$O(t) = I_1 * H_1^{eq}(t) + I_2 * H_2^{eq}(t) + I_3 * H_2^{eq}(t) + I_4 * H_2^{eq}(t) \tag{1}$$

where is $\mathrm{I}_i(t)$ the ith input (e.g., rainfall) and $H_i^{eq}(t)$ is the ith equivalent transfer function and equals the convolution of three elementary transfer functions:

$$H_i^{eq}(t) = K_i.\left(H_i^1 * H_i^2 * H_i^3(t)\right) \tag{2}$$

where is K_i a gain factor, and H_i^{α} is the αth transfer functions of the ith branch. Regarding transient systems, parallel branches may correspond to multiple simulta-

F. Bardolle (✉) · G. Porel
IC2MP, University of Poitiers, 4 rue Michel Brunet, 86022 Poitiers Cedex, France
e-mail: f.bardolle@brgm.fr

F. Bardolle · F. Bichot
BRGM Poitou-Charentes, 5 rue de la Goélette, 86280 Saint-Benoit, France

F. Delay
LHyGeS, University of Strasbourg, 1 rue Blessig, 67084 Strasbourg Cedex, France

N. Dörfliger
BRGM Water Environment and Ecotechnologies Division, 3 avenue Claude-Guillemin, 45060 Orléans Cedex 2, France

E. Pardo-Igúzquiza et al. (eds.), *Mathematics of Planet Earth*,
Lecture Notes in Earth System Sciences, DOI: 10.1007/978-3-642-32408-6_36,

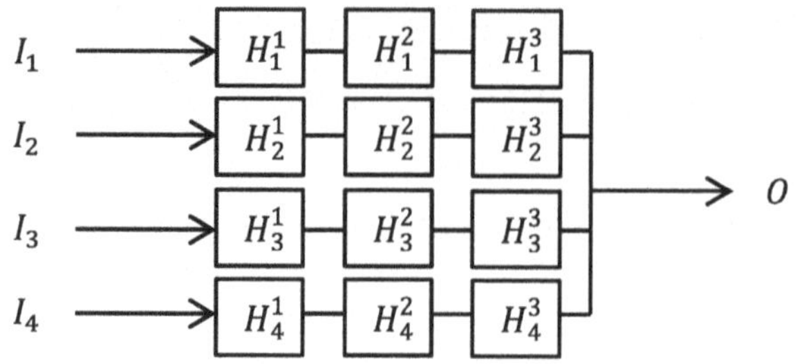

Fig. 1 Model architecture in series and parallel branches of transfer functions

Table 1 Three transfer functions from our function library

Distribution	Formula	Physical meaning	References
Gamma	$\mathrm{G(t)} = \frac{\mathrm{t^{b-1}.exp(-t/a)}}{\mathrm{a^b.\Gamma(b)}}$	Cascade of linear reservoirs	[1]
Normal	$\mathrm{N(t)} = \exp\left(\frac{-(\mathrm{t-a})^2}{\mathrm{b}^2}\right)$	Convection and dispersion	[2]
Power-law	$\mathrm{P(t)} = \frac{1}{(1+\mathrm{a.t})^{\mathrm{b}}}$	Diffusion	[3]

neous transfer mechanisms or to some kind of spatial variability, whereas a series would link successive phenomena.

The other novelty is the possibility to choose for each problem the number of inputs (branches), and the number and type of elementary transfer functions H_i^α. These functions are selected from a library of parametric probability density functions with some physical meaning regarding hydrological transfers (see examples in Table 1). Unused transfer functions are replaced by a Dirac function $\delta(t)$ corresponding to the identity element for convolution, i.e., $f(t) * \delta(t) = f(t)$.

With one parameter per branch (the gain factor K_i), and two parameters per elementary transfer function, the model has between 3 and 28 parameters.

2 Particle Swarm Optimization

Parameter estimation is performed by a particle swarm optimization (PSO) algorithm. Initially developed to simulate bird flocking [4], PSO consists in moving particles (vectors of size the number of parameters) in the parameter space to find the minimum of an objective function measuring a distance between simulated outputs and data. A particle i is moved iteratively and randomly in the parameter space while keeping for iteration $t+1$ the memory of its best experienced position p_i^t between $[0, t]$ and the memory of the best experienced position among all the individuals of the swarm g^t. After an initialization step assigning to each particle a random location and a velocity in the parameter space, each iteration calculates the value of the objective function for each particle and updates velocities (v_i^{t+1}) and positions (x_i^{t+1}) as follow:

$$v_i^{t+1} = w.v_i^t + c_1.r_1(g^t - x_i^t) + c_2.r_2(p_i^t - x_i^t) \tag{3}$$

$$x_i^{t+1} = x_i^t + v_i^{t+1} \tag{4}$$

where w is the inertia weight, c_1 and c_2 are acceleration constants, and r_1 and r_2 are random number uniformly distributed in (0,1). PSO is often considered as well-suited to ill-posed inverse problems because its global search through the parameter space and the exchange of information between individuals lowers the risks of converging toward a local minimum. The algorithm has been successfully applied to parameter estimation of rainfall-runoff models ([5] and [6]) and shows good convergence rates when compared to other global techniques [7].

3 Application

The systemic model and the PSO have been faced to data of river discharges (Sèvre Niortaise—France). The PSO seeks the parameters of transfer functions over a calibration period, from 08/1994 to 08/1996. The best inverse solution is then used to predict river discharge over two verification periods, from 08/1993 to 08/1994 and from 08/1996 to 08/1997 (Fig. 2). Only the first verification period contains discharges higher than 30 m^3/s, corresponding to flood events. The input data are a daily effective rainfall rate and a daily groundwater level (i.e., two branches in the model), and the output data is a daily discharge.

Five combinations of transfer functions were tested: G, G*N, P, P*N and G*P*N where the letters G, N and P stands respectively for Gamma distribution, Normal distribution and Power-Law distribution. The Nash-Sutcliffe efficiency (NSE) [8] is calculated to compare precision of simulations for the various combinations mentioned above. It ranges between $-\infty$ and 1, $NSE = 1$ being the optimal value.

NSEs reported in Table 2 show that each combination of transfer functions results in pretty good simulations for both calibration and prediction periods but sophisticated models like G*P*N do not necessarily produce better results than simple models like G. Moreover, uncertainty of input data may be responsible for the lower coefficients of the second prediction period.

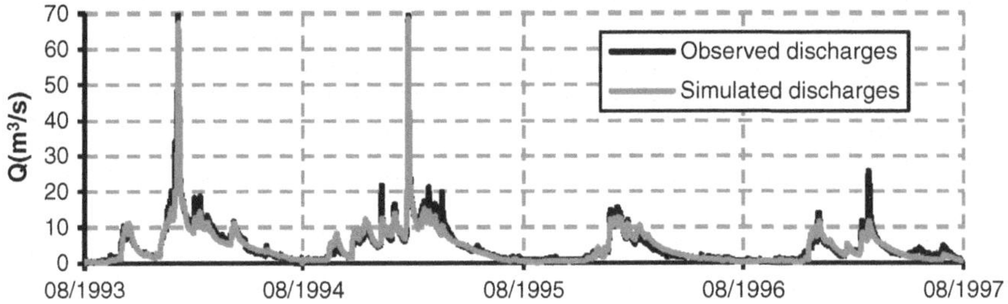

Fig. 2 Simulated and observed discharge for the gamma distribution Model

Table 2 Nash-Sutcliffe efficiency (NSE) for different transfer function combination

		G	G*N	P	P*N	G*P*N
Calibration	1994/95/96	0.945	0.944	0.840	0.911	0.945
Prediction	1993/94	0.912	0.912	0.882	0.931	0.911
	1996/97	0.747	0.743	0.748	0.771	0.737

Finally, standard deviations of the NSE of the ten best solutions are always inferior to 1E-3, attesting the good convergence of the PSO.

4 Conclusion

Multi-input versatile architecture and parametric transfer functions optimized by PSO form a model that produces satisfying results. For now, only the best solution is used but a stochastic model giving us an uncertainty measure of the simulation could be conceived with the other solutions found by the PSO algorithm.

References

1. Nash, J. E. (1957). The form of the instantaneous unit hydrograph. *IAHS Publication*, *45*, 114–121.
2. Pinault, J.-L., & Berthier, F. (2007). A methodological approach to characterize the resilience of aquatic ecosystems with application to Lake Annecy, France. *Water Resources Research*, *43*, W01418 1:15.
3. Drogue, C. (1972). Analyse statistique des hydrogrammes de decrues des sources karstiques. Statistical analysis of hydrographs of karstic springs. *Journal of Hydrology*, *15*, 49–68.
4. Kennedy, J., & Eberhart, R. (1995). Particle swarm optimization. *Proceedings of the IEEE International Conference on Neural Networks 1995* (Vol. 4).
5. Gill, M. K., Kaheil, Y. H., Khalil, A., McKee, M., & Bastidas, L. (2006). Multiobjective particle swarm optimization for parameter estimation in hydrology. *Water Resources Research*, *42*, W07417 1:14.
6. Chau, K. W. (2006). Particle swarm optimization training algorithm for ANNs in stage prediction of Shing Mun River. *Journal of Hydrology*, *329*, 363–367.
7. Fernández-Martínez, J. L., Mukerji, T., García-Gonzalo, E., & Fernández-Muñiz, Z. (2011). Uncertainty assessment for inverse problems in high dimensional spaces using particle swarm optimization and model reduction techniques. *Mathematical and Computer Modelling*, *54*, 2889–2899.
8. Nash, J. E., & Sutcliffe, J. V. (1970). River flow forecasting through conceptual models part I-A discussion of principles. *Journal of Hydrology*, *10*, 282–290.

Sequential Weights of Evidence as a Machine Learning Model for Mineral Deposits Prediction

Qiuming Cheng

1 Introduction

The idea of weighted training sample treatment was initially developed and used in AdaBoost which is an interesting algorithm developed in the mid-90s in machine learning and has been quickly adopted in many fields for combining the performance of multiple "weak" classifiers to produce a powerful "strong" classifier. Boosting was proposed in the literature of machine learning [1, 2] and has since received much attention and popularity not only in the computer science community for classification but also in statistics for optimization [3, 4]. The ordinary weights of evidence model with the assumption of conditional independence (CI) can be viewed as a special non-sequential boosting model. However, this assumption has limited its applicability in situations of noninsurance of CI [5, 6]. A new sequential weights of evidence model has been developed by the author as a general boosting model which can significantly reduce the influence of conditional dependency of evidential layers [7].

2 BoostWofE: A New Sequential Weights of Evidence

Assume a number of training points in two groups: mineral deposits denoted as $d_{\mathrm{i}} = 1, i = 1, \ldots, \mathrm{m}, (d_{\mathrm{i}} = 1) \in (\mathrm{D} = 1)$ and not mineral deposits denoted as $d_{\mathrm{j}} = 0, j = 1, \ldots, \mathrm{n}, (d_{\mathrm{j}} = 0) \in (\mathrm{D} = 0)$. Assume a number of variables

Q. Cheng (✉)
State Key Lab of Geological Processes and Mineral Resources, China University of Geosciences, Beijing 100083, China
e-mail: qiuming@yorku.ca

Q. Cheng
Department of Earth and Space Science and Engineering, Department of Geography, York University, Toronto M3J1P3, Canada

E. Pardo-Igúzquiza et al. (eds.), *Mathematics of Planet Earth*, Lecture Notes in Earth System Sciences, DOI: 10.1007/978-3-642-32408-6_37,

(evidences) of binary patterns $\mathrm{B_k}$ (k = 1, ..., N) are associated with the point event D. The new BoostWofE algorithm can be described as follows:

1. In the initial stage, an equal weight is assigned to each point,

$$w_1(d_i = 1) = 1/(n+m), w_1(d_j = 0) = 1/(n+m). \quad (1)$$

2. Repeat for evidence $\mathrm{B_k}$ with k = 1, 2, ..., N:
   (a) Calculate probabilities using the weighted training samples as

$$P_{w_k}(\mathrm{D}=1) = \sum_{i=1}^{m} w_k(d_i = 1),\ P_{w_k}(\mathrm{D}=0) = \sum_{j=1}^{n} w_k(d_j = 0) \quad (2)$$

$$\begin{aligned} P_{w_k}(\mathrm{B_kD}=1) &= \sum_{(d_i=1)\in(\mathrm{B_kD}=1)}^{m} w_k(d_i=1),\ P_{w_k}(\mathrm{B_kD}=0) \\ &= \sum_{(d_j=0)\in(\mathrm{B_kD}=0)}^{n} w_k(d_j=0) \end{aligned} \quad (3)$$

   where ($\mathrm{B_kD}$) stands for the intersect of sets $\mathrm{B_k}$ and D.
   (b) Compute the probabilities of misclassification of D after $\mathrm{B_k}$ applied

$$e_1 = P_{w_k}(\mathrm{B_k}=1\,|\,\mathrm{D}=0),\ e_2 = P_{w_k}(\mathrm{B_k}=0\,|\,\mathrm{D}=1) \quad (4)$$

   (c) Set the weights representing the relative importance of the two stages of classifier $\mathrm{B_k} = 1$ and $\mathrm{B_k} = 0$

$$W(\mathrm{B_k}=1) = \ln[(1-e_2)/e_1],\ W(\mathrm{B_k}=0) = \ln[e_2/(1-e_1)] \quad (5)$$

   (d) Update the weights of training samples for the next step

$$w_{k+1}(d) = \begin{cases} \dfrac{w_k(d=1)}{Z_1} \begin{cases} 1/(1-e_2),\ if\ (d_i=1)\in(\mathrm{B_k}=1\mathrm{D}=1) \\ 1/e_2,\ if\ (d_i=1)\in(\mathrm{B_k}=0\mathrm{D}=1) \end{cases} \\ \dfrac{w_k(d=0)}{Z_0} \begin{cases} 1/e_1,\ if\ (d_j=0)\in(\mathrm{B_k}=1\mathrm{D}=0) \\ 1/(1-e_1),\ if\ (d_j=0)\in(\mathrm{B_k}=0\mathrm{D}=0) \end{cases} \end{cases} \quad (6)$$

   where Z_1 and Z_0 are normalization factors for probability definition. The normalized weights of training points, $w_{k+1}(d)$, will be used in the next generation of boosting. Go back to the step 2 to implement the next generation. When all evidences are added and k = N, then go to the step 3.
3. Calculate posterior logit

$$Logit(\mathrm{D}\,|\,\mathrm{B_1}, \ldots, \mathrm{B_N}) = W_0 + W(\mathrm{B_1}) + \ldots + W(\mathrm{B_N}) \quad (7)$$

where $W_0 = \ln[P/(1 - P)]$ is the prior logit.

4. Calculate posterior probability with anti-logit transformation

$$P(D \mid B_1, \ldots, B_N) = \{1 + e^{-[W_0 + W^*(B_1) + \ldots + W^*(B_N)]}\}^{-1} \tag{8}$$

3 Application of BoostWofE for Prediction of Sn Mineral Deposit

For comparison purposes, the author reanalysis the dataset being used previously for prediction of Sn mineral deposits in Gejiu mineral district, China. The dataset includes 11 Sn mineral deposits as training points and five binary layers of geological features associated with occurrence of Sn mineral deposits in the same area in Cheng (Figs. 3–7)[8]: (1) Gejiu formation of carbonates (GF); (2) the 6 km buffer zone around the intersections of three groups of faults (DF); (3) local geochemical anomalies in stream sediments (LA); (4) regional geochemical anomalies in stream sediment samples (RA); and (5) buffer zones of 37 km around the center of intrusion (DI) interpreted as the center of buried intrusions by gravity data. These five binary layers are combined for calculating posterior probability maps by BoostWofE and WofE and for evaluation of the influence of CI.

A set of 113 points (102 regularly distributed non-mineral deposits and 11 known Sn mineral deposits) were used as training samples for implementation of both BoostWofE and WofE. In the calculation of weight, evidences are added in sequential. In each step, the training samples are assigned weights according to their spatial relationship with the evidence previously considered. Figure 1a shows the weights of training samples after the Gejiu Formation had been applied. These training samples with the new weights will be used for calculating the weight for the next evidence to be added. The results of weights calculated according to the order of patterns GF, DI, RA, LA and DF are shown in Fig. 1b. The weights obtained by the WofE are also calculated and shown in Fig. 1b. It can be seen from Fig. 1b that the two corresponding weights calculated for the first evidence by WofE and BoostWofE are equal, but the weights for the subsequent evidences are different. Other orders of these five evidences were also considered and compared. Based on the new weights obtained by BoostWofE and WofE we can calculate the posterior probabilities of each training points as shown in Fig. 1c. Further in Fig. 1d, it shows the observed numbers of mineral deposits and estimated numbers of mineral deposits from the posterior probabilities obtained by WofE and BoostWofE.

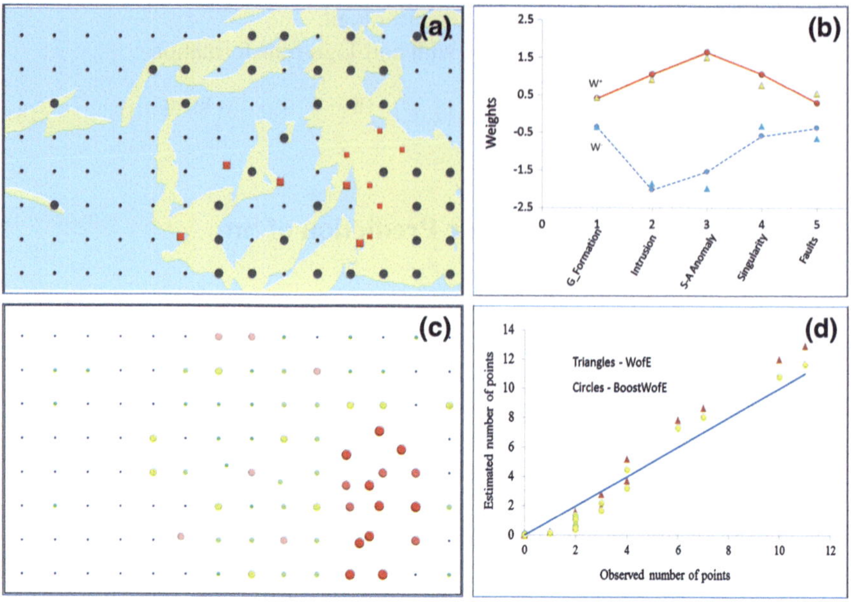

Fig. 1 **a** Map showing weighted training samples after applying the first binary classifier (GF). *Dots* represent non-mineral deposits and *squares* for Sn mineral deposits. The sizes of symbols corresponding to the weights calculated after GF being applied; **b** Plot showing comparisons of weights (W^+ and W^-) obtained by the BoostWofE (*triangles*) and WofE (*circles*); **c** Posterior probability calculated using the BoostWofE method. The size and color are corresponding to the probability; and **d** Comparison between observed mineral deposits (*solid line*) and estimated points by BoostWofE as *triangles* and by WofE as *circles*

4 Concluding Remarks

The case study has demonstrated that the estimated numbers of mineral deposits on the basis of posterior probability obtained by BoostWofE are generally in agreement with the observed results whereas the results obtained by WofE are notably overestimated. Further the results calculated using the Omnibus Test (OT) [5] by comparing the estimated number of points with the observed number of points, OT = 11/11.7 = 0.95 by BoostWof and OT = 11/12.9 = 0.85 by the ordinary WofE, also depict that the results obtained by BoostWofE show significant less effect of conditional dependency than those obtained by the ordinary WofE method. More case studies are needed to validate the effectiveness of BoostWofE.

References

1. Schapire, R. E. (1990). The strength of weak learnability. *Machine Learning*, *5*, 197–227.
2. Freund, Y. (1995). Boosting a weak learning algorithm by majority. *Information and Computation*, *121*(2), 256–285.
3. Friedman, J., Hastie, T., & Tibshirani, R. (2000). Additive logistic regression: A statistical view of boosting. *The Annals of Statistics*, *28*(2), 337–407.
4. Mease, D., & Wyner, A. (2008). Evidence contrary to the statistical view of boosting. *Journal of Machine Learning Research*, *9*, 131–156.
5. Bonham-Carter, G. F. (1994). *Geographic information system for geosciences: Modelling with GIS*. Oxford: Pergamon Press.
6. Agterberg, F. P. (2011). A modified weights-of-evidence method for regional mineral resource estimation. *Natural Resources Research*, *20*(2), 95–101.
7. Cheng, Q. (2012). *BoostWofE: A new weights of evidence model reducing the effect of conditional independency*. Math Geometry.
8. Cheng, Q. (2012). Multiplicative cascade processes and information integration for predictive mapping. *Nonlinear Processes in Geophysics*, *19*, 57–68.

Identification of Spatial Models of $\delta^{18}O$ in Precipitation of the Wet Season Over Spain Using Genetic Algorithms

Javier Heredia, Eulogio Pardo-Igúzquiza, Javier Rodríguez-Arévalo, Silvino Castaño, María F Díaz-Teijeiro, José E Capilla, Aantonio Prado and Lara Bardasano

1 Introduction

The water vapour in the clouds and water in precipitation alters their isotopic compositions because of the isotope fractionation: water vapour decreases its heavy isotopes ($^{18}O, ^{2}H$) while precipitation water has a relative increase in that value. The Global Network of Isotopes in Precipitation (GNIP) operates since 1961. Spain is part of GNIP since 2000 with the Red Española de Vigilancia Isótopos en Precipitación (REVIP) that has 14 stations in the Iberian Peninsula, one in Mallorca (Fig. 1) and another one in Canarias Islands.

J. Heredia (✉) · E. Pardo-Igúzquiza (✉) · S. Castaño
Instituto Geológico y Minero de España, Ríos Rosas 23, 28003 Madrid, Spain
e-mail: j.heredia@igme.es

E. Pardo-Igúzquiza
e-mail: e.pardo@igme.es

S. Castaño
e-mail: s.castano@igme.es

J. Rodríguez-Arévalo (✉) · M.F. Díaz-Teijeiro · A. Prado · L. Bardasano
Centro de Estudios y Experimentación de Obras Públicas, Alfonso XII, 3 y 5, 28005 Madrid, Spain
e-mail: javier.rodriguez@cedex.es

M. F. Díaz-Teijeiro
e-mail: marife.diaz@cedex.es

A. Prado
e-mail: antonio.prado@cedex.es

L. Bardasano
e-mail: lara.bardasano@cedex.es

J. E. Capilla
Universitat Politècnica de Valencia, Valencia, Spain
e-mail: jcapilla@upv.es

E. Pardo-Igúzquiza et al. (eds.), *Mathematics of Planet Earth*,
Lecture Notes in Earth System Sciences, DOI: 10.1007/978-3-642-32408-6_38,

Fig. 1 Monitoring network of isotopes in precipitation in Spain (modified of Google Earth)

In most of the Iberian Peninsula, the precipitation is concentrated between October and April: the wet season. To know the spatial trend of δ^{18}O in precipitation of the wet season allows to characterize the hydrogeological systems: recharge and residence times. The aim of this work is the statistical description of the spatial trend of the average value of δ^{18}O in precipitation of the wet season in the Iberian Peninsula. These regression models use different kind of variables: geographical (latitude, longitude, altitude, wind, distance to the sea) and climatic (temperature, precipitation, relative humidity, wind speed, atmospheric pressure and NAO index). The huge number of possible structures precludes an exhaustive search. Thus it was used a soft-computing method like genetic algorithms to indentify the optimal selection. A few geographical variables were used to model the global average annual of δ^{18}O in precipitation [1]. In Spain, the average annual of δ^{18}O in precipitation was modelling with geographical variables [2] and recently the time series of δ^{18}O in precipitation was modelling with climatological variables [3].

2 Methodology

Genetic algorithms (GA) are computer programs that simulate the process of biological evolution in order to solve optimization problems. They have been used to solve many problems in geosciences. The general scheme of a GA is the following: (i) To generate the starting population, where each individual is a solution to the stated problem. Each individual is represented by a chromosome and each chromosome is composed of genes where each gene represents some aspect of the solution; (ii) To evaluate the fitness of each individual, that is, the value of some objective function; (iii) To select the progenitors; (iv) To apply the operation of cross-over and mutation to the progenitors; (v) The descendents joint the progenitors to form a new generation; (vi) Go to step (ii) and repeat to advance another generation.

In the problem that we are dealing with, the optimization problem is a multiple least squares regression where the task is to show the combination of variables that best explain the spatial trend of δ^{18} O. There are 10 variables, plus their logarithms, their inverses, squares, cubes, square roots and the cross-products of any pair of the previous 60 variables. The maximum number of variables in the regression has been limited to 6. Thus, the chromosome has 24 genes, where four genes are used for describing each of the six allowed dependent terms in regression. Each population has been generated with 1000 individuals and it has been repeated for 1000 generations. The procedure of ranking has been used for selecting the progenitors, 500 out of the 1000, which produce 500 descendents in order to keep a constant size for the population. The fitness function has been the total sum of squares of the differences between experimental data and the value predicted by the model.

3 Results

The experimental data are the average monthly values of δ^{18} O measured at 15 locations (Fig. 1) of the Iberian Peninsula and for 7 months (from October to April, between 2000 and 2009) at each station. A calibration data set (53) and a validation data set (52) were selected from the total data set (105). The fitness function selected has been the adjusted squared correlation coefficient, R^2_{adj}. The R^2_{adj} takes into account: the goodness of fit of the model by the squared Pearson correlation coefficient, R^2, and the parsimony principle the number of the independent variables, m. The model validation was evaluated by least squares for each model. The results for calibration by the genetic algorithm and validation are shown in Fig. 2. The number of terms allowed has been changed from 1 to 6. There are two variables, NAO index and distance to the sea, that never appear in the best solution equations. The highest value of adjusted R^2_{adj} has been 0.855 for the case of six terms. Though, a similar value of 0.835 can be reached with tree terms. The best improvement can be seen when the number of allowed terms is larger than two. On the other hand, the validation results indicate that with more than 3 variables the models get worse.

4 Conclusions

The genetic algorithm has proven to be a very efficient method for best equation selection in multiple regression applied to the problem of determining isotopic content of δ^{18} O in precipitation from functionals of climatic and geographic variables.

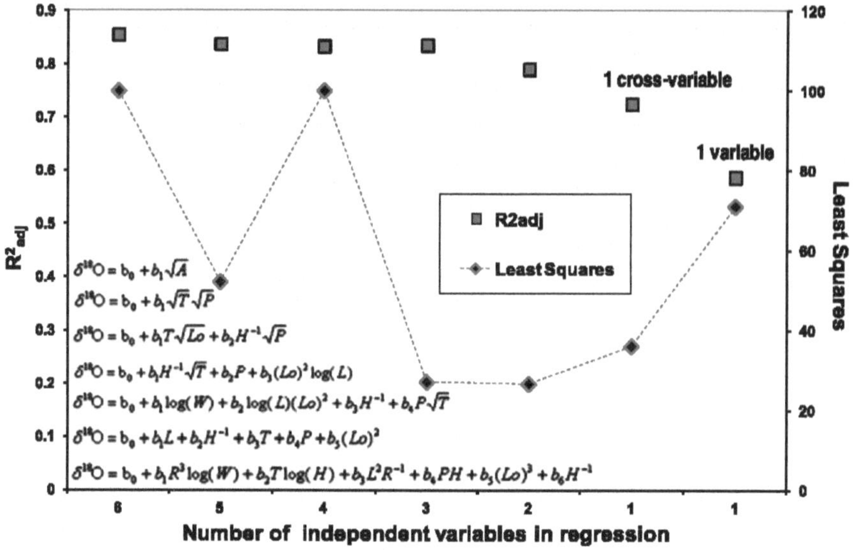

Fig. 2 R^2_{adj} and least squares for different optimal models found by the genetic algorithm. *T* temperature, *R* precipitation, *H* humidity, *W* wind speed, *P* atmospheric pressure, *D* distance to the sea, *L* latitude, *Lo* longitude, *A* altitude, *NAO* NAO index

The number of possibilities is too large to be examined exhaustively and the genetic algorithm offers an efficient solution in practical terms.

Acknowledgments This work has been developed using the results of the Project ISOTOPOSHIDRO (ref. CGL2009-12977) sponsored by Ministerio de Economía de España. The work of the second author has been supported by the Project KARSTINV (ref. CGL2010-15498). To the reviewers, who contributed to improve this paper.

References

1. Bowen, G.J., & Wilkinson, B. (2002). B.: Spatial distribution of $\delta^{18}O$ in meteoric precipitation. *Geology*, *30*(4), 315–318.
2. Rodríguez Arévalo, J., Castaño, S., & Díaz-Teijeiro, M.F. (2010). Contribution of the Spanish network for isotopes in precipitation and related modeling activities to characterize water bodies and their evolution. *European Groundwater Conference,* Madrid, Spain.
3. Heredia, J., Castaño, S., Díaz-Teijeiro, M.F., Rodríguez-Arévalo, J., Capilla, J., Sánchez, R., & Bardasano, L. (2012). Modelling the temporal variability of $\delta^{18}O$ values in precipitation in peninsular Spain and Balearic islands. 7th *AHPGG*, San Sebastián, Spain.

Space-Time Prediction of Extreme Events by Using Mathematical Logic Modeling of Cause-Effect Relations

Susanna Sirotinskaya

1 General Concept

Extreme events are effects of the unusual combination of factors (causes) which may be phenomena, processes or properties (structure, composition) of some objects. The extreme events are rare and are separated from each other in space or/and in time. As the environment is heterogeneous, manifestations (features) of the same factor may differ in different points of it. Therefore, identical extreme events may be linked to various feature sets related to the same combination of factors. These ideas can be expressed more formally as the two following sentences:

- An extreme event happens if and only if a certain combination of factors is present;
- The action of any factor from this combination may be manifested by different sets of non-alternative features, that is, the factor is present if and only if at least one feature from such sets is observed.

The tools of mathematical logic, in particular those of Propositional algebra, Predicate logic and Boolean algebra, allow expressing these two sentences in the form of a single Boolean equation which is used as theoretical model in methods of the cause-effect analysis. This equation is

$$F = (\vee^{1}{}_{1} \,\&\ldots\&\, \vee^{k}{}_{1}) \vee \ldots \vee (u^{1}{}_{J} \,\&\ldots\&\, u^{k}{}_{S}) \vee \ldots \vee (u^{1}{}_{t(1)} \,\&\ldots\&\, u^{k}{}_{t(k)}), \quad (1)$$

where F is the Boolean function characterizing the event under study; the variables in brackets are the function F arguments corresponding to some features related to different factors; k is the number of factors responsible for the event; $t(1), \ldots, t(k)$ are the numbers of features expressing these factors; &, $\vee$ are logical operations of conjunction and disjunction corresponding to "and" and non-alternative "or" in

S. Sirotinskaya (✉)
Rua Riachuelo, 535, ap. 301, Porto Alegre 90010-270, Brazil
e-mail: ascomputer@terra.com.br

E. Pardo-Igúzquiza et al. (eds.), *Mathematics of Planet Earth*,
Lecture Notes in Earth System Sciences, DOI: 10.1007/978-3-642-32408-6_39,

natural languages. The arguments of the function F are unknown parameters, and they are determined by processing of the given data. The values of the function F are 1 or 0, depending on the presence or absence of the event.

The solution of Eq. (1) is a formula of the same form in which the unknown arguments of the function F are substituted by variables corresponding to concrete features from the given data file. In accordance with the condition that only the most essential features should be taken in account, this formula includes the minimum number of variables (features) and the minimum number of variable conjunctions (combinations of features in brackets) in comparison with those in other possible formulas of the same form. The condition of minimizing both the number of individual variables and variable conjunctions is used as the optimization criteria and it is expressed mathematically by the requirement to minimize Disjunctive Normal Form (DNF) which is exactly the form of Eq. (1). This requirement, together with some conditions related to peculiarities of a study problem, makes possible the selection of the optimal solution corresponding to the most reliable cause-effect relation. Equation (1) may have several optimal solutions.

The features which are selected on the basis of solving Eq. (1) are used as predictive criteria. Combinations of these criteria in brackets represent favorable (unfavorable) settings in the case of space prediction and favorable (unfavorable, dangerous) situations in the case of time prediction. The prediction is performed by change of variables in the detected Boolean formula for the values 1 or 0, depending on the presence or absence of respective features. The result of prediction is expressed by values 1 (favorable) or 0 (unfavorable, dangerous) calculated by the derived formula.

2 Data Preparation and Processing

Primary data for performing the cause-effect analysis can be represented by conventional and contour maps as well as by numerical tables and text descriptions. In the case of the following construction of predictive maps, the specific range of original map scales is 1:25,000 to 1:15,00,000. In other cases, it may be anyone. Preparation of such data for computer processing includes three main procedures: creation of feature system, selection of sample objects and representation of data in the form of a binary table.

Creation of feature system represents the most laborious procedure of data preparation. This procedure differs for different types of primary data. In the case of conventional maps, a set of analyzable phenomena are given by subdivisions of map explanations, and a set of analyzable features is composed mainly by using units of explanations. In addition, some features created on the basis of map contents (e.g., direction of faults) may be introduced in the feature set.

In the case of contour maps, a map displays an individual phenomenon (property) which is described by quantitative characteristics in the form of contours. Therefore, intervals of characteristic values can be considered as phenomenon features. This idea is used just for the conversion of quantitative data into qualitative form [2], in particular, into the form of a binary table. The main condition for performing such conversion is the availability of two classes of sample objects, one of them being positive and the other negative relatively to the presence or possible occurrence of extreme events.

The same conversion is also used in the case of numerical tables. In the case of text descriptions, a feature system is created on the basis of professional knowledge.

The data preparation also includes the *selection of sample objects*. The performance of this procedure is closely related to the representation of final results. If the representation is a predictive map, the sample objects are selected on the basis of cartographic material. Then, they are areas of a standard size and shape enclosing the location of extreme events (e.g., mineral deposits, landslides). For example, a sample area may have a circle shape of the radius 2 km. If only the evaluation of some objects for the presence or possible occurrence of extreme events is required, the use of cartographic material for the selection of sample objects is not necessary.

The preparation of data is completed with the *construction of a binary table* in which the columns represent features and the rows correspond to sample objects. The values 1 and 0 (or omission) designate the presence and absence of a feature, respectively. A binary table can be converted further into compact form by the substitution of binary arrays for the strings of positive integers which are the codes of features with value of 1 in the corresponding binary array.

The use of the same mathematical model for the space, time and space-time prediction implies the identical way of data processing which includes detection of logical dependencies consistent with the given mathematical model in a data file; selection of optimal dependence; data analysis and data transformation (if necessary); introduction of knowledge of different categories in the data processing. These procedures are carried out with the help of an expert system with artificial intelligence capabilities [3]. Mathematical tools are Boolean algebra operations and transformations as well as some algorithms of the Boolean function minimization.

3 Modes of Prediction

Despite the possibility to use the same method and software both for space and time or space-time prediction, there is a difference in the mode of their application for different types of prediction. This difference is due to the relative importance given to the space and time parameters that, in its turn, depends on the research purpose.

The space prediction is the most suitable for study of phenomena having a low velocity of change. While performing this kind of prediction, the time is assumed fixed. Therefore, a data file does not include information referring to time. As study objects (e.g., mineral deposits) are distributed spatially, they can be shown by some symbol on original maps. In such case, the results of data processing, which are the Boolean formulas describing favorable (unfavorable) settings, can be expressed by predictive maps.

The time prediction is the most appropriate for study of phenomena having a relatively high velocity of change (e.g., weather events). In this case, a data table includes time points (e.g., years) of observation as objects. It is clear that all observations must be related to one fixed territory (e.g., the same station, region).

A possible application of time prediction consists in the detection of regularity in the manifestation of a phenomenon and identification of situations which deviate from this regularity being linked to the occurrence of extreme events.

In the space-time prediction, neither space, nor time is considered fixed (e.g., location and time points of earthquake). This prediction may be performed by using several cycles of space or time prediction. If the space prediction is performed, several data files for the same set of areas but for different time points are created (e.g., data file for one day before the earthquake, two days before it and so on). The purpose of this prediction mode is to detect the situations which are linked to time points preceding the extreme events. The identification of such situations in the environment is aimed at determining the exact time point of the extreme event occurrence. If the time prediction is used, several data files for the same range of time points but for different areas are processed. This approach seems to be appropriate for prediction of such extreme events as landslides.

Till now, the application of cause-effect analysis was limited to the space prediction [1, 3] and was mainly intended for the assessment of mineral deposits which apparently can be considered as extreme events. As regards to the time and space-time prediction, its practical application encounters significant difficulties. The main obstacle is the absence of data about the state of environment on the eve of extreme events. Obviously, in order to obtain such data the permanent monitoring of the environment in areas of possible extreme events is necessary.

References

1. Sirotinskaya S. V. (2011). Methods of mathematical logic in geosciences: Classification, pattern recognition, cause-effect analysis (p. 528). Porto Alegre: Editora UFRGS. (in Portuguese). (Susanna Sirotinskaya: Métodos de Lógica Matemática em Geociências: classificação, reconhecimento de padrões, análise de causa-efeito).
2. Sirotinskaya S. V., & Strieder A. J. (2008). Mathematical logic in data integration and in modeling: Basic elements (p. 282). Porto Alegre: Editora UFRGS. (in Portuguese) (Susanna Sirotinskaya & Adelir Jose Strieder: Lógica Matemática na Integração de Dados e na Modelagem: elementos básicos).
3. Sirotinskaya S. V. (2004). Cause-effect analysis in assessment of mineral resources. *Journal of Natural Resources Research*, *13*(1), 17–28.

Local Clustering in Spatio-Temporal Point Patterns

Jorge Mateu and Francisco J. Rodríguez-Cortés

1 Introduction and LISA Functions

LISA functions are built from local second-order characteristics of spatial point processes through product densities. The term LISA was coined by Anselin [1], but it was later when Collins and Cressie [2] developed theoretical properties, namely first- and second-order moments, of these functions. Our focus here is in extending the concept of LISA to the spatio-temporal context defining the LISTA functions. We define these new functions, present edge-corrected estimators, and develop their first theoretical moments.

A spatial point process X is a stochastic model governing the locations of events $\{\mathbf{u}_i = (x_i, y_i)\}$ in some Borel bounded set W of $\mathbb{R}^2$ [3]. Assume we have a planar point process X observed in a region $W \subset \mathbb{R}^2$ of area $|W| > 0$. Let $N(W)$ be the number of points of the process appearing in W, and define the spatial intensity ρ as the mean number of events per unit area. For a stationary and isotropic point process in the plane with intensity ρ, the K-function is defined by $\rho K(r) \equiv \mathbb{E}[N(B(\mathbf{0}, r) \setminus \{\mathbf{0}\})|\mathbf{0} \in X]$, with $r > 0$. The K-function provides an interpretable measure of clustering in a point process [4]. In particular, $\rho^2|W|K'(r)$ is the expected number of pairs of events in the observation region W with pairwise distance less than or equal to r. The K-function is a cumulative function and investigation of its derivative leads to another interpretable function, called the product density function, defined by $\rho^{(2)}(r) = \rho^2 K'(r)/2\pi r$ with $r > 0$. Denote its non-parametric kernel-based estimator by $\widehat{\rho_\varepsilon}^{(2)}(r)$, which depends on a bandwidth ε.

Now consider individual contributions to the estimated function that are analogous to the local statistics called *local indicators of spatial association* (LISA).

J. Mateu (✉) · F. J. Rodríguez-Cortés
Department of Mathematics, Universitat Jaume I, Castellón, Spain
e-mail: mateu@mat.uji.es

F. J. Rodríguez-Cortés
e-mail: cortesf@mat.uji.es

E. Pardo-Igúzquiza et al. (eds.), *Mathematics of Planet Earth*,
Lecture Notes in Earth System Sciences, DOI: 10.1007/978-3-642-32408-6_40,

An individual LISA product density function $\rho^{(2)i}(r)$ should reveal the extent of the contribution of the event $\mathbf{u}_i$ to the global estimate of $\rho^{(2)}(r)$, and may provide a further description of structure in the data. By analogy with the formation of the global product density estimate $\widehat{\rho_\varepsilon^{(2)}}(r)$, a localised version of the empirical product density function is given by

$$\widehat{\rho_\varepsilon^{(2)i}}(r) \equiv \frac{N(W)-1}{2\pi|W|}\sum_{j\neq i} w(\mathbf{u}_i,\mathbf{u}_j)\frac{\kappa_\varepsilon(||\mathbf{u}_i-\mathbf{u}_j||-r)}{||\mathbf{u}_i-\mathbf{u}_j||}, \qquad r>\varepsilon>0,$$

where $\{\mathbf{u}_1,\mathbf{u}_2,\ldots,\mathbf{u}_{N(W)}\}$ are locations observed in a window W, $w(\mathbf{u}_i,\mathbf{u}_j)$ is a spatial edge-correction factor [5], and κ_ε is a kernel function with bandwidth ε. This function provides a description of the local density of inter-event distances among the observed locations.

2 Spatio-Temporal Point Processes: LISTA Functions

Following Møller and Ghorbani [6], consider a spatio-temporal point process with no multiple points as a random countable subset X of $\mathbb{R}^2\times\mathbb{R}$, where a point $(\mathbf{u},s)\in X$ corresponds to an event at $\mathbf{u}\in\mathbb{R}^2$ occurring at time $s\in\mathbb{R}$. In practice, we observe n events $\{(\mathbf{u}_i,s_i)\}$ of X within a bounded spatio-temporal region $W\times T\subset\mathbb{R}^2\times\mathbb{R}$, with area $|W|>0$, and length $|T|>0$. In the sequel, $N(A)$ denotes the number of the events of the process falling in a bounded region $A\subset W\times T$. The intensity function of a spatio-temporal point process X can be defined as

$$\rho(\mathbf{u},s)=\lim_{|d\mathbf{u}\times ds|\to 0}\frac{\mathbb{E}[N(d\mathbf{u}\times ds)]}{|d\mathbf{u}\times ds|},$$

with $(\mathbf{u},s)\in\mathbb{R}^2\times\mathbb{R}$ and $d\mathbf{u}\times ds$ defining a small subregion around the event $(\mathbf{u},s)$ with volume $|d\mathbf{u}\times ds|$. Further, for $(\mathbf{u},s),(\mathbf{v},l)\in\mathbb{R}^2\times\mathbb{R}$ the product density is defined as

$$\rho^{(2)}((\mathbf{u},s),(\mathbf{v},l))=\lim_{(|d\mathbf{u}\times ds|,|d\mathbf{v}\times dl|)\to(0,0)}\frac{\mathbb{E}[N(d\mathbf{u}\times ds)N(d\mathbf{v}\times dl)]}{|d\mathbf{u}\times ds||d\mathbf{v}\times dl|}.$$

A process for which $\rho(\mathbf{u},s)=\rho$ for all $(\mathbf{u},s)$ is called homogeneous of first-order stationary.

For a second-order intensity-reweighted stationary (SOIRS) point process X, Gabriel and Diggle [7] and then Møller and Ghorbani [6] define the spatio-temporal inhomogeneous K-function by

$$K(r,t)=\int \mathbf{1}[||\mathbf{u}||\le r,|s|\le t]\,g_0(\mathbf{u},s)d(\mathbf{u},s), \qquad r>0,\quad t>0.$$

with $r > 0, t > 0$. For a Poisson process, $g_0 = 1$ and $K(r,t) = 2\pi r^2 t$. Both in the stationary and isotropic case and the SOIRS and isotropic case, the spatio-temporal pair correlation function is proportional to the derivative of $K(r,t)$ with respect to r and t. Thus for a SOIRS and isotropic spatio-temporal point process X

$$\rho^{(2)}((\mathbf{u},s),(\mathbf{v},l)) = \frac{\rho(\mathbf{u},s)\rho(\mathbf{v},l)}{4\pi r}\frac{\partial^2 K(r,t)}{\partial r \partial t}, \qquad r > 0, \quad t > 0.$$

Influenced by the idea of Fiksel [8], and under a SOIRS and isotropic point process X, we propose a general edge-corrected kernel estimate of the product density function by

$$\widehat{\rho}^{(2)}_{\varepsilon,\delta}(r,t) = \sum_{(\mathbf{u}_i,s_i),(\mathbf{u}_j,s_j)\in X}^{\neq} w(\mathbf{u}_i,\mathbf{u}_j)v(s_i,s_j)\frac{\kappa_{\varepsilon,\delta}(\|\mathbf{u}_i-\mathbf{u}_j\|-r,|s_i-s_j|-t)}{4\pi|W||T|\|\mathbf{u}_i-\mathbf{u}_j\|},$$

where κ is a multiplicative kernel, ε and δ are the spatial and temporal bandwidths, respectively, and $v(s_i,s_j)$ is a temporal edge-correction factor. Then κ can be written as

$$\kappa_{\varepsilon,\delta}(\|\mathbf{u}_i-\mathbf{u}_j\|-r,|s_i-s_j|-t) = \kappa_{1\varepsilon}\left(\|\mathbf{u}_i-\mathbf{u}_j\|-r\right)\kappa_{2\delta}\left(|s_i-s_j|-t\right).$$

A LISTA function can be constructed in the same manner as the global estimate. In this context, we begin by considering local features of the spatio-temporal K-function. Define for a spatio-temporal point process X

$$\{\rho K(r,t)\}^i \equiv \mathbb{E}\big[N(\mathbf{C}((\mathbf{u}_i,s_i),r,t)\setminus\{(\mathbf{u}_i,s_i)\})|(\mathbf{u}_i,s_i)\in X\big] \qquad r > 0, \quad t > 0,$$

as the number of extra events from $(\mathbf{u}_i, s_i)$ with spatial distance less than or equal to r, and time increment less than or equal to t. The expectation is conditional on observing $(\mathbf{u}_i, s_i) \in X$, and is calculated with respect to the reduced Palm measure. Note that $\mathbf{C}((\mathbf{u}_i, s_i), r, t)$ is a cylinder with center $(\mathbf{u}_i, s_i)$, radius r and height $2t$.

3 Conclusions

In this work we propose a general edge-corrected kernel estimate of a LISTA function, and develop close expressions for the theoretical first and second-order moments of the global and local spatio-temporal product density. The global density function detects large scale spatio-temporal interactions, while the local LISTA function is useful to highlight local clusters.

Acknowledgments Work partially funded by grant MTM2010-14961 from the Spanish Ministry of Science and Innovation.

References

1. Anselin, L. (1995). Local indicators of spatial association-LISA. *Geographical Analysis*, *27*, 93–115.
2. Collins, L., & Cresie, N. A. C. (2001). Analysis of spatial point patterns using bundles of product density LISA functions. *Journal of Agricultural, Biological, and Environmental Statistics*, *6*, 118–135.
3. Stoyan, D., Kendall, W. S., & Mecke, J. (1995). *Stochastic geometry and its applications* (2nd ed.). Chichester: Wiley.
4. Illian, J., Penttinen, A., Stoyan, H., & Stoyan, D. (2008). *Statistical analysis and modelling of spatial point patterns*. Chichester: Wiley.
5. Ripley, B. D. (1988). *Statistical inference for spatial processes*. Cambridge: Cambridge University Press.
6. Møller, J., & Ghorbani, M. (2012). Aspects of second-order analysis of structured inhomogeneous spatio-temporal point processes. *Statistica Neerlandica*, *66*, 472–491.
7. Gabriel, E., & Diggle, P.J. (2009). Second-order analysis of inhomogeneous spatiotemporal point process data. *Statistica Neerlandica*, *63*, 43–51.
8. Fiksel, T. (1988). Edge-corrected density estimators for point processes. *Statistics*, *19*, 67–75.

The Use of Remote Sensing Data in a Colombian Andean Basin for Risk Analysis

Olga Lucía Ocampo López and Jorge Julián Vélez Upegui

1 Introduction

Colombian Andean Basins are particularly vulnerable to climate change and climate variability; therefore, local studies are required for risk management. Understanding extreme events triggered by rainfall is a prerequisite for determining how climate events contribute to disaster occurrences, and for designing and implementing effective adaptation and disaster risk management strategies [1].

The estimation of hazardous event probabilities and associated return period at different intensity levels can be simulated by models that use historical data. The Intensity-Duration-Frequency (IDF) relationship has been a major focus of both theoretical and applied hydrology. Precipitation Area Duration Frequency (PADF) curves are used to model primary and secondary hazards such as non-hurricane rain events and floods. These curves are the representation of storm events characteristics that have occurred historically in a basin [2].

Without adequately accurate climate input information, it is extremely difficult to establish models [3]. Within many developing countries, as Colombia, stations are relatively sparse, the amount of missing data is high and the reliability of some data is low. Remotely sensed meteorological data sets can be employed to extract spatially explicit rainfall estimates to determine return period of events. Their power lies in the continuous acquisition of data at relatively short temporal intervals for large spatial extents at several spatial resolutions, which can therefore be used to support data generation for disaster risk modeling [2].

O. L. O. López · J. J. V. Upegui (✉)
Facultad de Ingeniería y Arquitectura, Universidad Nacional de Colombia Sede Manizales, Cra. 27 64-60 IDEA Campus Palogrande Manizales, Manizales, Colombia
e-mail: jjvelez@unal.edu.co

O. L. O. López
Universidad Autónoma de Manizales, Antigua Estación del Ferrocarril, Manizales, Colombia
e-mail: olocampol@unal.edu.co; olocampo@autonoma.edu.co

E. Pardo-Igúzquiza et al. (eds.), *Mathematics of Planet Earth*,
Lecture Notes in Earth System Sciences, DOI: 10.1007/978-3-642-32408-6_41,

This paper presents a comparative analysis of the maximum rainfall in a Colombian Andean Basin, using information from The Tropical Rainfall Measuring Mission (TRMM) satellite rainfall estimates and rain gauge registers, which are employed to generate PADF, in order to determine the return period of rain events.

2 Methodology

The Chinchiná River Basin has an approximate area of 1050 km^2, it is located on the western slopes at the central range in the Andes between 4°48′ and 5°12′N Latitude, at the south central region of Caldas, Colombia, (see Fig. 1). This Basin extends from the National Natural Park "Los Nevados" at 5400 m.a.s.l. to the Cauca River at 780 m.a.s.l.

The maximum rainfall analysis was carried out using rain gauge and remote sensing data. Daily rain gauge data was obtained from IDEAM (Institute of Hydrology, Meteorology and Environmental Studies of Colombia), CENICAFE (National Center for Coffee Research), CHEC (Caldas Hydroelectric) and Universidad Nacional de Colombia sede Manizales. Remote sensing daily data was extracted from the 3B42-TRMM algorithm (3B42-TRMM—Adjusted Merged-Infrared precipitation) which was taken from NASA [4]. The TRMM satellite provides space borne rain radar and microwave radiometric data that measures the vertical distribution of precipitation over the tropics, where the bulk of the Earth's rainfall occurs [2].

The Precipitation depth analysis was carried out by the incremental-isohyetal method. Spatially explicit rainfall estimates are represented using PADF curves which are determined by creating a relationship between the average maximum precipitation, the Area where this rainfall falls, the Duration of the precipitation event and the Frequency of the rainfall episode. The PADF curves creation procedure is shown in Fig. 2, which was suggested by Florida International University and its GIS Center [2]. The PADF curve generation software, developed by CAPRA [5], was used to obtain precipitation values for different areas and return periods. In this software, the frequency is conducted using a Gumbel distribution, which can be accepted as a good approach to rainfall distribution in Colombia.

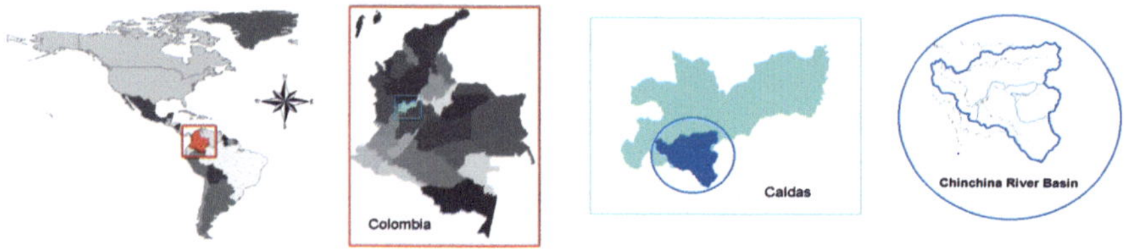

Fig. 1 Chinchiná River Basin location

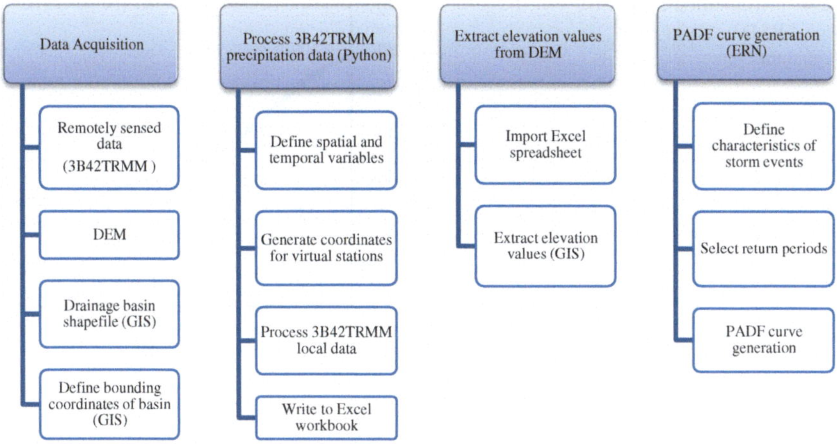

Fig. 2 Processing steps of TRMM precipitation data. Adapted [2]

3 Results

The reported extreme events are: 99 mm in Paramos, 150 mm in the upper Basin, 192 mm in the upper middle Basin, 170 mm in the middle Basin and 118 mm in the lower Basin. Regarding TRMM virtual stations, extreme precipitation fluctuates between 128 and 72 mm.

PADF curves were generated with remote sensing data, using the CAPRA model software; these curves, which are shown in Fig. 3, represent the probability of occurrence of maximum rainfall for different return periods and show the relationship between area and average depth.

PADF curves were also generated using rain gauge station data, with lower precipitation values being obtained in this case. The precipitation differences in PADF curves generated with remote sensing and rain gauge data, expressed in percentage, are shown in Fig. 4; these differences, with a range of 59 to 24 %, are 37 % average.

Fig. 3 PADF Curves 3B42-TRMM

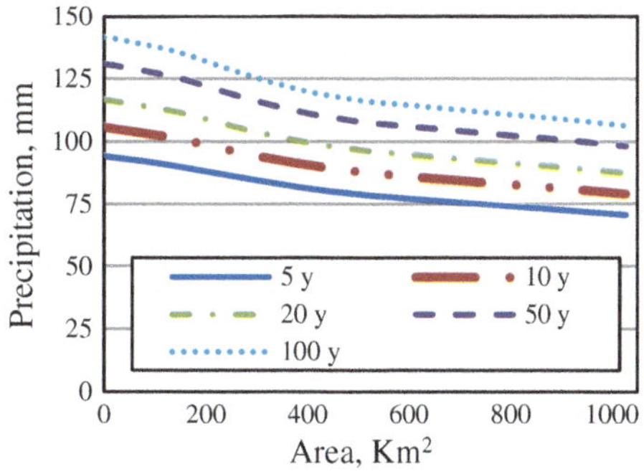

Fig. 4 Precipitation differences TRMM versus rain gauge data

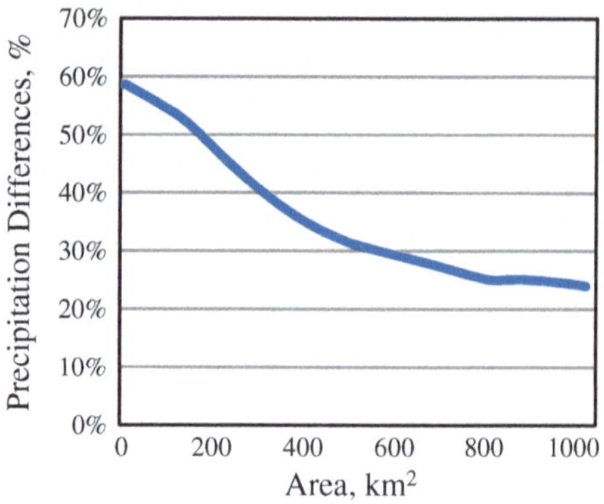

TRMM-satellite captures the spatial variability of rainfall in Chinchiná River Basin.

PADF curves for modeling primary and secondary risk in this Andean Basin should be generated from TRMM-satellite because better distribution curves are obtained with remote sensing data compared to rain gauge stations.

4 Conclusions

PADF curves are the representation of storm event that have occurred historically in a basin. There are significant differences between the PADF curves obtained by TRMM 3B42 daily rasters and rain gauge data in Chinchiná River Basin. Study results support the conclusion that it is advisable that remote sensing datasets are used for generating PADF curves for risk studies in this Colombian Andean Basin in order to perform a conservative design.

References

1. IPCC-Intergovernmental Panel on Climate Change. (2012). Managing the risk of extreme Events and Disasters to Advance Climate Change Adaptation.
2. Girs Center and Florida International University. (2011). Spatio-temporal precipitation estimates derived from remotely sensed data for use in the CAPRA model. Theory and Tutorial.
3. Hughes, D. A. (2006). Comparison of satellite rainfall data with observations from gauging station networks. *Journal of Hydrology*, *327*, 399–410.
4. National Aeronautics and Space Administration—NASA. ftp://disc3.nascom.nasa.gov/data/s4pa/TRMM_L3/TRMM_3B42_daily.
5. CAPRA. http://www.ecapra.org/.

Spatio-Temporal Analysis of Earthquake Occurrences Using a Multiresolution Approach

Orietta Nicolis

1 Introduction

The analysis and prediction of earthquake events is an important issue in environmental science. Space-time point-process models for earthquake analysis have been developed by [1–5]. In particular, the spatial-temporal Epidemic-Type Aftershock Sequence (ETAS) models introduced by [2] are commonly used to describe modern earthquake catalogs. Their basic assumption is that the risk of an earthquake occurrence at time t and location (x, y) consists in the contribution from the background rate μ and the contributions from each previous event. Then, the conditional intensity function can be written as

$$\lambda(x, y, t|\mathscr{H}_t) = \mu(x, y) + \sum_{i:t_i<t} g(x, y, t|x_i, y_i, t_i; m_i) \quad (1)$$

where $\mathscr{H}_t = \{(t_i, x_i, y_i, m_i); t_i < t\}$ is the history of occurrence times $\{t_i\}$ up to time t, with the corresponding epicenters $\{(x_i, y_i)\}$ and magnitudes m_i; $\mu(x, y)$ is the background seismicity rate, and the function g, called the triggering density, describes the rate of aftershocks (with magnitude $m > m_c$) at space-time coordinate (x, y, t) following the ith earthquake (x_i, y_i, t_i, m_i). Functional forms for the background rate $\mu(x, y)$ are not typically given. Nonparametric methods, such as splines and kernels, were proposed by [3, 4], and [5], among others. These studies suggest that smoothing the locations of past earthquakes provides a good estimate of the spatial distribution of future seismicity. [4] and [5] proposed a variable (adaptive) kernel estimator for spatially clustered point dataset. In particular, [5] found that declustering the data prior to smoothing epicenters resulted in larger probability gains than smoothing all quake locations. Alternatives to kernel estimation are the

O. Nicolis (✉)
Departamento de Estadistica, Universidad de Valparaiso,
Gran Bretaña 1111, Valparaiso, PA, Chile
e-mail: orietta.nicolis@uv.cl

E. Pardo-Igúzquiza et al. (eds.), *Mathematics of Planet Earth*,
Lecture Notes in Earth System Sciences, DOI: 10.1007/978-3-642-32408-6_42,

use of orthogonal expansions, and the use of splines (see, [6]). In this work we propose an estimator based on directional wavelets for identifying the rupture zones and estimating the background seismicity rate in Chile. Then, the resulting wavelet-based spatial intensity function will be included in the estimation of the space time ETAS model shown in Eq. (1).

2 Directional Wavelets for the Analysis of Point Processes

For $\mathbf{x} \in \mathbb{R}^2$, and any function $f(\mathbf{x}) \in \mathbb{L}_2(\mathbb{R}^2)$, the continuous directional wavelet transform for a scale a and an orientation θ is given by

$$W_f(a, \mathbf{b}, \theta) = \langle f, \psi_{a,\mathbf{b},\theta} \rangle = \int_{\mathbb{R}^2} f(\mathbf{x}) \overline{\psi_{a,\mathbf{b}}(\mathbf{x}, \theta)} d\mathbf{x}. \tag{2}$$

In literature, a variety of directional wavelets $\psi_{a,\mathbf{b}}(\mathbf{x}, \theta)$ have been proposed. In particular, [7] introduced a flexible function called fully-anisotropic directional Morlet wavelet, which is given by

$$\psi_{a,\mathbf{b}}(\mathbf{x}, \theta) = e^{i\mathbf{k}_0 \cdot \mathbf{C}\mathbf{x}} e^{1/2\mathbf{C}\mathbf{x} \cdot \mathbf{A}^T \mathbf{A}\mathbf{C}\mathbf{x}}, \tag{3}$$

where $\mathbf{k_0} = (0, k_0)$ is a wave vector with $k_0 \geq 5.5$, $\mathbf{A} = diag(D, 1)$ denotes a diagonal matrix, and D is the anisotropy ratio defined as the ratio of the length of the elliptical envelope in the y-direction to the length of the elliptical envelope in the x-direction. The matrix $\mathbf{C}$ is a linear transformation which rotates the entire wavelet through an angle θ defined as positive in the counterclockwise direction. Two examples of this fully-anisotropic wavelet for directions $\theta = 30, 90$ are shown in Fig. 1. In order to identify the behavior of the process in different directions, [8] introduced two new functions, $\eta(a, \theta)$ and $\zeta(a, \theta)$, given by $\eta(a, \theta) = \int |W_f(a, \mathbf{b}, \theta)|^2 d\mathbf{b}$ and $\zeta(a, \theta) = \frac{\eta(a,\theta)}{\int \eta(a,\theta) d\theta}$ where the component $|W_f(a, \mathbf{b}, \theta)|^2$, called *directional scalogram*, gives the distribution of energy at location $\mathbf{b}$, scale a and direction θ [9].

Since the method proposed by [7] can be used for detecting anisotropy in 2-D images, we extended the wavelet transforms of point processes proposed by [10] to spatial data. Then, the background seismicity rate of Eq. (1) can be written as $\mu(\mathbf{x}) = \sum_a \sum_{\mathbf{b}} W_f(a, \mathbf{b}, \theta) \psi_{a,\mathbf{b}}(\mathbf{x}, \theta)$, where $W_f(a, \mathbf{b}, \theta) = \int_{\mathbf{x}} \psi_{a,\mathbf{b}}(\mathbf{x}, \theta) dN(\mathbf{x})$ and $dN(\mathbf{x}) = 1$ if a point fall in $d\mathbf{x}$ and 0 otherwise.

3 Application to the Earthquake Catalogue of Chile

We consider the epicenter locations of 7249 earthquakes with a magnitude (M) 4.5 or larger for the period from 1903 through 2012, inside and around Chile, selected from the USGS web site, http://earthquake.usgs.gov/earthquakes/eqarchives/epic/.

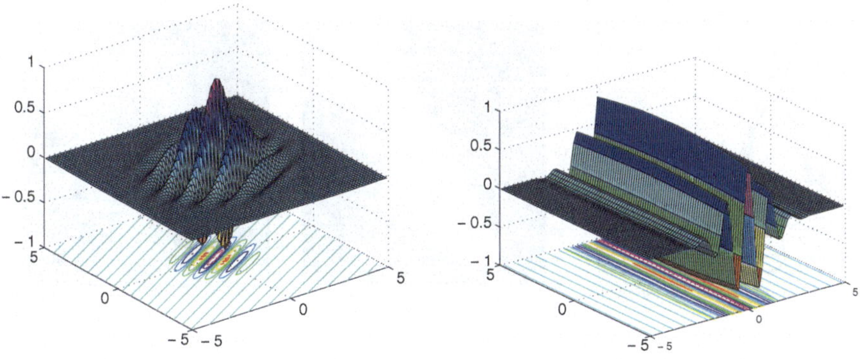

(a) (b)

Fig. 1 Fully-anisotropic directional Morlet wavelet with parameters: **a** $D = 0.8, k_0 = 5.5, \theta = 30$, **b** $D = 0.1, k_0 = 5.5, \theta = 90$

Figure 2 (left plot) show an example of the earthquake spatial distribution in Chile. By applying the directional wavelet transform to the epicenter data we have the scalogram shown in Fig. 3 (left plot) which identifies some dominant directions in the data. For example, the direction of 40 degree represents the main rupture zone given by the Nazca plate. The resulting hazard map using different directions are displayed in Fig. 2 (middle and right plots).

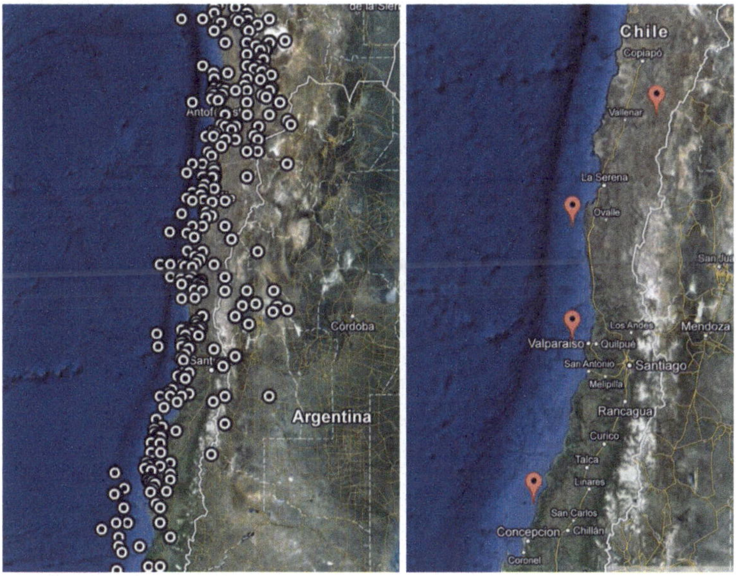

Fig. 2 Earthquakes in Chile from 1903 to 2012 with magnitude greater than 6 (*left plot*) and magnitude grater than 8 (*right plot*)

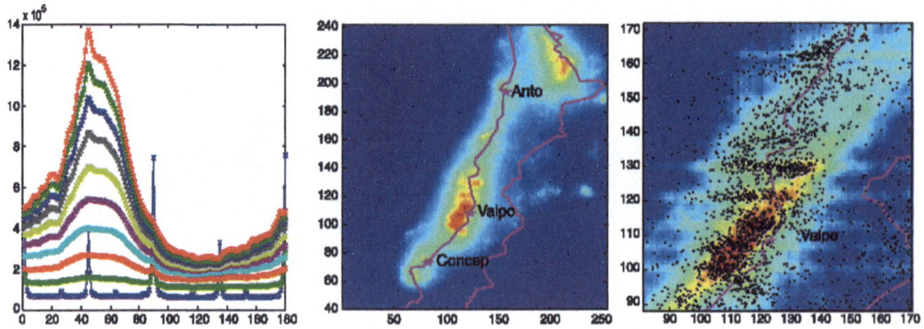

Fig. 3 Application of directional wavelets (with $D = 0.1$) to the earthquake catalogue of Chile with magnitude greater of 4.5 in the period 1903-2012: directional wavelet scalogram (*left plot*); wavelet-based estimated background rate using all directions (*middle plot*); wavelet-based estimated background rate of the Valparaiso area using angles of 40, 45, 90, 135, and 180° and earthquake epicenters (*black points*)

Acknowledgments The work has been partially supported by Fondecyt Regular 2013 (ID1131147).

References

1. Kagan, Y. (1991). Likelihood analysis of earthquake catalogues. *Journal of Geophysical Research. 106*, (Ser. B7), 135–148.
2. Ogata, Y. (1988). Statistical models for earthquake occurrences and residual analysis for point processes. *Journal of the American Statistical Association*, *83*, 9–27.
3. Ogata, Y. (1998). Space-time point process models for earthquake occurrences. *Annals of the Institute of Statistical Mathematic*, *50*, 379–402.
4. Zhuang, J., Ogata, Y., & Vere-Jones, D. (2002). Stochastic declustering of space-time earthquake occurrences. *Journal of the American Statistical Association*, *97*(3), 369–380.
5. Werner, M. J., Helmstetter, A., Jackson, D. D. & Kagan, Y. Y. (2011). High-resolution long-term and short-term earthquake forecasts for California. *Bulletin of the Seismological Society*. *101*, 1630–1648.
6. Vere-Jones, D. (1992). Statistical methods for the description and display of earthquake catalogues. In A. T. Walden & P. Guttorp (Eds.), *Statistics in the environmental and earth sciences* (pp. 220–246). London: Edward Arnold.
7. Neupauer, R. M., & Powell, K. L. (2005). A fully-anisotropic Morlet wavelet to identify dominant orientations in a porous medium. *Computers and Geosciences*, *31*, 465–471.
8. Kumar, P. (1995). A wavelet based methodology for scale-space anisotropic analysis. *Geophysical Research Letter*, *22*, 2777–2780.

9. Nicolis, O., & Mateu, J. (2012) Multiresolution analysis of spatial point processes for detecting linear patterns. GRASPA Working Papers (online), *44*. ISSN: 2037–7738.
10. Brillinger, D. (1998). Some wavelet analyses of point process data. 31st Asilomar conference on signals, systems and computers. *IEEE Computer Society*, 1087–1091.

Statistical Complexity Analysis of Spatiotemporal Dynamics

José M. Angulo and Francisco J. Esquivel

1 Entropy, Complexity and Multifractality

A variety of geophysical phenomena, among others, are characterized by certain forms of scaling behaviour intrinsic to spatiotemporal dynamics. Complexity and uncertainty inherent to such phenomena require the use of appropriate statistical tools for assessment.

Among other methodological approaches, information-theoretic related measures have been widely applied to quantify different structural aspects concerning evolution of a system. These include, for instance, classical Shannon entropy, and deformation parameter based generalizations such as Réyi and Tsallis entropies, as well as corresponding divergence and mutual information measures.

Formalization of measures quantifying complexity has received special attention in the last two decades, leading to a number of definitions also in an information-theoretic context, which have been applied in wide a variety of fields. Conceptually, complexity measures are aimed to reflect richness of behaviour in a system, with a balanced assessment between 'inner uncertainty' and 'departure from equilibrium' (see, for example, [1], and references therein).

As mentioned, dynamics of many natural phenomena is governed by scaling powerlaws related with self-similarity (fractality) of possibly different orders. State distributions associated to such systems are formally described in terms of multifractal measures, characterized by the presence of different local Hölder exponents. Generalized dimensions and multifractal spectrum, based on the scaling powerlaw

J. M. Angulo (✉)
Department of Statistics and Operations Research, University of Granada,
Campus de Fuente Nueva s/n, E-18071 Granada, Spain
e-mail: jmangulo@ugr.es

F. J. Esquivel
Andalusian Institute of Geophysics and Prevention of Seismic Disasters,
Campus de Cartuja s/n, E-18071 Granada, Spain
e-mail: jesquivel@ugr.es

E. Pardo-Igúzquiza et al. (eds.), *Mathematics of Planet Earth*,
Lecture Notes in Earth System Sciences, DOI: 10.1007/978-3-642-32408-6_43,

limiting behaviour of global and local Rényi entropies for the distributions associated to realizations at different resolution scales, constitute the main tools for multifractal analysis (see, for example, [2]).

Connections between complexity and multifractality concepts can be established based on the limiting behaviour according to scale of complexity measures. It can be proved, in particular, that LMC complexity with exponential entropy factor satisfies the limiting approximation in terms of the increment of generalized dimensions D_1 and D_2:

$$LMC_{\exp}(P_\varepsilon) = e^{H(P_\varepsilon)} \cdot D(P_\varepsilon) = e^{H(P_\varepsilon)} \cdot \left[e^{H_2(P_\varepsilon)} - \frac{1}{N_\varepsilon} \right] \sim \varepsilon^{-D_1} \varepsilon^{D_2} = \varepsilon^{D_2 - D_1}. \tag{1}$$

This result, which can be properly extended to some generalized forms of complexity, shows the potential use of variational properties of the generalized dimension curves for complexity assessment, in the multifractality context.

2 Application to Seismic Data

The above aspects are illustrated with application to a well known seismic data series occurred between October 8, 1987, and September 10, 1990, in the region surrounding Agrón village in the province of Granada, South of Spain. Three segments of

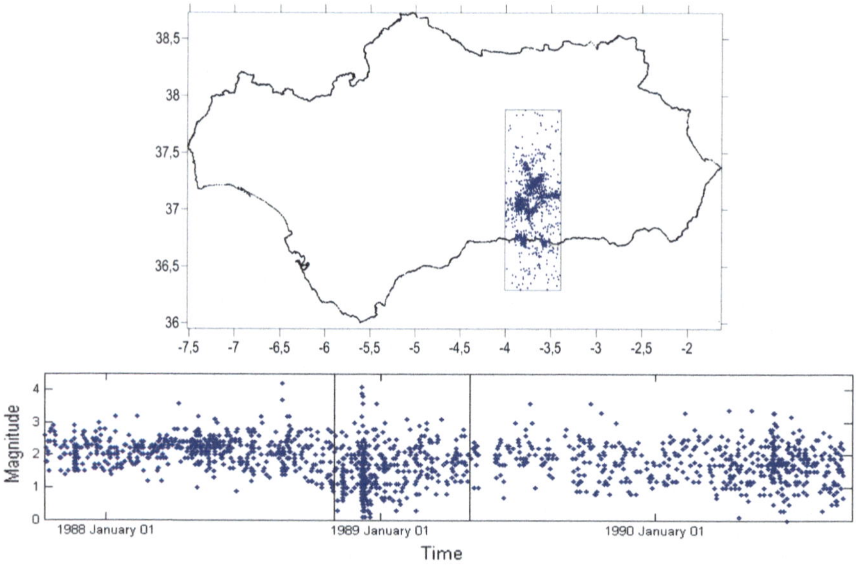

Fig. 1 Agrón data epicenters on Andalusian region (*top*), and magnitude-time plot (*bottom*)

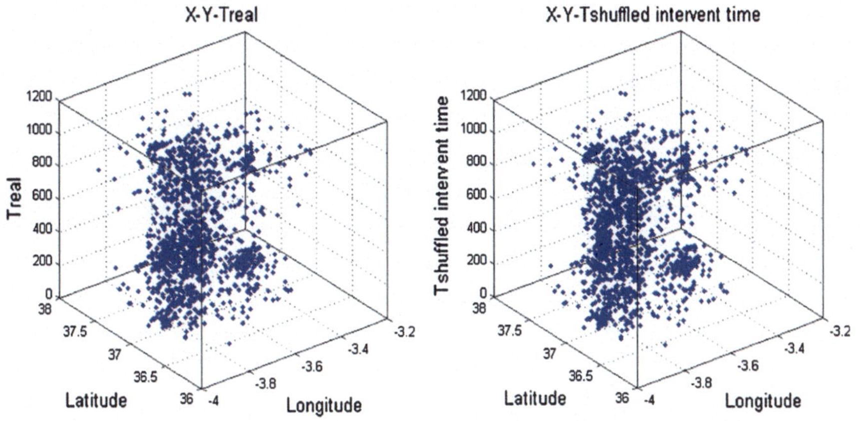

Fig. 2 Agrón data 3D scatter plots for (X, Y, T), corresponding to real and shuffled interevent time

equal size in number of events (551) are considered for comparative evolutionary assessment, referred as 'previous' (P), 'high activity' (HA), and 'subsequent' (S) subperiod, respectively (see Fig. 1). Figure 2 shows 3D scatter plots of (X, Y, T) components for real time and shuffled interevent time. Corresponding values of generalized dimensions D_0 (capacity dimension), D_1 (entropy dimension), D_2 (correlation dimension), and increments $D_1 - D_2$ and $D_{-\infty} - D_{\infty}$ (multifractal step), are given in Table 1 for the three subperiods, showing, among other significant differential aspects, a higher space-time structuring in the central HA subperiod. Finally, evolutionary changes of multifractal step and limiting LMC exponential complexity, also reflecting structural differences between the three subperiods, are displayed in Fig. 3 based on sliding windows, and with and without considering the counting effect of magnitude (weighting functions $\phi(m) = 1$ and $\phi(m) = \exp(m)$, resp.), showing a high intrinsical association of the latter with spatiotemporal dynamics.

Table 1 Values of generalized Rényi dimensions D_0, D_1, D_2, and increments $D_1 - D_2$ and $D_{-\infty} - D_{\infty}$, for X-Y-T, corresponding to real time and shuffled interevent time, for the three subperiods

Subperiod	Time	D_0	D_1	D_2	$D_1 - D_2$	$D_{-\infty} - D_{\infty}$
P	Real	2.6111	2.3895	2.1771	0.2124	1.6893
	shuffled IT	2.6122	2.3921	2.1952	0.1969	1.6721
HA	Real	2.4300	1.6363	1.0389	0.5974	2.7028
	shuffled IT	2.3636	1.8461	1.5193	0.3268	2.2375
S	Real	2.6399	2.4294	2.2091	0.2203	1.8299
	shuffled IT	2.6561	2.4684	2.2854	0.1830	1.6838

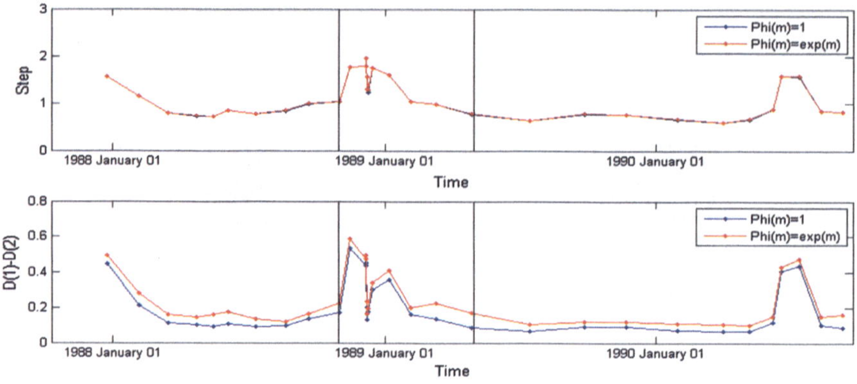

Fig. 3 Values of $D_{-\infty} - D_{\infty}$ (*top*) and $D_1 - D_2$ (*bottom*) for X-Y-Treal and weighting specifications $\phi(m) = 1$ and $\phi(m) = \exp(m)$, based on sliding windows of 100 events with 50 % overlap

3 Conclusion

Multifractal analysis provides a basis for complexity assessment, in terms of variational properties of generalized dimension curves, useful for evaluation of evolutionary structural changes, as well as comparison between systems.

Acknowledgments This work has been partially supported by Spanish grant MTM2012-32666, and Andalusian grant P08-FQM-3834.

References

1. Sen, K. D. (Ed.). (2011). *Statistical complexity: Applications in electronic structure*. New York: Springer.
2. Harte, D. (2001). *Multifractals: Theory and applications*. New York: Chapman and Hall.

Frequency Distributions and Scaling of Soil Texture and Hydraulic Properties in a Stratified Deep Vadose Zone Near Maricopa, Arizona

Alberto Guadagnini, Shlomo P. Neuman, Marcel G. Schaap and Monica Riva

1 Introduction

We analyze the distributional and scaling properties of soil texture data measured to a depth of 15 m over an area of 3600 m^2 in a vadose zone near Maricopa, Arizona, and of hydraulic properties estimated on the basis of these data with the Rosetta [1] neural network pedotransfer model. The sampling network at the site comprises a total of 1029 measurement locations distributed along several vertical wells and a horizontal transect (Fig. 1). A complete description of the available data set and details about the location of measurement points are found in [2].

Texture data include fraction f_i, $0 \leq f_i \leq 1$, of one of three texture categories i where $i = sa$, si or cl for sand, silt or clay, respectively. In addition we consider principal components of these fractions as calculated by Schaap [2] and logit transform of the fractions defined as $l_i = \ln[f_i/(1 - f_i)]$. Logit extends the domain of f_i from [0, 1] to the infinite domain $[-\infty, \infty]$. Hydraulic property estimates include porosity and its logit transform, saturated hydraulic conductivity and parameters of the van Genuchten-Mualem [3] constitutive moisture retention and relative hydraulic conductivity model. In this extended abstract we illustrate selected results corresponding

A. Guadagnini · M. Riva
Department of Civil and Environmental Engineering, Politecnico di Milano,
P.zza L. Da Vinci, 32, 20133 Milano, Italy
e-mail: alberto.guadagnini@polimi.it

M. Riva
e-mail: monica.riva@polimi.it

A. Guadagnini · S. P. Neuman (✉) · M. Riva
Department of Hydrology and Water Resources, University of Arizona, Tucson, AZ, USA
e-mail: neuman@hwr.arizona.edu

M. G. Schaap
Department of Soil, Water and Environmental Science, University of Arizona, Tucson, AZ, USA
e-mail: mschaap@ag.arizona.edu

E. Pardo-Igúzquiza et al. (eds.), *Mathematics of Planet Earth*,
Lecture Notes in Earth System Sciences, DOI: 10.1007/978-3-642-32408-6_44,

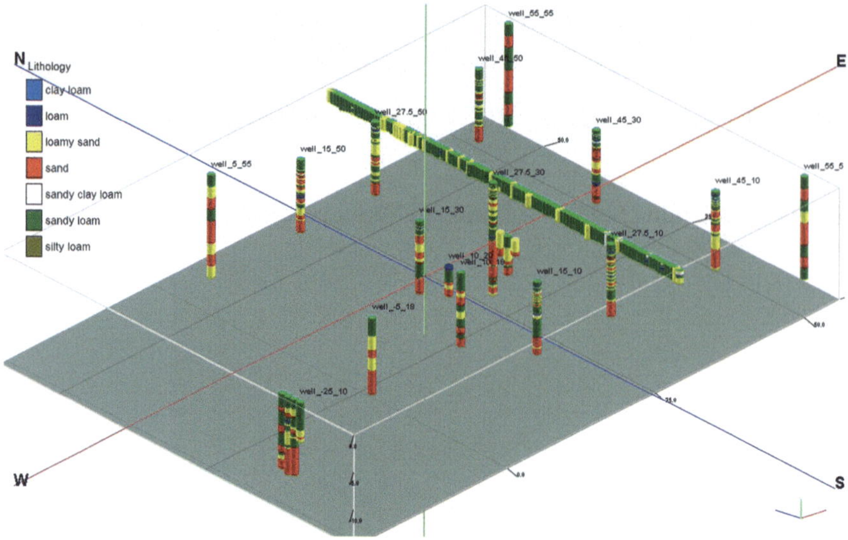

Fig. 1 Spatial distribution of sampling network at the Maricopa site (prepared by Yonggen Zhang)

to our analysis of some texture data and summarize our findings on the basis of all data.

2 Selected Results of Analyzing Texture Data

We start by fitting theoretical stable probability density (pdf) functions to frequency distributions of available f_i data using maximum likelihood (ML). All distributions exhibit slight asymmetry with tails heavier than those of the normal distribution. Few pass the stringent Shapiro-Wilk test of normality at a significance level of 0.05.

The manner in which vertical increments of the f_i data, at three different vertical separation distances or lags, s_V, vary spatially when arranged in a one-dimensional sequence is illustrated in Fig. 2. The variation is seen to be highly intermittent, as is that of logit transforms and principal components of the f_i data (not shown). Though the literature generally attributes such intermittency to multifractals, we suggest below that this is likely not the case here.

Vertical increments of all variables are near-symmetric at all lags; an example is shown for f_{sa} and l_{sa} in Fig. 3. They are generally heavy-tailed at small lags, with α values that tend toward 2 as lag increases. Such asymptotic behavior notwithstanding few of the increments pass Shapiro-Wilk tests of normality, at a significance level of 0.05, regardless of how close α is to 2.

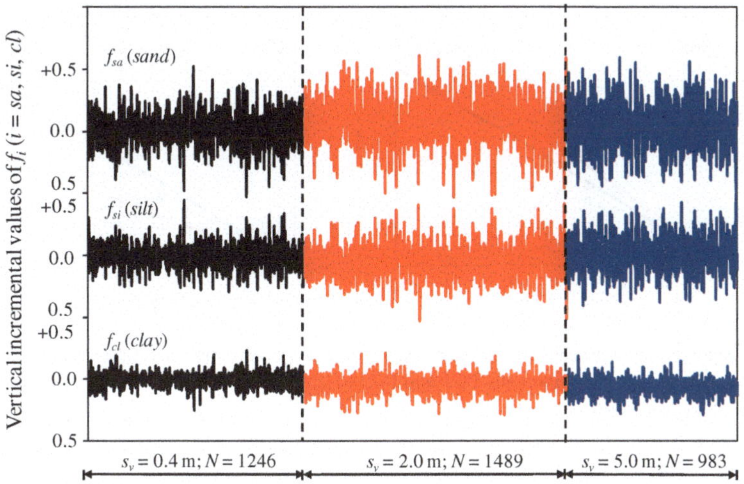

Fig. 2 Sequences of vertical f_i increments at lags $s_v = 0.4$, 2.0, and 5.0 m. N is sample size

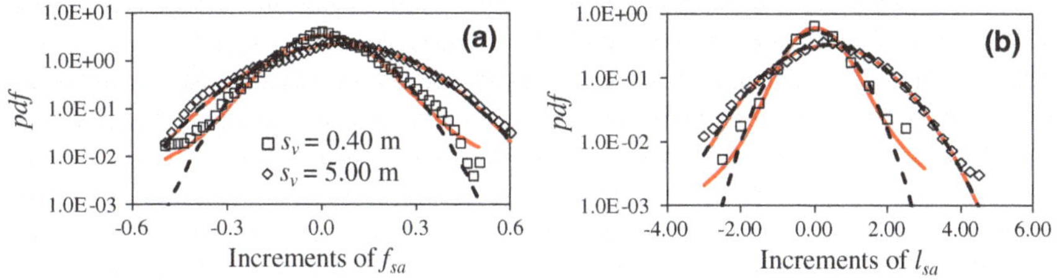

Fig. 3 Frequency distributions (*symbols*) of vertical increments of **a** f_{sa} and **b** l_{sa} at two lags and theoretical stable (*solid red*) and Gaussian (*dashed*) pdf's fitted to the increments by ML

We complete our illustrative example by noting that structure functions of all variables scale as a power of lag at intermediate lag values. As shown for some of the variables in Fig. 4, the corresponding exponent inferred by the method of moments varies in a nonlinear fashion with the order of the structure function. Though such nonlinear scaling is attributed in the literature to multifractals or fractional Laplace motions, our analysis of the Maricopa data suggests otherwise, as summarized below.

3 Summary of Findings

Vertical and horizontal spatial increments of both textural and hydraulic variables at Maricopa exhibit Gaussian or symmetric heavy-tailed distributions, nonlinear power-law scaling in a midrange of separation distances (lags), breakdown in power law scaling at small and large lags, extended power-law scaling at all lags, and various

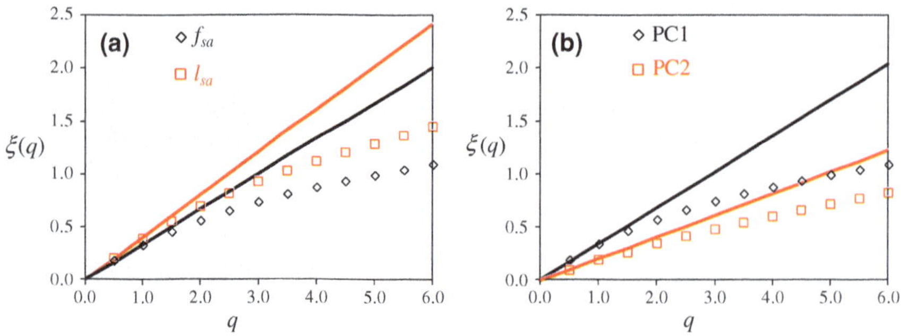

Fig. 4 Dependence of power-law exponent $\xi(q)$ on order q of structure function for **a** f_{sa} and l_{sa}; and **b** first two principal components PC1 and PC2 of f_i. *Straight lines* have slopes equal to $\xi(1)$, defining Hurst exponents

degrees of vertical to horizontal anisotropy. Such behavior is attributed in the literature to multifractals or fractional Laplace motions. We find it to be consistent with sub-Gaussian random fields subordinated to truncated (self-affine, monofractal) fractional Brownian motion (tfBm) or truncated fractional Gaussian noise (tfGn). Each variable is characterized by one Hurst exponent, H, that varies little with directions in most cases. All our estimates of $\alpha H/2$, except those corresponding to van Genuchten's [3] n parameter, are smaller than 0.5, indicating antipersistence.

Acknowledgments This work was supported in part through a contract between the University of Arizona and Vanderbilt University under the Consortium for Risk Evaluation with Stakeholder Participation (CRESP) III, funded by the U.S. Department of Energy.

References

1. Schaap, M. G., Leij, F. J., & van Genuchten, M Th. (2001). Rosetta: A Computer program for estimating soil hydraulic parameters with hierarchical pedotransfer functions. *Journal of Hydrology*, *251*, 163–176.
2. Schaap, M. G. (2013). Description, analysis and interpretation of an infiltration experiment in a semi-arid deep vadose zone. In P. K. Mishra & K. I. Kuhlman (Eds.), *Recent advances in hydrogeology* (pp. 159–183). New York: Springer (invited).
3. Van Genuchten, M. T. (1980). A closed-form equation for predicting the hydraulic conductivity of unsaturated soils. *Soil Science Society of American Journal*, *44*, 892–898.

Cascade of Proppant-Sandwiched Silt Blocks as a Double-Continuum: From Discovery to Mathematical Modeling

Anvar Kacimov, Said Al-Ismaily, Ali Al-Maktoumi, Hamed Al-Busaidi and Said Al-Saqri

1 Introduction

In the Barenblatt–Kazemi model of a double-porous compressible rock with regular blocks-fractures, containing oil, fluid is released to fractures-conduits in primary recovery schemes [1]. We studied the hydropedology of soils inside and in the vicinity of the Al-Khoud dam (area of 4 km^2, 23°37′ N latitude and 58°10′ E longitude, Muscat, Oman) using 33 pedons (with a textural analysis and lab-permeameter measurements of soil samples from different depths) and infiltration tests by double-ring and tension infiltrometers [2, 3]. A standard heterogeneity was detected in 31 of the excavated pedons. At two sites within the dam reservoir (ponded occasionally for several weeks per year) we found [2] an amazing soil (sedimentation) pattern geometrically resembling [1], i.e. silt blocks of sizes 30 × 20 × 20 cm and horizontal–vertical fractures of sandy apertures of 2–6 cm, sandwiched between the blocks. This is attributed to the "geotechnical" factor in soil formation (see [4] for regular soil genesis factors).

2 Mathematical Model of Infiltration

On a macroscale, a two-layered soil with a constant thickness d of the upper layer and the subjacent layer indefinitely extending in the z-direction (Fig. 1a) is subject to the Green–Ampt infiltration. The two layers in Fig. 1a are characterized by the triads (k_1, m_1, h_1) and (k_2, m_2, h_2) where $k_{1,2}$ are corresponding saturated hydraulic

A. Kacimov (✉) · S. Al-Ismaily · A. Al-Maktoumi · H. Al-Busaidi · S. Al-Saqri
Sultan Qaboos University, Al-Khod-123, PO Box 34, Muscat, Oman
e-mail: anvar@squ.edu.om

S. Al-Ismaily
e-mail: osman@squ.edu.om

E. Pardo-Igúzquiza et al. (eds.), *Mathematics of Planet Earth*,
Lecture Notes in Earth System Sciences, DOI: 10.1007/978-3-642-32408-6_45,

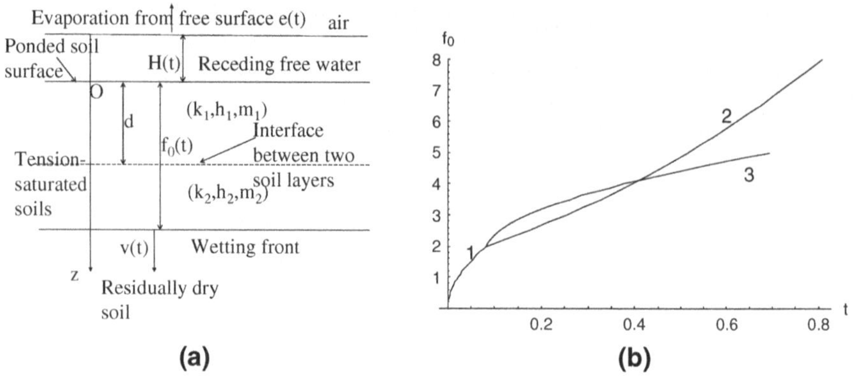

Fig. 1 Green–Ampt model in a two-layered soil and receding ponded water subject to infiltration and evaporation. **a** Vertical section, **b** position of wetting front in the upper layer (*1*) and lower layer (*2*–*3*) of two different capillary-hydraulic properties, H(t) = const

conductivities, $m_{1,2}$ are effective porosities and $h_{1,2}$ are water entry pressure heads. At time $t_d = 0$, the plane $z = 0$ is ponded at a given depth H_0. With time, the ponding depth, H(t), is either maintained constant or (if no extra water added as the case with the post-flash-flood events in Oman) decreases due to infiltration and evaporation of a rate $e(t_d)$ (retrieved from meteorological data). The wetting front position $f_0(t_d)$ is a target.

Unlike the Barenblatt–Kazemi model, in our case water and soil skeleton are incompressible and the hydraulic heads in the two layers are governed by two ODEs with the front kinematic conditions for the Darcian velocities $v_1(t_d) = m_1 df_0/dt$ and $v_2(t) = m_2 df_0/dt$, when the front is in the upper ($0 < t_d < T_d$) and lower ($T_d < t_d < T_{di}$) layers, correspondingly, where T_d is the instance of reaching the interface and T_{di} is the extinction time of ponding, if $H(t_d)$ naturally vanishes. At $t_d > T_{di}$ the second (drainage) front emerges [5].

We normalize all geometrical quantities to H_0 and introduce dimensionless time $t = t_d k_1/H_0$, $E = e/k_1$, $k_r = k_2/k_1$. Then we solve ODE for the front:

$$\begin{cases} m_1 \frac{d f_0(t)}{dt} = \frac{H(t)+h_1+f_0(t)}{f_0(t)}, & f_0(0) = 0, \quad 0 < t < T \\ m_2 \frac{d f_0(t)}{dt} = \frac{H(t)+h_2+f_0(t)}{d+(f_0(t)-d)/k_r}, & f_0(T) = d, \quad T < t < T_i \end{cases} \tag{1}$$

If no extra water is supplied to the reservoir we have an additional condition:

$$H(t) = 1 - E(t)t - m_1 d - m_2 f_0(t), \ t > T.$$

Equation 1 was integrated by the **NIntegrate** routine of Wolfram's *Mathematica*. Results for H = 1, dating back to Polubarinova–Kochina's explicit integration, are shown in Fig. 1b where branch 1 corresponds to $h_1 = 5$, $m_1 = 0.3$ and $d = 2$ (front in the upper layer), branch 2 matches this layer with $h_2 = 0.1$, $m_2 = 0.4$ and $k_r = 10$

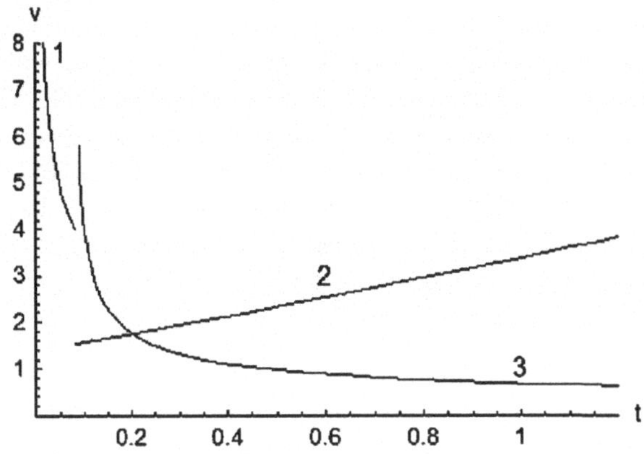

Fig. 2 Infiltration rate as a function of time, curves correspond to the data of Fig. 1b

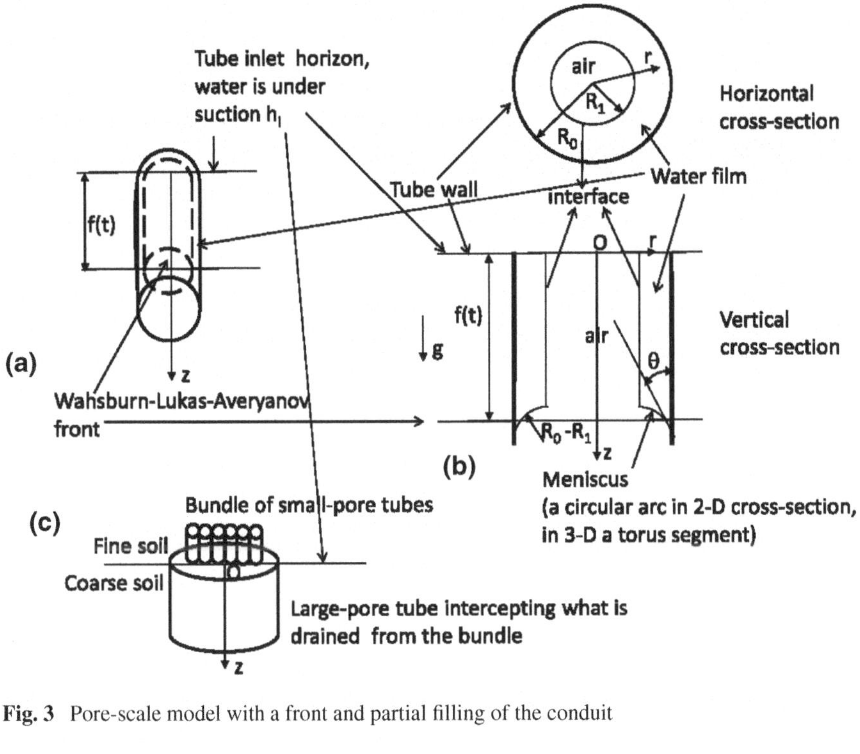

Fig. 3 Pore-scale model with a front and partial filling of the conduit

(coarser second layer) and branch 3—with $h_2 = 20$, $m_2 = 0.3$ and $k_r = 0.1$ (finer second layer). The infiltration rates are shown in Fig. 2. They are discontinuous functions at t = T. Curves 2–3 in Fig. 2 reach their horizontal asymptotes after a very large time, far beyond the duration of double-ring experiments and reservoir ponding in Oman. Clearly, if more than two layers are presented in the soil profile, then the infiltration rate will be a wiggling curve, with discontinuities smeared in experimental observations [6]. Our own infiltration tests concur with Ref. [6] i.e. with a wiggling shape of infiltration rates as functions of time and surmised lack of a horizontal asymptote of these curves in essentially heterogeneous soils.

In order to explain macro-scale results like ones in Fig. 2 or in Ref. [6], we developed a pore-scale model, which combines the Averyanov concept of a partially filled cylindrical pore channel [7] with the Washburn–Lucas meniscus-based front propagation (Fig. 3 depicts the basic model features).

References

1. Barenblatt, G. I., Entov, V. M., & Ryzhik, V. M. (1990). *Theory of fluid flows through natural rocks*. Dordrecht: Kluwer.
2. Al-Ismaily S. S., Al-Maktoumi A. K., Kacimov A. R., Al-Saqri S. M., Al-Busaidi H. A., Al-Haddabi M. H. (2013). A morphed block-crack preferential sedimentation: a smart design and evolution in nature. *Hydrological Sciences Journal*, In press.
3. Kacimov, A. R., Al-Ismaily, S., & Al-Maktoumi, A. (2010). Green-Ampt one-dimensional infiltration from a ponded surface into a heterogeneous soil. *Journal of Irrgation and Drainage Engineering*, *136*(1), 68–72.
4. Sandor, J., Burras, C. L., & Thompson, M. (2005). Factors of soil formation: human impacts. In D. Hillel (Ed.), *Encyclopedia of soils in the environment* (pp. 520–532). Oxford: Elsevier.
5. Kacimov, A. R., & Obnosov, Yu V. (2013). Pseudo-hysteretic double-front hiatus-stage soil water parcels supplying a plant-root continuum: the Green-Ampt-Youngs Model revisited. *Hydrological Sciences Journal*, *58*(1), 237–248.
6. Logsdon, S. D. (1997). Transient variation in the infiltration rate during measurement with tension infiltrometers. *Soil Science*, *162*(4), 233–241.
7. Kacimov, A. R., & Kayumov, I. R. (2002). Viscous flow through straight pore channels. *Journal of Porous Media*, *5*(3), 199–208.

Parameterization of Soil Thermal Diffusivity Versus Moisture Content Dependencies and Modeling Spatial Heterogeneity of Soil Temperature

Tatiana Arkhangelskaya

1 Objects and Methods

The arable soils of the paleocryogenic complexes of the East European Plain were studied in the Vladimir region (56°23′ N, 40°25′ E, 126 m AMSL) and in the Moscow region (54°20′ N, 37°37′ E, 177 m AMSL). The regular structure of the studied complexes was formed during the late Pleistocene periglacial period [1]. At the experimental plot in the Vladimir region the bulk density of the studied soils varied from 1250 to 1610 kg m^{-3} in the plow horizon and from 1100 to 1610 kg m^{-3} in the subsurface layer; the organic carbon content varied from 1.45 to 3.55 % in the plow horizon and from 0.40 to 4.65 % in the subsurface horizons [2]. Sand, silt and clay contents were 8–54, 30–74 and 14–35 % [3]. Soil thermal diffusivity was studied using the unsteady-state method [4]. The highest values ($1.93–5.90 \times 10^{-7}$ m^2 s^{-1}) were obtained for the EB horizons and the lowest ones ($1.69–3.37 \times 10^{-7}$ m^2 s^{-1}) for the Ah horizons [2].

At the experimental plot in the Moscow region the soil bulk density varied from 1190 to 1490 kg m^{-3} in the plow horizon and from 1340 to 1720 kg m^{-3} in the subsurface layer; the organic carbon content varied from 0.85 to 1.26 % in the plow horizon and from 0.30 to 1.22 % in the subsurface horizons. Sand, silt and clay contents were 6–42, 34–75 and 12–35 %. The topography of the studied 48×48 m^2 plot had no specific features in the areas occupied by soils with different morphological profiles [5].

T. Arkhangelskaya (✉)
Faculty of Soil Science, Moscow State University, Leninskie Gori, 1-12, Moscow, Russian Federation 119991
e-mail: arhangelskaia@gmail.com

E. Pardo-Igúzquiza et al. (eds.), *Mathematics of Planet Earth*, Lecture Notes in Earth System Sciences, DOI: 10.1007/978-3-642-32408-6_46,

2 Parameterization of Soil Thermal Diffusivity

The experimental $\kappa(\theta)$ curves, where κ is soil thermal diffusivity and θ is water content, were parameterized with the following approximating function [3]:

$$\kappa = \kappa_0 + a \exp\left[-0.5\left(\frac{\ln\left(\frac{\theta}{\theta_0}\right)}{b}\right)^2\right]. \tag{1}$$

The RMSE of approximation was 3.6 % for 33 undisturbed soil samples (Fig. 1b) and 6.4 % for two compacted soil samples with a pronounced sigmoid shape of the $\kappa(\theta)$ curve.

The suggested parameters can be interpreted quite clearly for the sigmoid-peak $\kappa(\theta)$ curves: κ_0 is the thermal diffusivity of dry soil, a is the difference between the highest thermal diffusivity at the optional water content θ_0 and the thermal diffusivity of dry soil, b is the half-width of the peak of the $\kappa(\theta)$ curve (Fig. 1a). Formula (1) can also be used to parameterize sigmoid experimental curves and even almost linear ones (Fig. 1b). For linear and sigmoid curves the interpretation of κ_0 parameter is the same: it is the thermal diffusivity of dry soil. Parameter a is a measure of κ growth with moisture, and b characterizes the width of moisture interval, where the main growth of κ occurs. Parameter θ_0 in the case of non-peak curves can be interpreted as the coordinate of "virtual maximum", which moves right as the $\kappa(\theta)$ curve becomes less sigmoid and more linear-like.

Parameters κ_0, a, θ_0 and b depend on soil properties and for loamy soils can be estimated from data on soil bulk density (ρ) and organic carbon content (C) [3]:

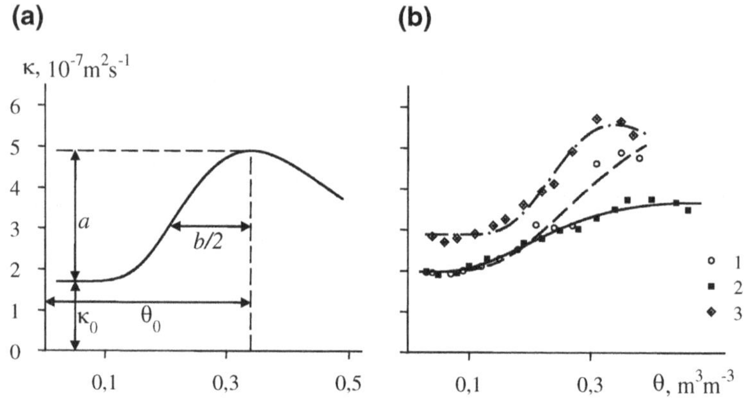

Fig. 1 **a** Graph of the function (1). **b** The examples of approximation of the experimental $\kappa(\theta)$ curves for the Ap (1), AE (2) and EB (3) horizons

$$\begin{aligned}
\kappa_0 &= -1.06 + 0.00198\,\rho + 0.20C,\\
a &= -0.58 + 0.00243\,\rho - 0.08C,\\
b &= 0.12 + 0.00012\,\rho + 0.12C,\\
\theta_0 &= -0.05 + 0.12a + 0.28b.
\end{aligned} \tag{2}$$

where ρ is in kg m^{-3}, C is in %, κ_0 and a are in 10^{-7} m^2 s^{-1}, θ_0 is in m^3 m^{-3}, b is dimensionless. Equations (1–2) allow to estimate thermal diffusivity of loamy soils with RMSE of 9–32 % [3].

3 Modeling Soil Temperature

The parameters of Eq. (1–2) were identified for soils from the Vladimir region and then were used to model soil temperature in the Moscow region. The input data were data on bulk density, organic carbon and soil moisture measured for the $8 \times 8\,\mathrm{m}^2$ grid within the $48 \times 48\,\mathrm{m}^2$ plot [5]. These data were used to model vertical distribution of soil thermal diffusivity in the gridpoints. The seasonal dynamics of soil temperature was modeled using 1D-approach, i.e., temperature was calculated independently for each gridpoint. The Fourier equation was used:

$$c_V(z,t)\frac{\partial T(z,t)}{\partial t} = \frac{\partial}{\partial z}\left(\lambda(z,t)\frac{\partial T(z,t)}{\partial z}\right), \tag{3}$$

where $T(z,t)$ is temperature, t is time, z is depth, $c_V(z,t)$ and $\lambda(z,t)$ are volumetric heat capacity and apparent thermal conductivity [6]. Thermal diffusivity $\kappa(z,t)$ is related with $\lambda(z,t)$ and $c_V(z,t)$ by the formula $\lambda(z,t) = \kappa(z,t) \times c_V(z,t)$.

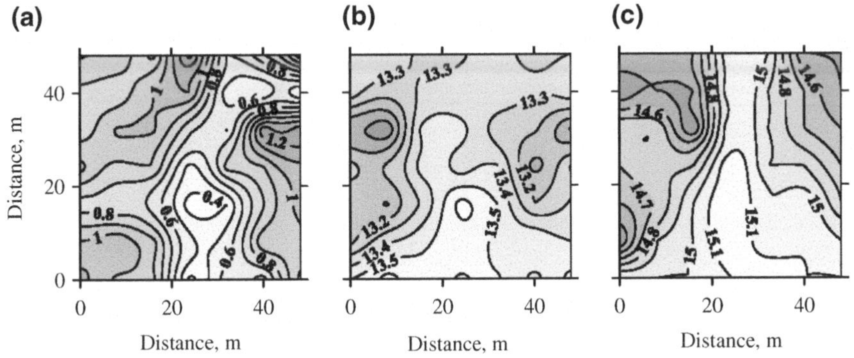

Fig. 2 Spatial distribution of organic carbon, %, at 30 cm depth (**a**), modeled (**b**) and measured (**c**) temperature, °C, at 50 cm depth on June 17, 2002 within the plot in the Moscow region

Soil temperature was modeled for the period from May 1 to June 17, 2002. The upper boundary condition was specified using data on daily average air temperatures at the Serpukhov weather station located within 14 km from the experimental site; the lower one was the condition of zero temperature gradient at 2 m depth. The initial soil temperature was set to 4 °C at all simulated depths in all gridpoints.

The pattern of simulated soil temperature spatial distribution was similar to the experimental one [5]. The coolest were the areas occupied by the soils with the AE subsurface horizon high in organic carbon (Fig. 2).

Acknowledgments Author thanks the Russian Fund for Fundamental Research for the financial support of the work (project No 13-04-01475).

References

1. Velichko, A. A., Morozova, T. D., Nechaev, V. P., & Porozhnyakova, O. M. (1996). Late-Pleistocene cryogenesis and modern soil formation in the Southern Taiga zone (with special reference to Vladimir Opol'e). *Eurasian Soil Science, 29*(9), 984–991.
2. Arhangelskaya, T. A. (2004). Thermal diffusivity of gray forest soils in the Vladimir Opolie region. *Eurasian Soil Science, 37*(3), 285–294.
3. Arkhangel'skaya, T. A. (2009). Parameterization and mathematical modeling of the dependence of soil thermal diffusivity on the water content. *Eurasian Soil Science, 42*(2), 162–172. doi:10.1134/S1064229309020070.
4. Parikh, R. J., Havens, J. A., & Scott, H. D. (1979). Thermal diffusivity and conductivity of moist porous media. *Soil Science Society of America Journal, 43*, 1050–1052.
5. Arkhangel'skaya, T. A., Khudiakov, O. I., Bedrina, T. N., & Mitusov, A. V. (2005). Lateral variability of agrophysical properties and heterogeneity of the hydrothermic field in the complex soil cover of the southern part of the Moscow region. *Moscow University Soil Science Bulletin, 60*(2), 8–15.
6. Campbell, G. S. (1985). *Soil physics with BASIC: Transport models for soil—plant systems*. Amsterdam: Elsevier.

A Packing Computational Method Relating Fractal Particle Size Distribution and Void Fraction in Granular Media

Carlos García-Gutiérrez, Miguel A. Martín, Francisco Muñoz Ortega, Miguel Reyes and Francisco J. Taguas

1 Introduction: From Particle Size Distribution to Packing

Particulate systems are of great interest in many fields of science and technology (see [1]). Packing of particles affects physical properties of the granular system. The observed and expected influence of particle size distribution on random packing structures increases the interest in relating both, either theoretically or by computational methods.

Modelling of PSD and the packing of particles of granular media produced by grinding has been jointly addressed in [2]. In [3] the effect of type of distributions on packing density is studied using dense random packing of sands, and in [4] a probabilistic approach is used to determine the distribution of the volume of voids between packed spheres once their size distribution is given. In [5] a three-dimensional Apollonian packing is used as a model for dense granular systems to investigate the fractal nature of packing. Packing of spheres with lognormal distribution of radii has been studied in [6] by means of computer simulations. Monte Carlo simulations are used in [7] to create the structure made of spherical particles of sizes obeying a given distribution.

Fractal models used for PSD of different granular media may be useful to study the packing structure. The goal of this work is to apply a version of the computational method used in [7] to the fractal PSD model proposed in [8]. In particular, we have been interested in studying the influence of the parameters defining of the fractal model on the resulting void fraction.

C. García-Gutiérrez · M. A. Martín (✉) · F. M. Ortega · M. Reyes · F. J. Taguas
Group of Fractals and Applications (PEDOFRACT), Technical University of Madrid, 28040 Madrid, Spain
e-mail: miguelangel.martin@upm.es

E. Pardo-Igúzquiza et al. (eds.), *Mathematics of Planet Earth*,
Lecture Notes in Earth System Sciences, DOI: 10.1007/978-3-642-32408-6_47,

2 The Particle Size Distribution Model

In practical terms the statistical similarity property for a grain distribution might be defined by the requirement that after sieving some amount of granular material with sieves of different size, the structure of the grain distribution of the material sieved is equivalent (in information terms) to the grain distribution of the initial one.

Let us suppose that the granular material is sieved retaining grains of size greater than r (think $r = 1/2$). Suppose that the fraction of material sieved is p_1 and the retained part $p_2 = 1 - p_1$

Let w_1 and w_2 be the linear transformations $w_1 = x/2$ and $w_2 = (x+1)/2$ which transform the interval [0, 1] into the subintervals [0, 1/2] and [1/2, 1] respectively.

Following [8], the invariant measure μ verifying

$$\mu(I) = \sum p_i w_i^{-1}(I) \tag{1}$$

is proposed as a model for PSD. The measure of any subinterval J is easily computable using Elton's theorem [9].

3 The Packing Algorithm

The computer simulation of packing has been done following the algorithm described in [7]. This algorithm, in a first step seeds (according to a uniform distribution) the initial positions of particles within a square domain of area equal to the sum of the areas of the particles multiplied by 4/3. In a second step, a process of rearrangement is iteratively repeating. In this process the relocation of each particle, i, is performed as follows: a search is made of all particles which overlap particle i, then a displacement vector is calculate for each particle j which overlaps i, following the equation:

$$R_{ij} = R_j + (R_i - R_j) \cdot \left(\frac{r_i + r_j}{d_{ij}}\right) \tag{2}$$

where R_i and R_j are the vectors of the centers of the particles i and j, the parameters r_i, r_j, are the radii of the particles i and j, and d_{ij} is the distance between the centers of the particles i, j. If there are more than one particle which overlaps the particle i, then the arithmetic mean of all R_{ij} is calculated.

The algorithm takes into account the cases when a particle is not in contact with another particle, then that particle is moved to contact its nearest neighbour.

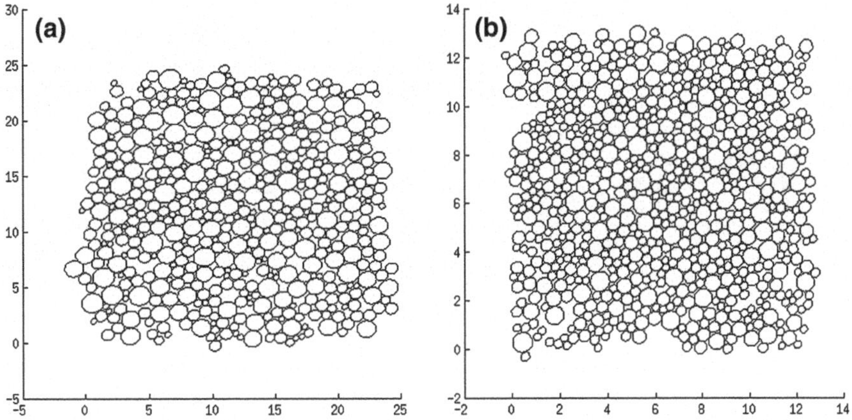

Fig. 1 Examples of the results of two simulations for **a** $P_l = 0.1$ and **b** $P_l = 0.7$

Table 1 Mean values and standard deviation of the porosity for the replications implemented for $P_l = 0.1$, $P_l = 0.4$ and $P_l = 0.7$

P1	Mean porosity	Standard deviation	Number of samples
0.1	0.283	0.0146	10
0.4	0.279	0.0130	10
0.7	0.259	0.0170	10

4 Results

Following Sect. 2, different PSD have been generated for different values of p_1 and p_2. The packing algorithm has been applied with the simulated PSD's with 10 replications for any of them. The influence of these parameters on the porosity have been studied. Figure 1 shows the results of the simulations for two different values of p_1. The variation of the mean porosity of replications is shown in Fig. 2 as a function of the sample size for three values of p_1. Mean values and the standard deviation values of the porosity measured over the replications implemented for the p_1 are summarized in Table 1.

A first conclusion is the dependence of the variance on the value p_1: for values showing a great variance, the number of particles used should be increased in order to ensure the reliability of computer simulations in obtaining the representative elemental volume of the porous medium. Results also seem to indicate the relation between porosity and the value p_1. Further simulations are planned.

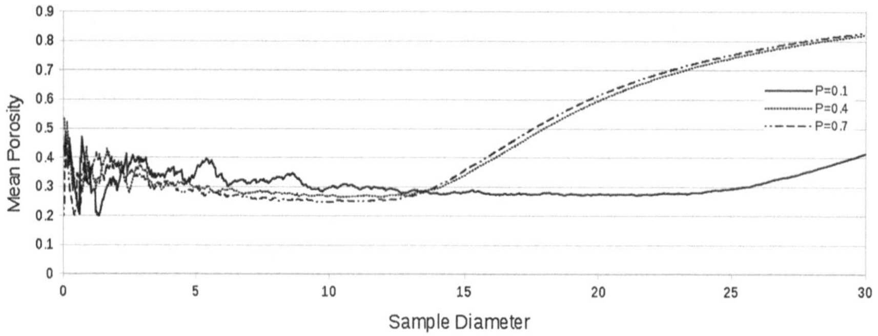

Fig. 2 The variation of the mean porosity as function of the sample diameter for $P_1 = 0.1$ (*continuous line*), $P_1 = 0.4$ (*dashed line*), $P_1 = 0.7$ (*dots* and *dashed line*). (the sample is a circle centered on the mean value of the coordinates of all particles and the unit of length is the same as for the particles)

Acknowledgments This research work was partially funded by Spain's Plan Nacional de Investigación Científica, Desarrollo e Innovación Tecnológica (I+D+I) under ref. AGL2011–25175.

References

1. Gray, W. A. (1968). *The packing of solid particles*. London: Chapman and Hall.
2. Andreasen, A. H. M., & Andersen, J. (1930). Uber die Beziehung zwischen Kornabstufung und Zwischenraum in Produkten aus losen Kornern. *Kolloid-Z, 50*, 217–228.
3. Sohn, H. Y., & Moreland, C. (1968). The effect of particle size distribution on packing density. *Canadian Journal of Chemical Engineering*, *46*, 162–167.
4. Rouault, Y., & Assouline, S. (1998). A probabilistic approach towards modeling the relationships between particle and pore size distributions: the multicomponent packed sphere case. *Powder Technology, 96*, 33–41.
5. Anishchik, S. V., & Medvedek, N. N. (1995). Three-dimensional apollonian packing as a model for dense granular systems. *Physical Review Letters*, *75*, 4314–4317
6. Nolan, G. T., & Kavanagh, P. E. (1993). Computer simulation of random packings of spheres with log-normal distributions. *Powder Technology*, *76*, 309–316.
7. He, D., Ekere, N. N., & Cai, L. (1999). Computer simulation of random packing of unequal particles. *Physical Review E*, *60*, 7098–7104.
8. Martín, M. A., & Taguas, F. J. (1998). Fractal modeling, characterization and simulation of particle-size distributions in soil. *Proceedings of the Royal Society of London A. Mathematical, Physical and Engineering Sciences*, *454*, 1457–1468.
9. Elton, J. (1987). An ergodic theorem for iterated maps. *Journal Ergodic Theory and Dynamical Systems*, *7*(4), 481–488.

Parallel Sets and Morphological Measurements of CT Images of Soil Pore Structure in a Vineyard

Fernando San José Martínez, Francisco J. Muñoz Ortega, Francisco J. Caniego and Fernando Peregrina

1 Introduction

The spatial arrangement of soil constituents—usually referred to as soil structure-controls important physical and biological processes in soil-plant-microbial systems, where microbial population dynamics, nutrient cycling, diffusion, mass flow and nutrient uptake by roots take place across many orders of magnitude in length scale. Therefore, a correct model of the geometry of soil pore structure is critical for understanding flow and transport processes in soils, creating synthetic soil pore space for hypothesis and model testing, and evaluating similarity of pore spaces of different soils.

Minkowski functionals provide computationally efficient means to measure four fundamental geometrical features of 3D objects such as pore space, i.e. the volume, the boundary surface, the mean boundary surface curvature and connectivity. Hadwiger's theorem confers a fundamental status to these geometrical features as morphological descriptors. Mecke [1] and Roth et al. [2] make use of the evolution of Minkowski functionals due to threshold variation to characterize 2D porous structures. Also, 2D porous structures were investigated by Mecke [3] and Vogel et al. [4]

F. San José Martínez (✉) · F. J. M. Ortega · F. J. Caniego
Departamento. Matemática Aplicada a la Ing. Agronómica, E.T.S.I. Agrónomos, Universidad Politécnica de Madrid, 28040 Madrid, Spain
e-mail: fernando.sanjose@upm.es

F. J. M. Ortega
e-mail: f.j.munoz.ortega@upm.es

F. J. Caniego
e-mail: j.caniego@upm.es

F. Peregrina
Instituto de Ciencias de la Vid y el Vino (CSIC-Universidad de La Rioja-Gobierno de la Rioja), Servicio de Investigación y Desarrollo Tecnológico Agroalimentario, La Rioja, 26076 Logroño, Spain
e-mail: fernandoperegrina@hotmail.com

E. Pardo-Igúzquiza et al. (eds.), *Mathematics of Planet Earth*,
Lecture Notes in Earth System Sciences, DOI: 10.1007/978-3-642-32408-6_48,

with Minkowski functions based on dilations and erosions. These functions account for the evolution of Minkowski functionals as the radius of dilation/erosion performed to the object varies. Arns et al. [5] analyzed these Minkowski functions to characterize 3D images of Fontainebleau sandstone. Vogel et al. [6] took advantage of Minkowski functions based on openings (i.e. erosion followed by dilation) to quantify soil 3D structure of arable soil and of repacked sand. Parallel sets, which can be understood in terms of dilations and erosion, were introduced by Mecke [3] to characterize and model 2D structures beyond two point correlation functions and predict percolations threshold in porous media. Arns et al. [5] also made use of parallel sets to determine the accuracy of the model they developed for the Fontainebleau sandstone.

Some preliminary results on the geometrical features captured by the evolution of Minkowski functionals of parallel sets of pore space will be presented here. We will also explore their suitability to characterize the influence of conventional tillage and permanent cover crop of resident vegetation on soil structure in a Spanish Mediterranean vineyard.

2 Mathematical Morphology of Pore Space

We will analyze binary 3D images of soil (black and white images) from X-ray computed tomography (CT) of intact soil columns. They contain two complementary phases: the phase of voids (pores) and the phase of soil matrix (mineral particles). Usually, a CT image of soil will be pictured as a set of points of a cube S where each point is part of pore space P or is part of soil matrix. These points are voxels of a three-dimensional image. The set P_r, also called parallel body of P at a distance r or r-parallel body to P , is the set of all points within a distance smaller than r from the object P. Roughly speaking, it is like a "skin" of thickness r was added to P. The parallel set P_r can also be described in terms of dilation of P by balls of radius r , as the union of all balls $B_r(\boldsymbol{x})$ of radius r centered at points $\boldsymbol{x}$ of P . The erosion of one phase is equivalent to the dilation of the complementary phase. Consequently, the erosion of $P$by a ball of radius r corresponds to the set of all positions of their centers $\boldsymbol{x}$ within P where $B_r(\boldsymbol{x})$fits completely into P . It is like a layer of thickness r was removed from P . We may generalize the notion of parallel body as follows, P_r is the dilation of P by balls of radius r when $r > 0$ and the erosion of P by balls of radius r when $r < 0$.

Minkowski functionals are a complete set of geometrical features as established by Hadwiger's theorem [7]. In simple terms this theorem states that any functional that assigns a number to any object and meets some very natural restrictions is a linear combination of the Minkowski functionals. In space there are four Minkowski functionals: volume, boundary surface area, mean breadth and Euler-Poincaré number. When the boundary surface of the object is smooth, mean breadth is the surface integral of mean curvature [7]. Points on the boundary surface of an object with positive curvatures settle on convex parts (protrusions) while points with negatives curvatures

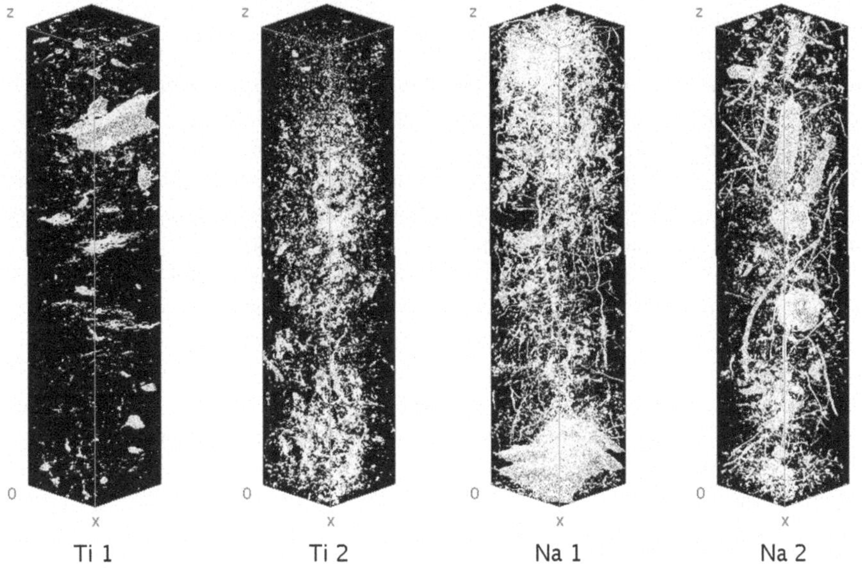

Fig. 1 View of 3D reconstructions of binary images of pore space (white) with voxel size of 50 μm in a box of 8.5 cm high (z-axis), 1.7 cm long (x-axis) and wide (y-axis) of samples from tillage soils (Ti) and natural cover crop (Na)

belong to concave parts (hollows). The Euler-Poincaré characteristic is an index of the topology of the pore phase and it quantifies pore connectivity. It can be computed as the sum of the number of connected components and the number of cavities of the object less the number of tunnels [1]. In this context a connected component of an object is any part of it whose points are connected to one another by curves of points contained in the object. Cavities are holes completely surrounded by the object, while tunnels are holes through the object connected with the exterior or background. For soils we may assume that pore space does not have cavities and, therefore, the connectivity number is the number of connected components subtracted by the number of tunnels. It is worth noting that the computation of Minkowski functionals reduces to their evaluation on cubes (or voxels) and their intersections (vertices, edges and faces) [9]. Finally, we have considered Minkowski functions build over parallel sets of the pores space. They provide a way to investigate the morphology of the pore space as it is dilated and eroded with balls of increasing radius.

3 Soil Samples and Image Acquisition

Soil columns were collected at the experimental farm "Finca La Grajera" property of La Rioja region government (northern Spain) at December 2010. The soil was classified as Fine-loamy, mixed, thermic Typic Haploxerepts according to the USDA soil classification. In this study we selected four columns collected between rows of a vineyard that was established in 1996. Two types of soil cover management in between rows were represented: (Ti) conventional tillage management between rows, which consisted in soil tillage of 15 cm depth by cultivator once every 4–6 weeks; (Na) natural soil with permanent cover crop of resident vegetation, which was dominated by annual grass and forbs common to La Rioja vineyards [9]. Columns were extracted vertically by percussion drilling between rows, within PVC cylinders of 7.5 cm interior diameter and 30 cm height from the upmost part of soil profile.

Columns were scanned with a PerkinElmer amorphous silicon (a-Si) detector with a Feinfocus FXE 225.51 microfocus beam source tube that was operated at 190 kV (53 uA) and 20 W. Only the upper half of the column was scanned and the region between 8 cm and 15 cm was selected. A stack of 1706 two-dimensional 16-bit grayscale images with a pixel size of 50 μm, 50 μm apart from one another were obtained for each sample. We selected a global method to segment images—the modes method—as we focused primarily on geometrical features evolutions and pattern analysis (Fig. 1).

4 Results and Discussion

Minkowski functionals were evaluated on five cubes per column. Cubes had 340 voxels per edge and they were centered on the axes of the column. Consecutive cubes shared a face, from top to bottom. The pore space P in cube c_i ($i = 1 \dots 5$) was eroded/dilated to yield parallel sets. Diameters of balls took nineteen different values for erosions and nineteen for dilation, as well; it was incremented from in steps of the voxel size (i.e. 50 μm). We considered densities of Minkowski functionals. Thus, we had volume fraction or image porosity, specific boundary surface area, specific mean curvature and specific Euler number.

Fig. 2 displays the evolution of these geometrical densities as function of erosion/dilation diameter. Differences between soil samples under natural resident vegetation cover (Na) and samples under conventional tillage (Ti) are noticeable even if samples Ti2 and Na2 have a similar evolution for dilations. Nevertheless, the evolution of image porosity and specific boundary surface with erosions diverges. It suggests that geometrical features of sample Ti2 are smaller than three voxels as they vanish with erosions of diameter smaller than that size.

Samples with natural resident vegetation cover (Na) show a greater amount of volume fraction and specific surface at any diameter of the balls used to erode/dilate as compared to samples form tillage soil (Ti). Similar results were reported by Peregrina et al. [9].

Evolution of specific mean curvature and connectivity seems to indicate that conventional tillage and resident vegetation cover produces two different pore structures; this difference is especially apparent when comparing samples Ti1 and Na1. Sample Na1 yields more specific mean curvature than sample Ti1 when dilated with balls smaller than seven voxels. In this range of diameters, mostly small convex

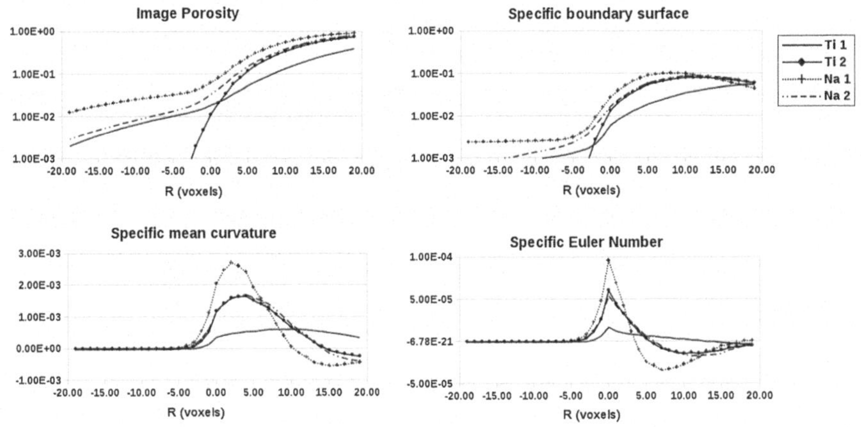

Fig. 2 Morphological functions of the pore space of soil columns: evolution of Minkowski functionals densities as a function of erosion/dilation diameter R. Negative values of R correspond to erosions while dilations correspond to positive values

voids should populate sample Na1 as compare to sample Ti1. High Euler numbers of sample Na1 at small diameters seem to suggest this behavior. But large diameters decrease specific mean curvature and Euler number of sample Na1 producing negative values while these geometrical measurements stabilize for sample Ti1 as diameter increases. This suggests that pore structure of sample Na1 contains a large number of small features as the number of small voids (i.e. connected components) overdue the number of soil materials tunnels through them; therefore, high values of the specific mean curvature from these small features of the Na1 pore space should be explained by the regularity of the surface that enclosed them. Na1 seems to display a richer structure as compared to sample Ti1. When the diameter of the ball used to dilate the object is seven both samples have the same positive specific mean curvature but sample Na1 has negative Euler number. Therefore, it suggests that geometrical features similar in size should dominate sample Ti1 while the dilations of sample Na1 show a more complex structure highly connected with tunnels through it as it seems to indicate negative Euler numbers. Low variation of specific mean curvature and Euler numbers of sample Ti1 is compatible with a pore structure made up with small irregular geometrical features of similar sizes that vanish as diameter of dilation increases and do not generate a complex and highly connected structure.

These results suggest that the evolution of morphological features with dilation/erosion is a suitable indicator of soil structure for cultivated soil and it seems to describe the influence of two different soil management practices (i.e. conventional tillage and natural cover crop) on soil structure in a Spanish Mediterranean vineyard. It is worth noting here how these results reflect the different pore structures as depicted by Fig. 1. The homogeneity of the pore space produced by tillage is obvious as compared to the heterogeneity of samples under resident vegetation crop. Similar geometrical features seem to dominate samples Ti2 and Na2; but, small structures discriminate between them and explain the behavior of specific image porosity and boundary surface when sample Ti2 is eroded.

These findings provide means to simulate natural porous media. The geometrical descriptors that seem to discriminate between these two types of samples could be used as inputs for morphological models of natural soil structures. Further investigations are needed to understand the role of morphological parameters in discrimination of a priori different natural soil structures and in determination of accurate models to mimic pore structure of natural soils.

Acknowledgments This work was partially supported by Plan Nacional de Investigación Científica, Desarrollo e Investigación Tecnológica (I+D+I) under ref. AGL2011/251675 and DGUI (Comunidad de Madrid) and UPM under ref. QM100245066.We thank the staff of the Servicio de Investigación y Desarrollo Tecnológico Agroalimentario (Gobierno La Rioja) for providing the experimental plots and helping with the field work and Thomas Redenbach at Fraunhofer ITWM where soil columns were imaged.

References

1. Mecke, K. R. (1998). Integral geometry and statistical physics. *International Journal of Modern Physics B*: *12*(9), 861–899.
2. Roth, R., Boike, J., & Vogel, H. J. (2005). Quantifying Permafrost Patterns using Minkowski Densities. *Permafrost and Periglacial Processes*, *16*, 277–290.
3. Mecke, K. R. (2002). The shape of parallel surfaces: porous media, fluctuating interfaces and complex fluids. *Physica A*, *314*, 655–662.
4. Vogel, H. J., Hoffmann, H., & Roth, K. (2005). Studies of crack dynamics in clay soil. I. Experimental methods, results, and morphological quantification. *Geoderma*, *125*, 203–211.
5. Arns, C.H., Knackstedt, M.A., Mecke, K.R.: Characterization of irregular spatial structures by parallel sets and integral geometric measures. Colloids and Surfaces A: Physicochemical and Engineering Aspects, 241, 351–372 (2004)
6. Vogel, J. H., Weller, U., & Schlüter, S. (2010). Quantification of soil structure based on Minkowski functions. *Computers and Geosciences*, *36*, 1236–1245.
7. Ohser, J., & Mücklich, F. (2000). *Statisitical analysis of microstructure in materials sciences*. Chichester: Wiley.
8. Michielsen, K., & De Raedt, H. (2001). Integral-geometry morphological image analysis. *Physics Reports*, *347*, 461–538.
9. Peregrina, F., Larrieta, C., Ibáñez, S., & García-Escudero, E. (2010). Labile organic matter, aggregates, and stratification ratios in a semiarid vineyard with cover crops. *Soil Science Society of America Journal*, *74*(6), 1–11.

Evaluation of Fractal Dimension in Karst Aquifers

Pedro A. Robledo-Ardila, Juan José Durán-Valsero and Eulogio Pardo-Igúzquiza

1 Introduction

The spatial heterogeneity of karst aquifers is widely recognized. The focus of this paper is to apply a 1-D fractal method as tool to obtain complementary information about internal structure of karst aquifers. The application of fractal geometry could provide substantial information on the network of conduits of the karst aquifers and may help us understand how the spatial distribution and temporal responses of their discharge [2, 4]. In so far as its application extends our degree of understanding of the internal organization of such aquifers and their hydrodynamic balance behaviour. Four examples have been taken in different carbonate massifs on the basis of flow data in springs with large discharges: the Tempul (Cabras range), La Cueva del Gato (Líbar range), The Villa (El Torcal de Antequera) and Villanueva del Rosario (Camarolos range) in Málaga and Cádiz provinces. For fractal dimension analysis the Scan Line method is used applying the equation $N \propto d^{-D}$ [2] using flow series (l/s) in a time sequence for the flow springs. Thus, the fractal dimension of a network of conduits of karst aquifer is a number such that 0–1 because the method is based in 1-D study of the system (Fig. 1). The higher the fractal dimension more developed is the stage of karst evolution. In this work we have simplified the analysis because de fractal equation method assumes all the possible variables to obtain a reliable result of the internal hierarchal of a karst aquifer. The differences of flows are considered here as

P. A. Robledo-Ardila (✉)
Spanish Geological Survey (IGME), C/ Ciudad de Queretaro s/n, 07007 Palma de Mallorca, Baleares, Spain
e-mail: pa.robledo@igme.es

J. J. Durán-Valsero · E. Pardo-Igúzquiza
Spanish Geological Survey (IGME), C/Ríos Rosas n°, 23, Madrid, Spain
e-mail: jj.duran@igme.es

E. Pardo-Igúzquiza
e-mail: e.pardo@igme.es

E. Pardo-Igúzquiza et al. (eds.), *Mathematics of Planet Earth*,
Lecture Notes in Earth System Sciences, DOI: 10.1007/978-3-642-32408-6_49,

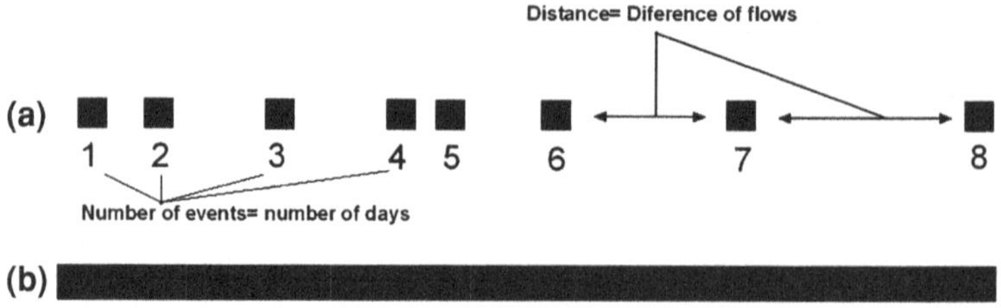

Fig. 1 Schematic example of the fractal dimension using the 1-D method. **a** No Fractal system. **b** Fractal system

the spatial distances en a line according the Hausdorff-Besicovitch dimension [3] and the days of the flow series is the number of events scaled with a minimum distance.

2 Materials and Methods

In the examples studied we opted for the Scan Line method (1-D study). The l/s data for each spring were considered in relation to the number of days. Fractal behaviour in a series of events (N) with a property (P) fulfils the following equation:

$$N \propto P^{-D} \tag{1}$$

where N is the number of objects, P is the property and D is an exponent related to the fractal dimension system, obtained from the above equation by the application of logarithms [2]. In the case of the aquifers studied, the temporal distribution of flows (series of daily flows expressed in l/s-day) was analyzed so as to ascertain whether or not their behaviour is fractal, and they may be classified by the fractal dimension of the distribution of distances between flows (D)d. The method is based on the measurement of daily flows in a 1-D cross-section. Daily flow was treated as a dimensional space between 0 and 1. The flow difference (N) scaled with a minimum distance (d) (which in this case is flow expressed in l/s) is hyperbolic (fractal) if it fulfils the equation (potential law) $N \propto d^{-D}$, where N is the number of days and d the flow difference. The D exponent in this case is related to the fractal system dimension (the volume of flows considered) and is obtained from the above equation by applying logarithms. In a double logarithmic graph with log (number of days sampled)-distance log (daily flow difference), the fractal dimension D(d) is the slope of the distribution curve fit.

3 Results

A gradation of flow and time in the discharge response for each of the springs is observed as regards the system fractal dimension. The Fig. 2 show the fractal dimensions of the flow discharged is <1 in each one (0.4448, 0.3999, 0.3067 and 0.3457, respectively). These results are not close to the theoretical value of 1 (fractal system).

The distribution trend line shows that in each of the systems analyzed there are various sets of flows spaced in a great number of intervals. La Cueva del Gato spring show a highly irregular gradation in their number of intervals. In Villanueva del Rosario spring the behaviour is similar though the log-log relationship shows a considerable decrease in the number of intervals, which are also more regular as to the number of events repeated in them. The Villa spring also exhibits a smaller number of intervals but similar to Villanueva del Rosario, though they are less homogenous in their log-log correlation. El Tempul spring show a great sets of spaced intervals. However, the relationship is highly regular, as the spacing shows few differences in its log-log correlation, with many sets of discharged flow intervals but with little difference between the flow volumes and the number of days on which these are repeated.

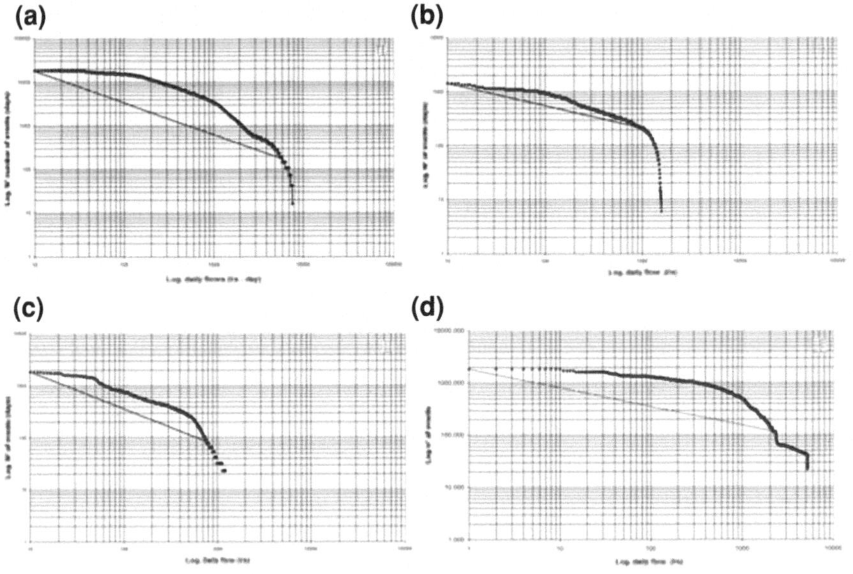

Fig. 2 Fractal dimension which is >1 in the four examples springs studied: **a** Cueva del Gato, **b** The Villa; **c** Villa Nueva Del Rosario: **d** The Tempul

4 Discussion

According the result of fractal dimension of flow discharge could be a related with their internal structure associated to a model in which some show a more inertial behaviour and others a more hierarchal (karstic). However, the trend line for the spatial distribution of flows show low values in all the cases (>0.5). If on the other hand we consider the intervals established in the aquifers in detail we find that there are specific sets of flows for which the steep slope of their distribution trend line is substantially modified, thus showing fractal behaviour as the fractal dimension of this particular set is notably close to 1 (self-similarity). Thus, the flow discharge analysed may be asserted to have multifractal behaviour, according to the set of intervals in which they are spaced, auto-scaling in their log-log correlation [1, 2]. The Fig. 3 shows the multifractal spectrum in the El Tempul spring. The method used has been 1D but in five different interval in the same data flow series.

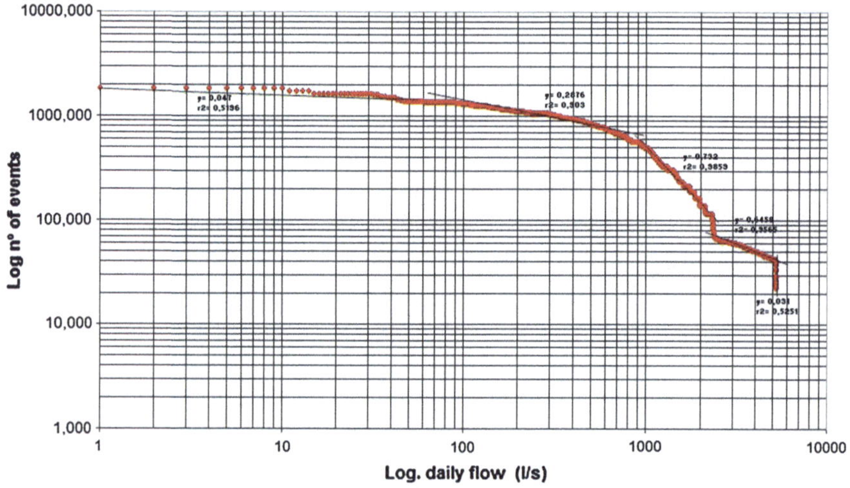

Fig. 3 Multifractal spectrum of the Tempul spring. Tow intervals are equal to >0.73 and >0.64, both close to 1 (good fractal system)

5 Conclusions

Fractal dimension method is a good complementary tool for its application in studies of karst aquifers. The results obtained indicate that the set of flow discharged is not highly fractal and this could be related with the internal organization degree of karst aquifers. However, some interval of flow shows a fractal dimension close to 1. For a more detailed study of fractal dimensions we should consider aspect such as, the conducts geometry and thickness, recharge or water balance in karst aquifers.

Acknowledgments This work has been developed under the UNESCO IGCP-598 and CGL2010 projects. We want to give thanks to the professor Bartolomé Andreo.

References

1. Bird, N., Cruz, M., Antonio, S., & Tarquis, A. (2006). Fractal and multifractal analysis of pore-scale images of soil. *Journal of Hydrology*, *322*, 211–219.
2. Gumiel, P., & Paredes, C. (1995). Interés de algunas aplicaciones de la geometría fractal en geología. *Tierra y Tecnología*, *14*, 12–21.
3. Henry, P. M. (1995). Hausdorff-Besicovitch dimension of Brownian motion paths. *Mathematical Journal*, *22*, 229–234.
4. Pardo-Igúzquiza Durán, J. J., & Rodríguez-Galiano, V. (2011). Morphometric analysis of three-dimensional networks of karst conduits. *Geomorphology*, *132*, 17–28.

Fractal Modelling of Karst Conduits

Eulogio Pardo-Igúzquiza, Juan José Durán, Pedro Robledo, Carolina Guardiola, Juan Antonio Luque and Sergio Martos

1 Introduction

A fractal is an object which in geometry can be described by a non-integer dimension. It was Mandelbrot [1] who conjectured the frequent occurrence of fractals in nature. Fractals have been used in karst research to model cave roughness [2, 3], cave distribution, fracture opening, fracture distribution, conduit distribution, conduit size, conduit morphology, conduit tortuosity, porosity and permeability [4]. The fractal dimension of karst conduits has been used as a morphometric parameter to describe three-dimensional karst conduit networks and it may be used to control stochastic simulations of such networks [5]. Nevertheless, another interesting application of fractal modelling is in the description of scale-invariant distributions and when extrapolating to smaller and larger scales beyond the measured range. This latter application is used here to calculate the conduit porosity and fracture porosity of a karst massif.

E. Pardo-Igúzquiza (✉) · J. J. Durán · C. Guardiola
Instituto Geológico y Minero de España (IGME), Ríos Rosas 23, 28003 Madrid, Spain
e-mail: e.pardo@igme.es

J. J. Durán
e-mail: jj.duran@imge.es

C. Guardiola
e-mail: c.guardiola@igme.es

P. Robledo
Unidad del IGME en Palma de Mallorca, Ciudad de Querétaro s/n, 07007 Palma de Mallorca, Spain
e-mail: pa.robledo@igme.es

J. A. Luque · & S. Martos
Unidad del IGME en Granada, Urbanización Alcazar del Genil, 4, 18006 Granada, Spain
e-mail: ja.luque@igme.es

S. Martos
e-mail: s.martos@igme.es

E. Pardo-Igúzquiza et al. (eds.), *Mathematics of Planet Earth*, Lecture Notes in Earth System Sciences, DOI: 10.1007/978-3-642-32408-6_50,

2 Methodology

Porosity in a karst system has a hierarchical structure involving matrix porosity, fracture porosity and conduit porosity. It can be hypothesized that its distribution follows a power law which is characteristic of parameters with fractal behaviour. Pardo-Igúzquiza et al. [5] provides an analysis of the distribution of conduits in a 3D volume. In this work, the volume of conduits in relation to the diameter of the conduit is considered. The hypothesis is that for conduits (*V*) that have a diameter greater than (*D*) the volume follows a power law:

$$V(D) \propto D^{-FD}, \tag{1}$$

where *FD* is the fractal dimension. When the experimental data are represented in a log-log plot, they should produce a straight line that defines the power law. This functional relationship can then be extrapolated to estimate the volume of conduits that are larger than 3 mm, which are considered for conduit flow, and the volume of conduits that are smaller than 3 mm that define the fracture porosity and are used when calculating Darcy flow. The fitting procedure is illustrated below.

3 Results

Measurements were taken for over a thousand sections from conduits in the Sierra de las Nieves aquifer in the province of Málaga in Southern Spain. The data were collected by the speleologists of the Grupo de Exploraciones Subterráneas (Sociedad Excursionista de Málaga). The speleologists use cave surveying to map conduits. A conduit is defined by profiles and a profile is a number of successive topographic stations. For each station the three-dimensional coordinates are known as well as the section of the conduit at that location. The section can be modelled as circular with a mean diameter. The spatial disposition of conduits is complex but can be characterized by its fractal dimension, which can be determined using the box-counting method as shown in Fig. 1. The resulting fractal dimension, 1.46, is similar to the typical dimension of 1.67 found for karst conduits [6].

Furthermore, the volume of the conduit can be calculated as the volume of a frustum of a right circular cone. In this way, for each diameter (*D*), it is possible to calculate the volume of conduits (*V*) with a diameter larger than *D*. The results have been plotted on a double-logarithmic scale in Fig. 2. The middle section of the plot shows a straight line and at the ends artifacts caused by sampling bias can be seen [4]. There are two kinds of sampling bias as a result of truncation and censoring. Truncation bias is introduced by the incomplete sampling of small features. There is a physical limit to cave exploration because the conduit must be large enough to allow the passage of the speleologists. On the other hand, truncation bias is introduced by the limited extent of the exploration, which could mean that some of the very large features have not been mapped.

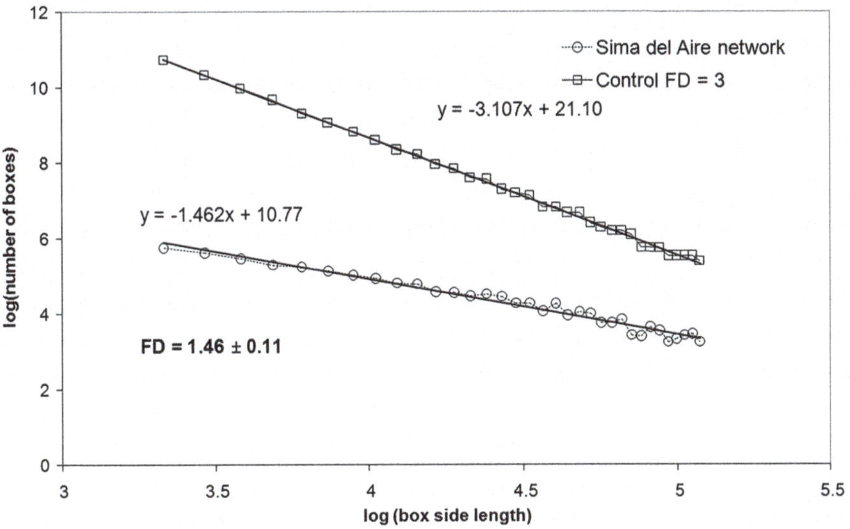

Fig. 1 Log-log representation of the spatial disposition of conduits using the box-counting method

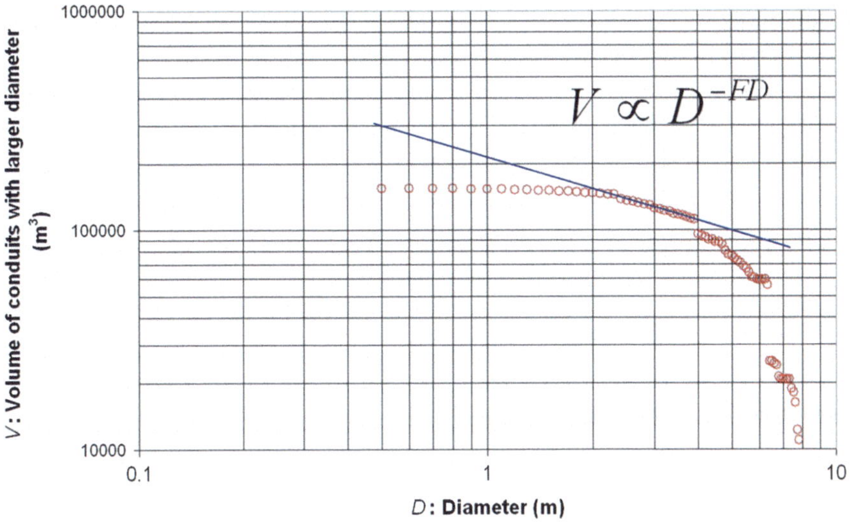

Fig. 2 Log-log representation of diameter of conduits versus accumulated volume of conduits with larger diameter

The linearity in Fig. 2 suggests that the data are scale invariant, or fractal, and can thus be extended beyond the limits of observed data. In doing so, it is possible to calculate that the volume of conduits larger than 3 mm is 3 10^6 m^3 and the volume of the karst massif where the conduits have been mapped is 1.2 10^9 m^3, giving a conduit porosity of 0.25 %. In the same way fracture porosity can be estimated as

0.25 %. The mean matrix porosity of the dolostone, which is the main constituent of the aquifer, is 2.66 %. The total porosity is therefore 3.16 %, very similar to the assumed mean porosity of karst aquifers in the Betic Mountains, which is 3 %.

4 Conclusions

Fitting a power law to experimental data of conduit sections can be used to obtain estimations of conduit porosity and fracture porosity in a karst massif. This technique provides quantitative measurements of otherwise difficult to evaluate parameters in karst hydrogeology. These values can be used for the mathematical modelling of flow in karst systems.

Acknowledgments This work was supported by research project CGL2010-15498 from the Ministerio de Economía y Competitividad of Spain. We would like to thank the speleologists of the Grupo de Exploraciones Subterráneas de la Sociedad Excursionista de Málaga.

References

1. Mandelbrot, B. B. (1967). How long is the coast of Britain? Statistical self-similarity and fractional dimension. *Science, 156*, 636–638.
2. Paredes, C., Mangin, A., & Eraso, A. (1996). El fenómeno de "scaling" en los sistemas kársticos. *Geogaceta, 20*(6), 1393–1396.
3. Curl, R. L. (1986). Fractal dimensions and geometries of caves. *Mathematical Geology, 18*(8), 765–783.
4. Mace, R. E., Marrett, R. A., & Hovorka S. D. (Eds.). (2005). Fractal scaling of secondary porosity in karstic exposures of the Edwards Aquifer. *Sinkholes and the Engineering and Environmental Impacts of Karst. ASCE, Geotechnical Special Publication 144, Proceeding of the Tenth Multidisciplinary Conference*, pp. 1–10.
5. Pardo-Igúzquiza, E., Durán-Valsero, J. J., & Rodriguez-Galiano, V. (2011). Morphometric analysis of three-dimensional networks of karst conduits. *Geomorphology, 132*, 17–28.
6. Jeannin, P.-Y., Groves, C., & Häuselaman, P. (2007). *Speleological investigations in Methods in Karst Hydrogeology edited by Nico Goldscheider and David Drew* (p. 264). London: Taylor & Francis.

Frequency–Area Distribution of Historical Landslides in the Sannio Apennine (Southern Italy)

Angelo Donnarumma, Paola Revellino and Francesco Maria Guadagno

1 Introduction

Landslides are examples of self-organized criticality (SOC) in nature [1]. In the context of SOC models a system exhibits a quasi stationary state (critical state). Moreover, the distribution of event sizes is scale invariant and the frequency distribution in time occurrence behaves as a 1/f noise [2]. First simple example of a SOC process adopted in landslide studies was the 'sandpile' model [3], based on the phenomenon that its slope keeps an angle of repose, but larger avalanche cannot be excluded in presence of a further addition of a single grain.

Several studies have shown that the probability density of landslide area $p(A_L)$ for many inventories exhibits a typical behavior (e.g. [4–7]): when plotted in bi-log graph, it increases linearly with the landslide size up to a maximum value, after which the $p(A_L)$ rapidly decreases with increasing landslide area, following a power law too. This behavior can be explained in terms of scale invariance, i.e. fractal statistics [8]. Different studies on frequency–area distributions of landslides agree on a power law tail behavior with exponent of -2.3 ± 0.6, behind the roll-over.

Accurate statistic analysis of the frequency-size (area or volume) distributions, as reported in many studies, is very important for several implications in the quantitative assessment of landslide hazard, because it allows to quantify the number of landslides that occur at different sizes but also the contribution of slope instabilities in landscape morpho-evolution.

A. Donnarumma (✉) · P. Revellino · F. M. Guadagno
Department of Sciences and Technologies, Via Port'Arsa, 11, 82100 Benevento, Italy
e-mail: angelo.donnarumma@unisannio.it

P. Revellino
e-mail: paola.revellino@unisannio.it

F. M. Guadagno
e-mail: guadagno@unisannio.it

E. Pardo-Igúzquiza et al. (eds.), *Mathematics of Planet Earth*,
Lecture Notes in Earth System Sciences, DOI: 10.1007/978-3-642-32408-6_51,

This paper aims at performing a statistic analysis of the frequency–area distribution of historical landslides in the province of Benevento (Southern Italy) by using a three-parameter 'inverse-gamma' Probability Density Function (PDF).

2 Landslide Inventory of the Study Area

The study area is the Benevento province (approximately 2,000 km^2), located in one of the most geologically complex sectors of mountain Apennine in Southern Italy. The morphological shape of slopes is controlled by recurring landslide events, in time and space [9, 10]. In the study area, 3,756 landslides were inventoried [11], covering an approximate area of 360 km^2, equal to about 18 % of the whole provincial surface. The inventory have been performed by means of interpretation of historical aerial photos of different time as from 1954, together with geological and field surveys carried out between 2001 and 2010. The landslide inventory is composed of the following types: 3,062 earth flows, 71 translational, rotational and composite slides, 596 falls and topples, 27 rock and debris avalanches, and debris flows.

3 Frequency-Area Statistics

In order to obtain the dependence of landslide frequency on landslide area, we use a landslide inventory.

The data binning was preliminarily made with 1-size, non zero intervals and was characterized by small dispersion. The $p(A_L)$ was obtained by dividing the area of each interval by its own amplitude. More specifically, we fixed 10^{-3} Km^2 as the starting point of the binning procedure and then we extracted the counts in each bin of a $5 \cdot 10^{-3}$ Km^2 amplitude.

Statistical distribution of landslide areas was obtained by using a probability density distribution $p(A_L)$ defined according to:

$$p(A_L) = \frac{1}{N_{LT}} \frac{\delta N_L}{\delta A_L}. \tag{1}$$

with the normalization condition

$$\int_0^{\infty} p(A_L) dA_L = 1. \tag{2}$$

where A_L is landslide area, N_{LT} is the total number of landslides in the inventory, and δN_L is the number of landslides with areas between A_L and $A_L + \delta A_L$. The PDF was computed by applying a three-parameter inverse-gamma distribution [12, 13] proposed by Malamud et al. [6] as general probability distribution for landslides:

$$p(A_L;\rho;a;s) = \frac{1}{a\Gamma(\rho)}\left[\frac{a}{A_L - s}\right]^{\rho+1} \exp\left[-\frac{a}{A_L - s}\right]. \tag{3}$$

where $\Gamma(\rho)$ is the gamma function of ρ; ρ controls the power law decay for medium and large landslide areas; a primarily controls the location of the maximum of the probability distribution; s primarily controls the exponential decay for small landslide areas. In our study, Maximum Likelihood iterative method were used to estimate inverse-gamma parameters for the landslide data set. The best fit corresponds to the following values: $\rho = 1.42$, $a = 9 \cdot 10^{-2}\mathrm{Km}^2$, $s = -1.38 \cdot 10^{-4}\mathrm{Km}^2$, $\Gamma(\rho) = 0.89$. Figure 1 shows the probability density, $p(A_L)$, obtained using Eq. 3. The observed distribution shows a distinct roll-over for landslide areas less than $10^{-1}\mathrm{Km}^2$. The linear portion of the curve with landslide area > $10^{-1}\mathrm{Km}^2$ shows a power-law decay with scaling exponent $\beta = (-\rho + 1)$ as 2.42 (determination coefficient $r^2 = 0.97$). Finally, in order to verify the SOC behavior of the real landslides, we performed a non-cumulative frequency–area distribution, for medium and large landslides, which satisfies a PL distribution with an exponent of $-0.92(N_L = A_L^{-0.92}, r^2 = 0.99)$.

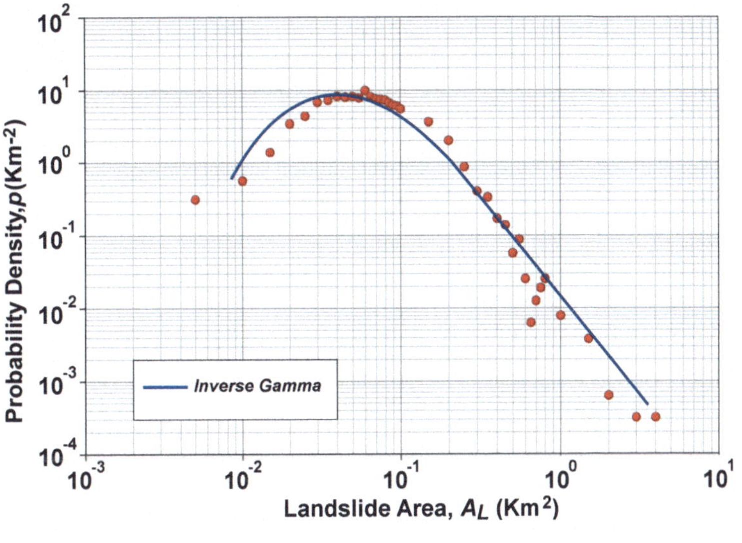

Fig. 1 Probability density of landslide area, $p(A_L)$, as a function of A_L, shown in logarithmic coordinates

4 Conclusions

Our study focused on the frequency distribution of A_L in the area of the Benevento province. Through a statistic analysis, we obtained the $p(A_L)$; its analysis indicates that the dataset exhibits negative power-law scaling of $p(A_L)$ as a function of A_L, for landslides exceeding a threshold area. The slope of the power law tail is about −2.4.

The analysis of landslide properties agrees with SOC model expectations, although a further step ($1/f^{\alpha}$ behavior) is needed in order to confirm the SOC system behavior. The full SOC modeling will be discussed in a forthcoming paper.

The obtained result can have important implications in quantitative landslide hazard assessments of the study area.

References

1. Malamud, B. D., & Turcotte, D. L. (1999). Self-organized criticality applied to natural hazards. *Natural Hazards*, *20*, 93–116.
2. Hergarten, S., & Neugebauer, H. (1998). Self organized criticality in a landslide model. *Geophysics Research Letters*, *25*, 801–804.
3. Bak, P., Tang, C., & Wiesenfield, K. (1988). Self-organized critically. *Physical Review A*, *38*, 364–374.
4. Guthrie, R. H., & Evans, S. G. (2004). Analysis of landslide frequencies and characteristics in a natural system, coastal British Columbia. *Earth Surface Processes and Landforms*, *29*, 1321–1339.
5. Guzzetti, F., Cardinali, M., Reichenbach, P., Cipolla, F., Sebastiani, C., Galli, M., et al. (2004). Landslides triggered by the 23 November 2000 rainfall event in the imperia province, Western Liguria, Italy. *Engineering Geology*, *73*(3–4), 229–245.
6. Malamud, B. D., Turcotte, D. L., Guzzetti, F., & Reichenbach, P. (2004). Landslide inventories and their statistical properties. *Earth Surface Processes and Landforms*, *29*, 687–711.
7. Van Den Eeckhaut, M., Poesen, J., Govers, G., Verstraeten, G., & Demoulin, A. (2007). Characteristics of the size distribution of recent and historical landslides in a populated hilly region. *Earth and Planetary Science Letters*, *256*, 588–603.
8. Chen, C. C., Telesca, L., Lee, C. T., Sun, Y. S. (2011). Statistical physics of landslides: New paradigm. *Europhysics Letters*, *95*(4), 49001, 3.
9. Donnarumma, A., Revellino, P., Grelle, G., & Guadagno, F. M. (2013). Failure analysis of shallow landslides using a three parameter Weibull distribution of slope angle. *Rendiconti Online Società Geologica Italiana*, *24*, 110–112.
10. Grelle, G., Revellino, P., Donnarumma, A., & Guadagno, F. M. (2011). Bedding control on landslides: A methodological approach for computer-aided mapping analysis. *Natural Hazards and Earth Systems Sciences*, *11*, 1395–1409 (2011). doi:10.5194/nhess-11-1395-2011.
11. Revellino, P., Grelle, G., Donnarumma, A., & Guadagno, F. M. (2010). Structurally-controlled earth flows of the Benevento Province (Southern Italy). *Bulletin of Engineering Geology and the Environment*, *69*(3), 487–500 (2010). doi:10.1007/s10064-010-0288-9.
12. Evans, M., Hastings, N., & Peacock, J. B. (2000). *Statistical Distributions* (3rd ed.). New York: John Wiley.
13. Johnson, N. L., & Kotz, S. (1970). *Continuous univariate distribution*. Boston: Houghton Mifflin.

A Spatiotemporal Remotely-Sensed Assessment of Peat Covered Areas Using Airborne Radiometrics

Jennifer M. McKinley, Antoinette Keaney and Alastair Ruffell

1 Introduction

Peat has a high proportion of soil carbon due to the relatively high carbon density of peat and organic-rich soils. For this reason it has become increasingly important to measure and model soil carbon stores and changes in peat stocks to facilitate the management of carbon changes over time. The aim of this research is to improve the assessment of carbon in organic-rich soil using remotely sensed airborne radiometric data of peat covered areas. The theory being applied is that saturated peat attenuates gamma-radiation from underlying rocks. This relationship can be used to estimate peat thickness, within certain limits. Irish peat is divided into blanket peatland (approximately 85 %) and raised peat bogs (approximately 15 % [1]). Raised bogs develop primarily in lowland areas from the accumulation of peat in fens. Blanket peat typically form on gentle slopes within upland regions. Two case studies, a raised bog and a blanket peat area, are presented to validate the use of airborne geophysical data to estimate peat thickness using the attenuation of bedrock geology radioactivity by superficial peat cover.

J. M. McKinley (✉) · A. Keaney · A. Ruffell
School of Geography, Archaeology and Palaeoecology, Queen's University Belfast, Belfast BT7 1NN, UK
e-mail: j.mckinley@qub.ac.uk

A. Keaney
e-mail: a.keaney@qub.ac.uk

A. Ruffell
e-mail: a.ruffell@qub.ac.uk

E. Pardo-Igúzquiza et al. (eds.), *Mathematics of Planet Earth*,
Lecture Notes in Earth System Sciences, DOI: 10.1007/978-3-642-32408-6_52,

1.1 Airborne Radiometrics

This research uses radiometric data generated by two airborne geophysical surveys. The Tellus Project [2] undertaken between 2004 and 2007 and the more recent EU-funded Tellus Border project (2011–2013), which was a continuation of the Tellus survey, covering the six bordering counties of the Republic of Ireland. As part of these airborne surveys, terrestrial radiation was sampled every second using a gamma-ray spectrometer (Explorium GR-820/3). Flight lines had a line spacing of 200 m and an average flight altitude of 56–60 m [2]. Total Count (TC) measurements, comprising a spectral summation including contributions from both natural and artificial radioactive sources, measured in counts per second (cps), were collected across the energy window from 0.41 to 2.81 MeV [3]. TC data provide a higher signal/noise ratio than individual spectral radioisotope levels and are used for the outputs shown since the signal/noise ratio is important when analysing attenuation and low count behaviour in the radiometric data [3].

2 Methods

2.1 Coregionalisation

Variability exists in peat depth and radiometric values across the raised peat bog (Fig. 1). Where the peat cover is thick the radiometric signal is reduced demonstrating the attenuation of the radiometric signal by saturated peat. The proximity of ground-based peat thickness measurements with airborne radiometric data at sufficient locations allowed the degree of spatial correlation between the datasets to be investigated. Coregionalization uses the degree of spatial correlation between measured peat depth and the attenuation of the radiometric signal to update a limited sampling regime of ground-based measurements with more densely remotely acquired data (Fig. 1a). Utilising the inverse spatial relationship (correlation coefficient r value = −0.49), cokriging uses the coefficients from cross variograms between ground-based peat thickness measurements and airborne radiometric data (Fig. 1b) to produce an updated cokriged map (Fig. 1c). The benefits of using the airborne survey data are twofold. The use of remote sensed data to produce a spatial map of peat thickness minimises destruction to a sensitive habitat. Secondly temporal changes to peat thickness can be assessed, in that areas where radiometric data do not correlate with previously measured peat depth may indicate drying out of the peat through extraction or encroachment by invasive vegetation.

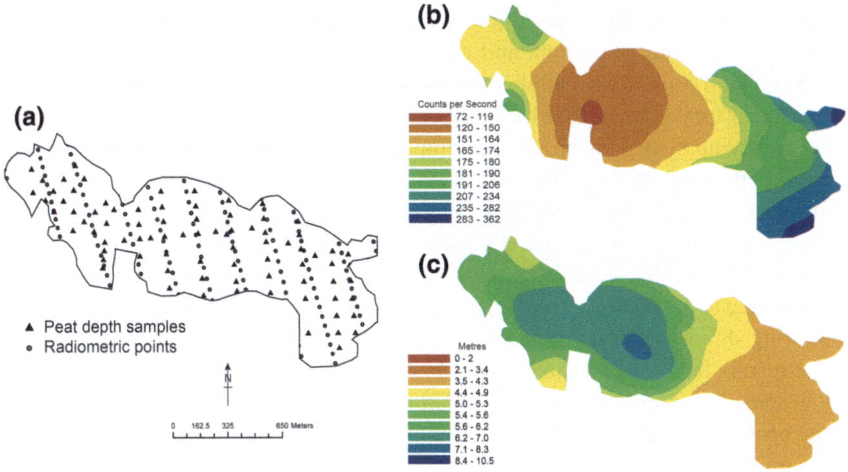

Fig. 1 **a** Measurement grids for peat depth and radiometrics; **b** Kriged map of total radiometrics (counts per second); **c** Cokriged output map of peat depth (metres) for the raised peat bog. Note the inverse relationship between the radiometric data and peat thickness demonstrating the attenuation of gamma-radiation by saturated peat (correlation coefficient r value = −0.49)

2.2 Using Remote Sensed Data to Update Peat Management Strategies

Blanket peat can cover kilometres of upland areas and although peat survey assessments may exist, peat depth measurements are limited. The use of remote sensed airborne radiometric data can provide a spatial estimate of peat thickness and an assessment of temporal changes in peat. Using parameters of fitted variogram models to the radiometric TC data (Fig. 2a), the kriged output map (Fig. 2b) illustrates

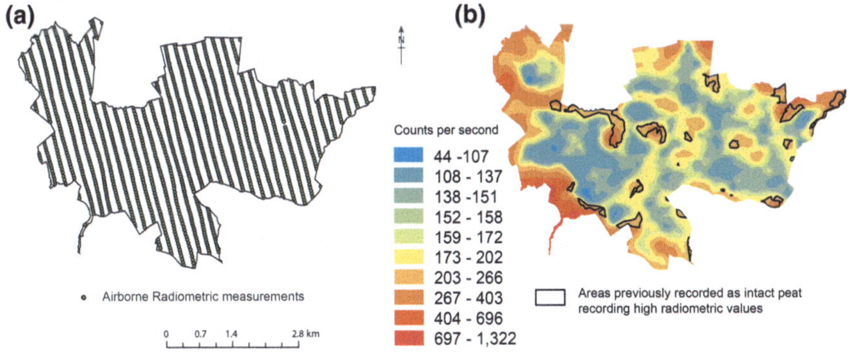

Fig. 2 **a** Measurement grid for airborne radiometrics; **b** Kriged map of total radiometrics (counts per second) for blanket peatland

the spatial variation of peat thickness across a blanket peatland using the inverse relationship between peat depth and airborne radiometric data. Where the peat is less thick, high radiometric total count values are expected and where the peat is thicker lower total counts values are expected as the radiometric signal is attenuated by the overlying peat. This provides a temporal assessment of peat changes for this blanket peatland when compared with previously conducted peat surveys. While the majority of the blanket peat remains unchanged (dark blue areas in Fig. 2b), areas are indicated with high radiometric values that previously were recorded as 'intact' thick peat. The implication is that these areas have experienced a reduction in peat thickness due to drying out, invasive species or illegal peat extraction. Ground-based monitoring should be focussed in these areas.

The results from this research have a broader significance to promote the use of remote sensing for spatial estimates of carbon stock, reduce the need for damaging on site peat monitoring and to record temporal changes in peat covered areas due to peat extraction and invading species.

Acknowledgments The Tellus project was funded by the Northern Ireland Department of Enterprise Trade and Investment and by the Rural Development Programme through the Northern Ireland Programme for Building Sustainable Prosperity. Tellus Border is supported by the EU INTERREG IVA programme, which is managed by the Special EU Programmes Body in Northern Ireland, the border Region of Ireland and western Scotland.

References

1. Tomlinson, R. W., & Milne, R.M. (2006). Soil carbon stocks and land cover in northern ireland from 1939 to 2000. *Applied Geography, 26*(1), 18–39.
2. Geological Survey Northern Ireland (GSNI): Tellus project overview. http://www.bgs.ac.uk./gsni/Tellus/index (2007).
3. Beamish, D., Cuss, R. J., Lahti, M., Scheib, C., & Tartaras, E. (2006). The Tellus airborne geophysical survey of Northern Ireland: final processing report 2006. Nottingham: British Geological Survey.

Monitoring of Urban-Damaging Landslides with Satellite Radar Missions: Arcos de la Frontera (Spain)

Guadalupe Bru, José Fernández, Pablo J. González and Kristy F. Tiampo

1 Introduction

In recent years, a number of newly constructed residential subdivisions located in the Arcos de la Frontera suburbs suffered severe structural damage due to slope instabilities which, in some cases, forced evacuation of the inhabitants. Wall cracks in buildings have been reported and, after an intense precipitation period in December 2009 and beginning 2010, slope failure caused the majority of damage in La Verbena neighborhood, not only affecting buildings but also public spaces and water supplies. Remediation efforts are still being carried out in order to stabilize the landslide. However, in the case of Pueblos Blancos subdivision, damages could be due to displacements of the earth fill on which it was constructed.

2 Geomorphological and Geological Setting

Arcos de la Frontera (Cádiz, southern Spain) is located in the Guadalete basin, between the Guadalquivir basin and the occidental border of the Subbetic System. The old city sits on top of a calcareous sandstone (commonly known as "caliza tosca") scarp with typical cross set dune structure. The substratum of the most recent urban area affected by the landslide is composed of siliciclastic sandstones and sandy

G. Bru (✉) · J. Fernández (✉)
Instituto de Geociencias CSIC, UCM, Madrid, Spain
e-mail: guadalupe.bru@igeo.ucm-csic.es

J. Fernández
e-mail: jft@mat.ucm.es

P. J. González · K. F. Tiampo
Department of Earth Sciences, University Western Ontario, London, ON, Canada

E. Pardo-Igúzquiza et al. (eds.), *Mathematics of Planet Earth*,
Lecture Notes in Earth System Sciences, DOI: 10.1007/978-3-642-32408-6_53,

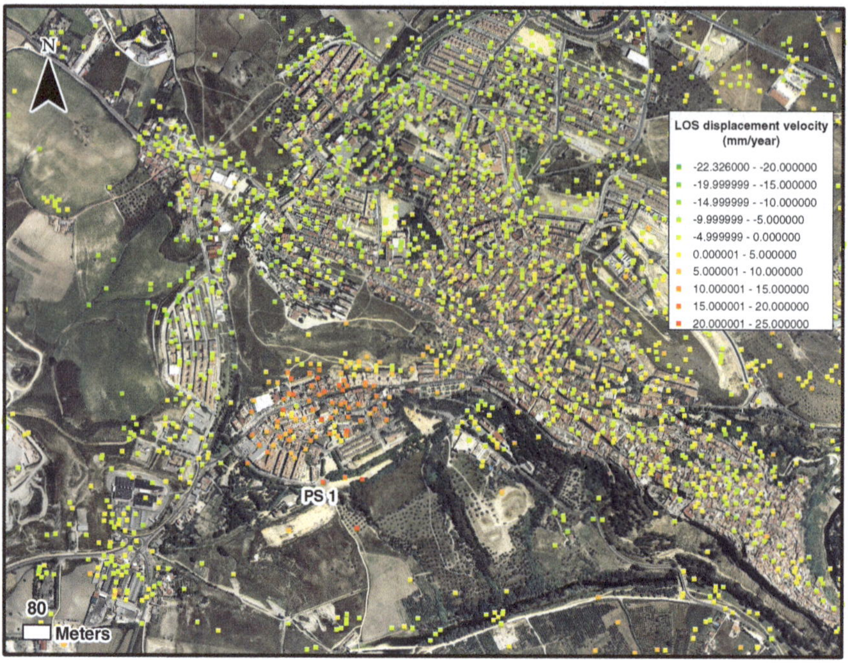

Fig. 1 Linear velocity map in LOS direction obtained by the processing of 10 SAR images for the period April 2011–January 2012

marlstones facies [3]. All the materials are Miocene-age and their genesis is pelagic, formed in an oceanic basin that actually constitutes the Guadalquivir depression. As mentioned earlier, Pueblos Blancos subdivision was constructed on earth fill.

The geomorphology of the area of interest is dominated by the meander-shaped channel of the Guadalete River. Both Verbena and Pueblos Blancos terrain slopes are gentle, the first one dipping towards the SE and the second to the W (Fig. 1).

3 InSAR: Principles, Data and Methodology

DInSAR techniques have become a powerful method to detect and monitor slow ground surfaces deformations at a relatively low cost. Although there are constraints, such as decorrelation, terrain slope topography and very rapid movements [4], its application to landslides studies is feasible [5–7]. On the one hand, classical differential interferometry can show displacements on the satellite Line of Sight (LOS) between two different images acquisitions. On the other, advanced techniques can compute displacement time series and velocity surface deformation using a stack of

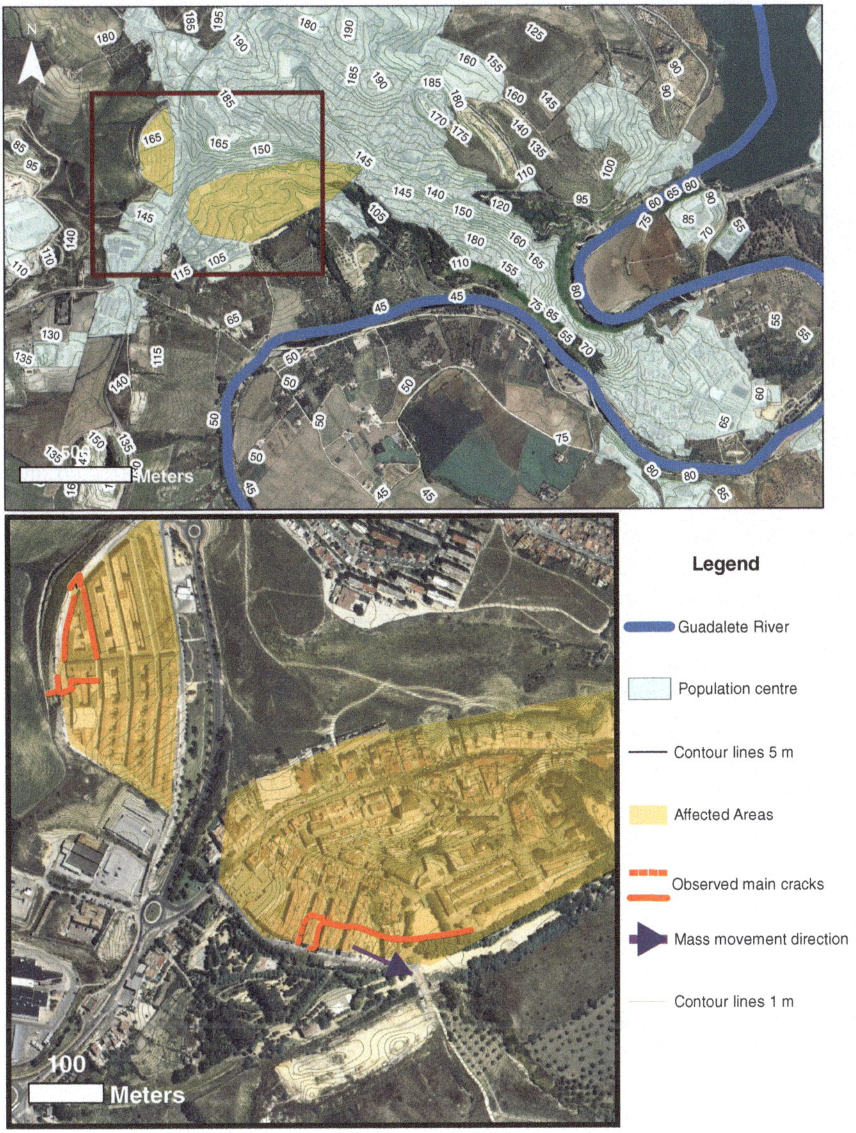

Fig. 2 Map of the study area, showing water streams, Arcos de la Frontera city, topography, main cracks observed in buildings and mass movement direction. The western affected area corresponds to Pueblos Blancos and the southern one includes the Verbena subdivision

SAR images, based on amplitude (full resolution) or coherence (multilook) criteria that are very useful to assess their activity.

A total of 22 descending ENVISAT SAR images, including the ENVISAT 2010+ mission, are processed with two techniques; the period covering March 2007–September 2010 (12 images) with the Coherent Pixel Technique, CPT and April 2011–January 2012 (10 images) with STAMPS software. We intend to correlate ground surface movements to intense precipitation periods and to evaluate the overall landslide behavior through displacement time series.

4 Results

The first linear velocity results obtained by the set of images from the period April 2011–January 2012 are shown in Fig. 1. It is seen that the southern affected area exhibits subsidence (ground surface to satellite range increase) in its north part, which according to the topography (Fig. 2), could correspond to the upper part of the landslide. However, Pueblos Blancos exhibits movement towards the satellite in the LOS direction. Daily precipitation data, obtained from the Spanish Meteorogical Agency, is compared to the displacement time series and show correlation (Fig. 3). On initial inspection, it could confirm our starting hypothesis (rainfall is correlated with landslide velocity). More detailed studies would be necessary.

Our results illustrate the benefits of including remote sensing techniques as a complementary tool for geotechnical engineering. In particular, they provide a significant advantage in structural monitoring in urban areas affected by slow landslides.

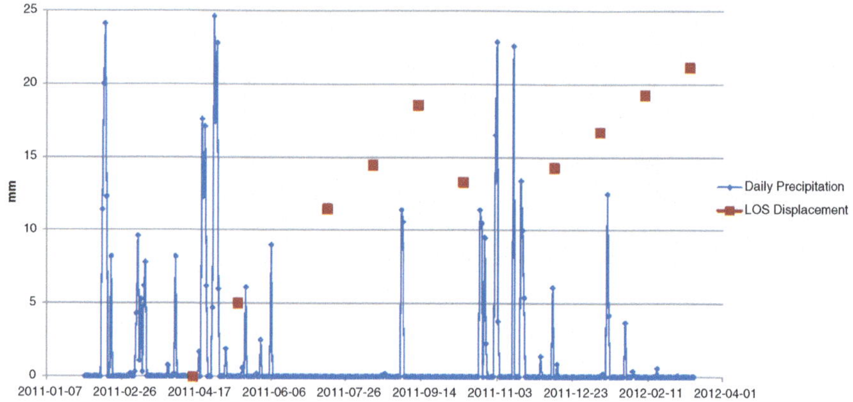

Fig. 3 Daily precipitation and LOS displacement of PS1 (labeled in Fig. 3)

Acknowledgments This research has been done in the frame of the MINECO research projects EOSLIDE (IPT-2011-1234-310000) and GEOSIR (AYA2010-17448). PJG acknowledges the Banting Postdoctoral Fellowship (Canadian Government). KFTs research is supported by an NSERC Discovery Grant. It is a contribution for the CEI Campus Moncloa.

References

1. Blanco-Sánchez, P., et al. (2009). *Pure and Applied Geophysics*, *165*, 1167–1193.
2. Hooper, A. (2008). *Geophysical Research Letters*, *35*, L16302. doi:10.1029/2008GL034654.
3. IGME: Hoja 1049–13-44: Arcos de la Frontera. Mapa geológico de España E:1:50000. Servicio de publicaciones del Ministerio de Industria. 54pp+map. (1990).
4. González, P. J., & Fernández, J. (2011). Error estimation in multitemporal InSAR deformation time series, with application to Lanzarote, Canary Islands. *Journal of Geophysical Research*, *116*, B10404. doi: 10.1029/2011JB008412.
5. Fruneau, B., Achache, J., & Delacourt, C. (1996). Observation and modelling of the Sant-E ' tienne-de-Tine'e landslide using SAR interferometry. *Tectonophysics*, *265*, 181–190.
6. Squarzoni, C., Delacourt, C., & Allemand, P. (2003). Nine years of spatial and temporal evolution of the La Vallette landslide observed by SAR interferometry. *Engineering Geology*, *68*, 53–66.
7. Herrera, G., Gutiérrez, F., García-Davadillo, J. C., Guerrero, J., Notti, D., Galve, J. P., et al. (2013). Multi-sensor advanced DInSAR monitoring of very slow landslides: The Tena valley case study (Central Spanish Pyrenees). *Remote Sensing of Enviroment*, *128*, 31–43.

Multivariate Variogram and Madogram: Tools for Quantifying Diversity/Dissimilarity in Spatiotemporal Data

Phaedon Kyriakidis, Dimitra Kitsiou and Dimitris Kavroudakis

1 Introduction

Multivariate measures of spatial association, such as the Mantel correlogram, are frequently used, especially in ecology, for quantifying (dis)similarity among multiple attributes as a function of neighborhood specification [5, 7]. In geostatistics, similar concepts are formulated under the umbrella term multivariable variogram, a distance-based measure of multivariate (dis)similarity linearly linked to classical auto- and cross-variograms [4]. The above statistics constitute average measures of pairwise attribute association among all neighboring units; hence the term global indices [6]. Local (site specific) univariate or bivariate indices of attribute association (Moran and Geary indices) have been developed in spatial econometrics [1], and are increasingly used to quantify spatially varying association in socio-economic and environmental data [5, 6]. In remote sensing applications, local multivariate variograms and madograms have been employed to quantify image texture, providing additional feature layers for improved thematic classification [2, 8]. Multivariate association indices, however, are not widely utilized in spatiotemporal settings. In this work, both global and local multivariate variograms and madograms are linked to Geary's and Gini's indices of association, and employed for elucidating spatiotemporal patterns in remotely sensed data within a marine science application context.

P. Kyriakidis (✉) · D. Kavroudakis
Department of Geography, University of the Aegean, Mytilene, Greece
e-mail: phkyriakidis@geo.aegean.gr

P. Kyriakidis
Department of Geography, University of California, Santa Barbara, CA, USA

D. Kitsiou
Department of Marine Sciences, University of the Aegean, Mytilene, Greece

E. Pardo-Igúzquiza et al. (eds.), *Mathematics of Planet Earth*,
Lecture Notes in Earth System Sciences, DOI: 10.1007/978-3-642-32408-6_54,

2 Materials and Methods

The objective is to quantify spatiotemporal patterns in remotely sensed data of chlorophyll-a (CHL) concentrations (in mg/m^3) and sea surface temperature (SST) values (in °C) acquired during the time period January 1998 through November 1999 over the North Aegean Sea in Greece. More specifically, CHL data were derived from the SeaWIFS sensor with nominal spatial resolution 10466 m, and were downloaded in processed form as monthly averaged concentrations from NASA's SeaWiFS Project site. SST data were derived from NOAAs AVHRR sensor with nominal spatial resolution 1600 m, and were download in processed form as monthly temperature values from the German Aerospace Agency. To place both data sets at a common spatial resolution, the SST data were interpolated (using only contemporaneous SST measurements) to the nearest grid node of the CHL raster. The time-averaged spatial distributions of SST and CHL are given in Fig. 1; the spatial extent of the study region in terms of UTM coordinates is approximately 635911–1033604 in the E-W direction and 4300642–4562281 in the S-N direction.

Multivariate measures of spatial association are constructed as a weighted averages of pairwise dissimilarities between data of multiple attributes measured at different locations or time instants; the weights are derived via an exogenous, e.g., distance-based, neighborhood specification as in classical variogram estimators. Pairwise multivariate attribute dissimilarity is quantified using some form of distance metric, e.g., Euclidean, Manhattan, or Mahalanobis [7]. When the Euclidean and Manhattan metrics are adopted, average dissimilarity between neighboring data pairs of a single attribute reduces to the geostatistical variogram and madogram measures, respectively [4]; these in turn correspond to Geary's and Gini's indices of association [3, 6]. Focused (location or time specific) versions of multivariate association indices are derived by restricting averaging of attribute dissimilarities to pairs of sites with a common origin (each location or time), thus facilitating the local characterization of spatiotemporal patterns from multivariate data.

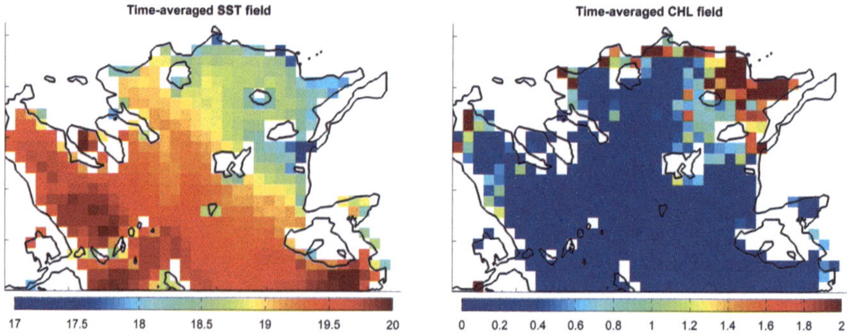

Fig. 1 Time-averaged (Jan 1998 through Nov 1999) spatial distribution of SST (*left*) and CHL (*right*) over the North Aegean Sea in Greece

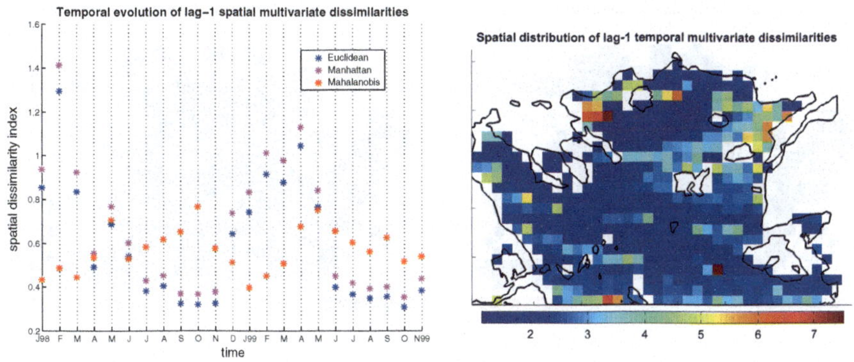

Fig. 2 Temporal evolution of different lag-1 spatial multivariate dissimilarities (*left*), and spatial distribution of lag-1 temporal multivariate dissimilarities based on the Manhattan distance metric (*right*) for SST and CHL data

3 Results

Figure 2 shows the temporal evolution of lag-1 spatial multivariate dissimilarities, based on the Euclidean, Manhattan and Mahalanobis distance metrics, as well as the spatial distribution of lag-1 temporal multivariate dissimilarities based on the Manhattan distance metric, for the SST and CHL data over the study region and period of interest. Here, a spatial lag of 1 is defined as the neighborhood of 8 nearest pixels around any central pixel. One can deduce that periodic temporal variability is more pronounced when multivariate association is quantified via the Euclidean and Manhattan distance metrics, since in those cases one is simply computing sums of univariate variograms and madograms, respectively. The spatial distribution of temporal lag-1 multivariate dissimilarities pinpoints areas of high variability corresponding to areas of increased ocean mixing in the region.

Figure 3 shows the spatial distribution of local lag-1 (with at least 4 informed pixels) spatial multivariate dissimilarities based on the Mahalanobis distance metric for January and February 1998. Areas of higher local dissimilarity indicate locally increased ocean mixing for the specific month.

4 Discussion

Global and local multivariate variograms and madograms were employed for characterizing the spatiotemporal distribution of satellite-derived sea surface temperature and chlorophyll-a concentration in the North Aegean Sea, Greece. The results indicate that such measures constitute useful tools for elucidating spatiotemporal patterns in remotely sensed data pertaining to multiple environmental variables.

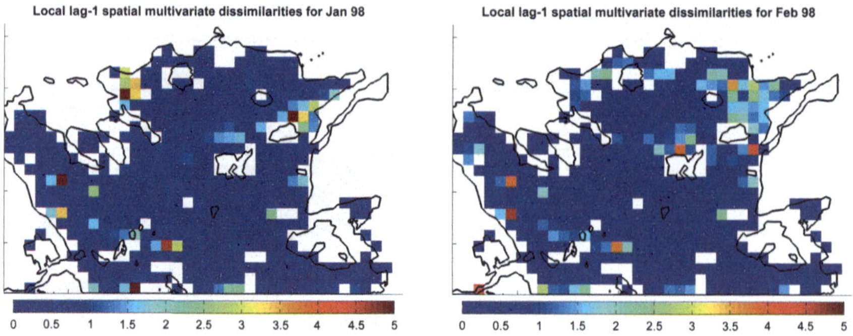

Fig. 3 Local lag-1 spatial multivariate dissimilarities for Jan (*left*) and Feb (*right*) 1998

Acknowledgments This work is part of the project "*Advances in Geostatistics for Environmental Characterization and Natural Resources Management*", implemented under the "*Aristeia*" Action of the "*Operational Programme Education and Lifelong Learning*", and co-funded by the European Social Fund (ESF) and Greek National Resources.

References

1. Anselin, L. (1995). Local indicators of spatial association. *Geographical analysis*, *27*(2), 93–115.
2. Atkinson, P., & Lewis, P. (2000). Geostatistical classification for remote sensing: An introduction. *Computers and Geosciences*, *26*(4), 361–371.
3. Biondi, F., & Qeadan, F. (2008). Inequality in paleorecords. *Ecology*, *89*(4), 1056–1067.
4. Bourgault, G., & Marcotte, D. (1991). Multivariable variogram and its application to the linear model of coregionalization. *Mathematical Geology*, *23*(7), 899–928.
5. Fortin, M. J., & Dale, M. (2005). *Spatial Analysis: A Guide for Ecologists*. Cambridge: Cambridge University Press.
6. Haining, R. (2003). *Spatial Data Analysis: Theory and Practice*. Cambridge: Cambridge University Press.
7. Legendre, P., & Legendre, L. (2012). *Numerical Ecology* (3rd ed.). Amsterdam: Elsevier
8. Li, P., Cheng, T., & Guo, J. (2009). Multivariate image texture by multivariate variogram for multispectral image classification. *Photogrammetric Engineering & Remote Sensing*, *75*(2), 147–157.

Spatiotemporal Interactions for Daily Mapping of PM_{10} with MODIS and Meteorological Data

Piero Campalani, Simone Mantovani and Peter Baumann

1 Aerosols Spatiotemporal Modelling

In recent years the modelling community has been increasingly exploiting spatiotemporal techniques for the Air Quality (AQ): despite the high spatial and temporal variability, it has been proved how aerosols can show a temporal autocorrelation, especially for secondary aerosols species (sulfate, nitrate, ammonium) [1].

Liu et al. (2009) successfully applied geostationary GOES AOT products in conjunction with land use and meteorological fields to feed a two-stage space-time additive model [2]; Gräler et al. in their recent report [3] showed that for the spatial interpolation of AQ, a prediction gain is achieved by including temporal correlation in the geostatistical (kriging) estimator.

The present paper—natural continuation of a previous study over the same area [4]—aims at exploring a spatiotemporal regression-based kriging technique to predict daily exposures of PM_{10}. Temporal correlation is evaluated by comparison of the same kriging estimator without time awareness.

Kriging represents a good trade-off between implementation ease, transferability, computational requirements and predictive power, making it a fascinating option with respect to chemical dispersion or dynamical coregionalization models [5]. It allows

P. Campalani (✉)
UNIFE, Via Saragat 1, Ferrara, Italy
e-mail: cmppri@unife.it; p.campalani@jacobs-university.de

P. Campalani · P. Baumann
JUB, Campus Ring 1, Bremen, Germany
e-mail: p.baumann@jacobs-university.de

S. Mantovani
MEEO Srl, Via Saragat 9, Ferrara, Italy

S. Mantovani
SISTEMA GmbH, Wäringerstraße 61, Wien, Austria
e-mail: mantovani@sistema.at

E. Pardo-Igúzquiza et al. (eds.), *Mathematics of Planet Earth*, Lecture Notes in Earth System Sciences, DOI: 10.1007/978-3-642-32408-6_55,

the inclusion of spaceborne imagery, whose usefulness for the AQ system is continuously experimented while still awaiting to reach the "promised land" [6].

2 The Daily Kriging Estimator

Although the annual average constraints on PM_{10} exceedances tend to take the precedence for being more restrictive, it is however over true that short-term peaks of PM_{10} can lead to equivalently dangerous effects on human health [7].

Validated [8] satellite Aerosols Optical Thickness (AOT) products from the polar-orbiting MODIS sensor at 1×1 km^2 of spatial resolution were used in conjunction with 27×27 km^2 meteorological fields to build a regression surface driving the trend of the kriging daily predictor. Spatiotemporal interactions were accounted via separable-covariance model, by means of the **R** *gstat* and *spacetime* packages.

Variogram fitting in time was done by selecting input data by month, this way trying to collect stable statistics for lags of at least 4–5 days. Spatial variograms were instead best chosen from the single daily observations, while usually a flattening of the semivariances was experienced when enlarging the time window (see Fig. 1).

Daily full maps of PM_{10} estimations and model errors were ingested in the *rasdaman* array database (http://www.rasdaman.org), whose open-standard OGC Web Coverage Processing Service (WCPS) and new spatiotemporal CRS concept [9]

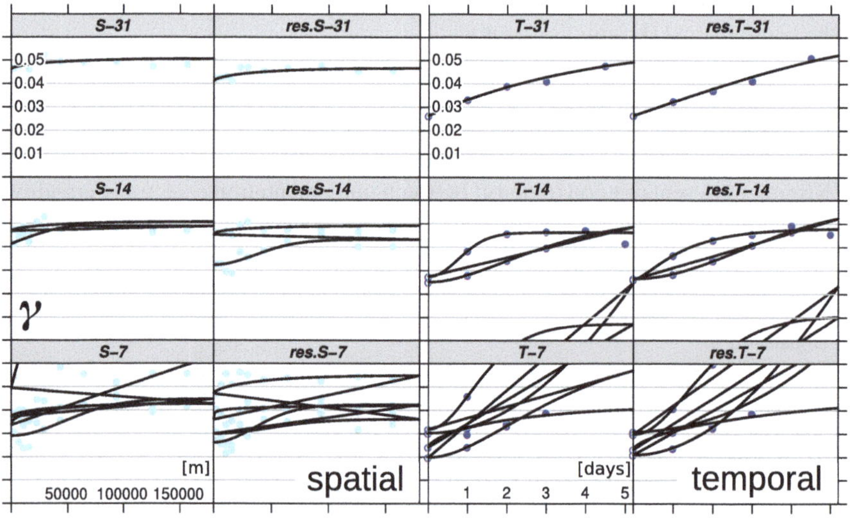

Fig. 1 Examples of variograms for both spatial and temporal pooling of scaled PM_{10} (Dec 2010). For each of the two tables: the first column is for the *direct* variogram of PM_{10} and *right* column is for the *residual* variograms after multiple regression; rows are respectively for the 1-month, 2-weeks and 1-week pooling intervals

allowed to go beyond the web-based raw visualization of maps, with on-the-fly analytics capabilities like spatial and temporal aggregation, traffic light indexing, or exceedances identification with arbitrary confidence interval.

3 Results

A preliminary session of regression analysis led to a separation of the available meteorological features in two subsets: pressure and wind components for the intermediate gap-filling of the AOT missing pixels; then humidity and mixing layer height—and the spatialised AOT (AOT^s)—for the PM_{10} mapping.

AOT regression observed an average adjusted goodness-of-fit (R^2) of 0.779 (0.096 σ) over the 3 years. Whereas pressure had a stable positive β (and an average Pearson's r of 0.37 with AOT), wind's role was highly variable.

PM_{10} regression showed a good adjusted R^2 too (0.785, 0.093 σ), although 0.05-significance was often not met. An ANOVA analysis revealed how AOT^s normalization via humidity and/or mixing layer height were not helping for a better fit, while higher AOT availability did not translate to more robust models.

The key results were visible from the cross-validation scores (Table 1): kriging paid off over mechanical inverse distance interpolation (9 % of error gain, not shown), but a modest though systematic worsening of the kriging precision and bias was observed when adding the temporal dimension to the model.

No relevant annual/seasonal variations occurred, except for a clear performance decrease during the cold season due to higher PM_{10} variance (Fig. 2).

Daily spatial variograms were finally sometimes not suitable over noisy days, in some cases with very short ranges (few km) or overly high nuggets. The monthly temporal variograms did not show strong profiles, thus, despite the very rugged topology of Austria, the complex wind profiles probably prevented the fine particles to keep a good memory over preceding days.

Table 1 Leave-one-out cross-validation median statistics using pure spatial External-Drift Kriging (KED), and separable covariance spatiotemporal (ST) KED (and relative % comparison)

	RMSE $[(\mu g/m^3)^2]$			ME $[(\mu g/m^3)]$		
year	KED	ST-KED	%	KED	ST-KED	%
2008	1.965	2.041	3.9	0.009	−0.059	556
2009	1.864	1.954	4.8	−0.013	−0.048	269
2010	1.835	1.927	5.0	0.002	−0.053	2550

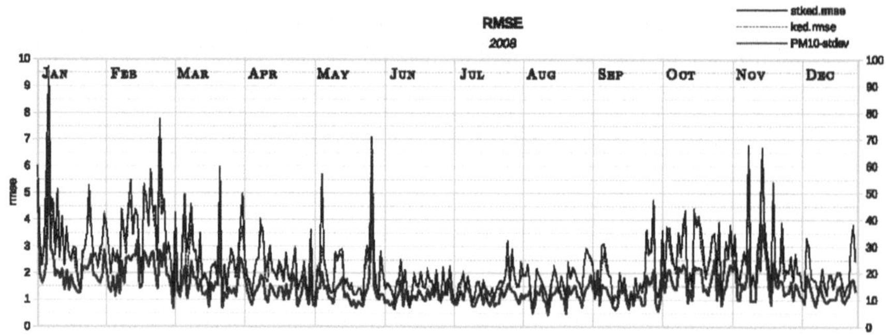

Fig. 2 Leave-one-out cross-validation chart of RMSE (2008) for ST-KED (—) and KED (- -) estimators (*left* Y scales, [$(\mu g/m^3)^2$]), and PM_{10} standard deviations (—), (*right* Y scales, [$\mu g/m^3$])

4 Conclusions and Outlook

The article presented an application of spatiotemporal kriging for the estimation of surface-level daily exposure of PM_{10} with covariates from MODIS satellite sensor and several meteorological fields.

The evidence of cross-validation scores again underlined the valuable contribution of a geostatistical kriging modelling over univariate mechanical interpolation for the PM_{10} estimation, as learnt from the preceding study over the same region [4].

Despite this, temporal correlation of the aerosols seemed to overly feed the predictor with uncertain data, which suffered a systematic performance loss when augmenting the input explanatory time frame from a single day to a set of days.

The relatively high uncertainty in the spaceborne pixels was forced to increase due to the mandatory gap-filling for the kriging external drift, and at the same time the pooled temporal variograms generally did not show a long memory of the aerosols observations, sometimes reaching the sill already at the first day lag.

Before the launch of new satellites for the AQ—like the upcoming geostationary NOAA GOES-R series—future models should probably try to avoid intermediate fillings in the input data with more complex geostatistics, e.g. Bayesian hierarchical space-time models; asymmetry in spatiotemporal covariance could also be addressed with more comprehensive models, e.g. with vine copulas [3].

Acknowledgments MEEO Srl and ZAMG for the provision of AOT and meteorological fields maps, respectively. The research leading to these results has received funding from the EU FP7 under GA n. 283610 "European Scalable Earth Science Service Environment (EarthServer)".

References

1. Turner, J., & Allen, D. (2008). Transport of atmospheric fine particulate patter: Part 2. *Journal of the Air & Waste Management Association*, *58*(2), 196–215.
2. Liu, Y., Paciorek, C. J., & Koutrakis, P. (2009). Estimating regional spatial and temporal variability of $PM_{2.5}$ concentrations using satellite data, meteorology, and land use information. *Environmental health perspectives, vol.*, *117*(6), 886.

3. Gräler, B., Gerharz, L., & Pebesma, E. (2012). *Spatio-temporal analysis and interpolation of* PM_{10} *measurements in Europe*. Technical report, ETC ACM.
4. Campalani, P., Mantovani, S., Hirtl, M., Caglienzi, M., Mazzini, G. (2012). Daily spatial prediction of PM_{10} mass concentrations with geostatistics: An Austrian case study. In: *geoENV*.
5. Fassò, A., Finazzi, F., D'Ariano, C. (2009). Integrating satellite and ground level data for air quality monitoring and dynamical mapping. http://hdl.handle.net/10446/954
6. Hoff, R., & Christopher, S. (2009). Remote sensing of particulate pollution from space: Have we reached the promised land. *J. Air & Waste Management Association*, *59*, 645–675.
7. Ezzati, M., Lopez, A., Rodgers, A., Murray, C. (2004). *Comparative quantification of health risks: Global and regional burden of diseases attributable to selected major risk factors*. Geneva: WHO.
8. Campalani, P., Nguyen, T. N. T., Mantovani, S., Bottoni, M., Mazzini, G. (2011). Validation of PM MAPPER aerosol optical thickness retrievals at 1×1 km^2 of spatial resolution. In: *19th International Conference on Software, Telecommunications and, Computer Networks*.
9. Baumann, P., Campalani, P., Misev, D., & Yu, J. (2012). Finding my CRS: A systematic way of identifying CRSs. In: *ACM SIGSPATIAL*.

Non Linear PS Time Series: Analysis and Post-Processing for Landslides Studies

Davide Notti, Claudia Meisina, Francesco Zucca and Alessio Colombo

1 Introduction

The development of SAR interferometry and in particular of the Persistent Scatterer Interferometry techniques (PSI) allowed mapping and monitoring extremely slow and very slow landslides [1] from regional to local scale [2–4]. Up to recent time PSI data was mainly exploited for spatial analysis of the movements. The temporal analysis of the movement (i.e. time series analysis—TS) was developed more recently [5, 6] when the non-linear processing algorithms were performed.

In this work we present a methodology to analyze the PS time series for the study of landslides. We apply the proposed procedure on a study area located in NW Italy (about 20000 km^2). It presents an heterogonous geological setting (Alps, Apennine, and Langhe Hills) and many landslides with a proper range of velocity (slow flows, deep seated gravitational slope deformations—DSGSD, complex landslides). Radarsat data were processed by T.R.E. s.r.l. with SqueeSAR algorithm [6] and cover the period 2003–2009. This allows us to have a wide number of time series of landslides to analyze.

D. Notti · C. Meisina (✉) · F. Zucca (✉)
Earth and Environmental science department, Pavia University, Pavia, Italy
e-mail: davidenotti@gmail.com

C. Meisina
e-mail: claudia.meisina@unipv.it

F. Zucca
e-mail: francesco.zucca@unipv.it

A. Colombo
ARPA Piemonte, Turin, Italy

E. Pardo-Igúzquiza et al. (eds.), *Mathematics of Planet Earth*,
Lecture Notes in Earth System Sciences, DOI: 10.1007/978-3-642-32408-6_56,

2 Post-Processing Tools

Detection of false trends. The time series may be affected by trends or anomalous measure not related to real ground movement of landslide but to other factors. It is quite easy detect these trends because usually they affecting the whole dataset. For instance, a problem may be related to a proper trend of the chosen reference point that can affect reversely all the dataset in this case it is better to change the reference point. Another problem may be related to an anomalous and spatially diffuse value of TS caused, for instance, by atmospheric events that can affect SAR image (Fig. 1a). In order to detect this kind of problems it is suggest to analyze very stable and high coherence TS on some stable areas, so if more than 1/3 of TS show a high dispersion value (> +/− 5 mm than standard TS) at the same date then it is an indication of probable noise of SAR image. It is recommended to remove the anomalous date from TS to avoid misunderstanding.

Detect and correct potential phase unwrapping mistakes. We propose a methodology to detect and correct potential phase unwrapping mistakes in the time series by only analyzing consequential displacement (Eq. 1). If the displacement (D) between two consecutive acquisitions (t_x, t_{x+1}) is major or equal than $+\lambda/4$ (the phase ambiguity) (e.g. 14 mm for C-Band) we correct the series by introducing a shift of $-\lambda/2$ (e.g. −28 mm for C-Band).

$$\text{IF} \quad D(t_{x+1}) - D(t_x) \geq \lambda/4\,\text{mm} \rightarrow D(t_{x+1}) - \lambda/2\,\text{mm} \tag{1}$$

This tool presents some limitations. In order to validate the unwrapping it is necessary to have other monitoring data available or a clear evidence from further sources (e.g. monitoring, rainfall, literature,…) that the landslides had relevant accelerations in the same period of potential phase unwrapping mistake (Fig. 1b).

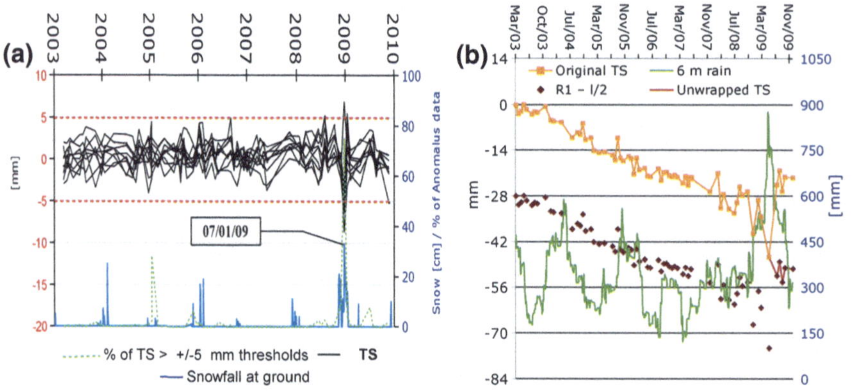

Fig. 1 **a** More than 50 % of time series of stable PS of this dataset show a variance > |5 mm| on 07/01/2009 related to snowfall in the same day of acquisition. **b** Planar slide in Langhe Hills: a time series potentially affected by phase unwrapping in April 2009

Averaging the time series for the areas with homogenous movement. In order to reduce the noise of TS and to better detect the general trends of deformations, we suggested averaging the time series. This can be made for the TS of PS over a landslide, with a same range of velocity and in located the same geomorphological setting.

Detect different trends on time series. The last step is to identify different trends in time series. A simple method to analyze the TS is to derive the time series of velocity. The velocity is the gradient of time series (dx/dt) calculated over an interval of at least 6 months to reduce variability noise effects. The method of deviation index (DI) [5] indicates how a trend after an event at time x, is different from previous trend, the index include variability corrections. The variability (or noise) represents the dispersion of the time series. The grater the DI, the more different are the trends. Other works [6] propose a classification of TS using models. In Fig. 2 b it is calculated the DI and the time series of velocity for a landslide in the Langhe Hills. The velocity increased up to 15 mm/yr from 3/5 mm/yr the Deviation Index is 5.

3 Landslides Time Series Interpretation

TS Validation. A further step is the comparison between the time series of PS data and time series of other instruments. This comparison it is important to make a cross validation to detect errors of both the monitoring systems. One of the advantages of PS time series is a sample frequency (e.g. 24 day for Radarsat) that is higher than other non-continuous monitoring systems like GPS or inclinometers (usually no more then 2–3 measures per year (Fig. 2). Another advantage is the possibility to compare time series that cover the same time intervals of for many landslides and so

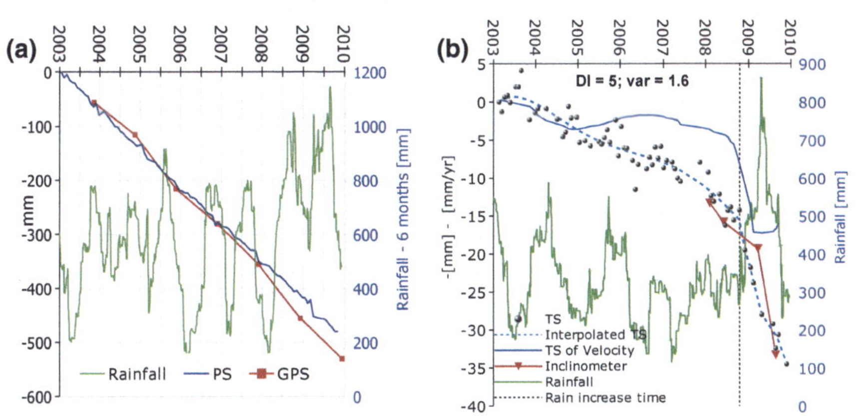

Fig. 2 Time series compared with 6 months cumulated rainfall and other monitoring instruments: **a** Deep seated landslide of Brenvetto (Turin), NW Alps; **b** A planar slide of Paroldo (Cuneo), Langhe Hills (**DI** is the Deviation Index; **var** is the variability)

it is easier to detect regional events and compare the different behaviour of landslides in relation with triggering events. The main con of time series of SAR data is related to the impossibility of increasing the number of measures during critical events of landslide accelerations. The other disadvantages are related to the small range of velocity detectable (up to 10 cm/yr) and to the already known PSI limitations [3].

First results in study areas. In our areas of study we observed good results in the description of landslides movements. It was possible to identify different behaviours of landslides. Overall we observed that DSGSD of the Alps have linear deformations with smooth changes related to rainfall or snow melting (Fig. 2a), wile the shallow landslides (especially in the Apennines and Langhe Hills) show time series with more abrupt change of trends (Fig. 2b). For instance, many landslides of Langhe hills are characterised by accelerations, confirmed by other monitoring systems in the winter and spring 2008–2009, in correspondence of a very rainy period.

Conclusions

The work provided to give some tools to analyze time series of PSI data for slow landslide analysis. The post-processing tools try to solve some time series problems in particular the detection of unwrapping errors and the identification of different trends. The time series analysis of case histories landslides show that the added value of time series is the good spatial and temporal sampling of PSI data.

References

1. Cruden, D. M., & Varnes, D. J. (1996). Landslide types and processes. In: A.K. Turner & R.L. Schuster (Eds.), *Landslides: 125 investigation and mitigation, Sp. Rep. 247, transportation research board, national research council* (pp 36–75). Washington DC: National 126 Academy Press.
2. Meisina, C., Zucca, F., Notti, D., Colombo, A., Cucchi, A., Savio, G., et al. (2008). Geological interpretation of PSInSAR data at regional scale. *Sensors*, *8*, 7469–7492.
3. Colesanti, C., & Wasowsky, J. (2006). Investigating landslides with space-borne Synthetic Aperture Radar (SAR) interferometry. Engineering Geology, *88*, 173–199
4. Herrera, G., Notti, D., García-Davalillo, J. C., Mora, O., Cooksley, G., Sánchez, M., et al. (2010). Analysis with C- and X-band satellite SAR data of the Portalet landslide area. *Landslides*. doi:10.1007/s10346-010-0239-3.
5. Cigna, F., Tapete, D., & Casagli, N. (2012). Semi-automated extraction of Deviation Indexes (DI) from satellite persistent scatterers time series: tests on sedimentary volcanism and tectonically-induced motions. *Nonlinear Processes in Geophysics*, *19*, 643–655.
6. Milone, G., & Scepi, G. (2011). A clustering approach for studying ground deformation trends in Campania region through the use of PS-InSAR time series analysis. *Journal of Applied Sciences Research*, *11*, 610–620.

Observation of the Mining-Induced Surface Deformations Using C and L SAR Bands: The Upper Silesian Coal Basin (Poland) Case Study

Marek Graniczny, Zbigniew Kowalski, Anna Piątkowska and Maria Przyłucka

1 Introduction

The Upper Silesian Coal Basin is located in the southern Poland and in the region of Ostrava-Karvina in the Czech Republic (Fig. 1). It covers an area of 7, 250 km^2. This is the major hard coal basin in Poland, and also one of the largest in Europe. Mining activities of Carboniferous hard coal deposits have been conducted over 200 years. The Upper Silesia is a big metropolitan area occupied by 37 towns with almost 3 million inhabitants.

The Upper Carboniferous coal-bearing formations are the most significant in the geological structure of the Upper Silesian Coal Basin. There are Precambrian, Cambrian, Devonian and Carboniferous formations in their basement. The Carboniferous overburden includes Quaternary, Miocene and Triassic rocks, and in the southernmost part also, rocks of the Carpathian overthrust; Permian and Jurassic deposits. The coal-bearing formations of the USCB include several lithostratigraphic series, reaching 8,500 m of depth. These series are featured by a gradual reduction of their thickness toward the east and southeast [1].

M. Graniczny (✉) · Z. Kowalski (✉) · A. Piątkowska (✉) · M. Przyłucka (✉)
Polish Geological Institute—National Research Institute, Rakowiecka 4, 00-975 Warsaw, Poland
e-mail: marek.graniczny@pgi.gov.pl

Z. Kowalski
e-mail: zbigniew.kowalski@pgi.gov.pl

A. Piątkowska
e-mail: anna.piatkowska@pgi.gov.pl

M. Przyłucka
e-mail: maria.przylucka@pgi.gov.pl

E. Pardo-Igúzquiza et al. (eds.), *Mathematics of Planet Earth*,
Lecture Notes in Earth System Sciences, DOI: 10.1007/978-3-642-32408-6_57,

1.1 Hazards

The main hazard is related to mining subsidence due to underground coal mining. It is estimated that in Upper Silesian Coal Basin about 600 km^2 already suffers from subsidence. The surface is characterized by numerous collapse cavities or basins with depths over a dozen of meters. They may remain dry or may be filled with water (inundation) depending on the local hydrogeological conditions. Subsidence areas can be inundated during floods [2]. However, mining subsidence also causes severe damage to linear structures like roads, railways, gas and water pipelines, electric power lines and sewage systems. Around Sosnowiec area houses strengthened with iron bars anchored in the walls are found to prevent further damage or collapse. However, even such reinforced buildings show cracks in the walls. Mining exploration also causes several risks and hazards related to surface and groundwater, which are mainly decrease of surface and groundwater quality due to anthropopression and risks resulting from induced seismic shocks [3].

There are different methods for evaluating hazards caused by deep hard coal mining. These methods include remote sensing techniques such as: very high resolution satellite imagery, aerial photos, thermal imagery, laser scanning, radar interferometry etc. and in situ conventional techniques such as geodetic surveying and GPS measurements. Additionally, it is necessary to collect systematic information on the extent of the mining works (including total thickness of exploited coal beds) as well as reported dames on buildings and infrastructures.

2 Interferometric Data Characteristic

In the presented research two types of processing were used: the traditional differential interferometry (DINSAR) and Permanent Scatterer Interferometry (PSI). Each of these methods has their advantages and limitations. Both allow covering a significant area of research and thus can be a better alternative to traditional geodetic measurement. Differential interferometry allows the registration of displacement with a size about decimetres occurred in the time between recordings of two scenes (usually several days). Changes on the surface are visualized as interference fringes. PSI method on the other hand allows to obtain a cloud of points, named permanent scatterers (PS). Each point contains information about the velocity of movement (mm/year) and is recorded on the basis of a set of scenes from the period of a few years. A more detailed description of the two methods can be found in the literature [4–6]. Both methods have been successfully used to monitor ground displacements associated with mining phenomena [7, 10], but also with earthquakes, volcanoes, landslides.

In this study three datasets were used: two PSI datasets and set of 5 differential interferograms. PSI dataset were processed by Tele-Rilevamento Europa (TRE) within DORIS project—the European Community's Seventh Framework Programme (FP7/2007–2013) Grant Agreement no. 242212 in 2011. The datasets consist of

70 ERS-1 and ERS-2 scenes registered between 17.05.1992 and 20.12.2001 (71709 PS points) and 31 ENVISAT scenes registered between 05.03.2003 and 29.09.2010 (32341 PS points). PSI datasets are characterized by average annual motion rate (VEL) in satellite Line of Sight (LOS) from $-40, 5$ to $+7, 41$ mm/year. Both datasets reflect the movement of ground caused by coal mining within the study area. PS points with negative VEL values are located on the edge of the area prone to vertical ground subsidence caused by hard coal mining. The areas of a great risk are not covered by PS, mainly due to very high values of subsidence exceeding the PSI technique detection threshold. Differential interferograms were received within Terrafirma project (ESRIN/Contract no. 17059/03/I-IW). The set contains five differential interferograms from ALOS-Palsar satellite (with time span: 22.02.2007–10.07.2007; 10.07.2007–25.08.2007; 25.08.2007–25.11.2007; 25.11.2007–25.02.2008 and 25.02.2008–27.05.2008) and were processed by Gamma Remote Sensing. The results show fringes exactly in the areas of the highest subsidence values caused by mining exploitation (where there are no PS points). Performing the combined analysis of PSI and DINSAR data the whole ground deformation patter of the mining subsidence area was detected.

3 SAR Data Interpretation

3.1 General Results Obtain in Upper Silesian Coal Basin

Processed SAR datasets were analyzed and compared in GIS and geostatistical software. In the first stage PSI datasets were compared with subsidence predictive models [8]. The model result consists of the map of areas at risk according to the category of mining areas. The map was made on the basis of recent and future mining activities foreseen until 2020. On the map, two main groups of areas and four categories have been distinguished (Fig. 1). First group (yellow color) includes category I (very small, harmless destruction of objects may occur) and category II (destruction of objects, easy to repair), with deformations that reach up to 3 m in 20 years' time. Second group (orange color) includes category III (big destruction of objects, protection is needed) and category IV (serious damage, objects are almost destroyed), with deformations varying from 3 to 12 m in 20 years' time. The comparison with both ERS and Envisat datasets indicates that the subsidence detected by PSI technique corresponds with subsidence predictive models. PS targets allow identifying the boundaries of the mining subsidence bowl areas, coinciding with those defined by Kowalski [8]. However it is observed a reduction of subsidence rate betweenERS and Envisat datasets,

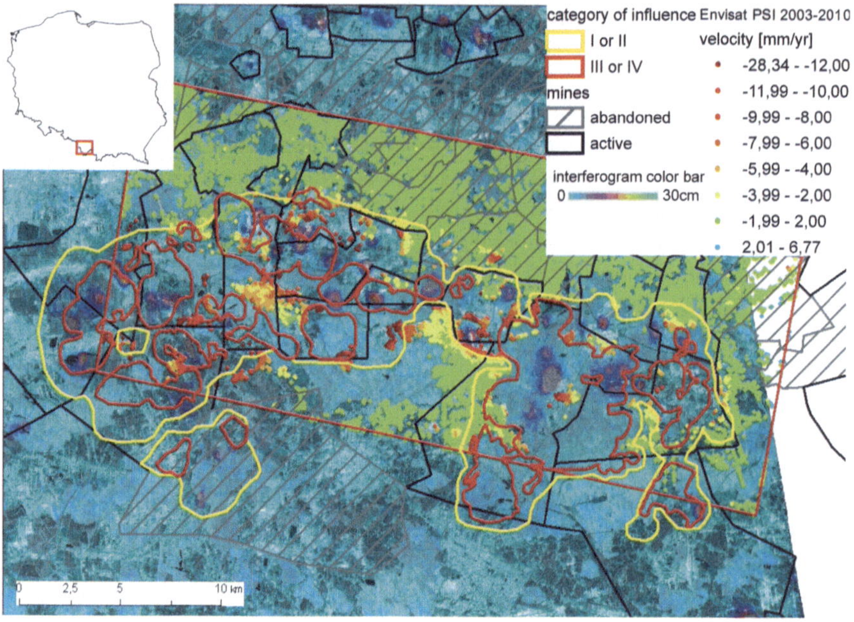

Fig. 1 Envisat PSI from period 05/03/2003–29/09/2010 on the background of ALOS-PALSAR interferogram 25/02/2008–27/05/2008

which could be related to a reduction in the USCB coal exploitation. Since 1994 coal mining exploitation has been reduced 50 % (in 1994 the production of hard coal was 134 million tons, whereas in 2010 the production is 77 million tons [9]). Additionally in areas of inactive mines a slight uplift is detected, which is probably connected with a rise of the groundwater level, previously pumped out during the exploitation time.

PSI data from ERS and ENVISAT satellites very well illustrate subsidence occurring in the active mining areas as long as the subsidence velocity does not exceed −40 mm/year. Unfortunately within the USBC mining area subsidence reaches up to 600 mm/year. This high subsidence rates explain why there are no PS points in the central part of the test area. On the other hand L-band differential interferograms allowed to detect deformation fringes that occurred between the acquisition of the used SAR images. Comparison of PSI data and differential interferograms reveal that they complement each other (Fig. 1). Fringes on the interferograms illustrate perfect changes on the surface happened too fast to be registered by PS points. Whereas PS points define boundaries of the area at risk and can be used to analyze subsidence with small rates, undetectable with the interferograms.

3.2 Interpretation of C and L-band SAR Data in Wujek Mine Case Study

"Wujek" coal mine is one of the companies that operate since the end of nineteenth century. In this area, five ALOS-PALSAR interferograms where compared with each other to analyze subsidence related with coal exploitation. The comparison is presented in Fig. 2.

Four areas of subsidence basins where defined and marked by numbers: 1—crossroad of Przodownikow and Szadoka Streets near Brynow railway station; 2—Tadeusz Kosciuszko Park in Brynow; 3—area near Katowice-Ligota railway station in Ligota; 4—crossroad of Brynowska and Tadeusza Kosciuszki Streets in Brynow. The analysis of the interferograms indicated that areas numbered 1 and 4 were strongly affected by subsidence between February and July 2007. Subsequently subsidence rate decreased and stopped in November 2011. Area number 2 was strongly affected by subsidence in July 2007, after that the subsidence decreased, but still was visible in May 2007. Area numbered 3 was not affected until November 2007 to May 2008.

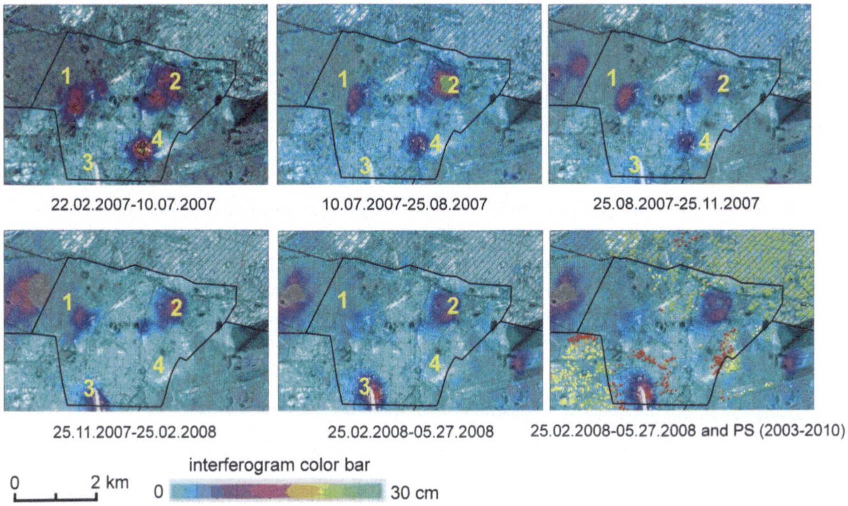

Fig. 2 Fringe on five ALOS-PALSAR interferograms illustrate the development of subsidence basins in the "Wujek" coal mine. The *sixth* picture shows the interferogram and Envisat PS points that complement each other

4 Conclusions

From the presented study few main conclusions can be made. In the area of active coal mining exploitation, differential interferograms processed from L-band data were successfully used to analyze and describe subsidence basins rate and movement. Singular C-band PSI datasets are insufficient to proper estimate subsidence values in areas with high movements rates. Interferograms complement this information. Data from the interferograms show subsidence in the range of decimeters per year, occurring in 46, 92 or 138 days periods. On the other hand PSI data provide PS velocity values measured in mm/year in 7–8 years periods. In the former deformation is derived from a raster where a significant error can be introduced in the interpretation, whereas in the latter the deformation measurements are provided point wise with a greater precision and accuracy. All these factors cause that the combination of PSI data and the interferograms may not be straight forward. C-band PSI datasets very well illustrate the trend of movements in the areas where these movements do not exceed −40 mm/year and can be used to define the boundaries of areas affected by subsidence. Whereas L-band interferograms detect subsidence with a rate decimeters per month.

Further study should be focused on comparison SAR data with geodetic levelling measurements and with maps of coal exploitation.

References

1. Volkmer, G. (2008). Coal deposits of Poland, including discussion about the degree of peat consolidation during lignite formation. In *Freiberg Geology, Oberseminar Geologie WS 2007/08*.
2. Graniczny, M., Czarnogorska, M., Kowalski, Z., Wegmuller, U. (2009). Using ALOS PALSAR, Envisat ASAR and ERS data for monitoring of subsidence and related ground failures in Upper Silesian Coal Basin, Southern Poland. In *EGU (European Geosciences Union) General Assembly 2009, Geophysical Research Abstracts*, 11, Vienna, Austria.
3. Czarnogorska, M., Graniczny, M., Kowalski, Z., & Wegmuller, U. (2008). Dynamika zmian powierzchni terenu na Gornym Slasku w okresie 10.07–25.08.2007 r.na podstawie danych interferometrycznych z satelity ALOS. *Przeglad Geologiczny*, *56*(7), 524–527.
4. Ferretti, A., Prati, C., & Rocca, F. (2001). Permanent scatterers in SAR interferometry. *IEEE Transactions on Geoscience and Remote Sensing*, *39*, 8–20.
5. Crosetto, M., Monserrat, O., Jungner, A., & Crippa, B. (2009). Persistent scatterer interferometry: potential and limits, ISPRS archives, vol. XXXVIII-1-4-7/W5. In *ISPRS workshop 2009 high-resolution earth imaging for geospatial information*, Hannover, Germany.
6. Carnec, C., Massonnet, D., Villan, J.P., King, C. (1994). Potential applications of differential SAR interferometry for monitoring the impact of underground mining. In *Proceedings of the 1st SAR interferometry workshop*, pp. 41–45, Tokyo, Japan.
7. Huanyin, Y., Guang, L., Perski, Z., Huadong, G. (2011). Satellite radar reveals land subsidence over coal mines. In *SPIE newsroom*, p. 3. doi: 101117/2.1201110.003898.
8. Kowalski, A. (2011). Wstępna informacja o działalności górniczej i aktywności deformacyjnej terenów górniczych i pogórniczych w obrębie Górnoślaskiego Zagłębia Węglowego. (unpublished, only in Polish).

9. Karbownik, A., & Turek, M. (2011). Zmiany w górnictwie węgla kamiennego: geneza, przebieg, efekty. *Przelglad górniczy*, *66(CVII)*7–8 (1064–1065), 11–18.
10. Herrera, G., Tomás, R., Lopez-Sanchez, J. M., Delgado, J., Mallorquí, J. J., Duque, S., et al. (2007). Advanced DInSAR analysis on mining areas: La Union case study (Murcia, SE Spain). *Engineering Geology*, *90*, 148–159.

Multi-Temporal Evaluation of Landslide-Induced Movements and Damage Assessment in San Fratello (Italy) by Means of C- and X-Band PSI Data

Silvia Bianchini, Deodato Tapete, Andrea Ciampalini, Federico Di Traglia, Chiara Del Ventisette, Sandro Moretti and Nicola Casagli

1 Introduction

Landslides can pose a serious threat to human lives and properties, especially in populated areas characterized by a significant cultural heritage, where the socio-economic losses and damages are stronger.

Detection of active ground movements on landslide-prone areas can greatly take advantage from advanced remote sensing techniques i.e. Persistent Scatterer Interferometry (PSI) [1]. Furthermore, the combined use of radar satellite data with traditional geomorphological tools like field surveys and *in-situ* observations can ease the mapping and monitoring of the impacts of such natural phenomena on buildings and manufactures of affected areas. C-band satellites provide long historical archives of motion rates and time series, covering wide areas at relatively low cost and medium spatial resolution. Whereas X band data with higher spatial/temporal resolution allow a more detailed investigation, even at the scale of a single building movement, in a shorter spanning time (e.g. some months) [2].

In this paper landslide effects in San Fratello area were investigated up to 2012 by exploiting an unprecedented availability of C- (ERS and RADARSAT) and X-band (Cosmo-SkyMed) data, combined with field survey. Estimation of ground motion rates and related impacts on cultural and social heritage led to a valuable and reliable damage assessment.

S. Bianchini (✉) · D. Tapete · A. Ciampalini · F. Di Traglia · Ch. Del Ventisette · S. Moretti · N. Casagli
Departement of Earth Sciences, University of Firenze, Via G. La Pira 4, 50121 Firenze, Italy
e-mail: silvia.bianchini@unifi.it

E. Pardo-Igúzquiza et al. (eds.), *Mathematics of Planet Earth*, Lecture Notes in Earth System Sciences, DOI: 10.1007/978-3-642-32408-6_58,

2 Study Area

San Fratello is located in the NE sector of Sicily Region (Italy) within Messina province on the Nebrodi Mountains. The rocks cropping out in the area consist of Cretaceous-Oligocene terrigenous-flyschoid to calcareous sedimentary sequences. Messina province is extensively affected by hydrogeological hazard. San Fratello is chronically affected by landslides, the most recent of which are the landslides recorded in 1922, on the NW sector of the village, and on 14 February 2010 on the SE slope, triggered by intense rainfall and causing huge damages and causalities. C-/X-band radar data, shown in Table 1, were used to successfully update the landslide inventory, enlarging the boundaries of most of the already mapped phenomena (Fig. 1). In particular, PSI results highlighted ground motions in historical (1992–2001), recent (2005–2010) and current (2011–2012) scenarios.

A projection of the measured satellite Line Of Sight velocities (V_{LOS}) along the most probable direction of real movement, i.e. the maximum slope direction, was carried out within each available PSI dataset obtaining V_{SLOPE}, in order to account more specifically for topographic and geomorphological slope conditions for local landslides analysis [3]. The directions of ground motion were subsequently confirmed by comparing *in-situ* observations undertaken in November 2012 and January 2013 with the available vectors of displacement along the steepest slope.

Table 1 PSI data stacks analyzed in this study

Parameters	ERS 1/2	RADARSAT 1	COSMO-SkyMed
Acquisition mode	Desc.	Asc. and Desc.	Desc.
Temporal span	1992–2001	2005–2010	2011–2012
PS density (PS/km^2)	2.25	112.73	400.62

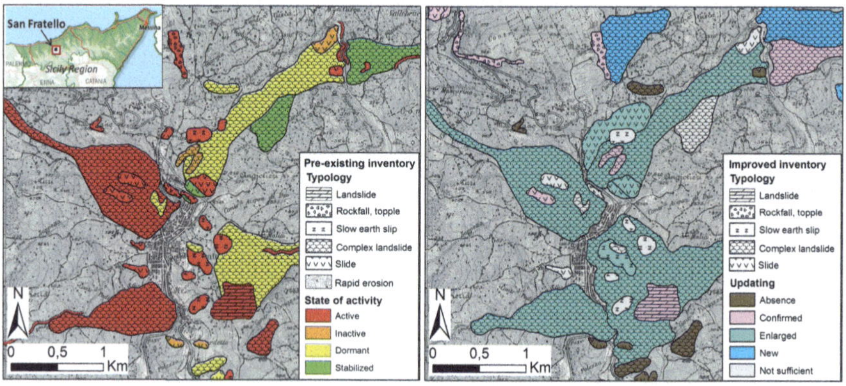

Fig. 1 Geographical location and landslide mapping of San Fratello area

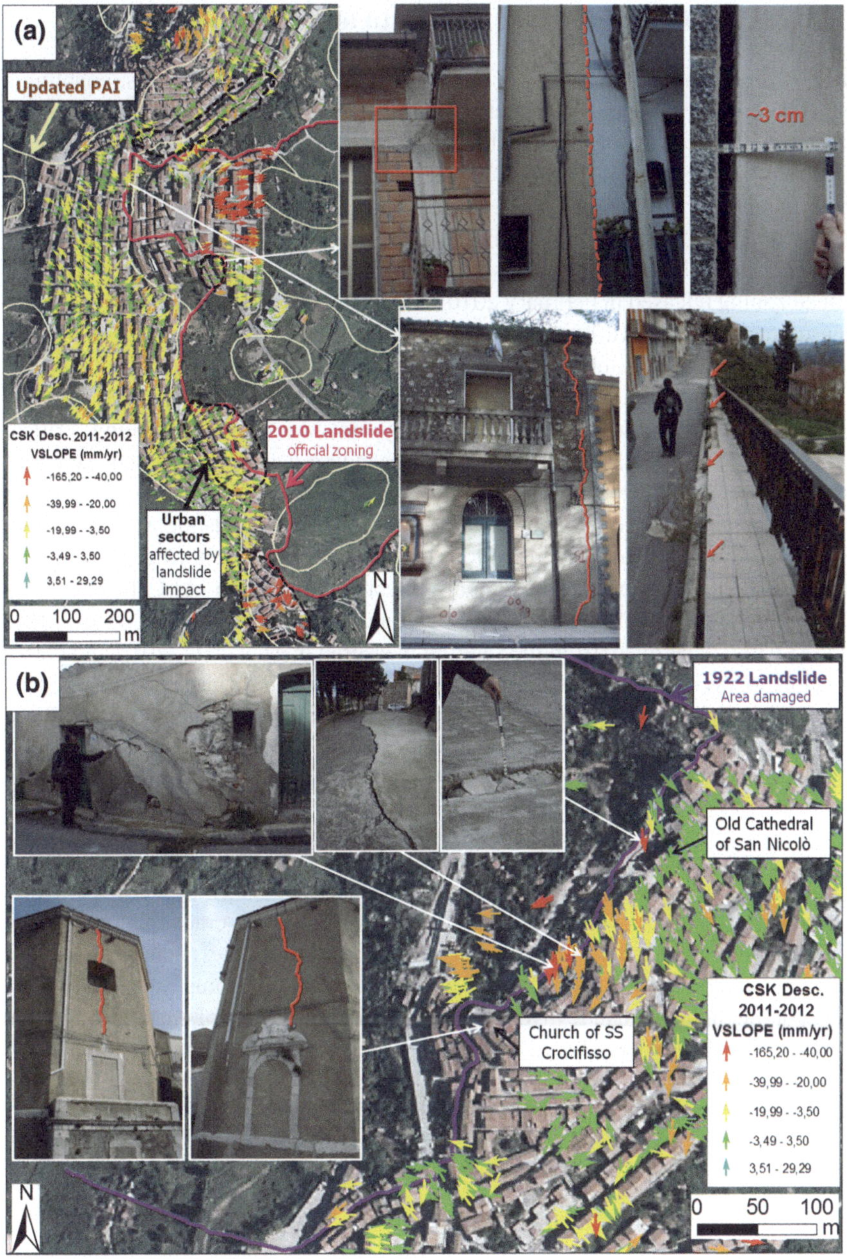

Fig. 2 Distribution of PS CSK V_{SLOPE} rates over 2010 (**a**) and 1922 (**b**) landslide areas

3 Cross-Validation Between PSI Data and Field Survey

Most of the PS targets show up on housetops and facades of buildings, enabling a highly detailed investigation of structural damages (Fig. 2a). The rapid radar mapping and subsequent zoning of the unstable urban sectors were carried out by following the object-oriented approach proposed by D. Tapete et al. [4] for applications on built environment. In particular we multi-layered and compared PSI data, local cadastre, the distribution of churches and sites of cultural interest, and the boundaries of the 1922 and 2010 landslides (Fig. 2). The latter were retrieved by combining the official zoning released in February 2010 and that drawn based on the interferometric measurements from a Ku-band Ground-Based SAR currently monitoring the south-eastern slope of San Fratello.

As the main outcome, the boundary of the 2010 landslide was updated, thereby including further urban quarters previously not considered as being critical (Fig. 2a). COSMO-SkyMed (2011–2012) descending data demonstrated their usefulness to precisely detect the single buildings affected by ground displacement. Four main areas located close to the official zoning of the 2010 landslide were definitely recognized as affected by ground motions in the last year of monitoring (black dotted lines in Fig. 2b). Unexpected movements detected close to the old Cathedral damaged by the landslide 1922 and the surrounding civil buildings (V_{LOS} up to -10.2 and -3.9 mm/year away from the satellite measured in May 2012–2013 and corresponding to V_{SLOPE} -50.8 and -14.5, respectively) found strong spatial correlation with the distribution and opening of the cracks along the pavement and wall surfaces (Fig. 2b). That definitely proved the current critical condition of this sector of San Fratello, thereby increasing the alert level for an area otherwise classified as stable.

4 Conclusions

The work provided a valuable cross-comparison between crack pattern survey focused on damage analysis of buildings, and PSI motion rates, achieving a complete and reliable investigation for an accurate impacts mapping and monitoring. The updating of the landslide inventory and the detailed zoning of unstable buildings supported local activities of civil protection and heritage conservation.

Acknowledgments This work was carried out within the DORIS project funded by the EC-GMES-FP7 initiative (Grant Agreement n°242212).

References

1. Ferretti, A., Prati, C., & Rocca, F. (2001). Permanent scatterers in SAR interferometry. *IEEE Transaction Geoscience Remote Sensors*, *39*(1), 8–20.
2. Bovenga, F., Wasowski, J., Nitti, D. O., Nutricato, R., & Chiaradia, M. T. (2012). Using COSMO-SkyMed X-band and ENVISAT C-band SAR interferometry for landslides analy-

sis. *Remote Sensing of Environment, 119*, 272–285.

3. Cigna F., Bianchini S., Casagli N (2012). How to assess landslide activity and intensity with Persistent Scatterer Interferometry (PSI): the PSI-based matrix approach. Landslides, 1–17.
4. Tapete, D., Fanti, R., Cecchi, R., Petrangeli, P., & Casagli, N. (2012). Satellite radar interferometry for monitoring and early-stage warning of structural instability in archaeological sites. *Journal of Geophysics and Engineering, 9*, S10–S25.

Integration of Earth Observation and Ground-Based HR Data in the Civil Protection Emergency Cycle: The Case of the DORIS Project

Paola Pagliara, Giuseppe Basile, Pierluigi Cara, Angelo Corazza, Andrea Duro, Bruno Manfrè, Roberta Onori, Chiara Proietti and Vincenzo Sansone

1 Introduction

The Italian national territory is exposed to a broader range of natural hazards, landslides included, that cause every year fatalities and a large economic damage. The vulnerability of its population and built environment is often severe and in some cases has been exacerbated by human activities. In this perspective, within the National Civil Protection Service the Department of Civil Protection (DPC) operates according to the following lines: prevention, forecast and assessment, early warning and alerting, emergency response and recovery from emergency.

The activities of DPC are daily supported by research efforts on the assessment of vulnerability and exposure to landslide and hydrogeological risk of population, buildings and critical infrastructures. A strong effort is also provided by European Earth Observation Programme Copernicus previously known as GMES (Global Monitoring for Environment and Security) [1]. In this framework the DORIS project (Ground Deformation Risk Scenario: an Advanced Assessment Service) is a preoperative advanced downstream service for the detection, mapping, monitoring and forecasting of ground deformations which integrates traditional and innovative

P. Pagliara (✉) · P. Cara · A. Corazza · A. Duro · R. Onori (✉) · Ch. Proietti (✉)
Italian Civil Protection Department, Roma, Italy
e-mail: paola.pagliara@protezionecivile.it

R. Onori
e-mail: roberta.onori@protezionecivile.it

Ch. Proietti
e-mail: chiara.proietti@protezionecivile.it

G. Basile · B. Manfrè
Regional Department of Civil Protection–Hydrogeological and Environmental Risks Service and Service for Messina Province, Messina, Italy

V. Sansone
Assessorato Regionale Territorio ed Ambiente–Dipartimento dell'Ambiente, Bologna, Italy

E. Pardo-Igúzquiza et al. (eds.), *Mathematics of Planet Earth*,
Lecture Notes in Earth System Sciences, DOI: 10.1007/978-3-642-32408-6_59,

Earth Observation (EO) and ground-based (non-EO) data and technologies to foster the ability of Environmental and Civil Protection authorities to manage the risks posed by ground deformations and improve the understanding of these complex phenomena [2]. The use of High Resolution (HR) data, both EO (optical and SAR) that non-EO to support Civil Protection Authorities in landslide risk management is well recognized [3–10]. In the DORIS project different sensor and data have been analyzed (COSMOSky-Med, TerraSAR-X, GB-InSAR, ERS and ENVISAT) and here is presented an operative integration and application of the different data and products in the whole emergency cycle.

2 The Test Site

The pre-operative service is tested in six study areas including two study areas in Italy, and one study area in each of the following European Countries: Poland, Spain, Hungary and Switzerland. The Messina Province, north-eastern part of Sicily, has been selected as test area for the project due to the events occurred in the period from October 2009 to March 2010, when the area was highly affected by more than one thousands of landslides that caused intense damages and casualties. The town of San Fratello in particular, on 14th of February 2010 a landslide of 20 millions of cubic meters occurred and caused severity damaged the eastern part of the town and 2,000 people were evacuated. Analysis and interventions for landslide risk mitigation have been performed since the first phase of the emergency and are still ongoing.

3 Data and Products

From the first phase of the emergency, the Civil Protection System has been activated in order to manage the landslide risk and many activities and analysis based on EO and non-EO data has been performed by the Civil Protection Departments, the Environmental Department, Regional Order of Geologists in Sicily and other local authorities and research centres.

The DORIS project took into account all the collected data and results to improve the products and their value added, not only in the emergency phase. In the DORIS project different monitoring techniques with HR data have been used, including remote sensing (optical and SAR) and in situ (GB-InSAR) data. In particular for SAR monitoring the images have been acquired from different satellites (ERS1/2, ENVISAT, Radarsat, TerraSAR-X, COSMOSky-Med). The availability of long time series of displacement obtained by combining ERS and ENVISAT data (1992–2010) [11] and different sensors (C and X band) give the possibility to exploit all the capability of the data referring to different types of landslide and to the different phases of the emergency. The analysis of the EO and non-EO data has been performed by the University of Florence-Earth Science Department and by the Research Institute of Geo-Hydrological Protection of the National Research Council (CNR-IRPI) as service providers of DORIS.

The DORIS products are the followings: Ground-deformation PSInSAR map and Ground-deformation PSInSAR time series, Ground-deformation velocity maps, event map, susceptibility map, landslide inventory map. The integration of in situ data (e.g. inclinometers, boreholes, geomorphological map, field surveys) provided by the Civil Protection Departments, the Environmental Department, Regional Order of Geologists in Sicily and other local authorities and research centres, allowed to better exploit the data and to improve the quality of the products.

4 Integration in the Civil Protection Emergency Cycle

The activities of landslide risk management are performed both in real time that in deferred time. The integration of the analysis performed in the first phase of the emergency and the activities performed in the DORIS project allowed to realize and

Table 1 Applications of DORIS products in the Civil Protection emergency cycle

Phase	Product	Action
Phase 1 Prevention and Preparedness	• Landslide Inventory Map; • Ground-deformation PSInSAR map • Ground-deformation PSInSAR time series; • Ground-deformation velocity maps; • Susceptibility Map	All products could be an input for the Hydrogeological Management Plan and for the assessment of landslide scenarios, which is part of the preparedness of civil protection emergency plans
Phase 2 Emergency	• Ground-deformation PSInSAR map; • Ground-deformation PSInSAR time series; • Ground-deformation velocity maps; • Event maps	Velocity maps for residual risk evaluation, monitoring and early warning; to define the preliminary stabilization measures based on the characterization of the landslides, event map to evaluate the magnitude of the event
Phase 3 Recovery	• Ground-deformation PSInSAR map; • Ground-deformation PSInSAR time series; • Ground-deformation velocity maps; • Event maps; • Landslide inventory maps; • Susceptibility Map	GB-InSAR and Satellite InSAR data for the evaluation of the stabilization measures and interventions. All products to update the pre-existing landslide inventory maps and to assess the geomorphological hazard for new settlements

test the pre-operative services and products that can be a valuable tool in different phases of the emergency cycle (Table 1).

5 Conclusions

The new and emerging technologies (EO, non-EO, aerial) integrated with the conventional analysis and methods, can greatly support the landslide risk management activities through all the phases of the emergency cycle, both in real that in deferred time. In addition, the validation of the products and the pre-operative application of the results have been made easier by the close collaboration between national, regional and local authorities in charge of landslide risk management.

DORIS is a pre-operative downstream service for landslide risk management and the products are still in the research phase and many test cases in the whole emergency cycle are mandatory, in order also to define an operative procedure taking into account the sustainability aspects.

References

1. Copernicus. http://copernicus.eu
2. DORIS Description of work document. GA n.o.: 242212
3. DORIS. http://www.doris-project.eu
4. Casagli, N., Farina, P., Leva, D., Nico, G., & Tarchi, D. (2003). Ground-based SAR interferometry as a tool for landslide monitoring during emergencies. In *Geoscience and Remote Sensing Symposium, Proceedings IGARSS '03. IEEE International*, (Vol. 4, 2924–2926).
5. Farina, P., Casagli, N., Ferretti, A. (2008). *Radar-interpretation of InSAR measurements for landslide investigations in civil protection practices*. First North American Landslide Conference, June 3–8, 2007 (pp. 272–283) Vail.
6. Cascini, L., Fornaro, G., & Peduto, D. (2010). Advanced low-and full-resolution DInSAR map generation for slow-moving landslide analysis at different scales. *Engineering Geology*, *112*, 29–42.
7. Cigna, F., Del Ventisette, C., Liguori, V., & Casagli, N. (2011). Advanced radar-interpretation of InSAR time series for mapping and characterization of geological processes. *Natural Hazards and Earth System Sciences*, *11*(3), 865–881. doi:10.5194/nhess-11-865-2011.
8. Mondini, A. C., Guzzetti, F., Reichenbach, P., Rossi, M., Cardinali, M., & Ardizzone, F. (2011). Semi-automatic recognition and mapping of rainfall induced shallow landslides using satellite optical images. *Remote Sensing of Environment*, *115*, 1743–1757. doi:10.1016/j.rse.2011.03.006.
9. Bianchini, S., Cigna, F., Righini, G., Proietti, C., & Casagli, N. (2012). Landslide hotspot mapping by means of persistent scatterer interferometry. *Environmental Earth Sciences*, *67*(4), 1155–1172. doi:10.1007/s12665-012-1559-5.
10. Guzzetti, F., Mondini, A. C., Cardinali, M., Fiorucci, F., Santangelo, M., & Chang, K. T. (2012). Landslide inventory maps: new tools for and old problem. *Earth Science Reviews*, *112*, 42–66. doi:10.1016/j.earscirev.2012.02.001.
11. Bonano, M., Calò, F., Manunta, M., Paglia, L., Zeni, G. (2011). *Long-term Analysis of Landslides via SBAS-DInSAR. In Proceedings of the Second World Landslide Forum*. Rome, October 3–7, 2011.

Geostatistical Analysis of PSI Radar Data: A Methodology to Assess Serviceable Limit State of Buildings

Margarita P. Sanabria, Carolina Guardiola-Albert, Roberto S. Tomás, Geraint Cooksley and Gerardo Herrera

1 Introduction

Subsidence caused by water withdrawal is a well-known phenomenon which affects worldwide areas. Structures built on these areas can suffer differential settlements and distortions since the subsidence is not always a homogeneous phenomenon. Monitoring subsidence is necessary in order to take remedial actions.

In the last decade Differential SAR Interferometry (DInSAR) has become an alternative to classical surveying techniques for measuring infrastructure displacements. DInSAR has shown a good capability for measuring small displacements of structures with subcentimetric accuracy [1–3]. Using geoestatistical tools, it is possible to create spatially exhaustive layers of displacements and the assessment of the corresponding accuracy and precision. With these derived subsidence maps, buildings serviceability limit state can be calculated. Serviceability limit states are those conditions which are not strength based but can make the structure unsuitable for its projected use. In foundations design, the most common serviceability limit states are differential settlements and angular distortions, which must be less than or equal to the corresponding limiting value stated for them.

M. P. Sanabria (✉)
Departamento de Infraestructura Geocientífica y Servicios, Instituto Geológico y Minero de España, Madrid, Spain
e-mail: m.sanabria@igme.es

C. Guardiola-Albert · G. Herrera
Departamento de Investigación y Prospectiva Geocientífica, Instituto Geológico y Minero de España, Madrid, Spain

R. S. Tomás
Departamento de Ingeniería Civil, Universidad de Alicante, Alicante, Spain

G. Cooksley
Altamira Information, Barcelona, Spain

E. Pardo-Igúzquiza et al. (eds.), *Mathematics of Planet Earth*, Lecture Notes in Earth System Sciences, DOI: 10.1007/978-3-642-32408-6_60,

The city of Orihuela is located in the Vega Baja of the Segura River (province of Alicante, SE Spain). According to [4] the geotechnical substratum is constituted by Permo-Triassic rocks (carbonates, dolomites and quartzites) and Tertiary sediments (marls, sandstones and conglomerates) that outcrop in the north and south of the city respectively. The late Pleistocene-Holocene sediments (sand, silt and clay), that fills the basin, are the most compressible sediments. In the past decades, a recent subsidence phenomenon is affecting the city of Orihuela. Subsidence is related to aquifer overexploitation during drought periods: 1993–1996 and 2006–2008 [5]. In the local press, subsidence related damages were reported in the western part city in 1995. Several heritage buildings were repaired since 90s.

Using the serviceability limit state, the methodology presented here aims to set buildings susceptible to suffer damage. Thanks to geostatistics, more specifically Conditional Sequential Gaussian Simulation (CSGS) [6], the uncertainty of estimated data can be associated. This information can be very useful for land management and also as a complementary tool for forensic analysis of structures.

2 Methodology

In the proposed methodology, the Stable Point Network (SPN) InSAR algorithm has been used to process SAR images (ERS-1/2 and Envisat ASAR) covering two different periods from July 1995 to December 2005 and from January 2004 to December 2008. The input data for the analysis were the persistent scatterers points (PSs) with the derived cumulative displacement along the satellite Line of Sight (LOS), a simplified geology layer, a compressible sediment thickness model and a polygon layer with 27 historical buildings of Orihuela City.

The first step is devoted to the statistical analysis of PSs and their spatial relationship with two of the conditioning factors of subsidence [5]: the geology and the compressible sediments thickness. Regarding the geology, it has been determined that it is a factor to be consider when performing an interpolation of PSs. Therefore, and considering the second step (interpolation of the PSs), the data has been divided into two populations based on geology. The first population include those PSs placed within the geotechnical substratum, and the second one those PSs placed on the compressible sediment areas. Regarding the compressible sediment thickness, it has been determined that the two variables (i.e. subsidence and soft soil thickness) are not directly correlated. Therefore, for the second step the thickness of compressible sediments has not been taken into account.

The second step is focused on the interpolation of the normalized cumulative displacement along the LOS of the four PSs data populations previously defined. The interpolation method was the CSGS [6] and 100 equally probable realizations were generated. The resulting products were percentiles maps (mean, 68th, 95th) and variance maps, with a resolutions of 20 m, that estimate the cumulative displacement in the city of Orihuela and characterizes the uncertainty of the estimation. The different

Table 1 Adopted β_{max} and δ_s criteria for the performed analysis

Expected damage level	Angular distortion (β_{max})	Maximum differential settlement (δ_s)
Low	$<1/3,000$	<25 mm
Medium	1/3,000–1/2,000	
High	$>1/2,000$	>25 mm

percentile maps were analysed. For the period 1995–2005 the mean was selected as the best estimator, while for the period 2004–2008 the 68th percentile was chosen.

The third step seeks to discriminate those buildings which can suffer damages due to subsidence, according to the serviceability limit state criterion. Maximum differential settlements (δ_s) and maximum angular distortions (β_{max}) have been calculated for each of the 27 historical buildings. Limiting criteria that classified the expected damage level are showed in Table 1.

3 Results and Discussions

Analysing the different directional variograms, anisotropy has been detected, matching the axis of the basin. This result, along with the analysis carried out in the first step, confirms that geology is a conditioning factor for subsidence and therefore it must be taken into account when performing the interpolation.

Furthermore, it appears that subsidence behaves as a different process in each of the analysed periods. Descriptive measurements show a larger subsidence during 2004–2008 period. Analysing the percentiles and variance maps, 2004–2008 period has less uncertainty in the estimations as well as more significant and continuous subsidence than during the period 1995–2005. This different behavior is related to the groundwater withdrawal during drought periods, like in 2004–2008, that caused a piezometric level drop, and hence a larger subsidence.

Applying serviceability limit states to the 27 historical buildings, it stands out that in most of them the expected damage level is medium or high. Considering maximum distortion (β_{max}) four hot spots were identified. The validation of this methodology has been carried out in one of the hot spots, the Church of Santa Justa and Rufina. In the last 10 years, there have been damages (cracks and ground settlement) in the NE and S of the church. The location of the two β_{max} calculated match with these damages. Moreover, the maximum differential settlement (δ_s) direction for both periods is NE-SW. This indicates that the church has tilted towards the SW. In this case, the data obtained with this methodology are consistent with the damages observed in the church.

References

1. Bru, G., Herrera, G., Tomás, R., Duro, J., De la Vega, R., Mulas, J. (2010). Control of deformation of buildings affected by subsidence using persistent scatterer interferometry. *Structure and Infrastructure Engineering*, *9*, 1–13.
2. Cascini, L., Ferlisi, S., Peduto, D., Fornaro, G., & Manunta, M. (2007). Analysis of a subsidence phenomenon via DInSAR data and geotechnical criteria. *Italian Geotechnical Journal*, *41*(4), 50–67.
3. Tomás, R., García-Barba, J., Cano, M., Sanabria, M. P., Ivorra, S., Duro, J., et al. (2012). Subsidence damage assessment of a gothic church using Differential Interferometry and field data. *Structural Health Monitoring*, *11*(6), 751–762.
4. Delgado, J., Alfaro, P., Andréu, J. M., et al. (2003). Engineering-geological model of the Segura River flood plain (SE Spain): A case study for engineering planning. *Engineering Geology*, *68*, 171–187.
5. Tomas, R., Herrera, G., Lopez-Sanchez, J. M., et al. (2010). Study of the land subsidence in the Orihuela city (SE Spain) using PSI data: Distribution, evolution and correlation with conditioning and triggering factors. *Engineering Geology*, *115*, 105–121.
6. Gómez-Hernández, J. J. & Journel, A. G. (1993). Joint sequential simulation of multiGaussian fields. In A. Soa-re (Ed.), *Geostatistics Tróia '92* (Vol. 1, PP. 85–94). Boston: Kluwer Academic Publishers.

Using Modern Sensor Data and Advanced Numerical Modelling for Slow Landslides Motion Forecasting

José A. Fernández-Merodo, Juan C. García-Davalillo and Gerardo Herrera

1 Introduction

Slow-moving landslides are a wide-spread type of active mass movement that cause severe damages to infrastructures and may be a precursor of sudden catastrophic slope failures. In this context, modelling slow-moving landslide behaviour is an important task in order to quantify and reduce the risk associated to this geological process. Two broad categories of models can be distinguished to predict landslide mobility: the phenomenological models and the physically based models. The first category employs empirical relationships, statistical or probabilistic approaches and artificial neural networks to relate soil movements and their causes. The latter provides relationships taking into account the mechanical soil behaviour.

Physically based models have been mostly used in practical cases to estimate landslide occurrence and stability conditions for a given scenario through a stability factor based on limit equilibrium analysis. Apart from earthquake studies, time dependent analysis is requested when: (i) hydrological conditions change as in the case of rainfall; (ii) resistant parameters are reduced as in the case of strain softening or weathering processes and (iii) creep behaviour is taken into account.

J. A. Fernández-Merodo (✉) · G. Herrera
Geological Hazards Group, Spanish Geological Survey, c/ Ríos Rosas, 23, 28003 Madrid, Spain
e-mail: jose.fernandez@igme.es

G. Herrera
e-mail: g.herrera@igme.es

J. A. Fernández-Merodo
M2i Mathematical Modelling in Engineering Group, ETSI de Caminos, Canales y Puertos, UPM. c/ Profesor Aranguren s/n, 28040 Madrid, Spain

J. C. García-Davalillo
Geological Hazards Group, Spanish Geological Survey, c/ Cirilo Amorós 42, 46004Valencia, Spain
e-mail: jc.garcia@igme.es

E. Pardo-Igúzquiza et al. (eds.), *Mathematics of Planet Earth*,
Lecture Notes in Earth System Sciences, DOI: 10.1007/978-3-642-32408-6_61,

Three numerical models have been recently proposed by the authors to reproduce the kinematics of the Portalet landslide (Central Spanish Pyrenees): (i) a simple 1D infinite slope visco-plastic model [1], (ii) a 2D elasto-plastic finite element model [2] and (iii) a 2D visco-plastic finite element model [3]. These models, ordered by increasing level of complexity, are compared by applying them to the Portalet case study.

2 The Portalet Landslide

The study area is located in the upper part of the Gállego River valley in the Central Spanish Pyrenees (Sallent de Gállego, Huesca) close to the Formigal ski resort. This is a structurally complex area, outcrops of Paleozoic material of Gavarnie mantle were affected by the Hercynian folding phases and the alpine tectonics. Pyrenean deglaciation and widespread structural relaxation shaped the landscape triggering complex landslides. We focus our work on the "Portalet landslide". It is a rotational slide earth flows, 30–50 m thick. The mobilized materials involve sands and gravels found within a clayey matrix with sandstone levels, greywackes and shales.

In summer 2004, the excavation of the foot of the slope carried out to build a parking reactivated the existing slide surfaces generating a new small earth slide called the "Parking landslide" 380 m long and 290 m wide. The occurrence of this new local landslide prevented the digging to be finished and affected the connection road to France. Constructive solutions were carried out to stabilize the hillside involving re-profiling of the landslide toe, building of small retaining walls and drainage systems. However, field observations indicate that the landslide is still moving.

An exhaustive and continuous monitoring campaign has been done since 2004, it includes inclinometers, piezometers, differential GPS, ground-based SAR measurements [1]. Advanced DInSAR processing of satellite SAR images from ERS and Envisat satellites (2001–2007) and TerraSAR-X satellites (2008) were also performed [4]. The monitoring reveals that the hillside moves following two patterns. The first one corresponds to a slow continuous motion of constant speed of about 100 mm/year, the second one corresponds to accelerations of the moving mass when water table rises during rain events. In the last eight years the cumulative surface displacement exceeds two meters.

3 Modelling Slow Landslides

Three numerical models have been recently proposed by the authors to reproduce the kinematics of the Portalet landslide: (i) a simple 1D infinite slope visco-plastic model, (ii) a 2D elasto-plastic finite element model and (iii) a 2D visco-plastic finite element model. Details of the proposed models can be found in [1–3].

The 1D infinite slope visco-plastic model [1] is a simple model, it gives a first approximation of the landslide kinematics. The model incorporates a simplified hydraulic model that approximate the water table evolution from the recorded daily rainfall intensity. The main advantage is that it takes very few parameters for the 1D model to be defined. In this way, after calibrating the unknown model parameters by back analysis in a fixed period of time, the simple model can be used to predict the landslide mobility in another period of time. There are two substantial shortcomings. The first one is that the model does not give spatial information, that means that dimensions of the landslide cannot be predicted. The second is that small variations in the input parameters produce large variations in the calculated solution, it is the case for instance for the initial water table depth or the viscosity parameter.

The 2D elasto-plastic finite element model [2] improves the 1D model since the geometry is better defined and the hydro-mechanical coupling is solved through Biot's equations. Nevertheless, the model is not able to reproduce the constant slow motion during dry periods. Moreover, position of the water table depth is still having a crucial role in the kinematics computation, if it reaches a level of 3.5 m below the surface, the slope becomes unstable causing the final collapse of the landslide.

The last model presented [3] overcomes these drawbacks through the visco-plastic constitutive relation of Perzyna's type. The movements remain "controlled" in an unstable situation. In the case of a safety factor less than one, deformations tend to infinity, but not instantaneously as in the case of elasto-plasticity but gradually, adjusting the deformation speed with the viscosity parameter. This model is able to reproduce the measured Portalet landslide kinematics [3].

Figure 1 compares the computed and measured mid-slope horizontal displacement and velocity from 05/10/2006 to 21/11/2006. Figure 2 shows the equivalent visco-plastic strain and displacement contours at 31/10/2011 using the 2D visco-plastic finite element model. It has to be pointed that a fairly well comparison between the computed ant the measured displacements/velocities in the monitoring campaigns has been checked in an extended period of time from 01/07/2004 to 31/10/2011 in [3]. Influence of the rainfall intensity on the landslide kinematics can also be verified in the original paper [3].

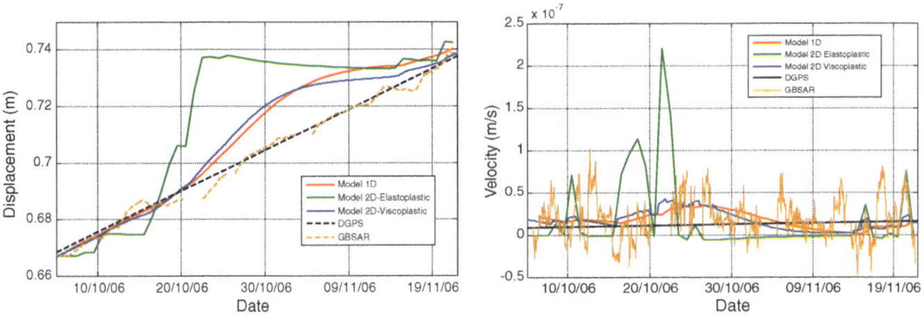

Fig. 1 Computed and measured mid-slope horizontal displacement and velocity from 05/10/2006 to 21/11/2006

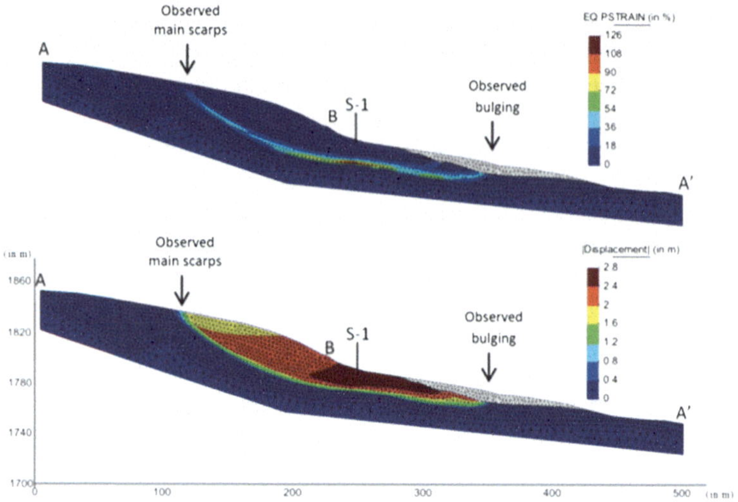

Fig. 2 Equivalent visco-plastic strain and displacement contours at 31/10/2011 [3]

4 Conclusions

The comparison made in this paper provides that after a careful and difficult calibration, the proposed models reproduce qualitatively and quantitatively, more or less accurately depending on the complexity of the model, the observed deformation patterns. These models can give successful short-term and medium-term predictions during stages of primary and secondary creep, i.e. at nearly constant strain rate. However, long-time predictions remain uncertain, stability depends strongly on the position of the water table depth and new failures during tertiary creep due to soil temporal micro-structural degradation are difficult to calibrate.

Acknowledgments This work has been partially funded by the Terrafirma program and the DO-SMS (SUDOE INTERREG IV B) and DORIS (EU-FP7-SPACE-2009-1 n° 242212) projects.

References

1. Herrera, G., Fernández-Merodo, J. A., Mulas, J., Pastor, M., Luzi, G., & Monserrat, O. (2009). A landslide forecasting model using ground based SAR data: The Portalet case study. *Engineering Geology*, *105*, 220–230.
2. Fernández-Merodo, J. A., et al. (2008). Modelling the portalet landslide mobility (iEMSs 2008): International congress on environmental modelling and software. In M. Sànchez-Marrè, J. Béjar, J. Comas, A. Rizzoli & G. Guariso (Eds.), *International Environmental Modelling and Software Society (iEMSs)*, Formigal, Spain.

3. Fernández-Merodo, J. A., García-Davalillo, J. C., Herrera, G., Mira, P., & Pastor, M. (2012). 2D visco-plastic finite element modelling of slow landslides: the Portalet case study (Spain). *Landslides*, doi:10.1007/s10346-012-0370-4.
4. Herrera, G., Notti, D., García-Davalillo, J. C., Mora, O., Cooksley, G., Sánchez, M., et al. (2011). Analysis with C- and X-band satellite SAR data of the Portalet landslide area. *Landslides*, *8*, 195–206.

Characterization of Underground Cellars in the Duero Basin by GNSS, LIDAR and GPR Techniques

Miguel Angel Conejo-Martín, Tomás Ramón Herrero-Tejedor, Enrique Pérez-Martín, Javier Lapazaran-Izargain, Jaime Otero-García, Juan Francisco Prieto-Morín and Jesús Velasco-Gómez

1 Introduction

The underground cellars that appear in different parts of Spain are part of an agricultural landscape dispersed, sometimes damaged, others at risk of disappearing.

This paper studies the measurement and display of a group of wineries located in Atauta (Soria), in the Duero River corridor. It is a unique architectural complex, facing rising, built on a smooth hillock as shown in Fig. 1.

These constructions are excavated in the ground. The cellars are nearly straight and arranged in parallel. The access to the cave or underground cellar has a shape of a narrow tube or down gallery. Immediately after, this space gets wider. There, wine is produced and stored [1].

M. A. Conejo-Martín(✉) · T. R. Herrero-Tejedor(✉) · E. Pérez-Martín(✉)
Departamento de Ingeniería Cartográfica, Geodesia y Fotogrametría—Expresión Gráfica, EUIT Agrícola, Universidad Politécnica de Madrid, Ciudad Universitaria S/N, 28040 Madrid, Spain
e-mail: miguelangel.conejo@upm.es

T. R. Herrero-Tejedor
e-mail: tomas.herrero.tejedor@upm.es

E. Pérez-Martín
e-mail: enrique.perez@upm.es

J. Lapazaran-Izargain(✉) · J. Otero-García(✉)
Departamento de Matemática Aplicada, ETSI de Telecomunicación, Universidad Politécnica de Madrid, Avenida Complutense 30, ES-28040 Madrid, Spain
e-mail: javier.lapazaran@upm.es

J. Otero-García
e-mail: jaime.otero@upm.es

J. F. Prieto-Morín(✉) · J. Velasco-Gómez(✉)
Departamento de Ingeniería Topográfica y Cartografía, ETSI Topografía, Geodesia y Cartografía, Politécnica de Madrid, Campus Sur, Carretera de Valencia, km 7, 28035 Madrid, Spain
e-mail: juanf.prieto@upm.es

J. Velasco-Gómez
e-mail: jesus.velasco@upm.es

E. Pardo-Igúzquiza et al. (eds.), *Mathematics of Planet Earth*,
Lecture Notes in Earth System Sciences, DOI: 10.1007/978-3-642-32408-6_62,

Fig. 1 Underground cellars. Atauta (Soria)

The measurement process and subsequent graph is addressed here and it aims to develop a new methodology according to the instruments used.

Observation and detection of the underground cellar, both on the outside and underground, it is essential to make an inventory of the rural patrimony [2]. The geo-detection is a noninvasive technique, adequate to accurately locate buried structures in the ground. Works undertaken include topographic work with the LIDAR techniques (Laser Imaging Detection and Ranging) and integration with data obtained by GNSS (Global Navigation Satellite System) and GPR (Ground Penetrating Radar). The results obtained are used to identify construction elements which form the underground cellar with an accuracy of centimeters. The graphic and cartographic results enable an optimal visualization of the study area and can be used in the reconstruction of the place.

2 Methods

In one of the cellars we made GPR prospecting from the surface and underground LIDAR scanning. The joint use of both techniques facilitates the determination and location of internal structures. LIDAR scanning provides the location of the internal visible structures of the cellar and its position related with surface. Through their fitting with the GPR profile, we estimate the Radio Wave Velocity (RWV) in the ground, required for locating the GPR detections of the hollow parts and old hidden structures.

The radar data were acquired using a Malå Ramac/GPR ProEx system equipped with unshielded antennae of 100 and 200 MHz, in order to compare the behaviour of different frequencies suitable for the area conditions and type of space [3]. Two profiles were done at each frequency, one along the selected cellar (itinerary 1, Fig. 2), and the second transversal to the selected and adjoining cellars (itinerary 2, Fig. 2).

The LIDAR used was a Faro Focus 3D. It was furnished with a telematic unit ambiguity and interval of 153.49 m and a range of 0.5–120 m outdoor with low ambient light. Point clouds registered by LIDAR and GPR where linked using GNSS techniques [4]. Also GNSS was used for georeferencing the entire survey [5].

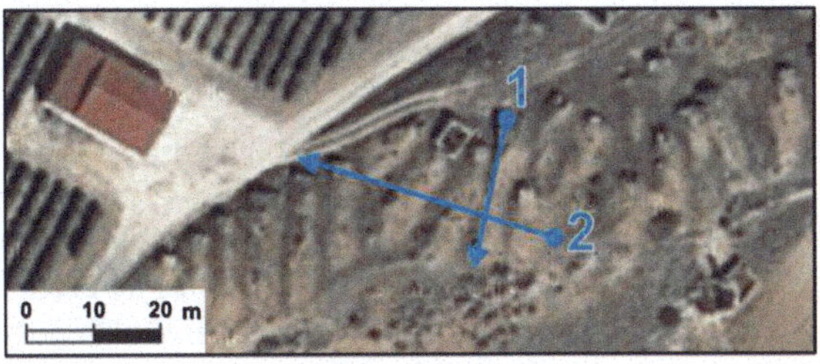

Fig. 2 Itineraries (1) and (2), both with GPR 100 and 200 MHz

3 Results

Figure 3 shows the 100 MHz GPR transparent profile obtained from the itinerary 1, superimposed on the LIDAR profile. The inner profile of the winery detected by LIDAR is represented in blue (hidden cavities are not detected). The GPR detection is shown in yellow and the road level in white.

We can observe that the dome rests on the road level. Other detectable structures are the chimney pipe (in red), a wide cavity around it, and a discontinuity over the structure that supports the roof stairs. The coupling of the results from GPR and LIDAR lets us estimate a value of 0.13 m/ns for the RWV in the medium (soil and rock), which is appropriate for a reasonably dry limestone. The radar detection profile does not match with the inner cavity detected by the LIDAR. This is due to the early GPR reflections in the hollow parts and old hidden structures, in addition to the limited resolution capability of the 100 MHz GPR, of ca 50 cm. A resonance effect appears in the staircase zone produced by multiple reflections between the stair treads and the ceiling. Over the cellar roof there may be a layer of fractured rock, which can produce a GPR reflection some centimetres above the roof.

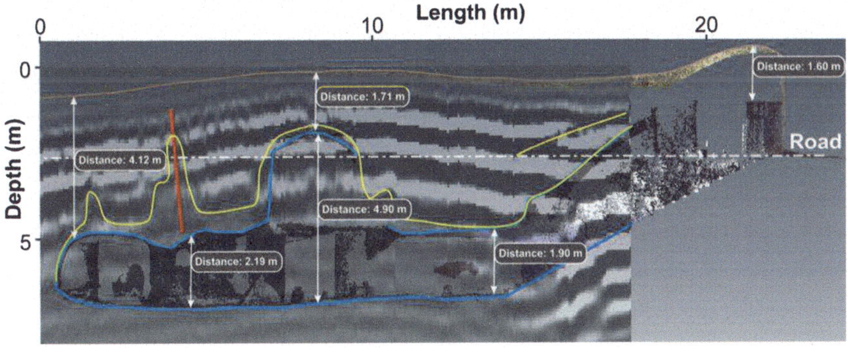

Fig. 3 GPR transparent profile superimposed on the LIDAR profile

4 Conclusions

The use of Geographic Information Technology allows better geovisualization of the rural heritage, as shown on this article.

Using 200 MHz GPR the penetration depth was scarce, while with 100 MHz we have obtained successfully results. It has detected the different cave-domes, the cave-ceiling and most of the cave-floors. It is also possible to detect the presence of other structures, as the entrance beam, the chimney or other close entrances.

The joint use of LIDAR and GPR techniques has revealed a faster method than conventional techniques, such as total station or photogrammetry. Also the RWV estimate is faster and more accurate than using only GPR. The accuracy obtained is centimetric, and GNSS technique makes feasible the combined use of LIDAR and GPR maintaining the accuracy and the survey speed.

The techniques described in this article are suitable to use on other natural cavities, archaeological cavities or multipurpose constructed underground spaces.

This project can help the underground cellars to be declared as Cultural Interest by the *Comisión de Patrimonio Cultural de Castilla y León–Junta de Castilla y León* (Heritage Department of the Regional Government of Castilla y León).

References

1. Martín Ocaña, S., & Cañas Guerrero, I. (2006). Comparison of analytical and on site temperature results on spanish traditional wine cellars. *Applied Thermal Engineering*, *26*(7), 700–708. doi:10.1016/j.applthermaleng.2005.09.004.
2. Pardo, J. M. F., & Guerrero, I. C. (2006). Subterranean wine cellars of central-Spain (Ribera de Duero): An underground built heritage to preserve. *Tunnelling and Underground Space Technology*, *21*(5), 475–484. doi:10.1016/j.tust.2005.07.004.
3. Lorenzo, E. (1996). Prospección geofísica de alta resolución mediante geo-radar. Aplicación a obras civiles. Monografías CEDEX, Madrid. ISBN-8477902569.
4. Han, J. Y., Perng, N. H., & Chen, N. J. (2013). LIDAR point cloud registration by image detection technique. *Geoscience and Remote Sensing Letters, IEEE*, *10*(4), 746–750.
5. Pérez-Martín, E., Herrero-Tejedor, T. R., Gómez-Elvira, M. Á., Rojas-Sola, J. I., & Conejo-Martin, M. A. (2011). Graphic study and geovisualization of the old windmills of La Mancha (Spain). *Applied Geography*, *31*(3), 941–949.

Study of Subsidence on Aquifers Having Undergone Extraction and Inactive Cycles

Rubén Martínez-Marín, Pablo Ezquerro-Martín, Juan Carlos Ibáñez-Carranza, Juan Gregorio Rejas-Ayuga and Maguel Marchamalo-Sacristán

1 Introduction

Water supply for human consumption is and will be a serious problem in the management of the urban development. In countries where rain patterns are very variable in in quantity and temporal distribution as well, the qualifier exploitation is a good possibility to be implanted. This management has two main aspects, extraction and inactive cycles when conditions allow them. During extraction cycles another problem to be considered is the affection produced on the ground surface due to the anthropic decline of the water table. Form the point of view of the civil engineering, the most relevant factor is the subsidence movements induced by the extraction on the surface. Additionally, it is important to know the velocity of those movements and their relationship with the piezometric level in the area. Studies of subsidence have been carried out, not only in Spain [1, 2] but in other countries [3], where overexploitation processes have been shown.

The 'Madrid basin' is a sedimentary basin with deposits from the Tertiary period. Inside it, two different lithology sets from Neogene can be observed: One of them with a high lacustrine character consisted of fine-grained sediments and sedimentary chemical rocks located in the central area of the basin (south of Madrid). The other set consists of detritic rocks on the highest edge of the basin, adjacent to the

R. Martínez-Marín (✉) · P. Ezquerro-Martín · J. G. Rejas-Ayuga · M. Marchamalo-Sacristán
Laboratorio de Topografia y Geomática, Technical University of Madrid, ETSI Caminos Canales y Puertpos C/Profesor Aranguren s/n, 28040 Madrid, Spain
e-mail: ruben.martinez@upm.es

J. C. Ibáñez-Carranza
Canal de Isabel II Gestión, C/José Abascal 10, 28003 Madrid, Spain
e-mail: jci@gestioncanal.es

J. G. Rejas-Ayuga
National Institude for Aerospace Technology INTA, Ctra.de Ajalvir km 4 s/n, 28850 Torrejon de Ardoz, Spain
e-mail: rejasaj@inta.es

E. Pardo-Igúzquiza et al. (eds.), *Mathematics of Planet Earth*,
Lecture Notes in Earth System Sciences, DOI: 10.1007/978-3-642-32408-6_63,

'Range of Guadarrama' (Fig. 1). This second set configures the area of interest, named 'Tertiary Detritic Aquifer of Madrid' (TDAM). From geotechnical and hydrogeological points of view, these units are porous and usually not consolidated, clearly anisotropic and heterogeneous with a thickness about 1,500 m. The most permeable levels are composed of an aggregation of sand and gravel, constituting interleaf layer with others less permeable, as clays, silts and clayey sands. Regarding the distance to the bedrock, from where those deposits are generated, a subdivision in three subfacies can be done:near, central and distant, being the central the most permeable (Table 1).

The well field of 'Fuencarral', located in the northwest area of Madrid, consists of nine wells to withdraw water from the 'TertiaryDetritic aquifer of Madrid'

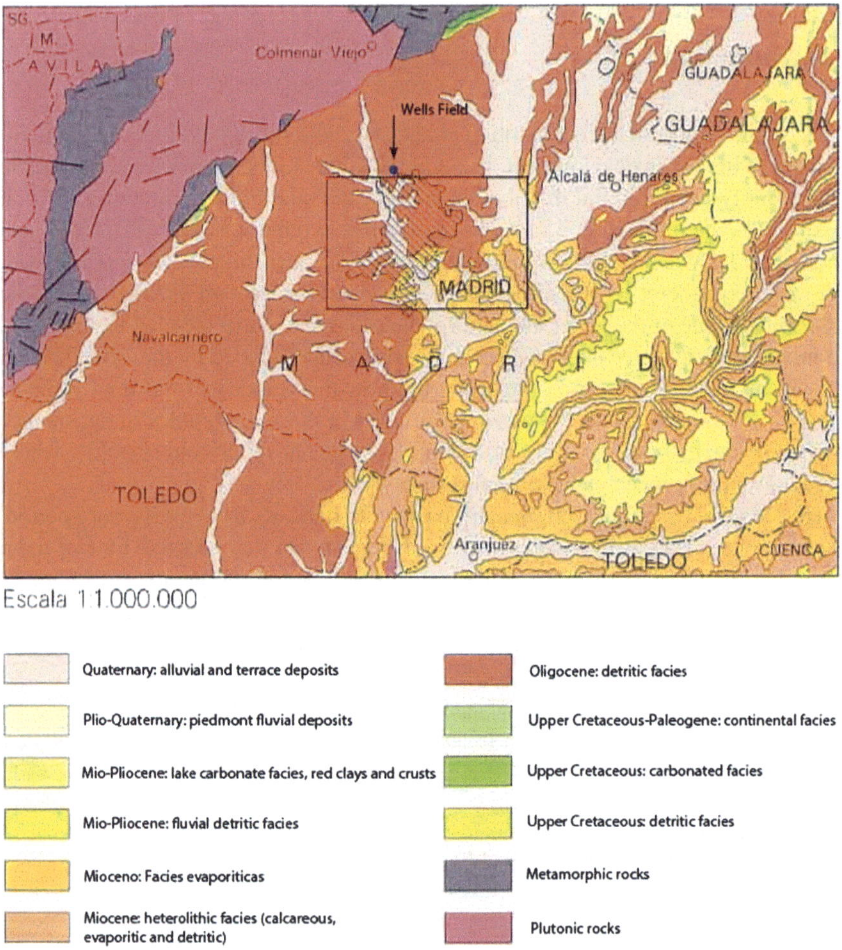

Fig. 1 Regional geology

Table 1 Hydraulics parameters

Parameter	Average value
Transmissivity	1–852 m^2/day
Vertical conductivity (k_v)	0.089–0.31 m/day
Horizontal conductivity (k_h)	0.9–10 m/day
porosity	1–20 %

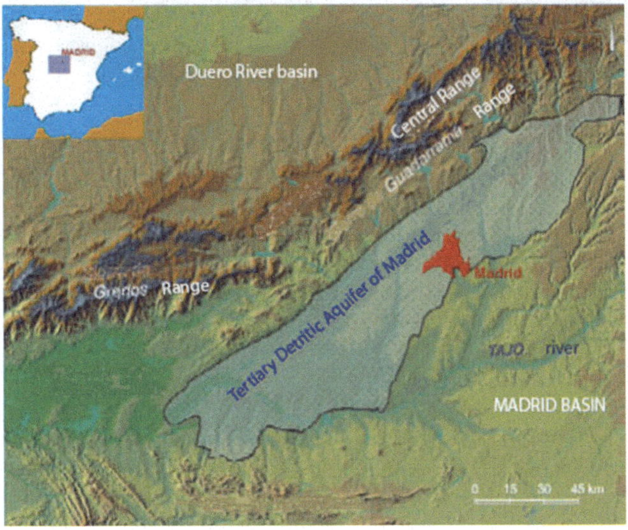

Fig. 2 General location. Tertiary Detritic Aquifer

(Fig. 2), which covers a total area of 2,500 km^2 and reneweable resources about 130–150 hm^3/year. Canal de Isabel II Gestion, the water management company of Madrid, has installed a network of 77 wells withdrawing in this aquifer, with depth from 350 to 700 m. Its supply capacity is about 69–78 hm^3 per one pumping year. Additionally to the withdrawls make by canal deIsabel II Gestion, this aquifer is used by industries, some municipal suppliers, residential areas and agricultural watering. They represent a consumption of 54 hm^3/year approximately (according to the official information from the local Government of Madrid).

2 Methods

A set of 26 images obtained by ERS-1 and ERS-2 satellites have been processedto calculate the vertical ground movement of the study area [4]. The images were acquired between January of 1997 and November of 2000, and cover a 100 × 100 km

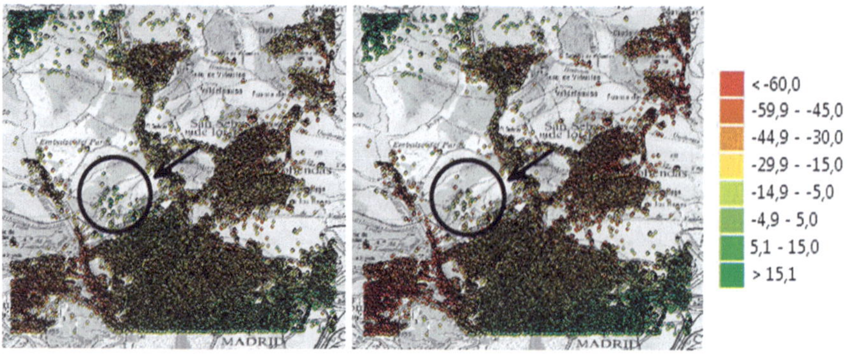

Fig. 3 PS distribution and PS values from the beginning (*left*, 04/02/1997) to the end (*right*, 18/10/2000) of the data series

area centered in Madrid capital. Once the images have been processed the result is a 'shapefile'(ESRI format) that content the PSs (Persistent Scatterers) determined by the PSP-IFSAR methodology (Persistent Scattered Pair—SAR Interferometry) [5], including:

- PSs Positions referred to the geographical Systerm based on WGS84.
- Average velocity of the movement for each PS, in mm/year for the study period.
- Displacement , in mm, referred to the origin image of the study period.

Based on the described data, investigation has been conducted under a combination of several factors:using absolute measurements, not only in movements but interest area (ring shapes with 500–1000 m radius), and pizometric measurements only from deep wells, which show, in a better way, the aquifer behavior Fig. 3.

3 Results and Discussion

Considering the whole period (1997–2000), Fig. 4 shows vertical displacements (mm) and the evolution of the piezometric level. Analysing that figure, three clear periods can be observed. The first one, from April 1997, to January 1999, is an inactive period, where the soil is recovering and the piezometric level as well. The second one, from February 1999 to March 2000, is an extraction period where the piezometric level is clearly decreasing. Finally, the third period, from April 2000 to December 2000, is another inactive period.

Several circles of influence have been defined and the correction coefficient between the piezometric level and the vertical movement, have been caculated. Analyzing the result we can confirm a strong relation between both studied variables. That relation decreases with distance, with correlation coefficients ranging from 0.97 to 0.62.

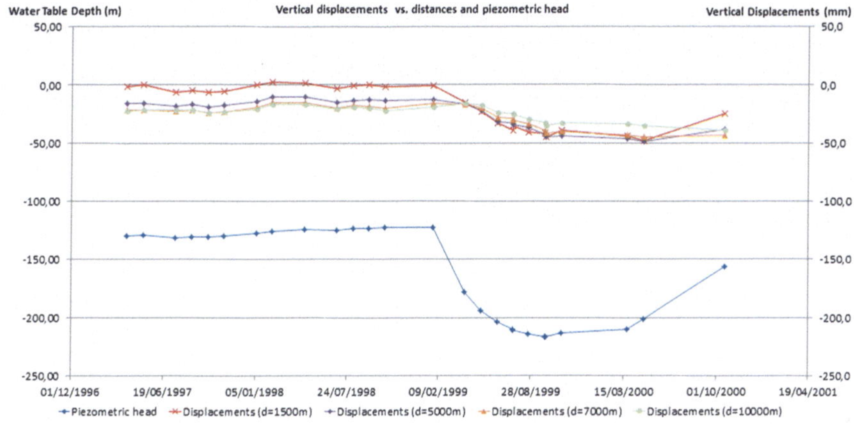

Fig. 4 Vertical displacements (mm) versus distances and piezometrtic head (1997–2000)

4 Conclusion

A DInSAR technique has been applied to determine vertical movements on the ground in 'a posteriori' process, since the studied period was from 1997 to 2000. This methodology is suitable to be applied in areas and periods was were no data are available. The study area is periurban zone with no other factors that can contribute to vertical movements, therefore all the movements of the ground are due to the water extraction. A strong relation between piezometric levels and vertical movements has been detected. That relation has been quantified calculating the correlation coefficients atseveral distances from the center of the study area.

Acknowledgments The authors would like to thank the canal de Isabel II Gestión (Madrid-Spain) for its contribution for the essential information and data for this reseach.

References

1. Peral, F., Rodriguez, A., & Mulas, J. (2004). Control of subsidence with borehole extensometers and surveying measurements in Murcia (Spain). *Proceedings ISC-2 on Geotechnical and Geophysical Site Characterization*. Rotterdam: Holland.
2. Tomas, R., Marquez, y., Lopez-Sanchez J. M., Delgado J., Blanco P., Mallorqui J., Martinez M., Herrera G., Mulas J. (2005). Mapping ground subsidence included by aquifer overexploitation using advanced differential SAR interferometry: Vega media of the Segura river (SE Spain) case study. *Remote Sensing of Environment*. *98*, 269–283.
3. Galloway, D., Jones, D.R. Ingebritsen, SE. (1999). Land subsidence in the United States Reston. p. 177. Virginia.

4. Hanssen, R.F. (2001). Radar interferometry. *Data interpretation and error analysis*. p. 308. Netherland: Kluwer Academic Publisher.
5. Contantini, M., Falco, S., Malvarosa, F., Minati, F. (2008). A new method for identification and analysis of persistent scatterers in series of SAR images. *Proceedings of the International Geoscience and Remote Sensing Symposium (IGARSS)*. p. 449–452, USA: Boston MA.

Advanced InSAR Techniques to Support Landslide Monitoring

Fernando Bellotti, Marco Bianchi, Davide Colombo, AlessandroFerretti and Andrea Tamburini

1 Introduction

Space-borne Synthetic Aperture Radar Interferometry (InSAR) is a widely used remote sensing technique to get surface deformation measurements with high precision over large areas. It is based on the phase comparison of SAR images, gathered simultaneously or at different times with slightly different looking angles from space or airborne platforms. The phase of a SAR image contains the superposition of many terms including the distance of the radar from the illuminated targets on the ground. The phase difference of two SAR images gathered at different times contains a phase term proportional to the target motion occurring along the sensor-target line-of-sight (LOS) direction during that time interval.

The measured phase difference shows an ambiguity cycle of 2π that corresponds to a 2-way travel path difference of λ (the used radar wavelength). Thus, in principle, InSAR has the potential to detect ground surface motion phenomena with the accuracy of a small fraction of the radar wavelength (usually from 3 to 24 cm) on large areas (thousands of km^2) with high spatial resolution (up to 1 m with the space-borne SARs of the last generation as the German TERRASAR-X and the Italian Cosmo-SKyMed–CSK).

F. Bellotti (✉) · M. Bianchi · D. Colombo · A. Ferretti · A. Tamburini
TRE, Ripa di Porta Ticinese, 79, 28003 Milano, MI, Italy
e-mail: fernando.bellotti@treuropa.com

M. Bianchi
e-mail: marco.bianchi@treuropa.com

D. Colombo
e-mail: davide.colombo@treuropa.com

A. Ferretti
e-mail: alessandro.ferretti@treuropa.com

A. Tamburini
e-mail: andrea.tamburini@treuropa.com

E. Pardo-Igúzquiza et al. (eds.), *Mathematics of Planet Earth*,
Lecture Notes in Earth System Sciences, DOI: 10.1007/978-3-642-32408-6_64,

However, apart from cycle ambiguity problems, the routine use of InSAR is limited by the so-called temporal and geometrical decorrelation, as well as to atmospheric artifacts. Temporal decorrelation appears where the electromagnetic signature of the targets changes with time (e.g. in presence of dense vegetation or water basins). Geometric decorrelation is generated by the target reflectivity variation as a function of the incidence angle. Finally, atmospheric heterogeneity superimposes on each SAR acquisition an atmospheric phase screen (APS) that could have the same spatial pattern (low-pass with typically 1 km decorrelation length) and amplitude of the surface deformation phenomenon under investigation.

2 Advanced InSAR Techniques

Persistent Scatterer Interferometry (PSI) is the collective term used inside the InSAR community to distinguish between single interferogram DInSAR and the second generation of InSAR technologies. The first of these to appear, in 1999, was the Permanent Scatterers (PSInSAR^{TM}) technique [1, 2, 4]. It overcomes the above said limitations of InSAR exploiting long temporal series of satellite radar data, acquired over the same area of interest at different times, to identify "natural radar targets" (Permanent Scatterers—PS), that return stable radar reflections over time back to the satellite, where displacement information can be retrieved.

SqueeSAR^{TM} interferometry, the latest evolution of PSInSAR^{TM} technology [3], developed by Tele-Rilevamento Europa and Politecnico of Milano, is one of today's most advanced technologies for ground deformation analysis and monitoring.

SqueeSAR^{TM} searches in the radar image set for targets that consistently reflect radar signals throughout the entire dataset of images, exploiting both permanently scattering ground targets (PS, e.g. buildings, pipelines, linear structures, open outcrops, etc.) and homogenous distributed scatterers (DS, e.g. homogeneous ground or scattered outcrops, uncultivated areas, debris covered areas).

Since its introduction, as the replacement to the widely accepted PSInSAR^{TM} algorithm, SqueeSAR^{TM} has challenged the industry standard by identifying many more ground points, increasing overall understanding of ground displacement occurring in an area of interest.

Thanks to their capability in detecting millimetre level displacements over long periods and large areas, PSInSAR^{TM}, and subsequently SqueeSAR^{TM} technique, have become a standard monitoring tool in a number of applications: natural hazards, geothermal, oil and gas, mining, urban and infrastructures monitoring.

3 Application to Landslide Monitoring

Thanks to its capability in detecting millimetre level displacements over long periods and large areas, SqueeSAR^{TM} analysis can be considered complementary to conventional geological and geomorphological studies in landslide detection and monitoring.

Several Italian Regions were studied with SqueeSAR™ in order to detect and monitor slope instability phenomena. The Valle d'Aosta Region (NW Italy) analysis represent one of the most successfully application. The study covered a time span from 1992 to 2011. Aim of the study was supporting the landslide inventory performed within the framework of the Italian Landslide Inventory (IFFI) Project. The integration of the conventional geological-geomorphological studies with the results of the SqueeSAR™ analysis, improved the results of the IFFI inventory, in terms of landslide areal extent, unmapped phenomena detection and activity assessment. The increasing interest of Italian authorities in the application of SqueeSAR™ resulted in the Piano Straordinario di Telerilevamento (PST) project, founded by the Ministry of the Environment. The aim of the project was to create the first national-scale database of interferometric information to map unstable areas. This was achieved by processing more than 12,000 ERS SAR scenes acquired over Italy and updating these results with the Envisat images available up to 2010.

SqueeSAR™ analysis is particularly suitable for the study of Deep-seated Gravitational Slope Deformations (DGSD), characterized by large areal extent and low surface displacement rate, ranging from few millimeters to tens of millimeters per year.

The SqueeSAR™ analysis realized over Huangtupo DGSD (China) provides quantitative information on displacements occurred (Fig. 1), reveals zones affected by

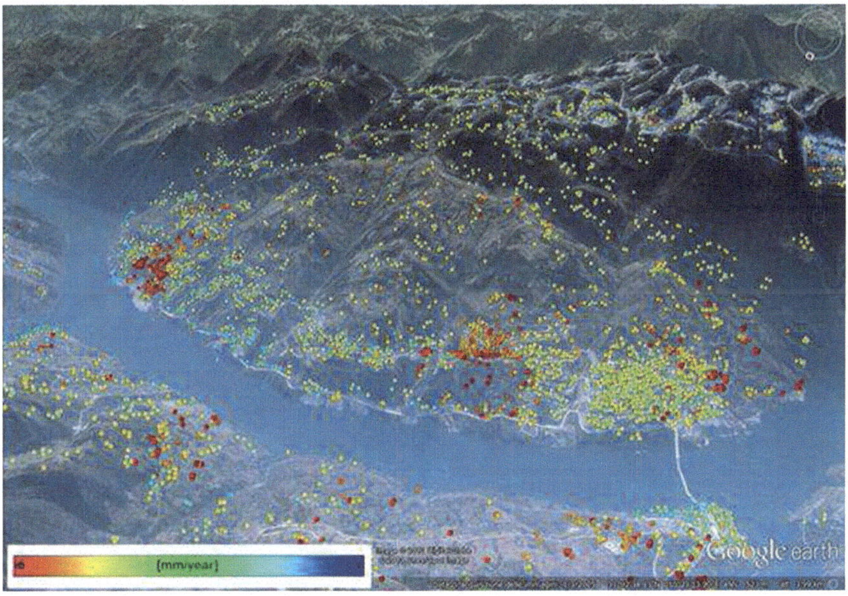

Fig. 1 Average velocity of the measurement points identified with the SqueeSAR™ over Huangtupo landslide

movements related to slope instability phenomena and identifies two different zones affected by velocities changes, consistent with landslide slip surfaces recognizes in previous works.

References

1. Ferretti, A., Prati, C., & Rocca, F. (2000). Nonlinear subsidence rate estimation using permanent scatterers in differential SAR interferometry. *IEEE Transactions on Geoscience and Remote Sensing*, *38*, 2202–2212.
2. Ferretti, A., Prati, C., & Rocca, F. (2001). Permanent scatterers in SAR interferometry. *IEEE Transactions on Geoscience and Remote Sensing*, *39*(1), 8–20.
3. Ferretti, A., Fumagalli, A., Novali, F., Prati, C., Rocca, F., & Rucci, A. (2011). A new algorithm for processing interferometric data-stacks: SqueeSAR^{TM}. *IEEE Transactions on Geoscience and Remote Sensing*, *49*(9), 3460–3470.
4. Kampes, B. (2006). *Radar interferometry: persistent scatterer technique*. Dordrecht: Springer.

Estimation of Information Loss When Masking Conditional Dependence and Categorizing Continuous Data: Further Experiments on a Database for Spatial Prediction Modelling in Northern Italy

Andrea G. Fabbri, Simone Poli, Antonio Patera, Angelo Cavallin and Chang-Jo Chung

1 Introduction

Prediction patterns are the result of modelling the spatial relationships between the distribution of natural events, e.g., the occurrence of landslides of a specific dynamic type, and that of map units that are assumed to represent the typical setting of the events. A spatial database is reanalysed that has been the object of differing analyses [2, 3]. Predictions are compared using either original categorical and categorized continuous supporting patterns or uncategorized continuous ones.

A direct supporting pattern, DSP, of the spatial distribution of 28 complex landslides, shown in Fig. 1, was previously used to obtain their spatial relationships with five categorical indirect supporting patterns, ISPs, representing the spatial context of the landslides: geology,**g**, land use, **l**, and permeability, **p** (with 7, 6 and 9 map units, respectively) in addition to internal relief, **i** and slope, **s** (categorized into 5 classes each). The five ISPs were selected, among a few more, to minimize the effects of conditional dependence. The same set of ISPs is reanalysed applying the Empirical Likelihood Ratio model in experiments that used also uncategorized continuous ISPs: namely aspect, **A**, curvature, **C**, digital elevation, **D**, in addition to internal relief, **I** and slope, **S**. The patterns of the database occupy a study area of 1,133,490 pixels of 20 m resolution, within a frame of 1291 pixels × 2045 lines. The scope of the comparison is to examine, using a spatial database constructed for landslide hazard in

A. G. Fabbri (✉) · A. Cavallin
DISAT, Università di Milano-Bicocca, 20126 Milan, Italy
e-mail: andrea.fabbri@unimib.it

S. Poli
ERM, Environmental Resourse Management, 20124 Milan, Italy

A. Patera
INGV, Istituto Nazionale di Geofisica e Vulcanologia, 00143 Rome, Italy

C.-J. Chung
Spatial Models Inc, Ottawa K1G 5K5, Canada

E. Pardo-Igúzquiza et al. (eds.), *Mathematics of Planet Earth*,
Lecture Notes in Earth System Sciences, DOI: 10.1007/978-3-642-32408-6_65,

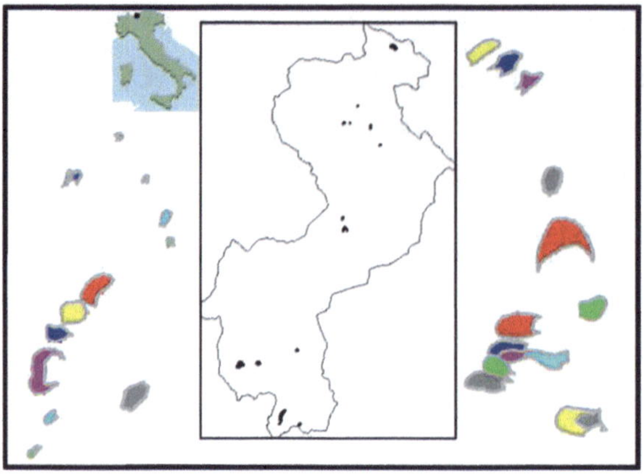

Fig. 1 Location of the Tirano study area in northern Italy and the distribution of the trigger areas of 28 complex landslides that occupy 1382 pixels. The trigger areas are surrounded by grey borders and are enlarged outside the study area frame

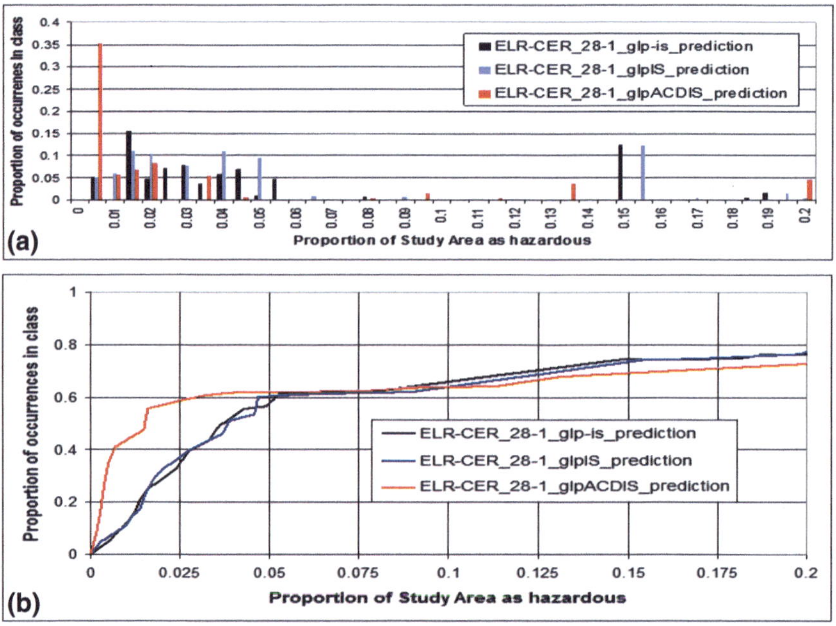

Fig. 2 ELR prediction-rate histogram (**a**) and cumulative curve (**b**) of the 20 % highest 0.5 % equal area classes in the Tirano study area obtained with 28-1 cross-validations with sets of ISPs: **glp-is** (*black*), **glpIS** (*blue*) and **glpACDIS** (*red*)

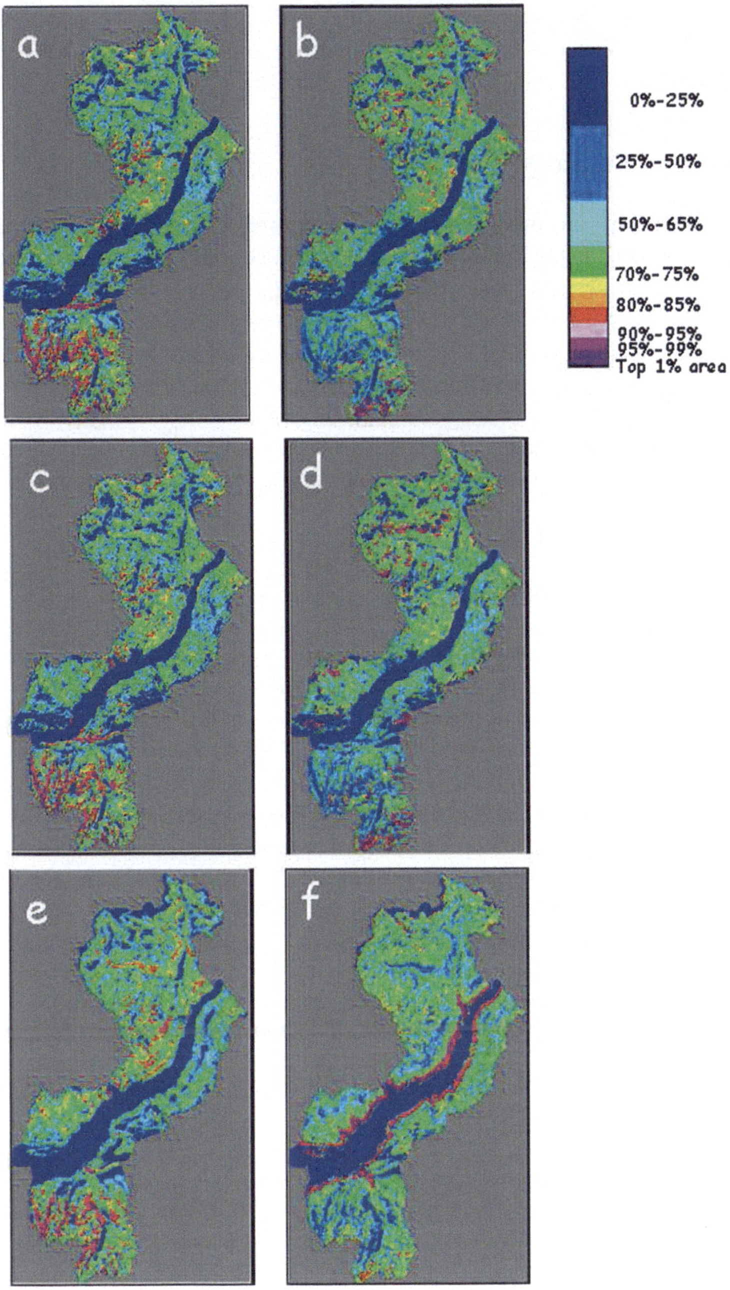

Fig. 3 Target and uncertainty patterns by ELR and the 28-1 cross-validation using ISPs: **glp-is** (**a**, **b**), **glpIS** (**c**, **d**) and **glpACDIS** (**e**, **f**). Ranked groups of classes are assigned pseudo-colours in the legend. A NE trending valley crosses the area

the Tirano study area in northern Italy, the effect of categorization of continuous data layers and the avoidance of conditional dependence implicitly existing between the spatial data layers.

2 Model and Strategy

Spatial relationships are modelled using an Empirical Likelihood Ratio function, ELR, based on the comparison of the distribution functions of all areas affected by landslide trigger zones and those of all areas not affected by landslides in the study area. Chung [1] discussed the use of ELR in prediction modelling.

The analytical strategy used to cross-validate the ELR prediction patterns is to sequentially iterate predictions using 28-1 landslides to predict the 28th. The resulting 28 prediction patterns are then analysed to obtain target and uncertainty patterns with rank-based statistics: the target pattern represents the median rank of the 28 predicted values for each pixel in the study area and the uncertainty pattern the corresponding range of ranks.

3 Results and Conclusion

The generation of target patterns provides measures of relative quality, reliability and comparability of spatial predictions. Figure 2 shows the 40 highest ranked 0.5 % equal area classes of study area predicted as hazardous in histogram and cumulative curve forms. Clearly using the continuous ISPs **ACDIS** provides classes with higher proportion of landslides within the highest 5 % of the study area. Comparing target and uncertainty patterns in Fig. 3, we can observe how the use of the continuous ISPs **IS** slightly improves the prediction with respect to using categorized **is**. Furthermore, using **ACDIS** improves it strongly and the uncertainty pattern indicates the effects of **D**, the digital elevation.

No particular advantage is observed either in categorizing continuous ISPs or in avoiding conditionally dependent ISPs, leaving out **ACD**, aspect, curvature and digital elevation. We encourage exploring the generality of such observations.

References

1. Chung, C. F. (2006). Using likelihood ratio functions for modelling the conditional probability of occurrence of future landslides for risk assessment. *Computers & Geosciences*, *32*, 1052–1065.
2. Fabbri, A. G., Poli, S., Sterlacchini, S., Cavallin, A., Chung, C. J. (2011). Uncertainty of class membership in spatial prediction modelling: follow-up study to an application to complex landslides. *Proceedings IAMG 2011, Peer-reviewed IAMG 2011 publication* (12 pp). doi:10.5242/iamg.2011.0241
3. Poli, S., & Sterlacchini, S. (2007). Landslide representation strategies of susceptibility studies using Weights-of-Evidence modelling technique. *Natural Resources Research*, *16*(2), 121–134.

3D-GIS Analysis for Mineral Resources Exploration in Luanchuan, China

Gongwen Wang, Shouting Zhang, Changhai Yan, Yaowu Song, Jianan Qu, Yanyan Zhu and Dong Li

1 Introduction

Three-dimensional (3D) geological modeling is an important method for understanding geological structures and exploring for mineral deposits. 3D–GIS can be used as interactive tool for mineral resources exploration [1–3]. The main advantages of 3D-GIS special analysis models are combining qualitative information of geological knowledge (e.g., metallogenic genesis/model) and quantitative data of geosciences to make decision of prospective targets in 3D space. Geoscience data can be derived metallogenic information for building more complete, more realistic geological and metallogenic models. Furthermore, these models can be modified at any time by adding new data from ongoing drilling and field surveys.

2 Data and Methodology

Luanchuan ore district is the most important Mo region in China. It has three large porphyry-skarn Mo deposits and more than 20 Pb–Zn deposits, ant it has measured reserves of molybdenum metal in excess of 3,000,000 tonnes in 2010. Although there are more than 50 years exploration data including geological, geophysical and geochemical data in this district, exploration depth is not more than 700 m. This paper presents 3D geological models visualization and spatial analysis methodology for molybdenum resources exploration by combing geological, geophysical, and

G. Wang (✉) · S. Zhang · C. Yan · J. Qu · Y. Zhu · D. Li
State Key Laboratory of Geological Processes and Mineral Resources,
China University of Geosciences, Beijing100083, China
e-mail: gwwang@cugb.edu.cn

C. Yan · Y. Song
Henan Institute of Geological Survey, Zhengzhou 450007, China

E. Pardo-Igúzquiza et al. (eds.), *Mathematics of Planet Earth*,
Lecture Notes in Earth System Sciences, DOI: 10.1007/978-3-642-32408-6_66,

geochemical data to identify high potential targets for mineralization at depth in Luanchuan district. Fig. 1 is the workflow of 3D-GIS for prospective targets and mineral resources assessment, and the main three steps are as follow:
Step1 A: 3D geological model of Luanchuan district (40.0 km × 20.0 km × 2.5 km) are constructed by one 1:10,000 scale geological and topographic map of good quality, nine 1:2000 scale cross-sections which contain most important detailed observations of geological objects, the logs of 358 boreholes, additional 978 litho-geochemical samples. 3D geological model was optimized using 1:25,000 scale gravity and magnetic data inversion combining geological and geochemical datasets at depth (the range is from 500 to 2000 m) for identifying main geological objects, e.g., the shapes of Jurassic granite (150 Ma) and Neoprpterozonic gabbro (743 Ma), the dips and strikes of ore-forming faults, and the thickness of metallogenic stratum. The 3D geological modeling progress includes geoscientific data acquisition, compilation, digitalizing and standardizing in a 3D coordinate system (x, y, z) [1]. Step1 B and C: On basis of the metallogenic information combining 1:2000 scale litho-geochemistry, 1:25,000 scale gravity and magnetic image, and 1:5000 scale Controlled Source Audiofrequency Magnetotelluric image in regional section Luanchuan district (Fig. 1 (Step1 A)), a new 3D model (17.0 km × 12.0 km × 2.5 km) (Fig. 1 Step1C) was derived for identifying prospective targets of porphyty-skarn Mo deposit at depth. Step1 D: Litho-geochemical information of Mo and Pb–Zn mineralization at the surface can delineate concealed porphyry-Mo orebody which is associated with Jurassic pluton in Luanchuan distric, the Step1 D section (Shibaogou, the location can be found in Step1 C) is derived by auto-cutting from 3D geological model of Luanchuan district.
Step2 A: 3D block model of geological objects in Luanchuan district can be applied to exploration targeting in Micromine software: (a) Extraction of zones within a defined distance to granite porphyry. (b) Identification of altered, fault, and gravity and magnetic anomaly zones favourable for Mo prospective target. Sept2 B and C: Geoscience data integrated using probability net neural and fractal methods to delineate prospective mapping in ArcGIS10.0 software with Spatial Data Modeller. The calculation steps of the above methodology can be found in [3]. Setp2 D: Borehole validation of prospective target within the above extracted zone.
Step3 A: Deposit-scale modeling for identifying new prospective targets of porphyry-skarn Mo deposit at depth. The 3D model of three large porphyry-skarn Mo deposits (4 km × 5.0 km × 2.5 km) are constructed by one 1:2000 scale geological and topographic map of good quality, the logs of 288 boreholes, additional 458 litho-geochemical samples. Step3 B and C: Step3 B is 3D model of the shapes of orebody, granite porphyry, and skarn rocks in 3D model of three large porphyry-skarn Mo deposits, and the orebody model is constructed by combining Mo grade ordinary kriging interpolation and fractal method. The 3D grade model was used to identify the centre or edge of Mo orebody and ore-forming fault, and the high grade value at bottom of orebody can indicate the deposit has a continuous Mo orebody at depth (Fig. 1). Step3 C and D: New Mo prospective targets identifying using queries in 3D-GIS environment. Nannihu and Shangfang porphyry-skarn Mo deposits have potential targets at depth (from 800 to 2000 m) by granite anomaly interpretation using gravity and magnetic data inversion. Deposit-scale borehole data were used

(a) to delineate probable mineralized zones by multifractal concentration-volume modeling of geochemical data and (b) to estimate metal resources by fractal modeling based on available orebody thickness and metal grade data.

3 Discussion and Conclusions

The prospective targets in Luanchuan district can be derived by three methods in the above steps: ① Step1 can be applied to recognize skarn-Mo deposit near to the Jurassic pluton (e.g., Shibaogou skarn-type prospective target around Shibaogou granite stock, Fig. 1 Step1D), and the target has no known deposit both at the surface and subsurface; ② Step2 can be applied to identify all potential targets of granite porphyry-skarn deposits which have multiple geoscience data; and ③ Step3 can be used to identify concealed orebody at depth of large porphyry Mo deposit, which show the known orebody has continuous at depth, and the prospective target are constrained by Jurassic pluton in the deposit (e.g., Nannihu porphyry-skarn Mo deposit, Fig. 1 Step3D). Based on the above prospective targets of luanchuan district, we conceptualized the metallogenic genesis in the Luanchuan ore region as follows. The Mo and Pb–Zn–Ag deposits/occurrences are controlled by the Jurassic tectonic-magmatic activity, such that skarn-Mo and hydrothermal vein Pb–Zn–Ag deposits are found mostly around granite (porphyry) pluton. Pb–Zn–Ag mineralization occurred in layered skarn or interlayer fracture zones which can be identified by litho-geochemical survey information. Therefore, a detailed 3D study of the concealed granite (porphyry) and the association with the host rock of marine carbonate-clastic strata (i.e., Nannihu, Meiyaogou and Sanchuan Formations) is considered for the construction of a 3D geological model and for prospective targets in luanchuan district.

The research results show that Luanchuan district has a complex geological setting, which is located in the East Qinling orogen. Magmatism is common throughout the geological evolution of the Qinling orogen. Neoprpterozonic gabbro intruded Luanchuan Group, Jurassic magmas intruded as small-scale stocks, dikes, and pipes, while the Cretaceous intrusions formed large-scale stocks and btholiths. The regional tectonic framework trends WNW-ESE, the later NE-trending tectonic cuts them and formed a network structure. The junctions between NW-trending and NE-trending faults control the distribution of small Jurassic plutons, which are important ore-bearing geological objects consisting of granodiorite, monzogranite and granite porphyry. All the large granite porphyry-skarn Mo deposits are located in Neoprpterozonic Luchan Group comprising dolomitic marble of Meiyaogou Formation, marble and schist of Nannihu Formation, metasandstone and biotite-bearing marble of Sanchuan Formation. Therefore, according to the porphyry-skarn genesis of Luanchuan district, the 3D models of ore-bearing plutons and the NE-trending and NW-trending faults are important information for identifying typical porphyry-skarn deposit in high potential targets. The gravity and magnetic data inversion, cross-sections and survey data (surface and borehole datasets) are integrated to construct potential targets at depth.

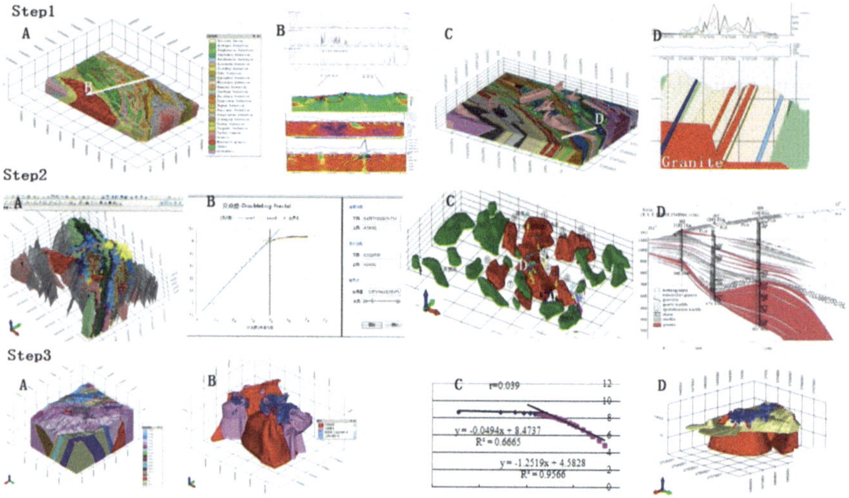

Fig. 1 Workflow of 3D-GIS for mineral resources exploration in Luanchuan district

References

1. Houlding, S. W. (1994). *3D geoscience modeling—computer techniques for geological characterization*. Berlin: Springer-Verlag.
2. De Kemp, E. A., Monecke, T., & Sheshpart, M. (2011). 3D GIS as a support for mineral discovery. *Geochemistry: Exploration, Environment, Analysis*, *11*, 117–128.
3. Wang, G., Zhang, S., Yan, C., Song, Y., Sun, Y., Li, D., et al. (2011). Mineral potential targeting and resource assessment based on 3D geological modeling in Luanchuan region, China. *Computers and Geosciences*, *37*, 1976–1988.

Animation of Groundwater Flow with STRING

Isabel Ostermann and Torsten Seidel

1 Introduction

In order to analyze 2D/3D simulation results for groundwater flow problems visualization is an essential resource. Most existing visualization techniques are easy to use and understand only for professionals. The intuitive representation of transient flow on planar and curved surfaces with the help of pathlets (comet-like shapes representing short parts of pathlines) which is used in STRING offers a novel tool to assist professionals as well as nonprofessionals in creating expressive animations of groundwater flow (see [3]). The length of a pathlet correlates with the local velocity and its direction corresponds to the direction of the velocity. As not only the velocity field is important but also its interaction with scalar fields (e.g. concentration, potential, pressure), scalar attributes can be optionally displayed in the background. The 'uniform' distribution of the pathlets is achieved by a novel, mesh-independent, intelligent, time-dependent seeding strategy based on the point management realized in the Finite Pointset Method (FPM) (cf. [2] for a recent survey on geometric flow visualization and, in particular, on seeding strategies). This innovative feature distinguishes our method from existing ones leading to a unique illustration quality.

I. Ostermann (✉)
Fraunhofer ITWM, Fraunhofer-Platz 1,
67663 Kaiserslautern, Germany
e-mail: isabel.ostermann@itwm.fraunhofer.de

T. Seidel
delta h Ingenieurgesellschaft mbH, Parkweg 67,
58453 Witten, Germany
e-mail: ts@delta-h.de

E. Pardo-Igúzquiza et al. (eds.), *Mathematics of Planet Earth*,
Lecture Notes in Earth System Sciences, DOI: 10.1007/978-3-642-32408-6_67,

2 Features of STRING

The pathlets move through the visualization plane with the flow velocity—Lagrangian formulation. A 'uniform' distribution of the pathlets during the whole animation is achieved by a novel, intelligent, time-dependent seeding strategy based on the point management realized in FPM. FPM is a meshfree generalized finite difference algorithm also employing a Lagrangian formulation with applications in continuum and fluid mechanics (see e.g. [4–6]). The advantages of the chosen seeding strategy are the independence of the seeding positions of the pathlets from the data mesh or point cloud as well as the avoidance of optical holes and accumulations. In a given geometrical 3D model the user can define horizontal as well as vertical 2D cuts as visualization planes. These planes can be additionally restricted by time-dependent clipping regions defined by polygonal lines, e.g. in case of saturated zones or large data sets. STRING offers the opportunity to illustrate time-dependent scalar fields in the background of the pathlet visualization either directly (continuous colormap or isosurfaces) or as contour lines. Rendered images can be saved as single frames together with geographic information as well as transparency. Afterwards, from these images an animation can be created with any external video editor. The technical features are: platform independence, graphical user interface for easy handling (default values are set), fast data access, possibility of parallelization, hardware acceleration by the use of standard OpenGL tools. The combination of the features described above, implemented in a market-ready, fast, user-friendly, flexible, and intuitive visualization tool, is novel and unique.

3 Visualization Process and Mathematical Tools

The visualization process—from given data set to resulting animation—consists of the following three steps: Loading and processing the data given by the established software system for groundwater and geothermal energy modeling called SPRING (see e.g. [1]), mapping the data to the geometrical representation, and finally rendering the geometrical representation of the data. Note that a comparison of STRING with traditional geometric visualization methods can be found in [3].

During the animation of the flow velocity we have the following iteration process: (i) Movement of the pathlets in one time step; (ii) Readjustment of the pathlet density through seeding and deletion; (iii) Saving the current position of the pathlets; (iv) Rendering the current state; (v) Go to (i). The movement of the pathlets is given by the solution of $d\mathbf{x}(t)/dt = \mathbf{v}(\mathbf{x}(t), t)$ with given initial condition $\mathbf{x}(0) = \hat{\mathbf{x}} \in \mathbb{R}^3$, where $\mathbf{v}(\mathbf{x}(t), t)$ is the velocity at pathlet position $\mathbf{x}(t) \in \mathbb{R}^3$ at time $t > 0$. In STRING the solution of this ordinary differential equation can be performed either based on the classical Euler method or the classical (fourth order) Runge Kutta method including an adaptive time step control. Velocities at positions not given by the data mesh or point cloud are interpolated with the help of the FPM-specific moving weighted least squares algorithm: For each velocity component $v_\star$ we are looking for a polynomial p satisfying the minimization problem $\min_{p \in \Pi^n} \sum_{i=1}^{I} \gamma\left(\frac{\|\mathbf{x}-\mathbf{x}_i\|}{h}\right) \|p(\mathbf{x}_i) - v_\star(\mathbf{x}_i)\|^2$. Π^n denotes the space of 1D polynomials of maximum order $n \in \{0, \ldots, 3\}$, $I \in \mathbb{N}$ denotes the number of available neighbors $\mathbf{x}_i$ of point $\mathbf{x}$, h denotes the averaged, scaled distance to the available neighbors, $\|\cdot\|$ denotes the Euclidean norm in 1D/3D, and γ denotes a weighting function. The order of the approximation is given by the number of available neighbors: at least 1 neighbor leads to order 0, at least 3 neighbors lead to order 1, at least 6 neighbors to order 2, and at least 10 neighbors to order 3. This approach is computationally efficient as the interpolation is done only locally and not based on all points. The results of the FPM-specific moving weighted least squares approximation of two neighboring time steps included in the original data set are used to linearly interpolate between the given time steps. The movement of the pathlets counteracts their 'uniform' distribution. In order to maintain the quality of the pathlet visualization, accumulations are collapsed and optical holes are filled. The filling process represents a novel, time-dependent seeding strategy for pathlets also directly applicable in 3D. It is characterized by a successive discrete hole search, where each pathlet is a possible source for a new one. Discrete search points around a 'source' are determined and, subsequently, search spheres around these points are checked for already existing pathlets. If there is none, the search point is added as a new pathlet.

4 Example

As an example we consider the 2D Elder problem to visualize transient convection cells with STRING (see [3] for further details). The simulation data set—a density driven flow in a porous, homogeneous, and fully saturated isotropic medium for a vertical profile—is generated with the water systems modeling software SPRING (see [1]). In Fig. 1 the transient flow is illustrated as pathlets in time with the concentration in the background. A series of transient convection cells is generated. The intelligent seeding strategy guarantees a 'uniform' distribution of the pathlets.

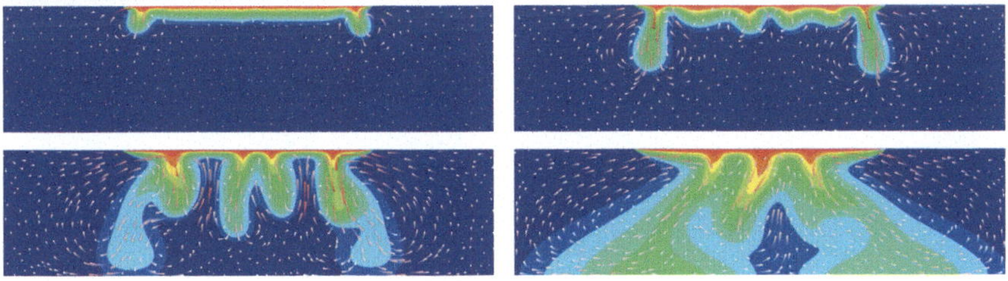

Fig. 1 Pathlet visualization (colored with a continuous gradient from high magnitude of the velocity in *red* to low magnitude of the velocity in *white*) and concentration (isosurfaces) after 1 (*top left*), 2 (*top right*), 4 (*bottom left*), and 8 (*bottom right*) years (see [3])

5 Conclusion

STRING offers a novel, unique, and intuitive way to animate transient flow problems. The main mathematical tools are 'uniformly' distributed pathlets illustrating the velocity on planar and curved surfaces of a 2D/3D model and the time-saving, FPM-specific moving weighted least squares method used to approximate values at positions not given by the original data set. Despite the original main focus, i.e. groundwater flow, this concept is generally suitable to visualize transient flow data in other application domains as well.

References

1. König, C., Becker, M., Diehl, A., Rosen, B., Rüber, O., Seidel, T., Werth, B., & Zimmermann, C. (2012). SPRING Manual. delta h Ingenieurgesellschaft mbH, Witten, 4.1 Edition, ISBN 978-3-00-040369-9.
2. McLoughlin, T., Laramee, R. S., Peikert, R., Post, F. H., & Chen, M. (2010). Over two decades of integration-based, geometric flow visualization. *Computer Graphics Forum, 29*(6), 1807–1829.
3. Seidel, T., König, C., Schäfer, M., Ostermann, I., Biedert, T., & Hietel, D. (2013). STRING—Intuitive Visualization of Transient Flow. In progress.
4. Tiwari, S., Antonov, S., Hietel, D., Kuhnert, J., Olawsky, F., Wegener, R. (2007). A meshfree method for simulations of interactions between fluids and flexible structures. In M. Griebel & M. A. Schweitzer (Eds.), *Meshfree Methods for Partial Differential Equations III* (pp. 249–264). New York: Springer.
5. Tiwari, S., & Kuhnert, J. (2005). A numerical scheme for solving incompressible and low Mach number flows by finite pointset method. In M. Griebel & M. A. Schweitzer (Eds.), *Meshfree Methods for Partial Differential Equations II* (Vol. 43, pp. 191–206). New York: Springer.
6. Tiwari, S., & Kuhnert, J. (2007). Modeling of two-phase flows with surface tension by finite pointset method (FPM). *Journal of Computational and Applied Mathematics, 203*(2), 376–386.

SHEE Program, a Tool for the Display, Analysis and Interpretation of Hydrological Processes in Watersheds

Jesús Mateo Lázaro, José Ángel Sánchez Navarro, Alejandro García Gil and Vanesa Edo Romero

1 Methodology

When the functionalities included in the SHEE package are combined together, the construction, manipulation, analysis and comparison of the hydrological processes which take place in a watershed are possible [1]. The program has been developed in the Department of Earth Sciences of The University of Zaragoza and its interface provides rapid and great quality OPENGL charts, in both RASTER and VECTOR formats. The software has numerous applications for either DEM management or hydrological processes simulation. Obtaining new coverages with the combination of DEM and simulated processes is also possible.

The DEM management is achieved using the GDAL (Geospatial Data Abstraction Library), which permits to import and export different archive formats and to make new coverages from multiple archives (e.g. the global DEM in Fig. 1). The program can combine coverages with different coordinate system thanks to the use of the PROJ4 library from the USGS. Thousands of terrestrial geodetic systems can be represented, transformed and converted between them. To do that, the program is able to obtain necessary Spatial Reference Organization parameters from the internet server transfer. Downloading information from WMS remote server is also possible. The program permits the use of available data sources during the hydrological processes

J. M. Lázaro (✉) · J. Á. Sánchez Navarro · A. García Gil · V. Edo Romero
Departamento de Ciencias de la Tierra, Universidad de Zaragoza, Pedro Cerbuna, 12, 50009 Saragossa, Spain
e-mail: jesmateo@unizar.es

J. Á. Sánchez Navarro
e-mail: joseange@unizar.es

A. García Gil
e-mail: agargil@unizar.es

V. Edo Romero
e-mail: vanesa@eid.es

E. Pardo-Igúzquiza et al. (eds.), *Mathematics of Planet Earth*,
Lecture Notes in Earth System Sciences, DOI: 10.1007/978-3-642-32408-6_68,

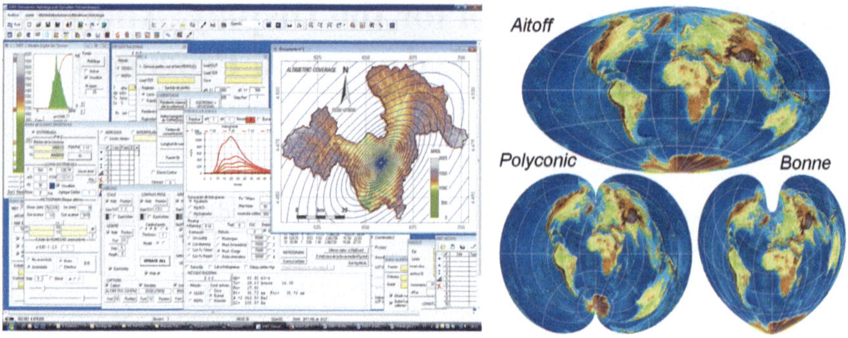

Fig. 1 Interface of the SHEE program and three projections examples (Aitoff, Polyconic and Bonne) of the SRTM 30 global DEM obtained with this package

modeling. Some of the data sources are the following: DEM's coming from different sources, rainfall frequency coverage, Curve Number coverage and real data. SAIH data stand out from all real episodes data because they provide a complete dataset of rainfall, discharge measurements and rain radar data.

A special case is the use of PLR (Parallel Linear Reservoir) models, where every DEM cell is considered as a reservoir combination in parallel. Water flows through each deposit at different flow rates [2]. These models are calibrated via the observation of recession curves of real hydrographs and they are very interesting because they allow the establishment of a precise water balance. The hydrological relations of a lineal deposit are driven by two ecuations: flow or deposit Eq. (1) and continuity or hydric balance Eq. (2). Their combination results in the runoff equation or discharge Eq. (3):

$$Q = \alpha \cdot S \tag{1}$$

$$R = Q + \frac{dS}{dt} \tag{2}$$

$$Q_2 = Q_1 \cdot \mathbf{e}^{-\alpha \Delta t} + R \cdot \left(1 - \mathbf{e}^{-\alpha \Delta t}\right) \tag{3}$$

Another feature of the model is that it allows making simulations with snowmelt models, introducing the runoff portion which comes from the snow into the recharge term (R). The reservoirs represent the different hydric deposit systems of the watershed (surface, subsurface and subterranean reservoirs, vegetation cover, snow, sheet flow, etc), whose evolution is presented in the simulated hydrographs.

2 Applications

2.1 Real Episodes Research

The PLR models have been applied to real episodes of several watersheds (a model of Bergantes river watershed in Fig. 2). A distributed model of total rainfall has been created for the study; reservoir parameters have been adjusted (Qo and α) starting from real hydrograph recession curves. The evolution of each reservoir during the flood is a relevant feature: the fast reservoir (reservoir 2, dashed line) increases and decreases rapidly, whereas the slow reservoir (reservoir 1, dotted line) takes part secondarily in the most intense stretch but it persists during all the recession.

2.2 Study of the Role of Snowmelt in Floods

A real episode occurred in January 2009 in the Valley of Roncal (Spain) is studied. In Fig. 3, the real hydrograph (blue) and the simulated hydrographs with and without snowmelt (red lines, solid and dashed respectively) are shown. The two model reservoirs with snowmelt are presented in gray color and the snowmelt contribution is observed in the hyetograph. When comparing the two simulations, the intervention of the watershed snow reservoir proves decisive for the flash flood occurrence.

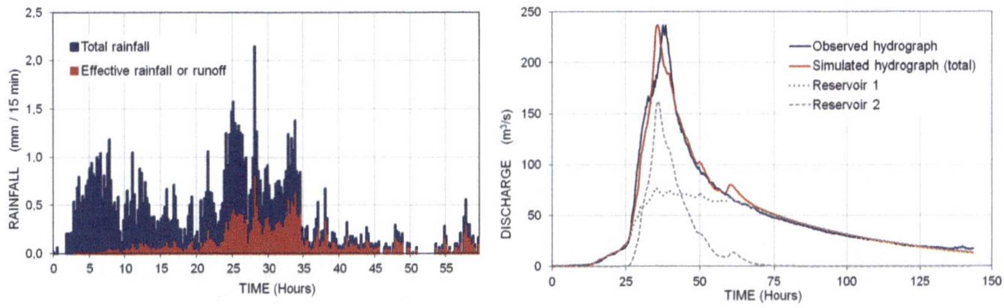

Fig. 2 Hyetographs and hydrographs for may 2003 episode in Bergantes watershed

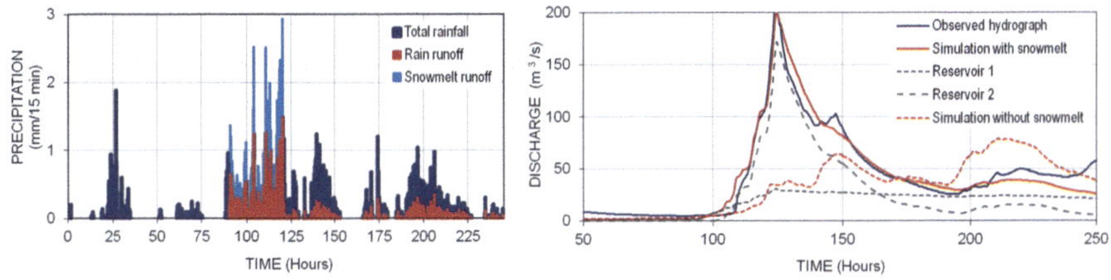

Fig. 3 Hyetographs and hydrographs for January 2009 episode in Roncal Valley

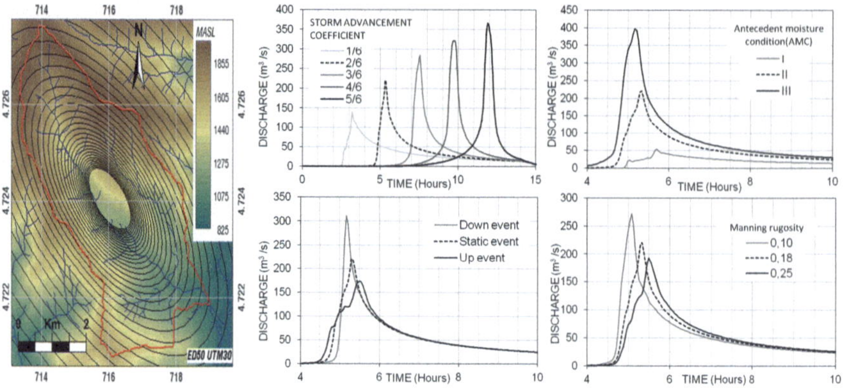

Fig. 4 Arás watershed with isohyets of standard event and hydrographs for process variation

2.3 Variability Analysis

Very severe flood flow occurred in the Arás watershed. Thanks to the simulations, a variability study was performed taking into account the changes in the main hydrological processes, such as the spatial and temporal rainfall distribution, the soil moisture conditions and the routing through drainage networks (Fig. 4). It is concluded that the soil moisture condition is the most determining process related to the magnitude of the flood flows produced by a similar rainfall amount.

3 Conclusions and Outlook

The SHEE program permits to configure multiple models of different hydrological processes such as spatial and temporal rainfall distribution, previous soil moisture state and water routing thorough the basin. The use of PLR models for the simulation of real episodes is very interesting for the works done. In the future, as more severe episodes analyses are assessed, it is clear that this software will permit to have a better knowledge of the responses of hydrological processes in extreme conditions.

References

1. Mateo-Lázaro, J., Sánchez-Navarro, J. A., García-Gil, A., & Edo-Romero, V. (2013). Developing and programming a watershed traversal algorithm (WTA) in GRID-DEM and adapting it to hydrological processes. *Computers & Geosciences*, *51*, 418–429.
2. Mateo-Lázaro, J., Sánchez-Navarro, J. A., García-Gil, A., Edo-Romero, V. (2012). Research the Snowmelt in Floods. (I) Adapting to the Linear Reservoir Models. (II) A real Event Assessment in a Gauged Catchment of the Pyrenean Mountains: Roncal Valley, Spain. 8.299-300 O communications. Geo-Temas, 13: VIII Geological Congress of Spain.

Mathematical Methods of the Data Analysis in a Prospecting Database for Geological Mapping

Oleg Mironov

1 Using Archive Data for Geological Mapping

Archives of geological prospecting data contain a large volume of geological information that should be re-used for various tasks.

In Moscow the project of geological and environmental mapping of the city territory at scale 1:10000 was fulfilled in 2007–2009 [1]. Twelve maps have been compiled for an area of more than 1000 km^2. The source information is the borehole database of geological archive that has been collected during the last 100 years. About 91000 boreholes logs have been used. The mapping used no new field investigations.

Before using archive data must be re-interpreted according to new challenges and modern geological concepts. This task is complicated because of following reasons.

1. Particular features in the geological structure admit an ambiguous stratigraphical definition of the layers' age even in contemporary investigations.
2. Borehole logs and other archive data expose a point of view of unknown specialists and there is a problem if we can trust it.
3. Usually engineering-geological reports describe lithological layers well but do not contain verified stratigraphical information.
4. Up to the last time the computer methods of data collecting and verification have not been exploited, therefore there are lacunas and banal errors in the source documents.

A large volume of data requires mathematical and computer methods for formal data verification. Such methods can find controversial points in data and show them to geologists who must take a final decision. It is not correct to use mathematics instead of the geological analysis and fix a database independently.

O. Mironov (✉)
Sergeev Institute of Environmental Geoscience, Russian Academy of Sciences,
Ulanski per. 13, P.O.box 145, Moscow 101000, Russia
e-mail: geoinf@geoenv.ru

E. Pardo-Igúzquiza et al. (eds.), *Mathematics of Planet Earth*,
Lecture Notes in Earth System Sciences, DOI: 10.1007/978-3-642-32408-6_69,

Mathematical methods are based on some formal rules that have their origin in theoretic concepts of the geological structure in considered area. Observed data deviations from those rules should be interpreted as:

1. A geological anomaly (e.g. a fault or a buried erosion valley).
2. An occasional deviation due to local specific (very rarely).
3. An interpretation mistake.
4. An error in the source data.

Two applications for Moscow borehole data are considered below. Moscow is situated in the centre of the Russian platform. The sediment formation was smooth, therefore concordant bedding is assumed as a primary approximation for the geological structure. There are no tectonic faults. The main geological features at the Moscow territory are erosion valleys of Carboniferous and pre-Quaternary (Jurassic) age which are fulfilled with posterior sediments. For example, at some places Holocene deposits are situated just over Jurassic or Carboniferous deposits.

At other regions analogous approaches may detect faults or unconformities.

2 The Analysis of the Correlation Diagrams for Borehole Logs

The formal definition of a correlation diagram for two boreholes is as follows [2]. Let x be the absolute height of the roof of some layer in the first borehole and y be the absolute height of the roof of the same layer in the second borehole. We attach a point (x,y) in the coordinate plane to this layer. The correlation diagram consists of all points attached to common layers in two boreholes. Layers that present in one of boreholes and miss in another one are displayed with vertical or horizontal lines.

Suppose the rate of the sediment formation at various points was the same or proportional. Then at a geological cross-section the layers marks are situated at the parallel or perspective lines (see Fig. 1a, b) and the points at the correlation diagram will lie on a straight line. The local fluctuations should be stochastic and small.

If one of boreholes is positioned in the buried erosion valley (Fig. 1c), then the upper point at the correlation diagram will deviate from a line drawn through other ponts. Upper layers from in another borehole may be absent in that borehole.

Visual analysis of correlation diagrams is straightforward. Various estimations may be used for straightness of the correlation diagram, e.g. a correlation coefficient between x and y or eccentricity of the first two principal components ellipse. Marks in boreholes are correlated; therefore thresholds for essential estimations should be established according analysis of practical data.

An example is shown at Fig. 2. The borehole 113459 is situated near the thalweg of a buried erosion valley. The mark of C_3rt layer roof in it is lower than predicted with preceding layers. Moreover, layers C_3nv and C_3pr are absent in this borehole. The correlation coefficient in this example is equal to 0.9.

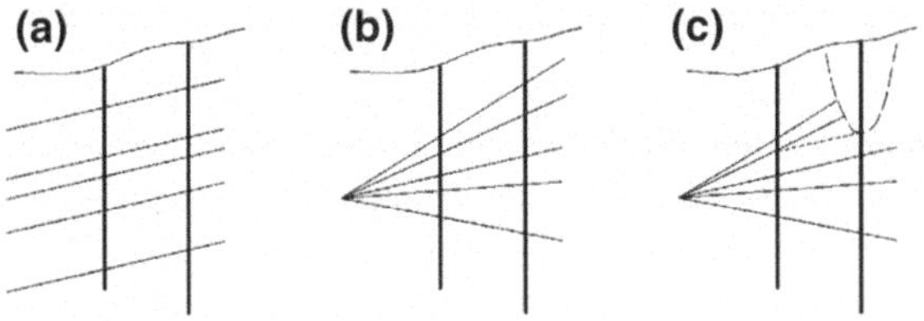

Fig. 1 Variants of layers position in boreholes

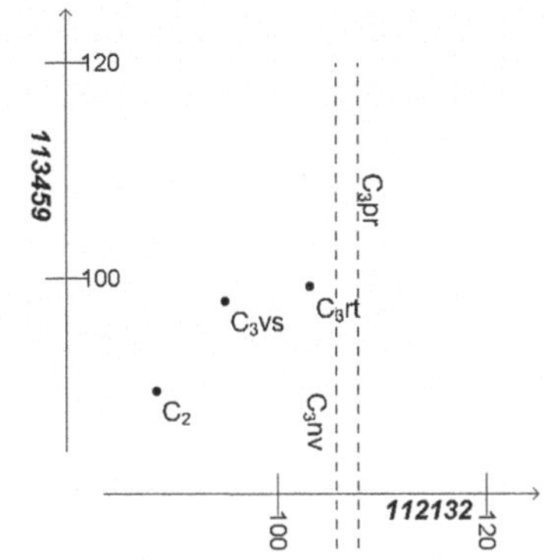

Fig. 2 Correlation diagram for boreholes 112132 and 113459

Correlation diagrams are useful to data verification. Pairs of boreholes with small values of correlation coefficients should be analyzed thoroughly, especially if the distance between boreholes is small too. It is reasonable to consider the scatterplot “distance—correlation coefficients”.

More than 3000 boreholes that exposed pre-Quaternary deposits were selected from the Moscow borehole database and correlation coefficients were calculated for pairs of boreholes with distance between them less than 1.5 km. About 85 % of pairs have correlation coefficients more than 0.95. Values less than 0.6 correspond to obvious data errors.

To test the hypothesis about the effect of geological anomalies 230 boreholes positioned in buried erosion valleys were selected. The minimum of their correlation coefficient with other boreholes situated at distance less than 500 m was calculated. Larger frequency of small values in this selection supports the hypothesis.

Statistics is purely descriptive. There is no statistical model for this problem yet.

3 The Trend Surface Analysis

We assume that the surface of deposits of fixed age in Moscow region was flat during the sediment formation. Tectonic movements may change its position in space but as a first approximation did not disturb the flatness. The fluctuations of flatness are caused by erosion processes but erosion zones occupy a small part of total area.

Suppose the trend plane for some deposits is found. Then large fluctuations of height marks in boreholes up from it should be interpreted as data errors or a wrong interpretation of drilling data. Large fluctuations of height marks in boreholes down from the trend surface may be interpreted as buried erosion valley.

Iteration process is used to find trend surface. The idea is to use the interpolated surface for the linear trend approximation only at those zones where a good quality of an interpolation can be guaranteed. At each step a set of boreholes is fixed (all data at the first step). The iteration consists of five steps.

1. Delaunay triangulation of source data point set is created.
2. "Large simplexes" are eliminated from the triangulation according the extended α-shape method [3, 4]. The result is called a shape.
   A "size" of a simplex is defined with a special function. The simplest geometry definition for a "size" is the minimal diameter of a disk that can cover the given simplex. Special "size" functions may be defined as a combination of geometry size of a simplex and variance of various kinds of data at its vertexes. Examples of shapes are demonstrated on Fig. 3.
3. Piecewise linear interpolation is used to build the surface over the selected shape.
4. The trend plane is the least-squares linear approximation for the surface built at step 3.
5. Differences between the trend plane and source height marks in the boreholes are calculated. Boreholes with large absolute values of differences are eliminated of the source data and investigated to explain fluctuations.

Steps 1–5 repeat until the trend plane stabilizes. In practice, 3 or 4 iterations are sufficient.

Algorithms described above are useful in the verification process for Moscow database. The majority of disputable data detected was explained later as a wrong data interpretation at the preceding stages of investigations or lack of information in the source documents.

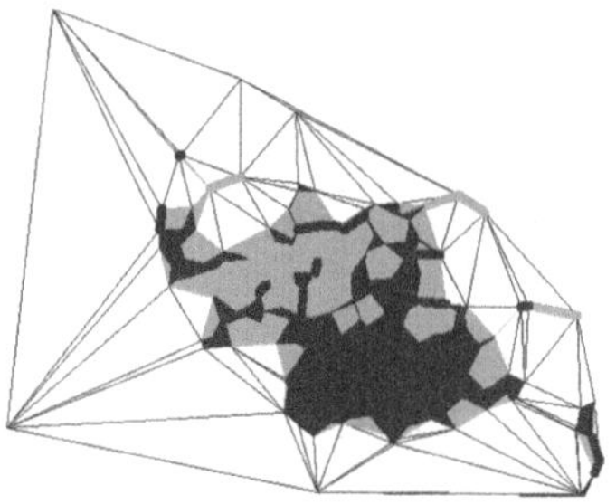

Fig. 3 The Delaunay triangulation of a point set and 2 α-shapes of this set (*shades* of a *gray color*)

References

1. Osipov, V. I., Burova, V. N., Zaikanov, V. G., Molodykh, I. I., Pyrchernko, V. A., & Savis'ko, I. S. (2012). A map of large-scale (detail) engineering geological zoning of Moscow territory. *Water Resources*, *39*(7), 737–746.
2. Guberman, Sh. I. (1987). *Informal data analysis in geology and geophysics*. Moscow: Nedra (in Russian).
3. Mironov, O.K. (2007). A shape of a point set. Applications to environmental and engineer geological cartography. Geoecologiya, inzhenernaya geologiya, hydrogeologiya, geocriologiya, 2, 173–179 (in Russian).
4. Edelsbrunner, H., & Mucke, E. P. (1994). Three-dimensional alpha shapes. *ACM Transactions on Graphics*, *13*, 43–72.

Geoprocessing Tool Regenerat: Characterization of Mineral Resource Quality of Renewable Sediment Deposits

Heinz Reitner, Sebastian Pfleiderer, Maria Heinrich, Irena Lipiarska, Piotr Lipiarski, Julia Rabeder, Thomas Untersweg and Ingeborg Wimmer-Frey

1 Introduction

In mountain regions, sand and gravel deposits of alluvial fans and debris cones can represent valuable mineral resources of aggregates if they constitute reasonable volumes and show good quality with regard to building material requirements. While quantity can often be estimated from the surface area of the deposit and thickness information from boreholes, the quality aspects are difficult to handle because sediment quality analyses are rarely available. Taking samples and analyzing them is often too expensive and time consuming, especially at a regional scale for a multitude of deposits.

Quality parameters of these deposits are largely controlled by morphology and geology of their provenance areas. The delineation of a certain provenance area and information about the morphology parameters can be derived utilizing digital elevation models (DEM). Information on the geology of the area can be found in existing geological maps of the study region, unfortunately the maps often do not describe the lithological properties of the sediment deposits. Therefore, by combining both information layers using a Geographic Information System (GIS) [1], an automated method was developed to provide a tool for deriving quality proxies.

The results of the GIS analysis were compared to the results of sedimentary analyses of samples at selected test sites [2].

H. Reitner (✉) · S. Pfleiderer · M. Heinrich · I. Lipiarska · P. Lipiarski · J. Rabeder · T. Untersweg and I. Wimmer-Frey
Geological Survey of Austria, Neulinggasse 38, Vienna, Austria
e-mail: heinz.reitner@geologie.ac.at

E. Pardo-Igúzquiza et al. (eds.), *Mathematics of Planet Earth*,
Lecture Notes in Earth System Sciences, DOI: 10.1007/978-3-642-32408-6_70,

2 Methods

To establish a connection between sedimentary deposits, provenance area morphology and geology, digital elevation models (DEMs) and geological maps were analyzed using a GIS. A geoprocessing tool was developed which proceeds as follows (Fig. 1):

After selection of sedimentary deposits in the digital geological map (1), the DEM was used to delineate the morphological catchment area of the sedimentary deposit (2) and to derive morphological parameters of the catchment area. Subsequently the geological units within the catchment area were selected (3) and their area percentage calculated (4).

For GIS analysis a raster based approach was chosen. The DEM was already available in raster format, but local depressions (so called sinks) had to be removed for the GIS algorithm to run smoothly. The DEM data then were used to generate

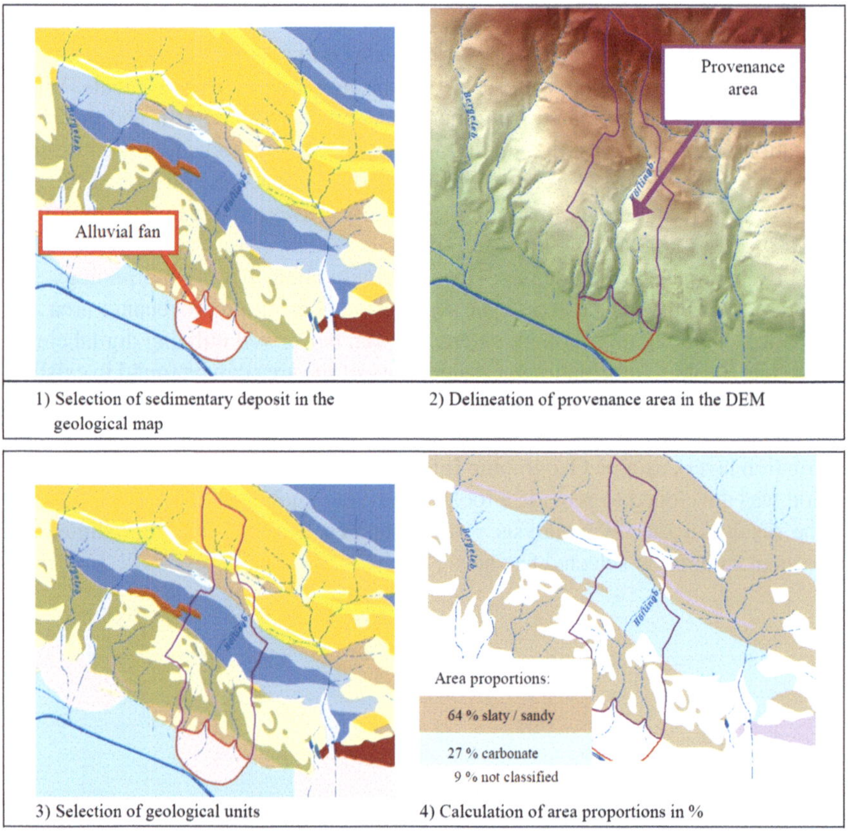

Fig. 1 Method of GIS analysis

flow accumulation and flow direction grids of the study area. For the geological maps a rasterization step was integrated into the geoprocessing tool.

For development the software ArcGis 10 [3] with its visual geoprocessing and programming environment ModelBuilder was selected. Within ModelBuilder a set of geoprocessing tools can be connected and the output of one geoprocessing tool can be used as input for the next tool. The ArcGis extensions “Arc Hydro Tools” and “Hydrology” provide geoprocessing routines for delineating a watershed by utilizing the raster calculation extension “Spatial Analyst”.

The ModelBuilder environment was extended using the programming language Python to provide additional functionality for accessing input parameters. A new graphical user interface was added to the ArcGis toolbox to serve as input form for geoprocessing parameters (Fig. 2).

Using ModelBuilder, the different steps of the analysis were combined into a single geoprocessing tool, called “Regenerat”.

The Regenerat geoprocessing tool can be applied to a multitude of selected sedimentary deposits at once. The tool processes the list of selected sedimentary deposits one by one, storing the resulting spatial objects and calculated attributes in a geodatabase for further evaluation.

For verification, the parameters were correlated with sediment properties derived from sediment analysis of samples taken in the field at selected sediment deposits [4].

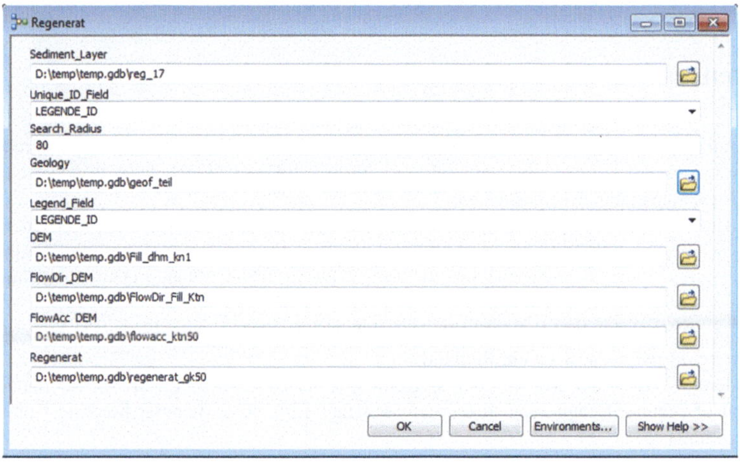

Fig. 2 Input form for geoprocessing parameters

3 Results

For each selected sedimentary deposit the provenance area and the according relevant parameters are calculated, such as surface area value, altitude range, mean and maximum values of flow length, minimum, mean and maximum of slope angle and the percentage of surface area of the lithostratigraphic units of the provenance area.

The results provide a quantification of the morphological and geological properties of the provenance area to support the evaluation of the potential quality of the deposit.

4 Discussion

If the morphology of the terrain exhibits a distinctive alpine relief, the geoprocessing tool is successful in the delineation of the provenance areas. Detailed geological maps of the study areas are essential for a successful interpretation of results in terms of sediment properties and resource quality. A regional study of aggregate resource quality in Austria is planned to further enhance the correlation of the GIS results with the results of conventional sediment analysis of samples taken in the field.

Acknowledgments The study was funded by the Austrian Academy of Sciences.

References

1. De Smith, M.J., Goodchild, M.F., & Longley, P.A. (2012). Geospatial analysis—a comprehensive guide to principles, techniques and software Tools. (4th ed.). Winchelsea: Winchelsea Press. http://www.spatialanalysisonline.com
2. Pfleiderer, S., Untersweg, T., Heinrich, M., Lipiarska, I., Lipiarski, P., Rabeder, J., Reitner, H., & Wimmer-Frey, I. (2012). Regenerat"—Methodenentwicklung rohstoffgeologische Evaluierung regenerativer Lockergesteinsvorkommen.—Final report, 32 p., Vienna: Geological Survey of Austria.
3. ESRI ENVIRONMENTAL RESEARCH INSTITUTE (2010). ArcGIS for Desktop 10.-GIS Software, http://www.esri.com, Redlands
4. Pfleiderer, S., Heinrich, M., Rabeder, J., Reitner, H., & Untersweg, T. (2013). Automated evaluation of renewable aggregate resources. Abstract 12th SGA Biennial Meeting Conference Proceedings.

Spatiotemporal Data Model for Multi-Factor Geological Process Analysis with Case Study

Gang Liu, Xiang Que, Xiaonan Hu, Shanjun Tian and Jiacheng Zhu

1 Introduction

A spatiotemporal database can presents spatial, temporal, and spatiotemporal database, it captures spatial, attribute and temporal aspects of data. All the individual spatial and temporal concepts must be considered to construct the data conceptual model [6], which can be used to support the corresponding spatial analysis [3]. Many GIS data models have been designed to incorporate temporal information into spatial databases, such as the Snapshot Model [1], the Space-Time Composite Model [4], Event-based Model [5] and Object-oriented Data Models [2].

Current event-based or event-driven spatiotemporal data model (ESTDM) can express causality of space-time changes during the spatiotemporal process. However, it cannot convey many interior factors that causing event changes or relationship between events, even hard to extract event types from specific process. A spatiotemporal data model for multi-factor geological process analysis is proposed to better support complex process simulation and analysis.

2 Spatiotemporal Data Model for Multi-factor Geological Process Analysis

The concept of process is used in many fields, here means the event (activity) sequences of a spatial object during its life cycle. Spatiotemporal process is the evolution course of a spatial object along with the time progress during a period. Geological spatiotemporal process is the evolution progress of geological object or

G. Liu (✉) · X. Que · X. Hu · S. Tian · J. Zhu
School of Computer Science, China University of Geosciences, Wuhan, China
e-mail: liugang67@163.com

E. Pardo-Igúzquiza et al. (eds.), *Mathematics of Planet Earth*,
Lecture Notes in Earth System Sciences, DOI: 10.1007/978-3-642-32408-6_71,

object set along with time, which is the process with multi-granularity responding to effective events.

Figure 1 shows the conceptual model of spatiotemporal data model with four layers considering the expression of geological spatiotemporal process.

For the design of spatiotemporal data model, it is composed of spatiotemporal class, geological object class, event class, status class, feature class, observation class and geological model class, corresponding to the expression of geometry, attribution, spatial relationship, semantics and behavior of geological spatial objects. The event is generated by the object regularly or irregularly according to observation data or other logical judgment.

The so-called extended spatiotemporal data model for geological process is Geological Event Multi-factor Driven Model, as depicted in Fig. 2.

Main operation mechanism of this GEMDM model is described as follows: (1) Application analysis model loaded into simulation model base by user;

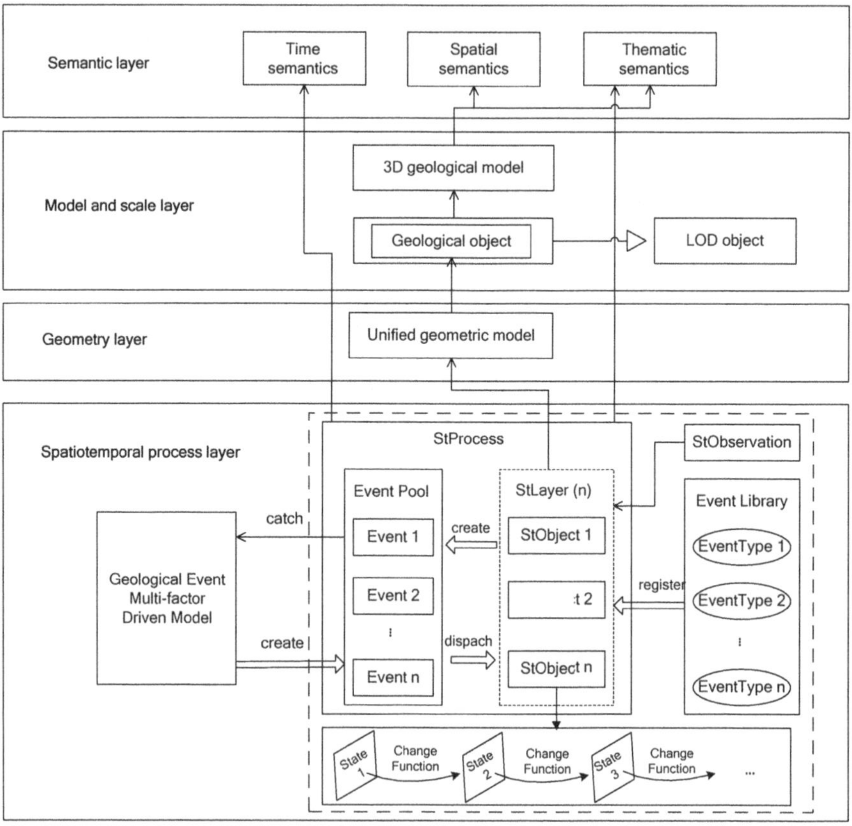

Fig. 1 Conceptual model of spatiotemporal data model with four layers considering geological spatiotemporal process (The basic model within gridlines is given by Wu Huayi from Wuhan University, China)

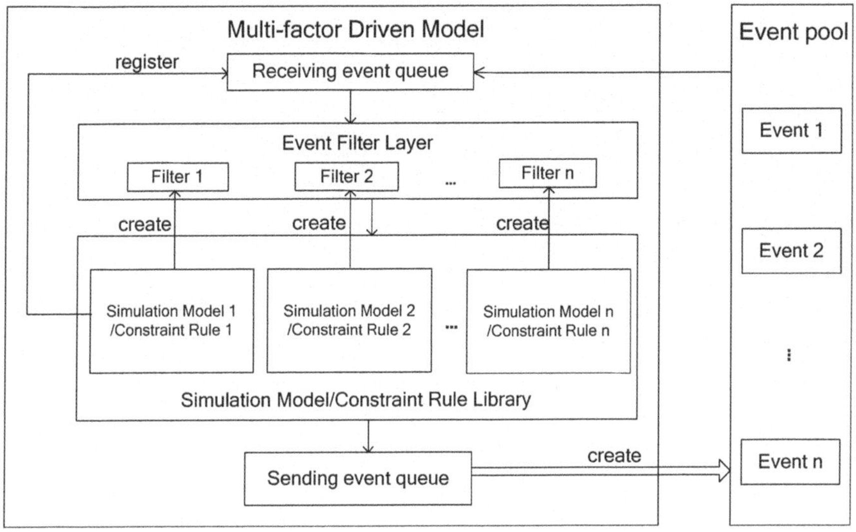

Fig. 2 Geological Event Multi-factor Driven Model (GEMDM)

(2) Creating corresponding event filter in the event filtering layer; (3) Registering related events to the receiving event queue; (4) When all the necessary events are received by event filters, corresponding events in the receiving queue are transferred to simulation model; (5) The simulation results are given out in the form of new events and registered into event pool for further processing.

3 Case Study

An example of the supervision and control on typical coal mine production of China is given to test the data model. Generally, there are more than 30 types sensor data included in coal mine monitoring in China, such as sensors about carbon monoxide, wind speed, wind pressure, temperature, gas, negative pressure, water level, coal level, flow, active power, reactive power, current, power factor, frequency, voltage, active energy, reactive energy, volume of production, Coal bunker full or empty, position, fault detection and so on. Most of these sensors are related to gas and device monitoring. However, some of them are associated with geological events, such as water level, tunnel roof pressure and working surface status.

Gas supervision plays a key role in the coal mine safety production. For the coal seam pressure prediction model, multi-factor driven model will receive following events: gas density change, gas gushing amount change, change and mining depth change. By means of model calculation and prediction, higher level events will be generated: normal gas density, abnormal gas density, normal gas gushing amount, abnormal gas gushing amount, normal gas pressure, abnormal gas pressure, etc. which can be used in subsequent processing. The time series process is shown in Fig. 3.

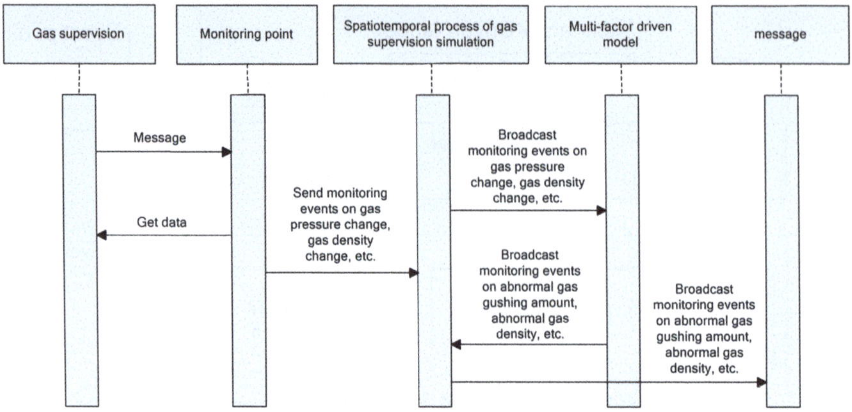

Fig. 3 Timing diagram of gas supervision and simulation process

Based on coal mine sensor network, system will receive many types of supervision events, such as tunnel extending event, mining surface change event, gas density change event, gas gushing amount change event and gas pressure change event. By the evaluation and analysis of multi-factor driven model, higher level events will be generated, such as multi-factor geological and environmental normal or abnormal events. According to the result of spatiotemporal process simulation, synthesize anomaly information can be used in mine safety production decision-making.

4 Conclusion

Geological process is a complex spatiotemporal evolution process. A spatiotemporal data model with extended multi-factor driven model is proposed in order to better support geological process analysis. Geological event multi-factor driven model is setting up event filter, constraint rule and simulation model to satisfy multi-level geological event processing. An example of coal mine supervision application focusing on gas supervision process is described to show the mechanism and feasibility of this model. Further researches include time granularity consistency and semantics model of geological phenomenon.

This work is supported by National High Technology Research and Development Program of China (No. 2012AA121401) and NSFC (No. 41172300).

References

1. Armstrong, M. P. (1988). Temporality in spatial databases. Proceedings: GIS/LIS'88, (vol. 2, pp. 880–889).
2. Worboys, M. F. (1995). *GIS: A computing perspective*. London: Taylor & Francis.
3. Yuan, M. (1996). Temporal GIS and spatio-temporal modeling, 3rd International Conference on Integrating GIS and Environmental Modeling. University of California. Santa Barbara-CA, USA. pp.21–26.
4. Langran, G., & Chrisman, N. R. (1988). A framework for temporal geographic information. *Cartographica, 25*(3), 1–14.
5. Peuquet, D. J., & Duan, N. (1995). An event-Based Spatiotemporal data model (ESTDM) for temporal analysis of geographical data. *International Journal of Geographical Information Systems, 9*, 7–24.
6. Koubarakis, M., & Sellis, T. (2003). *Spatio-temporal databases*. Germany: Springer.

Use of Variational Methods in Geological Mapping

Andrey N. Sidorov, Andrey G. Plavnik, Andrei A. Sidorov and Michail S. Shutov

1 Introduction

Splines have been used in geological mapping since 1970s, when some fundamental studies on the topic [1–3] were published. Initial use of spline-functions in geology [4] discovered a great potential in modeling of geological surfaces and fields. The spline-smoothing method became the most popular, because underlying geological data contains inherent 'noise'. Suppose that X, Y, Z are Gilbert spaces. An element $x^* \in X$ is called *smoothing spline* if:

$$x^* = \arg\min_{x \in X}(\|Tx\|_Y^2 + \rho\|Lx\text{-}z\|_Z^2). \tag{1}$$

L—being the linear operator, which describes a measurements model, z—being the measurements values, *T*—being the energy operator, which describes the physical nature of the mapping field and $\rho > 0$—being the smoothing parameter. For voluntary *T*—operator we are suggesting the approximal solution, based on Ritz-Galerkin method. By applying bicubic b-splines as the basis functions, we transform problem (1) into a system of linear algebraic equations with a dispersal symmetric matrix.

A. N. Sidorov · A. A. Sidorov · M. S. Shutov
Centre of Rational Use of Resources, 75 Malygina Street, 625026 Tymen , Russian Federation
e-mail: sidorov@crru.ru

A. A. Sidorov
e-mail: darth@crru.ru

M. S. Shutov
e-mail: shutov@crru.ru

A. G. Plavnik (✉)
West-Siberian Affiliate of Institute of Petroleum Geology and Geophysics SB RAS,
74 Taimyrskaya Street, 625026 Tymen , Russian Federation
e-mail: plavnik@ikz.ru

E. Pardo-Igúzquiza et al. (eds.), *Mathematics of Planet Earth*,
Lecture Notes in Earth System Sciences, DOI: 10.1007/978-3-642-32408-6_72,

This approach, applied in GST (Geo-Spline Technology) software, allows usage of not only values of geological field in the given points, in the mapping process, but also derivatives (gradients or curvatures), differential equations and integral values.

2 Regional Mapping Problems

Today, one of most topical problems in geological mapping is construction of detailed digital surface and property field models for large regions. One of the principal factors in mapping of large territories is reliance on raw data of varying type and quality. For example, it is in principle impossible to perform a new interpretation of all seismic surveying carried out in Western Siberia over the past 50 years, applying same methodology and parameters and achieve an accurate result. In reality we have to use the interpretation made during this period by different specialists applying different methods. For example, for some territories instead of a full seismic interpretation we only have fragments of time, depth or contour maps, whereas for other regions, only maps of different seismic horizons may be available. Our experience in surface modeling in Western Siberia region with the spline smoothing method is described in further detail below.

If we construct a grid model in an ordinary way using misaligned seismic interpolation, the result will be as shown in Fig. 1a. We have discovered that, the use of this horizon shape, i.e.—first derivatives (gradients) along the seismic line, in the mapping process, is more appropriate than the use of values of geological surface in shot points. A 'shape' is a more reliable parameter than t_0 values, as it is not dependent on a seismic phase or any adjustments that may be made.

In the suggested approach (1) L—operator can be modified in a way that the values in shot points should be considered as first derivatives along the seismic line. The mapping result made with this method is shown in Fig. 1b.

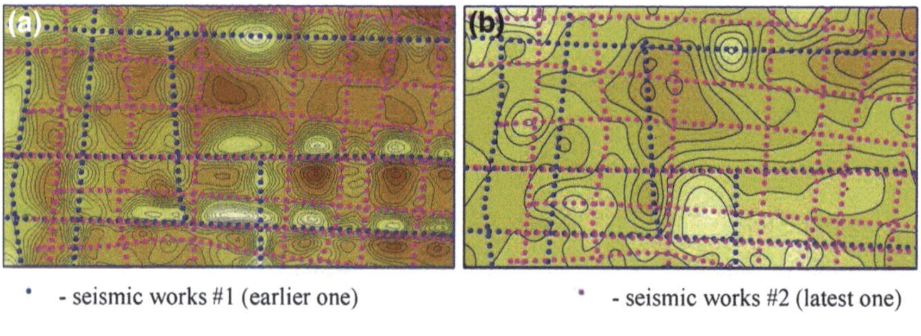

Fig. 1 **a** 'Structure noise' appears when seismic interpretation, made by different specialists is used in mapping process at ones in the ordinary way; **b** If the shape of horizon (first derivatives) along the seismic line is used, 'noise' caused by regular discrepancies in interpretation, disappears

We have used more than 2000 seismic surveys, approximately 10000 wells and other additional data (grids, contour maps etc.) to build a detailed structural map of the basis seismic horizon in Western Siberia that corresponds to the top of Jurassic deposits (Fig. 2).

If the seismic or well data used for the mapping of certain geological surfaces is insufficient, it is possible to use a relationship between the surface being mapped and an existing geological field. This relationship is described by T-operator and expressed as a set of differential equations: we call it—*model of knowledge*. In general there are inhomogeneous, up to second order, differential equations with coordinate depending coefficients.

The suggested approach makes it possible to construct maps of properties that correspond to the physical behavior of the mapping parameter. For example, we can modify T-operator as a Poisson equation to simulate a static fluid pressure field. We are using a simple model of knowledge to build inner Jurassic surfaces, which are not supported by detailed seismic interpretation:

$$\frac{\partial H_{a\text{-}t}}{\partial x} = a\frac{\partial H_{a\text{-}b}}{\partial x}, \quad \frac{\partial H_{a\text{-}t}}{\partial y} = a\frac{\partial H_{a\text{-}b}}{\partial y}. \tag{2}$$

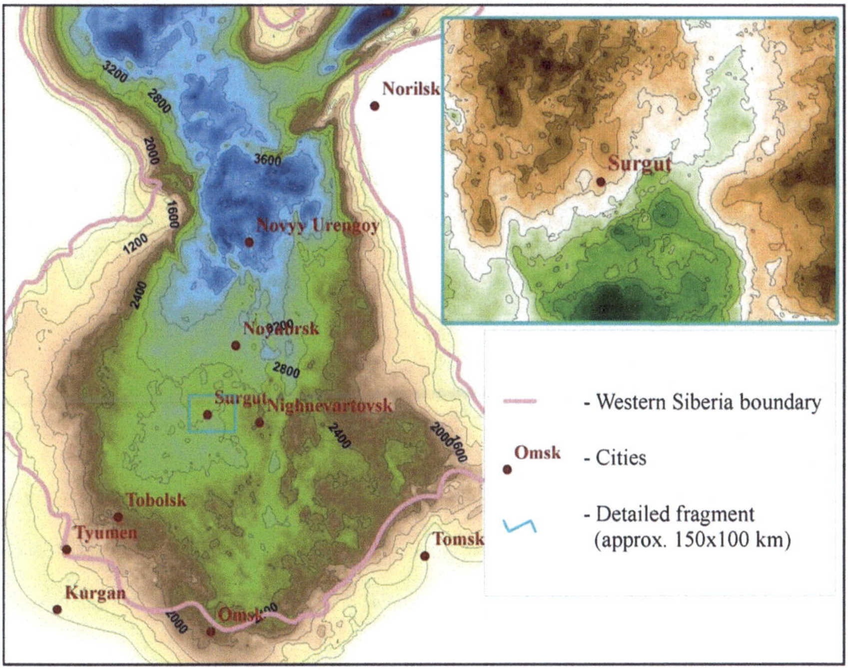

Fig. 2 Top of Jurassic deposits structure map for Western Siberia and detailed fragment of this map; Grid size is 1500 × 2000 km. approx., grid cell size—1 × 1 km

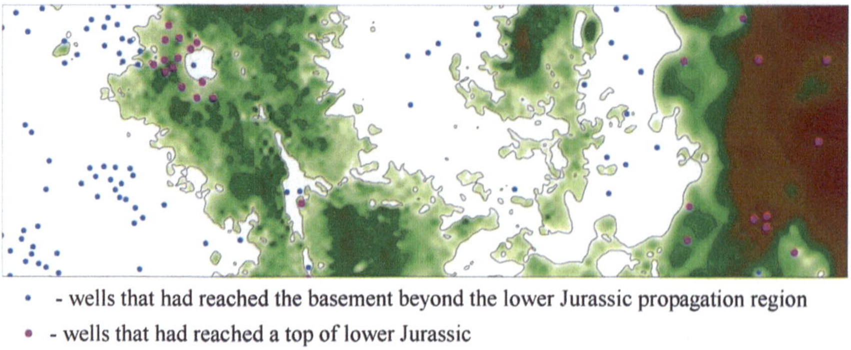

Fig. 3 Early Jurassic deposits thickness map; '*red*' wells were used in a mapping process as data, '*blue*' wells were used as test points; in more than 95 % of test points the model predicts zero early Jurassic thickness, as it should be

H_{a-b}—is a Jurassic deposits thickness map: we have built this detailed map using all available seismic interpretation and well data. H_{a-t}—is a thickness map between a mapping surface and bottom of Jurassic deposits, *a*—coefficient of conformity. When $a = 1$ the mapping surface has a shape as a top of Jurassic deposits, $a = 0$ when the inner horizon has a shape of a basement. This coefficient can be calculated during a mapping task solving: in this case it will be the most appropriate to satisfy all initial data. Despite its simplicity, model (2) is very reliable for mapping of Jurassic horizons in Western Siberia, because it demonstrates good predictability (Fig. 3).

Therefore, we can conclude that due to the *T*-operator, mapping is not a simple interpolation between data points, but a process of real mathematical modeling. In the simplest case *T*-operator can be transformed in to a well-known *minimum curvature* condition or if the mapping field can be described by a set of differential equations, a specialist may apply a more complicated model of knowledge.

References

1. De Boor, C. (1978). *A practical guide to splines*. New York: Springer.
2. Zavjalov, Y. S., Kvasov, B. I., & Miroshnichenko, V. L. (1980). *Methods of spline functions*. Moscow(Russia): Nauka.
3. Vasilenko, V. A. (1983). *Spline functions: theory, algorithms, programs*. Novosibirsk(Russia): Nauka.
4. Volkov, A. M. .(1980). *Solution of practical problems on computers*. Moscow(Russia): Nedra.

Comparison of Methods for Depth to Groundwater Calculation in Hard Rock Areas

Ronny Laehne, Dorothee Altenstein and Wolfgang Gossel

1 Introduction

There are several methods for depth to groundwater calculation based on different information. Information about geology or creeks are used to calculate the depth to groundwater in an easy way. On the one hand, there are some methods which directly calculate the depth to groundwater. On the other hand, many methods calculate the groundwater surface at a first step. These results got subtracted by the DEM information for calculating depth to groundwater.

Armbruster et al. [1] just used information about the geological units and set some fixed values for the depth to groundwater. For the depth to groundwater they defined a value for some aquifers like the Middle and Lower Bunter about 40 m (Fig. 1). This is a direct method for depth to groundwater calculation. The result correlates with the input data. Due to the simplified assumptions, the result looks non-realistic in some parts of the investigation. This method can be used to set these fixed values at a fist step of modeling.

In many other methods, an indirect way for calculating the depth to groundwater got used. Therefore at a first step, a groundwater surface is calculated [2, 3]. The following comparison describes the results of this first processing step. The chosen methods base on the assumption, that all creeks are connected to the groundwater.

R. Laehne (✉) · D. Altenstein · W. Gossel
Martin-Luther-University, von-Seckendorff-Platz 3, 06120, Halle, Germany
e-mail: ronny.laehne@geo.uni-halle.de

D. Altenstein
e-mail: dorothee.altenstein@student.uni-halle.de

W. Gossel
e-mail: wolfgang.gossel@geo.uni-halle.de

E. Pardo-Igúzquiza et al. (eds.), *Mathematics of Planet Earth*,
Lecture Notes in Earth System Sciences, DOI: 10.1007/978-3-642-32408-6_73,

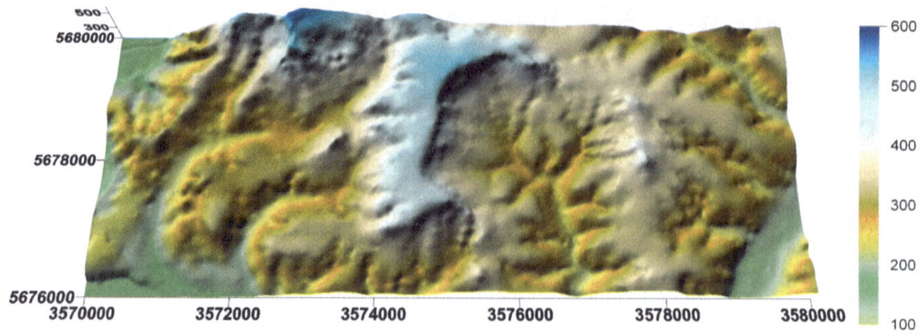

Fig. 1 Digital Elevation Model of the hard rock investigation area without any observation well (near Eschwege, Hesse, Germany)

2 Dataset

Based on one dataset, different opportunities to calculate the groundwater surface and the depth to groundwater were tested. The dataset just contains DEM information, shape of the creeks, creek direction and DEM slope. For the method of [1] additional information about the geology and the unit thicknesses are necessary. For the last mentioned method, fixed values for depth to groundwater are defined, e.g., for the unit Muschelkalk 20 m got considered. The result for the depth to groundwater calculation reflects the data input (Fig. 2). The method is a opportunity to get a fast overview about the depth to groundwater, but the result is not sufficient.

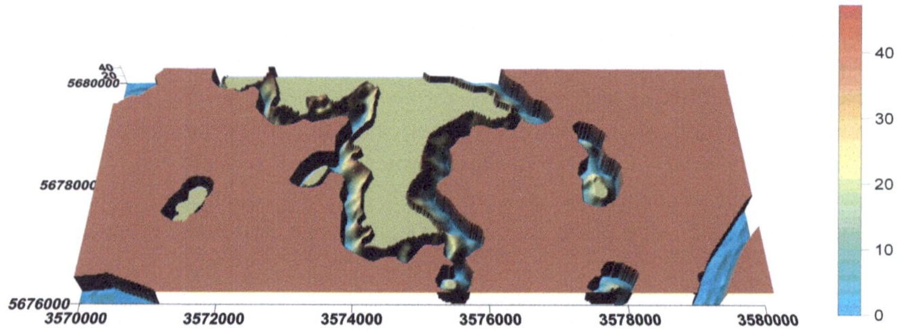

Fig. 2 Depth to groundwater, based on the method of [1]

3 Comparison of Different Methods

By comparing the indirect methods, the results mainly depend on the chosen interpolation method, in this case Nearest Neighbor interpolation, Triangulation, IDW and Kriging.

In a fast way, a Nearest Neighbor interpolation should be performed to generate the groundwater surface. [3] describes a method based on the creek dataset and later Triangulation. This method gives a non-realistic result. But it is useful to take this result as a first step in groundwater modeling e.g., as an input for some groundwater recharge calculations (Fig. 3). If it is necessary to create a more realistic result for the groundwater surface, interpolation methods like IDW or Kriging or the method Hydro-FaBer should be preferred.

If the dataset (DEM information, shape of creeks) is just interpolated by the IDW method, the result is more realistic, but there are some poor results in special parts of the investigation area. This method and the method based on the Kriging interpolation, don't take into account, that there can be made some assumptions for optimizing the result. Usually the creeks are under effluent conditions and the shape of groundwater surface depends on the DEM slope and the geology. The method Hydro-FaBer integrates the effluent conditions and the DEM slope to calculate the groundwater surface.

Based on these information and the assumptions a point based calculation for the depth to groundwater was done. The result (Fig. 4), e.g. the shape of the groundwater isolines is more realistic than the results of the other methods. A comparison to measured values was not possible due to the lack of these information.

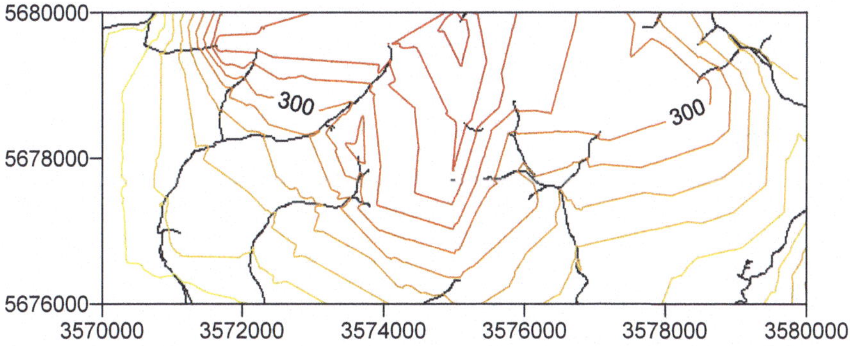

Fig. 3 Groundwater elevation, based on the method of [3]—TIN interpolation (dataset: creek elevation)

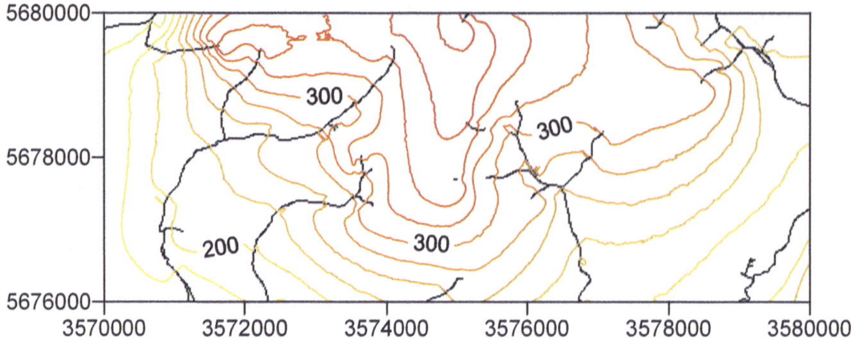

Fig. 4 Groundwater elevation, based on Hydro-FaBer method [2] and Kriging interpolation (dataset: creek elevation, creek direction, creek slope)

4 Conclusion

The method Hydro-FaBer can be used in areas with poor data amount about the groundwater surface. Thus, an overview and an input parameter for the hydrogeological modeling process got generated. Especially for assessing groundwater recharge, vulnerability mapping or as data input for numerical groundwater models in hard rock areas this method produces very good results for the groundwater surface.

For Hydro-FaBer only information about the creek slope, the creek direction and a DEM are needed. Therefore, the developed method studs for generating reasonable depth to groundwater, represented by a poor database. As a result, the comparison of different opportunities for calculation depth to groundwater shows the significant advantages of the method Hydro-FaBer.

References

1. Armbruster, V., Dudda, W., Hammer, J., Häffelin, H., Herre, H., Kilger, B.-M., et al. (2004). Hydrogeologische Erkundung Baden-Württemberg, Grundwasserdynamik, Grundwasserhaushalt, Grundwasserschutz, Enztal-Pforzheim, Mappe 3. - LGRB Baden-Württemberg.
2. Lähne, R., Gossel, W., & Wycisk, P. (2011). *Hydro-FaBer: A new tool for depth to groundwater Calculation*. Salzburg: IAMG. http://www.geologie.uni-halle.de/igw/umgeo/publikationen/Laehne_et_al_IAMG_2011_Hydro_faber.pdf.
3. Leppig, B. (2004). GIS gestützte Methoden zur räumlich-geometrischen Modellierung der Grundwasserdruckhöhe in Festgesteinen. *Mitt. Ing. Hydrogeol*, *88*, 11–25 (Aachen).

Line-Geometry-Based Inverse Distance Weighted Interpolation (L-IDW): Geoscientific Case Studies

Wolfgang Gossel and Michael Falkenhagen

1 Introduction

In geosciences parameter distributions and surfaces play an important role for the understanding of spatial processes. This information is, in most cases, represented by a point or cell data set in regular distances, a raster or grid. The values have to be interpolated to get this distribution. For certain topics, the basic information is given as isoline maps. This is advantageous compared to point data sets because additional expert knowledge can be inserted in the construction of the isolines. This method is used in geomorphology as well as in other geosciences. Topographical maps, geological structures, and contaminant distributions are drawn by experts not only based on point information but also on knowledge about coastlines, remote sensing images, the general directions of geological features (e.g. glacial channels or other erosive structures) or preferential groundwater flow directions.

Interpolation techniques can support the workflow from an isoline map to a quasicontinuous regular raster only to a certain extent. They are based on point data and use geometry based algorithms such as the triangulation method, the nearest neighbor method, the inverse distance weighted method or the spline method. Another approach is the use of spatial statistics: Geostatistical methods are used first to calculate the spatial correlation of the measured data, based on this parameter the interpolation is carried out. There are a few methods available to generate a raster directly

W. Gossel (✉)
Institute of Geosciences and Geography, Department of Hydrogeology and Environmental Geology, Martin- Luther University Halle,
Von-Seckendorff-Platz 3, D-06120 Halle, Germany
e-mail: wolfgang.gossel@geo.uni-halle.de

M. Falkenhagen
Institute for Applied Geosciences and Geography, Department of Remote Sensing and Cartography, Martin- Luther University Halle,
Von-Seckendorff-Platz 4, D-06120 Halle, Germany
e-mail: michael.falkenhagen@geo.uni-halle.de

E. Pardo-Igúzquiza et al. (eds.), *Mathematics of Planet Earth*, Lecture Notes in Earth System Sciences, DOI: 10.1007/978-3-642-32408-6_74,

based on isolines: (1) Triangulation takes the line parts as arms of the angle so that a fit to the lines is to a certain extent guaranteed. (2) Point interpolation methods can also be used by transferring lines to points but, in this case, line geometries are not persistent.

The utility of this method can be tested by applying it to certain processes such as flow paths, catchments as well as erosion, and sedimentation are processes that can be easily understood and estimated based on a visual analysis.

2 Methods

The L-IDW method is described in [1] in detail. In this method, the two nearest isolines are taken to calculate the value for a raster point or grid cell. In case of geometrical complications, other line parts are used for the estimation of the value as described in [1].

The L-IDW method provides the best interpolation results especially in "valleys" and "hills".

To test its practicality in geomorphology, flow path analysis derived from a digita l elevation model is used to compare the results, obtained with the L-IDW method with those achieved by some other approaches. The isolines of a topographical map of scale 1 : 25,000 have been digitized and interpolated with the L-IDW method to a 10 m raster. The midrange mountain area at the river Saale in Saxony-Anhalt, Germany, that was analyzed has heights between 98.75 m above sea level (masl) and 191.79 masl.

Another test has been focused on the outline of glacial channels. The erosional structures should be consistent in the interpolation result.

3 Results

In Fig. 1 a part of the resulting raster of the midrange mountain area is shown. The overlay of the flow paths is generated automatically by a GIS module. It is clearly visible that the flow paths meet the outline of the valley and also have nearly no breakdown in the current of the valley, as would be the case for flattened valleys in case of interpolation with triangulation.

In Fig. 2 the result for the base of the Quaternary in a part of Lower Saxony (Germany) is shown. One possibility for proving the result is the analysis of cross sections. The cross section perpendicular to the channel shows that there is no overshooting or undershooting with this method. The cross section along the glacial channel also shows the dominance of the isolines, that yields a smooth bottom of the valley without any "bulls eyes".

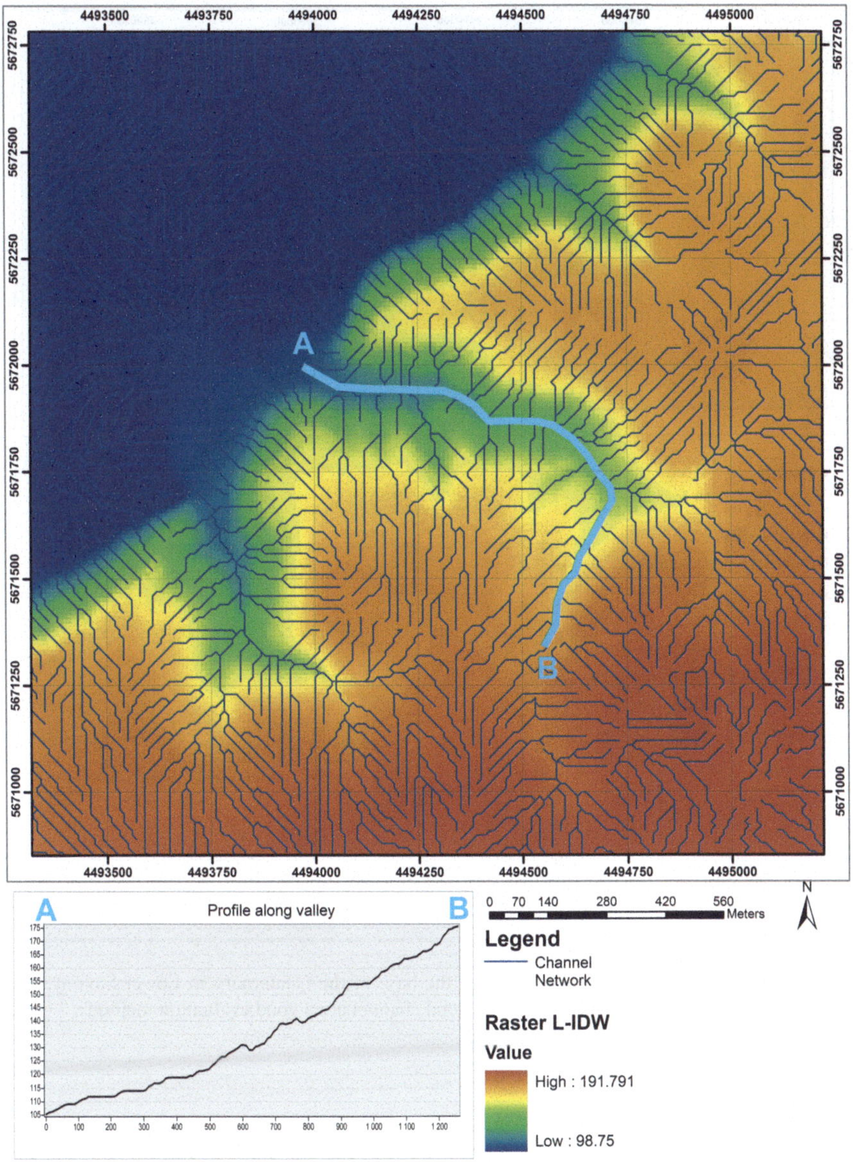

Fig. 1 Result of the L-IDW interpolation in a midrange mountain area. The flow paths were generated automatically

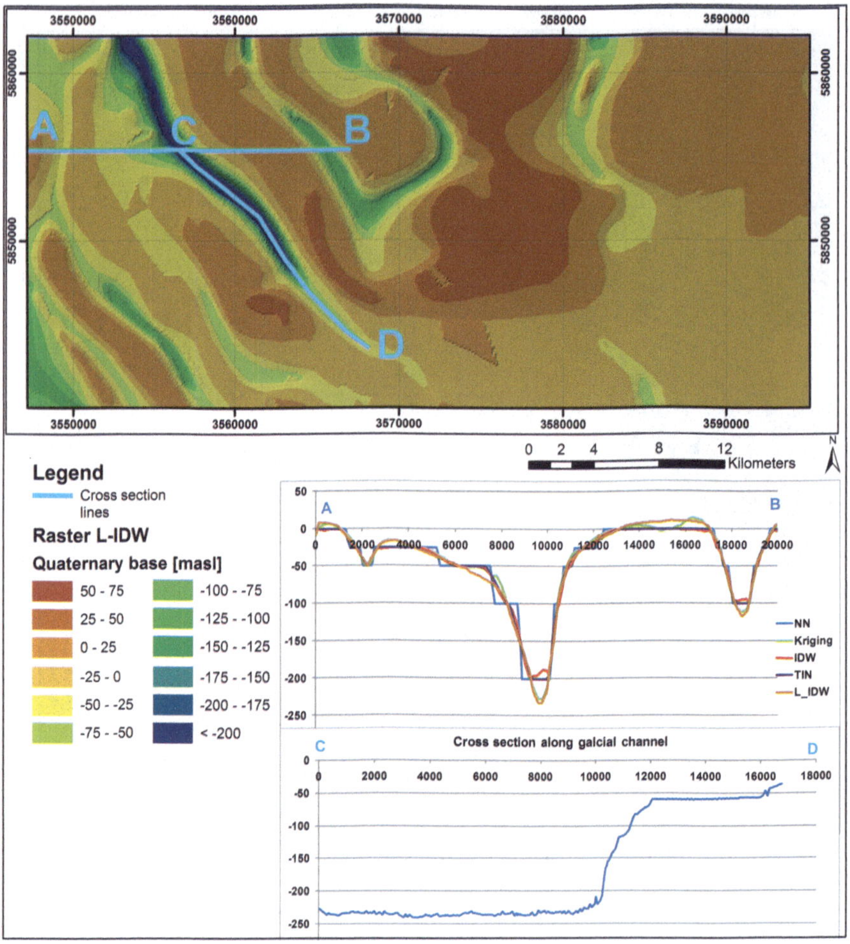

Fig. 2 Result of the L-IDW interpolation for the base of the Quaternary in Lower Saxony. The cross sections along and perpendicular to a glacial channel are a good evaluation method

4 Discussion

The results of the interpolation method for the two case studies are convincing. The comparison to the results of other interpolation methods shows a better fit to processes for the L-IDW method. The workflow for data input and visualization of results focuses common GIS data formats and the exchange via ASCII data. This makes data preparation not always easy but transparent and the result reliable.

Reference

1. Gossel, W., Chudy, T., & Falkenhagen, M. (2012). Interpolation based on isolines: line-geometrybased inverse distance weighted interpolation (L-IDW) with sample applications from the geosciences. *Zeitschr dt Ges Geowiss, 163*(4), 493–505.

Two-Dimensional Hydraulic Modeling and Analysis of Morphological Changes in the Palancia River (Spain) During a Severe Flood Event on October 2000

Beatriz Nácher-Rodríguez, Ignacio Andrés-Doménech, Carles Sanchis-Ibor, Francisca Segura-Beltrán, Francisco J. Vallés-Morán and Eduardo Albentosa Hernández

1 Introduction

In October 2000, the Palancia River underwent a large flood, which substantially modified the morphology of the river channel. These changes took place on a river bed previously altered by public works aimed at cleaning and channelizing the river section. The flood partially restored the original braided morphology, reconstructing bars and channels [1]. Moreover, river incision processes increased, particularly downstream of bridges. The aim of this work is to analyze the morphosedimentary changes through the development of a two-dimensional hydraulic model based on information of the recent flood event. The use of these tools is widely used for several purposes, such as urban flooding, environmental issues, or riparian vegetation or studies [2].

B. Nácher-Rodríguez(✉) · I. Andrés-Doménech · F. J. Vallés-Morán · E. Albentosa Hernández
Instituto Universitario de Ingeniería del Agua y Medio Ambiente, Universitat Politècnica de València, Camí de Vera s/n, 46022 Valencia, Spain
e-mail: beanacro@cam.upv.es

I. Andrés-Doménech
e-mail: igando@hma.upv.es

F. J. Vallés-Morán
e-mail: fvalmo@hma.upv.es

E. Albentosa Hernández
e-mail: ealbento@hma.upv.es

C. Sanchis-Ibor
Centro Valenciano de Estudios del Riego, Universitat Politècnica de València, Camí de Vera s/n, 46022 Valencia, Spain
e-mail: csanchis@hma.upv.es

F. Segura-Beltrán
Department de Geografia, Universitat de València, Avda. Blasco Ibáñez 28, 46010 València, Spain
e-mail: francisca.segura@igme.es

E. Pardo-Igúzquiza et al. (eds.), *Mathematics of Planet Earth*,
Lecture Notes in Earth System Sciences, DOI: 10.1007/978-3-642-32408-6_75,

2 Regional Settings

The Palancia River is 85 km long. The river basin (910 km^2) is regulated by the Regajo (6.6 hm^3) and Algar (6.3 hm^3) reservoirs. The mean annual flow is 1.3 m^3/s in the Regajo dam, but in the lower 25 km long reach the river bed is usually dry. The study encompasses the lowermost 6 km section of the river.

3 Materials and Methods

Channel morphology changes were analyzed comparing orthophotographs from dates previous and subsequent to the event (2000 and 2004), and oblique aerial photographs taken some days after the flood. Oblique images were rectified using ArcGIS TM version 9.3 (ESRI, Redlands, California, 2009), selecting between 10 and 12 ground control points (GCPs). The image distortion was assumed to be parabolic and, for this reason, a 2nd order polynomial was employed for georectifying. A DEM obtained from 2003 LiDAR data and field works contributed to improve the morphological analysis. River channel forms and textures were separately mapped.

The hydraulic numerical simulation was performed using GUAD 2D [3]. This software is a two-dimensional simulation model for the analysis of flood events. The water movement is governed by the laws of mass and momentum conservation under shallow water hypothesis. This implies that the variables are depth averaged, and a vertical hydrostatic pressure distribution is assumed.

In order to run the model, three basic inputs are needed. First of all, a $1 \times 1\,m^2$ Digital Elevation Model (DEM) of the study area was developed from vector topographic data. Then, surface resistance to the flow had to be estimated and introduced in the mathematical model by means of the Manning's roughness coefficients. The assessment of these coefficients was performed taking into account both grain size distribution in the river bed and vegetation features along the reach, with the help of historical ortophotographs, according to Cowan's estimation procedure [4]. Finally, the boundary and initial conditions were set. A hydrological study was performed to obtain a 35 hours long hydrograph (24–25 October 2000), which was assigned as the upstream boundary condition. Downstream, the water sea level was set.

4 Results

The main results obtained from the mathematical model are water depth and velocity in each cell of the model. These values can be used to perform a hydraulic-hydraulic-sedimentological analysis, in order to establish the mobilization zones of the bed material.

This kind of analysis requires not only the hydraulic variables, but also the grain size distribution of the river bed material, as the capacity of the flow to mobilize a granular material is closely bound to the average diameter of its particles (from 40 to 80 mm in the studied reach). Bedrock, concrete protections and roads were not included in this analysis.

Once the grain size was established, the hydraulic-sedimentological analysis was applied. This standard procedure is only valid outside the area of influence of bridge structures. It is based on the comparison of the shear stress produced by the flow, and the critical shear stress. Shear stress can be computed at each cell by means of the shear velocity (u^*) as shown in Eq. (1), while critical shear stress for a certain diameter (D) can be obtained applying Eq. (2). ρ is water density, γ and γ_S water and sediment specific weighs, respectively.

$$\tau_0 = \rho \, u_*^2 \tag{1}$$

$$\tau_c = 0.047 \, (\gamma_S - \gamma) \, D \tag{2}$$

These results were compared with field observations. The flood created a new braided pattern along the studied reach. Four different processes were identified: central bar and transverse bar formation (deposition), and chute cutoff and lobe dissection (erosion). Severe scour was also identified downstream two bridges.

5 Discussion

From the sedimentological analysis performed, two main conclusions arise. First, regarding the general behavior of the reach, the wider zones of mobilization coincide with those areas where the river has recovered the original braided pattern. On the other hand, it can be observed that the analysis does not reproduce accurately the scouring that occurred downstream the bridges. Indeed, severe erosion occurred at the Canet bridge (Fig. 1), but the model does not predict such mobilization of bed material. This is due to local tridimensional effects on the flow, like those related to the bridge piers, or also due to local changes in velocity as a consequence of variations in the bed material rigidity. The methodology presented in this paper does not take into account those effects, and further research is now being carried out in order to study local effects on scouring downstream bridges due to flow contraction and changes in the bed material rigidity [5].

6 Conclusions

The comparison of model results and channel morphological changes, reveals that these models constitute a powerful tool for the prediction of the river bed evolution. Further research is needed to improve calibration and to assess impact of the infrastructures.

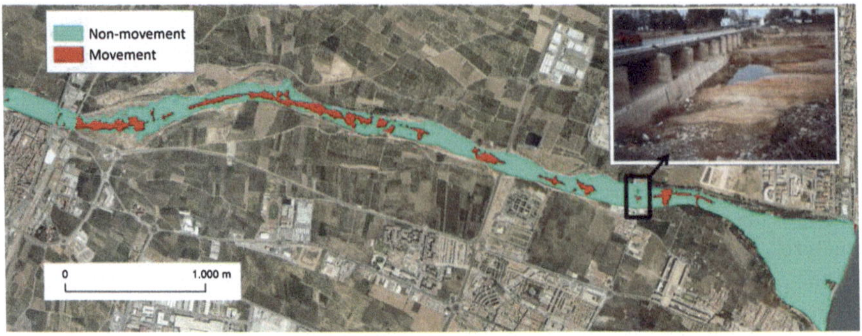

Fig. 1 Mobilized and non-mobilized areas in the Palancia river basin as predicted by the hydraulic-sedimentological analysis performed in the studied reach. The detailed view shows the effects of sediment scour downstream the Canet bridge

References

1. Segura-Beltrán, F., & Sanchis-Ibor, C. (2011). Efectos de una crecida en un cauce antropizado. La riada del Palancia de octubre de 2000. *Cuadernos de Geografía*, *90*, 145–167.
2. Néelz S., & Pender G. (2010). *Benchmarking of 2D Hydraulic Modelling*. Bristol: Environment Agency. ISBN: 978-1-84911-190-4.
3. Inclam SOFT (2012) Software modules Guad Creator (v1.1.1), Guad Gui (v1.2.0) and Guad View (v1.8.15), http://www.inclam.com/INCLAMSOFT/is_descargas.php
4. Cowan, W. L. (1956). Estimating hydraulic roughness coefficients. *Agricultural Engineering*, *37*(7), 473–475.
5. Nácher Rodríguez, B. (2012) Influencia de la contracción y de los cambios de rigidez del lecho en las condiciones críticas de inicio del movimiento aguas abajo de puentes sobre cauces. Tesina de Máster. Universitat Politècnica de Valencia. http://hdl.handle.net/10251/27858.

Stream Length-Gradient Index Mapping as a Tool for Landslides Identification

Jorge Pedro Galve, Daniela Piacentini, Francesco Troiani and Marta Della Seta

1 Introduction

Landsliding may have significant impact on the drainage network of montanious areas, where pronounced knickpoints can develop along the stream thalweg, influencing long-profile shape and trends. Moreover, marginal spillways, epigenetic gorges and terraces can be produced by subsequent stream adjustments. The effects of landslides over the hydrography are generally superimposed to other factors that control stream-profiles and landscape evolution, such as active tectonics, climate changes, lithology, and base-level variations. Hence, a better understanding of how landslides influence drainage networks and stream profiles is needed before either slope landforms or channel morphology can be used to infer the effects of external forcing on landscape evolution. Many quantitative geomorphic parameters like the Stream

J. P. Galve (✉)
Departamento de Ciencias de la Tierra, Universidad de Zaragoza, C/ Pedro Cerbuna 12, Zaragoza, Spain
e-mail: jpgalve@unizar.es

D. Piacentini
Dipartimento di Scienze Chimiche e Geologiche, Università di Modena e Reggio Emilia, L.go Sant'Eufemia, 19, 41121 Modena, Italy
e-mail: piacentini.daniela@unimore.it

F. Troiani
Dipartimento di Scienze della Terra, della Vita e dell'Ambiente, Università degli Studi di Urbino"Carlo Bo", Campus Scientifico Enrico Mattei, Via Cà Le Suore, 2/4, 61029 Urbino, Italy
e-mail: francesco.troiani@uniurb.it

M. Della Seta
Dipartimento di Scienze della Terra, Università degli Studi di Roma "La Sapienza", P.le Aldo Moro, 5, 00185 Rome, Italy
e-mail: Marta.dellaseta@uniroma1.it

E. Pardo-Igúzquiza et al. (eds.), *Mathematics of Planet Earth*,
Lecture Notes in Earth System Sciences, DOI: 10.1007/978-3-642-32408-6_76,

Length-Gradient (SL) index[1] [1], have been demonstrated to be suitable for detecting in homogeneous lithologies anomalies on stream-profiles related to active tectonics [2] and other sub-surface and surface processes [3]. A recent study conducted in a small catchment of the Central Apennines (Italy), [4] pointed out the influence of active deep-seated landslides on the occurrence of extreme anomalies of the SL index.

Nowadays, the availability of high-resolution Digital Terrain Models (DTMs) and the rising use of GIS software have led to a substantial improvement of the quantitative geomorphic methods to analyse large areas in a short time and cost-effectively. Therefore, it is fundamental to assess the sensitivity of geomorphic parameters of drainage networks and relief, such as the SL index, to suitably detect surface and sub-subsurface processes, which are, in turn, helpful to unravel the geomorphological evolution of a given area.

This work focuses on the spatial analysis of the SL Index within the piedmont area of the southern portion of the Emilia-Romagna Region (north Apennines, Italy) to test its suitability as a tool to support landslide detection.

2 Study Area

The study area covers ca. 2300 km^2 within the southern sector of the Emilia-Romagna Region in the Northern Apennines (Italy), including the mountain range, and the associated piedmont areas. The bedrock generally consists of terrigenous rocks belonging to the Miocene hemipelagic turbiditic sequence. The formation consists of arenites and pelites considerably varying in thickness and frequency within the area. The different rheological behaviour of the outcropping lithologies (arenites *vs* pelites), as well as the local litho-structural factors (i.e., downslope dipping of strata), favour the occurrence of frequent large deep-seated landslides (rock slides, earth flows, deep-seated gravitational slope deformations, and composite landslides). Large landslides within tributary basins, or reactivation of pre-existing landslides, often reach the streambeds and have generated knickpoints, epigenetic gorges, and in some cases stream-blockages.

3 Materials and Methods

The input dataset for the SL index computation derives from a 10 × 10 m gridded DTM. The point values have been interpolated with the Ordinary Kriging method to obtain the map of the areal distribution of the index. Successively, SL index anomaly maps have been elaborated and analyzed (Fig. 1a).

[1] The SL index derives from the Hack's equation: $SL = (dH/dL) \times L$, where: dH is the elevation difference between the extremes of a given stream-reach, dL is the length of the reach, and L is the distance from the mid-point of the reach upstream to the divide.

4 Results

The first result of the processing is a map of the areal distribution of the SL index. This map has been filtered to discard background values related to the different outcropping lithologies and, finally, the anomaly map of Fig. 1a has been obtained. This map and the profiles produced from it show the points/areas around the stream-segments where the extreme index values occur (Fig. 1b). At a regional scale, the alignments of this anomalies often coincide with the main active tectonic structures, highlighting knickpoints or knickzones related to the activity of the main faults or indirectly generated by the adjustment of streams to the local litho-structural setting. On the contrary, along tributary streams the main anomalies lie in correspondence with, or immediately downstream, large mass movements reported in the regional landslide inventory or in areas where the presence of landslides has been inferred by an initial aerial photo-interpretation (Fig. 1d). These points/areas have been checked in the field by a detailed geomorphological survey (Fig. 1c). After the survey, the 90 %

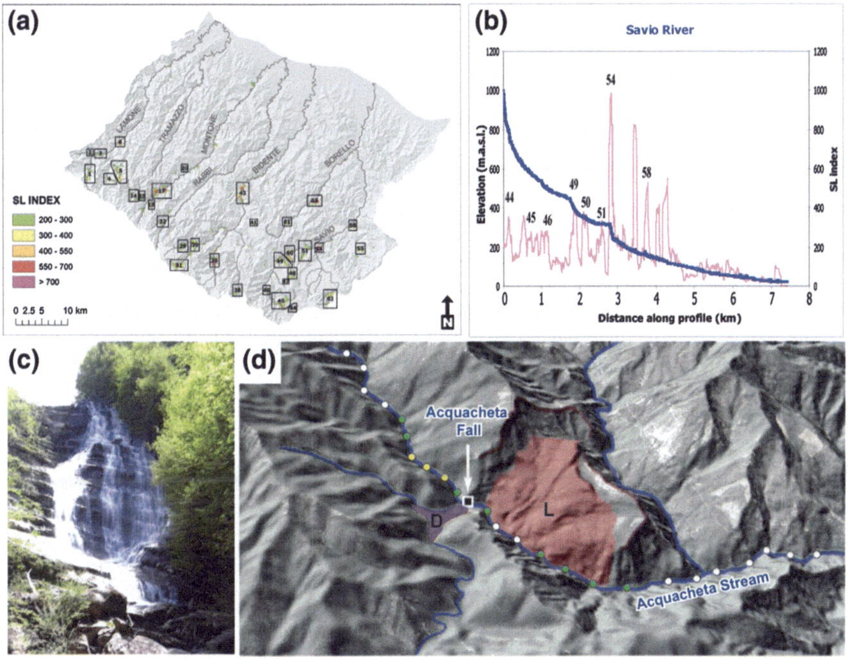

Fig. 1 Some examples of the aspects analyzed in the work. **a** SL index anomalies map indicating the knickpoints generated by landslides. **b** Example of one of the longitudinal and SL index profiles analyzed in the study area (Savio River). The number of anomalies related to landslides is marked. **c** Photo of a knickpoint generated by a landslide (Acquacheta Fall). **d** Geomorphological setting of the Acquacheta Fall and SL index points. The anomalous values are indicated by *green* and *yellow points*. L Landslide; D Lacustrine deposits accumulated because of the landslide dam responsible of the knickpoint

of the identified anomalies (27 out 30 anomalies) resulted to be connected to stream modifications due to active or partially reactivated large deep-seated landslides.

5 Discussion and Conclusions

This study shows that the stream-profile metrics may have a high potential in different geological studies when wide areas have to be studied with limited resources in a very short time. In particular, the application of the SL index mapping as a tool to detect landslides has demonstrated to be useful along tributary streams network, where the method provided the best results. On the contrary, in agreement with the current literature, along trunk valleys the SL index mapping was a useful tool to detect active tectonic structures. Finally, the SL index mapping can be a helpful method to detect those landslides that directly reached the streambeds and whose magnitude and activity was high enough to generate stream perturbations.

References

1. Burbank, D. W., & Anderson, R. S. (2001). *Tectonic geomorphology*. Oxford: Blackwell Science.
2. Hack, J. T. (1973). Stream-profile analysis and stream-gradient index. *U.S. Geological Survey Journal of Research*, *1*, 421–429.
3. Gonga-Saholiariliva, N., Gunnell, Y., Harbor, D., & Mering, C. (2011). An automated method for producing synoptic regional maps of river gradient variation: Procedure, accuracy tests, and comparison with other knickpoint mapping methods. *Geomorphology*, *134*, 394–407.
4. Troiani, F., & Della Seta, M. (2008). The use of the stream length-gradient index in morphotectonic analysis of small catchments: A case study from central Italy. *Geomorphology*, *102*, 159–168.

Quantitative Method on Historical Reconstruction of Coastal Geomorphological Change on Wave-Dominated Coast: A Case Study of the Pomeranian Bay, Southern Baltic Sea

Junjie Deng, Jan Harff and Joanna Dudzinska-Nowak

1 Introduction

Sea coasts are highly sensitive to the on-going global climate and environment changes, such as accelerated sea level rise and intensification of extreme storm events [1]. This is the case at the southern Baltic Sea coast, where eustatic change and glacio-isostatic land subsidence illustrated by [2] cause a relative sea level rise of up to 2 mm/y, and where strong storms events lead to continuous coastal retreat along the Polish coast [3]. Coastal protection and defense are required increasingly. Therefore, quantitative methods are needed to estimate coastal geomorphological changes including the reconstruction for the last centuries and the projection by 2100 AD (time span of climate modelling projection). The methods are applied at the Pomeranian Bay, southern Baltic Sea. One of selected key areas is shown in Fig. 2. This paper generally describes this methodological approach and its application.

2 Summary of Methodological Studies

The overview of the methodology is illustrated in Fig. 1 as a flow chart consisting of input data, methods and the outputs. There are three key methods including Geographical Information System, Dynamic Equilibrium Shore Model and complex long term morphodynamic model BS-LTMM that is described in detail in [4].

J. Deng (✉) · J. Harff · J. Dudzinska-Nowak
Institute of Marine and Coastal Sciences, University of Szczecin,
ul. Mickiewicza 18, 70-383 Szczecin, Poland
e-mail: junjie.deng@univ.szczecin.pl

J. Harff
e-mail: jan.harff@univ.szczecin.pl

J. Dudzinska-Nowak
e-mail: jotde@univ.szczecin.pl

E. Pardo-Igúzquiza et al. (eds.), *Mathematics of Planet Earth*,
Lecture Notes in Earth System Sciences, DOI: 10.1007/978-3-642-32408-6_77,

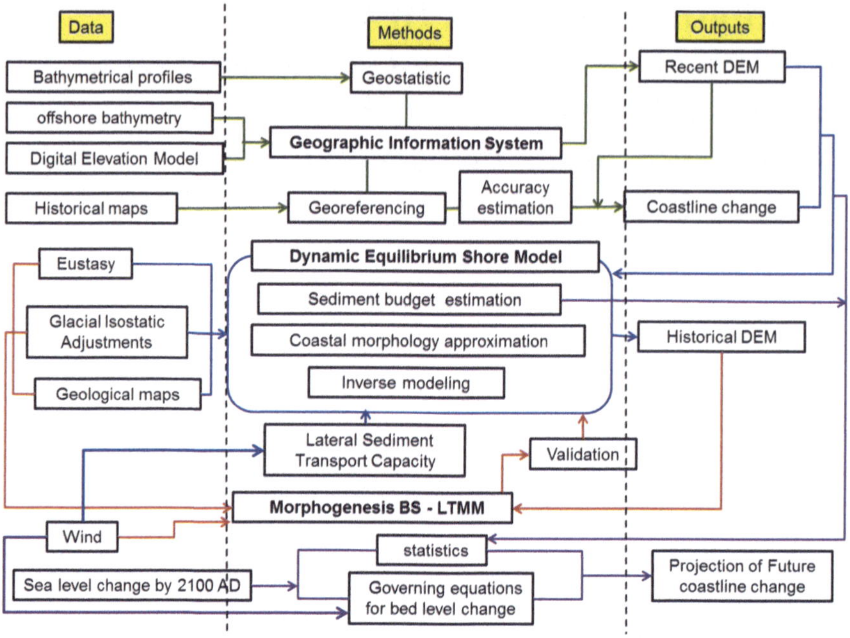

Fig. 1 The flow chart (of usage) of research methods

High resolution bathymetrical profiles are interpolated by Kriging that can provide Best Linear Unbiased Estimation. The accuracy of the historical maps (i.e. Messtischblatt maps) is quantitatively estimated by Root Mean Square Error of spatial differences of fixed points between historical maps and modern maps.

A mass balanced three dimensional source-to-sink model based on a generalization of Bruun rule model [5], named as Dynamic Equilibrium Shore Model (DESM), is elaborated. The main hypothesis of the model is that Lateral sediment source-to-sink transport dominates sediment flux in the Pomeranian Bay coastal system. The impact of rivers is limited at the vicinity of river mouth. The coast can thus be classified into two types in terms of sediment budget: semi-enclosed coast and open (boundary) coast. The research coastal area is subdivided into n zones with uniform width along the coast (Fig. 2). Based on recent DEM, for any time point $t \leq 0$, the depth of cross-shore profile representing each zone is described by the exponential function with the origin at the shoreline and the end point at closure depth:

$$y = a(1 - e^{(b_0 + \Delta b_t)(x + \Delta c_t)}) - s_t, t \leq 0. \quad (1)$$

where a is the exponential limit of the profile; b_0 is modern curvature coefficient; Δc_t is the known coastline displacement and S_t is known relative sea level change compared with recent time 2000 AD (t=0). The geomorphological change is thus described by the change of parameter Δb_t that is calculated by numerical iterations

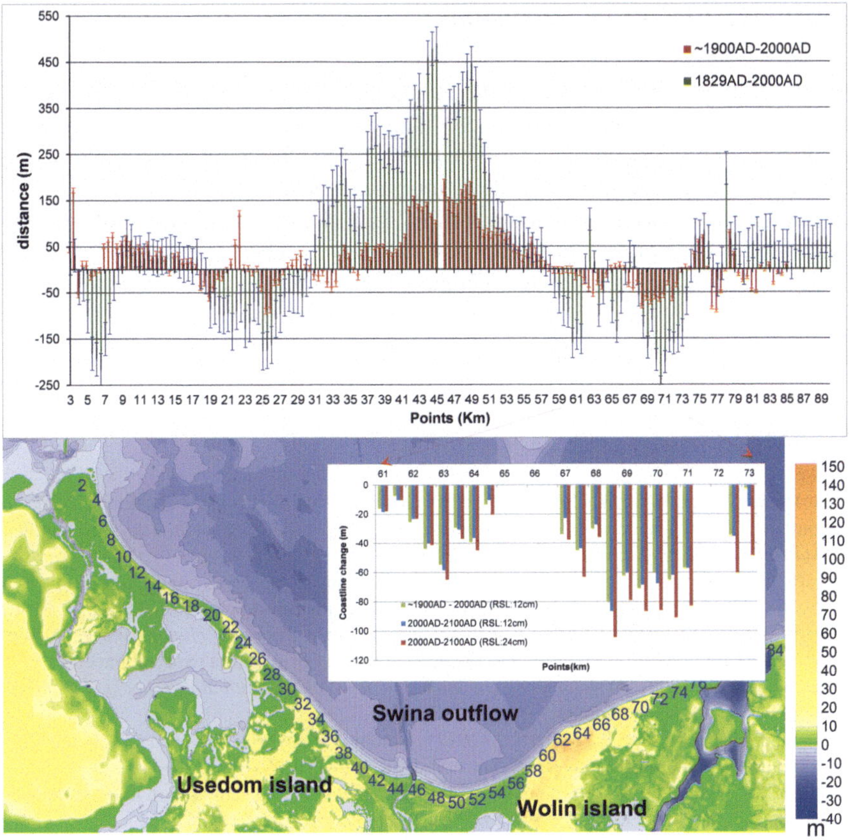

Fig. 2 Coastline changes with error bars in the past (reconstruction) and in the future (projection) with sea level rise scenarios of 12 and 24 cm by 2100AD

until the sediment budget of the research coast is balanced. The model DESM is validated by a comparison with long-term morphodynamic forward model (BS-LTMM) that starts with ~1900 AD DEM approximated by DESM, and by using nautical chart made in 1980s AD. The detail of description and validation of DESM is given in [6].

The sediment budget calculated by DESM can be simplified into a function of coastline change, sea level rise and other variables that can be obtained from recent DEM. The governing equations for bed level changes [7] connect the sediment budget functions with wind-wave climate driving lateral sediment flux. Therefore, the future coastline change can be projected when future wind-wave climate is stationary by 2100 AD.

3 First Results and Conclusions

The result at Swina Gate area of the Pomeranian Bay has led some preliminary conclusions. In Fig. 2, the historical coastline changes are quantified including the accuracy error bars. Besides, it also shows that DESM is also able to project (predict) reasonable coastline changes based on future sea level rise scenarios. The methodology proposed as a general flow chart in Fig. 1 can be a sufficient tool to study the historical geomorphological changes and the projection (prediction) of future changes, despite the fact that this methodology is limited at the sandy coast with continuously rising sea level and relatively stationary wind-wave climate.

References

1. Furmanczyk, K., Dudzinska-Nowak, J. (2009). Extreme storm impact to the coastline changes—South Baltic example. *Journal of Coastal Research, SI 56(Proceedings of the 10th International Coastal Symposium)*, 1637–1640.
2. Harff, J., Lemke, W., Lampe, R., Lüth, F., Lübke, R., Meyer, M., et al. (2007). The Baltic Sea Coast—a model of interrelations between geosphere, climate and anthroposphere. In J. Harff, W. W. Hay, & D. Tetzlaff (Eds.), *Coastline change—Interrelation of climate and geological processes* (Vol. 426, pp. 133–142). The Geological Society of America (Special Paper).
3. Zawadzka, E. (1999). *Development tendencies of the Polish south Baltic coast (in Polish)* (p. 147). Gdansk: GTN Gdansk.
4. Zhang, W. Y., Schneider, R., & Harff, J. (2012). A multi-scale hybrid long-term morphodynamic model for wave-dominated coasts. *Geomorphology, 149–150*, 49–61.
5. Bruun, P. (1988). The Bruun rule of erosion by sea-level rise: A discussion on large-scale two- and three-dimensional usages. *Journal of Coastal Research, 4*(4), 627–648.
6. Deng, J., Zhang, W., Harff, J., Schneider, R., Dudzinska-Nowak, J., Terefenko, P., Giza, A., Furmanczyk, K. A numerical approach to approximate historical morphology of wave-dominated coasts—a case study of the Pomeranian Bight, southern Baltic Sea. *Geomorphology* (Submitted).
7. Roelvink, D. Reniers, A. (2012). *A guide to modeling coastal morphology* (pp. 111–143). Singapore: World Scientific Publishing.

Linfo: A Visual Basic Program for Analysis of Spatial Properties of Lineaments

A. C. Dinesh, Vipin Joseph Markose and K. S. Jayappa

1 Introduction

Spatial properties of lineaments such as length, orientation, density and frequency have attracted research interest in the past few decades. Recent developments in remote sensing, geospatial technologies and computational algorithms will contribute to overcome the difficulties encountered in traditional lineament mapping. Major application of lineament studies are site selection for construction of dams, bridges, roads etc., as well as landslide risk assessment [1]; structural geological studies [2] and tectonic geomorphology [3]. Several computer programs such as SAL [4] and LINDENS [5] were developed for analyzing the spatial properties of lineaments. All these programs explained the detailed methodology for data processing and visualization. Geographical Information System (GIS) is the primary tool used for visualization of lineaments; hence a computer program which supports the GIS data formats will give an easy way for lineament analysis and interpretation. This paper describes a visual basic program called Linfo, which can be used for calculation of spatial properties of lineaments.

A. C. Dinesh (✉)
Marine and Coastal Survey Division, Geological Survey of India,
Mangalore 575001, India
e-mail: acdinesh@rediffmail.com

V. J. Markose (✉) · K. S. Jayappa (✉)
Department of Marine Geology, Mangalore University,
Mangalagangotri 574199, India
e-mail: vpnmrs@yahoo.co.in

K. S. Jayappa
e-mail: ksjayappa@yahoo.com

E. Pardo-Igúzquiza et al. (eds.), *Mathematics of Planet Earth*,
Lecture Notes in Earth System Sciences, DOI: 10.1007/978-3-642-32408-6_78,

2 Program Development

The program was developed in visual basic 6 which supports GIS platform and its structure. The primary aim of Linfo is to provide input for generating spatial maps of various lineament parameters such as length, density, frequency and intersection density. Direction of lineaments is usually represented by bearing, which divides direction (0–360°) into four quadrants of 90°. In this system, north and south are the dominant directions and measurements are determined in degrees from one of these directions. Figure 1 shows a typical lineaments in which (x_1, y_1), (x_2, y_2), (x_3, y_3) and (x_4, y_4) represent starting and end point coordinates of two lineaments. The following equation is applied for calculating the direction.

$$\theta = \text{Atan2}(\text{Sin}(x_2 - x_1).\text{Cos}(y_2), \text{Cos}(y_1).\text{Sin}(y_2) - \text{Sin}(y_1).\text{Cos}(y_2).\text{Cos}(x_2 - x_1)) \quad (1)$$

where, x_1, y_1 and x_2, y_2 are longitude and latitude of starting and end points, respectively.

In order to find the intersection points (X, Y) of lineaments; the following equation is applied (Fig. 1):

$$X = x_1 + ua(x_2 - x_1) \quad (2)$$

$$Y = y_1 + ua(y_2 - y_1) \quad (3)$$

$$ua = ((x_4 - x_3)(y_1 - y_3) - (y_4 - y_3)(x_1 - x_3))/((y_4 - y_3)(x_2 - x_1) - (x_4 - x_3)(y_2 - y_1)) \quad (4)$$

$$ub = ((x_2 - y_1)(y_2 - y_3) - (y_2 - y_1)(x_1 - x_3))/((y_4 - y_3)(x_2 - x_1) - (x_4 - x_3)(y_2 - y_1)) \quad (5)$$

If the denominators of the Eqs. (4) and (5) are 0, then the two lines are parallel. If the denominators and numerators of the equations for ua and ub are 0, then the two lines are coincident.

Fig. 1 Methodology for calculation of lineament orientation, length and intersection point (see text for explanation)

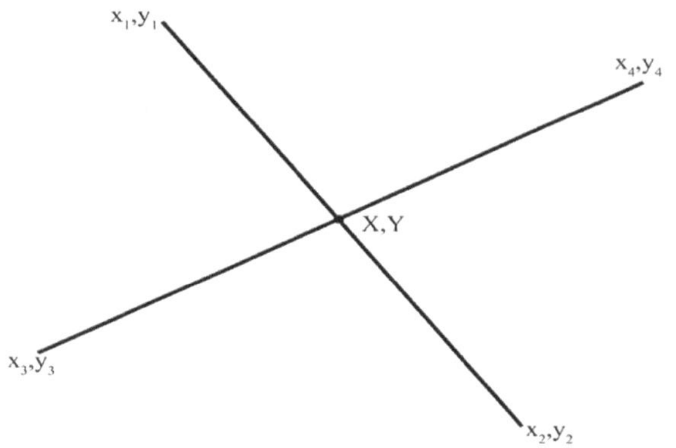

Lineament density is defined as the total length of lineaments per unit area. Frequency indicates the total number of lineaments per unit area. Intersection density is the total number of intersection points per unit area. The Linfo is designed to obtain unit areas by regular square grids defined by the user. After establishing the grids, the program calculates the total length, number of lineaments and number of intersection points within each grid and divides those values by the grid area. In order to check the efficiency and accuracy of Linfo, the results are validated by using lineament data of Kerala state, southwest India. A total of 175 lineaments were extracted from a lineament map of south India published by the Geological Survey of India and determined their spatial properties using Linfo. The lineament directions were determined and a rose diagram was generated using the program (Fig. 2a). A fixed

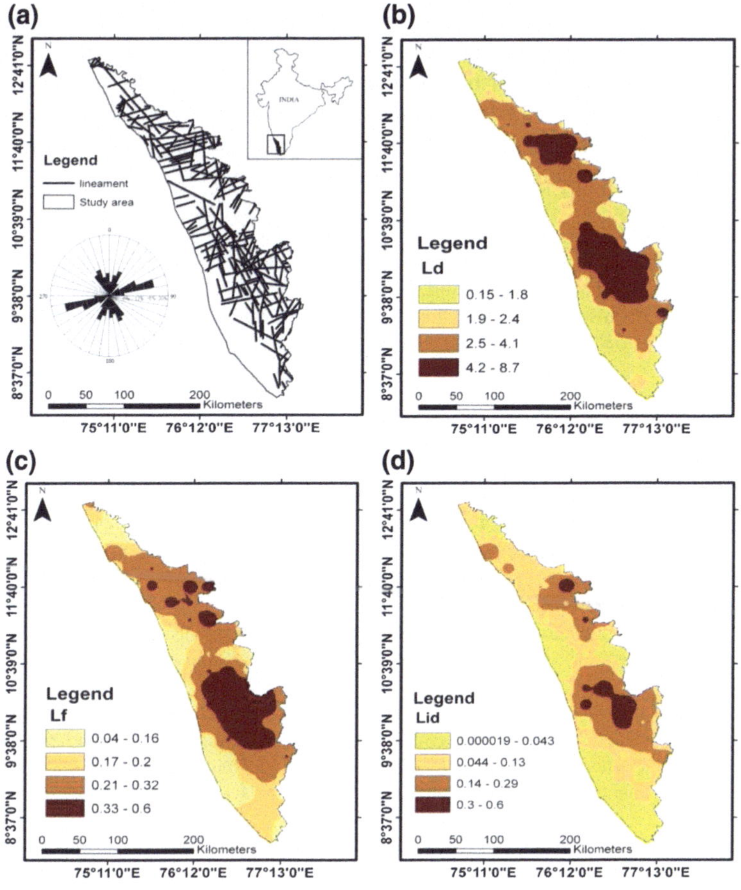

Fig. 2 **a** Location and lineament map of Kerala state. Spatial maps of lineament parameters: density **b**, frequency **c**, and intersection density **d**. These maps are prepared in ArcGIS 3D analyst by using the data calculated in grids covering 25 km^2 area in Linfo

grid size of 25 * 25 km is used to calculate the spatial properties of lineaments. All the three spatial maps (density, frequency and lineament intersection density), prepared by using Inverse Distance Weighted (IDW) interpolation method are shown in Fig. 2b, c, d.

3 Conclusion

An attempt has been made to develop the GIS utility program 'Linfo' that allows users to derive basic spatial properties of lineaments with greater accuracy. The lineament parameters such as orientation, length, density, frequency and intersection density can be easily calculated using the program. Linfo generates the regular square grids over lineament data and calculates the spatial properties in each grid. The program can handle large number of lineaments at a time and calculates the results faster than any other software. Testing of the program in a specific area shows that Linfo is effective in calculating lineament parameters such as density, frequency, intersection density, using the text data of ESRI line shapefile as an input. The program is a freeware that can be obtained from the corresponding author.

Acknowledgments The first author expresses gratitude to Shri. A. Sundaramoorthy, Director General, Geological Survey of India (GSI), Kolkata, Shri. R. C. Patra, Dy. Director General & HOD, Southern Region, GSI, Hyderabad and Shri. S. Kannan, Director-in-Charge, M&CSD, GSI, Mangalore for giving permission to publish this paper. The second author is thankful to Council of Scientific and Industrial Research (CSIR), New Delhi for providing financial support.

References

1. Stefouli, M., Angellopoulos, A., Perantonis, S., Vassilas, N., Ambazis, N., Charou, E. (1996). Integrated analysis and use of remotely sensed data for the seismic risk assessment of the southwest Peloponessus Greece. In *First congress of the Balkan Geophysical Society*, pp. 23–27 September, Athens, Greece.
2. Rahiman, T. I. H., & Pettinga, J. R. (2008). Analysis of lineaments and their relationship to Neogene fracturing, SE Viti Levu, Fiji. *Geological Society of America Bulletin*, *120*(11–12), 1544–1555.
3. Shahzad, F., Mahmood, S. A., & Gloaguen, R. (2009). Drainage network and lineament analysis: an approach for Potwar plateau (Northern Pakistan). *Journal of Mountain Sciences*, *6*(1), 14–24.
4. Casas, A. M., Cortes, A. L., Maestro, A., Soriano, M. A., Riaguas, A., & Bernal, J. (2000). LINDENS: a program for lineament length and density analysis. *Computers & Geosciences*, *26*, 1011–1022.
5. Ekneligoda, T. C., & Henkel, H. (2010). Interactive spatial analysis of lineaments. *Computers & Geosciences*, *36*, 1081–1090.

Numerical Karst: Spatio-Temporal Modelling of Karst Aquifer Systems

Eulogio Pardo-Igúzquiza, Sergio Martos, Juan Antonio Luque, Juan José Durán, Carolina Guardiola-Albert and Pedro Robledo

1 Introduction

Karst aquifers are very important as groundwater resources and as valuable environmental systems both on the surface and underground. However, their study requires approaches that go beyond those applied to typical detrital aquifers for two important reasons. First, karst aquifers usually correspond to positive relief features with limited or no pumping test data. Second, the karst aquifer is a very complex system where three kinds of porosity interact: matrix porosity, fracture porosity and conduit porosity. Hence, information is usually very scarce for such complex system. Because of this, the traditional approach has been to use black-box models where the main empirical data are the discharges from the karst springs. A transfer function is then created that relates rainfall to spring discharge. However, processes in karst systems are subject to significant spatio-temporal variability (rainfall, recharge, conduit and fracture density, unsaturated zone thickness, etc). The key of a transparent

E. Pardo-Igúzquiza (✉) · J. J. Durán · C. Guardiola
Instituto Geológico y Minero de España (IGME), Ríos Rosas 23, 28003 Madrid, Spain
e-mail: e.pardo@igme.es

J. J. Durán
e-mail: jj.duran@igme.es

C. Guardiola
e-mail: c.guardiola@igme.es

S. Martos · J. A. Luque
Unidad del IGME en Granada, Urbanización Alcazar del Genil, 4, 18006 Granada, Spain
e-mail: s.martos@igme.es

J. A. Luque
e-mail: ja.luque@igme.es

P. Robledo
Unidad del IGME en Palma de Mallorca, Ciudad de Querétaro s/n, 07007 Palma de Mallorca, Spain
e-mail: pa.robledo@igme.es

E. Pardo-Igúzquiza et al. (eds.), *Mathematics of Planet Earth*,
Lecture Notes in Earth System Sciences, DOI: 10.1007/978-3-642-32408-6_79,

box approach is the integration of all the available and reliable information, while the information that is uncertain can be simulated using an inverse procedure as described below.

2 Methodology

The inverse methodology for karst modelling is shown in Fig. 1. The main difficulty is the estimation of the location and density of conduits inside the karst massif. The only data available are provided by speleological exploration. However, this information is limited because not all the existing conduits have been explored or mapped and there is a limit to the size of conduits that can be explored. Nevertheless, they provide very valuable information that, together with a map of potholes and caves, can be used to create a map of conduit density. The sections that cannot be directly observed can be estimated through a procedure of karst conduit simulation that has been developed [1] to this end. However the conduit network is just one part of the karst system which has four additional important layers that must be estimated: the soil layer, the epikarst layer, the vadose layer and the saturated layer. The soil layer and the uppermost part of the epikarst layer must be determined for

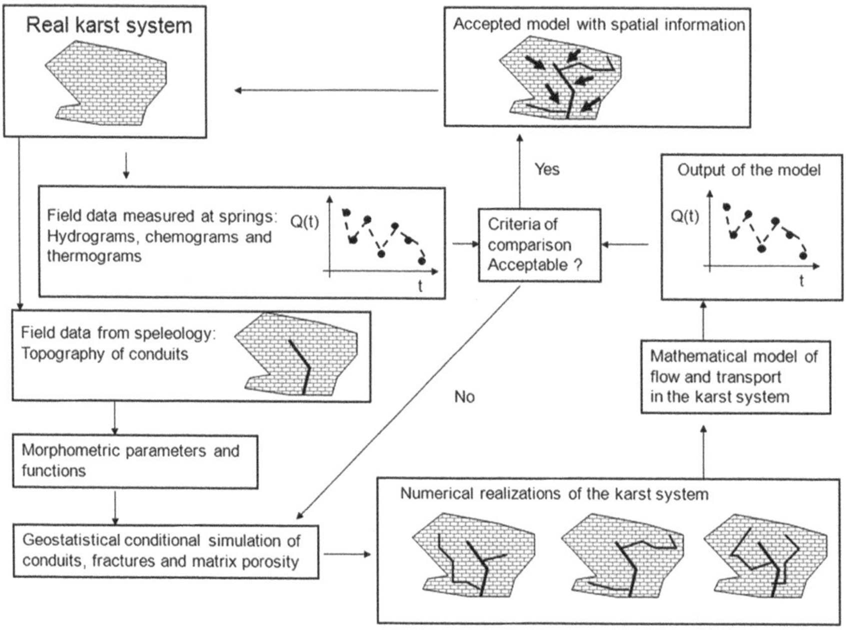

Fig. 1 Flowchart of the proposed methodology for a transparent box approach to the study of karst aquifers

water soil storage and recharge assessment. A method to estimate spatio-temporal recharge has been proposed in [2]. Furthermore, the epikarst may be well developed and may even be thick enough to form a perched aquifer. A method for mapping the epikarst using remote sensing and field data has been described in [3]. Another problem is the delineation of hydrogeological basins when the aquifer is drained by more than one important karst spring. Nevertheless, the most challenging problem is karst system flow simulation. First of all there are two kinds of flow. A fast flow that circulates through conduits that is responsible for peaks in the hydrogram after rainfall events. This flow can be simulated from the equations of open channel flow and pipe flow depending on the hydrodynamic conditions. Slow flow includes vertical flow through the vadose zone and flow along the saturated zone until the infiltrated water is discharged at the spring. There is also an interaction between the water from the pores and fractures and the conduits. Because the hydrograms of the springs are available, they can be used to calibrate the final spatio-temporal model of the karst system using an inverse procedure, whereby the characteristics of the different layers and conduits are changed until the simulated discharge fits the measured discharge.

3 Results

The Sierra de las Nieves is a high relief karst massif that hosts an important karst aquifer. It covers around 100 km^2 and it is drained by three main karst springs: Río Grande, Río Verde and Río Genal. The structure of the aquifer is composed of two blocks, one uplifted (Torrecilla block) with respect to the other (the Nava block), and separated by the Turquillas fault zone. Figure 2 shows recharge at one location in

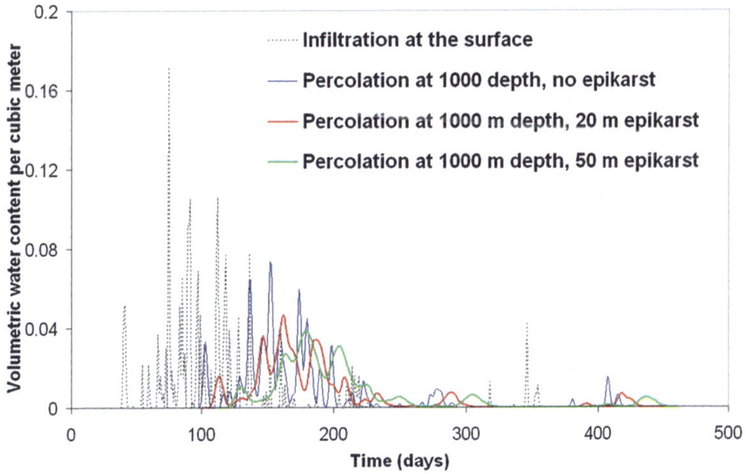

Fig. 2 Effect of the epikarst on percolation

the Torrecilla uplifted block and the effect of epikarst thickness on water percolation through a significant vadose zone of 1000 m (which is the case for the Sierra de las Nieves). The graph shows that the epikarst has the effect of delaying and attenuating the percolation wave. The routing of water from recharge to discharge is the main challenge of this project and, in order to address this issue, a spatially distributed description of the aquifer must be available. In order to achieve this, maps of spatially distributed recharge, epikarst development, vadose zone thickness, porosity and conduit density have been designed. The integration of all of the information together with the use of flow modelling (Richards' equation for unsaturated flow, Manning's equation for conduit flow, Darcy's equation for saturated flow, and the coupling between the different kinds of flow) is performed following the flow chart in Fig. 1, in order to gain understanding of the hydrogeological behaviour of the Sierra de las Nieves Aquifer.

4 Conclusions

Kart aquifers are complex systems and modelling them is a challenging task. This is even more complicated when the karst aquifer is in a high relief karst massif with a well developed network of conduits. Modelling a well-developed karst aquifer requires resolving several challenging problems, including creating a numerical model for the karst media itself and mathematical modelling of flow through the karst aquifer. The modelling of data layers using GIS technologies and an inverse approach are being used in order to provide a practical and cost-effective solution.

Acknowledgments This work was supported by research project CGL2010-15498 from the Ministerio de Economía y Competitividad of Spain. We would like to thank the reviewers for their constructive criticism.

References

1. Pardo-Igúzquiza, E., Dowd, P. A., Xu, C., Durán-Valsero, J. J., & Rodriguez-Galiano, V. (2012). Stochastic simulation of karst conduit networks. *Advances in Water Resources*, *35*, 141–150.
2. Pardo-Igúzquiza, E., Durán-Valsero, J. J., Dowd, P. A., Guardiola-Albert, C., Liñan-Baena, C., & Robledo-Ardila, P. A. (2012). Estimation of spatio-temporal recharge of aquifers in mountainous karst terrains: application to Sierra de las Nieves (Spain). *Journal of Hydrology*, *470*, 124–137.
3. Rodríguez-Galiano, V., Pardo-Igúzquiza, E., Durán, J. J., Chica-Olmo, M., Luque-Espinar, J. A., Guardiola-Albert, C., et al. (2012). Cartografía del epikarst integrando información de campo, geología e imágenes de satélite: caso de Sierra de las Nieves (Málaga). In J. Antonio López-Geta, R. F. Rubio, & D. L. Fernández (Eds.), *El agua en Andalucía: retos y avances en el inicio del milenio* (Vol. 2, pp. 1611–1620). Serie hidrogeología y aguas subterráneas No. 30. Madrid: Publicaciones del Instituto Geológico y Minero de España.

Surface Insights of Structural Relief Distribution Within the Madrid Cenozoic Basin from Fluvial and Terrain Morphometric Index

Julio Garrote and Guillermina Garzón

1 Introduction and Methods

In this paper we present the results of a morphometric analysis of the drainage network and relief of the Madrid Basin [1], aimed at checking whether they have been controlled by the distribution of lithologies or the stress field. The basin consists [2] of Late Miocene to Pliocene exorheic sediments overlying Oligocene to Middle Miocene deposits (Fig. 1). Some authors [3] have proposed that an extensional deformation has acted parallel to the NNW compression from the Miocene, not as a distinct extensional setting, but rather as result of flexural bending due to regional compression.

2 Results

The various indexes calculated (Fig. 2d–f) show a spatial distribution along the study area, which must be controlled either by the underlying lithology either by the presence of structural elements within the basin associated with recent tectonic stresses field; compressive stress field according to N-S to NW-SE and extensional stress field ac-cording to NW-SE [3]. The morphometric indexes results were compared first with the spatial distribution of lithologies within the basin, with the objective of determining possible relation-ships between the different variables. Two aspects can be highlighted, on the one hand the existence of a certain correlation between index

J. Garrote(✉) · G. Garzón
Department Geodinámica, Fac. Geología , Universidad Complutense de Madrid,
C/ José Antonio Novais, 2, 28040 Madrid, Spain
e-mail: juliog@ucm.es

G. Garzón
e-mail: minigar@ucm.es

E. Pardo-Igúzquiza et al. (eds.), *Mathematics of Planet Earth*,
Lecture Notes in Earth System Sciences, DOI: 10.1007/978-3-642-32408-6_80,

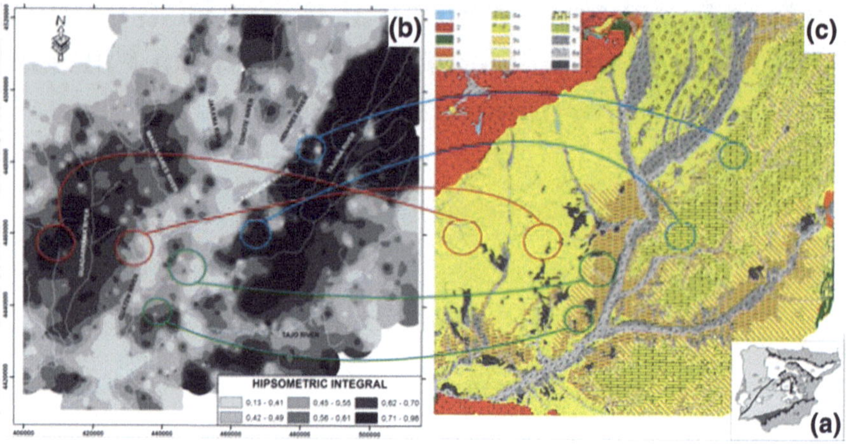

Fig. 1 Madrid Basin location map (**a**). HI map for the Madrid Basin (**b**) and its relationship with spatial distribution of lithologies (**c**). Example used lithologies legend: 5a: Limestones, marls and dolomites (Miocene–Pliocene); 5c: Gypsum and marl (Miocene); 5d: Arkosic sands (Neogene)

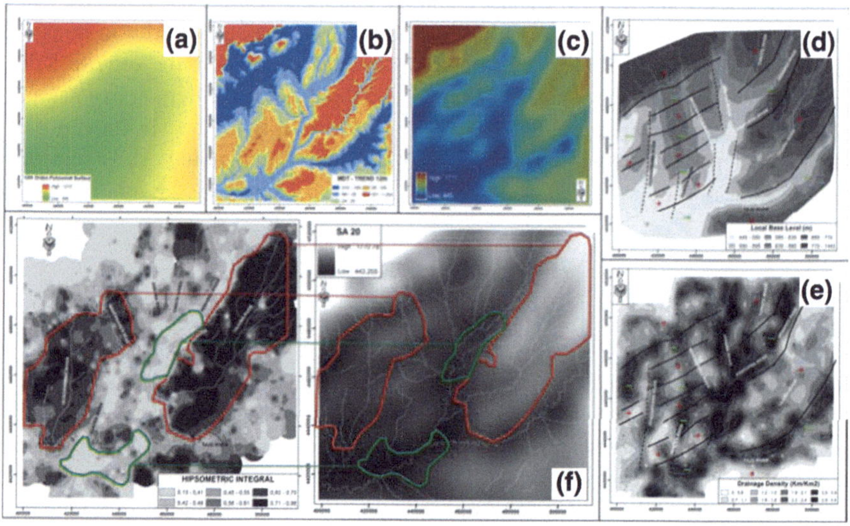

Fig. 2 Twelth order polynomial fit for the Madrid Basin (**a**) and residual map from DMT—Poli-nomial model (**b**). Twentieth order Harmonic model for the study area (**c**). Morphometric indexes of the basin (BL, Dd and HI) and relationship with high and low elevation areas of the harmonic model (**d**, **e** and **f**)

values and each lithologic unit defined. Furthermore, and more importantly, that the spatial distribution of morphometric index values is not adjusted to the distribution of lithologies, since there is a significant variability in the values within the same lithological unit, as reflected in Fig. 1.

The structural control on the relief of the basin has been previously pointed out [4] through the harmonic analysis of the relief. Here, we also show the results of a polynomial analysis (Fig. 2a) and its comparison with the DTM in order to separate the regional component of the relief (Polynomial model) and the map of residuals or local variations (Fig. 2b) [5, 6]. The latter shows a similar distribution of high value anomalous areas pointed to by the harmonic analysis (Fig. 2c). Based on these results, 20th order harmonic surface will be used as a model of deformation within the basin.

3 Discussion and Conclusions

After confirming that morphometric index results could not be satisfactorily ex- plained by the distribution of lithologies in the region, we have compared the differ- ent indexes with the basin deformation model obtained from the harmonic and polynomial analysis. In all cases (Dd, BL and HI), it can be observed the relationship between the uplifted and subsided areas of the harmonic model and the magnitude of the values that the morphometric indexes show (Fig. 2d–f). The uplifted areas of harmonic model are related with high BL and HI values, and low Dd values.

It can be also seen from the harmonic model results, the existence of lineaments in a NE-SW preferred direction [4], which are playing an important role in the configuration of the relief and deformation within the basin. However, the results shown by the morphometric analysis, point to the presence of other important directions that are not clearly reflected by the harmonic model. Following this idea, we derived the residual value map (Fig. 3a), which highlights the presence of lineaments within

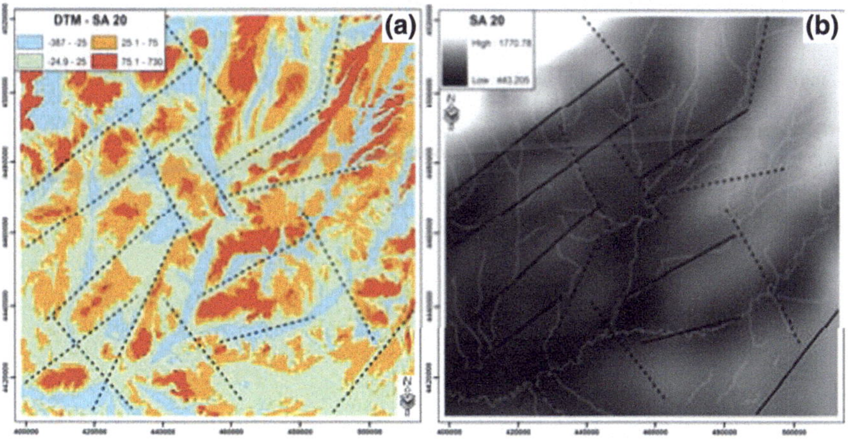

Fig. 3 Structural lineaments interpreted from Harmonic model residual map (**a**). Overlapped over the Harmonic model (**b**), three different situation can be identified for the lineaments

the basin, with a wider directional range than that shown directly by the harmonic model; and confirming what has been noted previously by morphometric indexes. This lineament scheme is more complex than previously proposed [1] from the T Factor index, even though some similar features and lineaments could be observed.

Lineament overlapping on the harmonic model (Fig. 3b) can be summarized in three situations: NE-SW lineaments observed clearly in the harmonic model; NE-SW lineaments which after interpretation in residual map could be easily defined over harmonic model; or NE-SW lineaments which have not clearly reflected in the harmonic model. In the latter two cases, it is necessary the presence of active structures responsible for some relief deformation beyond those defined from the harmonic model [6]. In summary, the morphometric indexes derived for several variables of the relief and drainage network of the Madrid Basin show the presence of a structural control on its evolution, which together with the erodability of sediments comprising the Tertiary basin fill, has been responsible for the present relief spatial distribution as well as for the shape and development of the main drainage system.

Acknowledgments This work has been funded by the project CGL2011-23857 of the Spanish R&D National Plan (FECYT).

References

1. Garrote, J., Heydt, G. G., & Cox, R. T. (2008). Multi-stream order analyses in basin asymmetry: A tool to discriminate the influence of neotectonics in fluvial landscape development (Ma- drid Basin, Central Spain). *Geomorphology*, *102*, 130–144.
2. Alonso-Zarza, A. M., Calvo, J. P., Silva, P. G., & Torres, T. (2004). Cuenca del Tajo. In J. A. Vera (Ed.), *Geologia de España* (pp. 556–561). Madrid: SGE-IGME.
3. De Vicente, G., & Muñoz-Martín, A. (2012). The madrid basin and the central system: A tectonostratigraphic analysis from 2D seismic lines. Tectonophysics. doi:10.1016/j.tecto.2012.04.003.
4. Tejero, R., González-Casado, J.M., Gómez-Ortiz, D., & Sánchez Serrano, F. (2006). Insights into the "tectonic topography" of the present-day landscape of the central Iberian Peninsula (Spain). *Geomorphology*, *76*, 280–294.
5. Grohmann, C. H. (2005). Trend-surface analysis of morphometric parameters: A case study in southeastern Brazil. *Computers & Geosciences*, *31*(8), 1007–1014.
6. Garrote, J. (2013). *La asimetría de la cuenca fluvial y otros índices morfométricos. Implicaciones para la Cuenca de Madrid*. PhD Thesis. UCM, Madrid.

Reinterpretation of Teide 2004–2005 Gravity Changes by 3D Line Segments Approximation

Peter Vajda, Ilya Prutkin and Jo Gottsmann

1 Data Reduction and Inversion

During the seismic unrest at the central volcanic complex (CVC) on Tenerife bulk gravity increase was recorded across a network at the CVC between May 2004 and July 2005 [5]. Here we aim at interpreting the gravity signal in terms of multiple sources using a non-linear inversion based on line segments approximation. The gravity changes were corrected for water table changes [5] and kriging-extrapolated onto an equidistant grid with a step of 500 m (Fig. 1). A trend, determined as a 2D harmonic function within the survey area, coinciding with the observed data on the boundary [6] was removed, in order to minimize the edge effects and truncation errors. Residual gravity changes after the removal of trend are shown in Fig. 2. Horizontal coordinates in the figures are UTM easting and northing, gravity is given in mGals (1 mGal = 10^{-5} m/s^2). The residual gravity changes (Fig. 2) were then best-fitted with the gravitational effect of line segments, each being defined by seven sought parameter, twice the triplet of coordinates and a line density [7]. These parameters were determined by minimizing residuals in L_2 norm by means of non-linear minimization procedure, using the gradient method of [4]. The line segments represent sources of temporal mass changes. With three line segments we achieve a fit of 3 μGal rms. Two of our segments, located to the NW of the Pico Viejo–Pico Teide (PV–PT) complex correlate with the NW seismogenic zone of [3].

P. Vajda (✉)
Geophysical Institute, Slovak Academy of Sciences, Bratislava, Slovakia
e-mail: Peter.Vajda@savba.sk

I. Prutkin
Institute of Geosciences, Jena University, Jena, Germany
e-mail: Ilya.Prutkin@uni-jena.de

J. Gottsmann
Department of Earth Sciences, University of Bristol, Bristol, UK
e-mail: j.gottsmann@bristol.ac.uk

E. Pardo-Igúzquiza et al. (eds.), *Mathematics of Planet Earth*,
Lecture Notes in Earth System Sciences, DOI: 10.1007/978-3-642-32408-6_81,

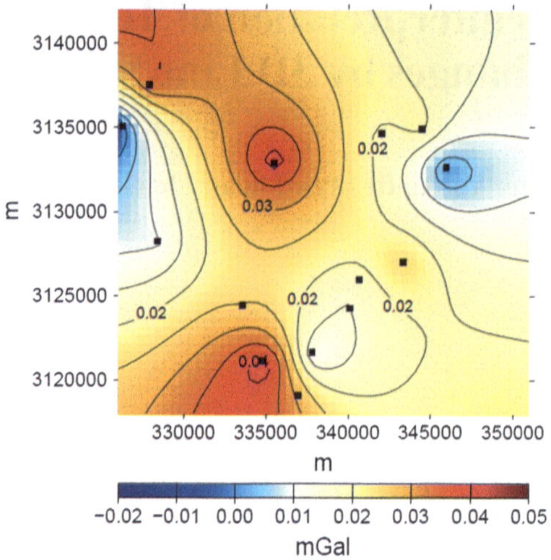

Fig. 1 Observed gravity changes (mGal)

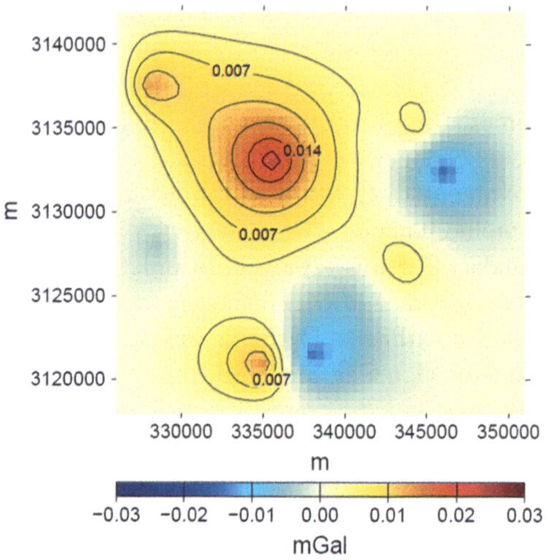

Fig. 2 Residual gravity changes (mGal)

They are located at depths 0–2 km b.s.l. The shallower SW line segment lies at the SW section of the caldera rim, and correlates with the SW seismogenic zone of [3].

We suspect that these line segments are composite sources representing shallower hydrothermal fluids and deeper magma injection. To test this hypothesis, we separate these composite sources into shallow and deep sources as follows. We decompose the residual gravity changes (Fig. 2) into a "shallow field" generated by presumed shallow sources and a "deep field" generated by deep sources. The division level of 4 km below sea level (b.s.l.) was chosen to match roughly the upper boundary of the two seismogenic zones of Ref. [3]. The decomposition procedure, based on stepwise

Fig. 3 Shallow field (mGal)

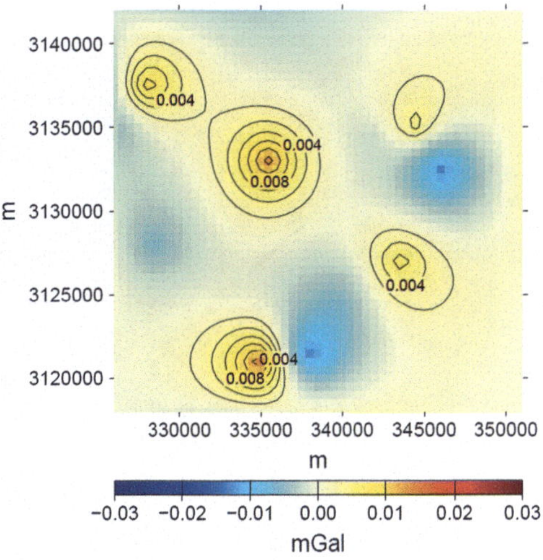

upward, downward and upward sequential harmonic continuations, is described in Ref. [6]. The shallow and deep fields resulting from the decomposition are presented in Figs. 3 and 4, respectively.

Again, both the shallow and deep fields are inverted in terms of their respective line segments. The inversion results in three shallow segments and two deep short and connected segments (one bent segment), see Fig. 5. The shallow and deep fields are approximated with rms of 2.3 and 0.7 μGal, respectively.

Fig. 4 Deep field (mGal)

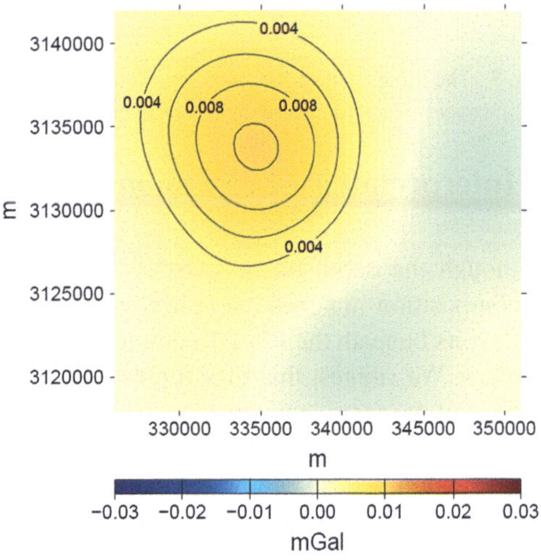

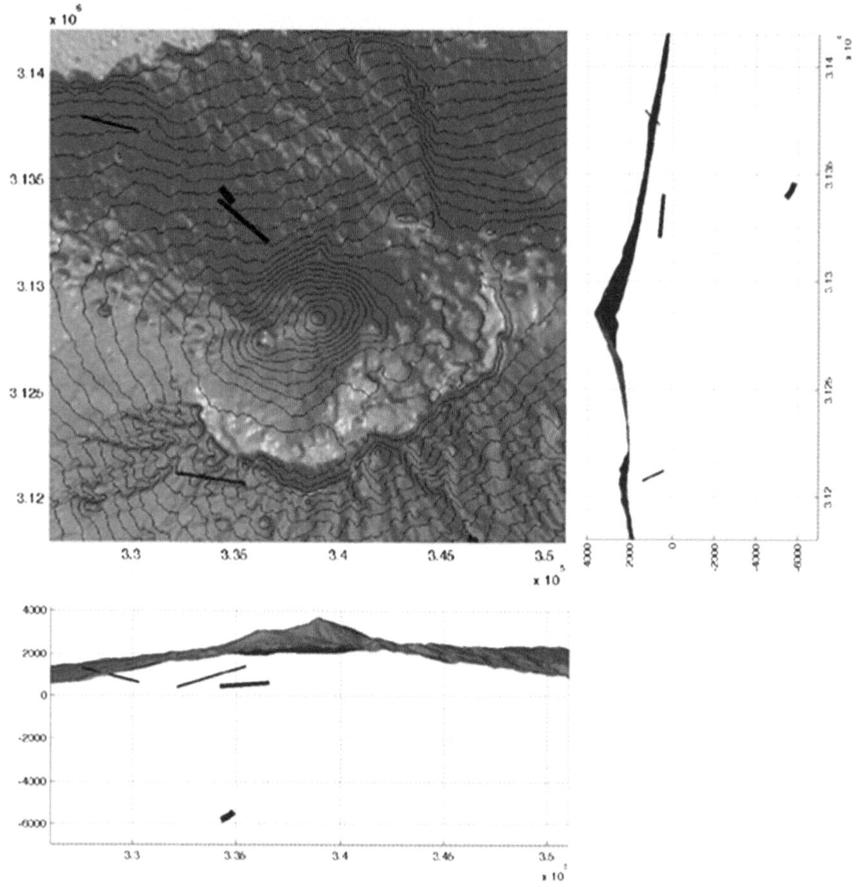

Fig. 5 The three shallow line segments and the short deep bent line segment

2 Interpretation of Inversion Results and Conclusion

Although the depth of the two NW segments found by inversion prior to signal decomposition matches the petrologically determined depth of phonolitic magma reservoirs beneath the PV–PT complex [1, 2], we do not interpret them as magmatic sources. We suggest that they represent a superposition of shallower hydrothermal water and deeper magma injection triggering the rise of the fluids. This is why we carried on with their decomposition into shallow and deep sources. Upon decomposition the inversion yields shallow and deep line segments. Due to the spatial distribution of the shallow segments and the position of the deep one, as well as their spatial correlation with the seismogenic zones of Ref. [3], we interpret the shallow segments as sources of hydrothermal fluids, while the short deep segment as a magma injection at a depth of about 6 km, within the NW zone of VT events swarm identified by

Ref. [3]. This hybrid nature of the observed unrest is best explained by the migration of hydrothermal fluids as a result of magma injection. Our inversion results and interpretation appear to be in fair agreement with the model of the tree-like magma intrusion proposed by Ref. [3].

Acknowledgments Jo Gottsmann acknowledges support by the Royal Society University Research Fellowship and EC FP7 project "VUELCO" (grant #: 282759). Peter Vajda was supported by the Slovak Research and Development Agency (contract No. APVV-0724-11) and by Vega grant agency (projects No. 2/0067/12 and 1/0095/12).

References

1. Ablay, G. J., & Martí, J. (2000). Stratigraphy, structure, and volcanic evolution of the Pico Teide–Pico Viejo formation, Tenerife, Canary Islands. *Journal of Volcanology and Geothermal Research, 103*(1–4), 175–208.
2. Andújar, J., Costa, F., & Martí, J. (2010). Magma storage conditions of the last eruption of Teide volcano (Canary Islands, Spain). *Bulletin of Volcanology, 72*, 381–395.
3. Cerdeña, D. I., del Fresno, C., & Rivera, L. (2011). New insight on the increasing seismicity during Tenerife's 2004 volcanic reactivation. *Journal of Volcanology and Geothermal Research, 206*, 15–29.
4. Fletcher, R., & Powell, M. J. D. (1963). A rapidly convergent descent method for minimization. *The Computer Journal, 6*(2), 163–168.
5. Gottsmann, J., Wooller, L., Martí, J., Fernández, J., Camacho, A. G., Gonzalez, P. J., et al. (2006). New evidence for the reawakening of Teide volcano. *Geophysical Research Letters, 33*, L20311.
6. Prutkin, I., Vajda, P., Tenzer, R., & Bielik, M. (2011). 3D inversion of gravity data by separation of sources and the method of local corrections: Kolarovo gravity high case study. *Journal of Apllied Geophysics, 75*(3), 472–478.
7. Vajda, P., Prutkin, I., Tenzer, R., & Jentzsch, G. (2012). Inversion of temporal gravity changes by the method of local corrections: A case study from Mayon volcano, Philippines. *Journal of Volcanology and Geothermal Research, 241—-242*, 13–20.

An Update GPS Velocity and Strain Rate Fields for the Iberian Region

Mimmo Palano, Pablo J. González and José Fernández

1 Introduction

The geodynamic setting of the Iberian Peninsula and the surrounding regions in the Mediterranean and the Atlantic is the result of the NW-SE directed convergent process between the Eurasian and African plates with simultaneous extension and formation of young oceanic basins in a back-arc setting due to the rapid roll-back of narrow slabs of subducting oceanic lithosphere [1]. One of these basins, the Alboran Sea is enclosed on three sides by the westernmost Alpine fold-and-thrust belt, the Betic Mountains to the N and NW and the Rif to W, SW and S. These mountain belts and the basin, together form the Gibraltar Orogenic arc, straddling the present-day Eurasia-Africa plate boundary (Fig. 1).

The region is marked by the occurrence of shallow depth ($h < 40$ km; Fig. 1) earthquakes of moderate magnitude. Along the Algerian margin, seismicity occurs on regional-scale WSW-ENE-trended structures with prevailing thrust focal mechanism solutions. Along the orogenic Arc and the Alboran Sea, seismicity is distributed over a wide area and occurs also at intermediate depth. Intermediate depth seismicity is focused along a N-S trend across the Alboran Sea and dipping southward from crustal depths beneath the Western Betics to a depth of $\sim$150 km beneath basin centre and occasionally deeper ($\sim$600 km) under the Central Betics. In the Rif, focal mechanism solutions depict complex seismotectonics with a mixture of solutions

M. Palano (✉)
INGV, Osservatorio Etneo—Sezione di Catania, P.zza Roma, 2, Catania 95123, Italy
e-mail: mimmo.palano@ct.ingv.it

P. J. González
Department of Earth Sciences, University of Western Ontario, London, ONN6A 5B7, Canada
e-mail: pgonzal4@uwo.ca

J. Fernández
Instituto de Geociencias (CSIC-UCM), Plza. Ciencias 3, 28040 Madrid, Spain
e-mail: jft@mat.ucm.es

E. Pardo-Igúzquiza et al. (eds.), *Mathematics of Planet Earth*,
Lecture Notes in Earth System Sciences, DOI: 10.1007/978-3-642-32408-6_82,

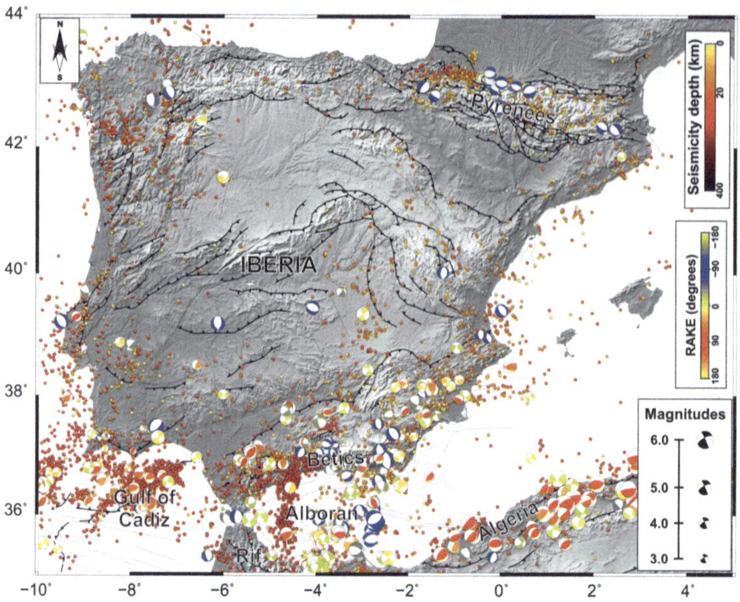

Fig. 1 Simplified tectonic map of Iberia. Crustal seismicity (M≥2.0) in the studied area since 2000 and focal mechanism solutions (since 1951, M≥3.0) are also reported. Focal mechanism solutions are colored according to rake (*red for pure thrust faulting, blue for pure normal faulting, and yellow for strike-slip faulting*)

including prevailing strike slip with subordinately reverse and few normal faulting solutions. In the Alboran Sea, crustal seismicity shows again prevailing strike-slip and subordinately normal faulting solutions. In the Betics, faulting is also complex with tectonic regime varying from normal (Central) to reverse (Eastern Betics) faulting along the mountain belt. Other intermediate-depth seismicity occurs SSW off Iberia (Gulf of Cadiz) where faulting is characterized by reverse solutions. Limited clusters of crustal seismicity can be recognized in NW Iberia and along the Pyrenees where focal mechanisms show normal faulting features.

2 GPS Data Processing, Velocity and Strain Rate Fields Computation

To study the crustal deformation pattern over Iberia region, we analyzed an extensive combination of GPS measurements carried out since 1999. In particular, we updated the results reported in [2] by extending the data processing up to the 2012.80 epoch and including into the processing additional data coming from networks developed mainly for mapping, engineering and cadastre purposes. GPS data were processed

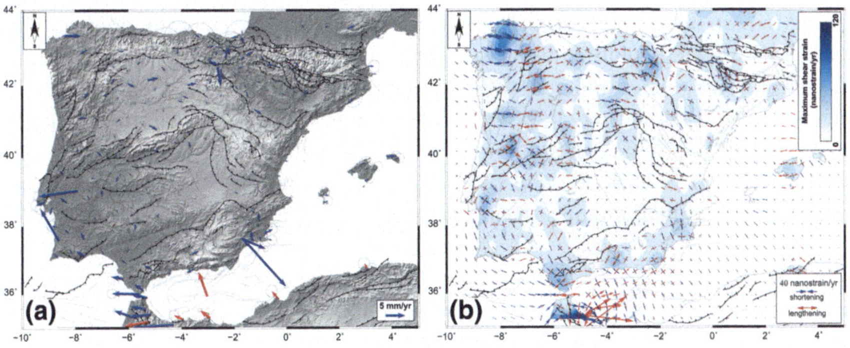

Fig. 2 **a** GPS velocities and 2-σ associated errors in a fixed Eurasian plate: blue arrows for sites analyzed in this study and red arrows for those coming from published solutions. **b** Geodetic strain-rate parameters: the color in background shows the magnitude of the maximum shear strain-rate, while arrows represent the greatest extensional (*red*) and contractional (*blue*) horizontal strain-rates

using GAMIT/GLOBK software [3, 4]. We estimated a consistent set of positions and velocities in an ITRF2008 frame [5] and in a fixed Eurasian reference frame [2].

To improve the detail of the geodetic velocity field for the studied area, we also performed a rigorous integration of our solutions with the ones reported in [6, 7] and [8]. In particular, since our solutions and the published ones share several common stations, we aligned their velocities to our fixed Eurasian reference frame by applying a Helmert transformation, obtained by finding the transformation that minimizes the RMS of differences between velocities of common sites. The average discrepancies are small, and the RMS for the 7–9 common stations is less than 0.4 mm/yr. The resulting velocity field, aligned to the Eurasian fixed reference frame is reported in Fig. 2a. In addition, by taking into account the observed GPS horizontal velocity field and associated covariance information we derived a continuous velocity gradient tensor on a regular $0.35 \times 0.35°$ grid using a "spline in tension" technique [9]. The velocity field was interpolated by removing from the computation all sites with fewer than 2.5 years of data and/or because of their suspicious movements with respect to nearby sites. As a final step, we computed the average 2D strain-rate tensor as derivative of the velocities at the nodes of each grid cell. The estimated strain-rates are shown in Fig. 2b.

3 Discussion and Conclusion

Geodetic data reveals that significant deformation occurs prevailing along the NW and SE margins of the Iberian Peninsula and along the Gibraltar arc, while on the inner parts of the peninsula, the crustal deformation occurs locally at rate <15 nanostrain/year. Along the NW margin, geodetic data evidences an E-W oriented contraction up to 55 nanostrain/year while along the SE margin a prevailing NNW-SSE-oriented contraction up to 30 nanostrain/year can be recognized. Along the Gibraltar arc, ~2–5 mm/year WSW motion can be detected. In particular, Al-Hoceima region shows a E-W-oriented elongation up to 90 nanostrain/year passing to a ca. E-W-oriented shortening of about −50 nanostrain/year in the Rif. The Alboran domain is characterized by elongation strain-rate axes WSW-ENE oriented coupled with shortening strain-rate axes, having the same magnitude (~25–40 nanostrain/year).

In conclusion, although on large areas of Iberia, the crustal deformation pattern is currently sampled or by few geodetic observations or by measurements hampered by large uncertainties (due to the short observation period), the observed framework is in good agreement with geological and seismological observations. The deployment of permanent GPS stations in the region will provide in the next years a better resolution of the crustal deformation pattern, providing new contributions to understanding the pattern of active deformation in this area exposed to a relevant seismic hazard.

Acknowledgments We thank all institutions that contributed to this work and for their efforts in providing good quality geodetic data, and research project AYA2010-17448. PJG acknowledges the Banting Postdoctoral Fellowship (Canadian Government). It is a contribution for the CEI Campus Moncloa.

References

1. Faccenna, C., Becker, T. W., Lucente, F. P., Jolivet, L., & Rossetti, F. (2001). History of subduction and back-arc extension in the central mediterranean. *Geophysical Journal International, 145*, 809–820.
2. Palano, M., González, P., & Fernández, J. (2013). Strain and stress fields along the Gibraltar Orogenic Arc: Constraints on active geodynamics. *Gondwana Research, 23*, 1071–1088.
3. Herring, T. A., King, R. W., McClusky, S. C. (2010). *Introduction to GAMIT/GLOBK, Release 10.4*. Cambridge: Massachusetts Institute of Technology.
4. Palano, M., Ferranti, L., Monaco, C., Mattia, M., Aloisi, M., Bruno, V., et al. (2012). GPS velocity and strain fields in Sicily and southern Calabria, Italy: Updated geodetic constraints on tectonic block interaction in the central Mediterranean. *Journal of Geophysical Research, 117*, B07401.
5. Altamimi, Z., Métivier, L., & Collilieux, X. (2012). ITRF2008 plate motion model. *Journal of Geophysical Research, 117*, B07402.
6. Serpelloni, E., Vannucci, G., Pondrelli, S., Argnani, A., Casula, C., Anzidei, M., Baldi, P., Gasperini P. (2007). Kinematics of theWestern Africa-Eurasia plate boundary from focal mechanisms and GPS data. *Geophysical Journal International, 169*, 1180-200.
7. Vernant, P., Fadil, A., Mourabit, T., Ouazar, D., Koulali, A., Davila, J. M., et al. (2010). Geodetic constraints on active tectonics of the Western Mediterranean: Implications for the kinematics and dynamics of the Nubia-Eurasia plate boundary zone. *Journal of Geodynamics, 49*, 123–129.
8. Koulali, A., Ouazar, D., Tahayt, A., King, R. W., Vernant, P., Reilinger, R. E., et al. (2011). New GPS constraints on active deformation along the Africa-Iberia plate boundary. *Earth and Planetary Science Letters, 308*, 211–217.
9. Wessel, P., & Bercovici, D. (1998). Interpolation with splines in tension: A Green's function approach. *Mathematical Geology, 30*, 77–93.

The 2012 Kilauea Volcano, Hawai'i, Slow-Slip Event Captured by cGPS and Satellite Radar Interferometry

Pablo J. González, Sergey V. Samsonov and Mimmo Palano

1 Introduction

Seismogenic fault systems accumulate stress along the locked portions of faults, which are subsequently or suddenly released during earthquakes. Slow-slip events (SSEs) are a complex processes occurring near the fault plane that release stress by producing atypical low slip-rates (lasting for days to months, or even years). SSEs have been detected along tectonic areas in subduction and strike-slip faults systems.

At subduction zones slow-slip phenomena have been detected at deep portions of the subduction megathrust zones (Cascadia, Guerrero in Mexico and Japan), but also at shallow depth (e.g., Nicoya in Costa Rica and the Hikurangi trench in New Zealand) [1]. Those two regions are assumed to bound the transition between different fault friction regimes, deep and shallow creep bounding the locked and seismogenic, stick-slip areas. However, slow slip phenomena seems to be not only an active process in subduction zones, but widely present at other fault systems [1], such as in deeper sections of major strike-slip fault systems such as the Parkfield segment of the San Andreas Fault in California.

Beneath the south flank of Kilauea volcano (Hawaii, USA) several SSEs have been detected using GPS and tiltmeter stations [2–4]. But, until now, differential radar interferometry has not been conclusive about the spatial pattern of vertical motion associated to these events, although GPS time series show a slight subsidence signal on the vertical component. Kilauea volcano is a very active volcano developed south

P. J. González (✉)
Department of Earth Sciences, Western University, London, ON, Canada
e-mail: pgonzal4@uwo.ca

S. V. Samsonov
Canada Centre for Remote Sensing, 588 Booth Street, Ottawa, ON, Canada
e-mail: sergey.samsonov@nrcan-rncan.gc.ca

M. Palano
INGV, Osservatorio Etneo—Sezione di Catania, P.zza Roma 2, Catania, CT, Italy
e-mail: mimmo.palano@ct.ingv.it

E. Pardo-Igúzquiza et al. (eds.), *Mathematics of Planet Earth*,
Lecture Notes in Earth System Sciences, DOI: 10.1007/978-3-642-32408-6_83,

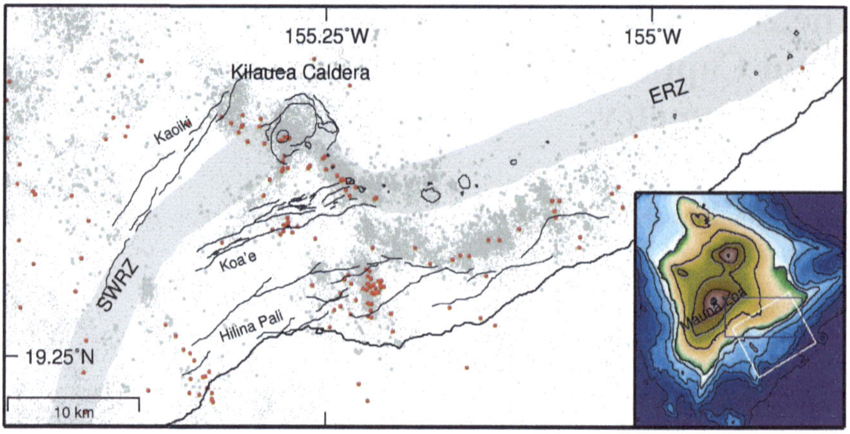

Fig. 1 Kilauea volcano map illustrating the major fault systems (*thin black lines*) and volcanic rift zones (*gray areas*). *Gray dots* represent the 2000–2012 located seismicity. *Red dots* are the epicenters of the earthquake during the 2012 SSE. *Lower inset* shows the Hawaii Island, *white rectangle* is the surface projection of the fault model

of Mauna Loa volcano in Hawaii Island. It rises above a ~5 km deep seafloor, and its southern flank shows continuous seaward deformation (Fig. 1). The seaward motion is assumed to occur due to slip at the interface of prevolcanics sediments laying over the oceanic crust and the volcanic succession [5]. This detachment surface has also produced large earthquakes and associated tsunamis such as the M7.2 1975 Kalapana earthquake [6].

In late May 2012, the most recent SSE began and it lasted until early June 2012. Following previous slow slip events, the SSE was accompanied by decollement (~8–10 km depth) earthquakes [4], but on June 5th, a relatively unusual very shallow earthquake swarm occurred across the Koa'e fault system, with two events having magnitudes around 3.5.

2 Ground Deformation Data Analysis and Modeling

A dense geodetic network of GPS stations is located in the Kilauea volcano for monitoring and research purposes. GPS data were processed using GAMIT/GLOBK software. We estimated a consistent set of positions and velocities in an ITRF2008 frame using additional far-field stations to define a local reference system using an established approach [7]. A network subset of 30 stations shows significant seaward displacements between May 27th and June 1st, 2012. GPS displacements range from ~5 cm near the coastline to almost no motion far from the coast (Fig. 2a). In addition, multiple different tracks from the Radarsat-2 satellite allowed us to generate nearly continuous ground deformation maps [8], around the summit caldera, south–west rift

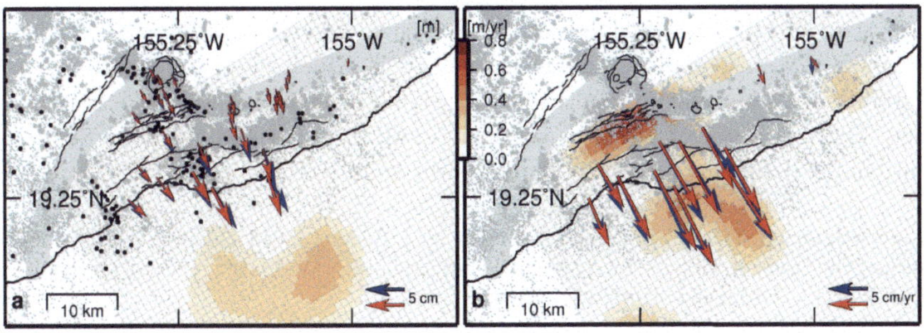

Fig. 2 GPS displacements (*arrows*) and fault slip distributions (*redish colored patches*). **a** 2012 SSE GPS observed (*blue*), and predicted (*red*) in cm, from the estimated slip [m]. **b** 2005–2012 estimate of the long-term flank motion: GPS observed (*blue*), and predicted (*red*) in cm/year, from the estimated fault slip-rate (m/year)

(SWR) and Koa'e fault system (Fig. 1). We detected a vertical (~2 cm, subsidence) signal along the coast is associated to the 2012 SSE.

Following, previous modeling approaches to understand the SSEs at Kilauea southern flank, we used a gently NNW-dipping thrust fault geometry corresponding to the decollement detachment structure [9]. We solved for the fault slip distribution occurred during the 2012 SSE by dividing the fault plane into 1 × 1 km dislocations. A fault slip model was obtained adopting a regularized linear inversion scheme constrained only by the observed GPS displacements (Fig. 2a). Spatial coverage of the interferograms is rather limited.

3 Discussion and Conclusions

The best-fitting fault slip model shows good agreement with the observed deformation (Fig. 2a). This model is consistent with a coast-parallel continuous patch of slip.

The model results present up to 40 cm of seaward displacement at depth ranging of 8–9 km and approximately 15 km off-shore beneath the southern flank. This slip pattern presented in Fig. 2a is in contrast with the long-term fault slip model obtained using the GPS stations with a linear and steady-state flank motion for the 2005–2012 period (Fig. 2b). Long-term flank motion seems to focus on two separated fault areas. The first fault segment depicts a narrow band with high slip rates (~70 cm/year) between the caldera and rift zones (ERZ and SWR) and the Hilina Pali fault system. A second segment is continuously sliding at a lower rate with max. 40 cm/year off the coast along the central Hilina Pali faults (Fig. 2b). The second fault segment only partially overlaps with the inferred SSE source area (Fig. 2a). Therefore, the 2012 SSE seems to have occurred along the fault plane at shallower depths, than long-term stable sliding segments that driven the flank motion.

These results illustrate that the seismo- and tsunamigenic decollement fault plane is segmented with spatially variable frictional areas, which varies from steady creeping to fully locked segments, and with patches of transient conditional friction properties. The definition of the fault segments with different sliding properties may contribute to define the seismic potential of the southern flank of Kilauea.

Acknowledgments This research was supported by a Banting Postdoctoral Fellowship of the Canadian Government. GPS data were obtained throughout UNAVCO data repository as courtesy of USGS-HVO, Hawaii and Stanford Universities.

References

1. Peng, Z., & Gomberg, J. (2010). An integrated perspective of the continuum between earthquakes and slow-slip phenomena. *Nature Geoscience, 3*, 599–607.
2. Cervelli, P., Segall, P., Johnson, K., Lisowski, M., & Miklius, A. (2002). Sudden aseismic fault slip on the south flank of Kilauea volcano. *Nature, 415*(6875), 1014–1018.
3. Brooks, B. A., Foster, J. H., Bevis, M., Frazer, L. N., Wolfe, C. J., & Behn, M. (2006). Periodic slow earthquakes on the flank of Kilauea volcano, Hawai'i. *Earth and Planetary Science Letters, 246*(3/4), 207–216.
4. Segall, P., Desmarais, E. K., Shelly, D., Miklius, A., & Cervelli, P. (2006). Earthquakes triggered by silent slip events on Kilauea volcano, Hawaii. *Nature, 442*, 71–74.
5. Owen, S., Segall, P., Lisowski, M., Miklius, A., Denlinger, R., & Sako, M. (2000). Rapid deformation of Kilauea volcano: global positioning system measurements between 1990 and 1996. *Journal of Geophysical Research, 105*, 18,983–18,998.
6. Lipman, P., Lockwood, J., Okamura, R., Swanson, D., & Yamashita, K. (1985). Ground deformation associated with the 1975 magnitude-7.2 earthquake and resulting changes in activity of Kilauea volcano, Hawaii. *U. S. Geological Survey Professional Paper, 1276*, 1–45.
7. Palano, M., et al. (2010). Etn@ref: a geodetic reference frame for Mt. Etna GPS networks. *Annals of Geophysics, 53*(4), 49–57.
8. Samsonov, S., & d'Oreye, N. (2012). Multidimensional time series analysis of ground deformation from multiple InSAR data sets applied to Virunga volcanic province. *Geophysical Journal International, 191*(3), 1095–1108.
9. Montgomery-Brown, E. K., Segall, P., & Miklius, A. (2009). Kilauea slow slip events: identification, source inversions, and relation to seismicity. *Journal of Geophysical Research, 114*, B00A03.

Spatiotemporal Analysis of Ground Deformation at Campi Flegrei and Mt Vesuvius, Italy, Observed by Envisat and Radarsat-2 InSAR During 2003–2013

Sergey V. Samsonov, Pablo J. González, Kristy F. Tiampo, Antonio G. Camacho and José Fernández

1 Introduction

Two active volcanoes, Campi Flegrei and Mt Vesuvius, are located in close proximity to the City of Naples, the third largest municipality in Italy with a population close to one million inhabitants. Previous studies have identified this region as one of the highest risk volcanic areas on Earth [1]. For monitoring volcanic hazards, surveillance networks are established to closely monitor changes in seismicity, gas emissions and active ground deformation as indicators of renewed volcanic/magmatic activities, potentially culminating in eruption. An alternative and complementary approach is to monitor ground deformation from space using InSAR [2, 3]. In this study we demonstrate a new advanced MSBAS InSAR processing technique that reveals signals at both Campi Flegrei and Mt Vesuvius. These signals are observed with a remarkable precision and resolution that have not been achieved before.

S. V. Samsonov (✉)
Canada Centre for Remote Sensing, 588 Booth Street, Ottawa, ON K1A0Y7, Canada
e-mail: sergey.samsonov@nrcan-rncan.gc.ca

P. J. González · K. F. Tiampo
Department of Earth Sciences, Western University, Biological and Geological Sciences Building, 1151 Richmond Street, London, ON N6A5B7, Canada
e-mail: pabloj.glez@gmail.com
e-mail: ktiampo@uwo.ca

A. G. Camacho · J. Fernández
Instituto de Geociencias, CSIC-UCM, Fac. Cc. Matemáticas, Plaza de Ciencias, 3, 28040 Madrid, Spain
e-mail: jft@mat.ucm.es

A. G. Camacho
e-mail: antonio_camacho@mat.ucm.es

E. Pardo-Igúzquiza et al. (eds.), *Mathematics of Planet Earth*,
Lecture Notes in Earth System Sciences, DOI: 10.1007/978-3-642-32408-6_84,

2 Data and Processing Methodology

For this study we collected and processed five independent InSAR data sets described in Table 1 with uninterrupted coverage from 2003 until 2013. We applied 2×10 multilooking to four standard beams and 4×5 multilooking to F6 fine beam and processed each data set independently with GAMMA software [4]. All possible interferometric pairs with perpendicular baseline less than 400 m were computed and the topographic component was removed using 90 m resolution SRTM DEM [5]. The orbital refinement was performed to remove residual orbital ramps and minor interpolation was applied for filling gaps in moderately coherent regions. The final filtered, unwrapped and interpolated interferograms were geocoded on a 90×90 m grid.

For the time series analysis we limited data to the Naples Bay area and re-sampled all interferograms to a common grid with GMT scripts [6]. Our final interferograms had dimensions of 724×603 pixels with the original resolution of about 90×90 m. We calculated the average interferometric coherence of the filtered products for each interferogram and selected only those with the average coherence above 0.5 for further processing. Over one thousand highly coherent inetrferograms were used in the MSBAS processing.

The MSBAS methodology was described in detail in [7, 8]. In the case of a single SAR set acquired by a sensor with an azimuth θ and an incidence angle ϕ, the time series of deformation can be reconstructed by applying the Small Baseline Subset method (SBAS) [9, 10]

$$AV_{los} = \Phi_{obs}, \quad V_{los} = A^{+}\Phi_{obs}, \quad d_{los}^{i+1} = d_{los}^{i} + V_{los}^{i+1}\Delta t^{i+1}, \qquad (1)$$

where A is a matrix constructed from the time intervals between consecutive SAR acquisitions, V_{los} is a vector of the unknown line-of-sight velocities, Φ_{obs} is a vector of observed interferogram values, A^{+} is a pseudo-inverse of matrix A found by applying the Singular Value Decomposition (SVD), and d_{los}^{i} is a line-of-sight displacement at the time t^{i}.

Table 1 Five InSAR data sets used in this study with uninterrupted coverage from 2003 until 2013

InSAR set	Orbit	Coverage	$\phi°$	$\theta°$	N	M
ENVISAT, Track 129	asc	2003–2012	23	354	55	276
ENVISAT, Track 036	dsc	2003–2012	23	196	58	196
R2, S3	asc	2009–2013	35	349	39	156
R2, S3	dsc	2009–2013	35	190	50	422
R2, F6	asc	2008–2013	48	351	48	407

Incidence ϕ and azimuth θ angles and number of available SAR images N and number of computed interferograms M is provided

In the case of K multiple SAR sets acquired by sensors with different orbital parameters (e.g. azimuth and incidence angles), the Eq. (1) can be rewritten in the following form for each set $k = 1, \ldots, K$

$$| \, S_N^k A \quad S_E^k A \quad S_U^k A \, | \cdot | \, V_N \quad V_E \quad V_U \, |^T = \Phi_{obs}^k \tag{2}$$

assuming that $V_{los} = \mathbf{SV} = S_N V_N + S_E V_E + S_U V_U$ and $\mathbf{S} = \{\sin\theta \sin\phi, -\cos\theta \sin\phi, \cos\phi\}$, where $\mathbf{S}$ is a line-of-sight unit vector with north, east and up components S_N, S_E, S_U, and $\mathbf{V}$ is a velocity (ground deformation rate) vector with components V_N, V_E, V_U.

Then the MSBAS method that includes all K sets of independently acquired SAR data can be presented in the following form

$$\begin{pmatrix} A^1 \\ A^2 \\ \ldots \\ A^K \end{pmatrix} \begin{pmatrix} V_N \\ V_E \\ V_U \end{pmatrix} = \begin{pmatrix} \Phi^1 \\ \Phi^2 \\ \ldots \\ \Phi^K \end{pmatrix} \quad or \quad \hat{A}\hat{V}_{los} = \hat{\Phi}_{obs} \tag{3}$$

where the new matrix $\hat{A}$ [as in (2)] has dimensions $3(\Sigma_{k=1}^K N^k - 1) \times \Sigma_{k=1}^K M^k$, the new vector V has dimensions $1 \times 3(\Sigma_{k=1}^K N^k - 1)$, and the new vector Φ_{obs} has dimensions $1 \times \Sigma_{k=1}^K M^k$, where M^k is a number of interferograms and N^k is a number of SAR images in k data set.

We further assumed that our data is not sensitive to the motion in the NS direction and the northern component was excluded from the computation. For smoothing the solution we applied Tikhonov regularization [11] with $\lambda = 0.25$.

3 Results

Results of the MSBAS processing are presented in Fig. 1. The linear deformation rate was calculated by fitting a straight line to the computed time series. The horizontal map (Fig. 1a) shows an expansion-like signal centered on the Campi Flegrei caldera with a rate in 2003–2013 of <1 cm/year and the slow westward motion in the eastern part of the region, east of Mt Vesuvius. The vertical map (Fig. 1b) shows well defined uplift at the Campi Flegrei caldera with the maximum 2003–2013 rate of 1 cm/year located near the coastal area. Subsidence with rate <1 cm/year is observed around and at Mt Vesuvius.

The time series of deformation show with remarkable precision the temporal evolution of ground displacement in both the vertical and horizontal directions. Two episodes of uplift occurred at Campi Flegrei during this time period (Figs. 1c–e), one in 2007 and the other which began in 2009–2010 and is currently ongoing. The second has reached a maximum observed rate of more than 10 cm/year in 2012–2013. The horizontal time series display steady motion in the E-W directions.

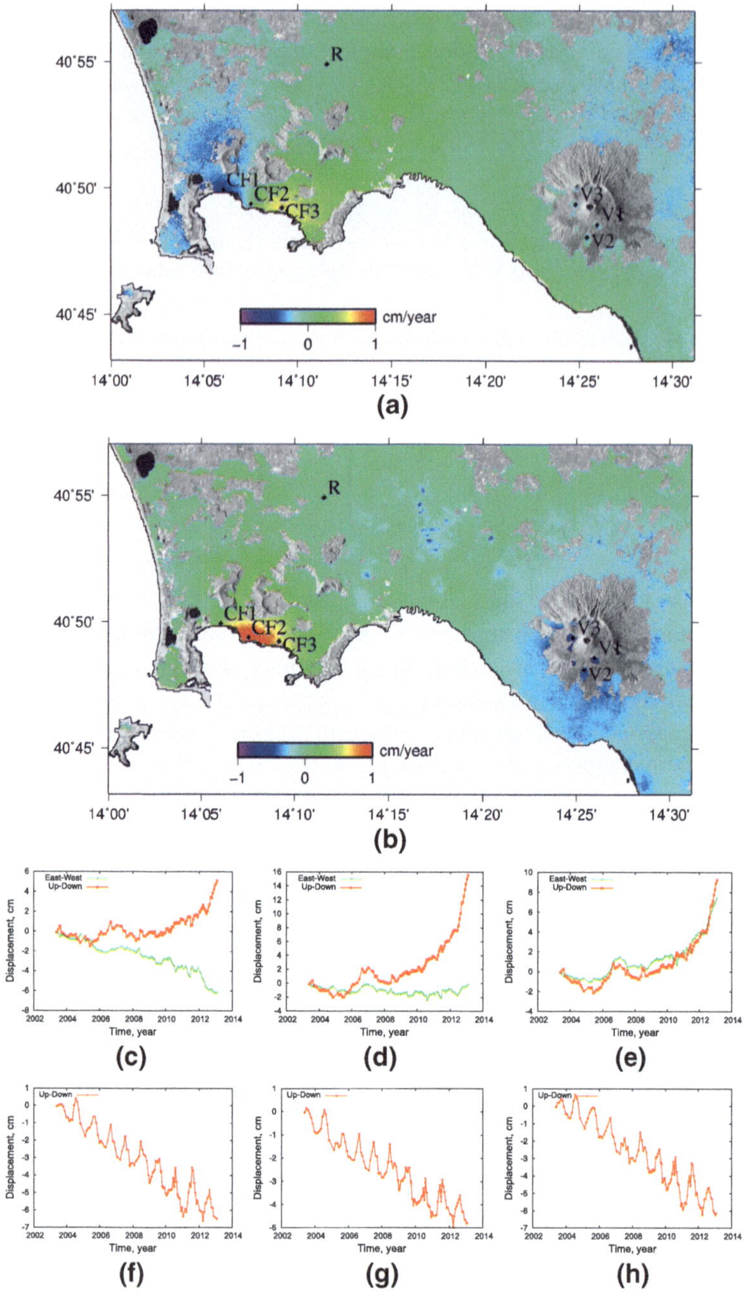

Fig. 1 Deformation rates and time series of ground deformation observed in Naples Bay area during 2003-2013. Horizontal east-west (**a**) and vertical up-down (**b**) deformation rates calculated by fitting linear model to computed time series. For points CF1-CF3 (**c–e**) located at Campi Flegrei caldera and V1-V3 (**f–h**) located at Mt Vesuvius we provide time series of ground deformation. Point R is reference assumed to be stable during observation period

A more unusual signal is observed at Mt. Vesuvius (Figs. 1f–h) that consist of steady subsidence with a sub-centimeter rate and seasonal oscillations with an amplitude of about 0.8 cm. These oscillations have a well-defined seasonal pattern with a maximum observed during the summer and a minimum observed during the winter. The amplitude and phase of those oscillations located within a few coherent patches at an altitude of about 700 m seems are nearly identical.

4 Discussion and Conclusion

In this work we applied for the first time the advanced MSBAS methodology to mapping ground deformation in the Naples Bay area that covers the Campi Flegrei caldera and Mt Vesuvius. We used ENVISAT and RADARSAT-2 data and achieved uninterrupted temporal coverage during 2003–2013.

Computed deformation maps show uplift at the Campi Flegrei caldera that started in 2009–2010 and continues until now. It is unknown if this uplift continues offshore but it seems likely. The temporal pattern of deformation demonstrates uplift with constantly increasing rate reaching over 10 cm/year presently. Due to its location, the proximity to the city of Naples, the more active monitoring for the signs of possible eruption is warranted.

The signal observed at Mt Vesuvius shows seasonal oscillations with an excellent temporal resolution and precision that has not been achieved before. Such performance can be attributed to the Singular Valued Decomposition (SVD) analysis that is capable of filtering the signal from noisy data. Our initial analysis suggests that these oscillations can be at least partially explained by the seasonal fluctuation in the water vapor pressure and other climate parameters.

The further work is under way to extend the temporal coverage to 1992-2013 by including ERS-1/2 data and also to include recent Cosmo-SkyMED data in order to achieve unprecedented temporal resolution for studying the observed signals. The source modeling using methodology proposed in [12] will be performed next in order to explain the origin of ground deformation observed at the Campi Flegrei.

Acknowledgments We thank the Canadian Space Agency (CSA) for providing RADARSAT-2 data and the European Space Agency (ESA) for providing ENVISAT data under projects C1P.7244 and C1P.12984. Figures were plotted with GMT and gnuplot software. The work of PG was supported by Banting Postdoctoral Fellowship. The work of KFT was supported by an NSERC Discovery Grant. This research was also supported by research project AYA2010-17448, a contribution for the CEI Campus Moncloa.

References

1. Isaia, R., Marianelli, P., & Sbrana, A. (2009). Caldera unrest prior intense volcanism in Campi Flegrei (Italy) at 4.0 ka B.P.: Implications for caldera dynamics and future eruptive scenarios. *Geophysical Research Letters*, *36*, L21303.

2. Massonnet, D., & Feigl, K. (1998). Radar interferometry and its application to changes in the Earth surface. *Reviews of Geophysics*, *36*(4), 441–500.
3. Rosen, P., Hensley, P., Joughin, I., Li, F., Madsen, S., Rodríguez, E., et al. (2000). Synthetic aperture radar interferometry. *Proceedings of the IEEE*, *88*(3), 333–382.
4. Wegmuller, U., & Werner, C. (1997). Gamma SAR processor and interferometry software. In *Third ERS symposium on space at the service of our environment*. Florence.
5. Farr, T., & Kobrick, M. (2000). Shuttle radar topography mission produces a wealth of data. *EOS Transactions, AGU*, *81*, 583–585.
6. Wessel, P., & Smith, W. (1998). New, improved version of the generic mapping tools released. *EOS Transactions, AGU*, *79*, 579.
7. Samsonov, S., & d'Oreye, N. (2012). Multidimensional time series analysis of ground deformation from multiple InSAR data sets applied to virunga volcanic province. *Geophysical Journal International*, *191*(3), 1095–1108.
8. Samsonov, S., d'Oreye, N., & Smets, B. (2013). Ground deformation associated with post-mining activity at the French-German border revealed by novel InSAR time series method. *International Journal of Applied Earth Observation and Geoinformation*, *23*, 142–154.
9. Berardino, P., Fornaro, G., & Lanari, R. (2002). A new algorithm for surface deformation monitoring based on small baseline differential SAR interferograms. *IEEE Transactions on Geoscience and Remote Sensing*, *40*(11), 2375–2383.
10. Usai, S. (2003). A least squares database approach for sar interferometric data. *IEEE Transactions on Geoscience and Remote Sensing*, *41*(4), 753–760.
11. Tikhonov, A., & Arsenin, V. (1977). *Solution of Ill-posed problems*. New York: Wiley.
12. A. Camacho, A., González, P., Fernández, J., & Berrino, G. (2011). Modelling ground deformation and gravity changes by means of 3d extended pressured bodies with free geometry. Application to deforming calderas. *Journal of Geophysical Research*, *116*(B10401).

Anthropogenic and Natural Ground Deformation Near Bologna, Italy Observed by Radarsat-2 InSAR During 2008-2013

Sergey V. Samsonov, Pablo J. González and Kristy F. Tiampo

1 Introduction

The over-exploitation of ground water resources in the Bologna region causes wide-spread ground subsidence. The rate of subsidence increased through time from 0.2–0.3 cm/year during 1897–1957 to 0.4–0.6 cm/year during 1957–2001 [1]. This area is also susceptible to seismic hazard. Recent 2012 Emilia seismic sequence with two M5.8 and M6.0 mainshocks caused casualties and damages to infrastructure and produced over 15 cm of ground uplift [2].

In this study we apply for the first time the advanced Multidimensional Small Baseline Subset (MSBAS) InSAR technique to map ground deformation during 2008–2013 period based on one hundred ascending and descending images acquired by the Radarsat-2 satellite. Produced deformation map and time series clearly map the spatial and temporal patterns of ground deformation and observe a further increase in subsidence rate up to -2.0 cm/year.

2 Data and Methodology

We collected and processed one ascending and one descending Radarsat-2 data sets described in Table 1 with coverage from 2008 until 2013. We applied 2×10 multilooking and processed each data set independently with GAMMA software [3].

S. V. Samsonov (✉)
Canada Centre for Remote Sensing, 588 Booth Street, Ottawa, ON K1A0Y7, Canada
e-mail: sergey.samsonov@nrcan-rncan.gc.ca

P. J. González · K. F. Tiampo
Department of Earth Sciences, Western University, Biological and Geological Sciences Building, 1151 Richmond Street, London, ON N6A5B7, Canada
e-mail: pabloj.glez@gmail.com

K. F. Tiampo
e-mail: ktiampo@uwo.ca

E. Pardo-Igúzquiza et al. (eds.), *Mathematics of Planet Earth*,
Lecture Notes in Earth System Sciences, DOI: 10.1007/978-3-642-32408-6_85,

Table 1 Two Radarsat-2 (beams Standard S3) InSAR data sets used in this study with coverage from 2008 until 2013

InSAR set	Orbit	Coverage	$\phi°$	$\theta°$	N	M
Radarsat-2, Standard 3	asc	2008–2013	35	349	59	127
Radarsat-2, Standard 3	dsc	2008–2013	35	190	38	90

Incidence ϕ and azimuth θ angles and number of available SAR images N and number of computed interferograms M is provided

All possible interferometric pairs with perpendicular baseline less than 400 m were computed and the topographic component was removed using 90 m resolution SRTM DEM [4]. The orbital refinement was performed to remove residual orbital ramps and minor interpolation was applied for filling gaps in moderately coherent regions. The final filtered, unwrapped and interpolated interferograms were geocoded on a 90×90 m grid.

For the time series analysis we re-sampled all interferograms to a common grid with GMT scripts [5]. Our final interferograms had dimensions of 2050×1734 pixels with the subsampled resolution of about 60×60 m. We calculated the average interferometric coherence of the filtered products for each interferogram and selected only those with the average coherence above 0.35 for further processing, which resulted in 127 ascending and 91 descending interferograms. We applied MSBAS methodology [6–8] that produced vertical and horizontal deformation rate maps and time series of deformation.

3 Results

Vertical deformation rate calculated with the MSBAS technique is shown in Fig. 1a. Wide spread subsidence is observed near the City of Bologna with rates up to -2 cm/year. Time series (Fig. 1b–g) show steady subsidence until early 2012. Since the second half of 2012 until present time series at points P3-P6 show a moderate reversal to uplift. A very slow westward motion is observed at points P4-P6. Since this methodology maps only relative to selected reference region motion a further analysis is needed in order to ensure its stability.

Co-seimic displacements produced by the 2012 Emilia sequence are observed NW of the epicenter area and shown as extent in the upper-left corner in Fig. 1. However, this region is severely decorrelated (due to dense vegetation), which does not allow for the automated phase unwrapping. The loss of fringes during the phase unwrapping produces incorrect estimation of co-seimic displacements (observed $\sim$6 vs. expected $\sim$15 cm).

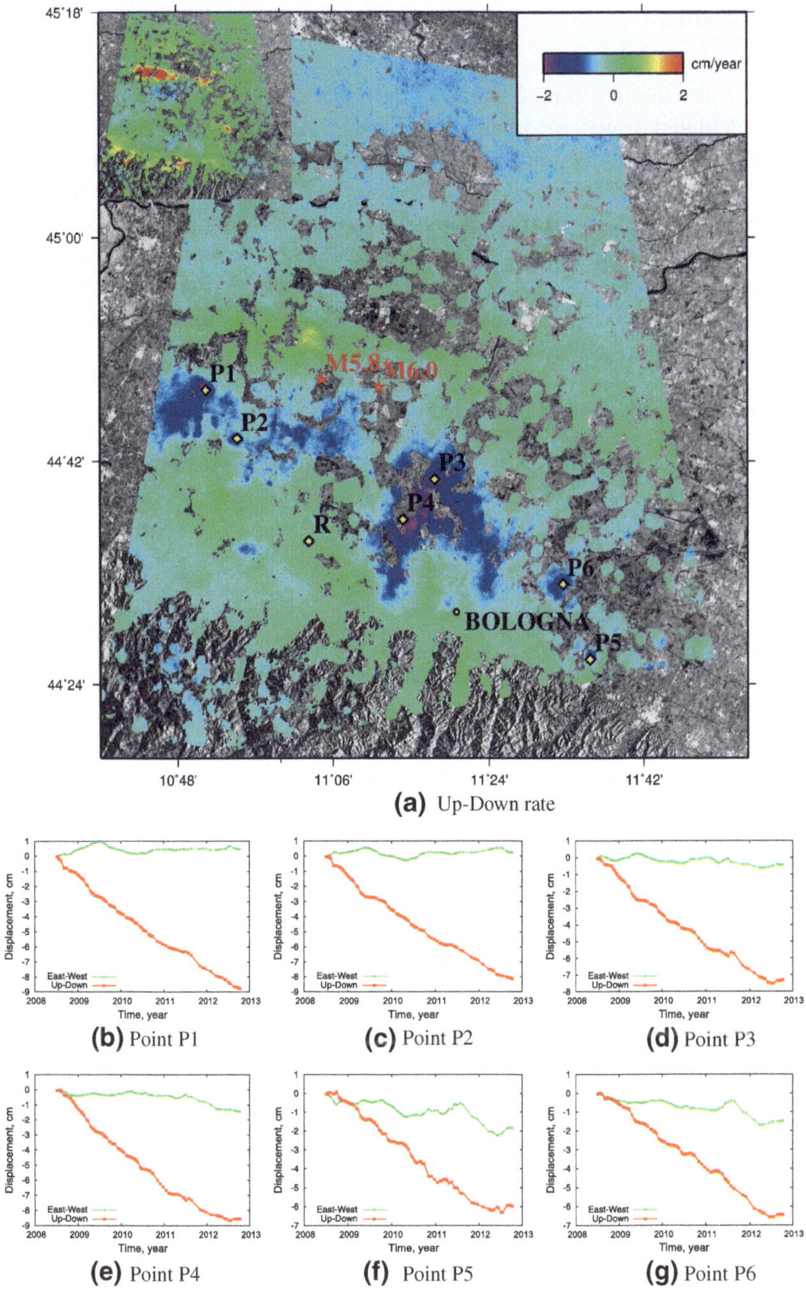

Fig. 1 2008-2013 vertical rate of deformation calculated with MSBAS method (**a**). For points P1-P6 we produce time series of vertical and horizontal deformation (**b–h**) that show steady subsidence and westward (relative to reference point R) motion. Locations of 2012 Emilia M5.8 and M6.0 mainshocks are shown as red stars. Vertical component of co-seismic displacements calculated by differencing 20121015 and 20120430 cumulative displacement maps is show as extent in *upper-left corner*

4 Conclusion

We present for the first time 2008-2013 deformation maps for the Bologna, Italy region calculated from one hundred Radarsat-2 images. We applied the fully-automated MSBAS methodology and produced vertical and horizontal deformation rates and time series. Observed signal corresponds to the anthropocentric ground subsidence near the City of Bologna and co-seimic deformation produced by the 2012 Emilia earthquake sequence. Due to severe decorrelation an accurate estimation of co-seismic surface displacements using the fully-automated approach was not possible but for urban regions located near the City of Bologna coherence was significantly larger resulting in a higher precision.

Acknowledgments We thank the Canadian Space Agency (CSA) for providing RADARSAT-2 data. Figures were plotted with GMT and gnuplot software. The work of PG was supported by Banting Postdoctoral Fellowship. The work of KFT was supported by an NSERC Discovery Grant.

References

1. Stramondo, S., Saroli, M., Tolomei, C., Moro, M., Doumaz, F., Pesci, A., et al. (2007). Surface movements in Bologna (Po Plain - Italy) detected by multitemporal DInSAR. *Remote Sensing of Environment*, *110*, 304–316.
2. Tizzani, P., Castaldo, R., Solaro, G., Pepe, S., Bonano, M., Casu, F., et al. (2013). New insights into the 2012 Emilia (Italy) seismic sequence through advanced numerical modeling of ground deformation InSAR measurements. *Geophysical Research Letters*. (in-press).
3. Wegmuller, U., & Werner, C. (1997). Gamma SAR processor and interferometry software. In *Third ERS Symposium on Space at the Service of our Environment*. Florence.
4. Farr, T., & Kobrick, M. (2000). Shuttle radar topography mission produces a wealth of data. *EOS Transactions, AGU*, *81*, 583–585.
5. Wessel, P., & Smith, W. (1998). New, improved version of the generic mapping tools released. *EOS Transactions, AGU*, *79*, 579.
6. Samsonov, S., & d'Oreye, N. (2012). Multidimensional time series analysis of ground deformation from multiple InSAR data sets applied to virunga volcanic province. *Geophysical Journal International*, *191*(3), 1095–1108.
7. Samsonov, S., d'Oreye, N., & Smets, B. (2013). Ground deformation associated with post-mining activity at the French-German border revealed by novel InSAR time series method. *International Journal of Applied Earth Observation and Geoinformation*, *23*, 142–154.
8. Samsonov, S., González, P., Tiampo, K., Camacho, A., & Fernández, J. (2013). Spatiotemporal analysis of ground deformation at Campi Flegrei and Mt Vesuvius, Italy, observed by Envisat and Radarsat-2 InSAR during 2003–2013. *Lectures Notes in Earth System Sciences*, 1–4.

Surface Displacements, Deformations and Gravity Changes Due to Underground Heat Source

Ladislav Brimich and Igor Kohut

1 Analytical Solution

The thermo-visco-elastic model presented in Hvoždara [4] was applied to estimate the volcanic ground deformations. Fundamental equations for the uncoupled thermo-visco-elastic problem for a point heat source located at depth ζ are given in Nowacki [6]. Thermo-visco-elastic gravity anomaly on the surface is given by Brimich [2].

Charco et al. [3] propose a simple method for including topographic effects in a 3D thermo-visco-elastic model that allows source depth to vary with the relief. On the Figs. 1, 2 and 3 are given horizontal, vertical thermo-viscoelastic displacement (in m) as well as the gravity changes computed for different relaxation times and the static value (thermoelastic case) considering (a) a flat surface, and (b)–(d) axis-symmetric volcanic cone with an average slope of the flanks of 15 °, 20 ° and 30 ° respectively. The influence of the topography is visible as the horizontal shift of maximal values of the both displacement components and particularly gravity changes.

2 Numerical Solution

To include the topography effect in the thermoelastic solution we have used the finite element method computation. The principles and basics of finite-element method are generally known and are described in numerous monographs (e.g., [1, 5]). All the computations are obtained by the COMSOL Multiphysics© software. Although numerical methods are time consuming, their results are more precise than the analytical approximate solutions since they allow to include structural characteristics of the medium as the topography.

L. Brimich (✉) · I. Kohut
Geophysical Institute of SAS, Bratislava, Slovakia
e-mail: geofbrim@savba.sk

E. Pardo-Igúzquiza et al. (eds.), *Mathematics of Planet Earth*,
Lecture Notes in Earth System Sciences, DOI: 10.1007/978-3-642-32408-6_86,

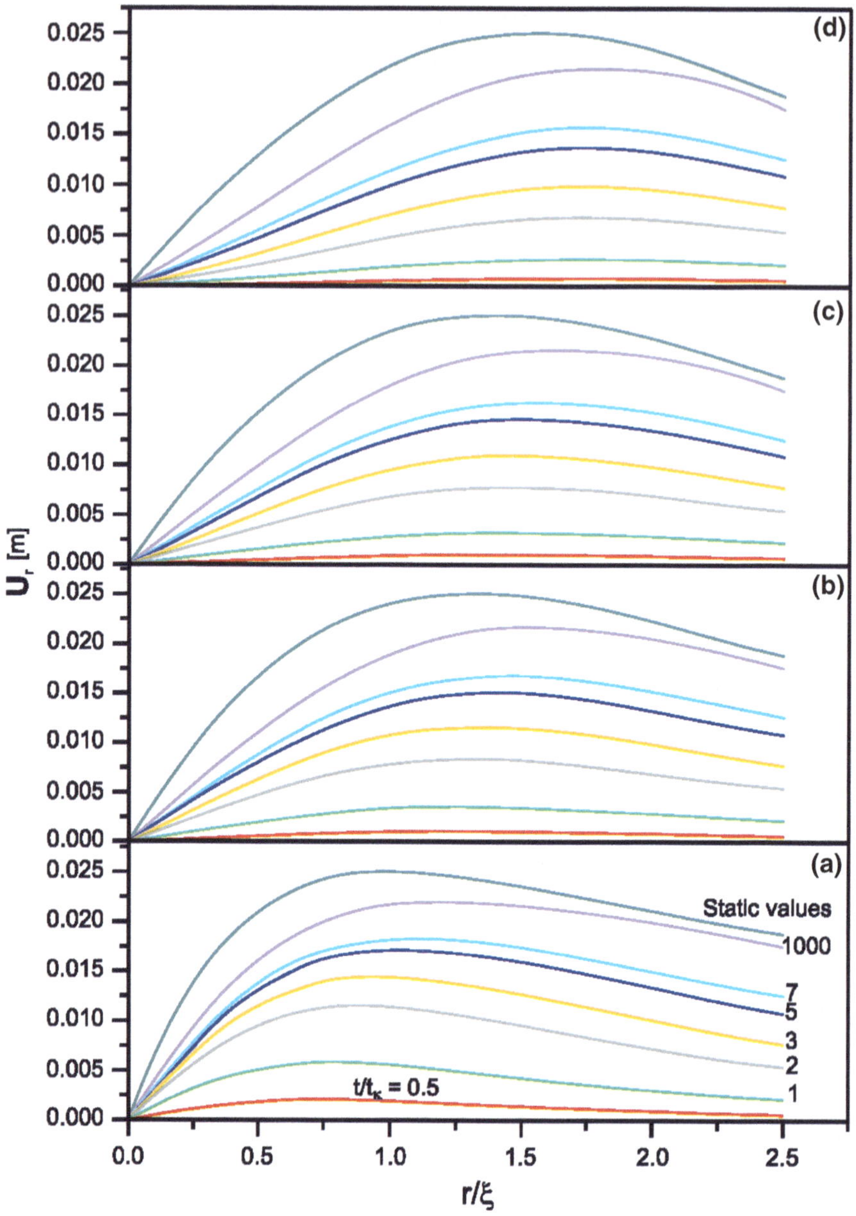

Fig. 1 Radial displacements

The models are homogeneous, isotropic, axi-symmetric with respect to vertical axis. In this way, the 3D rock massif has been modeled by an axi-symmetric section with respect to vertical axis passing through the heat source with 2 versions—with

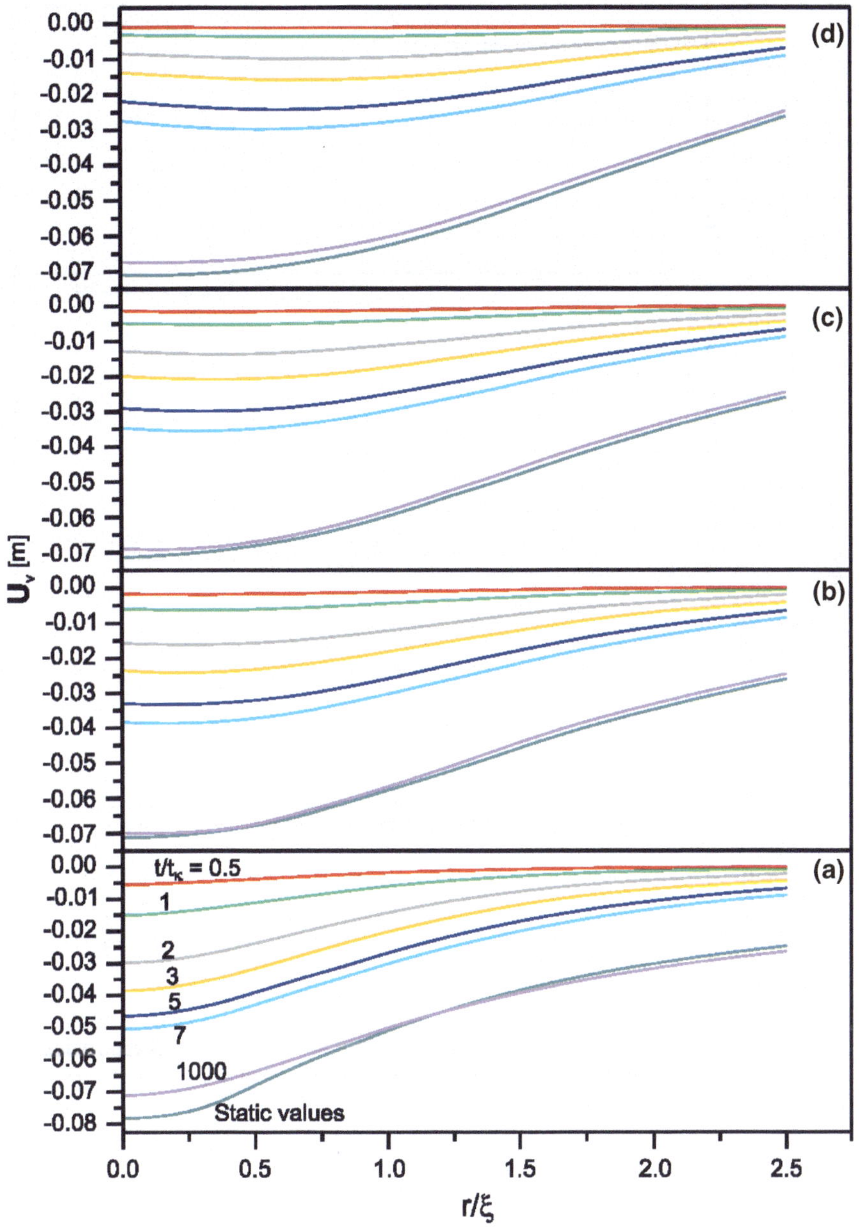

Fig. 2 Vertical displacements

and without topographical feature modeling the volcano cone (of 2 km height). The summit of the volcano is located over the thermal source. The domain horizontal length is 120 km and the vertical span is from +2 to −38 km in order to minimize

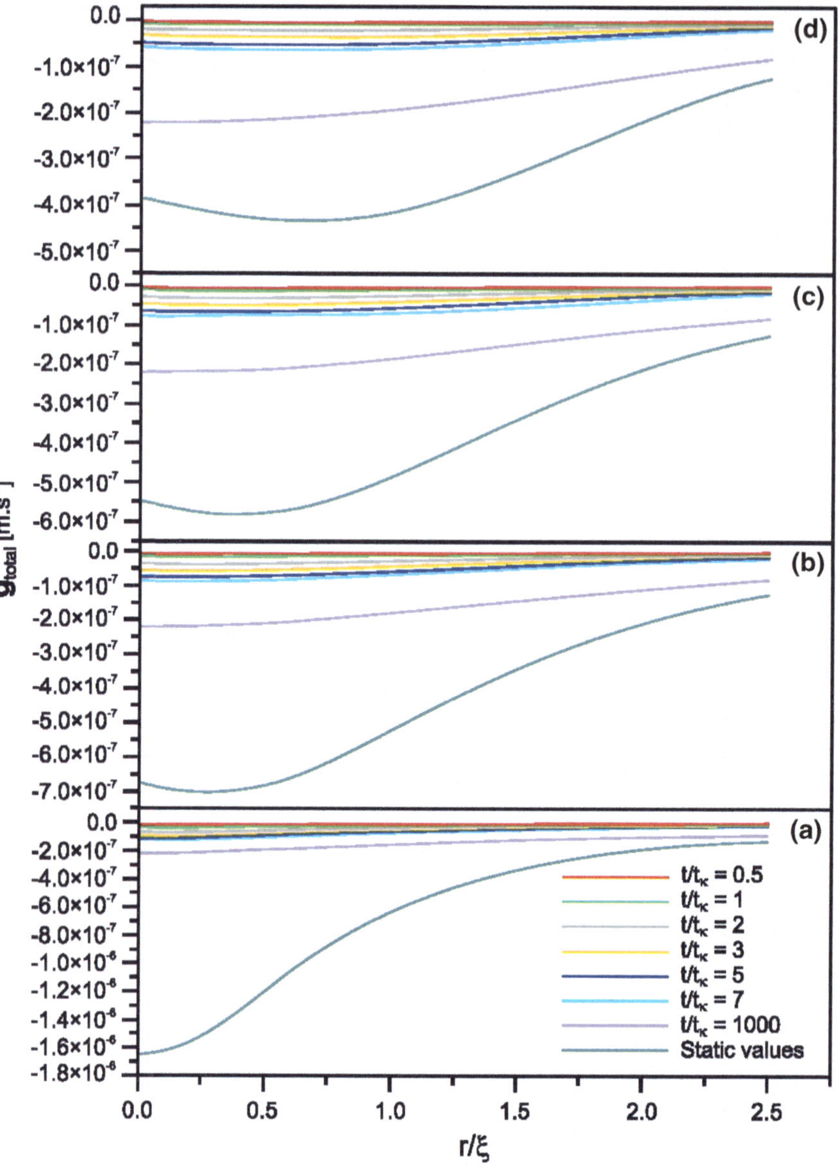

Fig. 3 Gravity changes

the influence of the external boundaries. The heat source is modeled by the spherical body in the 5 km depth. In the computation only the forces of thermal origin are considered.

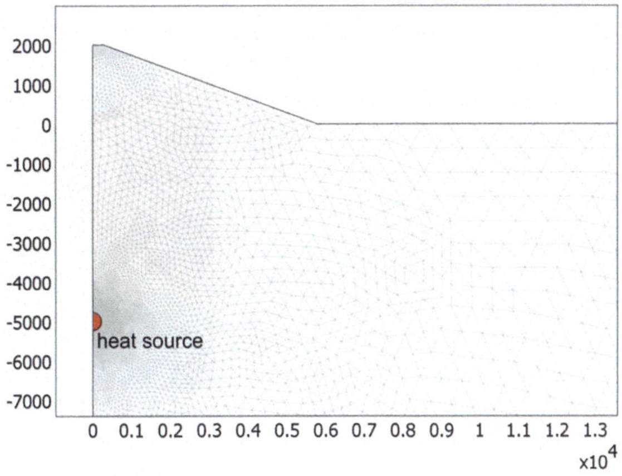

Fig. 4 Mesh with the position of the heat source

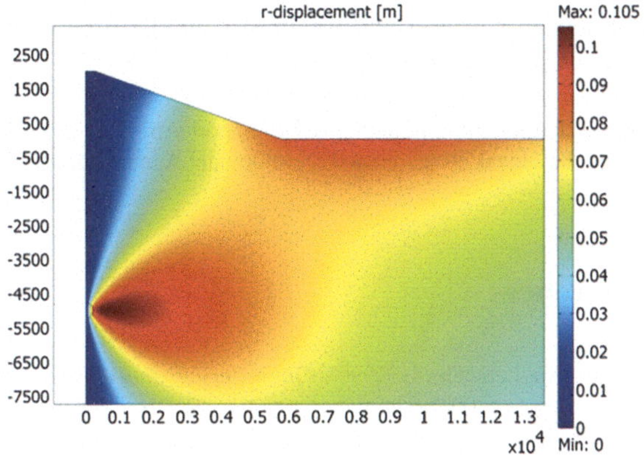

Fig. 5 Radial displacements

At first step, the domain was divided into finite elements. The mesh corresponding to each plane section is formed by 19568 triangular elements. In the neighborhood of the thermal source, the mesh is refined into smaller elements due to the large gradients of computed fields in this area (see Fig. 4). On the Figs. 5 and 6 are given radial and vertical displacements obtained using FEM.

3 Discussion

The methods described in this work can be very suitable to more complex models that consider sources of different geometries and allow elastic properties of the medium to vary with depth. While the analytical approximate methodology can be very attractive for solving the inverse problem, the numerical method described above may be used to include the topography when accurate solution is desired since it permits the

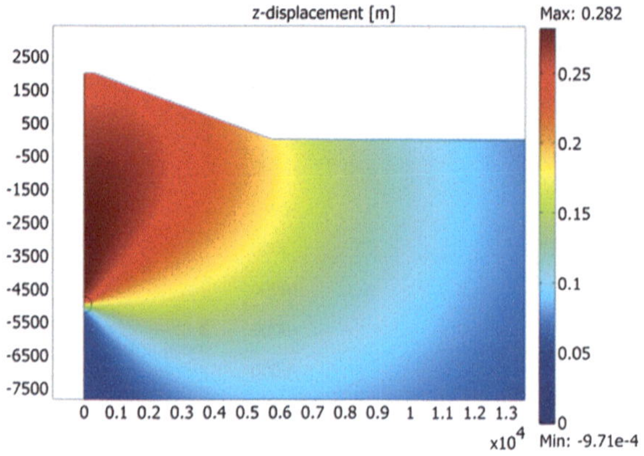

Fig. 6 Vertical displacements

consideration of non-uniform elastic and thermal properties of the medium and the local shape of the Earth's surface.

Acknowledgments The authors were supported by Vega Grant Agency under project No. 2/0067/12 and by the Slovak Research and Development Agency (contract No. APVV-0724-11).

References

1. Babuška, I., Szabo, B.(1990). *Finite element analysis*. New York, NY: Wiley.
2. Brimich, L. (2000). Thermoviscoelastic models of the deformations and gravity changes due to anomalous source of heat. *Acta Geodaetica et Geophysica Hungarica*, *35*, 37–48.
3. Charco, M., Brimich, L., & Fernández, J. (2002). Topography effects on dis-placements and gravity changes due to magma intrusions. *Geologica Carpathica*, *53*(4), 215–221.
4. Hvoždara, M. (1992). Thermo-viscoelastic deformation field due to a point source of heat in the halfspace. *Contributions of the Geophysical Institute of the Slovak Academy of Sciences*, *22*, 48–66.
5. Irons, B., & Ahmad, S. (1986). *Techniques of Finite Elements*. Chichester: Ellis Horwood.
6. Nowacki, W. (1962). *Thermoelasticity*. Oxford: Pergamon Press.

Polarization Persistent Scatterer InSAR Analysis on the Hayward Fault, CA, 2008–2011

Kristy F. Tiampo, Pablo J. González and Sergey S. Samsonov

1 Introduction

In northern California, relative motion of ~39 mm/year between the North American and Pacific is accommodated along a series of well-developed sub-parallel, primarily strike-slip faults including, but not limited to, the San Andreas, Hayward, Greenville, Concord and Rodgers Creek faults [1, 2]. Estimates suggest that while as much as 60 % of the relative motion is accommodated by the San Andreas alone [2], almost half of the strain accumulation is apportioned onto secondary faults. The Hayward fault, a right-lateral fault paralleling the San Andreas for nearly 100 km, accommodates much of the remaining motion and can be divided into two segments. The northern segment of the Hayward fault has not ruptured in approximately 300 years and the southern was responsible for the M $\sim$ 6.8, 1868 earthquake [3, 4]. Aseismic creep varies from 3 to 9 mm/year along the length of the Hayward, resulting in a slip deficit that will affect the magnitude of future large events [4, 5].

Earlier studies of the motion along the Hayward fault incorporated geodetic data from both continuous GPS and alinement arrays [2, 5, 6] differential InSAR (DInSAR) analyses that employed from 13 to 49 SAR images acquired over eight or more years [3, 4, 7, 8]. In addition, the DInSAR results were affected by the lack of coherence from the topographic relief on the eastern side of the fault.

Here we demonstrate the effectiveness of a new DInSAR technique [9] by estimating the deformation along the northern Hayward fault using fifteen quad polarization RADARSAT-2 SAR images acquired from April 2008 until March 2011. This new method selects persistent scatterers (PS) based on the polarization phase difference (PPD) in HH and VV polarized SAR images. PPD selects pixels with predominantly

K. F. Tiampo (✉) · P. J. González
Department of Earth Sciences, Western University, London, ON 6A 5B, Canada
e-mail: ktiampo@uwo.ca

S. S. Samsonov
Canada Centre for Remote Sensing, Ottava, ON K1A 0Y7, Canada

E. Pardo-Igúzquiza et al. (eds.), *Mathematics of Planet Earth*,
Lecture Notes in Earth System Sciences, DOI: 10.1007/978-3-642-32408-6_87,

even or odd bounce scattering properties, where odd bounce scattering occurs due to reflection from a flat, rough surface and even bounce scattering is from the interaction of the radar wave with standing objects. PPD requires a much smaller set of SAR acquisitions over a shorter time period than other PS techniques and is a powerful method for identification of PS candidates, producing a more complete spatial distribution than alternative DInSAR techniques.

Earlier work compared the results of the original PS technique with the PPD technique and showed that the PPD technique successfully identified a greater percentage of pixels in areas of low coherence [9]. Here the PPD method is used to reliably estimate the deformation along the Hayward fault with fewer acquisitions and increased spatial coverage over a shorter time than earlier DInSAR studies.

2 Data and Methods

PS analysis has emerged as an effective alternative where two-pass DInSAR methodology is limited by temporal decorrelation where pixels are selected based on identification of a predominant scatterering mechanism that is consistent over a long period of time [10]. PS candidates are selected based on their amplitude dispersion through time, $D = \sigma/A$, where σ is the standard deviation and A is the mean amplitude calculated for the coregistered set of SAR images. However, the selection of PS pixels with a high degree of accuracy requires at least thirty SAR images. It can be expensive and difficult to acquire large numbers of images and inevitably span longer time periods, making it difficult to identify short-term changes in deformation.

The PPD method does not require a large number of acquisitions for PS selection [9]. Using HH and VV polarized RADARSAT-2 SAR images, PS are selected based on their PPD, $\Delta\varphi = \varphi_{HH} - \varphi_{VV}$. φ_{HH} is the phase of a wave transmitted and received in horizontal polarization and φ_{VV} is the phase of a wave transmitted and received in vertical polarization, relative to Earth's surface. PPD values of zero and $\pm\pi$ correspond to scatterers with a dominant reflective mechanism, deterministic odd and even bounce scatterers. Generally, odd bounce scattering is caused by reflection from a flat, rough surface and even bounce scattering is produced by interaction of the wave with anthropogenic structures or other standing objects. $\Delta\varphi$ diverges from these extremes as the contribution from diffusive scattering increases [11, 12]. Diffusive scattering from vegetation results in values randomly distributed in $[-\pi, \pi]$.

For each pixel we can calculate a normalized average of absolute values of PPD, $\chi = \sum_{k=1}^{K} |\Delta\varphi_k| / K\pi$, where K is the number of SAR images, and select pixels with values close to 0 and 1 [9]. Threshold values of $\chi \leq 0.2$ and $\chi \geq 0.8$ are applied in order to select for those scatterers dominated by odd and even bounce mechanisms. If we investigate those pixels, after the removal of pixels over the water surface, approximately 131,000 pixels are recovered. 62.5 %, are the odd bounce scatterers and the remaining 37.5 %, are the even bounce scatterers.

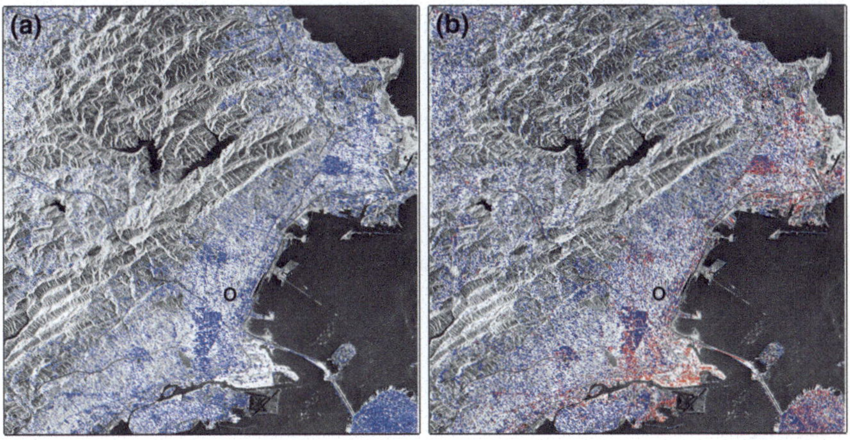

Fig. 1 **a** Pixels over the waterfront (O—Oakland) and the surrounding hills selected with a PS analysis (*blue*) and **b** pixels selected by the PPD analysis in the same region, where odd bounce pixels are *blue* and even bounce pixels are *red*. Modified from [14]

Figure 1 compares the PS and PPD method for a small region along the San Francisco waterfront and the surrounding hills. Pixels were selected by a PS analysis Fig. 1a and a PPD analysis Fig. 1b. In Fig. 1a the PS pixels are blue while in Fig. 1b the PPD odd bounce pixels are blue and the even bounce pixels are red. The total number of PS pixels is approximately 107,000, while the total number of PPD pixels is approximately 132,000, a significantly larger number. Visual inspection confirms that these correspond to roads, bridges, rocks and open surfaces. Surface displacements near the Hayward fault south from Point Pinole to just north of the city of Hayward, CA, were calculated using this PPD technique applied to 15 descending HH and VV polarization RADARSAT-2 SAR (FQ7) images acquired between April, 2008 and March, 2011. Processing techniques are detailed in [14].

3 Discussion and Conclusions

The final slip distribution associated with the surface deformation is in accordance with the broad velocity pattern obtained from regional continuous GPS and the alinement array [2, 14, 15]. Detailed sensitivity tests identify the importance of the additional information from the improved spatial coverage to the east obtained from the PPD analysis [14]. For example, deformation identified between ten and fifteen kilometers east of Point Pinole provided by that additional coverage significantly improves our understanding of slip at depth. For the first time, using this advanced PS technique, we map the displacement field on the vegetated hills to the east in addition to the urban are to the west of the fault. This additional information is obtained from only fifteen polarimetric acquisitions, less than half as many as previous DInSAR studies, at approximately half the cost. This can be used to estimate short-term shallow creep and with better spatial coverage on both sides of the fault [14]. Although the current availability of quad polarization data is significantly less than that of other acquisition modes of SAR data, the success of the PPD algorithm provides a new and innovative advance in the field of DInSAR.

Acknowledgments Images were acquired under SOAR-E project 5024 from the Canadian Space Agency. Research was funded by the NSERC and Aon Benfield/ICLR IRC in Earthquake Hazard Assessment, an NSERC Discovery Grant, and an Ontario Early Researcher Award. PJG also acknowledges support from a Banting postdoctoral fellowship. Figures were created with GMT software [16].

References

1. Argus, & D., Gordon, R. (2011). Present tectonic motion across the coast ranges and San Andreas Fault system in central California. *GSA Bulletin, 113*, doi:10.1130/0016-7606
2. d'Alessio, M. A., Johanson, I. A., Bürgmann, R., Schmidt, D. A., & Murray, M. H. (2005) Slicing up the San Francisco Bay Area: Block kinematics and fault slip rates from GPS-derived surface velocities. *Journal of Geophysical Research, 110*, B06403, doi:10.1029/2004JB003496
3. Schmidt, D., Bürgmann, R., Nadeau, R., & d'Alessio, M. (2005). Distribution of aseismic slip rate on the Hayward Fault inferred from seismic and geodetic data. *Journal of Geophysical Research, 110*, B08406.
4. Bürgmann, R., Schmidt, D., Nadeau, R., d'Alessio, F., Fielding, E., Manaker, D., et al. (2000). Earthquake potential along the Northern Hayward Fault, California. *Science, 289*, 1178–1182.
5. Lienkaemper, J. J., Galehouse, J. S., & Simpson, R. W. (2001). Long-term monitoring of creep rate along the Hayward Fault and evidence for a lasting creep response to 1989 Loma Prieta earthquake. *Geophysical Research Letters, 28*, 2265–2268.
6. Lienkamper, J. J., McFarland, F. S., Simpson, R. W., Bilham, R. G., Ponce, D. A., Boatwright, J. J., & Caskey, S. J. (2012). Long-term creep rates on the Hayward Fault: Evidence for controls on the size and frequency of large earthquakes. *Bulletin of the Seismological Society of America, 102*, doi:10.1785/0120110033
7. Bürgmann, R., Hilley, G., Ferretti, A., & Novali, F. (2006). Resolving vertical tectonics in the San Francisco Bay Area from permanent scatterer DInSAR and GPS analysis. *Geology, 34*, 221–224.
8. Lanari, R., Casu, F., Manzo, M., & Lundgren, P. (2007). Application of the SBAS-DInSAR technique to fault creep: A case study of the Hayward Fault. *CA. Remote Sensing of Environment, 109*, 20–28.
9. Samsonov, S., & Tiampo, K. (2011). Polarization phase difference analysis for selection of persistent scatterers in SAR interferometry. *IEEE Geoscience and Remote Sensing Letters, 8*, 331–335.
10. Ferretti, A., Prati, C., & Rocca, F. (2001). Permanent scatterers in SAR interferometry. *IEEE Transactions on Geoscience and Remote Sensing, 39*, 8–20.
11. Evans, D. L., Farr, T. G., van Zyl, J. J., & Zebker, H. (1988). Radar polarimetry: Analysis tools and applications. *IEEE Transactions on Geoscience and Remote Sensing, 26*, 748–789.
12. Ulaby, F. T., Held, D., Dobson, M. C., McDonald, K., & Senior, T. (1987). Relating polarization phase difference of SAR signals to scene properties. *IEEE Transactions on Geoscience and Remote Sensing, 25*, 83–92.
13. Goldstein, R. M., & Werner, C. L. (1998). Radar interferogram filtering for geophysical applications. *Geophysical Research Letters, 25*, 4035–4038.

14. Tiampo, K. F., González, P., & Samsonov, S. (2013). Results for aseismic creep on the Hayward fault using polarization persistent scatterer InSAR. *Earth and Planetary Science Letters*, doi:10.1016/j.epsl.2013.02.019
15. McFarland, F. S., Lienkaemper, J. J., & Caskey, S. J. (2009). Data from theodolite measurements of creep rates on San Francisco Bay Region faults, California, 1979–2010. U.S. Geol. Survey Open-File Report 2009-1119, p. 17 (2009, revised 2011).
16. Wessel, P., & Smith, W. H. F. (1998). New, improved version of generic mapping tools released. *Eos, Transactions American Geophysical Union*, *79*, 579.

Earth Models at the Low Andarax River Valley (SE Spain) by Means of Cross-Correlation of Ambient Noise

Abigail Jiménez, Antonio García-Jerez, Enrique Carmona, Francisco Sánchez-Martos and Francisco Luzón

1 Introduction

Surface wave tomography has proven to be very useful in imaging the crust and uppermost mantle on regional and global scales across much of the globe. Surface waves of different periods are sensitive to seismic shear wave speeds at different depths, with the longer period waves sampling the deepest structure. By measuring the dispersive character of surface waves, strong constraints can be placed on the shear wave velocity structure of the crust. Ambient noise cross-correlation can be regarded as a cheap and non-intrusive way of obtaining subsurface models. In this work, we want to improve our knowledge of the elastodynamic properties of the low andarax valley (LAV). The LAV is enclosed by the Sierra Alhamilla, with metapelitic outcrops, and by the Sierra de Gádor, which is a limestone-dolomite massif formation. The depression is filled by Miocene and Pliocene detrital deposits (marls, sandy silts, sands and conglomerates). Limestone–dolomite materials with a formation of phyllites quartzite form the basement of the depression (Fig. 1a) [1].

A. Jiménez (✉)
School of Environmental Sciences, University of Ulster, Northern Ireland, UK
e-mail: a.jimenez@ulster.ac.uk

A. García-Jerez · E. Carmona · F. Sánchez-Martos
Andalusian Institute of Geophysics, University of Granada, Granada, Spain

E. Carmona · F. Luzón (✉)
Department of Chemistry and Physics, University of Almería, Almería, Spain
e-mail: fluzon@ual.es

F. Sánchez-Martos
Department of Biology and Geology, University of Almería, Almería, Spain

E. Pardo-Igúzquiza et al. (eds.), *Mathematics of Planet Earth*,
Lecture Notes in Earth System Sciences, DOI: 10.1007/978-3-642-32408-6_88,

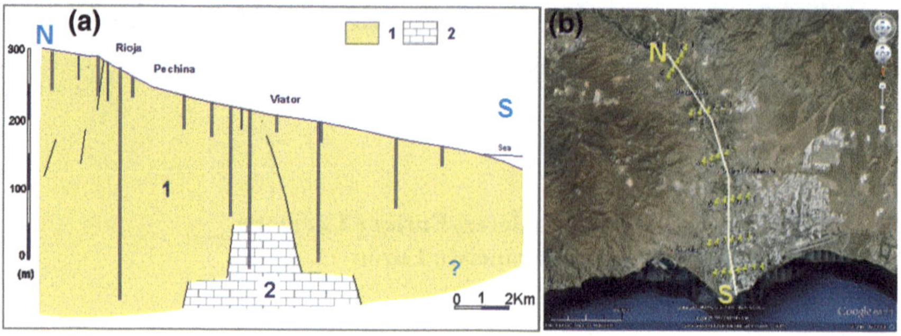

Fig. 1 **a** Geological section North/South across the low Andarax river valley completed by borehole data. *Legend 1* Marls, sandy silts, sands and conglomerates. *2* Limestones and dolomites. **b** Location geological cross-section and profiles from 0 (South) to 5 (North)

2 Data Analysis

The profiles we will study are perpendicular to the river bed (Fig. 1b), from the lower part (profile 0) to the upper part (profile 5). The data points are deployed every 500 m along the lines. We used CMG-3ESPD three-component broadband recorders manufactured by Guralp with reliable response bands from 120 s to 50 Hz, and Güralp CMG-6TD three-component broadband recorders with response bands from 30 s to 100 Hz, with independent GPS–based timing. The sampling rate was 100 samples/s. In this first analysis, the vertical components of motion have been used to invert the fundamental mode of the Rayleigh waves. For each pair of stations, we calculate the coherence (or SPAC) function. It has been shown (e.g., [2]) that:

$$\frac{\mathrm{Re}\left\lfloor \langle u_z(x,\omega)\rangle\langle u_z^*(y,\omega)\rangle \right\rfloor}{\sqrt{\langle |u_z(y,\omega)|^2\rangle}\sqrt{\langle |u_z(y,\omega)|^2\rangle}} = J_0(kr) \tag{1}$$

where the SPAC function [3] equals the Bessel function of zero order. The hypotheses considered are: (i) predominance of the fundamental mode for Rayleigh-wave propagation, and (ii) the medium can be locally approximated by a layered halfspace. Both hypotheses are reasonable and commonly assumed in velocity structure inversions based on surface waves. The experimental SPAC functions have been obtained as follows: first, we divided the traces into 1 min windows, and corrected them from baseline and trend. After that, we filtered the traces with a Butterworth band-pass filter with two poles between 0.1 and 5 Hz, and considered only 1 bit signals [4]. Then, we applied a 5 % cosine-tapered window, performed a temporal normalization for each trace [5], and calculate the FFT. Finally, the SPAC functions were computed between every pair of stations [2], and a Genetic Algorithm [6] was used in order to invert the S-wave structure. We applied GEOPSY [7] programs to calculate the theoretical dispersion curves, which are introduced in the argument of the Bessel function to obtain the theoretical SPAC coefficient. Evaluation of misfits was performed in terms of coherences.

3 Results and Conclusions

In Fig. 2 we represent the structural models obtained along the West-East profiles from this frequency-domain analysis. For each pair of consecutive points of the profile, all the couples of stations connected by paths passing through it were considered

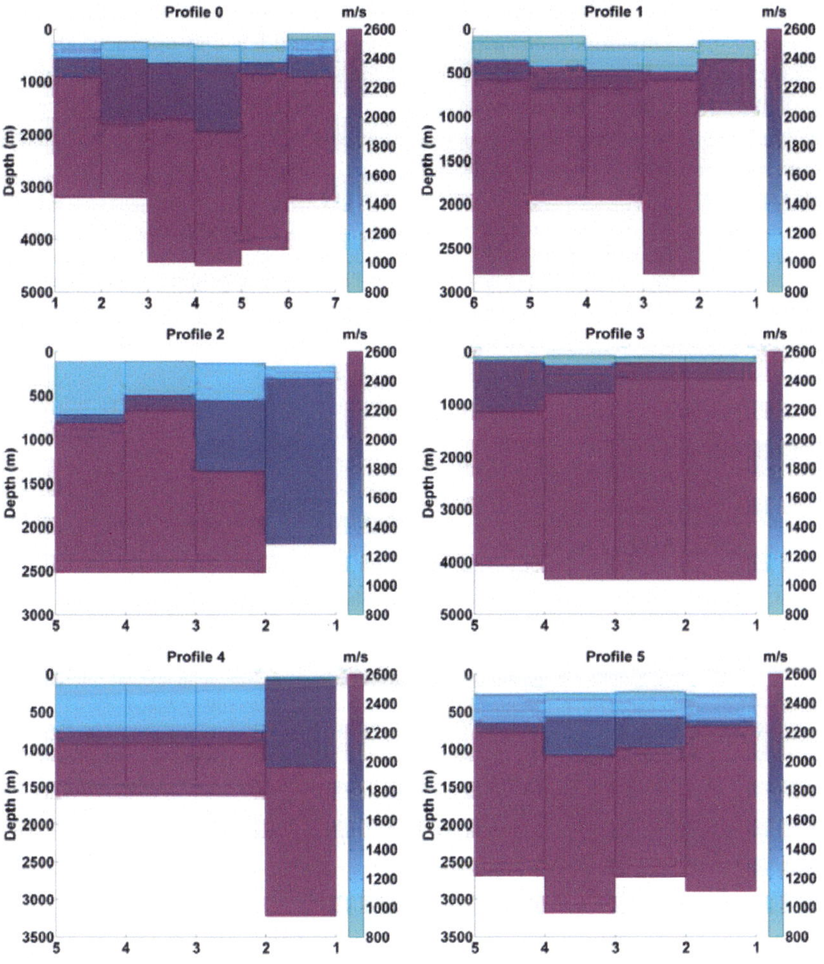

Fig. 2 Earth models, for profiles from 0 (South) to 5 (North). The *colorbar* represents the S wave velocities in m/s

in the ground model inversion. The influence of their respective SPAC curves, were weighted by using their fitness. The colored patches in Fig. 2 show the inverted models, where the maximum and minimum resolved depths have been calculated using the criterion of $\lambda/3$ penetration depth for Rayleigh waves. For profiles 0, 1, 2 and 5, a shallow soft layer with S wave velocity around 1 km/s has been prospected down to 500–700 m, overlying materials with significantly higher velocities. Although the computed depth of this layer reaches 800 m in profile 4, we could not get a large enough number of reasonably fitted paths to guarantee the reliability of this value. Our results at profile 3, which crosses the LAV 1 km upstream from Viator village, show a thickness of about 200–250 m for this soft layer. By comparison with a geological cross-section (see Fig. 1a), we observe that this depth approximately matches the rock basement formed of limestones and dolomites. The velocities of such materials are about 2 km/s. This important result lets us gain confidence in our methodology and makes possible to delineate the rock basement along the other profiles, taking those depths where these high S wave velocities are reached. Finally, the correspondence between the observed shallow detrital deposits in the LAV and the calculated soft layer with mean velocities $\sim$1 km/s should be remarked.

Acknowledgments This work was partially supported by the MICINN research project CGL2010-16250, Spain, by the EU with FEDER and by the research team RNM-194 of Junta de Andalucía, Spain. Thanks are given to A. Sánchez and S. Limonchi for their help in the field campaigns. The works by A. J. and A. G.-J. were supported by Juan de la Cierva grants, from the Spanish Government.

References

1. Sánchez-Martos, F., Pulido-Bosch, A., Molina-Sánchez, L., & Vallejos-Izquierdo, A. (2002). Identification of the origin of salinization in groundwater using minor ions (Lower Andarax, Southeast Spain). *The Science of the Total Environment, 29*, 43–58.
2. Luzón, F., Almendros, J., & García-Jerez, A. (2011). Shallow structure of Dception Island, Anctartica, from correlations of ambient seismic noise on a set of dense seismic arrays. *Geophysical Journal International, 185*, 737–748.
3. Aki, K. (1957). Space and time spectra of stationary stochastic waves, with special reference to microtremors. *Bulletin of the Earthquake Research Institute, University of Tokyo, 25*, 415–457.
4. Campillo, M., & Paul, A. (2003). Long-range correlations in the diffuse seismic coda. *Science, 299*, 547–549.
5. Bensen, G. D., Ritzwoller, M. H., Barmin, M. P., Levshin, A. L., Lin, F., Moschetti, M. P., et al. (2007). Processing seismic ambient noise data to obtain reliable broad-band surface wave dispersion measurements. *Geophysical Journal International, 169*, 1239–1260.
6. Jiménez, A., García, J. M., & Romacho, M. D. (2005). Simultaneous inversion of source parameters and attenuation factor using genetic algorithms. *BSSA, 95*, 1401–1411.
7. GEOPSY (2012). http://www.geopsy.org/

Some Insights About Volcano Deformation Interpretation

María Charco and Pedro Galán del Sastre

1 Introduction

In volcanic areas, it is expected that changes within magma system leading to eruption will result in precursory deformation measurable by geodetic techniques. In this way, deformation interpretation provides the link between the observed deformation and the inaccessible sources of such effects. The interpretation of geodetic data requires both modeling techniques and inversion approaches to characterize processes (pressure/volume changes), geometries and location at depth of magma chamber. Within the elastic frame, a variety of models have been proposed. The most commonly used is the Mogi model [1] that represents the simplest close analytical solution for a spherical expansion source embedded in a homogeneous isotropic elastic half-space with free surface. The computational simplicity of Mogi's model make it very suitable for geodetic data inversion. However, the internal structure of a volcano controls the details of the deformation field and in the case of fully 3D rheologies as well as complicated geometrical structures, a numerical method is required to simulate volcano deformation. The aim of this study is to provide a numerical tool for solving very fast and with appreciable degree of accuracy the inverse problem to characterize spherical sources. We show that the efficiency of this approach is more substantial when quantitative interpretation of volcanic deformation becomes of major importance in related geodetic monitoring efforts.

M. Charco (✉)
Facultad de CC. Matemáticas, Instituto de Geociencias (CSIC-UCM), Pza. de Ciencia, 3, Madrid, Spain
e-mail: mcharco@iag.csic.es

P. Galán del Sastre (✉)
Departamento de Matemática Aplicada a la Edificación, al Medio Ambiente y al Urbanismo, E.T.S. Arquitectura, UPM, Av. Juan de Herrera 4, Madrid, Spain
e-mail: pedro.galan@upm.es

E. Pardo-Igúzquiza et al. (eds.), *Mathematics of Planet Earth*,
Lecture Notes in Earth System Sciences, DOI: 10.1007/978-3-642-32408-6_89,

2 Problem Statement

Our intention is to study the response of an elastic media to an internal load that can reflect magmatic and/or hydrothermal processes at depth that results in strain. This problem can be formulated as a boundary value problem:

$$\begin{cases} \nabla \cdot \boldsymbol{\sigma} + \mathbf{f_P} = 0 & \text{in } \Omega \\ \sigma_{ij} = \lambda \sum_{k=1}^{3} \varepsilon_{kk}\delta_{ij} + 2\mu\varepsilon_{ij} & \text{in } \Omega \\ \varepsilon_{ij} = \frac{1}{2}\left(\frac{\partial u_i}{\partial x_j} + \frac{\partial u_j}{\partial x_i}\right) & \\ \mathbf{u} = \mathbf{0} & \text{on } \Gamma_2 \\ \boldsymbol{\sigma} \cdot \mathbf{n} = \mathbf{0} & \text{on } \Gamma_1 \end{cases} \tag{1}$$

where $\boldsymbol{\sigma}$ is the stress tensor; ε is the strain, that is related with the displacement field, $\mathbf{u}$; δ_{ij} is the Kronecker delta; $\mu = \mu(\mathbf{x})$ represents the shear modulus, also called the rigidity modulus or the second Lamé coefficient; $\lambda = \lambda(\mathbf{x})$ is the first Lamé coefficient; and $\mathbf{n}$ being the unit outward normal vector to the boundary $\Gamma = \overline{\Gamma_1 \cup \Gamma_2}$, $\Gamma_1 \cap \Gamma_2 = \emptyset$.

The inflation/deflation of magma reservoirs caused by pressure changes is usually modeled by considering a cavity with radius a inside the medium. In this case the problem domain depends on the cavity location. Thus, solving this problem via FEM implies mesh rebuilding when the source location or geometrical features of the source changes. The rebuilding of the mesh can be a limitation for solving the inverse problem through explorative schemes since they require to solve repeatedly problem (1) and therefore to change the source location and its geometric features on each realization. To overcome this limitation, we can take into account that the solution for the spherical cavity can also be obtained assuming three orthogonal force dipoles or center of dilatation, e.g., [2], when the source is much smaller that its depth. Thus, we propose the body force, $\mathbf{f_p}$, to be applied in (1) as the gradient of a Gaussian function whose variance is selected a priori depending on the size of the Finite Element discretization. Since this is an elastic system, the displacement field throughout the domain is a linear function of the source strength, $a^3\Delta P$.

Here, we use 3D models solved by FEM that simulate the volcano deformation as described above for the estimation of pressure changes, $a^3\Delta P$, and source location, (x_1', x_2', x_3'), that characterize volcano deformation sources. We refer to extensive literature for the formulation of the problem (1) via FEM [3]. To efficiently handle with non-linear inversions, the method we propose combines both a gradient based technique as least squares optimization and a sampling technique in an explorative scheme similar to the one proposed by [4].

3 Application

We study Teide stratovolcano (Tenerife, Canary Islands, Spain). For such as task, we perform two inversions to estimate the best-fitting sources for the synthetic data that could be observed at the GPS network stations that cover Tenerife island in case

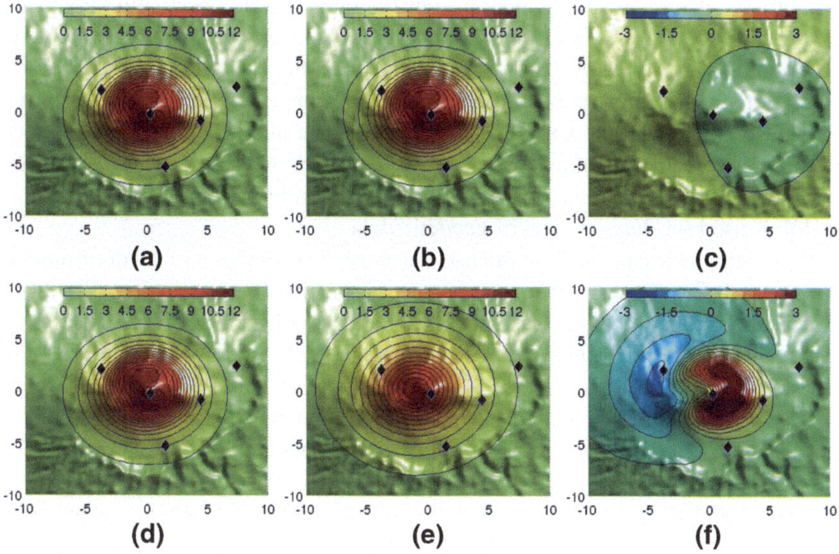

Fig. 1 **a**, **d** Synthetic vertical displacement, u_3, in cm (both graphics represent the same synthetic field); **b** u_3^{TenHeT} and **e** u_3^{TenHoF} considering the parameters obtained by inversion; model residuals: **c** $u_3 - u_3^{\mathrm{TenHeT}}$ and **f** $u_3 - u_3^{TenHoF}$. The *marks* indicate the location of GPS stations located around Teide

of intrusion beneath Teide summit. The synthetic data were obtained considering a heterogeneous domain with the real topography of the island that is contiminated with Gaussian noise of 1 cm standard deviation. The distribution of elastic parameters were estimated from previous gravimetric and seismic studies carried out in the island [5, 6].

One inversion is performed considering the structural model of the island used for the simulation of synthetic data (model TenHeT). The other one is performed with the same set of synthetic data but considering a flat free surface homogeneous domain and a reference elevation located at 3000 m above sea level (model TenHoF). Model TenHoF is representative of the analytical models, such as the Mogi model. While the first inversion allows to validate the inversiontechnique, the comparison between the results from model TenHeT and TenHoF reveals sensitivities to TenHoF assumptions.

Figure 1 shows the differences between the synthetic vertical displacement and the vertical displacement retrieved by TenHeT with the source parameters obtained from the inversion process. Predicted deformation from TenHeT (Fig. 1b) practically recovers the synthetic field (Fig. 1a). The differences are of order 10^{-5} cm (Fig. 1c). Considering the GPS precision attainable nowadays, we can conclude that the developed inversion methodology of geodetic deformation data observed at the GPS network of the Tenerife island is very accurate. Nevertheless, the quality of the inversion results is unreliable for TenHoF model since the influence of the medium assumptions retrieves residuals that are out the range of the observation precision with values around 3 cm (Fig. 1f). The magnitude and pattern of these systematic prediction differences suggest forward model assumptions should be carefully considered when designing a conceptual model of volcano deformation. Furthermore, the residuals show that the spatial distribution and the number of the GPS stations located nearby Teide volcano, could not be ideal for understanding the deformation field caused by a magma intrusionbeneath its summit when a typical model as TenHoF is used for the inversion of geodetic data.

Other authors have proposed a methodology for the estimation of the deformation source parameters taking into account structural medium heterogeneities [7]. They focus on the automation of the mesh generation process in response to perturbation of a spherical cavity position within the domain. We take advantage of our methodology and the total analysis time for 9,000 realizations of the forward model is about 17 h whereas [7] employees about 7 days in a similar machine (3 GHz quad core CPU) for 12,000 realizations. It is worth mentioning that their mesh contains 13,870 nodes versus the 108,990 nodes that integrates Tenerife mesh. Since our code is parallelized and the scalability is very good in this kind of problems were many independent forward models are computed, the CPU time can be reduced using as many processors as available. As an example we also perform an inversion in 4 quad core CPU and, as expected, the CPU time reduced by 4, i.e, we spent 4.5 h. Therefore, the inversion method with heterogeneous models could be considered as a semi-real time method of quantitative interpretation due to its scalability.

Acknowledgments The research of MC has been support by 200930I053 MICINN-CSIC Grant and by CGL2012-37222 Grant from MINECO. PGdS work has funded by grant CGL2007-66440-C04-01 from MICINN.

References

1. Mogi, K. (1958). Relations of eruptions of various volcanoes and deformation around them. *Bulletin of the Earthquake Research Institute, University of Tokyo*, *36*, 99–134.
2. Mindlin, R. D. (1936). Force at a point in the interior of a semi-infinite solid. *Physics*, *7*, 195–202.
3. Brenner, S. C., & Scott, R. (1994). *The mathetical theory of finite element methods* (p. 204). New York: Springer.
4. Camacho, A. G., et al. (2007). Interpretation of 1992–1994 gravity changes around Mayon volcano, Philippines, using point sources, *Pure and Applied Geophysics*, *164*, 733–749.
5. Watts, et al. (1997). A seismic study of lithospheric flexure in the vicinity of Tenerife, Canary islands. *Earth and Planetary Science Letters*, *146*(3–4), 431–447.
6. Camacho, et al. (2011). A new gravity inversion method for multiple subhorizontal discontinuity interfaces and shallow basins. *Journal of Geophysical Research*, *116*, B02413. doi:10.1029/2010JB008023.
7. Masterlark, T., et al. (2012). Nonlinear estimation of geometric parameters in FEMs of volcano deformation: integrating tomography models and geodetic data for Okmok volcano, Alaska. *Journal of Geophysical Research*, *117*, B02407.

Ensemble Kalman Filter Assimilation of Transient Groundwater Flow Data: Stochastic Moment Solution Versus Traditional Monte Carlo Approach

Marco Panzeri, Monica Riva, Alberto Guadagnini and Shlomo P. Neuman

1 Introduction

Assimilating temporal data with Kalman Filter (KF) related algorithms is a widespread inverse modeling technique in many areas of the earth sciences. The increasing popularity of this approach finds its origin in the simplicity of the associated first-order-based updating scheme, which is relatively easy to implement, computationally efficient and allows integrating dynamic data sequentially as soon as they become available (e.g., [1]). The ensemble Kalman filter (EnKF) is appropriate for large and nonlinear models of the kind required for realistic subsurface fluid flow simulations. It has traditionally entailed the use of a (numerical) Monte Carlo (MC) approach to generate a collection of interdependent random model representations. These are referred to in the literature as an ensemble, a term we use here in a more traditional stochastic context by relating it to ensemble moments of quantities of interest. The collection enables one to estimate the ensemble statistics (mean and covariances) of parameters and state variables that are used during the updating step. In practical problems, the computational burden associated with the MC framework requires keeping the number of MC realizations much smaller than the size of the system state and parameter covariance matrix. This often leads to spurious correlations and incorrect updates of the model parameters. Previously in Ref. [3] we proposed to circumvent the need for MC through a direct solution of approximate nonlocal moment

M. Panzeri (✉)
Department of Civil and Environmental Engineering, Politecnico dimilano, Piazza L. Da Vinci 32, 20133 Milano, MI, Italy
e-mail: marco.panzeri@polimi.it

M. Riva · A. Guadagnini · S. P. Neuman
Department of Hydrology and Water Resources, University of Arizona Tucson, AZ Tucson85721, USA
e-mail: neuman@hwr.arizona.edu

A. Guadagnini
e-mail: alberto.guadagnini@polimi.it

E. Pardo-Igúzquiza et al. (eds.), *Mathematics of Planet Earth*,
Lecture Notes in Earth System Sciences, DOI: 10.1007/978-3-642-32408-6_90,

equations (MEs) that govern the space-time evolution of conditional ensemble means and covariances of hydraulic heads and fluxes. Here, we compare the accuracy and computational efficiency of our newly proposed EnKF approach against the traditional MC framework. We do so for a scenario involving a pumping well operating in a two-dimensional randomly heterogeneous aquifer.

2 Numerical Model

We consider convergent flow to a well pumping at a constant rate $Q_P = 3$ (all quantities are given in consistent space-time units) from the center of a two-dimensional square domain measuring 40×40 and discretized into grid cells of uniform size 1×1. Deterministic head values of 1 and 0 are prescribed on the left and right boundaries while the top and bottom boundaries are impervious (Fig. 1). Storativity is set equal to a uniform value of 0.3.

The reference log-conductivity (Y) distribution (Fig. 1) has been generated using the sequential Gausian geostatistical simulation software SGSIM [2] as an unconditional statistically homogeneous and isotropic multivariate Gaussian field having zero mean and isotropic exponential covariance with variance 2 and integral scale 4. We solve the corresponding deterministic flow problem for a time period of 80 units to obtain a corresponding reference head distribution in space-time. We sample the reference Y field at 9 spatial locations and the reference head, h, at 20 locations (see Fig. 1) and 10 observation times ($T_k = 5 \times k$ with $k = 1, 2, \dots, 8$; $T_9 = 60$ and $T_{10} = 80$). The Y and h measurements are perturbed by adding white Gaussian noise having a standard deviation of 0.1 and 0.01, respectively. This renders, e.g.,

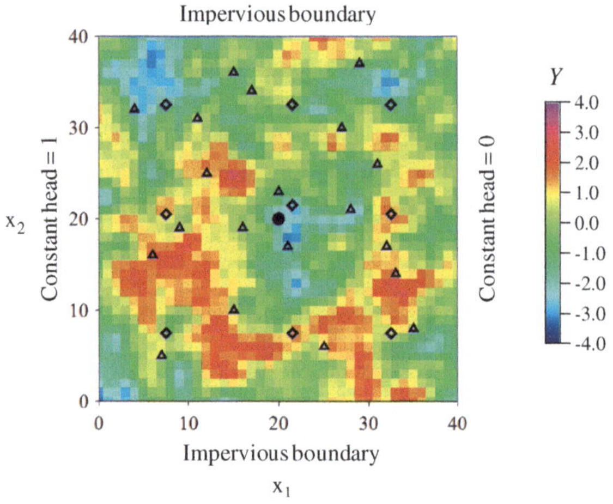

Fig. 1 Flow domain, boundary conditions, reference log-conductivity distribution, *pumping well* (•), *head* (Δ) and *log conductivity* (◇) measurement locations in the synthetic test case

perturbed Y values which are characterized by a relative difference ranging between 0 and 71 % with respect to their reference values at the nine Y sampling locations.

MC-based EnKF requires the initial generation of NMC realizations of equally likely Y fields conditioned on the noisy measurements. Initial head fields are obtained by solving the corresponding NMC deterministic steady state flow problems without pumping. The MEs approach is based on an initial mean Y field and corresponding covariance matrix which are constructed by projecting the nine Y noisy measurements onto all remaining grid elements via kriging. The steady state MEs are solved in the absence of pumping to obtain an initial mean h, corresponding h covariance, and cross-covariance between h and Y.

3 Results and Discussion

Figure 2 displays scatter plots of estimated mean ($\langle Y \rangle$) versus reference (Y_{ref}) log-conductivity values at the largest assimilation time $T_k = 80$ obtained through MEs- and MC-based approaches for different values of NMC. The results obtained with the two approaches tend to converge as NMC increases. This is also highlighted by the tendency to coincide of the linear regression lines fitted to the results.

As seen in Fig. 3a, slopes of regression lines fitted to the simulation results increase sharply with T_k at early assimilation times. Increasing NMC leads to improved performance of the updating scheme until the MC solution reaches convergence. This occurs at about $NMC \approx 10,000$ for the test case considered and no further significant improvement is noted for NMC up to 100,000 (Fig. 3). Figure 3a shows that MC-based slopes tend to their MEs-based counterpart with increasing NMC. The slopes of regression fits associated with MEs are generally larger than those related to MC. Figure 3b displays temporal dependence of the coefficients of determination (R^2) of the linear fits. Our results show that R^2 (a) is very low and does not improve

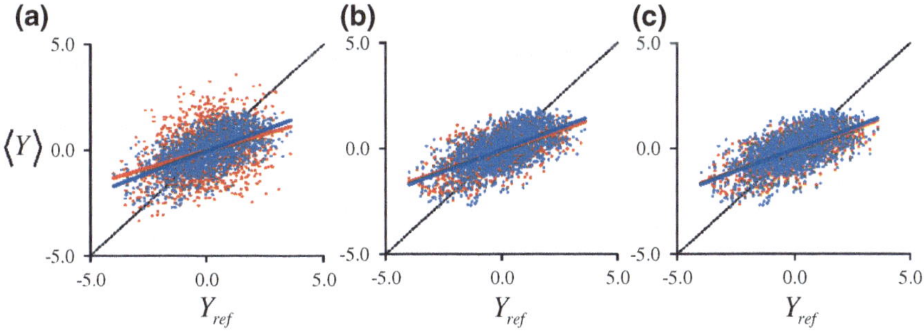

Fig. 2 Scatter plots of estimated mean ($\langle Y \rangle$) versus reference (Y_{ref}) Y values at $T_k = 80.0$ obtained by MEs (*blue symbols*) and MC (*red symbols*) approaches with NMC = **a** 100, **b** 1,000, **c** 10,000. Linear regression fits to the data (*blue* and *red lines* for MEs and MC approaches, respectively) are also shown

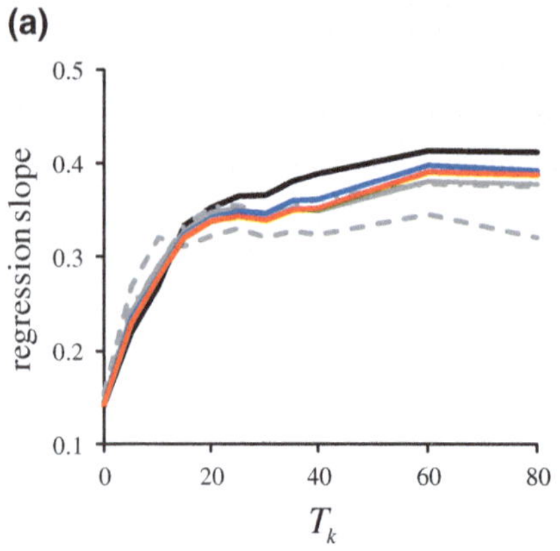

Fig. 3 Slopes (**a**) and coefficients of determination, R^2 (**b**) of *regression lines* fitted to *scatter plots* of estimated mean and reference Y values versus T_k for assimilations performed through ME (*solid black*) and MC approaches with $NMC = 100$ (*dashed*), 500 (*dashed-dotted*), 1,000 (*solid grey*), 10,000 (*solid blue*), and 100,000 (*solid red*)

with T_k when NMC is relatively small ($NMC = 100$), (b) it increases with NMC, and (c) its value is close to its MEs-based counterpart for $NMC \geq 1,000$.

Our reliance on MEs allows circumventing the drawbacks which are commonly observed in traditional (MC-based) applications of EnKF, such as the filter inbreeding and incorrect updates due to spurious covariances. In terms of computational efficiency of the two approaches, the CPU time required for each time interval on 10 standard 2.80 GHz Intel i7-860 processors working in parallel is 13,650 and $0.375 \times NMC$ seconds respectively for the MEs- and the MC- based approach.

References

1. Aanonsen, S. I., Nævdal, G., Oliver, D. S., Reynolds, A. C., & Vallès, B. (2009). Ensemble Kalman filter in reservoir engineering—A review. *SPE Journal*, *14*(3), 393–412. doi:10.2118/117274-PA.
2. Deutsch, C. V., & Journel, A. G. (1998). *GSLIB Geostatistical Software Library and User's Guide*. New York: Oxford University Press. 1998.
3. Panzeri, M., Riva, M., Guadagnini, A., & Neuman, S. P. (2013). Data assimilation and parameter estimation via ensemble Kalman filter coupled with stochastic moment equations of transient groundwater flow. *Water Resources Research*, *49*, 1–11. doi:10.1002/wrcr.20113.

New Analytical Solutions for Phreatic Darcian Flows Over Non-Planar Bedrocks

Anvar Kacimov, Yurii Obnosov and Osman Abdalla

1 Introduction

Groundwater flow in unconfined aquifers is characterized by a free (phreatic) surface and nonlinear boundary conditions there [1, 2]. Common catchment-scale reconnaissance models or regular annual assessment of aquifers' resources utilize a hydraulic Dupuit–Forchheimer (DF) approximation, which in steady regimes requires solving a boundary-value problem for a second-order ordinary differential equation. A more general, potential theory (PT), solves Laplace's equation, provided the aquifer is homogeneous. In the arid climate of Northern Oman, with a periodic occasional rainfalls of 200–300 mm/year in mountains (2–3 km high) and 100 m/year in the valley zones of catchments, which are several tens of kilometers long, recharge from the vadose zone to the water table can be neglected everywhere but the fractured rock in upper reaches of North Oman Mountains (NOM). The main factor controlling the shape and locus of the phreatic surface is the subjacent bedrock whose geometry is commonly inferred from geological data.

In the study area (Northern Oman), for which our model is developed, the geology ranges from the Precambrian basement rocks, mainly phyllites and slates, at the bottom of the succession occupying the core of NOM to karstified carbonate rocks (Hajar Supergroup, HSG) at the elevated areas to fractured ophiolitic sequence overlain by porous medium of Tertiary limestones and Quaternary alluvium gravel at the top of the geologic section. Recent monitoring of the water table, whose slope is steep in

A. Kacimov (✉) · O. Abdalla
Sultan Qaboos University, Al-Khod-123, PO Box 34, Muscat, Oman
e-mail: anvar@squ.edu.om

O. Abdalla
e-mail: osman@squ.edu.om

Y. Obnosov
Kazan Federal University, Kazan, Russia
e-mail: Yurii.obnosov@gmail.com

E. Pardo-Igúzquiza et al. (eds.), *Mathematics of Planet Earth*,
Lecture Notes in Earth System Sciences, DOI: 10.1007/978-3-642-32408-6_91,

the mountains and relatively mild in the valley part of the catchment, revealed a puzzling spatial variability, detected in direct borehole observations and reconstructed geophysical (mostly TDEM) surveys. The degree of this steepness, position of the water table and other aquifer characteristics are vital because groundwater is the main resource for agriculture and other sectors of Omani economy.

In standard DF or PT models the bedrock boundary of an unconfined aquifer is assumed to be planar [2]. In Refs. [3, 4] steep slopes of the free surface were attributed to a "groundwater fall" geometry of the bedrock, i.e. a non-planar aquifuge boundary making a step-down. In hillslope hydrology, both the DF and PT models are used but explicit closed-form solutions (like ours below) to phreatic-surface flow problems are rare. Here we extend the model of Ref. [4] and consider the following bedrock "anomalies": (a) an aquifer with an underlying aquifuge whose inclination changes abruptly from aquifer's upflow to downflow (Fig. 1) and (b) an aquifuge with a continuously varying slope. Correspondingly, we apply two different techniques: the hodograph method [4] and boundary-value problem method [5]. We assume a Darcian flow, ignore the capillary fringe, accretion or evapotranspiration to/from the vadose zone and any sinks-sources (e.g. pumping wells) in the flow domain.

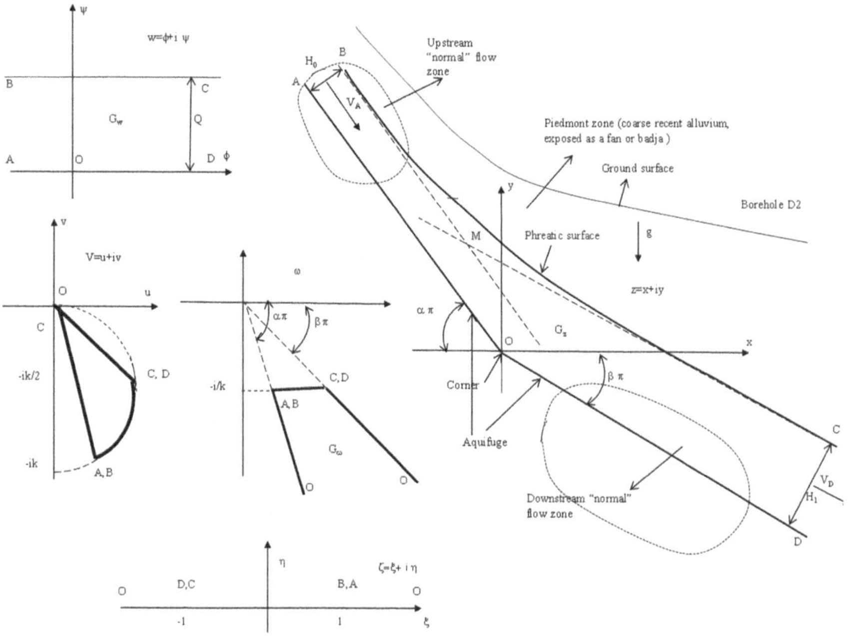

Fig. 1 Phreatic flow over a corner-shaped aquifuge; physical, complex potential, hodograph, inverted and auxiliary domains

2 Flow Over Non-Planar Aquifuge

The bedrock AOD makes a corner (Fig. 1). The origin of a Cartesian coordinate system coincides with the vertex O. The flanks of the wedge, OD and AO, dip at angles $\alpha\pi$ (counted from Ox positive clockwise) and $\beta\pi$ (positive counterclockwise), correspondingly. Without any loss of generality we consider here the "hillslope" case of $0 < \alpha = \text{const} < 1/2,\ 0 < \beta = \text{const} < 1/2$. If $\alpha > \beta$, flow decelerates downstream of the transition zone near point O, otherwise it accelerates. The flow rate (per unit length perpendicular to the plane of Fig. 1) is Q. A PT-based solution was obtained in Ref. [1] for $\alpha = -1/2,\ \beta = 1/2$; in Refs. [3, 4] the case of $\alpha = 0, \beta = 1/2$ was studied, in Ref. [2] winding seepage in domains with sharp-edged impermeabilities was considered.

If $\alpha > 0$ and $\beta < 1/2$, then BC far upflow and downflow of O is parallel to the bedrock i.e. a 1-D unidirectional flow is "normal" of saturated thicknesses H_0 and H_1 far above and below point O, respectively. The corresponding zones are schematically demarcated by dotted lines in Fig. 1. In these zones flow is aligned with the bedrock, the 1-D DF approximation works well and gives exactly the same solution as PT. In the conjugation zone of Fig. 1, the free surface BC is essentially non-straight. Dashed lines in Fig. 1 represent the "primitive" phreatic surface corresponding to two "normal" flows at constant slopes $\alpha\pi$ and $\beta\pi$, i.e. the straight lines $y = -\tan\alpha\pi x + H_0/\cos\alpha\pi$ and $y = -\tan\beta\pi x + H_1/\cos\beta\pi$. The "primitive" lines intersect at the point M and the corresponding "phreatic corner" BMC would be a simplistic Dupuit replica of AOD, translated. The angularity of AOD affects the shape of BMC in PT.

We introduce a complex physical coordinate $z = x+iy$, hydraulic head $h(x, y)$, Darcian velocity vector $\mathbf{V} = -k\nabla h$, velocity potential $\phi = -kh$, stream function ψ, complex potential $w = \phi + i\psi$ and complexified Darcian velocity $V = u + iv$. ϕ, ψ and h are harmonic functions. $\phi + ky = 0$ along BC. In the w-plane we have a strip G_w (Fig. 1) corresponding to the flow domain G_z. In the hodograph plane, we have a circular triangle G_V. In Fig. 1 the case of $\alpha > \beta$ is illustrated with O being a stagnation point. If $\alpha < \beta$ then $V_O = \infty$ i.e. the hodograph trigon is infinite. The magnitudes of velocities in the "normal" flow zones of Fig. 1 are $|V_A| = |V_B| = k\sin\alpha\pi$ and $|V_C| = |V_D| = k\sin\beta\pi$. From the mass balance, $Q = H_0|V_A| = H_1|V_D|$. We use the method of inversion [2] and invert G_V into a trigon G_ω where $\omega = dz/dw$ (Fig. 1). If $\alpha < \beta$ then G_ω in Fig. 1 is a standard triangle. We map conformally G_w onto G_ω via an auxiliary plane $\zeta = \xi + i\eta$ using the Schwarz–Christoffel formula twice. After some algebra we obtain equations of BC. Fig. 2 represents the results of computations for $\alpha = 0.35, \beta = 0.1$. This and computations for other tilt angles illustrate the accuracy of the DF approximation as compared with a full 2-D model.

The case of AOD in Fig. 1 as an arbitrary curve is tackled by the method from Ref. [5]: the G_w domain remains the same as in Fig. 1 but the shape of both boundaries of G_z is reconstructed by the so-called "inverse" technique. For this purpose, the pore pressure as a function of ξ in the auxiliary half-plane is given. Unlike the case of the wedge-shaped bedrock, the mathematical solution is written in terms of

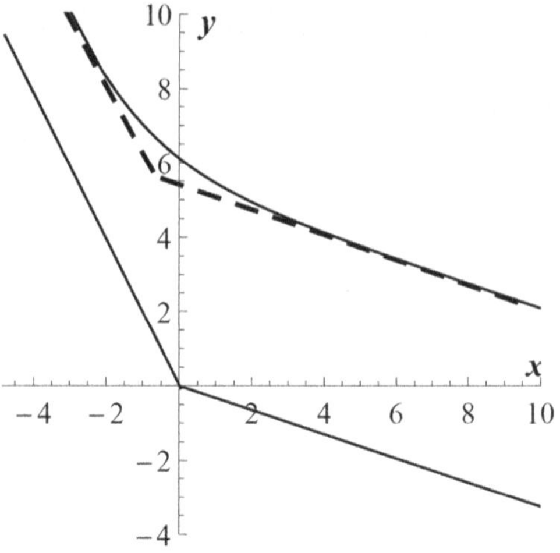

Fig. 2 Phreatic surface for $\alpha = 0.35, \beta = 0.1$ (*solid curve*) computed by PT, *dashed lines* are DF asymptotics

singular (Cauchy-type) integrals. Like on Fig. 2 the phreatic surfaces are plotted but the confining boundary also emerging as a part of solution.

Similarly to Ref. [6] our groundwater system if gravity-controlled. However, contrary to a humid climates (e.g. Canada), the water table in arid climates and catchments with a tick vadose zone is a "hydraulic" replica of the bedrock rather than of land topography.

Acknowledgments Support by the grant SR/SCI/ETHS/11/01, His Majesty Research Trust Fund (Oman) and by Russian Foundation of Basic Research grant N 12-01-97015-r_povolgh'e_a is appreciated.

References

1. Dachler, R. (1936). *Grundwasserstroming*. Wien (in German): Springer.
2. Strack, O. D. L. (1989). *Groundwater Mechanics*. Englewood Cliffs: Prentice Hall.
3. Li, W., Liu, Z., Guo, H., Li, N., & Kang, W. (2011). Simulation of a groundwater fall caused by geological discontinuities. *Hydrogeology Journal, 19*, 1121–1133.
4. Kacimov, A. R. (2012). Analytical solution for a phreatic groundwater fall: the Riesenkampf and Numerov solutions revisited. *Hydrogeology Journal, 20*, 1203–1209.
5. Kacimov, A. R., Klammler, H., Hatfield, K., & Ilyinsky, N. B. (2011). Constructal design of permeable reactive barriers: a groundwater hydraulics criterion. *Journal of Engineering Mathematics, 71*(4), 319–338.
6. Toth, J. (2009). *Gravitational systems of groundwater flow*. Cambridge: Cambridge University Press.

When Steady-State Is Not Enough

J. Jaime Gómez-Hernández, Teng Xu, Haiyan Zhou and Liangping Li

1 Introduction

It is important to give a proper characterization of subsurface hydrogeological properties for groundwater forecast, environmental risk assessment and subsurface resource management . However, in real word, a good characterization of subsurface hydrogeological properties is difficult to achieve from a sparse set of parameter observations. To solve this problem, stochastic inverse modeling methods are commonly used to improve the characterization making use of more abundant state data, such as piezometric heads. In order to complement a previous study [1] in which we show that enough transient piezometric head could be enough to give detailed characterization of the underlying hydraulic conductivity spatial heterogeneity, we have performed a study to show that steady-state piezometric information is not enough to properly characterize the conductivity field.

In this work, we will use a localized version of the normal-score ensemble Kalman filter (NS-EnKF) method developed by Zhou et al. [2] to study the power of time-invariant piezometric head observations in the characterization of a bimodal hydraulic conductivity field under steady-state flow conditions.

J. J. Gómez-Hernández (✉) · T. Xu
Research Institute of Water and Environmental Engineering, Universitat Politècnica de València, 46022Valencia, Spain
e-mail: jgomez@upv.es

T. Xu
e-mail: tenxu@upv.es

H. Zhou · L. Li
Center for Petroleum and Geosystems Engineering, University of Texas at Austin, Austin, TX, USA

E. Pardo-Igúzquiza et al. (eds.), *Mathematics of Planet Earth*,
Lecture Notes in Earth System Sciences, DOI: 10.1007/978-3-642-32408-6_92,

2 The Formulation of the Normal-Score Ensemble Kalman Filter for Steady-State Flow

The Normal-Score EnKF can be summarized as follows:

1. *Initialization step*. An ensemble of hydraulic conductivity fields is generated.
2. *Normal-Score transformation step*. A normal-score transformation is applied to the conductivity vector.
3. *Forecasting step*. Piezometric heads are calculated.
4. *Analysis step*. Normal-score transformed conductivities are updated by Kalman filtering.
5. *Back transformation step*. Back transform the updated normal-score transformed conductivities into conductivities using the inverse of the previously computed normal-score transform functions.

Besides, covariance localization is used to eliminate the effect of spurious correlations in the experimental covariance, and covariance inflation is used to avoid filter collapse [1].

In this work, the groundwater flow is assumed to be at steady-state. The governing equation for steady-state flow with external sinks/sources is

$$\nabla \cdot [k\nabla h] + q = 0 \tag{1}$$

where k is hydraulic conductivity $[LT^{-1}]$, h is hydraulic head $[L]$, q denotes sinks and sources per unit volume $[T^{-1}]$, and ∇ stands for $(\partial/\partial x,\ \partial/\partial y,\ \partial/\partial z)$.

2.1 Synthetic Example and Analysis

In this work, the construction of the synthetic bimodal confined aquifer, the boundary and initial conditions, conditions, the total simulation time and the positions of the 111 observation piezometers are the same as in the work by Xu et al. [1]. The reference log-conductivity field is shown in Fig. 1. The difference between the two works is that the observation piezometers in this work are time invariant. The flow simulator MODFLOW [3, 4] is used as the forward model.

In this work, three scenarios, shown in Table 1, are used to analysis the power of steady-state piezometric head in the characterization of a bimodal hydraulic conductivity field. The difference between the scenarios is in the use of the localization and variance inflation.

Figure 2 shows the ensemble mean (left column) and ensemble variance (right column) of the updated log-conductivity fields after 60 iterations in the assimilation step for the three scenarios.

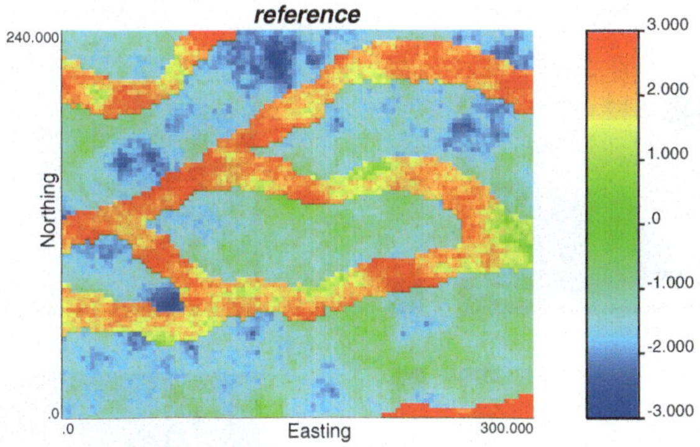

Fig. 1 Reference log-conductivity field

Table 1 Definition of scenarios

Scenario	Case 1	Case 2	Case 3
Observation piezometers (111)	√	√	√
Localization	√	√	
Variance inflation	√		

Comparing Fig. 2 with 1, we can see that steady-state piezometer heads fail to characterize the hydraulic conductivity field.

In contrast, in the work by Xu et al. [1], it has been proved that the transient piezometric heads carry enough information about the conductivity spatial heterogeneity, as long as a sufficiently large number of piezometers are available, that is, conditioning to transient piezometric head data could produce an ensemble of realizations that captures the main patterns of the non-Gaussian reference field, even when there is no prior information about the aquifer spatial heterogeneity patterns.

3 Summary

Steady-state piezometric head data are not enough to characterize the type of hydraulic conductivity heterogeneity displayed in Fig. 1.

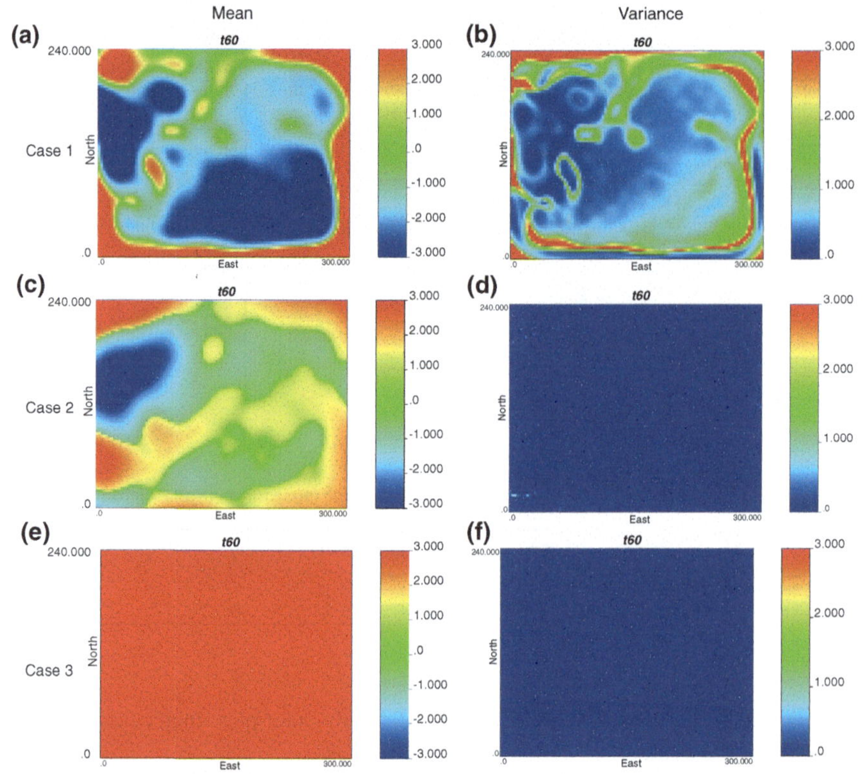

Fig. 2 Scenarios Case 1–3. Log-conductivity ensemble mean and variance computed after 60 iterations

References

1. Xu, T., Jaime Gómez-Hernández, J., Zhou, H., & Li, L. (2013). The power of transient piezometric head data in inverse modeling: an application of the localized normal-score enkf with covariance inflation in a heterogenous bimodal hydraulic conductivity field. *Advances in Water Resources*, *54*, 100–118.
2. Zhou, H., Gómez-Hernández, J., Hendricks Franssen, H., & Li, L. (2011). An approach to handling non-Gaussianity of parameters and state variables in ensemble Kalman filtering. *Advances in Water Resources*, *34*(7), 844–864.
3. McDonald, M., & Harbaugh, A. (1984). *A modular three-dimensional finite-difference groundwater flow model*. Washington: Scientific Publications Company.
4. Harbaugh, A., et al.(2000). MODFLOW-2000, the US Geological Survey modular ground-water model: user guide to modularization concepts and the ground-water flow process. US Geological Survey.

Hydrogeological and Thermal Modelling of an Underground Mining Reservoir

Clara Andrés Arias, Almudena Ordóñez Alonso and Rodrigo Álvarez García

1 Introduction

Since the profitability of using mine water as a geothermal resource has been demonstrated, many studies have been carried out in different countries to assess the potential of its exploitation.

Due to mining in the Central Coal Basin (CCB) of Asturias (Spain), original small multilayer aquifers in sandstones have given way to 'pseudoaquifers' with a behaviour similar to a karstic aquifer. Closed and inundated mining voids act as underground reservoirs, so they can be used as a geothermal source.

The efficient use of this type of resource can be optimized by applying thermal models. In this case, we chose the reservoir constituted by the Barredo and Figaredo flooded coal mines.

2 Area of Study

The area of study is located inside the CCB of Asturias (see Fig. 1a). The mean temperature in this area is 14 °C and the average yearly precipitation within the Barredo–Figaredo basin (Fig. 1a) is 874 mm.

C. A. Arias (✉) · A. Ordóñez Alonso (✉) · R. Álvarez García
Departamento de Explotación y Prospección de Minas, University of Oviedo, C/ Independencia, 13, 33004 Oviedo, Spain
e-mail: andresclara@uniovi.es

A. Ordóñez Alonso
e-mail: aoalonso@uniovi.es

R. Álvarez García
e-mail: alvarezrodrigo@uniovi.es

E. Pardo-Igúzquiza et al. (eds.), *Mathematics of Planet Earth*,
Lecture Notes in Earth System Sciences, DOI: 10.1007/978-3-642-32408-6_93,

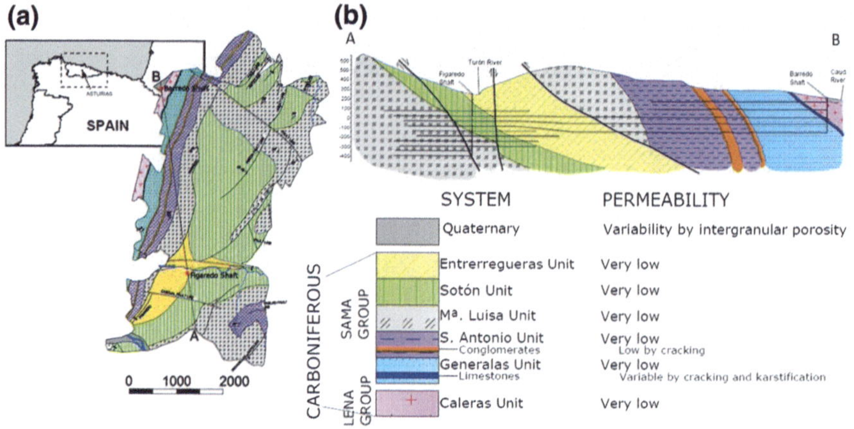

Fig. 1 Area of study: **a** Barredo–Figaredo basin; **b** Geological section

This basin was defined considering the superficial watershed and the extension of the mine voids that induce fractures and water filtration. It constitutes the unit that has been considered for modelling. It is crossed by the Turón River, which loses some of its flow by infiltration in the most mined area [1].

Figure 1b shows the existence of a synclinal structure (synclinal of Barredo). Hydrogeologically the materials are characterized by low porosity and permeability. The synclinal structure corresponds to a cyclic succession of parasequences formed by shales, sandstones and coal layers, with interbedded conglomerates and breccias [2]. Studies in the area show that the unaltered materials have permeabilities around 10^{-7} m/s [3].

Mining activity at the CCB started with "mountain mining" galleries and it continued by extraction wells to exploit the layers below the valley. At the studied area, when both shafts finished their mining activity, the water level was at −180 m.a.s.l. Then, pumping was stopped for 295 days, allowing the water level to increase up to +150 m.a.s.l. Now, pumping is resumed and water level is maintained at that height. Currently, water pumped from Barredo shaft is being geothermically used, with an average temperature of 20 °C at a depth of 100 m.a.s.l. For more information of this system consult the PhD thesis [4].

3 Numerical Model Description

For the simulation of the example presented in this paper, FEFLOW [5] was used. The 2D model developed is based on a cross-section according to the line that joins Barredo and Figaredo shafts, considering the mine works (Fig. 1b).

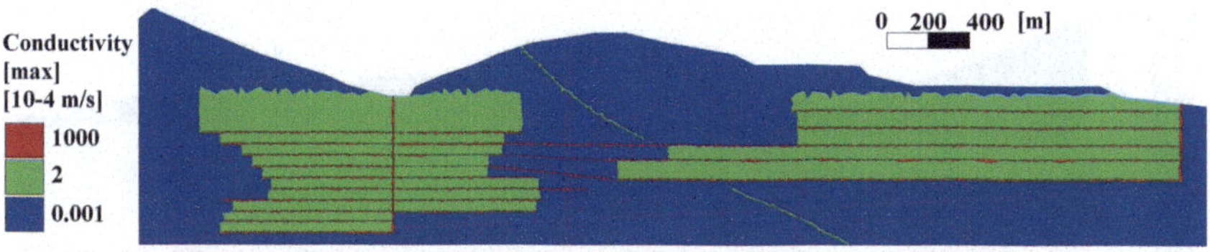

Fig. 2 Material hydraulic conductivities defined after calibration

The conceptual model is composed of isotropic, heterogeneous, porous and unconfined medium, and simulations were performed with a saturated/variably saturated model. As the materials in the study area are not permeable, lateral inflow was neglected. A Type 2 Neumann boundary condition was used for rainfall infiltration (variable flow, obtained from a previous hydrological study of the fraction of the effective rainfall which infiltrates into the basin [1]). The recharge coming from the river was modelled by injection wells with an average infiltration flow from Turón River in the stretch intercepted for the modelled section.

A previous calibration of the hydrogeological parameters (hydraulic conductivity, porosity and retention coefficient) was performed, comparing the water level measured during the flooding process of the mine with the simulated values.

The thermal model starts after the flooding period, when pumping (3.5 m^3/day) is re-established in the reservoir to keep a constant flooded level. Extractions have been considered in both wells, maintaining a zero water flow balance in the system (inflow = outflow).

Additionally, the temperature of the infiltrating water was used as boundary condition, and at the bottom edge a constant geothermal heat flow of 65 mW/m^2 was defined. As initial condition, the temperature of the geological massive varies between 14 °C at the surface and 25 °C in the deepest area of the model, considering the Earth's geothermal gradient of 0.03 °C/m.

Once the model was defined and calibrated (Fig. 2), several scenarios of functioning of the geothermal system were tested to assess the long-term viability of it: (i) 30 years of simulation without reinjection, (ii) 60 years of simulation without reinjection and (iii) 30 years of simulation with reinjection of used water (1.4 m^3/day) in Figaredo shaft.

4 Results

The calibration was considered valid once the Pearson correlation coefficient between the simulated and the real water level values recorded during the flooding period reached a value of 0.75.

Figure 3 shows a 2D vertical cross section of the model and the calculated temperatures after 30 years without reinjection of water (case i). The water mainly flows through the galleries which remain colder, whereas a more constant temperature is

Fig. 3 Temperature in the reservoir after 30 years of simulation, without reinjection

maintained in the massif. After 30 years of simulation, there is a similar distribution of temperatures for the third scenario (with reinjection), reaching a temperature of 18 °C in Barredo shaft at a depth of 100 m.a.s.l. This value is a bit lower than the current average temperature, but still within the limits of the natural fluctuations of the system throughout a year. For longer simulations (case ii) there are not significant changes. Once the equilibrium is reached, colder water remains mainly on the top of the reservoir, because it is constantly renewed by the pumping at Figaredo and Barredo shafts. In the lower areas of the system, water is kept at higher temperatures.

5 Conclusions

The defined model efficiently represents the temporal evolution of the water table in the mine reservoir during the flooding period with a correlation coefficient of 0.75.

Regardless of the simulation time and the different scenarios studied, the evolution of the underground reservoir temperature always follows the same patterns. Therefore, it can be concluded that the geothermal system under the current demand and environmental conditions, would be able to supply water at the required temperature during more than 30 years.

Acknowledgments The authors thank the company HUNOSA the information provided for the realization of this work and also thank at DHI-WASY for its help and support using the information package FEFLOW. (Grant: UNOV-09-BECDOC-S).

References

1. Ordóñez, A., Andrés, C., Álvarez, R., Jardón, S. (2010). Aprovechamiento de las Aguas Subterráneas como Recurso Hídrico y Energético. In *Seguridad y Medio Ambiente*, vol 28 (pp. 46–60). Spain: Fundación Mapfre.
2. García-Loygorri, A., Ortuño, G., Caride de Liñán, C., Gervilla, M., Greber, C.H., Feys, R. (1971). El Carbonífero de la Cuenca Central Asturiana. Trabajos de Geología, Universidad de Oviedo, vol 3, pp. 101–150.

3. Fandos, P., Rodríguez, F., Gutiérrez, A. M., Álvarez, J.J. (2004). El Yacimiento de HUNOSA en la Cuenca Carbonífera Central. Servicios de Geología del Caudal y del Nalón.
4. Jardón, S. (2010). Aprovechamiento de las Aguas de Mina en la Cuenca Central Asturiana como Recurso Energético. Aplicación al Embalse Minero Barredo-FIgaredo. PhD Thesis.
5. Diersch, H.-J. G. (2005). *FEFLOW finite element subsurface flow and transport simulation system. Reference Manual*. Berlin: WASY, Institute for Water Resources Planning and Systems Research.

Effect of Entrapped Gas Below the Phreatic Surface on Pressure Propagation and Soil Deformation

Héctor Montenegro, Oliver Stelzer and Bernhard Odenwald

1 Introduction

An essential characteristic in transient groundwater flow problems is the phenomenon of storage capacity. The water storage is governed by three processes: change in void volume of the skeleton, change in fluid saturation and change in fluid density. Due to the negligible compressibility of water the storage in (fully saturated) aquifers is assumed to be dominated merely by soil deformation. The presence of entrapped gas below the phreatic surface, for example as a result of natural water level fluctuations, in combination with "rapid" external changes in head and/or stress can substantially affect pore pressure propagation dynamics [2, 3]. The expression "rapid" must be considered in regard to the soil's hydraulic conductivity and involves accordingly a wide range of time scales [3].

The altered dissipation dynamics is the result of the increased storage volume due to the gaseous phase compressibility [2]. Retarded excess pressure enhances the occurrence of large hydraulic gradients and seepage forces which in turn may affect the effective stress, leading to soil deformation and eventually to a reduction of soil stability [2].

H. Montenegro (✉) · O. Stelzer · B. Odenwald
BAW Federal Waterways Engineering and Research Institute,
Kussmaulstr.17, D-76187 Karlsruhe, Germany
e-mail: hector.montenergo@baw.de

O. Stelzer
e-mail: oliver.stelzer@baw.de

B. Odenwald
e-mail: bernhard.odenwald@baw.de

E. Pardo-Igúzquiza et al. (eds.), *Mathematics of Planet Earth*,
Lecture Notes in Earth System Sciences, DOI: 10.1007/978-3-642-32408-6_94,

2 Gas Entrapment and Storage Capacity

Assuming the porous medium as rigid, the entrapped pore gas as immobile and any gas volume change being followed by storage or release of pore water the storage capacity C [1/L] is numerically equivalent to a volume of soil gas dV_g [L^3] compressed or expanded (i.e., water stored or released) by a unit volume of aquifer V_0 [L^3] under a unit head dh [L] (or likewise pore pressure du_w [F/L^2]) decline.

$$C = \frac{1}{V_o} \cdot \frac{dV_g}{dh} = \frac{1}{V_o} \cdot \frac{dV_g}{du_w} \cdot \frac{du_w}{dh} = \frac{1}{V_o} \cdot \frac{dV_g}{du_w} \cdot \gamma_w = \frac{d\theta_g}{du_w} \cdot \gamma_w. \tag{1}$$

Hilf [1] related a change in gas content $d\theta_g = dV_g/V_0$ [-] to a gas pressure change du_g [F/L^2] considering the porosity n [-], the initial (i.e., at atmospheric conditions) water saturation S_0 [-], the atmospheric pressure u_a [F/L^2] and the volumetric coefficient of solubility h* [-] quantifying the dissolved gas:

$$d\theta_g = \frac{dV_g}{V_o} = \frac{du_g}{(u_a + du_g)} \cdot (1 - S_o + h^* \cdot S_o) \cdot n. \tag{2}$$

Assuming equilibrium between gas u_g and water pressure u_w below the phreatic surface (i.e., $u_g = u_w$) and substituting $d\theta_g$ in Eq. 1 yields the storage capacity term C as the derivative of the gas content- pressure head curve (see Fig. 1, left) which, in contrast to the specific storage S_s is pore pressure dependent [3]:

$$C = \frac{d\theta_g}{du_w} \cdot \gamma_w = \frac{(1 - S_o + h^* \cdot S_o) \cdot n \cdot \gamma_w}{u_a + du_w} \tag{3}$$

Figure 1 shows the gas content θ_g (left) and the capacity C (right) with increasing pressure head u_w/γ_w (or likewise depth below phreatic surface for hydrostatic conditions) for an initial gas content $\theta_{g0} = 0.02$ vol %. The right graph reveals the huge specific storage and the non-linear pressure head dependence in the presence of even small volumes of entrapped gas (customarily used specific storage $S_s < 1 \cdot 10^{-5}$ 1/m).

3 Coupled Flow-Deformation Analysis Considering Gas Entrapment

To examine the effect of entrapped gas on excess pore pressure dynamics in a deformable soil a coupled groundwater flow-deformation (consolidation) analysis was performed with Plaxis FE-Software version 2D 2012. Gas entrapment was considered by assigning a saturation-pressure relationship according to Fig. 1 (left). The configuration of the numerical 1D column experiments is given in Fig. 2 and considers two different loading situations. In the first case a static load at the top of the

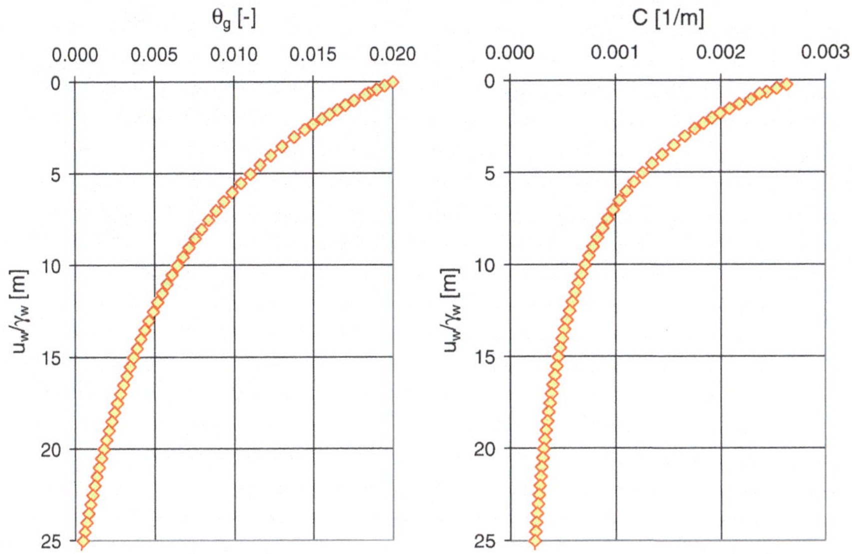

Fig. 1 Gas content (*left*) and storage capacity C (*right*) as function of pressure head u_w/γ_w

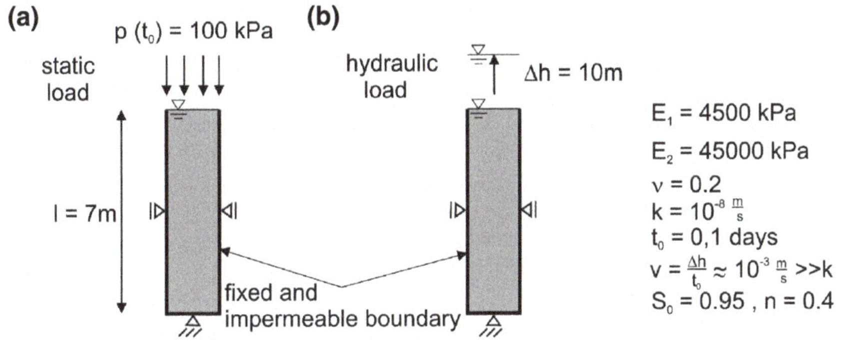

Fig. 2 Experimental set up for the fully coupled consolidation analysis

column is increased to $\Delta p = 100\,\text{kPa}$ during 0.1 day. In the second case the water level is increased by $\Delta h = 10\,\text{m}$ during the same time period.

Calculations for two linear elastic soils with different stiffness values E_1 and E_2 were performed. The results are presented in Fig. 3. In the static loading case the change in total stress results in a rise of pore pressure. Without entrapped gas the excess pore pressure immediately ($t < t_0$) rises to the corresponding load Δp and dissipates with time accompanied by settlements. Considering gas entrapment the initial excess pressure is lower and settlements already start during loading.
In the hydraulic loading case without gas entrapment the pore pressure reacts instantaneously on the pressure head change Δh showing only negligible displacements as

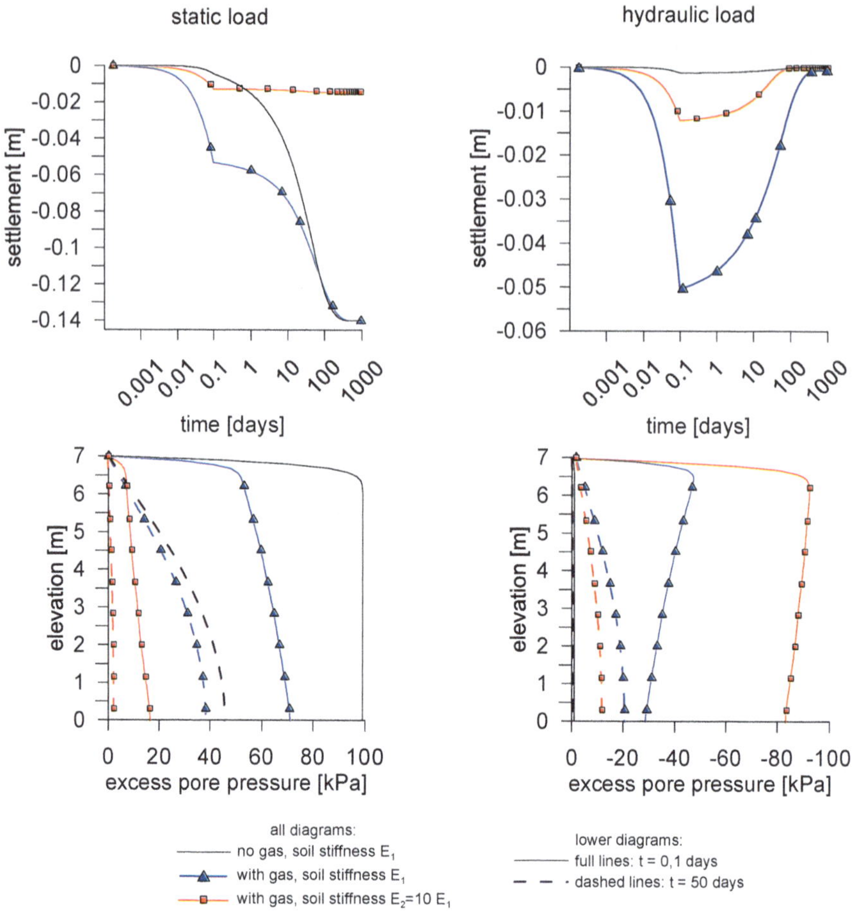

Fig. 3 Calculated settlements and excess pore pressures for static (*left*) and hydraulic loading

the effective stresses remain essentially unaffected. Entrapped gas bubbles enhance storage which retards the propagation of the increased pressure head producing seepage forces at the top which in turn induce an instant settlement. In the course of time the initial excess pore pressures dissipate until reaching steady state and the elastic deformation, in contrast to the static loading case, is completely reversed. The magnitude of initial excess pore pressure and settlement during loading depends on the soil stiffness. The implications of altered pore pressure dissipation due to gas entrapment will be displayed with examples from engineering practice in the presentation.

References

1. Fredlund, D. G., Rahardjo, H., & Fredlund, M. D. (2012). *Unsaturated Soil mechanics in engineering practice*. New York: John Wiley & Sons, Inc.
2. Köhler, H.-J., Feddersen, I., & Schwab, R. (1999). Soil and structure deformations due to reconstruction of an old lock built on unsaturated submerged clay. *In: Proceedings of the* 2nd *International Symposium on Pre-failure Deformation Characteristics of Geomaterials, Torino. Italy,* september 26–30, 1999.
3. Montenegro, H., Köhler, H.-J., & Holfelder, T. (2004). Inspection of excess pressure propagation in the zone of gas entrapment below the capillary fringe. *In: Proceedings, International Conference From Experimental Evidence towards Numerical Modelling of Unsaturated Soils, Weimar.* Germany, Vol. 2, September 2003.

Time Series Analysis of Groundwater Hydrographs: Case Study From a Hardrock Area

Wolfgang Gossel and Ronny Laehne

1 Introduction

Time series analyses are used manifold in sciences: Registrations of starobservations with a telescope in astronomy as well as temperature values in meteorology and water levels in hydrology and hydrogeology are only a few examples for applications of analyses of trend, periods and autocorrelation. The method itself is generally an a posteriori method to estimate parameters, search for correlations etc. Time series analysis of groundwater hydrographs is a challenging task. Data loggers avoid the former problems of irregular registration and computers speed up every year so that the analysis is improved. The question of finding, proving and interpreting trends, periods and autocorrelation in hydrographs, hydrochemical data sets and meteorological observations is of high interest for hydrogeologists. Based on groundwater level measurements in a porphyry aquifer in Halle (Germany) the time series analysis with the proposed methods of trend identification and period scanning method is compared to the Fast Fourier Transformation (FFT). Therefore an evenly sampled hydrograph (see Fig. 1) is used although period scanning is capable of analyzing unevenly sampled time series. The results for an extended unevenly sampled data set will be presented in an additional publication. First steps of this analysis are presented in [1].

In cases of gaps in the hydrograph the analysis was carried out by interpolation of values in the past. The methods for this step are manifold and not only linear interpolation between two measurements. After this step the analyses of trend and periods are applied.

W. Gossel (✉) · R. Laehne
Institute of Geosciences and Geography, Department of Hydrogeology and Environmental Geology, Martin—Luther University Halle, Von-Seckendorff-Platz 3, D-06120 Halle, Germany
e-mail: wolfgang.gossel@geo.uni-halle.de

Ronny Laehne
e-mail: ronny.laehne@geo.uni-halle.de

E. Pardo-Igúzquiza et al. (eds.), *Mathematics of Planet Earth*,
Lecture Notes in Earth System Sciences, DOI: 10.1007/978-3-642-32408-6_95,

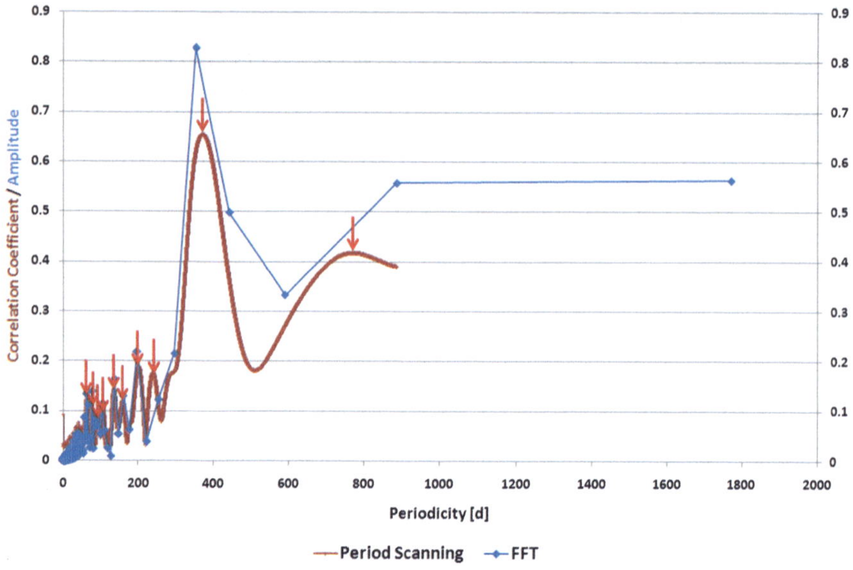

Fig. 1 Hydrograph and results of trend analysis and a superposition of the 10 highest correlating detected periods marked in Fig. 2

2 Methods

The unevenly recording of groundwater data with gaps and periods with high resolution measurements makes the application of the classical time series analysis ineffective, as it is described in [2] and [3]. Therefore a new set of methods was developed that consists of two of the three steps, the analyses of trend and periods, which seem to be more important than the autocorrelation.

The trend analysis is simply carried out via a linear regression line. The advantage is that an information about the coefficient of determination and therefore the correlation For the identification of periods in a time series the period scanning was developed. Here the measured time series is (after elimination of a significant trend) correlated to synthetic periodic signals of plain cosine functions. For the correlation the Pearson correlation coefficient is used:

$$r_{XY} = \frac{Cov(X,Y)}{\sigma(X)\sigma(Y)}$$

$$with$$

$$r_{XY} = Correlation\,Coefficient$$

$$Cov(XY) = Covariance(XY) = \sum\nolimits_{i=n}^{n} (x_i - \overline{x})(y_i - \overline{y})$$

$$\sigma(X) = Standard\,deviation(X) = \sqrt{\sum\nolimits_{i=n}^{n}(x_i - \overline{x})^2}$$

$$\sigma(Y) = Standard\,deviation(Y) = \sqrt{\sum\nolimits_{i=n}^{n}(y_i - \overline{y})^2}$$

Compared to the Fast Fourier Transformation (FFT) it has three advantages: (1) All possible periods in a time series can be identified. (2) The correlation coefficients instead of intensities are calculated and therefore the significance of a certain period can directly be outlined (two sided significance T-test). (3) Time series with temporary unevenly sampled data can be analyzed. Disadvantages are the low velocity of the analysis and the possible non-uniqueness of the identified periods.

The tested frequencies have to be set by the user and should be between the Nyquist frequency (double of the average sampling rate) and the statistically necessary frequency (half of the total time range of the measurements).

3 Results

The results of the analysis of a 4 years hourly registered groundwater hydrograph show a highly significant trend with a very small gradient. The equation of the trend is given by $y = -0.00043 \times +0.104$.

In Fig. 2 the results of the period scanning as well as with an FFT are shown. The period scanning shows results also in shorter periods than the FFT. The advantage of the period scanning is very clear therefore. The significance level for the 95 % confidence interval is 0.00797 so that several short periods can be identified clearly. The challenge is additionally to find out the independence of the periods from longer ones and in the case of the two indicated periods this can be proven.

4 Discussion

The application of the new methods to the groundwater hydrograph analysis is a clear improvement. The comparison of the period scanning with the FFT shows at certain ranges of the periodogram a higher resolution of the period scanning method. The identification of the significance prevents from misinterpretation. This is already helpful for the trend analysis, but especially for the analysis of periodicities.

More practical advantages are obvious for the period scanning method compared to the results of a FFT: If the hydrograph has gaps of a few days due to changes of the data loggers etc. there is no problem to work with the whole unevenly sampled time series and not only with Short Time Fourier Transformation or wavelet analysis as described in [4]. Another case that has been proven at other hydrographs is the measurement in shorter time steps within the time series of low resolution [5]. The

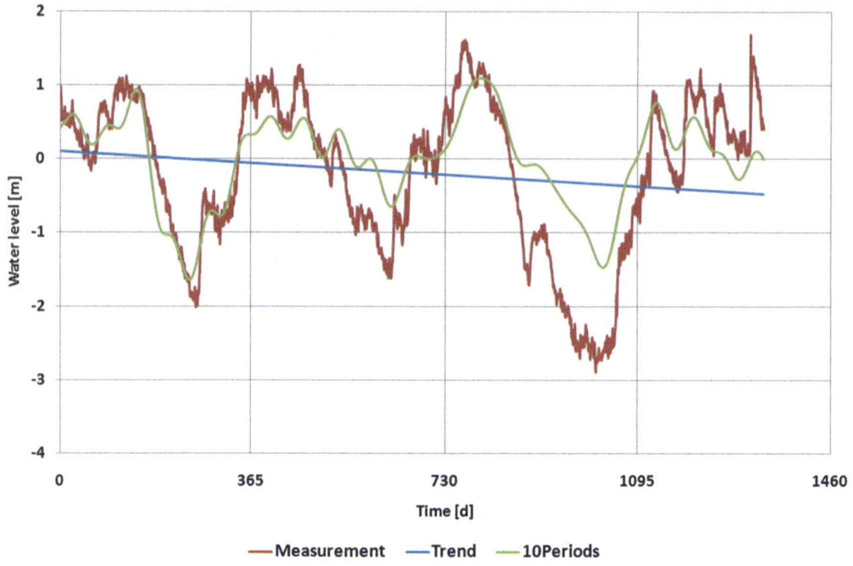

Fig. 2 Result of the period scanning and the FFT. The correlation coefficients on the y-axis indicate the significance (limit of the 95 % confidence interval: 0.008)

correlation coefficients help also in this method to identify the significant periods and separate them from insignificant periods.

The workflow takes in general more time than the classical time series analysis, especially the period scanning, but with a high optimization and the use of parallel processing the calculation times are also reduced and amount to a few minutes for 50,000 measurements.

The source code of the calculating kernel is OpenSource and can be downloaded via www.mo2geo.org together with the manual.

References

1. Gossel, W., Lähne, R., Rienäcker, J., Krauss, G., & Wycisk, P. (2011). Auswertung von Grundwasserstandsdaten auf dem Weinberg-Campus der Martin-Luther-Universität Halle-Wittenberg.- Hallesches Jahrbuch 32/33: 129–141.
2. Langguth, H. R., & Voigt, R. (1980). *Hydrogeologische methoden*. Springer: New York. 486pp.
3. Panagopoulos, G., & Lambrakis, N. (2006). The contribution of time series analysis to the study of the hydrodynamic characteristics of the karst systems: Application on two typical karst aquifers of Greece. *Journal of Hydrology*, *329*, 368–376.
4. Muttiah, R. S., Harme, R. D., & Richardson, C. W. (2005). Discharge and sedimentation periodicities in small sized watersheds. *Catena*, *61*, 241–255.
5. Gossel, W. (1999). Hydrogeologie und Grundwasserhaushalt ausgewählter anthropogen wenig beeinflusster Grundwassereinzugsgebiete in Berlin.- Berliner Geowiss. Abh. (A): 1–170.

Coupled Hydrogeophysical Simulation of a Pumping Test in an Unconfined Aquifer and its Associated Gravimetric Anomaly

Andrés González-Quirós and José Paulino Fernández-Álvarez

1 Introduction

The difficulty to measure subsurface water content and hydrologic properties has led to a growing number of geophysical methods. It is presented here an application of time-lapse gravity for hydrological purposes, a field known as hydrogravimetry. This technique provides a direct field measure since a change in water content causes a change in terrain bulk density, leading to variations in local gravitational acceleration.

This relation was highlighted first by [1]. The development of new instruments with higher resolutions ($<1\,\mu$Gal) and precisions and ($<5\,\mu$Gal), has increased the interest of gravity measurements for hydrological purposes. [2] showed how a pumping test in an unconfined aquifer could be monitored with gravity measurements and [3] performed a coupled hydrogeophysical inversion, extended by [4] to more complex conditions. All of them, however, used different simulation codes for the hydrogeological and geophysical forward problems. A coupled hydrogravimetric problem of a pumping test in a confined aquifer with a unique Finite Element Code was performed by [5].

The same code, COMSOL Multiphysics, is used here to reproduce a coupled hydrogravimetric problem of a pumping test in an unconfined aquifer. We illustrate how a gravity profile on the surface can help to estimate a transient flow parameter.

Given only two observation points of the stationary phreatic level, we show how the corresponding coupled gravity simulation provides information about the specific yield with no need to perform transient flow tests.

A. González-Quirós (✉)
Hydro-Geophysics and NDT Modelling Unit, Polytechnic School of Mieres (University of Oviedo), C/Gonzalo Gutiérrez Quirós S/N, 33600 Mieres, Spain
e-mail: andres@hydrogeophysicsndt.com

J. P. Fernández-Álvarez (✉)
Department of Mining Exploitation and Prospecting, Polytechnic School of Mieres (University of Oviedo), C/Gonzalo Gutiérrez Quirós S/N, 33600 Mieres, Spain
e-mail: pauli@uniovi.es

E. Pardo-Igúzquiza et al. (eds.), *Mathematics of Planet Earth*, Lecture Notes in Earth System Sciences, DOI: 10.1007/978-3-642-32408-6_96,

2 Methodology

2.1 Relation of Gravity with Aquifer Storage Changes

In an aquifer, changes in hydraulic head lead to mass changes. If the aquifer is unconfined the coefficient of storage is called specific yield, S_y[-]. The change in gravity (Δg_{gw}) due to changes in storage could be approximated to a slab (h [m] is the water table elevation, ρ [kg/m^3] the density of water, and, t_1 and t_2 [s] the beginning and the end of the interval):

$$\Delta g_{gw} = 2\pi G\rho S_y \Delta h = 2\pi G\rho S_y \left(h\left(t_2\right) - h\left(t_1\right)\right). \tag{1}$$

The shape of the water table is here different from a slab and the gravitational response of the drawdown cone is calculated with a superposition of finite disks [6].

2.2 Simulating the Gravitational Response of a Pumping Test

2.2.1 Hydrogeological Models

Two hydrogeological models have been performed. A 2D horizontal model of a pump in an unconfined aquifer was developed under Dupuit assumptions of horizontal flow following the methodology proposed by [7].

In order to simulate fluxes both in horizontal and vertical directions, we used the Arbitrary Lagrangian-Eulerian (*ALE*) method explained in [8]. In ALE mode the whole domain (2D-axisymmetric) is free to deform (*Free Deformation* option in COMSOL) and the water table boundary moves matching the atmospheric pressure.

2.2.2 Modelling a Gravimetric Anomaly

COMSOL does not have a proper module for gravity calculation. However, a gravity module can be created from an electrostatics one since both processes are governed by Poisson's equation [9].

2.2.3 Coupling the Gravity to the Hydrogeological Models

In the 2D hydrogeological problem a variable called "equivalent density" has been created (change in mass per unit surface of aquifer ($kg \cdot m^{-2}$)), which relates the drawdown with the change in mass:

$$\Delta m_u = \rho \cdot S_y \cdot \Delta h. \tag{2}$$

The solution of the hydrogeological model is coupled with a *linear extrusion* operation to a 3D domain for the gravity calculations. The variable "equivalent density" is implemented in the *surface change density* submenu in COMSOL.

In ALE model the "equivalent density" is change in mass per unit of volume ($kg \cdot m^{-3}$):

$$\Delta m_u = \rho \cdot S_y \cdot \tag{3}$$

The gravity is here directly calculated over the hydrological model with the "equivalent density" implemented via the *space change density* submenu.

3 Results, Discussion and Conclusions

Two different methods for the calculation of the gravimetric anomaly related with water extraction in steady-state conditions have been successfully implemented in a unique coupled simulation code.

The results have been checked satisfactory with analytical solutions. The resulting gravity profiles (Fig. 1) represent microgravity measurements over the surface.

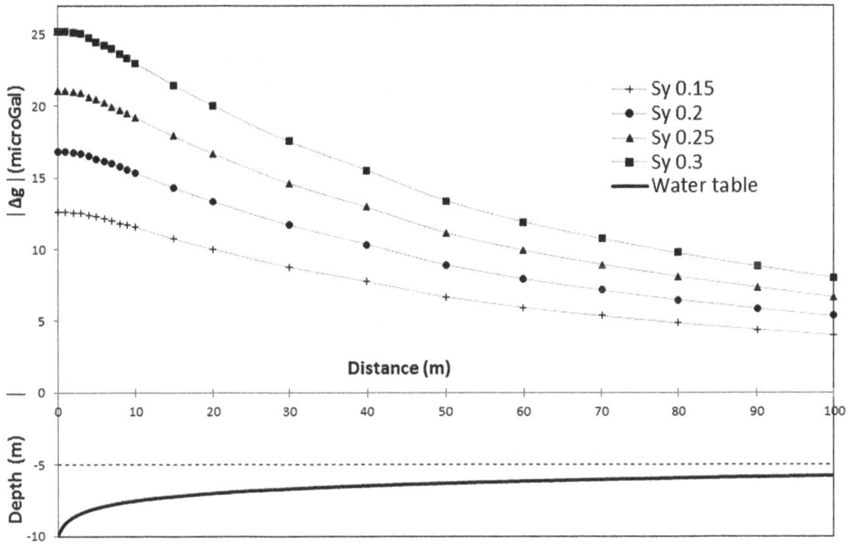

Fig. 1 Change in gravity (in absolute value) for different values of specific yield (Sy) obtained with the hydrogeological model (at the *bottom*). Initial water table was 5 m depth from the surface (*dashed line*)

The gravity anomalies vary between zero (beyond the radius of influence) and the maximum value near the pumping well.

It has been observed that, under adequate circumstances (depth of water table, enough drawdown cone) changes in position of hydraulic head and in value of specific yield cause gravity anomalies that are higher than the resolution and precision of actual micro-gravimeters (see point 1).

Under steady-state pumping conditions, and having enough observation wells, it is possible to know the shape and volume of the drawdown cone. Given the gravity data of mass change, and given a known water density, it is possible to estimate a specific yield value.

The use of a unique simulation code allows the method to be used for more complex situations and is a first step to perform a complete hydrogeophysical inversion.

Acknowledgments This work has been supported by the PCTI Asturias 2006–2009, and partially supported with an 80 % within the main priority axis 1 from the FEDER Operative Program from the Principality of Asturias 2007–2013.

References

1. Montgomery, E. L. (1971). Determination of coefficient of storage by use of gravity measurements. Ph.D. thesis, University of Arizona
2. Damiata, B. N., & Lee, T. C. (2006). Simulated gravitational response to hydraulic testing of unconfined aquifers. *Journal of Hydrology*, *318*(1–4), 348–359.
3. Blainey, J. B., Ferré, T. P. A., & Cordova, J. T. (2007). Assessing the likely value of gravity and drawdown measurements to constrain estimates of hydraulic conductivity and specific yield during unconfined aquifer testing. *Water Resources Research*, *43*(12), W12408.
4. Herckenrath, D., Auken, E., Christiansen, L., Behroozmand, A. A., & Bauer-Gottwein, P. (2012). Coupled hydrogeophysical inversion using time-lapse magnetic resonance sounding and time-lapse gravity data for hydraulic aquifer testing: Will it work in practice? *Water Resources Research*, *48*, W01539.
5. Fernández Álvarez, & J. A., Sánchez Morán, A. (2012). Contribución a la mejora de las capacidades de la microgravimetría: simulación numérica de la extracción de un fluido en un acuífero y de su anomalía gravimétrica asociada. 7ª AHPGG, San Sebastián
6. Telford, W. M., Geldart, L. P., & Sheriff, R. E. (1990). *Applied Geophysics* (p. 770). New York: Cambridge University Press.
7. Kitanidis, P. K. (2008). Depth-averaged modeling of groundwater flow and transport. COMSOL conference, Boston
8. Jin Y., Holzbecher, E., & Ebneth, S. (2012). Simulation of pumping induced groundwater flow in unconfined aquifer using arbitrary Lagrangian-Eulerian method, *Proceedings of the 2012 COMSOL Conference, Milan*
9. Butler, S. L., & Sinha, G. (2012). Forward modeling of applied geophysics methods using Comsol and comparison with analytical and laboratory analog models. *Computers and Geosciences*, *42*, 168–176.

Towards a More Efficient Simulation of Surface-Groundwater Interaction in Conjunctive Use Systems: Selective Compression and Modal Masking in the Eigenvalue Method

Oscar David Álvarez-Villa, Eduardo Cassiraga and Andrés Sahuquillo

1 Introduction

During the last decades, some mathematical techniques have been widely used to reduce large groundwater flow models. In general, three main families of reduction approaches have been recognized: driven by data, driven by model and combined. In driven by data reduction, the main objective is to reproduce some available information of the aquifer's flow variables using a black box reduced model. Combined reduction uses some specific model results or empirical data to define the patterns and build an orthogonal subspace that reduces the size of the original model. Proper orthogonal decomposition has been the most widely used approach to achieve a combined groundwater reduction [1]. On the other hand, the main goal in model oriented reduction is to build an orthogonal subspace that captures the mathematical structure of the groundwater flow model. This can be done via the eigenvalue method [2, 3]. In this paper, we present a new reduction methodology called eigenvalue method with selective compression and modal masking (EVM-SCMM) as a tool to execute a modal reduction of oversized groundwater flow models. The mathematical framework to solve the groundwater flow PDE via EVM-SCMM is presented and some relevant results of applying EVM-SCMM on rectangular aquifers are shown.

O. D. Álvarez-Villa (✉) · E. Cassiraga · A. Sahuquillo
Universitat Politècnica de València, Camino de Vera, s/n, 46022 València, Spain
e-mail: osdaalvi@gmail.com

E. Cassiraga
e-mail: efc@dihma.upv.es

A. Sahuquillo
e-mail: asahuq@hma.upv.es

O. D. Álvarez-Villa
GOTTA Ingeniería SAS, Cra 78A No 48-91 Int. 201, Medellín, Colombia

E. Pardo-Igúzquiza et al. (eds.), *Mathematics of Planet Earth*,
Lecture Notes in Earth System Sciences, DOI: 10.1007/978-3-642-32408-6_97,

2 Mathematical Framework of the EVM-SCMM

Let's assume a linear, time invariant aquifer, whose spatial domain has been discretized in n active nodes using finite differences (FD) to solve the groundwater flow PDE. This assumption implies that aquifer's hydraulic parameters and boundary conditions don't change in time and it allows to use the superposition principle. Also, let's define $\mathbf{h}(t) \in \mathrm{R}^n$ as the vector of piezometric heads in the active nodes [L]. These heads can be written as $\mathbf{h}(t) = \mathbf{u} + \mathbf{w}(t)$, defining that $\mathbf{u} \in \mathrm{R}^n$are the steady state heads subjected to the boundary conditions imposed to the original groundwater model and $\mathbf{w}(t) \in \mathrm{R}^n$are transient heads that satisfy the following equation:

$$\mathbf{A}\mathbf{w}(t) + \boldsymbol{\Psi}\mathbf{r}(t) = \mathbf{S}\frac{d\mathbf{w}(t)}{dt} \tag{1}$$

where $\mathbf{A} \in \mathrm{R}^{n\times n}$ is the conductances matrix [L^2/T], $\mathbf{S} \in \mathrm{R}^{n\times n}$is the matrix of storages [L^2], $\boldsymbol{\Psi} \in \mathrm{R}^{n\times n_a}$ is the matrix of time invariant elemental excitations, n_a is the number of external actions (EA) and $\mathbf{r}(t) \in R^{n_a}$ is the vector of EA intensities [L^3/T]. The boundary conditions for Eq. (1) are zero and its initial condition is $\mathbf{w}(0) = \mathbf{h}(0) - \mathbf{u}$ [2]. The EVM-SCMM's reduction procedure acts over (1) as follows. First, let's assume that the intensities of each EA are zero and $\mathbf{w}(t) = \mathbf{V}\Phi(t)$, where is $\mathbf{V} \in \mathrm{R}^{n\times n}$ an orthogonal eigenvectors matrix $\Phi(t) \in \mathrm{R}^n$ and is a vector of time dependent continuous functions. It can be shown that the solution for $\mathbf{w}(t)$ is [2]:

$$\mathbf{w}(t) = \mathbf{VEV}^{*}\mathbf{Sw}(0) + \mathbf{v}\left(\frac{\mathbf{I}-\mathbf{E}}{\boldsymbol{\Lambda}}\right)\mathbf{V}^{*}\boldsymbol{\Psi}\mathbf{r}(t) \tag{2}$$

where $\boldsymbol{\Lambda} \in \mathrm{R}^{n\times n}$ is the diagonal matrix of eigenvalues. $\boldsymbol{\Lambda}$ and $\mathbf{V}$ can be obtained using general eigenvalue problem's solvers and the general solution for the transient heads is $\mathbf{w}(t) = \mathbf{VE}\Phi(0)$, where $\mathbf{E}$ is a diagonal exponential matrix, whose diagonal components are $e_{ii}(t) = \varphi(0)e^{-\lambda_i t}$. Assuming that the simulations are carried out in discrete time intervals of duration Δt, from Eq. (2), the states of the aquifer are:

$$\Phi(t) = \mathbf{E}|_{\Delta t}\Phi(t-1) + \left(\frac{I - \mathbf{E}|_{\Delta t}}{\Lambda\mathbf{F}}\right)\mathbf{Br}(\tau) \tag{3}$$

where $\mathbf{F} \in \mathrm{R}^{n\times n}$ is a diagonal matrix that contains the volumes enclosed by the eigenvectors and $\mathbf{B} = \mathbf{FV}^{*}\boldsymbol{\Psi} \in \mathrm{R}^{n\times n_a}$is the matrix of the allocation coefficients.

According to (3), the allocation coefficients, $b_{i,j}$, represent the direct influence of each EA on the changes of the aquifer's dynamic. Hence, if the modal allocation coefficient corresponding to any given EA is small, then the changes on the aquifer's states, due to such excitations, are also small. Based on the above mentioned concept, for any EA, let's define that an *effective mode* contributes effectively to the changes in the aquifer's states. Also, the changes on the aquifer's states due to EA acting on a *residual mode* are negligible. Accordingly, the selection of effective modes is based on the concept of *participation limit*, b_{lim}, using the following criteria:

$$\begin{cases} \text{If } b_{i,j} \geq b_{lim} \ (\lambda_i, \mathbf{v}_i) \text{ is effective for EA } j \\ \text{If } b_{i,j} < b_{lim} \ (\lambda_i, \mathbf{v}_i) \text{ is residual for EA } j \end{cases} \tag{4}$$

which allows to build the *effective states mask*, $\mathbf{Z} \in \mathrm{N}^{n \times n_a}$. $\mathbf{Z}$ is a boolean matrix that indicates which modes are excited by any given EA. The localization of the $z_{i,j} =$ falseelements is defined in accordance to the indexes contained in the masking set $\mathrm{PC}\left\{(i, j) : b_{i,j} < b_{\text{lim}}\right\} \forall i = 1, \ldots, n$ and $j = 1, \ldots, n_a$. Applying (4) on (3), the size of the matrixes involved on the aquifer's state are reduced because only $p+1$ modes are effective for, at least, one of the EA considered. Therefore:

$$\widehat{\Phi}(t) = \widehat{\mathbf{E}}|_{\Delta\mathbf{t}} \widehat{\Phi}(t-1) + \left(\frac{\widehat{\mathbf{I}} - \widehat{\mathbf{E}}|_{\Delta\mathbf{t}}}{\widehat{\Lambda \mathbf{F}}}\right) \widehat{\mathbf{B}}\mathbf{r}(\tau) \tag{5}$$

are the reduced or effective states of the aquifer, where $\widehat{\mathbf{F}}, \widehat{\Lambda}, \widehat{\mathbf{I}}, \widehat{\mathbf{E}}|_{\Delta\mathbf{t}} \in \mathrm{R}^{(p+1)\times(p+1)}$ are the reduced diagonal matrices defined for Eq. (3), $\mathbf{B} \in \mathrm{R}^{(p+1)\times n_a}$ is the reduced matrix of allocation coefficients and $\widehat{\Phi}(t) \in \mathrm{R}^{p+1}$is the vector containing the aquifer's effective states. The last effective state, $\varphi_{p+1}(t)$, contains the conservation mode defined in [3] and [4] to preserve the equation of continuity on the aquifer. The mask of effective modes, $\mathbf{Z} \in \mathbf{N}^{(p+1)\times n_a}$, acts decreasing the number of arithmetic calculations on the matrix operation $\widehat{\mathbf{B}}\mathbf{r}(\tau)$ because: (*i*) $\widehat{\mathbf{B}}$ can be stored using a sparse matrix format, saving virtual memory, (*ii*) operations are performed only regarding to the effective modes and (*iii*) the matrix-vector multiplication has to be done using sparse algorithms. This scheme of calculation improves the EVM-SCMM's performance when many EA are acting. Finally, the control parameters (heads, volumes, surface-ground water fluxes, etc.) required by the conjunctive use model are calculated using the expressions proposed by [4].

The main limitation of the EVM-SCMM is that the efficient generation of effective modes demands the use of iterative algorithms to solve the generalized eigenvalue problem. However, the computational and storage requirements in such generation can be diminished using a physically based stopping criterion in the right way. For instance, the modal accumulated allocation factor can be used. These factors represent how accurately the reduced model preserves the conservation equation and allows to avoid unnecessary iterations during the modal generation.

3 Application of the EVM -SCMM on Rectangular Aquifers

The EVM-SCMM was applied to reduce groundwater flow models on homogeneous rectangular aquifers subjected to uniform recharge and punctual pumping. The effects of changing the parameterization of the river-aquifer connection on the reduction's scheme of the EVM-SCMM were analyzed. The results of the reduced model are compared to those obtained using classical FD simulation, discretizing time and space. To test the accuracy of those reduced simulations, the root squared mean error (**RSM**), the Nash's efficiency (**E**$_2$) and the bias coefficient (**BIAS**) have been estimated. The estimations for punctual and distributed EA show that, when the number of effective modes increases, **RMS** and **BIAS** decreases to zero and **E**$_2$ approximates to 100 %. It has been found that decreasing the riverbed conductance improves the EVM-SCMM's performance because fewer modes are needed to obtain more accurate results. The efficiency of the flow simulation via EVM-SCMM is similar to the observed using embedded multi-reservoir models [5]. Considering either homogeneous or heterogeneous rectangular aquifers, the accuracies exhibited by the simulations via EVM-SCMM were comparable to those obtained via the EVM or DF, but its execution was much faster. To simulate the interactions between a river and a rectangular aquifer discretized in 75000 blocks during 5000 daily intervals, classical DF took around 1 h while EVM-SCMM finished in less than 2 min.

4 Concluding Remarks

The main advantage of the EVM-SCMM is that, increasing the number of cells in the FD representation of the aquifer's spatial domain, it is possible to consider a more detailed description of the spatial variability of the aquifer's hydraulic parameters and EA in the groundwater flow modeling. It also increases the computational efficiency of the simulations, especially when a conjunctive use system is modeled.

References

1. Vermeulen, P. T. M., Heemink, A. W., & Te Stroet, C. M. B. (2004). Low dimensional modeling of numerical groundwater flows. *Hydrological Processes*, *18*(8), 1487–1504.
2. Sahuquillo, A. (1983). An eigenvalue numerical technique for solving unsteady linear groundwater models continuously in time. *Water Resources Research*, *19*(1), 93–97.
3. Pulido-Velázquez, D., Sahuquillo, A., Andreu, J., & Pulido-Velázquez, M. (2007). An efficient conceptual model to simulate surface water body-aquifer interaction in conjunctive use management models. *Water Resources Research*, *43*, W07407.
4. Andreu, J., & Sahuquillo, A. (1989). Efficient aquifer simulation in complex systems. *Journal of Water Resources Planning and Management*, *113*(1), 110–129.
5. Pulido-Velázquez, M., Sahuquillo, A., Ochoa-Rivera, J. C., & Pulido-Velázquez, D. (2005). Modeling of stream-aquifer interaction: the embedded multi-reservoir model. *Journal of Hydrology*, *313*, 166–181.

Rational Lanczos Reduction of Groundwater Flow Models to Perform Efficient Simulations of Surface-Ground Water Interaction in Conjunctive Use Systems

Oscar David Álvarez-Villa, Eduardo Cassiraga and Andrés Sahuquillo

1 Introduction

The Rational Lanczos method (RLANRM) [1] has been used in the last 20 years to reduce groundwater flow models of several kinds of aquifers. Generally, those aquifers have been spatially discretized via finite elements [2], but in this paper we present an adaptation of RLANRM to simulate surface-ground water relationships via finite differences (FD) in conjunctive use systems. First, the mathematical framework to solve the partial differential equation (PDE) of groundwater flow via RLANRM is presented. Later, the algorithm for the reduced simulation is discussed. Finally, relevant results of applying RLANRM on rectangular aquifers are shown.

2 Mathematical Framework of RLANRM

Let's consider a linear, time invariant aquifer, whose spatial domain has been discretized in n active FD nodes to solve the groundwater flow PDE. Accordingly, the aquifer's hydraulic parameters and boundary conditions do not change in time and the superposition principle applies. Also, let $\mathbf{h}(t) \in \mathrm{R}^n$ be the vector of piezometric heads in each active FD node of the model [L]. By superposition, it can be written

O. D. Álvarez-Villa (✉) · E. Cassiraga · A. Sahuquillo
Universitat Politècnica de València, Camino de Vera, s/n, 46022 València, Spain
e-mail: osdaalvi@gmail.com

E. Cassiraga
e-mail: efc@dihma.upv.es

A. Sahuquillo
e-mail: asahuq@hma.upv.es

O. D. Álvarez-Villa
GOTTA Ingeniería SAS, Cra 78A No 48-91 Int. 201, Medellín, Colombia

E. Pardo-Igúzquiza et al. (eds.), *Mathematics of Planet Earth*,
Lecture Notes in Earth System Sciences, DOI: 10.1007/978-3-642-32408-6_98,

that $\mathbf{h}(t) = \mathbf{u} + \mathbf{w}(t)$, where $\mathbf{u} \in \mathrm{R}^n$ is the steady state solution subjected to the boundary conditions imposed to the original groundwater model and $\mathbf{w}(t) \in \mathrm{R}^n$ is the transient solution for the following dynamical system:

$$\mathbf{A}\mathbf{w}(t) + \boldsymbol{\Psi}\mathbf{r}(t) = \mathbf{S}\frac{d\mathbf{w}(t)}{dt} \qquad (1)$$

where $\mathbf{A} \in \mathrm{R}^{n\times n}$ is the conductances matrix $[\mathrm{L}^2/\mathrm{T}]$, $\mathbf{S} \in \mathrm{R}^{n\times n}$ is the matrix of storages $[\mathrm{L}^2]$, $\boldsymbol{\Psi} \in \mathrm{R}^{n\times n_a}$ is the matrix of time invariant elemental excitations acting over the aquifer, n_a is the number of external actions (EA) and $\mathbf{r}(t) \in \mathrm{R}^{n_a}$ is the vector of intensity for those AE $[\mathrm{L}^3/\mathrm{T}]$. The boundary conditions imposed to (1) are zero and its initial condition is $\mathbf{w}(0) = \mathbf{h}(0) - \mathbf{u}$ [3]. Assuming that the intensities of the EA are zero and $\mathbf{w}(t) = \mathbf{V}\Phi(t)$, where $\mathbf{V} \in \mathrm{R}^{n\times n}$ is an orthogonal projection matrix and $\Phi(t) \in \mathrm{R}^n$ is a vector of states, substituting $\mathbf{w}(t)$ in (1) and applying the variable's separation technique, two problems are obtained [3]: (*i*) a generalized eigenvalue problem and (*ii*) a time dependent first order differential equation. Applying the inverse transformation combined with a spectral shift [2, 4], the previously mentioned generalized eigenvalue problem can be expressed as:

$$(\mathrm{A} - \sigma\mathbf{S})^{-1}\mathbf{S}\mathbf{V} = \boldsymbol{\Omega}\mathbf{V} \qquad (2)$$

where $\boldsymbol{\Omega} \in \mathrm{R}^{nxn}$ is the diagonal matrix of shifted eigenvalues. Executing m steps of the rational Lanczos iteration [4], the Eq. (2) can be expressed as follows [2]:

$$\mathbf{X}^*\mathbf{S}(\mathbf{A} - \sigma\mathbf{S})^{-1}\mathbf{S}\mathbf{X} = \boldsymbol{\Pi};\ \mathbf{X}^*\mathbf{S}\mathbf{X} = \mathbf{I} \qquad (3)$$

where $\mathbf{X} \in \mathrm{R}^{nxm}$ is the matrix of Lanczos vectors, $\boldsymbol{\Pi} \in \mathrm{R}^{mxm}$ is the tri-diagonal matrix of Lanczos coefficients. The m eigenvalues of $\boldsymbol{\Pi}$ are good approximations of m eigenvalues of (2). The participation factor for the ith Lanczos vector is defined as the proportion which contributes to the aggregate volume of the EA, as follows:

$$p_{i,j} = \left(\sum_{k=1}^{n} s_{k,k}x_{k,i}\right)\mathrm{x}_i^*\Psi_j \qquad (4)$$

with $i = 1, \ldots, m$ and $j = 1, \ldots, n_a$, x and s are the elements of $\mathbf{X}$ and $\mathbf{S}$, $\mathbf{x}_i$, ψ_j are the ith and jth column vectors of $\mathbf{X}$ and $\boldsymbol{\Psi}$, respectively. The accumulated participation factors, pa_j with $j = 1, \ldots, n_a$, are calculated as criteria to stop the Lanczos iteration, because they have unitary upper limits. Thus, when pa_j is close enough to one for all EA acting over the aquifer, the generation of the Krylov's reduction subspace is stopped.

Replacing (3) in (2), defining that $\mathbf{A}_\sigma = \mathbf{A} - \sigma\mathbf{S}$ and using the $\mathbf{S}$-orthonormality of Lanczos vectors, the equation for the aquifer's states $(\Phi(t)) \in \mathrm{R}^m$ is obtained [2]:

$$\boldsymbol{\Pi}\frac{d\Phi(t)}{dt} - \mathbf{G}\mathbf{r}(t) = (\sigma\boldsymbol{\Pi} - \mathbf{I})\Phi(t) \qquad (5)$$

where $\mathbf{G} = \mathbf{X}^*\mathbf{S}\mathbf{A}_\sigma\boldsymbol{\Psi} \in \mathrm{R}^{mxn_a}$ and $\mathbf{w}(t) = \mathbf{X}\Phi(t)$ is the approximated solution for transient piezometric heads. The Lanczos states are also approximated using an implicit FD for the time dependent derivative. Therefore, assuming that transient simulations are performed using time intervals of equal duration Δt and the EA's intensities are constant during those intervals, the following expression is obtained:

$$\left(\frac{\boldsymbol{\Pi}}{\Delta t} + \mathbf{I} + \sigma\boldsymbol{\Pi}\right)\Phi^{t+1} = \frac{\boldsymbol{\Pi}}{\Delta t}\Phi^{t} + \mathbf{G}\mathbf{r}(\tau) \tag{6}$$

which is a tri-diagonal system of linear equations where the vector of Lanczos states at time $t+1$ is unknown. Equation (6) is solved using bi-conjugate gradient solvers.

Finally, to calculate the integrated volumes of surface-ground water interaction, the time integration of Lanczos' states is executed numerically via Simpson's rule. During such process, the τth interval is divided in h subintervals of equal duration, with h even. Hence, the integrated Lanczos states are computed as follows:

$$\boldsymbol{\Theta}(\tau) = \int_{t=t_0}^{t+1=t_f} \Phi(t)dt \approx \frac{\Delta t}{3h}\left[\Phi^{t_0} + 2\sum_{j=1}^{h/2-1}\Phi^{2j} + 4\sum_{j=1}^{h/2}\Phi^{2j-1} + \Phi^{t_f}\right] \tag{7}$$

where $\boldsymbol{\Theta}(t) \in \mathrm{R}^m$ is a vector containing the integrated Lanczos states of the aquifer. Finally, the control parameters (piezometric heads, accumulated volumes, ground-surface water fluxes and volumes, etc.) required by the conjunctive use model can be obtained efficiently using the expressions proposed by [5].

3 Application of the RLANRM on Rectangular Aquifers

The RLANRM was applied to reduce groundwater flow models on rectangular aquifers. The effect of changing the riverbed's conductance (RC) on the RLANRM reduction scheme was quantified comparing the results from the reduced models with those obtained using classical FD simulation. To test the accuracy of the reduced simulations, root squared mean error (**RSM**), Nash's efficiency ($\mathbf{E}_2$) and bias coefficient (**BIAS**) have been estimated. Figure 1 exhibits the estimated $\mathbf{E}_2$ for the surface-ground water interactions simulated with RMLANRM in an homogeneous rectangular aquifer for recharge and punctual pumping, considering six RC values.

Results show that as the size of the Krylov subspaces increases, $\mathbf{E}_2$ approximates to 100 % and, simultaneously, **RSM** and **BIAS** decrease to zero. Thus, when the RC is low, the RLANRM's performance improves because fewer Lanczos vectors are needed to obtain more accurate simulations. On the contrary, when river-aquifer connection is almost perfect, the reduced model's accuracy decreases and the cost of building the Lanczos' vectors augments. Numerical experiments have shown that the classical eigenvalue method (EMV) [3] is more efficient than RLANRM to perform the simulation of the transient river-aquifer interactions because it possesses a simple

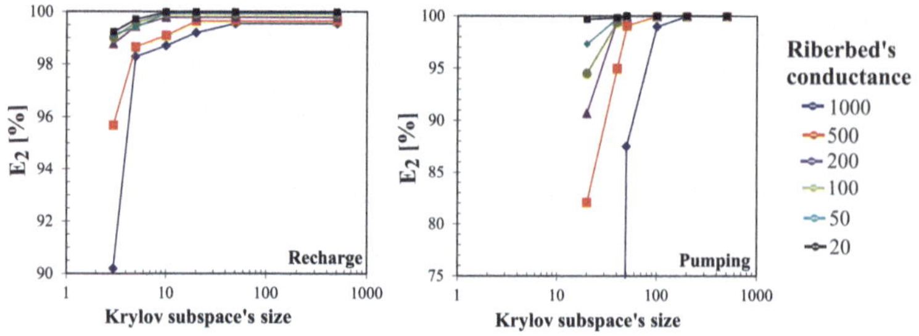

Fig. 1 E_2 estimated for the surface-ground water relationships simulated with RMLANRM in homogeneous rectangular aquifer for recharge and pumping

explicit state equation. This disadvantage is more evident for monthly simulation because the integration of the Lanczos' states using Eq. (7) is more demanding.

4 Concluding Remarks

The RLANRM is more efficient than classical FD to simulate groundwater flow in complex conjunctive use systems, but its performance is slightly lower than the exhibited by the EVM. It also allows to deal with very large groundwater flow models, when many cells are used in the FD representation of the aquifer's spatial domain. Consequently, it is possible to consider more detailed descriptions on the spatial variability of the aquifer's hydraulic parameters and the imposed EA.

References

1. Dunbar, W. S., & Woodbury, A. D. (1989). Application of the Lanczos algorithm to the solution of the groundwater flow equation. *Water Resources Research*, *25*(3), 551–558.
2. Zhang, K. (2000). *The modal reduction method for simulation of groundwater flow and multispecies contaminant transport in fractured porous media*. Doctoral Dissertation: The University of Manitoba.
3. Sahuquillo, A. (1981). An eigenvalue numerical technique for solving unsteady linear groundwater models continuously in time. *Water Resources Research*, *19*(1), 93–97.
4. Nour-Omid, B., Parlett, B. N., Ericsson, T., & Jensen, P. S. (1987). How to implement the spectral transformation. *Mathematics of Computation*, *48*, 663–673.
5. Andreu, J., & Sahuquillo, A. (1989). Efficient aquifer simulation in complex systems. *Journal of Water Resources Planning and Management*, *113*(1), 110–129.

Efficient Generation of Effective Modes and Spectral Masks to Build Reduced Groundwater Flow Models Using the Eigenvalue Method with Selective Compression and Modal Masking

Oscar David Álvarez-Villa, Eduardo Cassiraga and Andrés Sahuquillo

1 Introduction

The most computer demanding step in the modal reduction of a groundwater flow model via the eigenvalue approach with selective compression and modal masking (EVM-SCMM) is the generation of the effective modes. In this paper we propose: (i) physically based criteria to the selection of those effective modes, (ii) efficient algorithms for the generation of those effective modes and (iii) physically based criteria to stop the modal generation. The proposed generators are based on iterative methods to solve a symmetric generalized problem of eigenvalues (SGEVP).

2 EVM-SCMM and Physical Criteria to Modal Generation

Assuming that the reduced groundwater flow simulations are carried out in discrete time intervals of equal duration Δt, the EVM-SCMM obtains the following expression for the effective states of the aquifer (recall that $\mathbf{w}(t) = \mathbf{V}\hat{\Phi}(t)$, $\mathbf{w}(t) \in \mathrm{R}^n$ are transient piezometric heads and $\mathbf{V} \in \mathrm{R}^{n\times n}$ is the eigenvector matrix) [1, 2]:

O. D. Álvarez-Villa (✉) · E. Cassiraga · A. Sahuquillo
Universitat Politècnica de València, Camino de Vera, s/n, 46022 València, Spain
e-mail: osdaalvi@gmail.com

E. Cassiraga
e-mail: efc@dihma.upv.es

A. Sahuquillo
e-mail: asahuq@hma.upv.es

O. D. Álvarez-Villa
GOTTA Ingeniería SAS, Cra 78A No 48-91 Int. 201, Medellín, Colombia

E. Pardo-Igúzquiza et al. (eds.), *Mathematics of Planet Earth*,
Lecture Notes in Earth System Sciences, DOI: 10.1007/978-3-642-32408-6_99,

$$\hat{\Phi}(t) = \hat{\mathbf{E}}|_{\Delta t}\hat{\Phi}(t-1) + \left(\frac{\hat{\mathbf{I}} - \hat{\mathbf{E}}|_{\Delta t}}{\hat{\mathbf{\Lambda}}\hat{\mathbf{F}}}\right)\hat{\mathbf{B}}\mathbf{r}(\tau) \tag{1}$$

where $\hat{\mathbf{\Lambda}} \in \mathbf{R}^{(p+1)\times(p+1)}$ is the diagonal matrix of effective eigenvalues, $\mathbf{r}(\tau) \in \mathbf{R}^{n_a}$ is the vector of intensity for the external actions (EA) in the τth interval of simulation $\hat{\mathbf{E}} \in \mathrm{R}^{(p+1)\times(p+1)}$, is an effective diagonal exponential matrix, $\hat{\mathbf{F}} \in \mathrm{R}^{(p+1)\times(p+1)}$ is a diagonal matrix containing the volumes enclosed by the effective eigenvectors, $\hat{\mathbf{B}} = \hat{\mathbf{F}}\hat{\mathbf{V}} * \psi \in \mathrm{R}^{(p+1)\times n_a}$ is the matrix of the effective allocation coefficients and $\hat{\Phi}(t) \in \mathrm{R}^{p+1}$ is the vector of the aquifer's effective states. As defined by [1], the last effective state, $\varphi_{p+1}(t)$, is used to preserve the equation of continuity on the aquifer.

According to Eq. (1), the allocation coefficients, $b_{i,j}$, represent the direct influence of the jth EA on the changes of the aquifer's flow dynamic. Consequently, if the allocation factor of the ith mode corresponding to the jth EA is small, the changes of the aquifer's states due to such EA are also small. Therefor, let's define that: (i) an *effective mode* contributes effectively over the changes in aquifer's states, (ii) the changes on the aquifer's states due to EA acting on a *residual mode* are negligible and (iii) defining that $\mathbf{v}_i$ is the ith eigenvector, the selection of effective modes is based on the concept of *participation limit*, b_{lim}, using the following criteria:

$$\text{If } b_{i,j} \geq b_{lim}(\lambda_i, \mathbf{V}_i) \text{ is effective for EA } j \tag{2}$$

which also allows to build the *effective modes mask*, $\mathbf{Z} \in \mathrm{R}^{(p+1)\times n_a}$. $\mathbf{Z}$ is defined as a boolean matrix that indicates which modes are excited by any given EA and its $z_{i,j}$ = false elements are defined according to the indices contained on the masking set $\mathrm{PC}\{(i,j) : b_{ij} < b_{lim}\} \forall i = 1, \ldots, p+1$ and $j = 1, \ldots, n_a$. Thus, $\mathbf{Z}$ indicates the location of the effective and residual modes in the matrix $\mathbf{B}$. $\mathbf{Z}$ also diminishes the computational charge of $\hat{\mathbf{B}}\mathbf{r}(\tau)$ because the operations can be done using sparse algorithms. Now, the accumulated allocation factors for the jth EA are defined as:

$$b_{a_j} = \sum_i b_{i,j} \tag{3}$$

where ba_j has an upper unitary limit and it represents the percentage of mass conservation on the aquifer. The stop of the effective modes generation process is evaluated calculating the accumulated allocation coefficients and comparing them with a lower limit imposed to the reduced model, called *level of continuity*, ba_{lim}.

Incorporating the previously defined physically based criteria on iterative methods to solve a SGEVP, it is possible to generate efficiently effective modes for modal reduction via the EVM-SCMM. The advantage of using iterative methods is that the sparse structure of the groundwater flow model is exploited to save operations and storage. Two iterative sparse generators of effective modes have been implemented.

The first generator is based on a modification of the algorithm proposed in [3], it uses ILU preconditioning to accelerate the vector iterations and conjugate gradient to minimize the Rayleigh quotient. This vector generation works very fast when a small amount of the lesser magnitude effective modes are needed. The algorithm generates sequentially each mode, calculates the allocation coefficients to check if it is effective and, finally, computes the accumulated allocation coefficients and checks if $ba_{i,j} > ba_{lim}$ for all EA to stop the generation. The second one is based on the rational Lanczos iteration with explicit restart [4] to calculate sequentially sets of modes. The explicit restart deflates a new initial Lanczos vector to prevent the convergence to one of the already available eigenvectors. Once each restart has built a new set of modes, the generator calculates the allocation coefficients for all the modes to check which ones are effective. This process is repeated for all obtained modes. Then, the accumulated allocation coefficients are calculated and the generator checks if $ba_{i,j} > ba_{lim}$ for all EA to stop the generation. If the stopping criterion is not achieved, another restart is performed using a new spectral shift to improve the convergence.

3 Application to Heterogeneous Aquifers

Results of several numerical experiments have shown that the rational Lanczos generator is very efficient, even for large scale groundwater flow models where the spatial domain of the aquifer has been discretized using tens of thousands of nodes belonging to a finite differences (FD) network and many EA are imposed. For instance, an heterogeneous irregular aquifer, connected with a sinuous river has been analyzed. Its geometry, hydraulic parameters and EA are presented in Fig. 1.

The aquifer's domain has been discretized via FD using 21850 active nodes. For the heterogeneous aquifer shown in Fig. 1, the Rayleigh minimization generator wasn't efficient enough because a lot of effective modes were needed. As stated in [3]

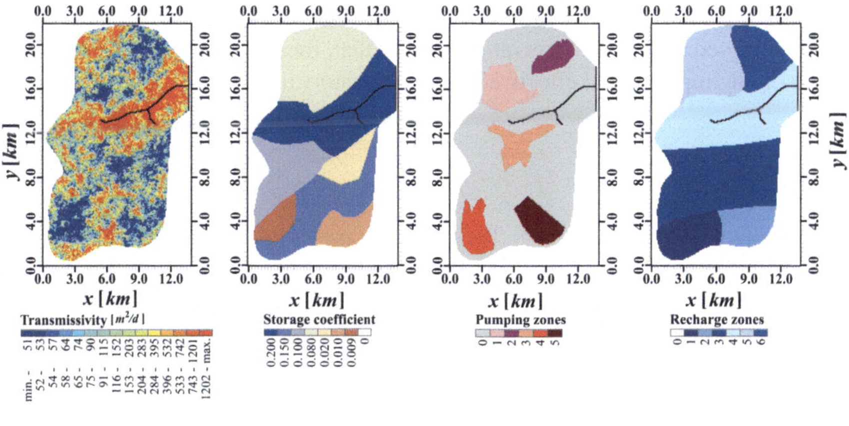

Fig. 1 From *left* to *right*: transmissivity field, storage coefficient field, regions of homogenous pumping and zones of homogenous recharge for the heterogeneous irregular aquifer

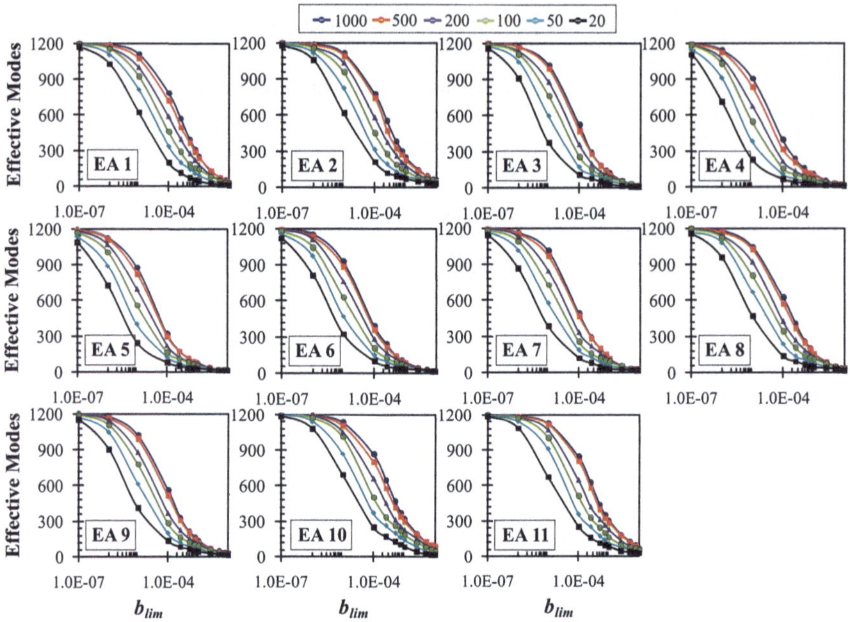

Fig. 2 Effective modes generated by Lanczos' iteration on a highly heterogeneous aquifer

this method has the problem that the cost of the vector deflation augments as long as the execution advances. Consequently, the effective modes were obtained applying the rational Lanczos generator. Such effective modes are presented in Fig. 2 for each EA and different values of riverbed's conductance (RC). The results show that the iterative generation of modes, combined with physical criteria, is more efficient than calculating the complete spectra of the eigenproblems to apply the classical modal truncation [1] and it allows to reduce efficiently large groundwater flow models. It is also clear that high values of RC augments the number of required effective modes for any imposed b_{lim}. This observation implies that, for high RC, the computational charge of the implemented effective modes generators increases.

References

1. Sahuquillo, A., & Andreu, J. (1988). The eigenvalues approach for solving linear groundwater flow problems. In: Custodio, E. et al. (Eds.) *Groundwater flow and quality models* (pp. 151–164). D. Reidel Publishing Company, Guilin.
2. Álvarez-Villa, O. D. (2013). Efficient simulation of surface-ground water relationships in conjunctive use systems using reduction techniques on linear, time invariant flow models. Doctoral dissertation, Technical University of Valencia (in Spanish).
3. Gambolati, G. (1993). On time integration of groundwater flow equation by spectral methods. *Water Resources Research*, *29*(4), 1257–1267.
4. Grimes, R. G., Lewis, J. G., & Simon, H. D. (1991). *A shifted block Lanczos algorithm for solving sparse symmetric generalized eigenproblems: Rep RNR-91-012*. NASA—Ames Research Center.

Application of Eigenvalue Method with Selective Compression and Modal Masking and Rational Lanczos Reduction Method to the Efficient Simulation of Surface–Ground Water Interactions in Heterogeneous Aquifers

Oscar David Álvarez-Villa, Eduardo Cassiraga and Andrés Sahuquillo

1 Introduction

We present an application of the eigenvalue method with selective compression and modal masking (EVM-SCMM) and rational Lanczos method (RLANRM) to reduce efficiently the groundwater flow models of two highly discretized linear heterogeneous aquifers. We have performed a sensitivity analysis about the influence of changing the riverbed's conductance (RC) and the reduction parameters on a more accurate representation of the surface-ground water interactions. In this analysis, the modal participation limits for the EVM-SCMM and the size of the Krylov's subspaces for the RLANRM have been the sensitivity parameters of study.

2 Short Descriptions of the Reduction Methods

The EVM-SCMM and RLANRM are model oriented reduction techniques; it means that, in both methods, orthogonal subspaces that capture the mathematical structure of the groundwater model are built. Once the reduction subspace is available, the groundwater flow model is projected, its size is largely reduced and the transient simulation is more efficient. The main differences between the EVM-SCMM and

O. D. Álvarez-Villa (✉) · E. Cassiraga · A. Sahuquillo
Universitat Politècnica de València, Camino de Vera, s/n,46022 València, Spain
e-mail: osdaalvi@gmail.com

O. D. Álvarez-Villa
GOTTA Ingeniería SAS, Cra 78A No 48-91 Int. 201,Medellín, Colombia

E. Cassiraga
e-mail: efc@dihma.upv.es

A. Sahuquillo
e-mail: asahuq@hma.upv.es

E. Pardo-Igúzquiza et al. (eds.), *Mathematics of Planet Earth*,
Lecture Notes in Earth System Sciences, DOI: 10.1007/978-3-642-32408-6_100,

RLANRM are the following. First, EVM-SCMM uses a subspace generated by the eigenvectors of the model, while RLANRM uses an easy to obtain Krylov's subspace that implicitly approximates such eigenvectors. Second, the EVM-SCMM's equation of state is explicit, while the RLANRM's states have to be obtained numerically. Consequently, the simulation via EVM-SCMM is faster than using RLANRM.

3 Configuration of the Aquifers

A rectangular aquifer with three bands of homogenous hydraulic properties is connected to a straight river. The length between the river and the opposite impervious wall is 5000 m and the distance between the impervious walls perpendicular to the river is 7500 m. The transmissivities in the bands of the aquifer vary between 2500 and 250 m^2/d and their storage coefficients vary from 0.2 to 0.07. The RC decreases from 5000 to 20 m^2/d. The aquifer is subjected to: (*i*) an uniform, time varying recharge and (*ii*) time varying punctual pumping in five locations. The aquifer's spatial domain has been discretized via finite differences (FD) using 3285 active nodes, the block's length is 100 m. and the simulation horizon is 1000 days.

Also, a highly heterogeneous aquifer with irregular boundaries connected with a sinuous sloped river has been analyzed. Its geometric configuration, hydraulic parameters and the external actions (EA) are presented in Fig. 1. The integrated intensities of those EA vary on time according to Fig. 2. 21,850 active nodes were used to discretize the aquifer's spatial domain via FD, the length of each square block's side was 100 m and the simulation horizon was 5100 days. The corresponding transmissivity field was obtained via Gaussian sequential stochastic co-simulation [1] using a sample extracted from the Walker Lake data set [2]. In both aquifers, the control

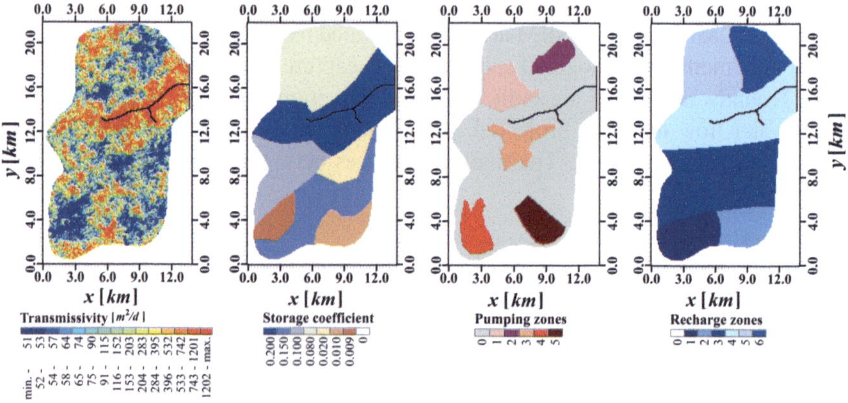

Fig. 1 From *left* to *right*: transmissivity field, storage coefficient field, homogeneous zones of recharge and homogenous regions of pumping for the heterogeneous irregular aquifer

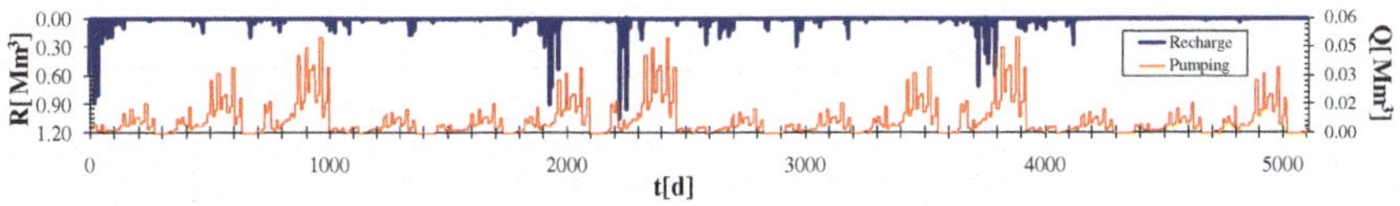

Fig. 2 Integrated volumes of recharge and pumping for the heterogeneous aquifer

parameters consisted of surface-ground water interchange fluxes on the drainage network. The obtained results are compared to the analog simulations performed using classical FD models, in which time and space were discretized.

4 Simulation's Results and Analysis

To solve the problem of eigenvalues for the rectangular aquifer, a Rayleigh quotient minimization generator [3] has been used. Applying such generator, the first 30 effective modes were calculated for four RC values. Also, the eigenproblems for the highly heterogeneous aquifer was solved using a rational Lanczos generator [4]. Using such generator, the first 2,000 effective modes were calculated for six RC's values. The rational Lanczos iteration was applied to obtain Krylov subspaces formed by 200 and 2,000 Lanczos vectors for the rectangular and the highly heterogeneous aquifers, respectively. Then, using those effective modes and Krylov subspaces, the transient simulations of surface–ground water relationships have been performed via EVM-SCMM and RLANRM. To test the accuracy of the reduced simulations, root squared mean errors (**RSM**), Nash's efficiencies (**E**$_2$) and bias coefficients (**BIAS**) were estimated using FD simulated reference hydrographs.

The estimated **RSM**, **BIAS** and **E**$_2$ for the rectangular aquifer are presented in Table 1. In general, the reduced simulations were more accurate for the lowest RC. Nevertheless, the simulations via EVM-SCMM have been more accurate than those obtained applying RMLANRM, particularly for the highest imposed RC values.

The results from the sensitivity analysis for the surface–ground water interactions on the highly heterogeneous aquifer, simulated with EVM-SCMM and RMLANRM are shown in the Fig. 3. These graphs exhibit the behaviors of estimated **RSM**, **BIAS**

Table 1 Performance indexes for surface–ground water fluxes in a rectangular aquifer

	RC = 5000 m^2/d		RC = 200 m^2/d		RC = 100 m^2/d		RC = 20 m^2/d	
Index (%)	EVM-SCMM	RM-LANRM	EVM-SCMM	RM-LANRM	EVM-SCMM	RM-LANRM	EVM-SCMM	RM-LANRM
RMS	0.56	1.57	0.22	1.09	0.13	0.51	0.29	0.071
E$_2$	99.86	98.84	99.95	99.03	99.98	99.68	99.99	99.96
BIAS	0.20	0.26	0.153	0.203	0.125	0.11	0.09	0.02

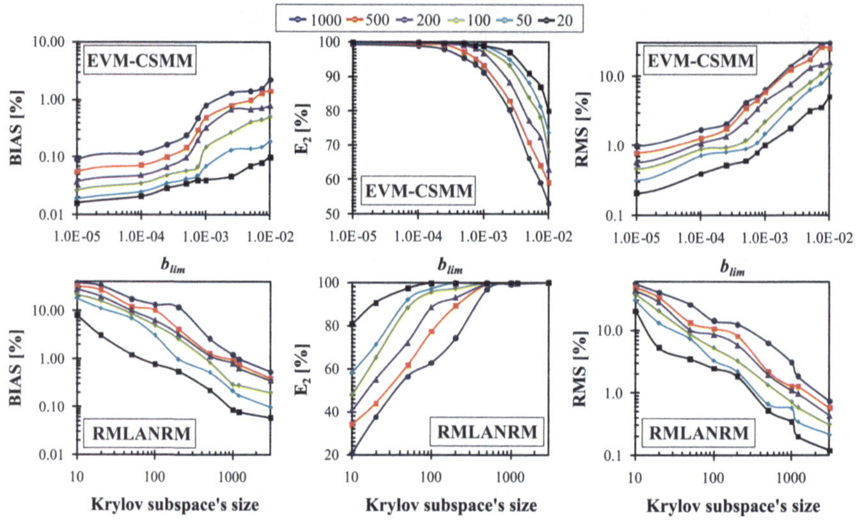

Fig. 3 BIAS, $\mathbf{E}_2$ and **RSM** estimated for the surface–ground water relationships simulated with EVM-SCMM and RMLANRM in the highly heterogeneous aquifer

and $\mathbf{E}_2$ as a function of: (*i*) b_{lim} is the participation limit, a parameter that allows to select the effective modes in the EVM-SCMM (if b_{lim} is low, a bigger amount of effective eigenvectors are used in the reduction, therefore the size of the reduced model is bigger), (*ii*) the size of the Krylov's subspace or, equivalently, the amount of Lanczos' vectors used by the RMLANRM and (*iii*) six different RC values. In general, the estimations diminish when the larger reduced models are used. These results prove the power of both techniques to reduce efficiently groundwater flow models and to obtain accurate simulations. Although EVM-SCMM obtains accurate aggregated surface–ground water interactions more efficiently, RMLANRM calculates more accurate the aquifer's distributed variables. Numerical experiments have shown that **RSM**, **BIAS** and $\mathbf{E}_2$ estimated for piezometric heads and partial water balances over fractions of the drainage network are more optimal for RMLANRM than for EVM-SCMM. In summary, the choice of the reduction methodology depends on the desired control parameters to be better represented.

References

1. Gómez-Hernández, J. J., Journel, A. (1993). Joint sequential simulation of multi-Gaussian fields. In V. I. A. Soares (Ed.), *Geostatistics Tróia 92*. Dordrecht: Kluwer Academic Publishers.
2. Isaaks E. H., Srivastava, R. M. (1989). *Applied geostatistics*. New York: Oxford University Press.

3. Gambolati, G. (1993). On time integration of groundwater flow equation by spectral methods. *Water Resources Research*, *29*(4), 1257–1267.
4. Grimes R. G., Lewis J. G., Simon H. D. (1991). A shifted block Lanczos algorithm for solving sparse symmetric generalized eigenproblems. Rep RNR-91-012. NASA-AMES R C.

A Multipoint Flux Domain Decomposition Method for Transient Flow Modeling in Complex Porous Media

Andrés Arrarás, Laura Portero and Ivan Yotov

1 Introduction

Let us consider a single-phase Darcy flow on a domain $\Omega \subset \mathbb{R}^2$, i.e.,

$$\begin{aligned} p_t + \nabla \cdot \mathbf{u} &= g \qquad \text{in } \Omega \times (0, T], \\ \mathbf{u} &= -K \nabla p \qquad \text{in } \Omega \times (0, T], \end{aligned} \tag{1}$$

with initial condition $p = p_0$ in $\Omega \times \{0\}$ and boundary condition $p = 0$ on $\partial\Omega \times (0, T]$, where $\partial\Omega$ denotes the boundary of Ω. Here, $\mathbf{u}$ is the Darcy velocity, p is the pressure, and K is a symmetric and positive definite tensor representing the rock permeability divided by the fluid viscosity. Homogeneous pressure boundary conditions are chosen for the sake of simplicity in the exposition, but alternative boundary conditions can also be considered.

In this work, we present a combined multipoint flux domain decomposition technique for solving problem (1). More precisely, we extend the so-called multipoint flux mixed finite element (MFMFE) method (cf. [1]) to evolutionary diffusion problems by introducing a domain decomposition time-splitting scheme (cf. [2]). In the sequel, we describe in detail how the method is constructed and provide a numerical illustration of its performance in a practical setting.

A. Arrarás (✉) · L. Portero (✉)
Departamento de Ingeniería Matemática e Informática, Universidad Pública de Navarra, Edificio de Las Encinas, Campus de Arrosadía, 31006 Pamplona, Spain
e-mail: andres.arraras@unavarra.es

L. Portero
e-mail: laura.portero@unavarra.es

I. Yotov
Department of Mathematics, University of Pittsburgh, 301 Thackeray Hall, Pittsburgh, PA 15260, USA
e-mail: yotov@math.pitt.edu

E. Pardo-Igúzquiza et al. (eds.), *Mathematics of Planet Earth*,
Lecture Notes in Earth System Sciences, DOI: 10.1007/978-3-642-32408-6_101,

2 The Multipoint Flux Mixed Finite Element Method

Let $\mathcal{T}_h$ be a conforming, shape-regular and quasi-uniform partition of Ω into quadrilateral elements, where $h = \max_{E\in\mathcal{T}_h} \operatorname{diam}(E)$. The velocity and pressure finite element spaces are denoted by V_h and W_h, respectively. On any element E, they are defined via a suitable transformation of the lowest order Brezzi–Douglas–Marini spaces on the reference element (cf. [3]). As a result, the MFMFE approximation to (1) reads: *Find* $(\mathbf{u}_h, p_h) : [0, T] \to V_h \times W_h$ *such that*

$$
\begin{aligned}
(\dot{p}_h, w) + (\nabla \cdot \mathbf{u}_h, w) &= (g, w) \qquad &&\forall\, w \in W_h,\\
(K^{-1}\mathbf{u}_h, \mathbf{v})_Q &= (p_h, \nabla \cdot \mathbf{v}) &&\forall\, \mathbf{v} \in V_h,
\end{aligned} \tag{2}
$$

with a suitable $p_h(0)$ to be specified below. The inner product $(K^{-1}\cdot, \cdot)_Q$ represents the trapezoidal quadrature rule, which is defined as the sum over all $E \in \mathcal{T}_h$ of

$$(K^{-1}\mathbf{q}, \mathbf{v})_{Q,E} = \tfrac{1}{2}\textstyle\sum_{i=1}^{4} |T_i| K^{-1}(\mathbf{r}_i)\mathbf{q}(\mathbf{r}_i) \cdot \mathbf{v}(\mathbf{r}_i),$$

where $\{\mathbf{r}_i\}_{i=1}^4$ are the vertices of E, and each $|T_i|$ is the area of the triangle formed by the two edges sharing $\mathbf{r}_i$.

Let us define $P_h = (P_{h,1}, P_{h,2}, \ldots, P_{h,N_e})^T$, where $P_{h,i} = p_h(\mathbf{c}_i)$, $\mathbf{c}_i$ is the centroid of E_i, and N_e denotes the number of elements in $\mathcal{T}_h$. Hence, the MFMFE method (2) can be reduced to a cell-centered pressure system of the form (cf. [1])

$$\dot{P}_h(t) + A_h P_h(t) = G_h(t) \qquad t \in (0, T], \tag{3}$$

with $P_h(0) = P_h^0$, where each component $(P_h^0)_i = \frac{1}{|E_i|}\int_{E_i} p_0(\mathbf{x})\, d\mathbf{x}$ and $|E_i|$ is the area of E_i. In (3), the discrete diffusion operator is given by $A_h = D^{-1}BM^{-1}B^T$. If we denote by $\{\mathbf{v}_i\}_{i=1}^{2N_f}$ and $\{w_j\}_{j=1}^{N_e}$ the basis functions of V_h and W_h, respectively, then $(B)_{ij} = -(\nabla \cdot \mathbf{v}_j, w_i)$ and $(M)_{ij} = (K^{-1}\mathbf{v}_j, \mathbf{v}_i)_Q$, with N_f being the number of edges in $\mathcal{T}_h$. Furthermore, $D = \operatorname{diag}(|E_1|, |E_2|, \ldots, |E_{N_e}|)$ and $(G_h)_i = \frac{1}{|E_i|}(g, w_i)$.

3 The Domain Decomposition Splitting Method

Let Ω be decomposed into two overlapping subdomains Ω_1 and Ω_2 such that $\Omega = \Omega_1 \cup \Omega_2$. Each subdomain $\Omega_k \subset \Omega$ is chosen to be an open disconnected set involving m_k connected components, i.e., $\Omega_k = \cup_{l=1}^{m_k} \Omega_{kl}$, for $k = 1, 2$. Such components are pairwise disjoint, thus satisfying $\Omega_{ki} \cap \Omega_{kj} = \emptyset$, for $i \neq j$. Under this decomposition, we construct a smooth partition of unity consisting of two non-negative functions $\{\rho_k(\mathbf{x})\}_{k=1}^2$. These functions fulfill, for any $\mathbf{x} \in \overline{\Omega}$, the conditions: (a) $\mathrm{supp}(\rho_k(\mathbf{x})) \subset \overline{\Omega}_k$; (b) $0 \leq \rho_k(\mathbf{x}) \leq 1$; and (c) $\rho_1(\mathbf{x}) + \rho_2(\mathbf{x}) = 1$.

In this framework, recalling the semidiscrete problem (3), we introduce the operator splittings $A_h = A_{h,1} + A_{h,2}$ and $G_h = G_{h,1} + G_{h,2}$. The split terms are given by $A_{h,k} = D^{-1} B \Gamma_k M^{-1} B^T$ and $G_{h,k} = \hat{\Gamma}_k G_h$, for $k = 1, 2$, where Γ_k and $\hat{\Gamma}_k$ are two diagonal matrices of the form $\Gamma_k = \mathrm{diag}(\rho_k(\mathbf{r}_1) I_{f_1}, \rho_k(\mathbf{r}_2) I_{f_2}, \ldots, \rho_k(\mathbf{r}_{N_v}) I_{f_{N_v}})$ and $\hat{\Gamma}_k = \mathrm{diag}(\rho_k(\mathbf{c}_1), \rho_k(\mathbf{c}_2), \ldots, \rho_k(\mathbf{c}_{N_e}))$, respectively. In this case, f_i is the number of edges that share the ith vertex point, I_{f_i} denotes the identity matrix of order f_i, and N_v is the number of vertices in $\mathcal{T}_h$. Note that $2N_f = \sum_{i=1}^{N_v} f_i$.

In order to solve system (3), we consider a domain decomposition variant of the classical Peaceman–Rachford alternating direction implicit method (cf. [4]). To this end, the time interval $[0, T]$ is partitioned as $t_n = n\tau$, for $n = 0, 1, \ldots, N_T$, where τ is the (constant) time step and $N_T = [T/\tau]$. The fully discrete scheme is thus given, for $n = 0, 1, \ldots, N_T - 1$, by

$$\begin{aligned} P_h^{n+1/2} &= P_h^n + \tfrac{1}{2}\tau F_{h,1}(t_n, P_h^n) + \tfrac{1}{2}\tau F_{h,2}(t_{n+1/2}, P_h^{n+1/2}), \\ P_h^{n+1} &= P_h^{n+1/2} + \tfrac{1}{2}\tau\, F_{h,1}(t_{n+1}, P_h^{n+1}) + \tfrac{1}{2}\tau F_{h,2}(t_{n+1/2}, P_h^{n+1/2}), \end{aligned} \tag{4}$$

where $F_{h,k}(t_n, P_h^n) = -A_{h,k} P_h^n + G_{h,k}(t_n)$, for $k = 1, 2$. Note that P_h^n is an approximation to the solution $P_h(t)$ of (3) at time $t = t_n$. The linear system to solve at the kth internal stage of (4) involves the system matrix $(I + \frac{1}{2}\tau A_{h,k})$, for $k = 1, 2$. Recalling that $\mathrm{supp}(\rho_k(\mathbf{x})) \subset \overline{\Omega}_k$, the elements of $A_{h,k}$ are null outside $\overline{\Omega}_k$. Since Ω_k further consists of m_k disjoint connected components, the preceding linear system is indeed a collection of m_k uncoupled linear subsystems which can be solved in parallel.

4 Numerical Illustration

Let us consider a single-phase flow through a porous medium containing a low-permeability streak. In particular, (1) is posed on $[0, 1] \times [0, 1] \times [0, 2]$, with $g(\mathbf{x}, t) = 0$ and $p_0(\mathbf{x}) = \cos(\pi x/2)$. Pressure is specified to be equal to 1 on the boundary $\{0\} \times (0, 1)$ and equal to 0 on $\{1\} \times (0, 1)$. In turn, zero flux is set on $[0, 1] \times \{0, 1\}$. The flow domain contains a low-permeability region which is delimited by the bold lines in Fig. 1. The permeability throughout the domain is uniform and isotropic ($K = I_2$), except in the low-permeability streak, where the parallel and normal components to the local streak orientation are equal to 0.1 and 0.001, respectively. A stationary version of this problem is studied in [5]. Here, we use a strip-wise decomposition of Ω, considering two subdomains with two disjoint components each.

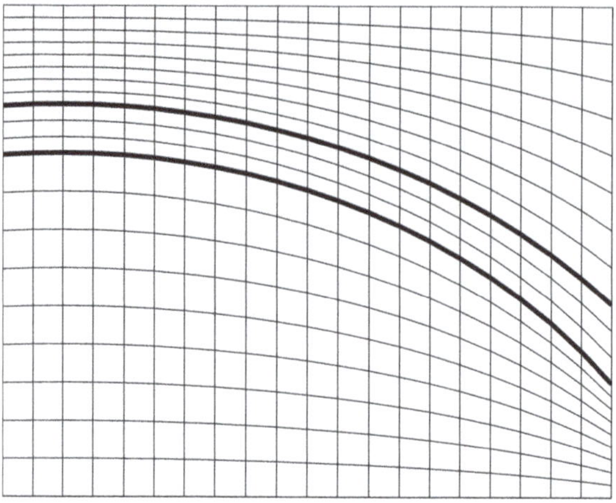

Fig. 1 Geometry of the flow domain and logically rectangular grid

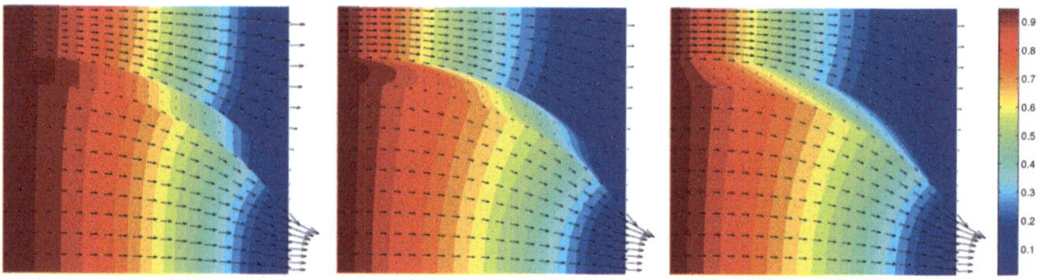

Fig. 2 Calculated pressure distributions (contour maps) and velocity fields at times $t = 0.02$ (*left*), $t = 0.2$ (*center*), and $t = 2$ (*right*)

In Fig. 2, we show the pressure distribution and velocity field obtained at different times. The method considers the logically rectangular grid displayed on Fig. 1, with 200×200 elements, and a time step $\tau = 5\text{E-}03$. Note that, as time evolves, the pressure gradient above the streak increases near the boundary $x = 0$, while decreasing as we approach $x = 1$. As a result, the velocity gets larger in the upper-left corner of the flow domain, and diminishes as we move towards the right end. The change in the pressure gradient—and, thus, in the velocity—is much more subtle below the streak. As expected from the physical configuration, no flow enters the streak.

References

1. Wheeler, M. F., and, Yotov, I. (2006). A multipoint flux mixed finite element method. *SIAM Journal on Numerical Analysis*, *44*, 2082–2106.
2. Mathew, T. P. A. (2008). Domain decomposition methods for the numerical solution of partial differential equations. Lecture Notes in Computational Science and Engineering., (Vol. 61). Berlin: Springer, p 764.
3. Brezzi, F., Douglas, J, Jr, & Marini, L. D. (1985). Two families of mixed finite elements for second order elliptic problems. *Numerische Mathematik*, *47*, 217–235.
4. Peaceman, D. W., & Rachford, H. H, Jr. (1955). The numerical solution of parabolic and elliptic differential equations. *Journal of the Society for Industrial and Applied Mathematics*, *3*, 28–41.
5. Durlofsky, L. J. (1993). A triangle based mixed finite element-finite volume technique for modeling two phase flow through porous media. *Journal of Computational Physics*, *105*, 252–266.

Optimal Reconstruction of 3D Fracture Networks (FEBEX Field Test, Grimsel Site, Swiss Alps)

Israel Cañamón, Rachid Ababou and F. Javier Elorza

1 Introduction

Many studies have developed methodologies to infer fracture size distributions out of planar [4] or more recently curved [2] trace maps, although there are some precautions to be taken when doing that. We found some works dealing with the estimation of trace length statistical parameters in cylindrical boreholes sampling, or with the comparison between analytical and simulated trace length histograms for various disk or elliptical fractures synthetic distributions. In this paper, however, we describe a methodology to use the statistical information of the traces left on a cylindrical gallery to simulate a real fractured network. The granitic medium reconstructed in this paper is located at the Grimsel Test Site (GTS, in Switzerland) in the southern part of the Central Aar Massif, around 400 m below the surface. The cylindrical gallery from which fracture trace data have been collected corresponds to the FEBEX experiment, and has been excavated in the northern part of the Laboratory tunnel of the GTS.

I. Cañamón (✉) · F. J. Elorza
Escuela Técnica Superior de Ingenieros de Minas y Energía,Universidad Politécnica de Madrid,C. /Ríos Rosas 21,28003 Madrid, Spain
e-mail: israel.canamon@upm.es

R. Ababou
Porous Media Group, Institut de Mécanique des Fluides de Toulouse,University of Toulouse, 1 Allée Camille Soula,Toulouse 31400, France
e-mail: franciscojavier.elorza@upm.es

E. Pardo-Igúzquiza et al. (eds.), *Mathematics of Planet Earth*,
Lecture Notes in Earth System Sciences, DOI: 10.1007/978-3-642-32408-6_102,

2 Geomorphological Data

Some of the parameters necessary to generate the fracture network have come directly from field data available in the literature [3]. Other ones have been computed indirectly, and the remaining ones are to be fitted by the optimization algorithm:

- *Fracture orientation*. Two exploratory boreholes (FEBEX-95001 and FEBEX-95002) were drilled and intersecting fractures were measured [3] and classified according to the different discontinuity systems identified in [6]. To simplify the generation process, genetically and morphologically similar fractured systems have been grouped in the stereonet 'space' (Fig. 1a). Thus, only 4 fracture families were simulated, with uniform distributions for dip and plunge angles.
- *Fracture aperture*. Measurements in exploratory boreholes only distinguish qualitative aperture data: 'filled', 'open' and 'open-wet' fractures (Fig. 1b). For the network generation, arbitrarily increasing apertures have been assigned according to this classification, but they have been adjusted at a later stage to fit field hydraulic measurements. Final aperture values were 3.5E-9, 6.65E-6 and 1.65E-5 m for 'filled', 'open' and 'open-wet' fractures respectively.

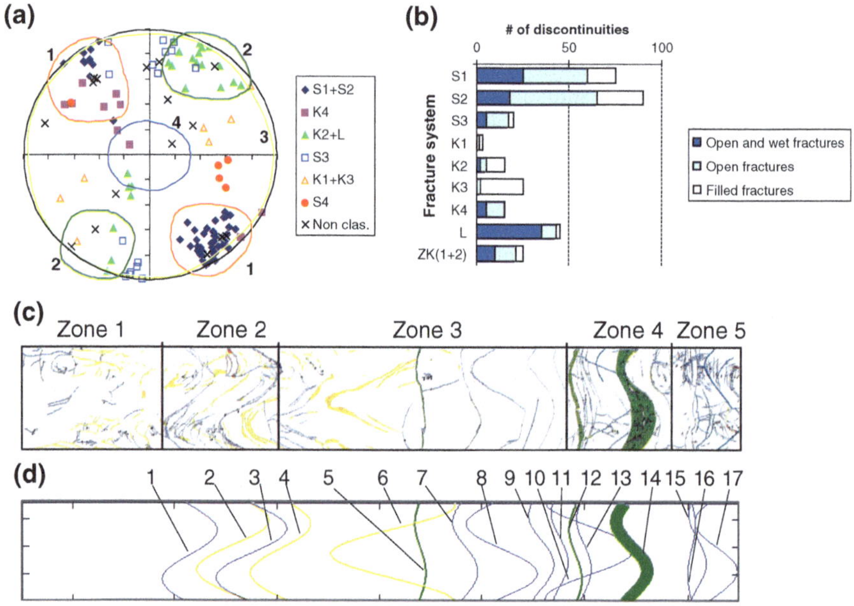

Fig. 1 Geomorphological data from the exploratory boreholes and the drift wall: **a** Stereonet of the fractures intersecting the exploratory boreholes (Fracture systems described in [6]: ductile (S1, S2 and S3), brittle (S4/K4, K2/L, K1, K3 and S5) and tensile (ZK1 and ZK2)) and the 4 families used in the simulation; **b** Number of discontinuities 'filled', 'open' and 'open-wet' from each fracture system in the exploratory boreholes; **c** Measured trace map of the drift wall with the division into five different density zones; and **d** Seventeen main fractures from deterministic simulation

- *Fracture density*. We have used as a calibration parameters the p_{21} density (trace length/intersecting plane surface) coming from the cylindrical trace map of the FEBEX drift. Moreover, an inhomogeneous p_{21} density has been reproduced indirectly by considering five different zones in the trace map (Fig. 1c).
- *Fracture location*. A homogeneous Poisson process has been used to define the center coordinates (x_{cf}, y_{cf}, z_{cf}) for each fracture 'f' within the bounds of the generation domain. This distribution has been locally modified by a specific algorithm to adjust the inhomogeneity of fracture density along the gallery.
- *Fracture size*. The power law distribution (Pareto PDF) was adjusted for fracture size. From the three parameters of this distribution (R_{MIN}, R_{MAX}, and exponent b), only one was fixed. Indeed, the maximum radius R_{MAX} was set to 100 m to ensure the existence of large 'domain-crossing' fractures. The other two parameters were fitted by the optimization algorithm. The main fractures observed in the near field (those with full traces in the gallery wall trace map) were simulated deterministically to reproduce the hydraulic behavior in the tunnel (Fig. 1d).

3 Optimization Method and Results

A Monte-Carlo algorithm has been implemented to reconstruct stochastically the synthetic fractured medium. An optimization procedure based on simulated annealing [5] has been used to adjust fracture size distribution so as to minimize the discrepancy between the synthetic fractured medium and the real one, according to the geologic data available. A new approach for adapting the search interval for possible new points which takes into account number of rejected and accepted points over a number of iterations has been implemented [1]. This approach obtains global minima with a lower computational cost (less number of objective function evaluations).

The objective function minimizes the χ^2 error measure of the discrepancy between the observed and simulated tunnel trace histograms (both trace lengths and 3D chord lengths). Due to the stochastic nature of the objective function, we have averaged the statistics of 3 realizations (fracture networks) to evaluate each value of the objective function. The final simulated fracture network has a total of 2 906 474 fractures, and has been obtained as one of the best realizations out of 750 generated with the optimal parameters found in the optimization process. Figure 2 shows the trace map of the simulated fractured medium, in good agreement with the measured one.

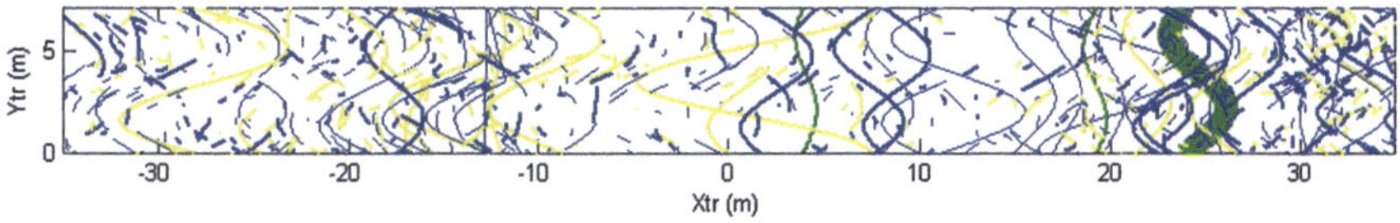

Fig. 2 Trace map of the optimally reconstructed fractured medium

4 Conclusions

We have developed a methodology to simulate a 3D fractured network that optimally fits the cylindrical tunnel trace map and other geomorphological data. The optimization procedure is based on the Simulated Annealing technique and optimizes histograms with both 2D and 3D information of the measured trace map. A good agreement between the simulated and the real fractured medium was obtained.

References

1. Goffe, W. L., Ferrier, G. D., & Rogers, J. (1994). Global optimization of statistical functions with simulated annealing. *Journal of Econometrics*, *60*(1/2), 65–100.
2. Gupta, A. K., & Adler, P. M. (2006). Stereological analysis of fracture networks along cylindrical galleries. *Mathematical Geology*, *38*(3), 233–267.
3. Keusen, H. R., Ganguin, J., Schuler, P., & Buletti, M. (1989). *Grimsel Test Site: Geology*. Nagra Technical Report, NTR 87-14E.
4. La Pointe, P. R. (2002). Derivation of parent fracture population statistics from trace length measurements of fractal fracture populations. *International Journal of Rock Mechanics and Mining Sciences*, *39*, 381–388.
5. Metropolis, N., Rosenbluth, A., Rosenbluth, M., Teller, A., & Teller, E. (1953). Equation of state calculations by fast computing machines. *Journal of Chemical Physics*, *21*, 1087–1092.
6. Pardillo, J., Campos, R., & Guimerá, J. (1997). Caracterización Geológica de la Zona de Ensayo FEBEX (Grimsel - Suiza). CIEMAT, Technical Report 70-IMA-M- 2-01.

Numerical Analysis of Stream-Groundwater Exchange in a Floodplain Following a Dike Opening

Héctor Montenegro

1 Introduction

For over a decade dike opening has been discussed as a measure to improve the connectivity between river and floodplain. Dike opening initiates intense changes in floodplain hydraulics (velocity, ponding duration etc.). Despite the importance of stream-groundwater interactions for matter and energy fluxes their dynamics are yet not well understood [1]. Their study demands for surface-subsurface observations of heads, exchange volumes before, during and after flooding events at a relevant scale.

2 Lenzen Dike Opening

The dike opening near the village of Lenzen at the River Elbe is among the largest floodplain restoration schemes in Germany. In 2009 more than 4.2 km^2 of cultivated land till then protected from inundation by levees were reconnected to the stream by a series of breaches and flood chutes cut in the existing dike. With flood rise at 900 m^3/s the stream steps from its bed and starts inundating the floodplain. Beyond a discharge of 1200 m^3/s through flow along the whole floodplain up to the new built levee occurs. An Array of 12 groundwater and 8 surface water monitoring stations were installed (see Fig. 1). Large surface flow velocities and water level changes during flooding imposed particular robust monitoring stations.

H. Montenegro (✉)
BAW Federal Waterways Engineering and Research Institute, Kussmaulstr. 17, 76187 Karlsruhe, Germany
e-mail: hector.montenegro@baw.de

E. Pardo-Igúzquiza et al. (eds.), *Mathematics of Planet Earth*,
Lecture Notes in Earth System Sciences, DOI: 10.1007/978-3-642-32408-6_103,

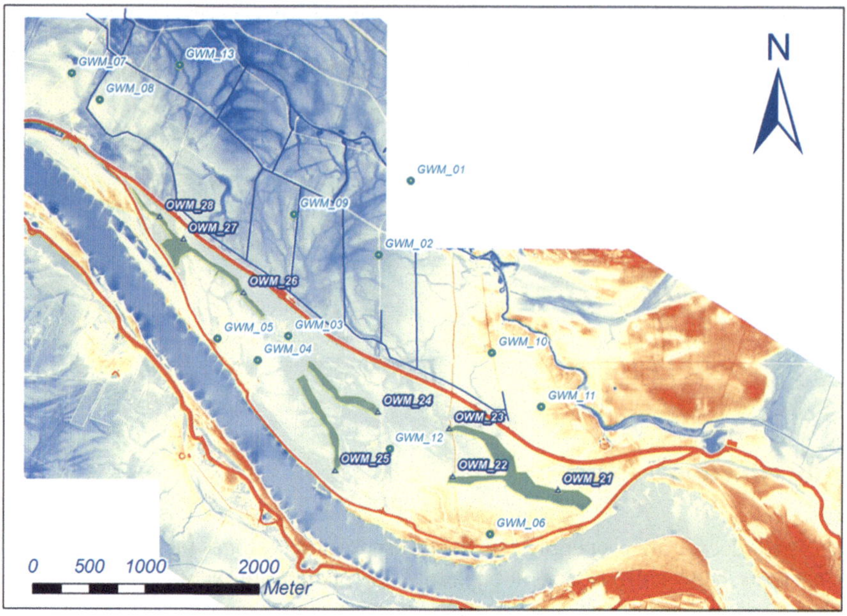

Fig. 1 Elevations (*blue*: low, *red*: high), flood chutes (*green*) and *arrays* of groundwater (GWM) and surface water (OWM) monitoring stations in the study area

2.1 Specific Floodplain Hydraulic Features

For centuries the fertile floodplains were protected from periodic inundation by dikes. An arrangement of irrigation and drainage ditches supported the balance between water drainage and water retention needs. The good hydraulic link between the surface and subsurface water bodies due to the permeable gravel and sand fluvial aquifer the river bed is cut into is typical in the study area. Another feature are the loamy fluvial deposits on top of the permeable sediments in the floodplains. Fluctuating river stages in such a layered system generate transitions from confined to unconfined groundwater flow conditions depending on the stage rise and loamy layer thickness. These shifts have an effect on groundwater flow dynamics. The reach of groundwater fluctuations in response to river stage rises, and their propagation velocity (or lag time) from the shore into the adjoining fluvial aquifer depend on the aquifer's current condition and thus varies in the curse of time depending on the actual river stage (Fig. 2).

Fig. 2 Aerial view of the study site during floodplain inundation

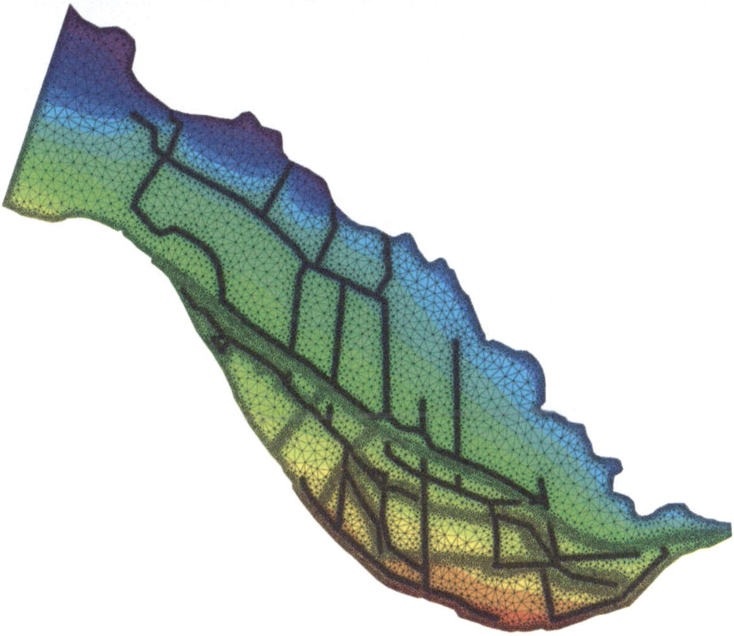

Fig. 3 Finite element discretization of the study area

3 Groundwater FE Modelling

A transient 2D groundwater flow model was set up to interpret the surface/subsurface head measurements in terms of water exchange dynamics. Exchange processes were limited to the river shores, the flood chutes and the ditch system, due to the limited permeability of the loamy layer. Based on a digital elevation model (see Fig. 1) the study site was discretized by triangular finite elements with an spatial resolution of approx. 10–100 m permitting a precise representation of the complex geometry of shoreline, flood chutes and the ditch system (see Fig. 3). The representation of the transition between confined and unconfined conditions proved essential. A suitable consideration required the spatial distribution of the loam layer thickness which was interpolated from existing soil boring logs.

Conductivity and storage properties as well as leakage coefficients were estimated based on observations from a period with no surface water inflows. Based on that parameter set a flooding event was simulated. The event consisted of a stage with increased groundwater exfiltration to the chutes/ditches (rising flood) followed by a surface inflow from the Elbe (wave peak) and a final infiltration phase from the chutes/ditches to the aquifer. The surface inflow from the Elbe was treated as an external boundary condition according to the observed surface water heads [2]. Despite considerable simplifications the model was able to reproduce the general groundwater and exchange dynamics reasonably.

References

1. Friese, K., Witter, B., Miehlich, G., & Rode, M. (2000). *Stoffhaushalt von Auenökosystemen Böden und Hydrologie, Schadstoffe Bewertungen*. Berlin: Springer.
2. Montenegro, H., & Litwora, A. (2011). Untersuchung des Wirkungszusammenhangs zwischen Abflussdynamik und Grundwasser in Flussauen. Retrieved from http://vzb.baw.de/publikationen.php?file=kolloquien/3/Tagungsband_Geohydraulische%20Erkundungsverfahren_221111.pdf.

Models of Parallel Linear Reservoirs (PLR) with Watershed Traversal Algorithm (WTA) in Behaviour Research of Hydrological Processes in Catchments

Jesús Mateo Lázaro, José Ángel Sánchez Navarro, Vanesa Edo Romero and Alejandro García Gil

1 Introduction

Here we present a computer procedure of identifying and routing of basins in digital elevation models, whether grid or raster type. The method can be useful in hydrology and geomorphology applications, among other disciplines, because inside its general structure, it supports models of hydrological processes such as PLR models, allowing an accurate investigation of actual episodes.

2 Methodology

The WTA algorithm intends to cover all cells in the DEM in an ordered way. These cells within a given catchment differentiate or distinguish the watershed and its cells from the rest of the DEM. The flow chart in Fig. 1 shows the structure of the algorithm (in [1] there is a further explanation). For a given root node, all nodes in its catchment must be run orderly. In Fig. 1, for the node 1, the algorithm follows the path indicated by the arrows, blue arrows for up direction and red arrows for down direction.

J. M. Lázaro (✉) · J. Á. S. Navarro · V. E. Romero · A. G. Gil
Departamento de Ciencias de la Tierra, Universidad de Zaragoza,
Pedro Cerbuna, 12, 50009 España, Zaragoza, Spain
e-mail: jesmateo@unizar.es

J. Á. S. Navarro
e-mail: joseange@unizar.es

V. E. Romero
e-mail: vanesa@eid.es

A. G. Gil
e-mail: agargil@unizar.es

E. Pardo-Igúzquiza et al. (eds.), *Mathematics of Planet Earth*,
Lecture Notes in Earth System Sciences, DOI: 10.1007/978-3-642-32408-6_104,

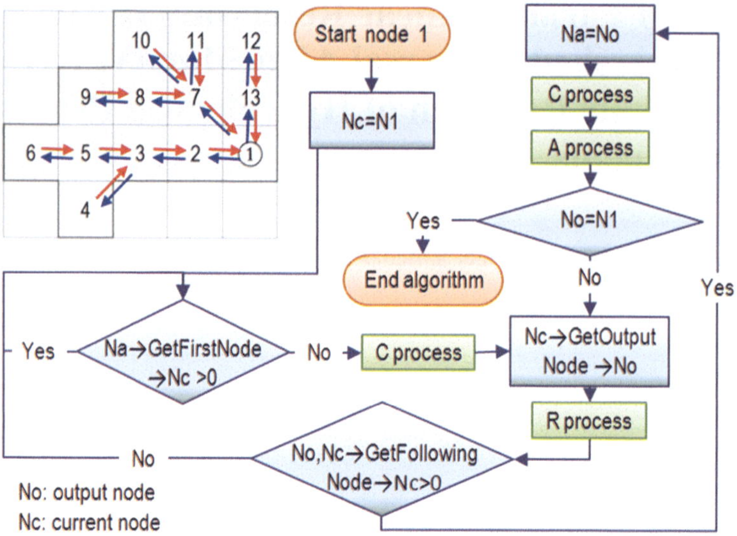

Fig. 1 Graphic routing of a watershed and its flow chart (*blue boxes*) are represented. Inside the structure, models of hydrological processes are introduced (*green boxes*). The C process is the curve number model, whereas A represents the PLR model, both applied to the grid cells. R is a routing model (e.g. Muskingum), applied to the nodes connections

With this procedure, all cells have been covered following a logic sequence with the following qualities: (1) Other algorithms can be created by introducing models, placing them in the right places. (2) These new models will be able to obtain the necessary data from the right places and provide data to other models. (3) There are two traversal directions, upslope and downslope. Depending on the type of calculation which performs a particular procedure, one direction or another will be the most representative of the hydrologic processes. (4) When a node is passed for the last time in a downward direction, all nodes in the watershed which were located above it will be covered; when applying measures or models, the influence of all nodes is thus perceived. (5) When a node is reached in an upward direction, the influence of all nodes of the main flow to the first point in the watershed is perceived.
A special case is the use of PLR (Parallel Linear Reservoir) models, where every DEM cell is considered as a reservoir combination in parallel. Water flows through each deposit at different flow rates [2]. These models are calibrated by observing the recession curves of real hydrographs and they are very interesting because they allow the establishment of a precise water balance. The hydrological relations of a lineal deposit are driven by two equations: flow or deposit equation and continuity or water balance equation. Their combination results in the runoff equation or discharge Eq. (1):

$$Q_2 = Q_1 \cdot e^{-\alpha\Delta t} + \mathbf{R} \cdot \left(1 - e^{-\alpha\Delta t}\right) \tag{1}$$

With regard to DEM characteristics, the used program, SHEE [1], can manage any format, size, accuracy and reference system. E.g., Global DEM has been used like STRM30 (3.6 Gb and 30″), MDT5 of spanish territory (120 Gb and 5 m), Lidar, etc.

3 Application to Actual Events in the Bergantes River

The Bergantes watershed has been selected because of its data availability and its natural state of conservation. It is a Mediterranean 1,050 km^2 basin, with a 84 km long main channel. A GRID size of 20 m was used.

The investigation of real episodes followed these steps: (1) creation of a total rainfall distributed model, (2) determination of the reservoir parameters, (3) determination of the effective rainfall, which can be deduced from rainfall and streamflow measurements of the rainfall and flow registers, (4) establishment of the water balance, (5) deduction of the Curve Number model, (6) generation of effective rainfall hyetographs, and (7) obtaining of simulated hydrographs. Figure 2 shows hyetographs and hydrographs for two episodes.

The fast reservoir (dash line) increases and decreases quickly, whereas the slow one (dotted line) is secondarily involved in the most intense stretch but it persists during the whole recession.

The slow reservoirs in the stretch with higher discharge have a significant influence. The subsurface and groundwater flows and other characteristics of the basin are relatively quickly activated and produce a significant contribution to the discharge. For example, there are cases where one-third of the discharge is produced during the peak flow.

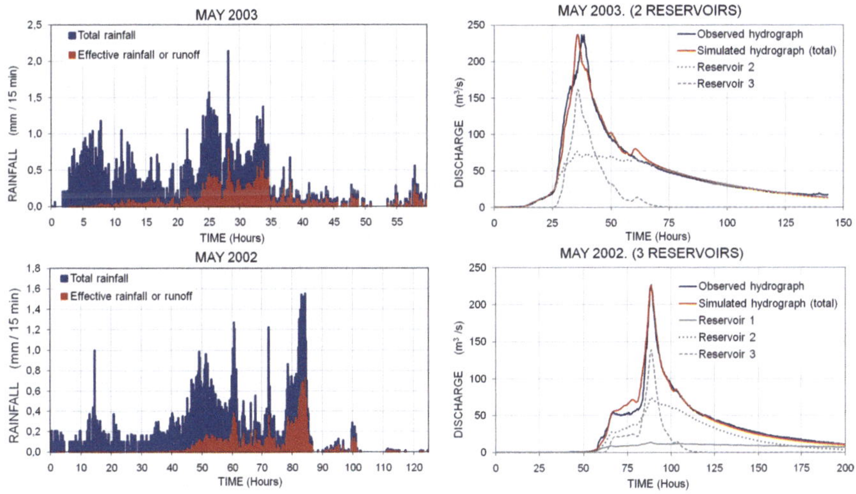

Fig. 2 Hyetographs of total and effective rainfall, and real and simulated hydrographs for two real events in Bergantes watershed

4 Conclusions

The main tool presented runs the entire drainage network of a watershed in both directions, upwards and downwards, which is suitable for entering inside its structure, models of hydrological processes that occur in basins or assessing its characteristics. The main attribute of this method is that information about hydrological processes and properties is transmitted during the routing from one area to another of the basin. The simulation using PLR models represents and divides the different hydrological processes (e.g., surface and groundwater flow, snow reservoirs), which allows an observation of the different contributions from each reservoir. This type of observation makes it possible to define a partial hydrograph for each reservoir in addition to the total hydrograph. The partial hydrograph analysis and its relation to the total hydrograph provide relevant information relating to the hydrological behaviour of the watersheds and their processes.

References

1. Mateo-Lázaro, J., Sánchez-Navarro, J. A., García-Gil, A., & Edo-Romero, V. (2013). Developing and programming a watershed traversal algorithm (WTA) in GRID-DEM and adapting it to hydrological processes. *Computers and Geosciences*, *51*, 418–429.
2. Mateo-Lázaro, J., Sánchez-Navarro, J. A., García-Gil, A., & Edo-Romero, V. (2012). Research the Snowmelt in Floods. (I) Adapting to the Linear Reservoir Models. (II) A real Event Assessment in a Gauged Catchment of the Pyrenean Mountains: Roncal Valley, Spain. 8.299-300 O communications. Geo-Temas, 13: VIII Geological Congress of Spain.

An Analytical Solution of Tide-Induced Head Fluctuations in an Inhomogeneous Coastal Aquifer

Luis Guarracino and Leonardo Monachesi

1 Introduction

The interaction between groundwater and seawater induced by sea tides has been extensively analyzed through analytical solutions. Since the 1950s, many analytical solutions have been obtained for both single confined aquifers and two-layer aquifer systems (e.g. [1, 2]). However, the analysis of heterogeneity has been limited to piecewise constant values of hydraulic parameters (e.g. [3, 4]). In a recent work by Monachesi and Guarracino [5], exact and approximate analytical solutions are derived for a confined aquifer with hydraulic conductivity that linearly increases with the distance to the coastline. The continuous increase of hydraulic conductivity can be useful for studying alluvial coastal aquifers. In this type of aquifer, progressively finer sediments are usually deposited on the downstream part of the depositional zone, giving as a result a continuous increase of hydraulic conductivity in the upstream direction. To the authors' knowledge, these analytical solutions are the only ones that consider a continuous variation of hydraulic properties with distance.

2 Mathematical Model

Consider a confined aquifer that ends at the coastline and extends landward infinitely as shown in Fig. 1. In order to derive an analytical solution, the following assumptions are made: the datum of the induced head fluctuation is the mean water level; water

L. Guarracino (✉) · L. Monachesi
CONICET, Facultad de Cs. Astronómicas y Geofísicas, Universidad Nacional de La Plata, Paseo del Bosque s/n, 1900 La Plata, Argentina
e-mail: luisg@fcaglp.unlp.edu.ar; lguarracino@hotmail.com

L. Monachesi
e-mail: lmonachesi@fcaglp.unlp.edu.ar

E. Pardo-Igúzquiza et al. (eds.), *Mathematics of Planet Earth*,
Lecture Notes in Earth System Sciences, DOI: 10.1007/978-3-642-32408-6_105,

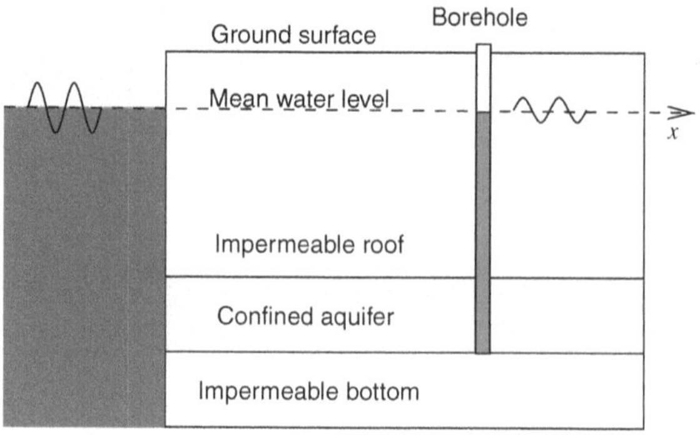

Fig. 1 Schematic representation of the confined coastal aquifer. The x-axis is horizontal and positive landward with its origin at the coastline

flow is horizontal and obeys Darcy's law; the effect of density variations on water flow is neglected; the hydraulic conductivity K increases quadratically with the distance; the specific storativity S_s is constant.

According to the above assumptions, tide-induced head fluctuations are described by the following boundary-value problem:

$$\frac{\partial}{\partial x}\left(K(x)\frac{\partial h}{\partial x}\right) = S_s\frac{\partial h}{\partial t}, \quad x \in [0, \infty), \quad t \in (-\infty, \infty) \tag{1}$$

$$h(0, t) = A\cos(\omega t), \quad t \in (-\infty, \infty) \tag{2}$$

$$\lim_{x\to\infty} K(x)\frac{\partial h}{\partial x} = 0, \quad t \in (-\infty, \infty). \tag{3}$$

where $h(x, t)$ is the water head, A the amplitude, and ω the angular frequency of the sea tide. Note that only the periodic solution is considered, so no initial condition is needed to solve the differential problem (1)–(3). The hydraulic conductivity is assumed to have the following expression:

$$K(x) = K_0\,(1 + bx)^2 \tag{4}$$

where K_0 is the hydraulic conductivity at the coastline and b is the rate of increase.

3 Analytical Solution

Let $H(x, t)$ be a complex function that satisfies (1)–(3) after the boundary condition (2) is replaced by $H(0, t) = Ae^{i\omega t}$. Because $h(x, t)$ is the solution of (1)–(3), it follows that $h(x, t) = \Re\,[H(x, t)]$, where $\Re$ denotes the real part of the complex expression.

Substituting $H(x,t) = Ay(x)e^{i\omega t}$ into (1)–(3) yields

$$(1+bx)^2 \frac{d^2y}{dx^2} + 2b(1+bx)\frac{dy}{dx} - i2a^2y = 0 \quad x \in [0,\infty) \tag{5}$$

$$y(0) = 1 \tag{6}$$

$$\lim_{x\to\infty} K_0 (1+bx)^2 \frac{dy}{dx} = 0 \tag{7}$$

where $y(x)$ is a complex function and $a = \sqrt{\frac{\omega S_s}{2K_0}}$ is the tidal propagation parameter.

The solution of the differential problem (5)–(7) leads to the solution of the original differential problem (1)–(3):

$$h(x,t) = Ae^{-0.5(p+1)\ln(1+bx)} \cos(0.5q\ln(1+bx) - \omega t) \tag{8}$$

where $p = \frac{1}{\sqrt{2}}\sqrt{\sqrt{1+2^6\left(\frac{a}{b}\right)^4}+1}$ and $q = \frac{1}{\sqrt{2}}\sqrt{\sqrt{1+2^6\left(\frac{a}{b}\right)^4}-1}$.

Note that when $b \to 0$ the analytical solution (8) becomes

$$h(x,t) = Ae^{-ax}\cos(ax - \omega t) \tag{9}$$

which is the classical solution for a homogeneous aquifer obtained by Jacob [1].

Figures 2 and 3 show, respectively, the amplitudes and time-lags of (8) for different rates of increase of hydraulic conductivity b for a hypothetical example. In comparison with the homogeneous model, the quadratic heterogeneity produce more dampened amplitudes near the coastline. The dampening effect is almost negligible

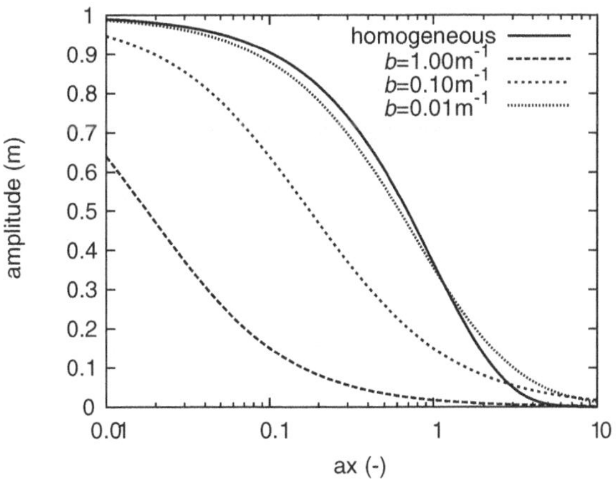

Fig. 2 Amplitude versus dimensionless distance ax for different rates of increase of hydraulic conductivity. The sea tide is considered semi-diurnal (period of 12.4 h) with an amplitude $A = 1$ m. The hydraulic parameters of the aquifer are assumed to be $S_s = 10^{-3}$ m^{-1}, $K_0 = 0.8$ m/h

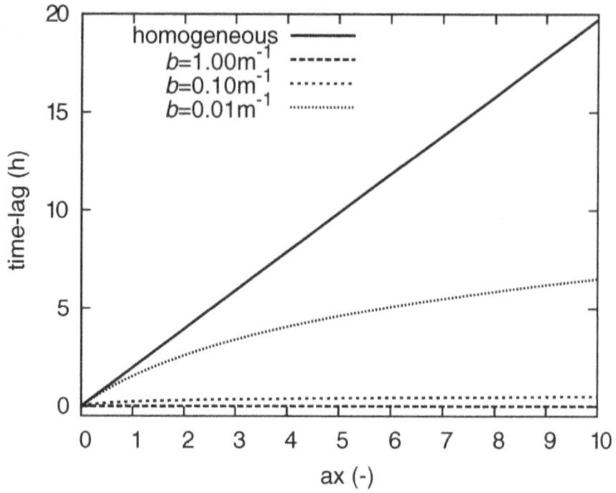

Fig. 3 Time-lag versus dimensionless distance ax for different rates of increase of hydraulic conductivity. The sea tide is considered semi-diurnal (period of 12.4 h) with an amplitude $A = 1$ m. The hydraulic parameters of the aquifer are assumed to be $S_s = 10^{-3}$ m^{-1}, $K_0 = 0.8$ m/h

for small values of b. The time-lags are smaller that the time-lag of the homogeneous aquifer, which shows a linear increase with the distance.

The proposed solution is used to estimate the hydraulic conductivity field in a coastal aquifer on Sapelo Island, Georgia [6]. The dataset consists of amplitude and time-lag values estimated from water level fluctuations recorded in 8 monitoring wells. The parameters of the hydraulic conductivity model (4) are estimated by fitting the amplitude and time-lag of (8) to the field dataset using an exhaustive search method. The estimated parameters are $K_0 = 0.3632$ m/h and $b = 0.9572$ m^{-1}, which are consistent with the high degree of heterogeneity reported in the study site.

4 Conclusions

A new analytical solution that predicts tide-induced head fluctuations in a heterogeneous coastal aquifer is presented. The analytical solution provides closed-form expression for both the amplitude and time-lag. The conceptual model assumes a hydraulic conductivity field that increases quadratically with the distance to the coastline. This type of heterogeneity produces dampened amplitudes near the coastline and a faster transmission of induced fluctuations into the aquifer. The applicability of the proposed solution is demonstrated using field data from literature.

References

1. Jacob, C. E. (1950). Flow of groundwater. In H. Rouse (ed.), *Engineering hydraulics* (pp. 321–386). New York: Wiley.
2. Li, H. L., & Jiao, J. J. (2002). Analytical solutions of tidal groundwater flow in coastal two-aquifer system. *Advances in Water Resources*, *25*, 417–426.

3. Guo, H., Jiao, J. J., & Li, H. (2010). Groundwater response to tidal fluctuations in a two-zone aquifer. *Journal of Hydrology*, *381*, 364–371.
4. Chuang, M. H., Huang, C. S., Li, G. H., & Yeh, H. D. (2010). Groundwater fluctuations in heterogeneous coastal leaky aquifer systems. *Hydrology and Earth System Sciences*, *14*, 1819–1826.
5. Monachesi, L. B., & Guarracino, L. (2011). Exact and approximate analytical solutions of groundwater response to tidal fluctuations in a theoretical inhomogeneous coastal confined aquifer. *Hydrogeology Journal*, *19*, 1443–1449.
6. Schultz, G., & Ruppel, C. (2002). Constrains on hydraulic parameters and implications for groundwater flux across the upland-estuary interface. *Journal of Hydrology*, *260*, 255–269.

Mapping of Flood-Plain by Processing of Elevation Data from Remote Sensing

Viviana Aguilar Muñoz and Márcio de Morisson Valeriano

1 Introduction

The relief is a key component in studies of the physical world, since this is both an indicator and a condition of environmental processes (geological, hydrological, climatic, etc.), besides influencing the availability and distribution of resources necessary for survival (soil nutrients, water, etc.) [1]. Thus, the contribution of mathematical modeling of the topographic surface for application in environmental research can be valued in a wider sense than merely geometric [2]. Relief metric descriptors like slope and vertical curvature have been reported in the literature as topographic variables [3]. These can be obtained by computational manipulation of elevation data, represented in Digital Elevation Models (DEM), by means of geomorphometric techniques [4].

A specific methodology was developed to compute topographic variables related to soil distribution: dissection deepness, dominancy, internal relief and topographic height [5], which can be adapted to calculate the relief-elevation index [6]. The pur-pose of this experiment is to apply that methodology for relief mathematical model-ing of a Brazilian watershed where every year are recorded numerous losses from floods, and to evaluate the potential of each of these variables for mapping of floodplain areas.

V. A. Muñoz(✉) · M. de Morisson Valeriano
National Institute For Space Research—INPE, Av. dos Astronautas 1758, São José dos Campos, CEP 12201-970, Brazil
e-mail: viaguila@dsr.inpe.br

V. Aguilar Muñoz
PhD CAPES/CNPq—IEL National—Brazil, São José dos Campos, Brazil

M. de Morisson Valeriano (✉)
CNPq Researcher, São José dos Campos, Brazil
e-mail: valerian@dsr.inpe.br

E. Pardo-Igúzquiza et al. (eds.), *Mathematics of Planet Earth*,
Lecture Notes in Earth System Sciences, DOI: 10.1007/978-3-642-32408-6_106,

2 Materials and Method

This experiment was developed on Itajaí River basin, located in the Brazilian South Atlantic hydrographic region. This site is bounded within the coordinates from 50°21'17"W to 48°39'04"W and from 26°22'44"S to 27°51'02"S. The basin's re-lief is characterized by steep hills at the south, west and northwest and plains bound-ed by the Atlantic Ocean at the northeast. Elevation ranges between zero and approximately 1700 m.a.s.l. DEM from Topodata database [7] was taken as input. This model represents a geo-statistic refinement (90–30 m) of SRTM-C, available to Brazilian territory [8].

The methodology for computing the topographic variables [6] using Geographic Information Systems (GIS) tools consisted on applying differential filters, followed by arithmetic and logical operations on the DEM. New elevation surfaces were reconstructed from maximum, medium and minimum quotas identified over the DEM by mathematical inequalities of elevation quotas, selected within radial distances of approximately 810 m (27 × 27 pixels windows mobile). The arithmetic combination of these information layers (surfaces) follows the equation set of the Table 1.

3 Results and Discussion

Figure 1 shows an excerpt of topographic variables of the Itajaí basin. It is noted that each of these descriptors enhances some particular characteristic of relief. Floodplains were recognized in this work as lower, deeper and wider lands, with low dominance and less volume of available material for erosion.

Table 1 Mathematical models for computing the topographical variables of this work

Variable	Description	Equation	
Dissection deepness	Vertical distance from the elevation maximal surface to a point on the topography surface	$D_p = Z_{max} - Z$	(1)
Dominancy	Vertical distance from a point on the topography surface to the average elevation surface	$P = Z - Z_{med}$	(2)
Internal relief	Vertical distance from a point on the elevation maximal surface to a point on the elevation minimal surface	$I = Z_{max} - Z_{min}$	(3)
Relief-elevation index	Ratio between the vertical distance from the average elevation surface to the minimal and the internal relief	$E = \frac{Z_{med} - Z_{min}}{Z_{max} - Z_{min}}$	(4)
Topographic height	Vertical distance from a point on the topography surface to elevation minimal surface	$H = Z - Z_{min}$	(5)

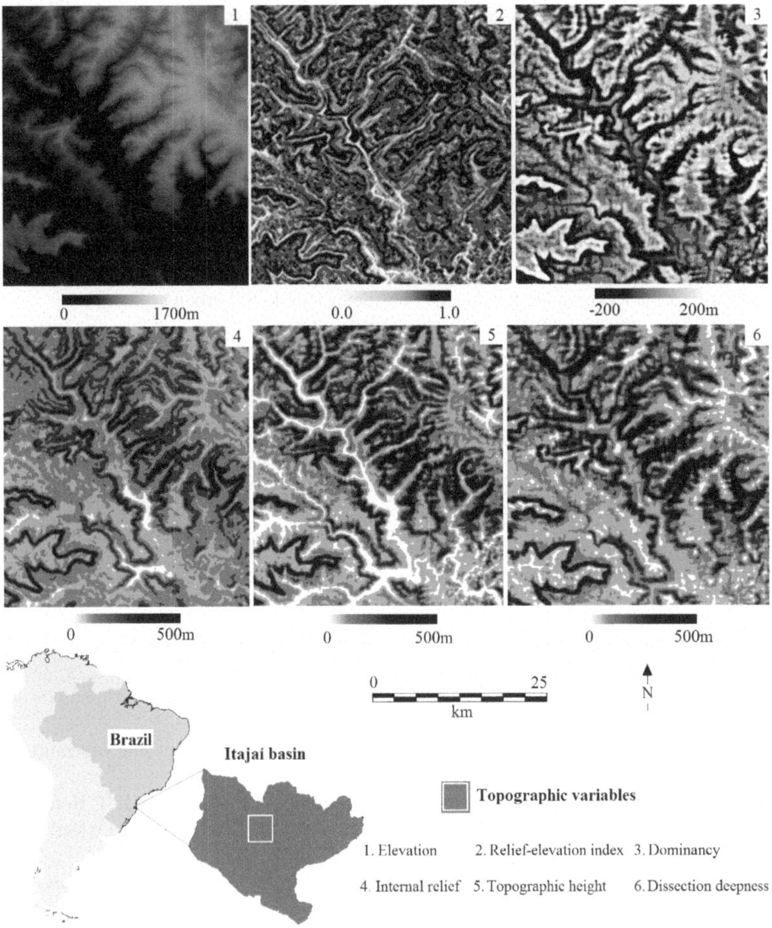

Fig. 1 Excerpt of topographic variables extracted from elevation data of Itajaí basin, Brazil

Figure 2 shows an excerpt of the topographic height besides an urban grid within the basin. It is possible to observe that Río do Sul has been developing along a floodplain. This city is frequently affected by floods, and for this reason the Civil Defense has declared public emergency several times [9].

4 Conclusions

It was possible to extract new information from relief, important for environmental studies, by the manipulation of elevation data through algorithms based on relatively simple mathematical models. Its observation allows us to conclude that the

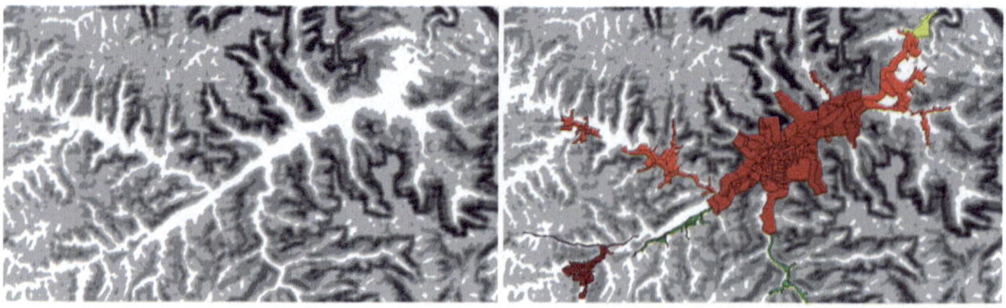

Fig. 2 Excerpt of the topographic height (*left*) beside of Rio do Sul urban grid (*right*)

topographicBreak height highlights the floodplains better than the other variables, although each contributes to its mapping. Development of urban areas on the floodplain basin indicates a high degree of public exposure to flooding in the Itajaí basin.

References

1. Guisan, A., & Zimmermann, N. E. (2000). Predictive habitat distribution models in ecology. *Ecological Modelling*, *135*, 147–186.
2. Shary, P. A., Sharaya, L. S., & Mitusov, A. V. (2002). Fundamental quantitative methods of land surface analysis. *Geoderma*, *107*, 1–32.
3. Florinsky, I. V., & Kuriakova, G. A. (1996). Influence of topography on some vegetation cover properties. *Catena*, *27*, 123–141.
4. Pike, R. J., Evans, I. S., & Hengl, T. (2009). Geomorphometry: a brief guide. In T. Hengl & H. I. Reuter (Eds.), *Geomorphometry: Concepts, Software, Applications* (pp. 3–30). Oxford: Elsevier B. V.
5. Muñoz, V. A. (2009). *Análise geomorfométrica de dados SRTM aplicada ao estudo das relações solo-relevo*. Dissertação (Mestrado em Sensoriamento Remoto) - Instituto Nacional de Pesquisas Espaciais: São José dos Campos.
6. Pike, R. J., & Wilson, S. E. (1971). Elevation-relief ratio, hypsometric integral, and geomorphic area-altitude analysis. *Geological Society of America Bulletin*, *82*, 1079–1084.
7. Topodata: Banco de Dados Geomorfométricos do Brasil. http://www.dsr.inpe.br/topodata/index.php; http://www.webmapit.com.br/inpe/topodata/.
8. Valeriano, M. M., & Rossetti, D. F. (2012). Topodata: Brazilian full coverage refinement of SRTM data. *Applied Geography*, *32*, 300–309.
9. Defesa Civil em Santa Catarina: http://www.defesacivil.sc.gov.br/.

Effect of Spatial Heterogeneity on Rate of Sedimentary O_2 Consumption Reaction

Tanushree Dutta and Simonetta Rubol

1 Introduction

Quantification of soil hotspots and hot-moments is one of the main challenges to understand the variability encountered in soils in terms of respiration activity and trace gas emissions [1]. At the current state of the art, determination of the link between the biological and soil heterogeneity is a challenge [2]. The latter is also a function of the scale at which the phenomena are observed [3]. However this target is difficult given the lack of instrument to measure spatial variability in bacteria activity at high frequency. In order to fill this gap and to improve the current knowledge, our work aim in studying the temporal evolution of oxygen (O_2) consumption maps resulting from the addition of DI water, humic acid and glucose. The overall objective of this study is to investigate the link between the spatial average consumption rate and land use and also to trace the spatial variability of O_2 for a given time point.

2 Materials and Methods

2.1 Study Sites

For this study, fresh sediments were collected from sites under contrasting land use namely forest (three, old, well developed and recent forests), cropland (two, conventional and non-conventionally treated), swamp, riverbed and a fallow land

T. Dutta (✉) · S. Rubol
Hydrogeology Group, Departmen of Geotechnical Engineering and Geosciences, Universitat Politècnica de Catalunya-BarcelonaTech, 08034 Barcelona, Spain
e-mail: tanushree.dutta@upc.edu

S. Rubol
Department of DICA, University of Trento, via Mesiano 77, 38123 Trento, Italy
e-mail: simonetta.rubol@upc.edu

E. Pardo-Igúzquiza et al. (eds.), *Mathematics of Planet Earth*,
Lecture Notes in Earth System Sciences, DOI: 10.1007/978-3-642-32408-6_107,

(with weeds) adjacent to one of the croplands. In addition to land use, the sites differ in their location, vegetation and soil properties. Details of the location of the sites, vegetation and management are not shown here (available on request).

2.2 Sediment Characterizations and Sample Preparation

Fresh samples were brought to the laboratory and stored at 4 °C. For analysis, fresh un-sieved sediments were used to avoid the soil disturbances caused by sieving. The sediments were screened for roots and gravels and homogenized to obtain replicate samples for analysis. A sub-sample was oven-dried (105°) for 24 h for the measurement of gravimetric moisture content. Another subset of samples were airdried and sieved (<2 mm) for analysis of organic C, total N, pH, cations and anions . For ionic analysis, soils were extracted by shaking a 3 g (fresh wt) subsample with 30 mL of 2M KCl for 1 h and filtering through Whatman no. 1 filter paper following the method described in [4].

2.3 Preparation of Substrate Media

Substrate media capable of providing 100 mg $NO_3^- - Nkg^{-1}$ and 40 mg glucose or humic acid kg^{-1} soil were prepared using DI water. The concentration of the substrates corresponded to the substrate concentration used for measuring the maximum activity of the biomass of denitrification enzymes [5]. DI water with no substrate was used as control treatment for comparison.

2.4 Experimental Set Up

Visisens sensor foils were glued on the bottom of the BD Falcon™ six-well plates (area = 9.6cm^2), which were then filled with 3 g of fresh homogenized (un-sieved) soil. Each well was placed on the top of the Visisens (Presens) optical sensor for monitoring O_2 consumption. Soil O_2 maps (1 pixel = 9.2 μm) were obtained using the time drive measurement feature of the Visisens analytical software every 30 s for 40 mins after the addition of substrate. Replicated were also performed with oven-dried soil, to check that no O_2 consumption was detected by the Visisens instrument (data available upon request). Each sediment type (representing different land use) was replicated three times for each substrate. Substrate media was added homogeneously to the surface of the soils into the wells using pipette.

3 Results

3.1 O_2 Spatial Variability for Different Substrate and Land Use

Oxygen consumption was in general highly variable among replicates, indicating the fundamental role of spatial variability across the field site. However, for all the replicate, the O_2 consumption following glucose addition resulted in a greater reduction (80 %) of the initial O_2 values than the one measured after DI and humic acid addition. The latter resulted respectively in no reduction and 20 % reduction of the initial value. In addition, for each treatments, the O_2 consumption for the cropland (soil E) and the fallow land (soil G) were significantly higher than the ones measured in river sediment, forest and swamp, although the initial oxygen content was in general higher in the forest and the river sediments (about 100 %) than in the other soil types (around 80 %). Fig. 1 illustrates the spatial variability measured across land use after adding glucose, where each points of the curves has been obtained by averaging the values of the O_2 map recorded every 20 s for 40 mins.

3.2 O_2 Maps and Soil Heterogeneity

Figure 2 shows the oxygen maps measured at t = 0, 10, 20, 30 and 40 mins after glucose addition for all the soil types. The figure clearly shows that the oxygen is consumed non-uniformly in the area of soil investigated, with hotspots in which the oxygen values span along a wide range, thus permitting the coexistence of aerobic and anaerobic condition within a small portion of the soils. Since the diameter of the hotspots is in general bigger than the size of bacteria (i.e. 0.1–1 μm), our results suggest that this O_2 gradient may indicate the presence of different colonies which can be tolerant to rapid change in O_2 values. To the best of our knowledge, this is the first work, which is able to investigate the spatial variability of O_2 in sediments at

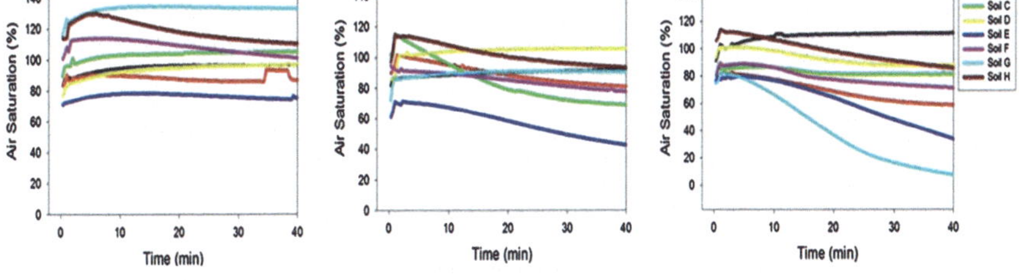

Fig. 1 Temporal evolution of spatial average O_2 map consumption following, **a** no substrate addition, **b** humic acid addition and **c** glucose addition

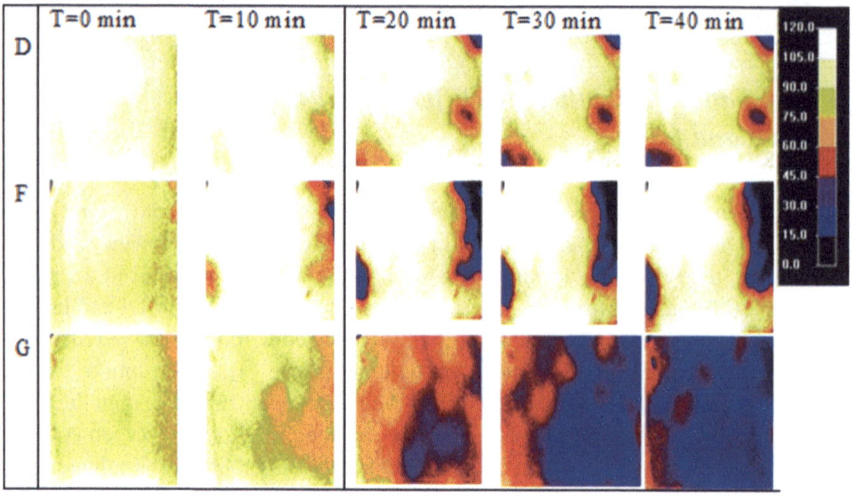

Fig. 2 O_2 maps showing spatial variability and hotspots of O_2 consumption following glucose addition at time (T) = 0, 10, 20, 30 and 40 mins

high frequency. The data suggest that even on a very small spatial scale the O_2 may vary in the soil. Determining a model to predict the soil variability is challenging, however the averaging operation in modeling has to be done with caution, since in a small portion of soil (e.g. 1 cm^2), respiration rate may span on a wide range of values. We believe that these types of analysis may provide the range of variability across soil type. In our specific analysis, we found that the cropland soil presented more homogenous oxygen consumption along the soil surface with respect to the other soils. This may indicate the presence of a more homogeneous soil, where the substrates diffuse more uniformly.

Acknowledgments SR acknowledges the support of the Provincia Autonoma di Trento and the European Commission within the 7° Programma Quadro–Azioni Marie Curie Cofund, PAT.

References

1. Rubol, S., Silver, W. L., & Bellin, A. (2012). Hydrologic control on redox and nitrogen dynamics in a peatland soil. *The Science of the total environment*, *432*, 37–46.
2. Groffman, P. M., Butterbach-Bahl, K., Fulweiler, R. W., Gold, A. J., Morse, J. L., Stander, E. K., et al. (2009). Challenges to incorporating spatially and temporally explicit phenomena (hotspots and hot moments) in denitrification models. *Biogeochemistry*, *54*, 49–77.
3. Sanchez-Vila, X., Rubol, S., Carles-Brangaria, A., & Fernàndez-Garcia, D. (2013). An analytical solution to study substrate-microbial dynamics in soils. *Advances in Water Resources*, *54*, 181–190.

4. Bremner, J. M., & Keeney, D. R. (1966). Determination and isotope-ratio analysis of different forms of nitrogen in soils: 3. Exchangeable ammonium, nitrate and nitrite by extraction ± distillation methods. *Soil Science Society American Proceedings*, *30*, 504–507.
5. Groffman, P. M., Holland, E. A., Myrold, D. D., Robertson, G. P., & Zou, X. (1999). Denitrification. In G. P. Robertson, D. C. Coleman, C. S. Bledsoe, & P. Sollins (Eds.), *Standard soil methods for long term ecological research* (pp. 272–288). New York: Oxford University Press.

Optimal Development of Regional Rain Network Using Entropy and Geostatistics

Hadi Mahmoudi-Meimand, Sara Nazif and Hasan-Ali Faraji-Sabokbar

1 Introduction

Rainfall is hydro-meteorological variable with drastic spatial and temporal variations. In recent years, more attention is paid to application of entropy theory and geostatistical method in rain gauge network design.

Karamouz et al. [1], utilized the measure of transinformation in entropy theory for selecting the best water quality monitoring stations from a set of potential monitoring sites along a river. Yeh et al. [2] introduced a model composed of kriging and entropy with probability distribution function to relocate the rainfall network and to obtain the optimal design with the minimum number of rain gauges. Chen et al. [3] proposed a method composed of kriging and entropy that can determine the optimum number and spatial distribution of rain gauge stations in catchments.

In this study the entropy and geostatical methods are used in combination to improve the rain gauge network density.

2 Materials and Methods

In this paper, the optimal locations of rain gauges development in a monitoring network are determined using the entropy theory by considering the maximum uncertainty (minimum redundant information in the system) and the maximum rainfall estimation error based on Kriging method.

H. Mahmoudi-Meimand (✉) · H.-A. Faraji-Sabokbar
Department of Cartography, University of Tehran, Tehran, Iran
e-mail: Mahmoodi.gis@gmail.com

H.-A. Faraji-Sabokbar
e-mail: hafaraji@gmail.com

S. Nazif
School of Civil Engineering, College of Engineering, University of Tehran, Tehran, Iran
e-mail: snazif@ut.ac.ir

E. Pardo-Igúzquiza et al. (eds.), *Mathematics of Planet Earth*,
Lecture Notes in Earth System Sciences, DOI: 10.1007/978-3-642-32408-6_108,

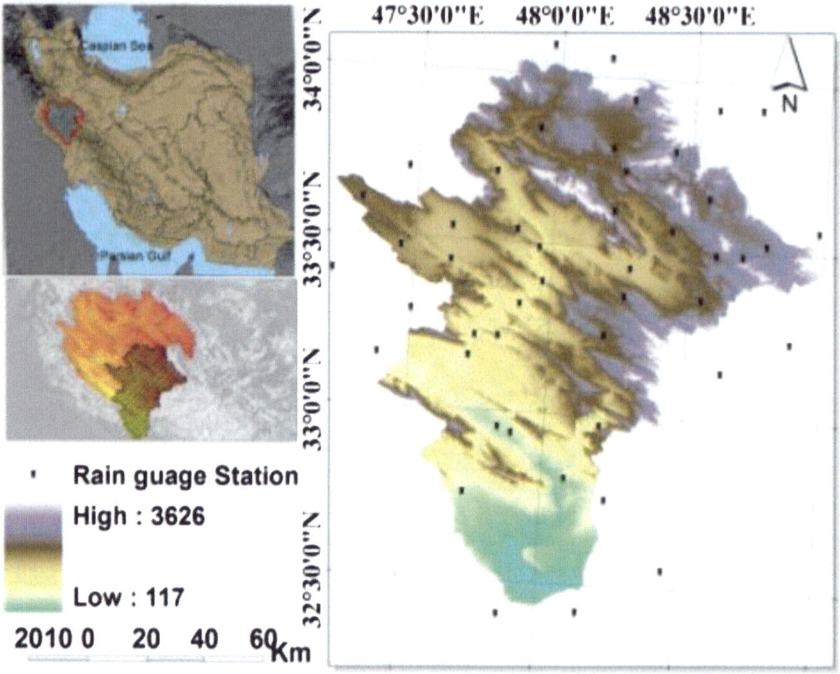

Fig. 1 Location and rain gauge of the study area

2.1 Calculating the Rainfall Estimation Variance Using Kriging and the Transformation Entropy

In this study, the rainfall is spatially interpolated using ordinary kriging. In kriging the error associated with the estimates in points with no measurements are calculated. Rainfall Kriging Variance is calculated by the following formula:

$$\sigma_{\tilde{z}}^{2} = \mu + \sum_{i=1}^{N} \lambda_i \gamma_{i0}. \tag{1}$$

where $\sigma_{\tilde{z}}^2$ is the kriging variance, which provides a measure of the error associated with the kriging estimator, μ is the Lagrange parameter and γ_{i0} is Variogram values i between the i sample and estimated spot and λ_i is the weight of rain gauge i. Due to high variability of rainfall in different months, the weighted average of monthly rainfall is estimated. The final layer of rainfall estimation error at the basin for the maximum kriging error as a measure of choice in network structure optimization model for locating new stations rain guage used.

The transferable information T(X,Y) of two rain gauge stations X and Y is the mutual information or overlapped information of X and Y, where f(x, y) is the joint

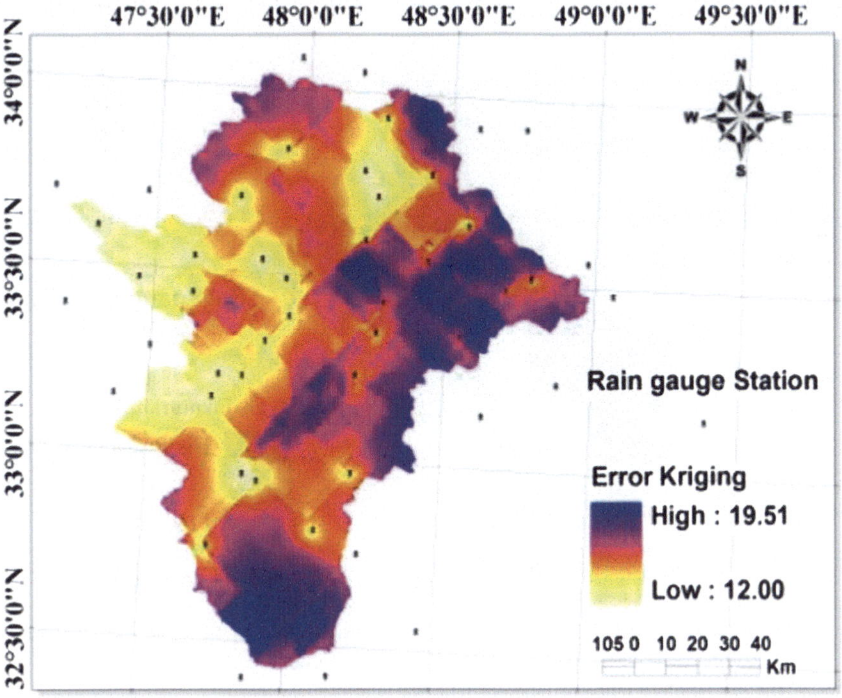

Fig. 2 The final layer of the variance estimates at the basin

entropy of x and y, f(x) and f(y) are the probability density function of the variables x and y, respectively.

Transinformation (T(x, y)) in the discrete form can be expressed as follows:

$$T(x, y) = T(y, x) = -\int_{-\infty}^{\infty}\int_{-\infty}^{\infty} f(x, y)\mathrm{Ln}\left(\frac{f(x, y)}{f(x)f(y)}\right). \qquad (2)$$

Equation (2) for variables with normal distribution can be simplified as follows [4]:

$$T(x, y) = -\frac{1}{2}\mathrm{Ln}\left(1 - r_{xy}^2\right) \qquad (3)$$

where r_{xy} is the correlation coefficient between x and y. The final layer of entropy transfer at the basin for the minimum transformation entropy as a measure of choice in network structure optimization model for locating new stations rain guage used.

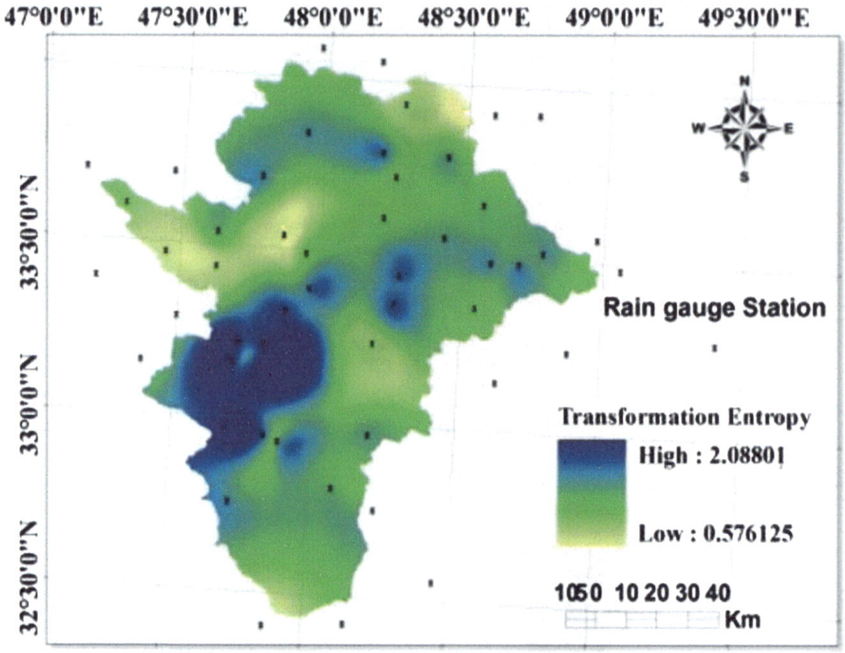

Fig. 3 The final layer of entropy transfers at the basin level

3 Rain Gauge Network Optimization Model

The objective of this model is combination of maximizing the minimum transinformation entropy and minimizing the maximum Error variance which is quantified as follows:

$$\text{Min } Z = \sum_{j=1}^{n} \sum_{i=1}^{m} a_{ij} \left\{ \alpha_1 \left(\text{Errs}_{ij} - \text{Errs}_{max} \right)^2 + \alpha_2 \left(\text{Ents}_{ij} - \text{Ents}_{min} \right)^2 \right\}. \tag{4}$$

The main constraints are as follows:

$$\begin{cases} a_{ij} = b_{ij} \times c_{ij}, b_{ij} > \frac{r_1 - \text{Ents}_{ij}}{r_1}, b_{ij} < \frac{r_1}{\text{Ents}_{ij}}, c_{ij} > \frac{\text{Errs}_{ij} - r_2}{r_2}, c_{ij} < \frac{\text{Errs}_{ij}}{r_2}, \sum_{i=1}^{2} \alpha_i = 1, \\ a_{ij}, b_{ij}, c_{ij} \in [0, 1]. \end{cases} \tag{5}$$

To standardize these criteria, the following relations are used:

$$\text{Ents}_{ij} = \frac{\text{Ent}_{ij} - \text{Ent}_{min}}{\text{Ent}_{max} - \text{Ent}_{min}} \qquad \text{Errs}_{ij} = \frac{\text{Err}_{ij} - \text{Err}_{min}}{\text{Err}_{max} - \text{Err}_{min}}$$

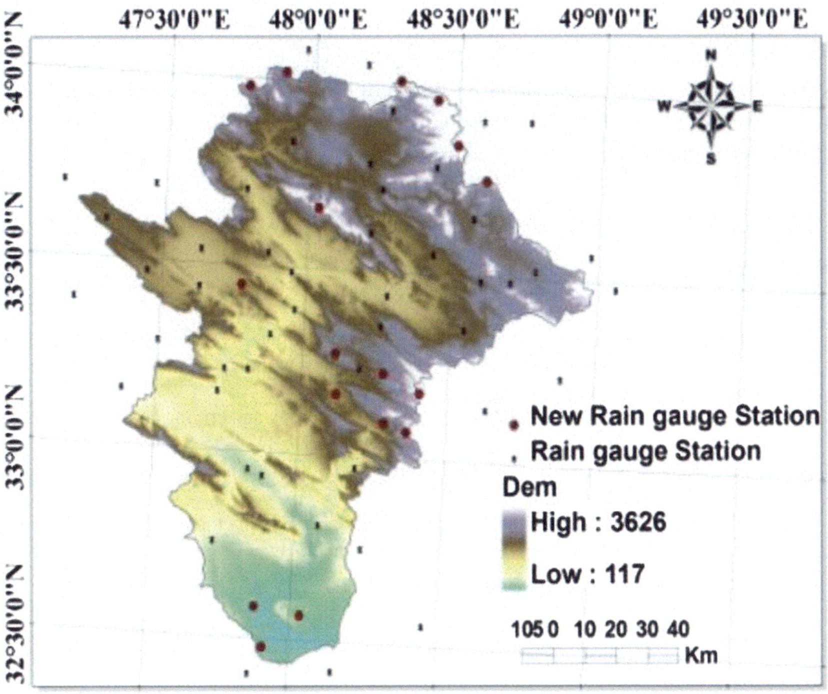

Fig. 4 Location of 17 selected rain gauge stations of the study area

where Ent_{ij} and Err_{ij} are the real value of entropy and error kriging, respectively. $Err_{max} = \max(Err_{ij})$ and $Ent_{min} = \min(Ent_{ij})$. $Errs_{max} = \max(Errs_{ij})$, $Ents_{min} = \min(Ents_{ij})$.

r_1 is the threshold value for the selected range of data entropy and r_2 is the threshold value for the selected range of data error kriging. a_{ij}, b_{ij} and c_{ij} are auxiliary binary variables. α_1 and α_2 coefficients are between 0 and 1. Their values are determined by the decision maker.

4 Results and Discussion

Rainfall data of the 49 stations in the Karkhe located at the south western part of Iran (Fig. 1) in months October to April calculated for each month. The developed layers of rainfall estimation variance are weighted based on monthly rainfall and overlaid. The final layer of the overlapping layers are obtained and as a measure of the objective function be considered (Fig. 2). The transformation entropy layer is developed in the similar way (Fig. 3). The estimation variance and entropy layers are combined based on the objective function and the optimal points of rain gauges development are determined based on the optimization model. In the considered case study, the application of the proposed method to an existing rain network over the Karkhe catchment region under a minimum transformation entropy of 30 % and maximum Kriging error of 60 % resulted in 17 new rain stations to be added to the original network (Fig. 4).

References

1. Karamouz, M., Nokhandan, A. K., Kerachian, R., & Maksimovic, C. (2009). Design of on-line river water quality monitoring systems using the entropy theory: A case study. *Journal of Environmental Monitoring and Assessment, 155*, 63–81.
2. Yeh, H. C., Chen, Y. C., Wei, C., & Chen, R. H. (2011). Entropy and kriging approach to rainfall network design. *Journal of Paddy and Water Environment, 9*, 343–355.
3. Chen, Y.-C., Wei, C., & Yeh, H.-C. (2008). Rainfall network design using kriging and entropy. *Journal of Hydrological processes. 22*, 340–346.
4. Mogheir, Y. (2003). Assessment and redesign of groundwater quality monitoring networks using the entropy theory-Gaza strip case study. PhD thesis, University of Coimbra, Coimbra, Portugal.

Application of Drastic Methodology for Evaluation of Guarani Aquifer Vulnerability: Study Case in Ribeirão Bonito, SP, Brazil

Rafael Gonçalves Santos, Mara Lúcia Marques, Gabriela Trigo Ferreira and José Ricardo Sturaro

1 Introduction

The Guarani Aquifer System (GAS) consists in one of the biggest freshwater reserves in the world, comprising parts of Brazil, Argentina, Paraguay and Uruguay, being not only a source for supply to the population but also a means in order to develop economic activities and leisure. Areas where the aquifer is outcropped require studies that aim to a better protection and spatial planning, once they are intimately connected to direct recharge of the aquifer.

This study presents a Vulnerability mapping through the application of DRASTIC method. Today the DRASTIC method is a standardized system for evaluating groundwater pollution potential. DRASTIC has been widely used in many countries because the inputs required for its application are generally available or easy to obtain from public agencies [1].

1.1 Study Area

The study area is a mosaic of topographic charts in scale 1:10,000, covering the entire city of Ribeirão Bonito and parts of other neighboring municipalities located in the

R. G. Santos (✉) · M. L. Marques · G. T. Ferreira · J. R. Sturaro
UNESP, IGCE, Av. 24A, Bela Vista 1515, Rio Claro, SP , Brazil
e-mail: faelgeo89@hotmail.com

M. L. Marques
e-mail: maraluciamarques@hotmail.com

G. T. Ferreira
e-mail: gabitrigoferreira@hotmail.com

J. R. Sturaro
e-mail: sturaro@unesp.br

E. Pardo-Igúzquiza et al. (eds.), *Mathematics of Planet Earth*,
Lecture Notes in Earth System Sciences, DOI: 10.1007/978-3-642-32408-6_109,

center of São Paulo state. Most extension is comprised by outcrop and recharge areas of Guarani Aquifer, which presenting a layered system composed of sandstones of the Triassic period (Pirambóia Formation) and Jurassic (Botucatu Formation) over there aquifer outcropped. A minor extension is covered to Serra Geral (basaltic) and Adamantina Formations (fine sandstones and very fine) from the Cretaceous period, both structures are located above the aquifer and confined or semi-confined the Guarani Aquifer [2].

2 Methodology

DRASTIC is an established aquifer vulnerability mapping methodology, developed by the U.S. Environmental Protection Agency (EPA) [3]. This method defines seven parameters that contribute to intrinsic aquifer vulnerability. The DRASTIC PESTICIDE system is designed to be used where the activity of concern is the application of pesticides to an area. The pesticide DRASTIC differs from general DRASTIC in the assignment of weights (Table 1).

The seven parameters were assembled and combined using a weighted sum equation to produce the final intrinsic vulnerability map:

$$\text{DRASTIC} = \text{D}r^*\text{D}w + \text{R}r^*\text{R}w + \text{A}r^*\text{A}w + \text{S}r^*\text{S}w + \text{T}r^*\text{T}w + \text{I}r^*\text{I}w + \text{C}r^*\text{C}w \quad (1)$$

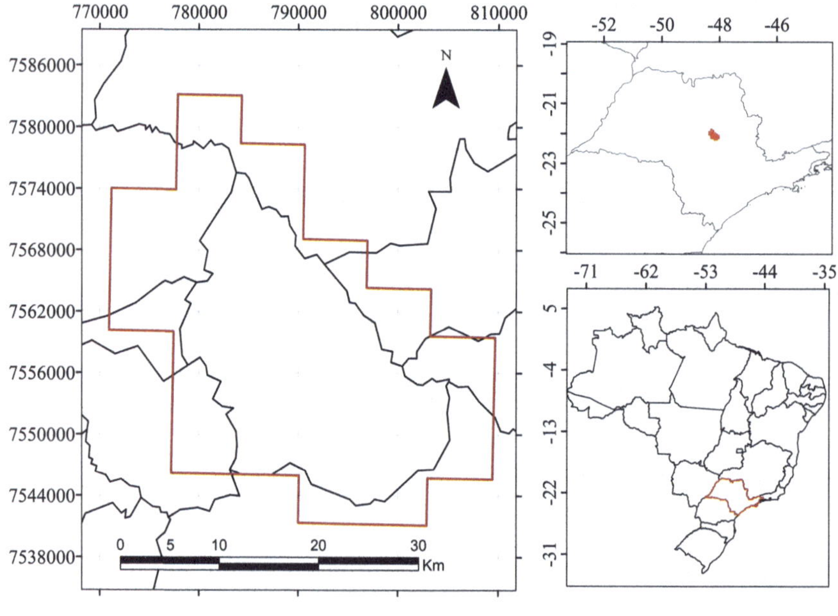

Fig. 1 Location of the study area in Ribeirão Bonito, SP, Brazil

Table 1 The DRASTIC Parameters and their relative weights [1]

Parameters (experimental data)	General weight	Pesticide weight
D: Depth to water [4]	5	5
R: Recharge [2]	4	4
A: Aquifer media [4]	3	3
S: Soil media [2]	2	5
T: Topography	1	3
I: Impact of the vadose zone [4]	5	4
C: Conductivity (hydraulic) of the aquifer [4]	3	2

where r is the rating for the parameter and w is an assigned weight for the parameter.

3 Results and Discussions

The analysis results of the mapping and vulnerability Index of the Guarani Aquifer in the region of Ribeirão Bonito are shown in Fig. 2.

The results prove that the DRASTIC method presented a more homogeneous area than the DRASTIC PESTICIDE. This fact is due to the greater weight of the soil and

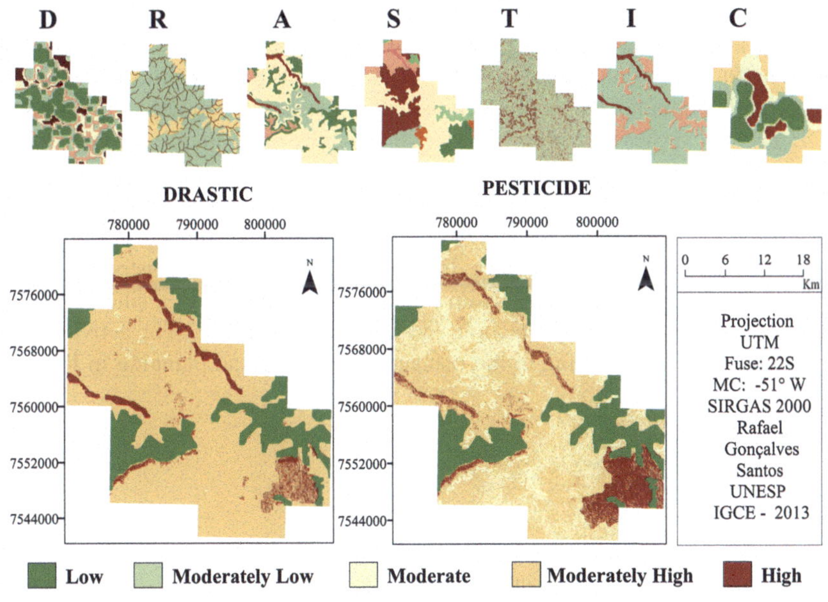

Fig. 2 Vulnerability of the Guarani Aquifer through the drastic and drastic pesticide methods

topography variable in the final summation. The areas of greatest vulnerability are concentrated where the aquifer is outcropped (Pirambóia and Botucatu Formations) and along watercourses (alluvial sediments) that make direct recharging the aquifer.

Areas of sedimentary rocks consisting mainly of sandstones and sandy soils, also presented higher vulnerability. Areas with low vulnerability are located where the aquifer presents a confining layer with the presence of igneous rocks (Serra Geral Formation). The results indicate that the majority of the study area, showed a moderately high vulnerability DRASTIC index; the index has presented the results PESTICIDE moderately high and moderate.

4 Conclusion

The GAS vulnerability map, obtained through the application of DRASTIC method, demonstrates the fragility of recharge areas and aquifer outcrop, enabling the identification of those areas with most vulnerability index to contamination from potential agents of pollution. The vulnerability mapping also allows electing the most appropriate areas for installation of economic activities with high potential to contaminate water resources. The DRASTIC method demonstrates to be an important tool for the management of water resources in the study area and other areas located under the same lithology.

The index DRASTIC and DRASTIC PESTICIDE presented similarities in vulnerability mapping, both pointed that most vulnerable areas are mostly located on sandstone formations and sandy soils. The study area is an agricultural region typically with large monoculture production and, in this sense, the DRASTIC PESTICIDE method exposes the vulnerable areas that require more attention due to the use of agricultural inputs.

The results indicate that most vulnerable areas are located on sandstone formations, also highlight the need for protection and management along the watercourses, since they supply the aquifer directly. The DRASTIC method demonstrated to be an efficient tool in planning and analyzing the intrinsic vulnerability, since it proved vulnerable areas requiring protection and areas in which different activities can be installed, thus contributing to the management and spatial planning in Ribeirão Bonito.

References

1. Jovanovic, N. Z., Adams S., Thomas A., Fey M., Beekman H. E., Campbell R., Saayman I. & Conrad J. (2006). Improved DRASTIC method for assessment of groundwater vulnerability to generic aqueousphase contaminants, WIT Transactions on Ecology and the Environment, Vol 92, waste Management and the Environment III, pp. 393–402.

2. Rabelo, J. L., Quaresma, J. E., & Wendland, E. (2004). *Conceptual model of outcrop area of the Guarani Aquifer in the central region of São Paulo*. In: XIII XIII Brazilian Congress of Groundwater, Cuiabá-MT.
3. Aller, L., Bennett, T., Lehr, J. H., Petty, R. J., & Hackett, G. (1987). *DRASTIC: A standardized system for evaluating groundwater pollution potential using hydrogeologic settings*. Dublin: USEPA.
4. Queluz, J. G. T., & Sturaro, J. R. (2011). Spatialization of the vulnerability of the Guarani aquifer in Ribeirão Bonito - SP. In: II Symposium of Applied Geostatistics on Agricultural Sciences, Vol. 01. pp. 01–05. Botucatu: UNESP.

A Case Study of Geometric Modelling via 3-D Point Interpolation for the Bathymetry of the Rabasa Lakes (Alicante, Spain)

África de la Hera, Enrique López-Pamo, Esther Santofimia, Guillermo Gallego, Raquel Morales, Juan J. Durán and José M. Murillo-Díaz

1 Introduction

In a partnership between the Geological Survey of Spain and the Regional Government of Alicante, a study of the wetlands in the province is being carried out. The Rabasa Lakes, located North-East from the city of Alicante, have been formed as a consequence of the rising of the groundwater level in a quarry (Fig. 1) that was abandoned in the 1970s. There are three lakes of size 0.5, 1.5 and 3 ha. The largest lake has been selected to carry out a bathymetric study.

The methods to obtain bathymetric maps of rivers, lakes, ocean floors, etc. are diverse. Acoustic, visual and computer technology, as well as Global Positioning Systems (GPS) are at the foundations of modern bathymetry systems. Single beam echo sounders (SBES), on ships infer the topography of the sea floor by emitting an

A. Hera (✉) · E. López-Pamo · E. Santofimia · R. Morales · J. J. Durán-Valsero · J. M. Murillo-Díaz
Geological Survey of Spain, C/ Ríos Rosas, 23, 28003 Madrid, Spain
e-mail: a.delahera@igme.es

E. López-Pamo
e-mail: e.lopez@igme.es

E. Santofimia
e-mail: e.santofimia@igme.es

R. Morales
e-mail: r.morales@igme.es

J. J. Durán-Valsero
e-mail: jj.duran@igme.es

J. M. Murillo-Díaz
e-mail: jm.murillo@igme.es

G. Gallego
Universidad Politécnica de Madrid, Avda. Complutense 30, 28040 Madrid, Spain
e-mail: ggb@gti.ssr.upm.es

E. Pardo-Igúzquiza et al. (eds.), *Mathematics of Planet Earth*,
Lecture Notes in Earth System Sciences, DOI: 10.1007/978-3-642-32408-6_110,

Fig. 1 Location and view of the Rabasa Lakes in the province of Alicante, Spain

acoustic pulse and measuring the travel time it takes the wave to return to the ship after being reflected by the sea floor. Multibeam echo sounders [1] (MBES) provide higher accuracy and angular resolution by correcting for the movements of the ship. Bathymetry in shallow waters (depth $< 50\,\mathrm{m}$) can be measured from aircrafts by means of airborne laser radar (LIDAR) [2]. Such systems infer the water depth by sensing the echo of emitted laser light pulses that are reflected by the water surface and the water bottom. Optical bathymetry methods are usually cost effective and efficient over large areas of clean water bodies.

2 Methodology

In this work, a bathymetry technique based on acoustic technology is presented. The estimation of the topography of the Rabasa lake bed consists of two stages. First, in situ measurements are obtained by means of sonar and GPS equipment. A Garmin 160C Fishfinder sonar probe and a Garmin 76S GPS receiver working in differential mode were used to collect a total of 335 points measurements spatially distributed throughout the lake. In a second stage, the data is processed by scientific software [3] that, by means of the method of ordinary kriging, returns and estimate of the wetland bathymetry.

Kriging [4, 5] is widely used in hydrogeology as the preferred method for constructing datasets, usually over unstructured 2-D grids, suitable for contouring. It is a method closely related to the best linear unbiased estimation (BLUE) in Statistics. Given the observed depths $\mathrm{Z}(\mathrm{x}_\mathrm{i})$ at locations $\mathrm{x}_\mathrm{i} \in \mathrm{R}^2, i = 1, \ldots, N$, the kriging method interpolates the depth at an unobserved location x_0 using a linear combination of the known depths: $\check{\mathrm{Z}}(\mathrm{x}_0) = \sum_{\mathrm{i}=1}^{\mathrm{N}} \mathrm{w}_\mathrm{i}(\mathrm{x}_0)\mathrm{Z}(\mathrm{x}_\mathrm{i})$. The weights $\mathrm{w}_\mathrm{i}\,(\mathrm{x}_0)$ depend on the distance from x_0 to the known locations x_i, i.e., $\|\mathrm{x}_0 - \mathrm{x}_\mathrm{i}\|$, and are calculated based on hypothesis of the ordinary kriging method.

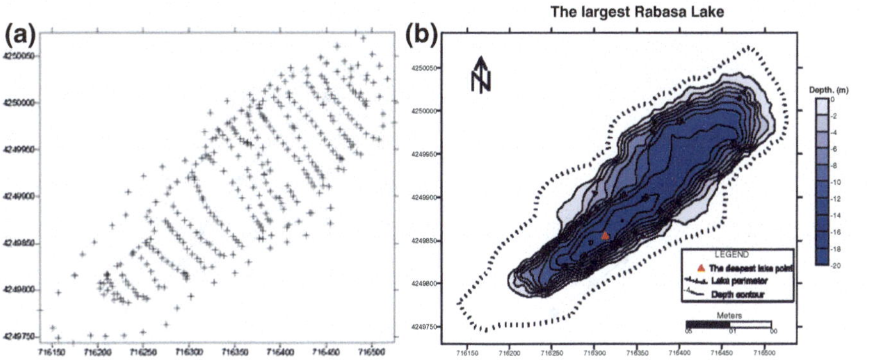

Fig. 2 Water depth is measured in meters Coordinate axes are in UTM units.**a** Field points measured of the largest Rabasa Lakes. **b** Corresponding bathymetric chart (i.e., contour map) obtained by ordinary kriging

3 Results

The geometric modelling of a lake basin by means of an interpolated surface through a set of scattered 3-D points allows us to analyze the lake bed as a whole. For the largest Rabasa Lakes 335 points were measured, out of which 217 belong to the flooded area. Figure 2a shows the location of the collected field points. The minimum, median and maximum water depths are 0, 8.4 and 18 m, respectively. The mean water depth and the standard deviation are 8.49 and 5.33 m, respectively. The average distance between nearest neighbor points is 5.93 m. The corresponding topographic map estimated using ordinary kriging is shown in Fig. 2b. The lake shows an elliptical morphology with an almost flat lake bottom. The distance between adjacent points of the generated grid is 2.93 m in each axis. This is approximately half the average spacing of the input data. The estimated minimum depth by kriging is 18.16 m, which is very close to the observed depth.

The knowledge of the morphology of the lake allows us to establish aquifer-lake connections (in case the groundwater level is known) and to compute the storage capacity of the lake. Other statistics, such as the area and volume of the basin at different depths (Fig. 3) can be computed. The resulting values are paramount in hydrological studies to characterize and forecast the behaviour of the lake depending on other environmental variables. For example, they can be used in balance equations to predict the response of the water stored in the wetland to several input and output water flows such as rain, springs, runoff, evaporation, water pumping, etc. According to the fourth report of the Intergovernmental Panel on Climate Change [6] forecast, considered climate change scenarios suggest a decrease of the water inputs due to fewer precipitations and lower groundwater level, and possible runoff due to isolated rainfalls.

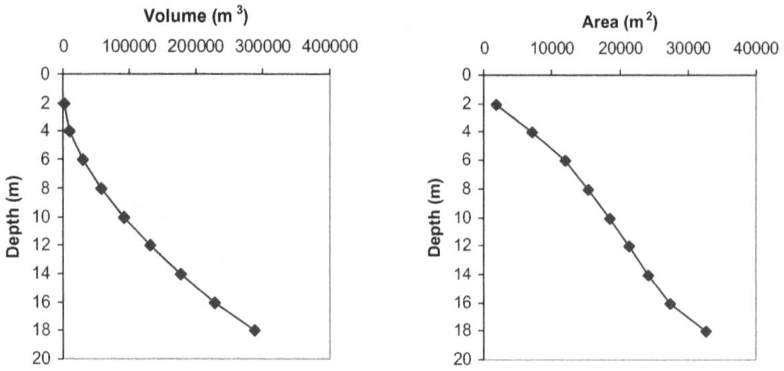

Fig. 3 Depth-volume and depth-area (*hypsometric*) curves of the largest Rabasa Lakes. Depth is measured from the surface of the lake

4 Conclusion

In this work, a case study of the application of least squares interpolation of scattered 3-D points (kriging) has been presented for a hydrological characterization of a lake. The sonar technology used to acquire the physical measurements to be interpolated was selected based on technical considerations (lake size, accuracy, etc.) as well as economical ones. The proposed methodology is mathematically well founded and reliably describes the bathymetry of the lake bed in agreement with the observations.

Acknowledgments Authors acknowledge the Regional Government of Alicante and the IGME for the partnership and cofunding related to this project. We thank Prof. A. Valdés (Complutense University of Madrid) for his useful comments.

References

1. Geng, X., & Zielinski, A. (1999). Precise Multibeam Acoustic Bathymetry. *Marine Geodesy*, *22*(3), 157–167.
2. Whitman, E. C. (1996). Laser airborne bathymetry - Lifting the littoral. *Sea Technology*, *37*(8), 95–98.
3. SURFER Software, v. 8.01 (2002). Scientific Software Group. http://www.ssgsurfer.com/.
4. Chiles, J. P., & Delfiner, P. (1999). *Geostatistics*. Wiley Series in Probability and statistics: Modeling Spatial Uncertainty.
5. Tonkin, M. J., & Larson, S. P. (2002). Kriging water levels with a regional-linear and point-logarithmic drift. *Ground Water*, *40*(2), 185–193.
6. Pachauri, R. K., & Reisinger, A. (Eds.). (2007). *Contribution of Working Groups I*. IPCC, Geneva, Switzerland: II and III to the Fourth Assessment Report of the Intergovernmental Panel on Climate Change.

Analysis of Groundwater Monitoring Data Sets with Non-Detect Observations: Application to the Plana de Sagunto (Valencia, Spain) Groundwater Body

Juan Grima, Juan Antonio Luque-Espinar, Juan Ángel Mejía-Gómez and Ramiro Rodríguez

1 Introduction

European countries are facing the problem of significant pollution of soil and groundwater resources [1]. The current legislative framework, i.e. the Water Framework Directive (WFD) and the Groundwater Directive (GD) require Member States (MS) to establish groundwater monitoring networks and design a monitoring plan as part of a programme of measures for achieving WFD environmental objectives. Groundwater data frequently evidence that many non-commonly monitored synthetic or naturally occurring chemicals, are present in aquifers. In other cases a fraction of the monitoring data shows no detection of such contaminants, what does not necessarily means that the contaminant is not present at any level. In order to address the uncertainty associated with the unknown mean concentration of data sets with observations qualified as non-detects, statistical methods for handling left censored data have to be applied.

J. Grima (✉) · J. A. Luque-Espinar
Spanish Geological Survey, C/Ríos Rosas, 23, 28003 Madrid, Spain
e-mail: j.grima@igme.es

J. A. Luque-Espinar
e-mail: ja.luque@igme.es

J. A. Mejía-Gómez
Campus Irapuato-Salamanca, División Ciencias de la Vida. Departamento de Ciencias Ambientales, Universidad de Guanajuato, Guanajuato, Mexico
e-mail: jangelmg@gmail.com

R. Rodríguez
Instituto de Geofísica, UNAM, Ciudad Universitaria, 04510 México, D.F., Mexico
e-mail: ramiro@geofisica.unam.mx

E. Pardo-Igúzquiza et al. (eds.), *Mathematics of Planet Earth*,
Lecture Notes in Earth System Sciences, DOI: 10.1007/978-3-642-32408-6_111,

1.1 Study Area and Data

In the framework of the Common Implementation Strategy (CIS) of the WFD all Member Satates and Norway agreed to develop a number of Pilot River Basins to check the various issues developed in Guidance Documents (GDs). In Spain, the GDs were evaluated in the Júcar River Basin District (JRBD), aiming to provide tools for the management of its hydraulic resources [2]. The main pressures identified are water abstraction, point source and diffuse pollution and saline intrusion [3]. Plana de Sagunto Groundwater Body (Fig. 1) has been selected as a case study due to the fact that is at risk of not achieving good chemical status. All the information about pollutants concentration has been downloaded from the web page of the JRBD Authority by means of its direct download utility. In Plana de Sagunto there are 13 stations with values above the threshold value for nitrates (50 mg/L). Historical data have been compared until 2010.

2 Methodology and Data Analysis Techniques

The first step to take in the process of time series analysis is to make a visual representation of data [4] to identify patterns, trends or seasonal variation. If observations are independent and identically distributed random variables, linear regression can be used to model the relationship among a dependent variable and several explanatory variables. However, applying linear models to groundwater data is not suitable, because of its intrinsic characteristics. In fact, many qualitative groundwater variables neither are normally distributed nor have even a symmetric distribution [5]. To avoid such a difficulty [6], local regression (LOESS) or locally weighted scatterplot smoothing (LOWESS) are used. In Fig. 2 time series (with 8 values at least) for nitrates of Plana de Sagunto are displayed jointly with its smoothed line. An upward trend is clear at station JUIG003940.

To deal with censored data various methods have been developed. The simplest one is substituting a constant for all observations below the detection limit. Distributional methods depend on the assumption that the data fit a given distribution while non-parametric alternatives are based upon the survival function, commonly used in medical research. It does not require knowing the distribution of data and can handle multiple detection limits. In this paper three substitution methods (assigning zero, half of the detection limit and the detection limit respectively to the censored values), two distributional methods (Maximum likelihood estimation and Robust regression or ROS) and one non-parametric approach (Kaplan-Meier) have been tested to estimate the distribution parameters for nitrates in Plana de Sagunto Groundwater Body. Different censoring degrees have been artificially imposed to data sets and confidence intervals calculated afterwards. Resampling techniques (bootstrap) are used to set confidence intervals for the original data and for ROS estimates. To check compliance with a standard the lower confidence limit is used. All results (Table 1) have

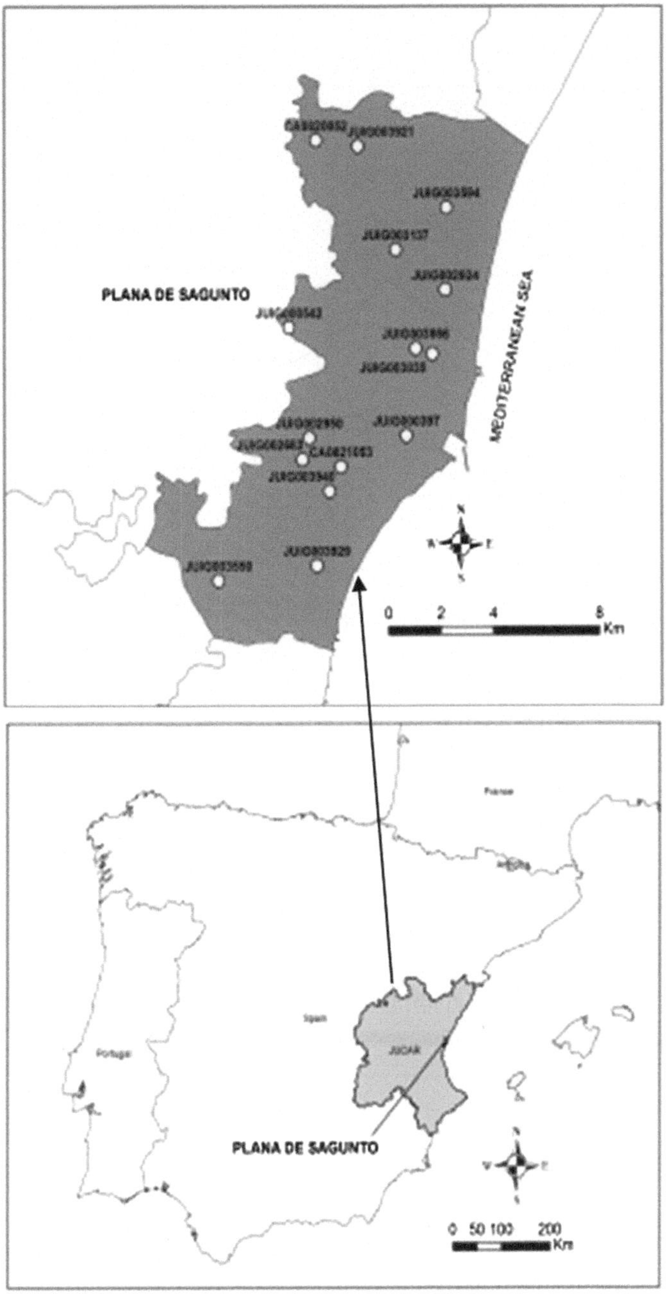

Fig. 1 Location of Jucar River Basin and Plana de Sagunto GWB

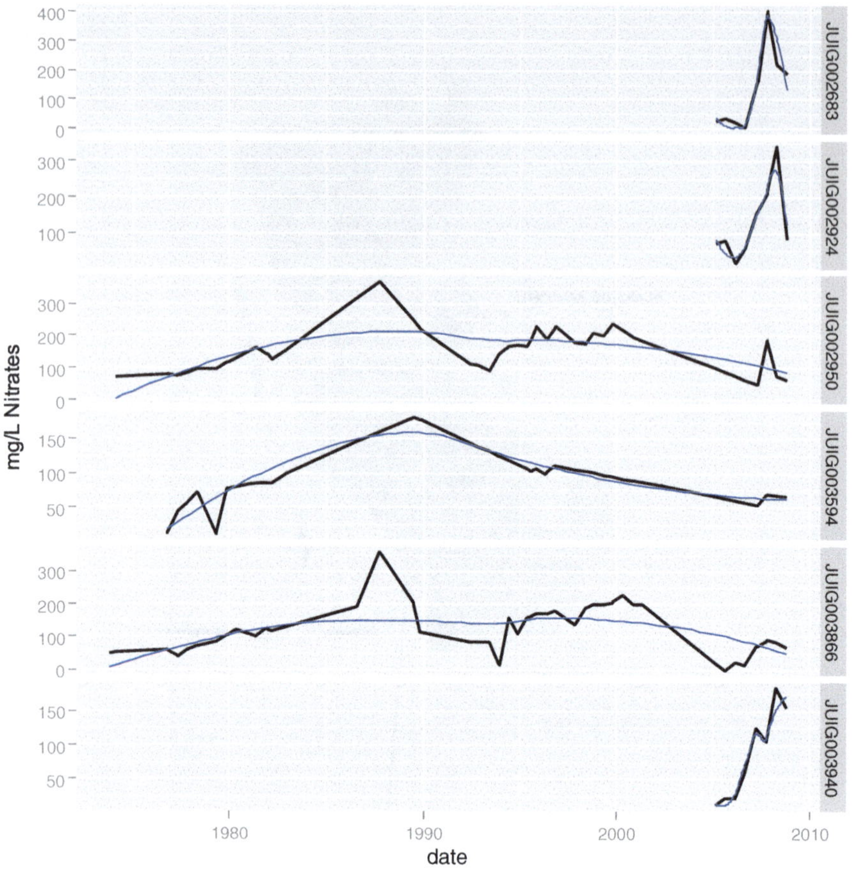

Fig. 2 Time series and smoothed lines at Plana de Sagunto GWB

been elaborated within the framework of the R software environment for statistical computing [7].

3 Results and Discussion

Sample size has great impact on confidence intervals. With small sample sizes (less than 10) maximum likelihood estimation performs poorly as a consequence of assuming distributional assumptions (provides larger confidence intervals relative to other procedures). Substitution methods (half the detection limit) can provide an indication of compliance (although further sampling is definitely required to assess chemical status). On the other hand, when there is more than one detection limit (or it is expected as a consequence of changing analytical precision over time) it is necessary

Table 1 Confidence limits of the mean using selected methods under different censoring degrees

LCL – UCL = upper-lower confidence limit Station Code	Mean confidence interval								
	Original Data	Censored data							
		% data censored	Values above detection limit	Substitution			Maximum likelihood estimation	Robust regression on order statistics	Kaplan-Meier
				0	0.5 detection limit	Detection limit			
JUIG002683	N = 8	12.5	7	50.1–230.8	49.1–232.2	53.2–229.7	39.8–1455.2	49.51–231.2	39.6–231.2
mean = 133	LCL = 47.5	25	6	46.2–228.1	50.6–231.7	52.9–231.2	33.7–2629.5	47.7–236.23	40.7–233.1
sd = 139.3	UCL = 230.3	50	4	27.6–223.5	62.1–234.3	80.1–238.4	17.6–29791.1	–	136.5–265.4
JUIG002924	N = 8	12.5	7	63.3–202.5	67.3–202.3	71.6–203.8	64.3–317.4	73.07–202.03	64.2–205.1
mean = 129.4	LCL = 67.7	25	6	50.9–199.7	65.4–200.2	77.9–204.9	57.3–366.7	71.71–207.31	69.1–207.7
sd = 106.1	UCL = 203.9	50	4	26.6–192	59–198.4	94.7–213.9	37.3–604.1	–	73.8–214
JUIG002950	N = 34	11.7	32	129–178.2	132.7–180.4	135.5–180.4	133.2–187.3	135.7–180.3	135.2–180.6
mean = 157	LCL = 135.1	23.5	16	64.3–139.1	123–171.5	179.9–206	101.9–191.1	143–193.1	177.7–204.9
sd = 68.9	UCL = 180	47.1	12	43.5–119.8	120–166.5	194.6–216.7	84.4–181.7	132.3–198.1	192.5–215.7
JUIG003594	N = 17	15.3	12	44.1–92.8	60.8–97.6	76.1–102.9	50.5–147.5	69–100.9	74.4–102.6
mean = 79.5	LCL = 61.6	23.1	11	39.4–89.8	59.7–96	78.1–104	47.7–149.8	67.4–101.4	75.7–103.3
sd = 40.2	UCL = 98.7	46.1	7	18.6–75.5	54.7–91.6	88.8–110.4	32.8–163.2	49.8–101.8	85.3–109.1
JUIG003866	N = 40	12.8	33	94–143.1	99.2–145.9	105.1–148.1	148–1012.7	107–148.9	108.3–149.8
mean = 122.4	LCL = 99.8	25.6	29	85.6–139.1	99.5–145.7	113.1–152.1	143.4–1276.6	106.9–152.6	111.9–151.8
sd = 75.1	UCL = 145.5	48.7	20	81.2–136.3	100.1–145.8	118.3–155	140.1–1456	110.7–162.5	116.9–154.6
JUIG003940	N = 8	12.5	7	439.9–128.1	3.9–130.1	46.3–130.5	38.9–256.3	42.4–130.6	41.6–132.8
mean = 85.9	LCL = 42.9	37.5	5	30.5–127.9	38–129.2	48.5–130.3	29.1–508.1	50.9–145.9	64.5–137.6
sd = 66.6	UCL = 130	50	4	19.5–124	50.1–130.5	78.9–136.9	21.6–802.7	–	101.7–147.1

to use particular techniques to meet these requirements. In summary, ROS and K-M are the recommended methods, as they provide more precision and are capable of dealing with several detection limits. Furthermore, a minimum of 10-15 samples above detection limit is suggested to obtain accurate results. Upward trends must be tested before any statistical analysis, removing its influence.

When values qualified as non-detected exceed 50 % results are not accurate and other alternatives must be explored. The assessment has led to the conclusion that Plana de Sagunto GWB shows poor chemical status as a result of nitrate pollution.

References

1. Grima, J., Ballesteros, B., López, J., & García, O. (2002). Sustentabilidad de las técnicas de recuperación de acuíferos y suelos contaminados y necesidades de investigación en la Unión Europea. In J. A. López Geta, J. Grima Olmedo & L. Rodríguez Hernández (Eds.), Investigación, Gestión y recuperación de acuíferos contaminados. Ponencias.
2. Estrela, T., Maestu, J., Fidalgo, A., Pérez, M. A., Fullana, J., & Pujante, A. M. (2004). Jucar Pilot River Basin. Provisional Article 5 Report.
3. Ministerio de Medio Ambiente y Medio Rural y Marino. (2007). Estudio General sobre la Demarcación Hidrográfica del Júcar.
4. Luque, J. A., Grima, J., Varela, M., & Sánchez, I. Searching for trends in data from groundwater monitoring networks through exploratory data analysis. European Groundwater Conference, Poster Session and Proceedings, Madrid, May 20–21, 2010.
5. Montgomery, H., Loftis, J. C., & Harris, J. (1986). Statistical characteristics of ground-water quality variables. *Ground Water*, *25*, 176–184 .
6. Helsel, D. R., & Hirsch. R. M. (2002). Techniques of Water-Resources Investigations of the United States Geological Survey. Book 4, Hydrologic Analysis and Interpretation. Statistical Methods in Water Resources. *Environmental Science and Pollution Research International*, *14*(5), 297–307.
7. R Development Core Team. R (2011). A language and environment for statistical computing. R Foundation for Statistical Computin, Vienna, Austria. ISBN 3-900051-07-0.

The Mesh Optimization of the Environmental Investigation Applied to the Diagnosis of the Quality of the Basement (Algerian Experience)

Abderrahim Gheris

1 Introduction

Potentially polluted site is made to be sampled at different stages of its characterization or remediation, to meet various objectives. In all cases, the problem of the sampling strategy in terms of "how to reach a sufficient level of knowledge relative to study objectives, with a minimum data positioned better?" Speaking of sampling strategy means then also take into account financial constraints or deadlines, decide of the sample sizes appropriate to the objectives of study, to program the sampling phases in order to best resolve uncertainties or to integrate all existing data to determine the number and location of additional samples. Currently, sampling strategies most often result of a reflection of practitioners based on their experience, knowledge of the site as well as common sense. The objective of this work is to show what can be the contribution of geostatistics to rationalize sampling of polluted sites. Through the use of database collected during a project of research estimation and evaluation of the risk of spreading the pollution of ground water, emitted by a plant of paintings [2]. The Algerian regulations do not provide specific recommendations on the sampling methodology. The Algerian methodological guide recalls the different approaches of the points location and refers to the recommendations developed by the U.S. EPA in 1991 [5] and by the european standardization [1]. The site in question has a multiple pollution (three pollutants) on an area of about 5 ha of land. It is a plant of paintings located in north eastern Algeria, where various chemicals containing heavy metals were dumped for many years from the 80 s. Between the years 2008 and 2010 a study was carried out on this site to locate and study the special distribution of soil contamination by heavy metals. For this purpose a systematic sampling campaign (soil and water) has been executed, the results of this campaign

A. Gheris (✉)
INFRARES Laboratory, University Mohamed Chérif Messadia, BP1553, 41000 Souk Ahras, Algeria
e-mail: agheris@hotmail.fr

E. Pardo-Igúzquiza et al. (eds.), *Mathematics of Planet Earth*,
Lecture Notes in Earth System Sciences, DOI: 10.1007/978-3-642-32408-6_112,

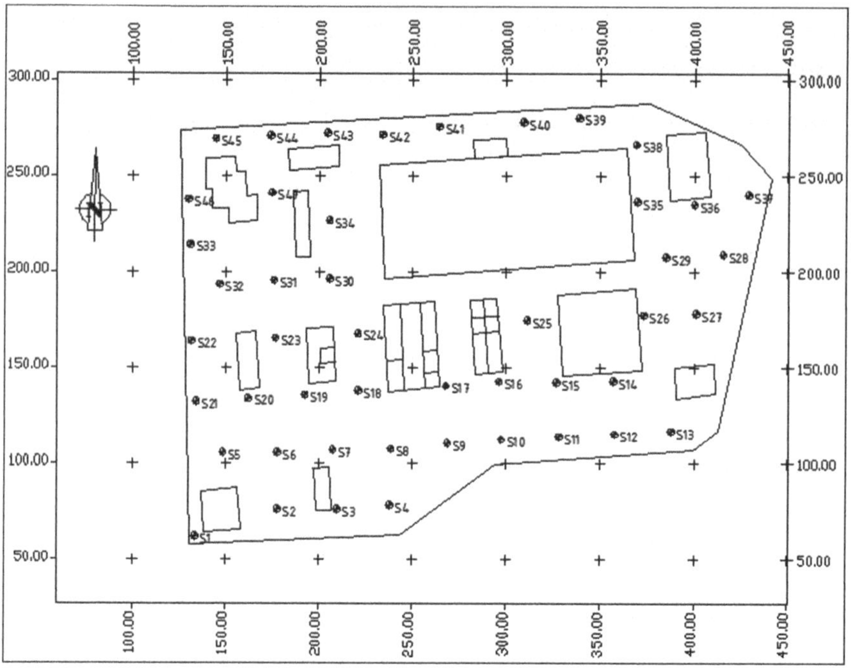

Fig. 1 Location on the site plan of sampling (36°17′19, 89″N; 7°56′20, 71″E)

were mustered in a database [2]. Therefore, over a period of 2 years, successive campaigns have resulted to the realization of surveys using a mechanical auger. From the beginning the sampling strategy was to perform systematic surveys with an advancing from south to north, the boreholes are shown in Fig. 1. Laboratory analysis of 47 soil samples collected during this campaign showed high levels of three substances: lead, cadmium, and chromium [3]. The groundwater level is located at about 10 m depth at the right site. The sands clay aquifer is a very productive aquifer (permeability greater than 10^{-3} m/s). The basement consists of yellowish sandy clay, silty sand with pebbles, yellowish clay and limestone concretion.

1.1 Production of Maps of Contamination

The criterion of rationalization of the selected sample assumed to estimate, from the available data on each pollutant, a 3D map of the probability of exceeding its guideline value. Several geostatistical methods allow estimating a probability of exceedance. Include the estimation by kriging, usually after data transformation (eg., log transformation log, Gaussian) [4]. According to the study [2, 3] and for each of the three pollutants, it has been deduced from the 3D map of probability of exceeding

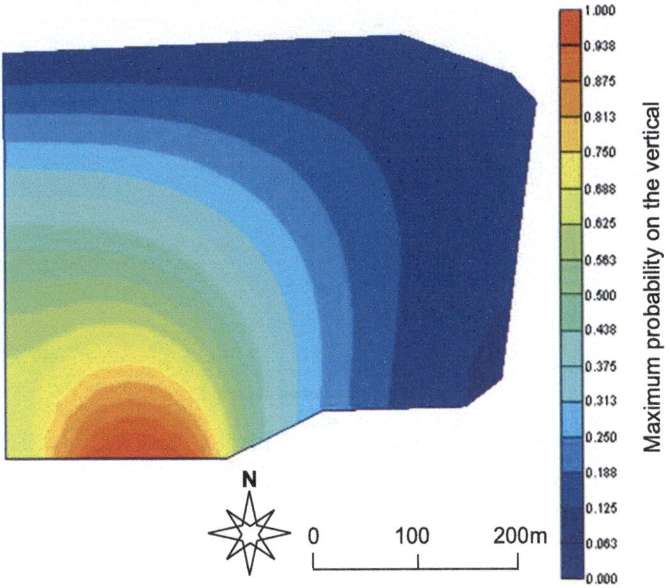

Fig. 2 Probability map derived from existing data

a 2D map synthetic for the highest probabilities calculated on the vertical (from 0 to 6.0 m depth). For example, Fig. 2 shows the resulting map for lead. As expected, the first polls in the region of higher probability showed high soil that impregnated. As and progresses to the northward the impregnations were less important. The area of highest probability corresponds to an area incineration of packaging waste for raw materials; she occupies about 8 % of the total surface. Therefore the map of probabilities reflects the spatial representation of the spread of pollution.

2 Probabilistic Simulation and Results

Now using the database of test results in the laboratory [2], we performed a series of simulation under various scenarios, to test the performance of different types of sampling strategies used to detect circular source occupying 8 % of the total site area (area highlighted by the previous study). The first simulation scenario is to consider only a certain number of polls (between 15 and 40) disposed according to a layout random, the second scenario, activation of the same points, but according to a layout stratified random, and finally a transverse and longitudinal profile which forms a fish bone. The results of the performance of these themes are shown in the graph in Fig. 3.

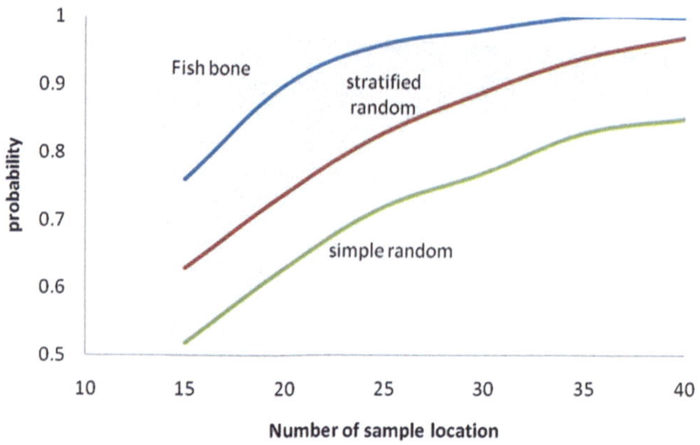

Fig. 3 Performance of the different strategies

3 Conclusion

We can say that the least costly strategy in financial term is the diagnostic by profiling. Then comes the geostatistical analysis exploratory, which allows developing through the integration of all the information collected a risk mapping. Additional sampling is recommended to minimize the uncertainty about the levels of pollutants in the uncertain zones, for example using methods of linear or non-linear kriging.

References

1. AFNOR. Norme NF ISO 10381–1 (X 31–008-1). (2003). Qualité du sol - Échantillonnage-Partie 1: Lignes directrices pour l'établissement des programmes d'échantillonnage.
2. Gheris, A. (2011). Les sols contaminés, modélisation et prédiction du transfert des particules dans un aquifère (cas de la ville de Souk Ahras). Geotechnical PhD Thesis, University of Guelma, Algeria.
3. Gheris, A., & Mekssaouine, M. (2010). 3D numerical simulation of the impact of the activity of a production unit of paintings on the groundwater of the city of Souk Ahras. First international water conference, CIEAU 2010, University of Annaba, Algeria.
4. Pellet, M. & Laville-Timsit, L. (1993). Échantillonnage des Sols par Caractérisation d'une Pollution: Guide Méthodologique. Édition du BRGM, rapport n° 37865 ENV 4S 93. 80 p.
5. US Environmental Protection Agency. (1991). Superfund program-respresentative sampling guidance-volume 1: soil. OSWER Directive 9360.4–10, EPA 540/R-95/141.

Data Archives: Development and Application in Environmental Management

Konstantin Alexeev

1 Introduction to the Problem

The presented technology was conceived because of existing problems of information support of geological and nature-use management in the territory of Russia. Today large volumes of geodata aren't systematized and aren't provided to the uniform standard, that significantly complicates their use, because of huge territory and disunity of various departments and the working groups. Unlike existing GIS-projects, the described technology unites all types of the data, which are storing in different, not connected with each other directly, systems, in uniform metabase. And, as the independent system, it allows and assumes integration with existing (Russian and international) information systems for providing the unified access to data.

Rapid development of geology in Soviet period led to accumulation of a great volume of the data obtained during carrying out field researches [1]. At the same time this information in most cases is not structured and even not transferred to a digital form. Natural consequence of the described situation is the dissociation of the information resources which is strongly complicating processing and use of the saved-up scientific potential.

For the solution of this problem development of the structured data archive is offered. The scheme of carrying out offered works is defined by the following stages:

- Organization of structure of storage
- Definition of a necessary minimum of the description of object
- Expansion of signs of search and description of objects if necessary
- Creation and introduction of regulations of granting the signs given on the basis of the created list
- Collecting and specification of information on stored data
- Metadatabase creation

K. Alexeev (✉)
State Scientific Centre VNII geosystem, Varshavskoye chausse, 8, 117105 Moscow, Russia
e-mail: pro920@bk.ru

E. Pardo-Igúzquiza et al. (eds.), *Mathematics of Planet Earth*,
Lecture Notes in Earth System Sciences, DOI: 10.1007/978-3-642-32408-6_113,

- Control of a condition of data (planned and on demand)
- Gradual copying of data from various sources on a local resource in the ordered form

2 Process and Main Problems of the Structured Data Archive Development

The main problem of the organization of structure of data storage in environmental management is ignorance, and often and misunderstanding by users of information fullness of available materials and inability, and as a result–difficulty of allocation from all volume of the presented data of really necessary sets of data.

The impossibility of structurization following from here is explained by absence of data as those.

Usually, the array of information used in work becomes data to the end of performance of work and can be presented in the structured set only at the last stage–a stage of formation and representation of reporting materials. Unfortunately, at this moment data any more aren't of interest to further use and further are stored in a type of backup copies on remote resources, without the distinct description of the executed structurization and an essence of the carried-out works. For this reason the subsequent works performed on the basis of these data, begin again with processing of an initial material.

Due to the problems of association of resources described above and elaboration of uniform vision of an information field in environmental management at the first stage of creation of archive the database comprising information on processed objects, will comprise only two fields of records. The first field will comprise an object name which, as a rule, is a file name. Thus, the first field will define at the same time location of object. The second field will contain the text description of object in any form. On the basis of this initial database by the analysis and specification of the text description information base of archive and the rule of management will be created by it. Further the quantity of fields will be increased but in view of the fact that objects will actually not be stored in base, its volume won't be catastrophically great. Also possibility of change of structure of base as addition of new fields for objects again arriving in archive, and base recalculation as a whole is supposed. Full recalculation and recreation of archive is possible both at the direction, and in the course of planned procedural works.

The specified way of collecting and storage of information has many advantages. Let's list the main:

- lack of need for the big massif of means of storage at physical level and, as a result, the small cost of the project at the initial stage and at further development;
- possibility of participation in archive practically all interested the organization and, probably, individuals in view of the fact that for providing data needs only exis-

tence constant (or the archive regulated by means) connections to the international Internet network;
- overcoming of a psychological problem of some organization to "lost" the acquired material. In fact, all materials remain with "owner".

3 Examples of Applications

Now presented technology of data archives creation are used for systematization and gathering of cartographic data using during scientific researches (geological, geophysical and etc.) in SSC VNIIgeosystem. The data types, collected in the archives, are as follows

- rasters of thematic maps (source and referenced)
- rasters of geological and seismic profiles (source and referenced)
- thematic vector layers (source and resulted)
- GIS-projects collecting source or/and resulting data
- text documents–reports, describing results of scientific projects

Main steps of development of the application and loading process included :

1. Verification of the register of cartographic materials, correction of mistakes, reduction to the set requirements
2. Verification of the register of digital materials, correction of mistakes, reduction to the set requirements
3. Creation of the uniform array of the spatial data displaying a geographical binding of digital materials
4. Coordination among themselves register of cartographic materials, register of digital materials and data on their geographical position.
5. Loading of the created interconnected data file in Internet GIS, control of a workplace of the user, debugging.

The project providing to the authorized user (as showing at Fig. 1):

- viewing of materials of the geologic-geophysical contents, including bibliographic data and technological stages of their processing,
- search of necessary materials both in spatial criteria, and in keywords,
- downloading of a demanded set of materials on its local computer.

As technological platform of the system we used multifunctional geoinformation server—technological platform for compilation of web-GIS application [2].

In an ideal, all data on which the link metabase have to be moved on a local resource. Thus, the backup copy of archive as a whole will be created. But the assessment of possibility of realization of this stage will be possible only after some time after project start.

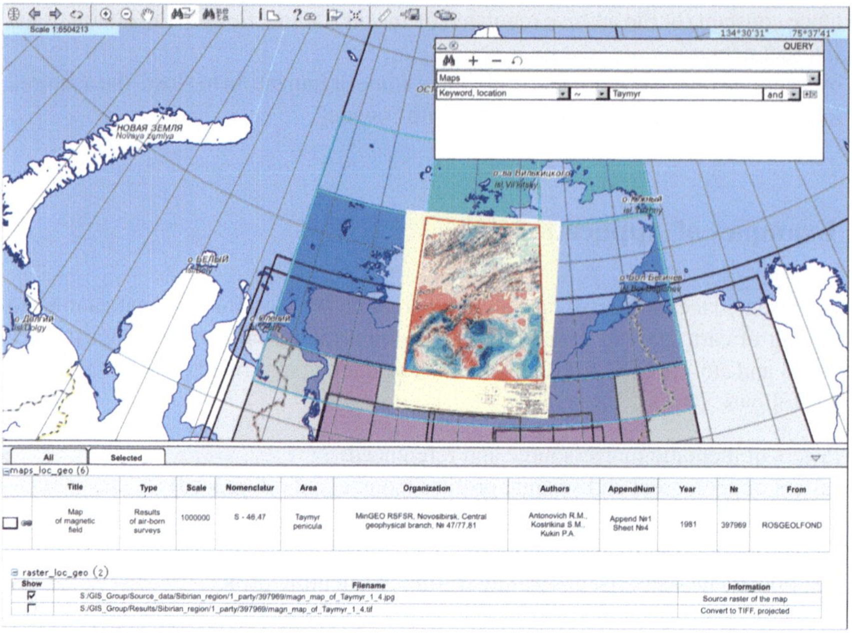

Fig. 1 Example of the window shown in the database

Using of the presented system will allow to significantly reduce of time of search and preparation of data for carry out of various works in the field of environmental management. Now the project is at a development and testing stage. We are planning to finish approbation and to make of the full-function version of archive of geodata to the middle of 2014.

References

1. Alexeev K.V., & Lyubimova A.V. (2012). Problems of integration of information resources during scientific researches. *Proceeding of International Conference IT for Geosciences EAGE*, Dubna.
2. Cheremisina E.N., Lyubimova A.V., & Soukhanov M.G. (2009). *MSG—tools for spatial data integration, processing and presentation*. 4th International Conferenace Earth from Space—The Most Effective Solution, Moscow, Russia.

Ecological Remediation Volume (ERV) in Coastal Aquifers Affected by Seawater Intrusion. Methodology and Application in the Oropesa-Torreblanca Plain (Spain)

Arianna Renau-Pruñonosa, Ignacio Morell, David Pulido and Jorge Mateu

1 Introduction

The coastal aquifers present hydrodynamic and hidrogeochemical peculiarities that add complexity to the correct groundwater resources management [9]. In them, the balance water could be modified (sometimes very notably) on a natural or anthropic way, due to variations produced in the water balance components. That process would produce the advance of the seawater intrusion [1, 4].

In the last years in Spain, and following the application of the Water Framework Directive [7, 8], research on groundwater management and sustainable use of groundwater has increased [2, 3, 5]. Such reports tend to define and quantify the submarine groundwater discharge of coastal aquifers and present interesting methodological frameworks to determine it. However, it is well known that the submarine groundwater flow is the result of all the actions taken on an aquifer. Therefore, the study of marine intrusion should not be solely focused on determining the optimal groundwater discharge, but also on quantifying, and rectifying if possible, the actions that led to the problem. The objective of this research is to define a new methodology to determine the *Ecological Remediation Volume (ERV)*. *ERV* is the volume of pumping

A. Renau-Pruñonosa (✉) · I. Morell · J. Mateu
Coastal Aquifers Research Group, Jaume I University, Castellón (Spain). IUPA, Av. Sos Baynat s/n, 12071 Castelló de la Plana, Spain
e-mail: arenau@uji.es

I. Morell
e-mail: morell@uji.es

J. Mateu
e-mail: mateu@mat.uji.es

D. Pulido
Coastal Aquifers Research Group, Spanish Geological Survey, Urb. Alcázar de Genil, 4, Edif. Zulema, 18006 Bajo, Granada, Spain
e-mail: d.pulido@igme.es

E. Pardo-Igúzquiza et al. (eds.), *Mathematics of Planet Earth*, Lecture Notes in Earth System Sciences, DOI: 10.1007/978-3-642-32408-6_114,

reduction that has to be reached in an aquifer, or a sector of it, to yield an intrusion flow lesser or equal to the estimated intrusion for the system's natural state (reference intrusion).

2 Methodology

The methodological proposal for the calculation of *ERV* stems from a constant-density mathematical flow model calibrated on a steady-state (MODFLOW). It consists in the iterative reduction of pumping involved in the intrusion process, until the water table yields an intrusion equal or lower than the estimated value for the system's natural state (reference intrusion). The following steps should be followed:

- Phase 1: Modelling the system's natural state to establish the reference intrusion. The resulting model will be named *reference model in a natural state* (*NS model*).
- Phase 2: Delimiting the areas affected by marine intrusion and trying to reduce their surface by iteratively reducing the pumping involved in the process, until the water table yields flows from the constant head level (intrusion flows coming from the sea) equal or lower than the estimated value for the system's natural state. The initial model in this phase will be named *reference model in an influenced state* (*IS model*).
- Phase 3: Calculating the *ERV* from the *IS model* and the last scenario of Phase 2 (the one associated to a flow intrusion equal or lower than the estimated value for the natural state).

The use of a constant-density model is due to their simplicity. They require less information. Nevertheless, the use of variable-density codes might give more exact results with a higher degree of complexity.

3 *ERV* of Plana de Oropesa-Torreblanca Aquifer

The Oropesa-Torreblanca Plain, a pliocuaternary coastal aquifer, is located in Eastern Spain (Fig. 1). The Plain is approximately of 80 km^2. It supports seawater intrusion due to overexploitation from the last decades of the last century (Fig. 2).

The mathematical model, used to calculate *ERV*, is a constant-density mathematical flow model calibrated on a steady-state (MODFLOW). The aquifer was built with two active layers that represent the Plain and the cretaceous aquifer (underlying). The grid was regularly spaced, 500 m $\times$ 500 m. The hydraulic conductivity and storage coefficient vary from 5 to 200 m/day, and from 1 to 10 %, respectively. The Plain borders with aquifers in the northern, western and southern part. They present hydraulic connections with the study area, which will be simulated a constant head boundary condition (CHD). The Mediterranean Sea is located to the east and it will be approached with a constant head boundary condition (CHD). The humid zone will

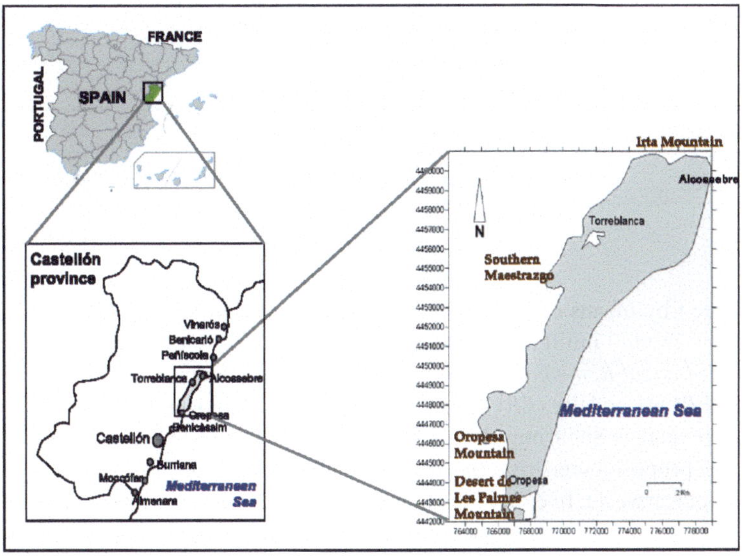

Fig. 1 Location map of Oropesa-Torreblanca Plain (Eastern Spain)

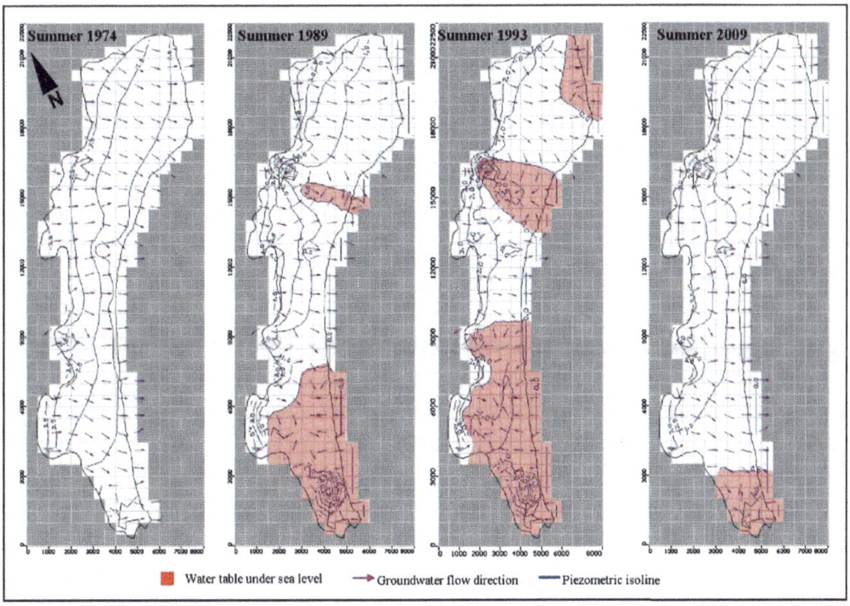

Fig. 2 Piezometric evolution of Oropesa-Torreblanca Plain according to the simulation of the groundwater flow (MODFLOW) calibrated in transitory-state (1973–2009) [6]

Table 1 *ERVh* , *ERVm* and *ERVs* Plana de Oropesa-Torreblanca aquifer

	Pumping (hm^3/year)		*ERV* (hm^3/year)
	IS model	Last scenario (Phase 2)	
Humid year *(h)*	34.5	33.5	1.0
Medium year *(m)*	34.5	32.8	1.7
Dry year *(d)*	34.5	30.8	3.7

be simulated by means of drain boundary conditions (DRN). The recharge (RCH) comes from rainfall infiltration (66.4 mm/year) and irrigation returns (20 mm/year).

The *Ecological Remediation Volume (ERV)* can be calculated for humid (*ERVh*), medium (*ERVm*), and dry (*ERVd*) years. The mean rainfall in humid *(h)*, medium *(m)* and dry *(d)* years is 884.6 mm/year, 520.3 mm/year and 337.5 mm/year respectively. In the three proposed scenarios, the *NS model* indicates that the intrusion of reference is inexistent (phase 1). In case of the humid year, the *ISh model* (phase 2) shows that the marine intrusion affects on the Southern sector and there are nine wells involved in the process. To reach there reference intrusion it is necessary to reduce 1 hm^3/year (*ERVh*) the pumpings in this sector (Table 1). The *IS model* corresponding to the medium year (*ISm model*) shows that the marine intrusion, like in the previous case, affects to the Southern sector, but in this case there are 14 wells involved in the advance of the seawater intrusion. The *ERVm* (medium year) is of 1.6 hm^3/year (Table 1). Finally, the study of a dry year *ERV* reveals that the marine intrusion (*ISs model* of the phase 2) concerns the Southern and the Central sectors, where the wells involved in that process are 24 and 7, respectively. In the phase 3, the *ERV* for a dry year (*ERVs*) is of 3.7 hm^3/year, 2.8 hm^3/year correspond to the Southern sector and 0.9 hm^3/year to the Central sector (Table 1).

4 Conclusions and Suggestions

The *Ecological Remediation Volume (ERV)* can be calculated for humid (*ERVh*), medium (*ERVm*), and dry (*ERVd*) years. The application of this methodology in the Oropesa-Torreblanca Plain shows that the advance in seawater intrusion (with respects to the natural state intrusion), in any of the three scenarios (humid, medium or dry years) is mainly due to pumping. On average, in humid years the sector affected by a higher intrusion is the Southern sector, and the corresponding *ERV* is 1 hm^3/year. In medium years, the *ERVm* is 1.7 hm^3/year and the process affects exclusively the Southern sector. In dry years the process affects the Southern and Central sectors; in this case, the *ERVd* would be 2.8 and 0.9 hm^3/year respectively. The *EVRs* total is 3.7 hm^3/year.

For the groundwater management, it proposes to use, in any case, the maximum restriction, the *ERVs*, to guarantee goundwater discharge to the sea that it would prevent the seawater intrusion and it would preserve the sustainability of the resource.

Alternatively, it would apply the reduction corresponding to the *ERVm* and to use the higher pumping reductions (*ERVs*) in case of periods abnormally dry. Following these recommendations, it proposes in case of Oropesa-Torreblanca Plain to apply the maximum restriction (*ERVs*) being this of 2.8 hm^3/year in the Southern sector and 0.9 hm^3/year in the Central sector. In case of applying the reduction corresponding to the *ERVm*, in the Southern sector the pumping reduction will be of 1.6 hm^3/year and in case of dry periods it would be necessary to apply the restrictions corresponding to the*ERVs*.

References

1. Arslan, H., & Demir, Y. (2013). Impacts of seawater intrusion on soil salinity and alkalinity in Bafra Plain, Turkey. *Environmental Monitoring and Assessment185*(2), 1027–1040.
2. Gómez, J.D., Díaz, J.A., & Meléndez, M. (2012). Cálculo de Descargas Ambientales al Mar en Acuíferos Detríticos Costeros Mediante Modelos Matemáticos de Flujo. Aplicaciones en la Demarcación del Júcar. In: New Contributions to the Knowledge of Coastal Aquifers (Vol. 1, pp. 127–234).Alicante: Instituto Geológico y Minero de España TIAC'12.
3. Instituto Geológico y Minero de España; Universitat Jaume I de Castelló (IGME-UJI): Descargas Ambientales al Mar en las Masas de Agua Subterránea Costeras de la Provincia de Castellón (Cuenca del Júcar) (2009). Madrid: Instituto Geológico y Minero de España y Dirección General del Agua.
4. Kaman, H., Kurunc, A., Sonmez, N. K., Cetin, M., Uz, B. Y., & Aslan, G. E. (2011). Preliminary investigation of seawater intrusion into Inland through Acisu Creek in Antalya, Turkey. *Journal of Food, Agriculture and Environment*, *9*(2), 612–617.
5. Meléndez, M.; Díaz, J. A., & Gómez, J.D. (2012) Estimación de las Descargas Ambientales al Mar y el Estado de la Intrusión Marina en Acuíferos Costeros de la Zona Central de la Cornisa Cantábrica de España (Asturias y Cantabria). In: New Contributions to the Knowledge of Coastal Aquifers (Vol. 1, pp. 489–499). TIAC'12. Alicante: Instituto Geológico y Minero de España.
6. Renau-Pruñonosa, A. (2013). Nueva Herramienta para la Gestión de las Aguas Subterráneas en Acuíferos Costeros. Volumen Ecológico de Remediación (VER). Metodología y Aplicación a la Plana de Oropesa-Torreblanca (MASub080.110). *Doctoral Thesis*. Castellón: Jaume I University. ISBN: 978-84-695-7809-4.
7. Unión Europea (UE-DMA) (2000): Directiva Marco del Agua, Directiva 2000/60/CE del Parlamento Europeo y del Consejo de 23 de Octubre de 2000. Diario Oficial de las Comunidades Europeas de 22/12/2000. L 327/1-327/32.
8. Unión Europea (UE): Directiva 2006/118/CE del Parlamento Europeo y del Consejo de 12 de Diciembre de 2006 (2006). Diario Oficial de la Unión Europea de 27/12/2006. L 372/19-372/31.
9. Werner, A. D., Bakker, M., Post, V. E. Vandenbohede, A., Lu, CH., Ataie-Ashtiani, B., Simmons, C.T., & Barry, D.A. (2013). Seawater intrusion processes, investigation and management: Recent advances and futures challenges. *Advances in Water Resources*, *51*, 3–26.

A Radon Risk Map of Germany Based on the Geogenic Radon Potential

Peter Bossew

1 Introduction

Indoor radon (Rn) is believed an important risk to human health. Therefore Rn monitoring programs have been initiated in many countries. Currently the European Commission is discussing Basic Safety Standards for protection against ionizing radiation, requiring inter alia to establish radon action plans and to identify so-called radon prone areas. Risk is commonly defined as probability that indoor Rn concentration exceeds a threshold.

In Germany datasets of Rn concentration in soil air, of soil permeability and of indoor Rn concentrations exist linked to geology. However indoor Rn has not been surveyed representatively, thus the data cannot be used directly for risk assessment. Instead it was decided to use geogenic radon as predictor in a probabilistic "transfer" model soil-indoor Rn. Indoor data shall be used to "calibrate" the model. Known geogenic Rn shall allow estimating the mean Rn risk in rooms and buildings with given physical characteristic as regards Rn infiltration and accumulation.

2 Materials and Methods

Data are taken from the German GIS-based Rn database which includes indoor Rn concentrations together with information on relevant house-related control parameters (floor level, year of construction, etc.), soil radon, soil permeability and geology. This study restricts to indoor Rn values of dwellings in ground floor levels of houses with basement.

P. Bossew (✉)
German Federal Office for Radiation Protection, Köpenicker Allee 120–130, 10318 Berlin, Germany
e-mail: pbossew@bfs.de
e-mail: peter.bossew@reflex.at

E. Pardo-Igúzquiza et al. (eds.), *Mathematics of Planet Earth*, Lecture Notes in Earth System Sciences, DOI: 10.1007/978-3-642-32408-6_115,

Radon potential is defined RP := $C_S/(-{}_{10}\log(k)-10)$, based on a concept suggested by [1]; C_S = Rn concentration in soil air (kBq/m^3), k = permeability (m^2). For geological classification a simplified geological map is used (after [2]).

Local probability that indoor Rn concentration C exceeds a threshold is derived from geogenic Rn, quantified by the RP. As first step, a map of the RP is generated; Second, a "transfer" model $RP(x) \rightarrow prob[C(x)>c]$ is established. Here the fact that RP and C are not observed at the same locations poses a particular challenge. For the transfer model a Gumbel copula model of the bivariate joint distribution of RP and C is estimated.

3 Results

3.1 The predictor Map

Geology is considered as a categorical predictor. This was done as in [3] as a simplified regression kriging scheme, but with Gaussian sequential simulation instead, in order to recover local conditional distributions. The raw values RP_i are transformed into $Y_i := \ln(RP_i / RP_G)$, with RP_G the mean RP over geological class G. Y_i is subjected to spatial modelling, which results in realizations of the simulation algorithm. Each is back-transformed which yields the wanted conditional distributions $F_{RP}(x)(rp)$. The resulting map is shown in Fig. 1. (Empty cells: no geological unit covered by observations could be assigned to the centre of the cell.)

This method is simple but has drawbacks: (1) RP_G are simple means which may be biased because usually the observations are not uniformly (e.g. randomly) distributed within each geological class, but may be clustered; (2) The variogram is estimated for the variable Y, although no continuity of Y(x) or RP(x) across geological borders can be anticipated.

3.2 Model of Bivariate Distribution

We want to estimate the local exceedance probability $p(x) = prob[C(x) > c|RP(x) = rp]$ which can be derived from the bivariate joint distribution $F_{C,RP}$. Introducing a copula $\Psi_{C,RP}$, we find $1-p = \partial\Psi_{C,RP}(F_C(c),F_{RP}(rp))/\partial F_{RP}(rp)$. The marginals F_C and F_{RP} were determined by spatial de-clustering of the data.

As copula model a Gumbel type is chosen for several reasons. (1) It seems appropriate to strongly right-skew data, as typical for Rn related quantities. (2) Its parameter ϑ is related to Kendall τ by $\tau = 1-1/\vartheta$, and thus easy to estimate from data. (3) It allows upper tail dependency, other than binormal copulae, which do not allow conditional prediction of high extremes. Alternatives will be explored in the future.

Inserting the Gumbel copula leads to an analytic expression of p in dependence of ϑ and the marginals $F_{RP}(rp)$ and $F_C(c)$ which can be evaluated once ϑ is known.

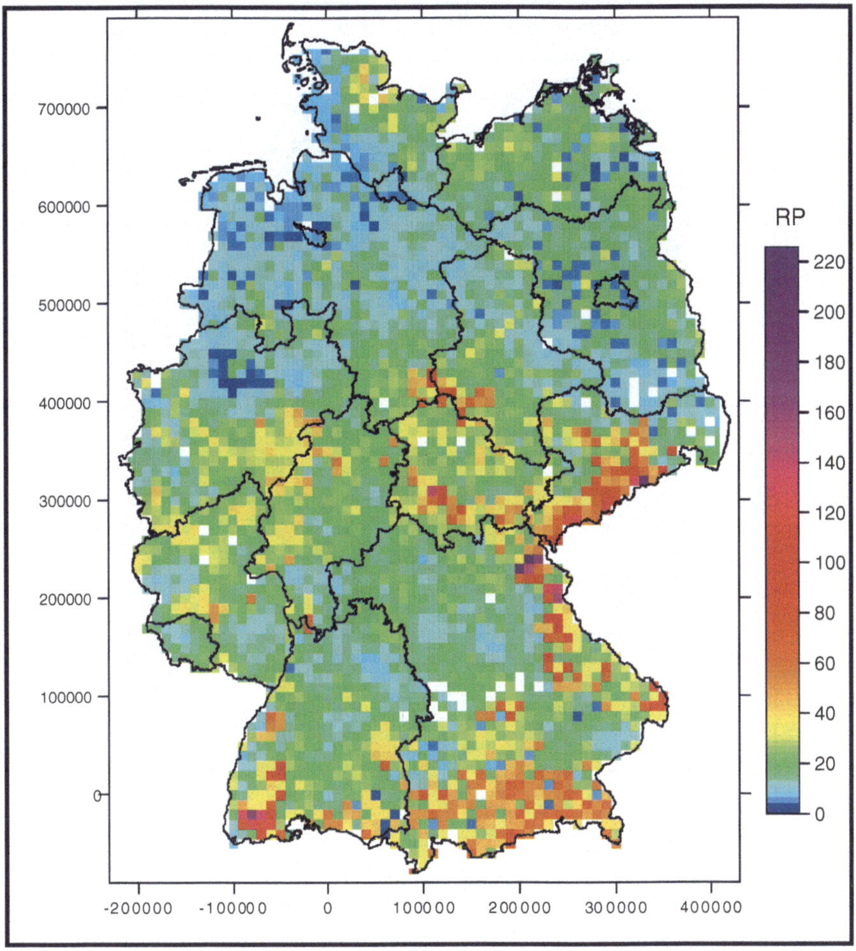

Fig. 1 Estimated expectation of the radonpotential, based on 100 realizations of sequential Gaussian simulation. Domain: territory of germany; axis units: m

3.3 Lagged Correlation

The lagged Kendall correlation $\tau(h)$ for lag h is computed for subsets of the data allowing estimation of uncertainty. The wanted value for $h = 0$ is estimated by extrapolation of $\tau(h)$ towards $h = 0$. (The method corresponds to estimating a nugget from an empirical variogram.) For the Rn data we find $\tau(0) = 0.16$ which is a rather weak correlation; this reflects the fact the indoor Rn (C) is controlled also by other factors than soil Rn (RP). The value was confirmed by correlating a collocated dataset generated by estimating (ordinary kriging) RP at the locations of C.

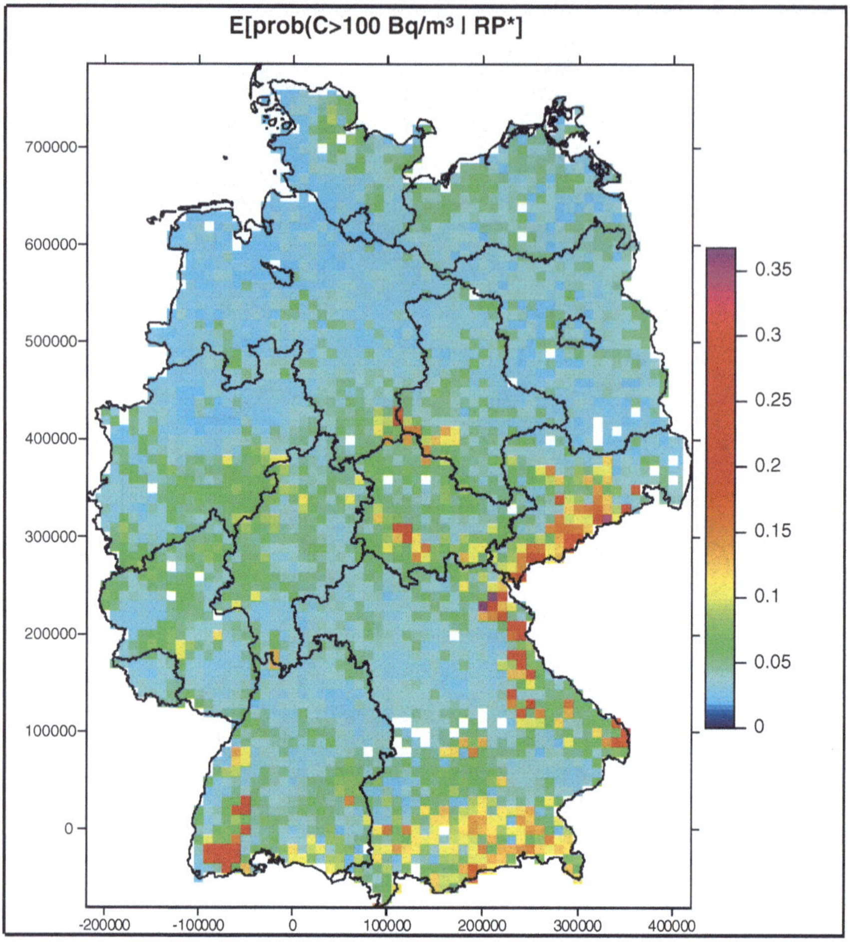

Fig. 2 Probability that indoor Rn concentration exceeds 100 Bq/m^3, estimated from the radon potential; expectation over realizations of the RP

3.4 The Rn Risk Map

The local (at unsampled x*, a grid node) expectation of the wanted exceedance probability p is calculated as

$$E_{\text{spat}}[\text{prob}(\text{C}(\text{x}^*) > \text{c})|\text{RP}(\text{x}^*)] = \int_{\omega\in\Omega} \text{d}W_{\text{RP(x)}}(\text{rp}(\omega, \text{x}^*))\text{p}(\text{rp}(\omega, \text{x}^*)),$$

where $W_{\text{RP}(x^*)}$ denotes the local conditional distribution of predictor RP (result of Sect. 3.1), rp(ω, x*) its ω-th realization at x* and Ω the sample space of the simulation, whose realizations are indexed by ω and p(rp) the exceedance probability as function of the RP (from the model Sect. 3.2). In practice this is simply the mean over realizations. The resulting map is shown in Fig. 2. Similarly the local expectation of C at x^* can be calculated, involving a second integration over $F(c|RP(x^*) = rp(\omega,x^*))$, with $F = 1-p$, to be performed numerically.

Since the transform $RP \rightarrow p(RP)$ is not linear, $E_{spat}[prob(C(x^*) > c)|RP(x^*)] \neq p(C(x^*) > c|E_{spat}(RP)(x)]$ which would be easier to calculate, but contains a bias.

4 Discussion and Conclusions

The transfer model is global in that the same model for the joint distribution of indoor and soil Rn is assumed over the domain (the territory of Germany). This may be considered an analogue to second-order stationarity in RF theory. However factors which control indoor Rn which have a regional trend may invalidate the strict assumption and render the proposed solution an approximation.

Validation of the probabilities p(x) is not easy because only relatively few cells with sufficient data are available. For those cells with higher mean RP (i.e. stronger geogenic control) the empirical probabilities lie well within a 95 % confidence interval (also recovered from the realizations, as above) of the estimated p(x).

The results turn out sensitive against choice of the marginal distributions F_{RP} and F_C resulting from data de-clustering while sensitivity against the Gumbel ϑ is relatively low.

"Complete" bivariate geostatistical modelling, e.g. by cokriging or co-simulation, has not been performed because (1) a relatively simple (in comparison) transfer model geogenic → indoor Rn was requested, and (2) because co-variography has been considered prohibitive. Simplifications such as the "Markov M1" assumption suffer from non-collocated data.

Among issues to be addressed in future are inclusion of further predictors such as geochemical quantities, extension to other categories of indoor Rn data and higher spatial resolution.

References

1. Neznal, M., Neznal, M., Matolin, M., Barnet, I., & Miksova, J. (2004). The new method for assessing the radon risk of building sites. Czech Geol. Survey Special Papers, 16, Czech Geol. Survey, Prague, p. 47. Retrieved March 29, 2013 from http://www.radon-vos.cz/pdf/metodika.pdf
2. Kemski, J., Siehl, A., Stegemann, R., & Valdivia-Manchego, M. (2001). Mapping the geogenic radon potential in Germany. *The Science of the Total Environment*, *272*(1–3), 217–230.
3. Bossew, P., Dubois, G., & Tollefsen, T. (2008). Investigations on indoor radon in Austria, part 2: Geological classes as categorical external drift for spatial modelling of the radon potential. *Journal of Environmental Radioactivity*, *99*(1), 81–97.

Quantitative Risk Management of Groundwater Contamination by Nitrates Using Indicator Geostatistics

Mario Chica-Olmo, Eulogio Pardo-Igúzquiza, Antonio Luque-Espinar, Víctor Rodríguez-Galiano and Lucía Chica-Rivas

1 Introduction

Groundwater Daughter Directive of the European Union (Directive 2006/118/EC, 2006) recognizes that groundwater is a valuable natural resource, and as such should be protected from deterioration and chemical pollution. Mainly because of agricultural practices, and also demographic growth and economical development, the amount of nitrogen released to the subsurface from fertilizers, sewage and animal wastes has increased during the last decade. Due to the high solubility and mobility of nitrate, it leaches through the soil and vadose zones to reach the water table. Thus there is an important threat that groundwater resources will become contaminated by nitrates. This is the case of the Vega de Granada aquifer in Southern Spain; a detritic aquifer that has suffered from the increase of urbanization and intensive agriculture and cattle raising. Then it is important to assess the risk of contamination of the aquifer by drawing maps of the content of nitrates in groundwater. However, the groundwater is only accessible at a finite number of wells and furthermore no

M. Chica-Olmo (✉) · V. Rodríguez-Galiano
Departamento de Geodinámica, Facultad de Ciencias, Universidad de Granada,
Avenida Fuentenueva s/n,18071 Granada, Spain
e-mail: mchica@ugr.es

E. Pardo-Igúzquiza
Instituto Geológico y Minero de España (IGME),
Ríos Rosas 23, 28003 Madrid, Spain
e-mail: e.pardo@igme.es

J. A. Luque-Espinar
Unidad del IGME en Granada,
Urbanización Alcazar del Genil, 4,18006 Granada, Spain

L. Chica-Rivas
Facultad de Ciencias de la Salud, Universidad de Granada,
Avda. Madrid s/n., 18071 Granada, Spain
e-mail: luciachicarivas@hotmail.com

E. Pardo-Igúzquiza et al. (eds.), *Mathematics of Planet Earth*,
Lecture Notes in Earth System Sciences, DOI: 10.1007/978-3-642-32408-6_116,

all those wells are monitored for nitrate content. As a result, the practical problem becomes to draw maps of the probability of exceed given thresholds of nitrate content like the drinking water nitrate regulatory limit using that limited experimental information only. This assessment of where and to what extent the groundwater is at risk of contamination by nitrates will help the environmental managers to adopt the appropriate strategies for aquifer protection [1, 2] . Geostatistics provides the right tools of spatial statistics for estimating those probabilities taking into account the spatial correlation of data. Particularly there are various forms of indicator kriging (e.g. median indicator kriging, multiple indicator kriging and various forms of indicator cokriging) that allow to estimate the probability of exceeding a threshold. In Europe the normative establishes a limit value of 50 mg/l for drinking water and a value of 37.5 mg/l for starting points for trend reversal initiatives.

2 Methodology

Giving a continuous variable like nitrate content $Z(u)$ at a particular spatial location, the indicator variable $I(u)$ at the same location is defined for a given threshold z_c as:

$$I(u) = \begin{cases} 1 & \text{if} \quad Z(u) \leq z_c \\ 0 & \text{otherwise} \end{cases} \tag{1}$$

Given an experimental sample of n data of nitrate contents $\{Z(u_1), Z(u_2), \ldots, Z(u_n)\}$ by applying (1) it is possible to have the n data of indicators $\{I(u_1), I(u_2), \ldots, I(u_n)\}$. At a location u_0 where the nitrate content has not been monitored, the nitrate content $Z(u_0)$ is obviously unknown. For risk evaluation and the design of managements strategies, more relevant than estimating the value of $Z(u_0)$, it is to estimate the probability of exceeding the threshold of interest z_c,which can be evaluated by estimating the corresponding indicator value $I^*(u_0)$ that can be interpreted as an estimate of the probability $\mathrm{P}(Z(u_0) \leq z_c\}$ and thus the probability of excess being $1 - I^*(u_0)$ and where the star as superscript means estimated value. The value $I^*(u_0)$ may be obtained by using different kriging approaches like: median indicator kriging (Me-IK), multiple indicator kriging (Mul-IK), median indicator cokriging (Me-ICoK) and linear coregionalization model indicator cokriging (LCM-ICoK). Indicator kriging uses only the threshold of interest z_c while indicator cokriging uses several thresholds (10, 37.5, 50, 75 and 100 ppm in our case study) including the one of interest. Median indicator kriging uses the variogram estimated for the median threshold as the variogram for any threshold appropriately scaled for variance (variogram sill) while LCM-ICoK uses the linear model of coregionalization, [3].

3 Results

There are 198 monitoring wells homogeneously distributed across the Vega de Granada aquifer where nitrate was measured during the year 2003. This is a detritic aquifer in Southern Spain [4] which is highly nitrate vulnerable zone (NVZ) because of a very important agricultural land use which implies diffuse contamination. Others sources of nitrate are related to a high density of population and to industrial activities. The indicator maps for thresholds of 10, 37.5, 50, 75 and 100 ppm where estimated by the four kriging methods and the results for the threshold of 37.5 are shown in Fig. 1. It may be seen how Mul-IK gives the smoother map of probabilities. Me-IK and Me-CoIK give similar maps and with a high contrast for high probabilities. Finally LCM-CoIK gives a high contrast in the low probabilities. In order to compare the methods quantitatively, a validation procedure has been performed in which ten data are taken away from the experimental data and the indicators are estimated are those locations. The validation procedure is repeated 10 times. The four methods were ranked according to their performance (mean squared error, MSE) and a scoring of 3, 2, 1 and 0 were given to each method depending on its ranking in MSE. The total score was 60 for Me-IK, 100 for Mul-IK, 50 for Me-CoIK and 87 for LCM-CoIK. Thus Mul-IK is the best. The methods that use the median indicator give the worse results, with Me-CoIK being the worse. In fact observing the map of

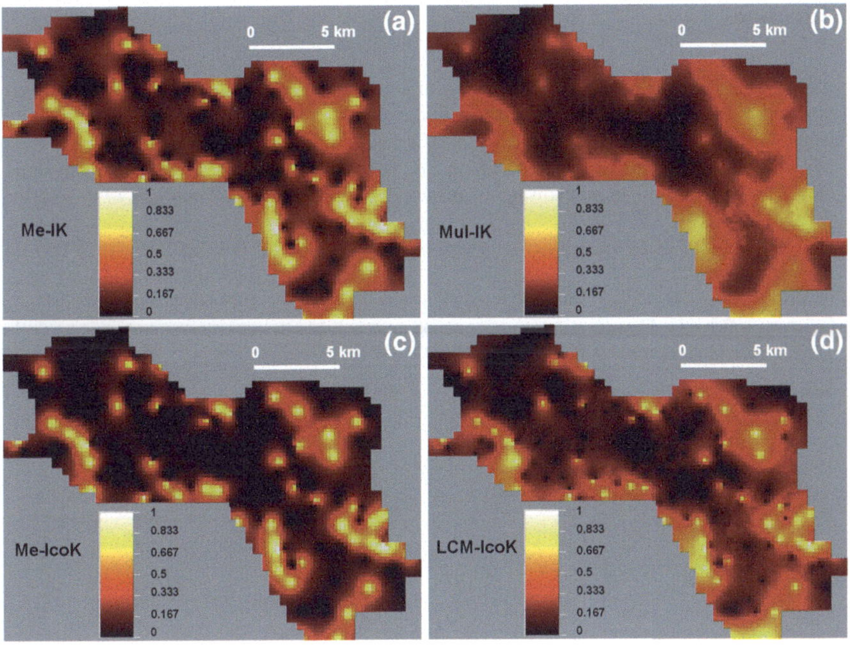

Fig. 1 Indicator probability for a threshold of 37.5 ppm of nitrate

Fig. 1b, it is possible to see the clear spatial anisotropy of nitrates in groundwater with larger continuity along the direction of the groundwater flow and with a good correlation with the permeability of the soil.

4 Conclusions

Different kriging approaches are available to estimate the probability that the content of nitrate at an unsampled location will exceed a threshold of interest. Using a validation technique it has been possible to rank the methods according to their performance, Mul-IK has given the best results and the map produced by Mul-IK is more easily interpretable. The second best method is LCM-CoIK which from a theoretical point of view should be the best because it takes information from several thresholds. However, LCM-CoIK has almost double order violations than Mul-IK and further research is needed in order to have a conclusive proof.

Acknowledgments This work was supported by research projects CGL2010-15498 and CGL2010-17629 from the Ministerio de Economia y Competitividad of Spain. We would like to thank the reviewers by their constructive criticism.

References

1. Assaf, H., & Saadeh, M. (2009). Geostatistical assessment of groundwater nitrate contamination with reflection on DRASTIC vulnerability assessment: The case of the upper litani basin, lebanon. *Water Resources Management*, *23*, 775–796.
2. D'Agostino, V., Greene, E. A., Passarella, G., & Vurro, M. (1998). Spatial and temporal study of nitrate concentration in groundwater by means of coregionalization. *Environmental Geology*, *36*(3–4), 285–295.
3. Goovaerts, P. (1997). *Geostatistics for natural resources evaluation* (483 pp). New York: Oxford University Press.
4. Pardo-Igúzquiza, E., Chica Olmo M., Rigol-Sánchez, J. P., Luque-Espinar, J. A., & y Rodríguez Galiano, V. (2011). Una revisión de las nuevas aplicaciones del cokrigeaje en Ciencias de la Tierra. *Boletín Geológico y Minero*, *122*(4), 497–516.

Geologically Constrained Groundwater Monitoring Network Optimization in Halle (Germany): A Case Study with Saltwater Intrusion Along a Fault System

Thomas Horschig, Ronny Lähne, Michael Falkenhagen and Wolfgang Gossel

1 Introduction

The foundation of Halle in 806 A.D. was based on geological phenomena, i.e. brine springs. The study region of Halle (Saale) covers 135 km^2 in the urban region. A major fault, the Halle-fault, separates two large-scale tectonic units: the Merseburg-block and the Halle-Wittenberg-block. Both blocks have a completely different structure belonging to different regional tectonic blocks. The Merseburg-block is built of units from the Zechstein (Upper Permian), the Bunter (Lower Triassic) and the Muschelkalk (Middle Triassic). The Halle-Wittenberg-block is built of permo-carbon sediments and vulcanits [3]. In the Cretaceous around the area of the halle-fault the earth crust began to fracture. The Halle-Wittenberg-Block was uplifted several hundred meters and therefore the Halle-fault enables the upcoming of high mineralized deep water. The improvement of groundwater monitoring in the urban area of Halle according to higher ranked structures is the content of this work.

2 Methods

The spatial optimization in the present work has been done using Geostatistical Temporal Spatial algorithm (GTS) [1] and the MSN Spatial Sampling Optimization software [4]. The MSN Spatial Sampling Optimization software can be used to increase the accuracy of a long term monitoring network, determining a number of points that can be added to the existing grid. In contrast GTS can be used to determine redundant wells while maintaining an appropriate level of accuracy as well as determining new sites for monitoring wells. The data set is part of the long

T. Horschig (✉) · R. Lähne · M. Falkenhagen · W. Gossel
Institute of Geosciences and Geography, Department of Hydrogeology and Environmental Geology, Martin-Luther University Halle, Von-Seckendorff-Platz 3, 06120 Halle, Germany
e-mail: thomas.horschig@student.uni-halle.de

E. Pardo-Igúzquiza et al. (eds.), *Mathematics of Planet Earth*,
Lecture Notes in Earth System Sciences, DOI: 10.1007/978-3-642-32408-6_117,

time monitoring network of Halle (Saale). In the study area 282 wells were observed in different temporal periods while measuring different parameter. For the spatial optimization the electric conductivity was chosen to be the observed parameter.

2.1 Spatial Optimization with GTS

To optimize an existing monitoring network means to sustain the information and data quality, maybe improve the quality, from the monitoring network but to reduce the cost and effort. The aim is a balance between cost saving and a loss of accuracy. GTS can be used for long time monitoring optimization (LTMO). It works with a decision-logic framework arriving an optimal design by using known semi-objective statistical and geostatistical techniques in a novel manner [1]. Those techniques are i.e Kriging and the Locally Weighted Quadratic Regression (LWQR).

2.2 Spatial Optimization with MSN

Another optimization method bases on the Mean of Surface with Nonhomogenity (MSN) theory [4]. This method combines the qualities of spatial stratification and Kriging Variance. The MSN-method decomposes a nonhomogeneous surface into smaller subareas that are locally homogeneous in the mean and calculates the spatial mean of the surface and its variance afterwards. Hu and Wang [2] developed a software package for spatial sampling optimization using the MSN theory. It can determine if a monitoring network is set up at the optimum or if additional monitoring wells are needed. To accelerate the optimization progress a Monte Carlo and Particle Swarm Optimization algorithm is implemented. The software package is freely available to download.

3 Application of the Methods

3.1 MSN: Spatial Sampling Optimization Software (MSN-SSO)

The dataset consists of 282 wells that monitored the electric conductivity. The Mean is 2392 ± 177.40 with $\alpha = 0.05$ and a standard error of the mean of 90.52. The software estimates the adding points for an optimal sampling design. The MSN-SSO software suggests adding two new monitoring wells to the network. The result is mapped in Fig. 1.

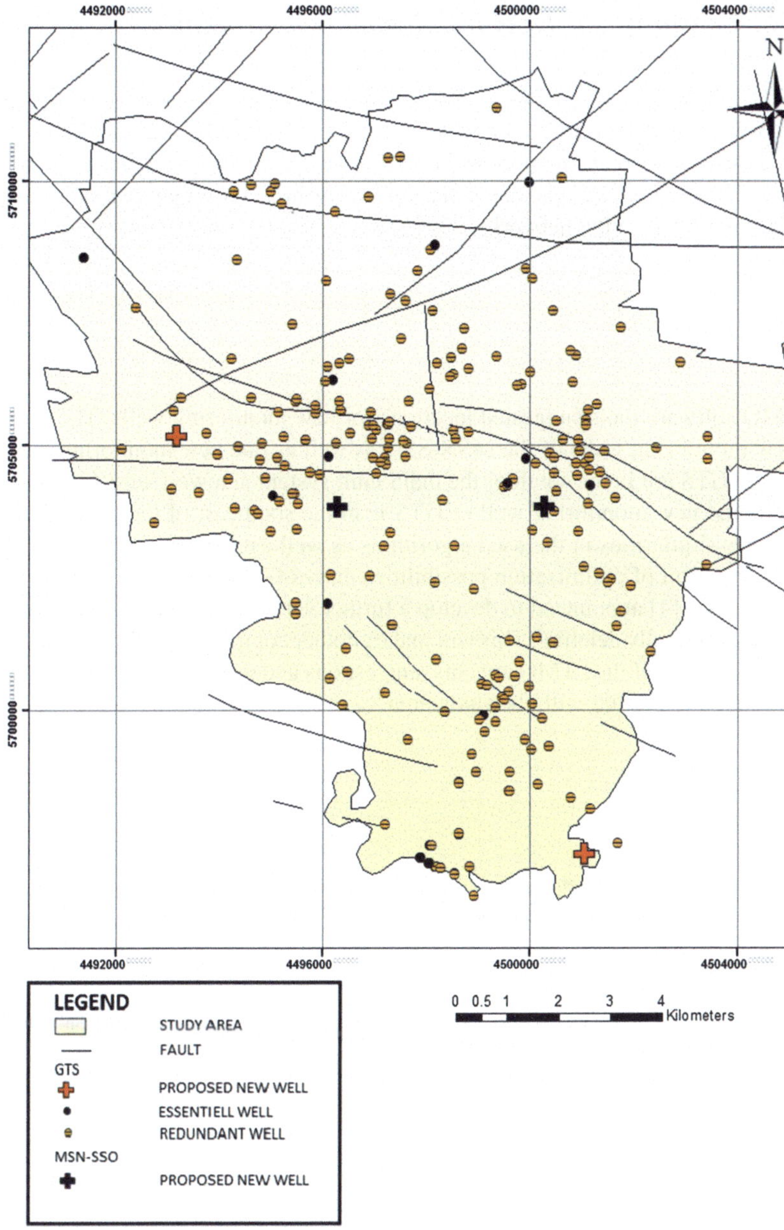

Fig. 1 LTMO Optimization result of study area

3.2 GTS: Geostatistical Temporal Spatial Optimization

While 282 wells were monitored the optimization result suggests that the monitoring of only 16 wells is essential. This conclusion results from only one variable, the electric conductivity. Choosing more variables for the optimization will decrease the number of redundant wells. Like the MSN-SSO software GTS suggests adding two new monitoring wells to the network.

4 Conclusion

Both LTMO software tools suggested locations for new monitoring wells. The suggested new monitoring wells of the MSN-SSO as well as one new monitoring well suggested by GTS are in the south of the main fault system at a west–east-line. The other suggested new monitoring well of GTS is in the southeast of the study area. That indicates similarities in the used algorithms as well as significant differences. The wider spectrum of optimization possibilities makes GTS to the more complete optimization tool. [4] announced to develop a further version that will include more functions like optimally deleting. This will make another comparison more balanced. Of course the results of the LTMO are only suggestions and would have to be proved by experts that are familiar with the study area.

References

1. Cameron, K., & Hunter, P. (2010). Optimization of LTM networks using GTS: Statistical approaches to spatial and temporal redundancy, Air Force Center for Environmental Excellence (AFCEE), pp. 1–16.
2. Hu, M.-G., & Wang, J.-F. (2010). A spatial sampling optimization package using MSN theory. *Environmental Modelling and Software*, *26*(4), 546–548.
3. Schwab, M., Herold, U., Rappsilber, I., & Thomae, M. (2006). Geologischer Überblick zur Halle-Störung. *Mitt. Geol. Sachsen-Anhalt, Band*, *10*, 9–12.
4. Wang, J.-F., & Hu, M.-G. (2009). Modeling spatial means of surfaces with stratified nonhomogeneity. *IEEE Transactions on Geoscience and Remote Sensing*, *47*(12), S.4167–4174.

Informative Trace-Element Features of Cassiterite from Tin Mineralized Zones and Breccias in the Russian Far East: Application of Logic-Informational Analysis

Nina Gorelikova, Irina Chizhova and Filipp Balashov

1 Introduction

An important informative feature of ore minerals is the association of their trace elements (Be, Ta, Nb, In, Ga, Sc, W, Fe, Mn, Ti, and Cr) that depends on the geological situation, the genetic type of the deposit, and the productivity of the mineral assemblages. The statistical analysis of trace elements in cassiterite by techniques of multivariate statistics testifies that the trace-element composition of cassiterite is controlled by such factors as the regional metallogeny, the character of magmatism, the lithology of the host rocks, and the genetic type of the deposits [1]. The analysis of geochemical associations of minerals in magmatic rocks and ores makes it possible to solve a broad range of geological problems.

2 Geology of the Vysokogorskoe Deposit and Associations of Trace Elements in Its Minerals

The Vysokogorskoe tin deposit in accretionary folded system of the Sikhote Alin was produced within a broad time span from the Cretaceous through Paleogene in relation to multiple alternating environments of Andean-type suprasubduction and a Californian-type transform continental margin [1]. Along with mineralized zones

N. Gorelikova (✉) · I. Chizhova · F. Balashov
Institute of Geology of Ore Deposits, Petrography, Mineralogy and Geochemistry of Russian Academy of Sciences, Staromonetny per., 35, Moscow, Russia
e-mail: ngor@igem.ru

I. Chizhova
e-mail: tchijova@igem.ru

F. Balashov
e-mail: balash@igem.ru

E. Pardo-Igúzquiza et al. (eds.), *Mathematics of Planet Earth*,
Lecture Notes in Earth System Sciences, DOI: 10.1007/978-3-642-32408-6_118,

and veins, tin is concentrated in economic dyke-shaped orebodies of fluid-magmatic breccias. These are extended veins and mineralized breccia zones. The veins are replaced by mineralized breccias at a depth of ~500 m. Our investigation of trace elements in cassiterite from the Vysokogorskoe deposit was conducted to clarify their distribution in the ore veins and breccias and to use these elements as indicators of ore-forming processes. No consensus is reached so far on the distribution character of trace elements in cassiterite. Our study was aimed at elucidating the distribution character of trace elements in the mineral depending on the type and composition of the ore zones and at identifying indicative elements that can characterize similarities of and differences between the ore zones and fluid-magmatic breccias and eventually provide insight into the specifics of the mineral-forming processes.

3 Mathematical Processing

In order to reveal trace-element differences inherent in cassiterite from the ore veins and breccias, we applied logical-information analysis, which is underlain by a specialized algorithm of the pattern recognition method based on studying variation sequences of objects and the similarity concept. In its core, this method is underlain by distinguishing various groups of datasets with the aim of describing objects to be tested and to estimate their significance and similarity. The identification of the informative indicators is based on the principle of the common features of objects belonging to the same class. To distinguish indicators that are informative for describing class A_k and to separate it from other classes, we use a function based on the probability that indicator j is present at the sites of class A_k and is not present at the sites of any other class, and the probability that indicator j is not present at the sites of class A_k but is present at the sites of other classes.

Specialized functions are used to develop a decision rule for site identification: a site belongs to a class that has the maximum value of the function. As a result, we have a relation between coefficients for any examined site that characterizes the degree of similarity of this site with other classes.

Our database comprises eleven-element compositions of 37 cassiterite samples from ore veins and fluid-magmatic breccias collected from levels 179 m up to 435 m. The parameters used herein are In, Sc, W, Nb, V, Cr, Be, Ti, Fe, and Mn.

4 Results of Mathematical Processing

Our study was focused on the following two major tasks: (1) to conduct comparative analysis of the mineralized zones in order to reveal their similarities and distinctions; (2) to identify representative distinctive signs of the ore veins and ore breccias. In solving the former task, we revealed ranges of indicators of the primary meanings of signs (elements) that characterize every class (zone) and are based on distinguishing

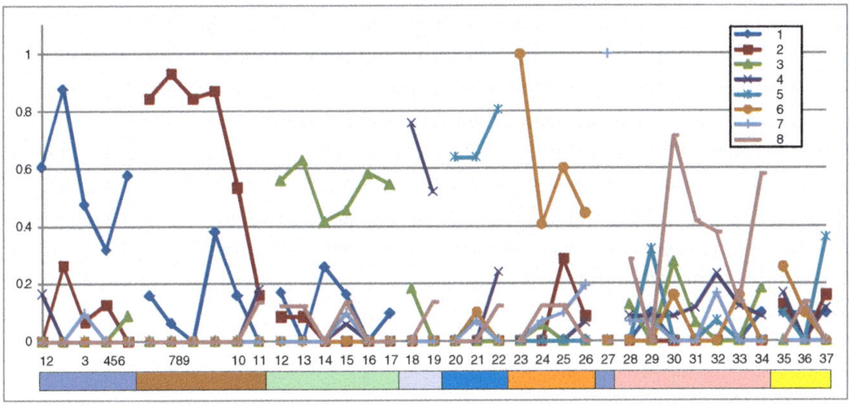

Fig. 1 Identification of objects from standard samples (*1–34*) and recognition objects based on the obtained estimations. Legend. Zones: *1–5* Kulisnaya; *12–17* Vostok 8; *18*, *19–58* Tectonicheskaya; *20–22* Severnaya; *23–26* Explozivnaya; *27* Aiskaya; *28–34* Turmalinovaya

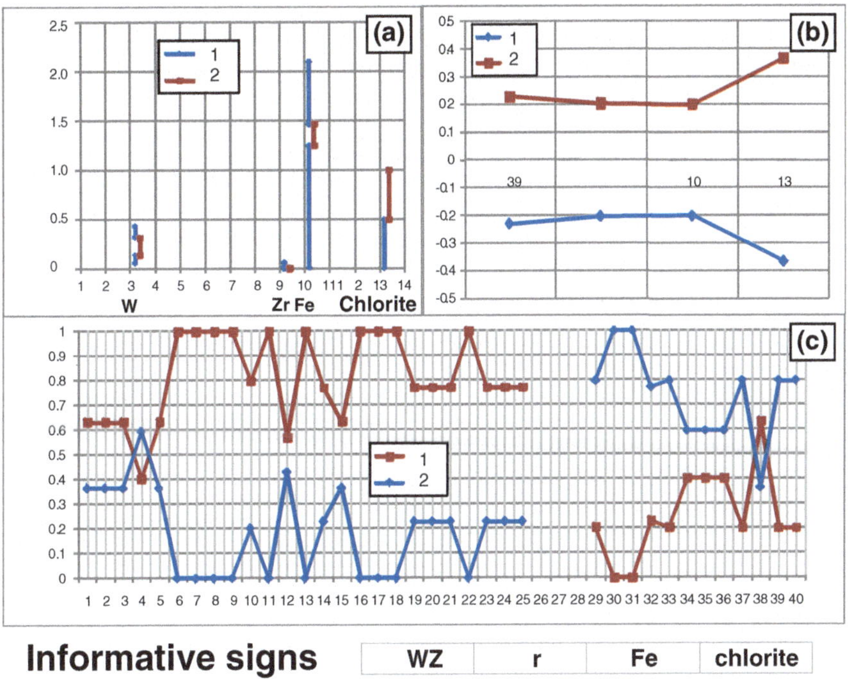

Fig. 2 Identification of **c** ore veins and fluid-magmatic breccias on the basis of **a** interval-indicators and **b** weights of signs. Legend: *1* ore veins; *2* breccias

the intervals and informative weights of signs; we conducted the recognition of the studied objects (Fig. 1). Every mineralized zone shows unique features of its ore-forming processes, and all of the zones compose a homogeneous family. The most informative signs for the recognition are W, Fe, Mn, and V.

When solving the second task, we found signs characterizing ore veins (class I) and fluid-magmatic breccias (class II). Intervals-indicators and informative weights are established for every class, and two classes are then identified on this ground. The informative signs for the recognition of the two classes are W, Zr, Fe, and also the content of chlorite in the ores (Fig. 2).

5 Conclusions

Our comparative mathematical analysis revealed significant differences between the studied classes of the database, with these differences manifested in specific interval-indicators of elements (signs). The established intervals-indicators and informative weights of signs permitted us to conduct the identification of the objects. The differences in the cassiterite composition from ore veins and breccias testify that the mineral deposition parameters (T°, redox-potential, pH and Eh, etc.) were different. This method can be applied to assay ore veins and conduct predictive metallogenic analysis of unexplored areas.

Acknowledgments The present research was executed under financial support of the Russian Foundation for Basic Research (projects 11-05-00006 and 11-05-00942) and Earth Science Devision (program 2).

References

1. Gorelikova, et al. (2006). Discriminating geodynamic regimes of tin ore formation using trace element composition of cassiterite: the Sikhote Alin case (Far Eastern Russia). Composition data analysis in geosciences. Geological Society, London, pp. 43–57.
2. Chizhova, I. A. (2010). Logic-informational modeling while forecasting metallogenic analysis of prospective areas. In *Works of IGEM RAS*. Issue 4 (1995). Modern problems of ore geology, petrology, mineralogy and geochemistry (pp. 59–84). Moscow: IGEM RAS (in Russian).

A Spatial statistical Approach for Sedimentary Gold Exploration: A Portuguese Case Study

Pierre Goovaerts, Teresa Albuquerque and Margarida Antunes

1 Introduction

In Portugal, ore extraction and processing has been an important economic activity and mines were actively developed until the early 1970s. The recent spike in gold prices and technologic development make extraction and processing more effective, leading to a regain interest in abandoned gold mining areas with a few experimental explorations in Alentejo, southern Portugal.

Geochemical cartography is a goal in mining prospection since the late 1920's. Recent development of analytical methods and computational resources facilitates the implementation of geochemical mapping and its use in natural resources management [1]. Geochemical modeling in environmental applications is mostly oriented to the recognition and quantification of anthropogenic impacts. An accurate characterization of the natural background values is an essential and unavoidable step to evaluate the influence of mining activities on the environment.

The aim of this manuscript is the development and application of a spatial statistical approach for sedimentary gold exploration in an old abandoned mining region—Monfortinho region. The analysis focuses on heavy metals (Fe, Ba, Cu, Cr, B, Zn, Sb, Pb, Sn, Ni, Mn, Be, Mo, Co, Y, Cd, Ag, V, As, W, Nb and U) and is based on a point support dataset of 376 stream sediment samples.

P. Goovaerts (✉)
BioMedware, Inc., 121 W. Washington St., 4th Floor-TBC, Ann Arbor, MI 48104, USA
e-mail: goovaerts@biomedware.com

T. Albuquerque · M. Antunes
CIGAR—Geo-Environmental and Resources Research Center, FEUP,
Oporto and Polytechnic Institute of Castelo Branco, 6001-909 Castelo Branco, Portugal

E. Pardo-Igúzquiza et al. (eds.), *Mathematics of Planet Earth*,
Lecture Notes in Earth System Sciences, DOI: 10.1007/978-3-642-32408-6_119,

2 Methodology

2.1 Sampling

Around mine tailing sites, the mineralogical content of the material exploited consists of inert materials from the gangue constituent's mineralization or mineral constituents of rocks [2]. The presence of anomalies in some chemical elements in the surrounding areas of tailings or mineralized areas indicate the action of dominant wind and transport of fine dust from the superficial layers of the heap [3]. The stream sediments resulting from the alteration of rocks by various physical and chemical processes are mobilized, transported and deposited along the water lines.

The geochemical composition of stream sediments and their spatial distribution in the study area, were characterized using a total of 376 representative samples, collected in a narrow region ranging from 50 m upstream to 100 m downstream the streams' confluences [4] (Fig. 1). Almost all water lines correspond to open valleys, so our point-support stream sediments samples correspond to incipient and not evolved soils. All the samples were collected on schist and were prepared through reduction, drying and grinding. Total concentration of As, B, Cu, Ba, Pb, Zn, Ni, Sb, Mn, Be, Mo, V, Co, Y, Cd, Nb, Fe, Cr, Ag, Au, W and U were analyzed by ICP-AES, with a precision of 20 ppm for As and 10 ppm for the other elements [4]. Tin and W were analyzed by X-ray fluorescence spectrometry and plasma emission with a precision of 10 ppm [4].

Gold was determined only for the 12 samples collected inside the old mine area. Its value was predicted at other locations using linear regression ($R^2 = 0.46$) and four

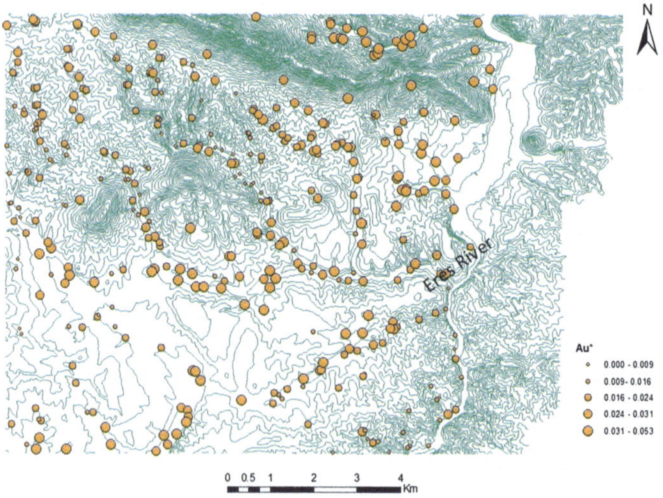

Fig. 1 Location map of sediment samples. The size of *yellow dots* is proportional to the gold concentration

metals (Fe, As, Mn and W) which are known to be mostly associated with the local gold's paragenesis. Fig. 1 shows the location of samples with the gold content that was either measured directly or estimated by regression. In the subsequent analysis, these values were assumed to be known with certainty. Indeed, the main objective of the present study is the visualization and delineation of potential zones of low and high values for future prospections instead of the accurate estimation of gold content.

2.2 *Geostatistical Methodology*

The delineation of zones of high and low contents in gold was conducted through the application of local cluster analysis (LCA) [5]. The basic idea is to compute at each grid node a local indicator of spatial autocorrelation (LISA) and test whether this statistic is significantly positive, indicating the existence of an aggregate of grid nodes with similar gold content. Since the data are not gridded, the first step is the derivation of gold content at all grid nodes. Kriging is not recommended for this step because its smoothing effect will cause the detection of artificial clusters. Following Goovaerts [6], gold content was first simulated using sequential Gaussian simulation and each of the 100 realizations underwent a LCA to identify each grid node that belongs to a cluster of small gold content (low value surrounded by low values) or a cluster of large gold content (high value surrounded by high values). The one hundred classified maps were then processed to derive the most likely classification of each node and the associated likelihood (i.e. frequency of occurrence of that class). This approach has the advantage of incorporating the uncertainty attached to the gold map through the local cluster analysis.

3 Results and Conclusions

Figure 2b shows the average of 100 simulated maps of gold content which indicates higher values along the Erges River and downstream the abandoned gold mines. Each

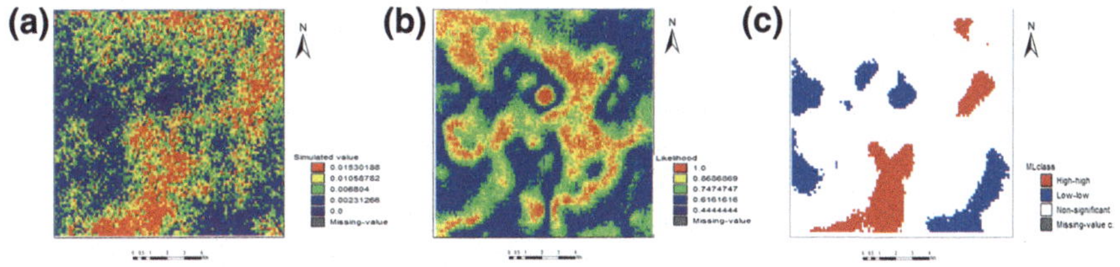

Fig. 2 **a** Average simulated map of gold content; **b** Likelihood map for the classification into clusters of low or high values; **c** Classes with the highest frequency of occurrence

simulated map underwent a LCA, leading to the allocation of some grid nodes to clusters of low gold content (Low-low) or high gold content (High-high), whereas the LISA statistic was not significantly different from zero at other grid nodes. Figure 2b shows the frequency of the most likely class that is displayed in Fig. 2c. Lower frequencies indicate zones of changes (i.e. boundaries) or transitions between classes of values. The location of hot and cold spots shows a clear Au enrichment along the Erges River downstream the old abandoned sedimentary mineralization.

Acknowledgments We are grateful to Instituto Geológico e Mineiro (Portugal) for the data on sediments.

References

1. Antunes, I. M. H. R., & Albuquerque, M. T. D. (2013). Using indicator kriging for the evaluation of arsenic potential contamination in an abandoned mining area (Portugal). *Science of the Total Environment*, *442*, 545–552.
2. Maroto, A. G., Navarrete, J., & Jimenez, R. A. (1997). Concentraciones de metales pesados en la vegetación autoctona desarrollada sobre los suelos del entorno de una mina abandonada. *Boletín Geologico y Minero*, *108–1*, 67–74.
3. Santos Oliveira, J. M., Pedrosa, M. Y., Canto Machado, M. J., & Rochas Silva, J. (1998). Impacte ambiental provocado pela actividade mineira. Caracterização da situação junto da Mina de Jales, avaliação dos riscos e medidas de reabilitação. Actas do V Cong. Nac. Geol. *84/2*, E74–E77.
4. Instituto Geológico e Mineiro (1988). Reports from a Prospecting project for tungsten, tin and associated minerals from Góis-Segura. Metalic Mineral Prospecting Section. Porto, Portugal (pp. 10).
5. Anselin, L. (1995). Local indicators of spatial association-LISA. *Geographical Analysis*, *27*, 93–115.
6. Goovaerts, P. (2006). Geostatistical analysis of disease data: visualization and propagation of spatial uncertainty in cancer mortality risk using Poisson kriging and p-field simulation. *International Journal of Health Geographics*, *5*, 7.

Coalbed Methane Production Analysis and Filter Simulation for Quantifying Gas Drainage from Coal Seams

C. Özgen Karacan and Ricardo A. Olea

1 Introduction

Coal-seam degasification, or coalbed methane (CBM) production, prior to coal mining is an indispensable practice for reducing gas-in-place (GIP) in the coal for controlling methane emissions during underground mining, for controlling greenhouse gas emissions from coal seams and also for using produced methane as an unconventional energy source [1]. In order to be able to take full advantage of the benefits of CBM production, however, properties of coal, heterogeneities in the field and the coal-seam reservoir properties should be well understood.

Gas and water production rate analysis of CBM wells help determining dynamic reservoir properties of the coal seam, which change during degasification, at a given well location. Therefore, the ability to determine coal reservoir properties at spatial well locations is important for predicting high-flow-capacity areas of the reservoir and for estimating GIP and its change between particular production periods. Moreover, geostatistics can be used to determine the values of GIP at intervening spaces between these well locations for evaluating gas drainage efficiency spatially, and for benefits to miner safety, greenhouse gas reduction and energy production.

This paper presents an application that combine filter simulation with time-varied reservoir properties obtained from history matching of gas and water production from 86 degasification wells in a 12,900-acre field in Alabama. GIP in the Jagger seam of the Mary Lee coal group was simulated between 1987 (initial time) and 2011 to estimate degasification efficiency in this coal and its uncertainty.

C. Ö. Karacan (✉)
Office of Mine Safety and Health Research, NIOSH, 626 Cochrans Mill Road,
Pittsburgh, PA 15236, USA
e-mail: cok6@cdc.gov

R. A. Olea
Eastern Energy Resources, USGS, 12201 Sunrise Valley Dr., MS 956,
Reston, VA 20192, USA
e-mail: rolea@usgs.gov

E. Pardo-Igúzquiza et al. (eds.), *Mathematics of Planet Earth*,
Lecture Notes in Earth System Sciences, DOI: 10.1007/978-3-642-32408-6_120,

2 Technical Work and Results

This work has three main phases; (1) Production history matching of gas and water rates from 86 vertical CBM wells to determine time-dependent reservoir properties, (2) Calculation of volumetric GIP at spatial well locations, and (3) Modeling of GIP in the coal seam using filter simulation technique. Production history matching analyses were completed using Fekete's F.A.S.T. CBM^{TM} software version 4.7. The data obtained from history-matching results were used to calculate volumetric GIP values and CBM production at each well location [2]. For geostatistical simulations of this data, Stanford Geostatistical Modeling Software (SGeMS) [3] implementation of the filter simulation, FILTERSIM, was employed [4, 5].

Methane GIP for the Jagger seam was calculated using reservoir data obtained from production history matching. The GIP values at wellbore grids (0.92 acre each) were calculated for grids corresponding to each of the 86 degasification wells. Table 1 provides basic statistics of these data (in million cubic feet (MMscf) per 0.92 acre). The values given in Table 1 show that initial mean GIP of Jagger seam, at borehole locations, was 1.29 MMscf, which decreased to 0.81 MMscf in 2011, as a result of degasification in the study area between start of degasification and 2011.

The univariate statistical GIP data given in Table 1 refer to the potential amount of methane capacity of Jagger seam at a given date from a 0.92 acre area, if GIP is assumed to be constant throughout the study area. Although point-wise data and evaluation of GIP can be helpful, this approach is a simple average and does not account for spatial differences between data locations. Therefore, geostatistical simulation was used to assess spatial properties of CBM production efficiency.

The spatial data locations for FILTERSIM are shown in Fig. 1. For modeling, 86 data points were used in simulation grids that had 115 and122 columns and rows, respectively. Each grid cell was 200 ft in x- and y-directions to give a grid area of 0.92 acre. Thus, each simulation grid had 14,030 grid cells and represented a total area of 12,900 acres.

FILTERSIM aims to capture patterns or structures from training images (TI) and condition them to local data. There are no strict rules regarding generation of TIs. In this work, we aimed for statistical and spatial representation of data and the presence of geological discontinuities in the TIs. For this purpose, first order Voronoi decomposition (nearest neighbor) was employed by placing faults into grids as discontinuities based on their spatial locations. Figure 1 shows initial GIP and its TI generated for the Jagger seam. The TIs for each case were further examined using

Table 1 Basic statistics of GIP values calculated at 86 wellbore locations for Jagger seam.

Time	Min.	Max.	Mean	Std. D.	Q5	Q50	Q95
Initial	0.821	1.688	1.292	0.200	0.979	1.284	1.635
2011	0.256	1.589	0.810	0.274	0.387	0.787	1.312

Q values are at the specific quantiles of the distribution

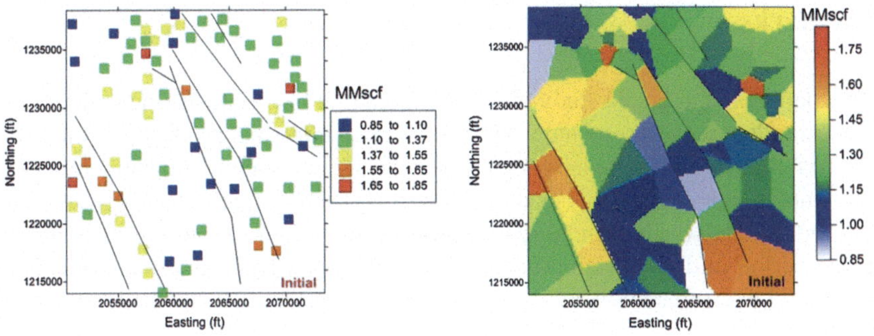

Fig. 1 Initial GIP values of Jagger seam calculated for the CBM well locations and the TI representing this data. *Black lines* show the faults in the area

Q–Q plots and semivariograms for statistical and spatial representativeness before using them in FILTERSIM.

The FILTERSIM simulations were conditioned to hard data only by optimizing simulation parameters and by checking the Q50 realizations against TIs. These plots were also compared with the Q–Q plots of hard data and TI and their statistics. Eventually, 100 realizations were generated for quantifying initial and the 2011 GIP values of the Jagger seam in the study area. Figure 2 shows the Q50 realizations of GIP for these two periods, and also GIP histograms from all realizations.

Simulation results and their distribution indicate that Jagger seam had an initial methane quantity between 16 and 18 Bscf. Due to degasification activities in the area, methane quantity in the coal seam decreased by ~6 Bscf to values between 10.5 and 12 Bscf (~35 % decrease) in 2011. Furthermore, realizations spatially show where the most efficient degasification was achieved. Based on Q50 realization shown in Fig. 2, most of the methane was produced from Northern margin of the area, as well as regions in the northeast outside of the faults.

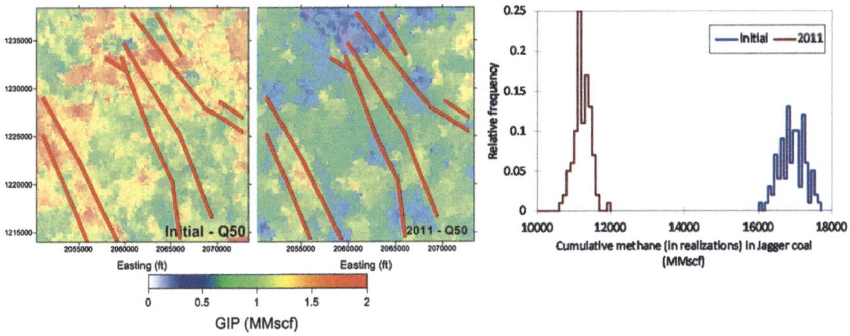

Fig. 2 Q50 (50 % quantile) realizations of initial and 2011 GIPs, as well as GIP histograms from all 100 realizations. *Red lines* show the faults in the area

3 Remarks

In this paper, an integrated study of CBM production history matching and geostatistical simulation was presented to estimate GIP quantity and production efficiency from coal seams. An example was given for the Jagger seam of Alabama by using filter simulation as the geostatistical method. It was shown that statistically and spatially representative TIs were able to reproduce both the patterns and the data for a given period in degasification history. The data obtained from these simulations can be used to estimate GIP, its change in time, and the uncertainty associated with predictions.

References

1. Karacan, C. Ö., Ruiz, F. A., Cote, M., & Phipps, S. (2011). Coal mine methane: a review of capture and utilization practices with benefits to mining safety and to greenhouse gas reduction. *International Journal of Coal Geology*, *86*, 121–156.
2. Karacan, C. Ö., Olea, R. A., & Goodman, G. (2012). Geostatistical modeling of gas emission zone and its in-place gas content for pittsburgh seam mines using sequential gaussian simulation. *International Journal of Coal Geology*, *90–91*, 50–71.
3. Remy, N., Boucher, A., & Wu, J. (2009). *Applied geostatistics with SGeMS*. Cambridge: A User's Guide. Cambridge University Press.
4. Zhang, T., Switzer, P., & Journel, A. G. (2006). Filter-based classification of training image patterns for spatial simulation. *Mathematical Geology*, *38*, 63–80.
5. Wu, J., Zhang, T., & Journel, A. G. (2008). Fast FILTERSIM simulation with score-based distance. *Mathematical Geosciences*, *40*, 773–788.

The Relationship Between Skin and Apparent Wellbore Radius in Fractal Drainage Areas

Tom Aage Jelmert

Nomenclature

- B : Fluid volume factor
- k_w : Permeability at the reference length
- p : Pressure
- q : Production rate
- μ : Viscocity
- D : Fractal dimension
- h : Thickness
- p'_w : Ideal pressure
- S : Skin factor
- θ : Connectivity index

1 Theory

Barker [1] proposed a generalized radial flow model for flow in a sparse fracture network. Chang and Yortsos [2] proposed an alternative model based on the additional assumption of fractal geometry. We use the fractal model. Predictions based on this model give a single answer, but the model is not deterministic. Many realizations are possible. A prediction may be regarded as an average or the expected value for known fractal dimension and conductivity index.

Consider steady state flow in a fractal drainage area with Euclidian dimension $d = 2$, which corresponds to cylindrical geometry. Then, the flow coefficient [2] is:

$$\beta = D - \theta - 1. \tag{1}$$

T. A. Jelmert (✉)
Norwegian University of Science and Technology, NTNU, n-7491 Trondheim, Norway
e-mail: tom.aage.jelmert@ntnu.no

E. Pardo-Igúzquiza et al. (eds.), *Mathematics of Planet Earth*,
Lecture Notes in Earth System Sciences, DOI: 10.1007/978-3-642-32408-6_121,

The permeability- and porosity functions are: $k(r) = k_w r_D^{\beta-1}$ and $\varphi(r) = \varphi_w r_D^{D-2}$. A homogeneous reservoir is characterized by $\beta = 1$, $D = 2$ and $\theta = 0$. The radial flow equation for a fractal reservoir, Jelmert [3], is:

$$p(r) = p'_w + \frac{q\mu B}{2\pi k_w h}\left(\frac{1}{1-\beta}\left(r_D^{1-\beta} - 1\right)\right). \tag{2}$$

The model has extensions to high velocity flow [4].

The value $\beta = 1$is not included in Eq. (2). This case must be solved for separately. Then, $D - \theta = 2$. Use of L'Hospital's rule leads to the conventional (logarithmic) result, see Eq. (5). There is a continuous transition over the singularity. The limit is akin to the homogenous reservoir solution. If in addition, $\theta = 0$, then the limiting behavior is the homogeneous reservoir solution.

The apparent wellbore radius may be thought about as the fictitious wellbore radius which makes the ideal wellbore pressure (without skin) equal to that of an equivalent actual well with skin, p_w. See Fig. 1.

The left and right hand triangles have the same slope, m, hence:

$$\frac{\Delta p_S}{1 - r_{aD}^{1-\beta}} = \frac{\Delta p_e}{r_{eD}^{1-\beta} - 1}. \tag{3}$$

Radial distances r_a and r_e are the apparent wellbore radius and distance to the external boundary respectively. After a short calculation, we obtain:

$$S = -\frac{1}{1-\beta}\left(r_{aD}^{1-\beta} - 1\right). \tag{4}$$

The dimensionless distance is: $r_D = r/r_w$. Note that $r_a = r_w$ for $S = 0$

The traditional logarithmic relationship is included in Eq. (4) as limiting behavior.

Fig. 1 Pressure versus radial distance profile

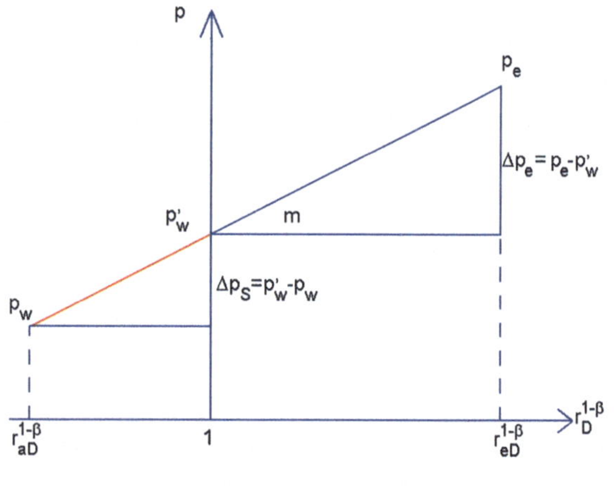

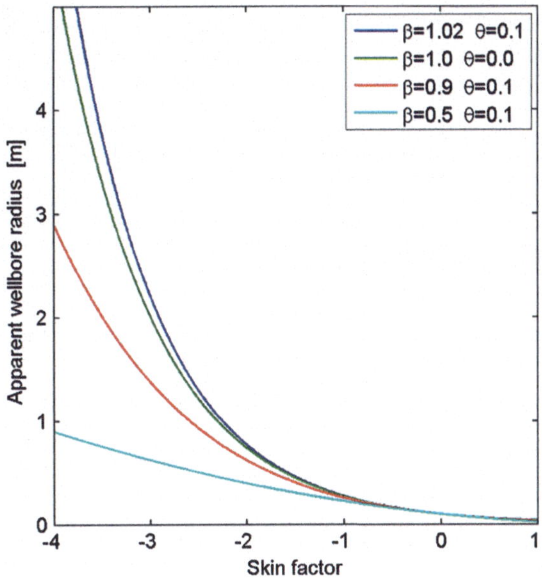

Fig. 2 Apparent radius versus the skin factor

$$S_{\beta\to 1} = -\lim_{\beta\to 1}\left(\frac{1}{1-\beta}\left(r_{aD}^{1-\beta}-1\right)\right) = -\ln \mathrm{r}_{\mathrm{aD}}. \tag{5}$$

2 Discussion

The skin factor, which may be obtained by well testing, has influence on the productivity of wells in fractal drainage areas [3, 5]. For a negative skin factor, the wellbore pressure, p_w, is higher than on the reservoir side of the skin, p'_w. Such behavior is not a natural phenomenon. It is problematic to include a negative skin directly into an analytical well test simulator. Use of an apparent wellbore radius avoids this problem.

Mathematical model equations for well test interpretation can be solved for a limited number of unknowns. These depend on many variables. Hence, information or estimates from other sources are necessary. A numerical estimate can be plugged into the equation or used as check on a value obtained by well test interpretation.

The reference length for the permeability, k_w, is arbitrary. We recommend the wellbore radius, r_w. Then, it may be possible to estimate k_w by core analysis and well logging.

There is a theoretical lower limit for the skin factor. The apparent wellbore radius cannot by larger than the external radius, r_e. Then:

$$S_{\min} = -\frac{1}{1-\beta}\left(\mathrm{r}_{\mathrm{eD}}^{1-\beta}-1\right). \tag{6}$$

The skin factor may assume large positive values when $\beta \geq 1$. There is a theoretical maximum value for $\beta < 1$, which is given by the condition $r_a \geq 0$, which yields:

$$S_{max} = -\frac{1}{1-\beta}. \tag{7}$$

3 Conclusions

The traditional (logarithmic) relationship between the apparent wellbore radius and skin is included in the generalized equation as limiting behavior.

The relationship between the dimensionless apparent wellbore radius and skin depends on the flow coefficient, β.

The traditional (logarithmic) pressure profile is included in the generalized equation as limiting behavior.

A fractal drainage area gives rise to pressure behavior like a homogeneous reservoir when $D - \theta = 2$.

The reference length, r_w is arbitrary. We recommend use of the wellbore radius since the rock properties may be estimated from core analysis and logs.

The generalized apparent wellbore radius may be used to generate well test type curves for negative values of the skin factor.

References

1. Barker, J. A. (1988). A generalized radial flow model for hydraulic tests in fractured rock. *Journal of Water Resources Research*, *24*(10), 1796–1804.
2. Chang, J., & Yortsos, Y. C. (1990). Pressure transient analysis of fractal reservoirs. *Journal of SPE Formation Evaluation*, *5*(1), 31–38.
3. Jelmert, T.A. (2009). Productivity of fractal reservoirs. *SPE 128080, Annual Technical Conference and Exhibition*, AlKhobobar, May 9–11, 2009.
4. Jelmert, T.A. (2010). High velocity flow in a fractal reservoir, IAMG Annals of Mtg, Budapest, Hungary, 29 August–2nd September.
5. Jelmert, T.A. (2011). Productivity of bounded drainage areas with fractal rock properties. *IAMG Annals Mtg*. Salzburg, 5–9 September.

Parameterization of Channelized Training Images: A Novel Approach for Multiple-Point Simulations of Fluvial Reservoirs

Mohamed M. Fadlelmula, Serhat Akin and Sebnem Duzgun

1 Introduction

The scarcity of geological subsurface information is one of the most significant sources of uncertainty in a reservoir modeling process. Thus, stochastic simulation techniques are developed to systematically take such kind of uncertainty into account. These techniques generate multiple equiprobable reservoir models while maintaining the geological structure. The traditional two-point based geostatistical modeling techniques, such as the sequential indicator simulation, fail to capture complex geological structures [1, 2]. In contrast, the multiple-point simulations (MPS) technique, which is based on a training image (TI) concept, is proven to be a powerful technique in modeling such complex structures [3]. The simulation carried out by the MPS technique is taking the patterns from the training image and anchoring them to the actual hard data [4].

Selection of the right TI that represents the target spatial phenomenon is crucial for an effective use of MPS technique [5], but it involves uncertainty. In their works Boisvert et al. [5] and Arslan [6] addressed this type of uncertainty. However, the construction of TIs involves uncertainty too. Some of the parameters that are responsible for such type of uncertainty are investigated by Strebelle [1] where he studied the sensitivity of MPS to the size of the TI, width of the structure displayed by the TI, and the rotation of the TI. Nevertheless, the literature addressing the issue of

M. M. Fadlelmula (✉) · S. Akin
Department of Petroleum and Natural Gas Engineering,
Middle East Technical University, Ankara, Turkey
e-mail: mfmining@yahoo.com

S. Akin
e-mail: serhat@metu.edu.tr

S. Duzgun
Department of Mining Engineering, Middle East Technical University,
Ankara, Turkey
e-mail: duzgun@metu.edu.tr

E. Pardo-Igúzquiza et al. (eds.), *Mathematics of Planet Earth*,
Lecture Notes in Earth System Sciences, DOI: 10.1007/978-3-642-32408-6_122,

TI construction and the uncertainty associated is very limited. Thus, this paper was developed with the aim of analyzing the impact of such uncertainty on simulated reservoir models.

2 Methodology

To analyze the impact of TI uncertainty on reservoir modeling, a synthetic fluvial channel reservoir model (original reservoir) is generated utilizing the dataset of Stanford VI project of Remy et al. [7]. The modeling is performed with a proposed MPS methodology. This methodology is summarized as follows:

1. Generate a reservoir facies realization using the facies hard data and MPS algorithm; the Single Normal Equation Simulation (SNESIM) of Stanford Geostatistical Modeling Software (SGeMS)
2. Calculate the experimental variogram (in 3D) of the generated realization
3. Fit a variogram model to the experimental variogram
4. Generate the corresponding porosity file (in 3D) using the Sequential Gaussian Simulation (SGSIM) of SGeMS
5. Extract the porosity data out of SGeMS
6. Upscale the data
7. Calculate the corresponding permeability values using defined logarithmic relations
8. Construct a reservoir model using the Builder tool of CMG software then input the porosity and permeability data calculated into the model
9. Run the model by CMG's black oil simulator IMEX
10. Extract OOIP and ROIP values.

After modeling the original reservoir, the channel structure of the original TI is represented mathematically in two dimensions (2D) with a Sine function. The parameters of the sine function (i.e. amplitude and phase) are modified to generate different 2D TIs. Next, the number of channels in each generated TI is modified to generate more 2D TIs. Then, the third dimension (Z-direction) slices are added to generate three dimensional TIs of the 2D ones. Note that, although the Z-direction slice is a gridding parameter it also controls the size of the structure (i.e. channel) since the added slices are identical to the first one (i.e. 2D TI). As a result, a TI becomes a function of four parameters, namely, the number of channels, the number of waves in each channel which is controlled by the phase value, the amplitude value of waves, and the number of Z-direction slices.

In this study 368 TIs are generated. The values of parameters considered in the generation of these TIs are summarized in Table 1. The generated TIs are then used to simulate the original reservoir model utilizing the proposed MPS methodology. Next, the uncertainty impact of the TI parameters on the cumulative oil production, the original and recoverable oil in place (OOIP and ROIP) are analyzed.

Table 1 The values of the TI's parameters that are used in the generation of the 368 TIs

Number of channels	Number of waves	Number of Z slices	Amplitude level (value in pixels)	Number of TIs generated
1, 2, and 3	1	1, 2, …, 7, and 8	High (75)	$3 \times 1 \times 8 = 24$
1 and 2	2	1, 2, …, 7, and 8	High (75)	$2 \times 1 \times 8 = 16$
1 and 2	3	1, 2, …, 7, and 8	High (75)	$2 \times 1 \times 8 = 16$
1, 2, 3, 4, and 5	1, 2, and 3	1, 2, …, 7, and 8	Medium (57)	$5 \times 3 \times 8 = 120$
1, 2, …, 7, and 8	1, 2, and 3	1, 2, …, 7, and 8	Low (30)	$8 \times 3 \times 8 = 192$

3 Results and Discussions

Based on the analysis of the parameters' impact on the cumulative oil production it is found that TIs having less than 5 Z-direction slices (Fig. 1), three channels, and two waves in each channel, produce erroneous models. Thus, these values are set as thresholds for accurate modeling practices. In addition, the results show that as the number of channels and waves in a TI increase the uncertainty in the cumulative oil production of the simulated reservoir decreases. For example, the uncertainty range decreases from 435 MSTB to 70 MSTB when the number of channels having medium-amplitude waves increases from 1 to 3 in an 8 Z slices TI. However, the increase in the number of Z slices from 5 to 8 and in the amplitude level have no noticeable effect on the uncertainty of the reservoir. So, it is unnecessary to generate TIs with more than 5 Z slices when modeling 3D reservoirs. Furthermore, it is noticed that the number of channels in a TI, the amplitude level of their waves, and the number of the Z slices in the TI have no defined effect on the OOIP and ROIP. In other words, their effect is random. However, the amount of the OOIP and ROIP are directly proportional to the number of waves in the channels of the TI.

Based on aforementioned observations thresholds are defined as 5 Z-direction slices, three channels and two medium or low amplitudes waves in each channel. To verify that the defined thresholds provide the most representative reservoir models the uncertainty range of the OOIP prediction is considered. The OOIP values of the reservoirs modeled using all the generated TIs are plotted and their uncertainty range is found to be between +1.549 and 0.345 %. However, this range is reduced to 0.382 – 0.104 % when only the OOIP of the reservoirs developed with TI that meet the defined thresholds are plotted.

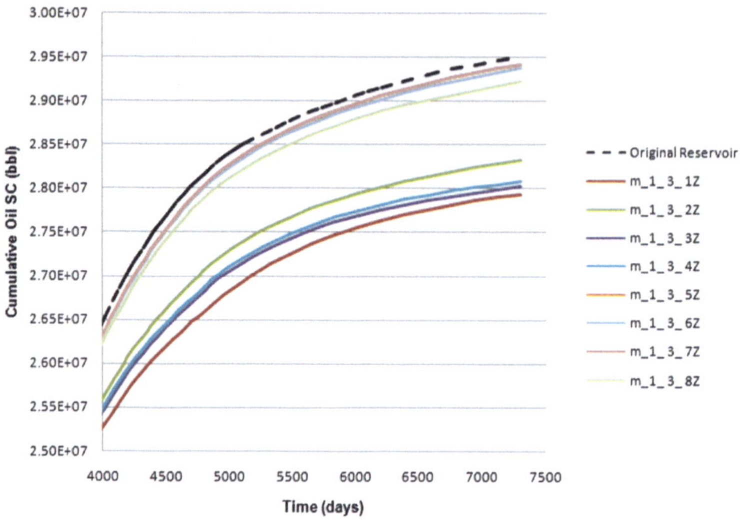

Fig. 1 Comparison of the cumulative oil productions between the original reservoir and reservoirs generated with TIs having different numbers of Z-direction slices (i.e. 1Z, 2Z, 3Z …, 8Z). The TIs used are those having three channels, each with a medium amplitude wave. Where, (*m*): medium amplitude, (*1*): one wave, (*3*): three channels and (*1Z*): one Z slice

Acknowledgments The first author wishes to acknowledge the Scientific and Technological Research Council of Turkey (TUBITAK) for providing a doctoral research fellowship.

References

1. Strebelle, S.B. (2000). *Sequential Simulation Drawing Structures from Training Images*. Ph.D. thesis, Stanford University, California.
2. Strebelle, S., and Journel, A.G. (2001). Reservoir modeling using multiple-point statistics. *In: Society of Petroleum Engineers Annual Technical Conference and Exhibition*, SPE Paper 71324, New Orleans.
3. Journel, A. G. (2005). Beyond covariance: the advent of multiple-point geostatistics, geostatistics banff 2004. *Quantitative Geology and Geostatistics*, *14*, 225–233.
4. Feyen, L., & Caers, J. (2006). Quantifying geological uncertainty for flow and transport modeling In Multi-modal heterogeneous formations. *Advances in Water Resources*, *29*, 912–929.
5. Boisvert, J., Pyrcz, M., & Deutsch, C. (2007). Multiple-point statistics for training image selection. *Natural Resources Research*, *16*, 313–321.
6. Arslan, I. (2005). *Structural model selection*. California: Stanford UniversityStanford University.
7. Remy, N., Boucher, A., & Wu, J. (2009). *Applied geostatistics with SGeMS: A user's guide*. New York: Cambridge.

Mathematical and Physical Models for the Estimation of Wind-Wave Power Potential in the Eastern Mediterranean Sea

George Galanis, George Emmanouil, Christina Kalogeri and George Kallos

1 Introduction

In the last decades, the estimation and monitoring of renewable energy resources are in the front line of interest of the scientific and technical community worldwide since a variety of important issues that surpass the borders of pure scientific research are affected. In the present work a high resolution study for the wind and wave energy potential is presented for the area of Eastern Mediterranean Sea focusing especially in the Levantine Basin and the Aegean Sea area. An integrated atmospheric-wave modeling system is utilized and ten year (2001–2010) simulation runs are performed. Observations from different sources (meteorological stations, buoys and satellites) are assimilated into the models for improving the initial conditions used [3, 6]. The relevant outcomes are analyzed by means of a complete statistical system employing conventional statistical measures, probability distribution fitting tools, as well as postprocessing modules based on Kalman filters for the elimination of possible systematic biases. For the latter, new advances obtained from the Information Geometry framework are utilized providing new ideas on the way that the discrepancies between model and observed data should be measured.

G. Galanis (✉) · G. Emmanouil · C. Kalogeri
Hellenic Naval Academy, Section of Mathematics, Xatzikyriakion,
18539 Piraeus, Greece
e-mail: ggalanis@snd.edu.gr

G. Galanis · C. Kalogeri · G. Kallos
Department of Physics, Atmospheric Modeling and Weather Forecasting Group,
University Campus, Bldg. PHYS-V, University of Athens, 15784 Athens, Greece

E. Pardo-Igúzquiza et al. (eds.), *Mathematics of Planet Earth*,
Lecture Notes in Earth System Sciences, DOI: 10.1007/978-3-642-32408-6_123,

2 The Modeling Systems Employed

Two state of the art atmospheric models are employed performing simulations for a ten year period (2001–2010): The Skiron/Eta regional forecasting system [5] and the open source atmospheric model WRF [9]. The former has been developed at the University of Athens by the Atmospheric Modeling and Weather Forecasting Group. It is based on the Eta/NCEP model and consists of a full physics non-hydrostatic model with sophisticated convective, turbulence and surface energy budget scheme. The horizontal resolution used for the present study was 0.05 x 0.05 degrees covering the whole Mediterranean region while 45 vertical levels up to 50 hPa have been utilized. Initial and boundary conditions are obtained from a high-resolution reanalysis (15 x 15 Km). On the other hand, the WRF model has been a collaborative partnership between NCAR, NOAA-NCEP, US military laboratories and Universities. It is a numerical weather prediction system suitable for a broad spectrum of applications from meters to thousands of kilometers. For the present study a horizontal resolution 0.02 x 0.02 degrees over the area of interest have been applied with 35 vertical levels up to 50 hPa. Initial and boundary conditions are obtained from the GFS high-resolution reanalysis (15 x 15 Km).

The wave parameters that are necessary for the estimation of wave power monitoring are simulated by the wave model WAM [7]. This is a third generation wave system that solves the wave transport equation with-out any presumptions on the shape of the 2d (frequencies/angular) wave spectrum. For the present study, the ECMWF version, CY33R1 [2], has been adopted. The horizontal resolution used for the wave model's simulations were 1/60 x 1/60 degrees, while the wave spectrum was discretized in 25 frequencies and 24 directions.

The estimation of the wind and wave power potential over the areas of interest is based on the following formulas:

$$\mathrm{P_{wind}} = \frac{1}{2}\mathrm{pv}^3 \quad [W] \quad \mathrm{P_{wave}} = \rho g \int_0^{2\pi}\int_0^{\infty} \mathrm{f}^{-1}\mathrm{E}(\mathrm{f},\theta)\mathrm{dfd}\theta = \frac{\rho g^2}{64\pi}\mathrm{H_s^2T_e}[\mathrm{W/m}] \tag{1}$$

[8] where p stands for the wind and ρ the water density, v for the wind speed, $E(f,\theta)$ denotes the two dimension wave spectrum, Hs the significant wave height, Te the wave period and g the gravity acceleration.

In addition to the above modeling systems and targeting to the elimination of biases that numerical simulation systems usually exhibit when focusing on local applications, a postprocess module has been developed. It is based on a non-linear version of Kalman filter [3] that estimates the bias of the modeled data as a polynomial of the modeled parameter:

$$y_i = a_{0,i} + a_{1,i}\cdot m_i + a_{2,i}\cdot m_i^2 + a_{3,i}\cdot m_i^3 + \ldots + \varepsilon_i \tag{2}$$

where m_i denotes the direct output of the model and y_i the corresponding bias. The estimation of the involved covariance matrices [3, 6] is based on new techniques that have been developed within the framework of Information Geometry. The latter recognizes the distributions of modeled and corresponding recorded data as statistical manifolds categorized in non-Euclidean spaces [1]. In this context, the underlying geometry may differ significantly from the classical one and the distances between the data sets—the discrepancies of the models—are estimated by minimum distance curves (geodesics) avoiding simplifications adopted by the classically used least square methods. A detailed presentation of this approach can be found in [4].

3 Statistical Analysis: Results

The statistical analysis of the modeled results for the wind/wave power outputs and the environmental parameters that affect their evolution, is based on a variety of statistical measures including average values, deviation and asymmetry measures as well as the kurtosis (the fourth moment of the data), providing information for the most frequent values, their expected deviation but also for the exposure of the data under study to extreme events. On the other hand, the model outputs and the corresponding observations have been studied by a probability density function (pdf) point of view: This approach gives the full package of information and is further exploited in order to estimate the discrepancies between modeled and observed data using techniques of Information Geometry, as mentioned in the previous section. Some indicative results are presented in Figs. 1, 2. The former depicts the 10-year main wave statistical parameters over the Cyprus sea areas, while the latter provides the full package of the statistical analysis at a point in Aegean Sea (Cyclades islands).

The south-western sea area of Cyprus seems to keep the primary role in wind and wave power potential being exposed to long period waves (swell) that favor rather smooth evolution of the wave power with low exposure to extreme events and available amounts of energy even in the absence of winds. On the other hand, the

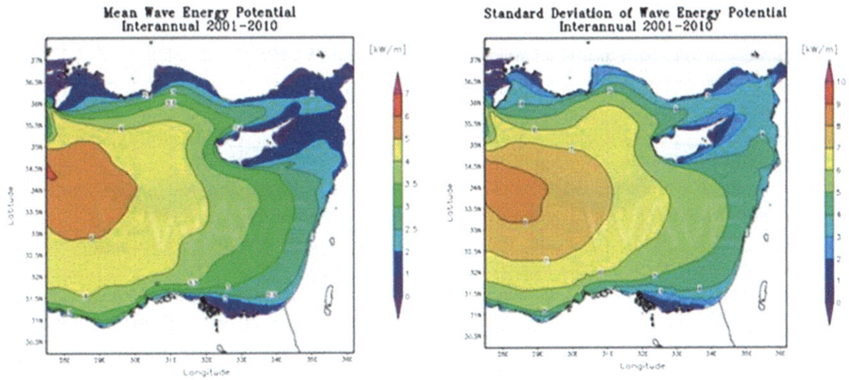

Fig. 1 Mean values and deviation of wave energy potential in the Levantine Basin

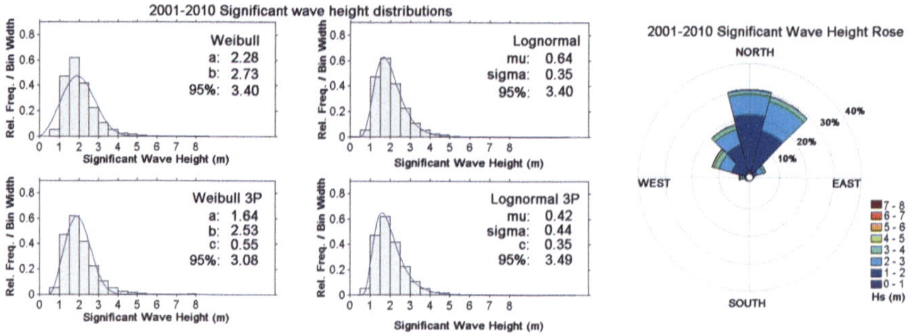

Fig. 2 The analysis of the significant wave data for a central Aegean point

Aegean Sea is a wind dominated area where the wind and wave power potential are closely related.

Acknowledgments This work has been supported by the Research Program "Development and application of new mathematical and physical models for monitoring the wind and sea wave energy potential" funded by the Greek Secretariat of Research and Technology.

References

1. Arwini K, Dodson CTJ (2008) Information geometry: Near randomness and near independence. *Lecture Notes in Mathematics 1953*, Berlin: Springer.
2. Bidlot, J., Janssen, P., Abdalla, S., & Hersbach, H. (2007). *A revised formulation of ocean wave dissipation and its model impact. ECMWF Tech. Memo. 509* (p. 27). Reading, United Kingdom: ECMWF.
3. Galanis, G., Emmanouil, G., Kallos, G., & Chu, P. C. (2009). A new methodology for the extension of the impact in sea wave assimilation systems. *Ocean Dynamics*, *59*(3), 523–535.
4. Galanis, G., Chu, P. C., Kallos, G., Kuo, Y. H., & Dodson, C. T. J. (2012). Wave height characteristics in the north atlantic ocean: A new approach based on statistical and geometrical techniques. *Stoch Environ Res Risk Assess*, *26*, 83–103.
5. Kallos G (1997) The regional weather forecasting system SKIRON. *Proceedings, Symposium on regional weather prediction on parallel computer Environments*, 15–17 October 1997, (p. 9), Athens, Greece.
6. Kalnay, E. (2002). *Atmospheric Modeling*. Data assimilation and predictability: Cambridge University Press. 341.
7. Komen, G., Cavaleri, L., Donelan, M., Hasselmann, K., Hasselmann, S., Janssen, P. (1994) *Dynamics and Modelling of ocean waves*, Cambridge: Cambridge University Press.
8. Pontes, M. T. (1998). Assessing the european wave energy resource. *Transactions of the American Meteorological Society*, *120*, 226–231.
9. Skamarock, W. C., Klemp, J. B., Dudhia, J., Gill, D. O., Barker, D. M., Wang, W. & Powers, J. G. (2005) A description of the Advanced Research WRF Version 2. NCAR Tech Notes-468+STR.

Geostatistical AVO Direct Facies Inversion

Leonardo Azevedo, Pedro Correia, Rúben Nunes and Amílcar Soares

1 Introduction

One of the main challenges in geophysics is retrieving high resolution seismic inverse elastic models (e.g. acoustic and elastic impedance modes) with high reliability along with their intrinsic uncertainty. Seismic inverse problems are nonlinear with non-unique solutions due to their intrinsic properties: the limited bandwidth of the seismic reflection data, noise and assumptions on physical models. For these reasons the degree of uncertainty related with the inverted best-fit elastic models may be considerably high and it should be integrated on the solution.

Due to the nature of seismic inverse problems, a probabilistic framework is a very valid approach to pose the inverse problem. Stochastic, or geostatistical, seismic inversions are a common practice in any seismic reservoir characterization playing a key role in the reservoirs' uncertainty and risk assessment. The most common inversion algorithms allow one to retrieve reliable acoustic impedance (AI) and elastic impedance (EI) models along with their intrinsic uncertainty. Herein, we present a new iterative geostatistical methodology for seismic AVO inversion based on sto-

L. Azevedo (✉) · P. Correia · R. Nunes · A. Soares
Center for Modeling Petroleum Reservoirs, CERENA/DECivil, Instituto Superior Técnico, Universidade Técnica de Lisba, Lisbon, Portugal
e-mail: leonardo.azevedo@ua.pt

P. Correia
e-mail: pedrocorreia1984@gmail.com

R. Nunes
e-mail: nunesrfm@gmail.com

A. Soares
e-mail: asoares@ist.utl.pt

L. Azevedo
CESAM and Departamento de Geociências,
Universidade de Aveiro, Campus Universitário de Santiago,
3810-193 Aveiro, Portugal

E. Pardo-Igúzquiza et al. (eds.), *Mathematics of Planet Earth*,
Lecture Notes in Earth System Sciences, DOI: 10.1007/978-3-642-32408-6_124,

chastic simulations and co-simulations. From the proposed inverse methodology one is able to retrieve at once density, p-wave velocity (Vp), s-wave velocity (Vs) and facies models while the uncertainty space associated with each model is still considerably explored. In addition to the elastic models, and as part of the solution, we can also derive the corresponding R-zero and Gradient volumes [1]. The great advantage of this new algorithm is the ability to retrieve, directly from the inversion process a facies model, which was previously defined from the available well-log data, conditioned by both the available well-log and seismic data.

We successfully applied this new inverse methodology to a pre-stack synthetic seismic dataset, built based on a real dataset and denominated from now one as real seismic data, where random noise was added. The final best fit inverse elastic models of Vp, Vs, and density produced a synthetic pre-stack seismic that has a correlation coefficient with the observed pre-stack seismic data of about 0.8

2 Methodology

The direct AVO facies inversion workflow (Fig. 1) is an iterative and convergent process based on two key main ideas: on each iterative step, the density, Vp, Vs, and facies models are perturbed towards a desired match between the real and the synthetic pre-stack seismic data is reached. At each iteration, the match between real and inverted data is evaluated for the entire set of simulated models. Then, the best parts of each are selected recurring to a genetic algorithm and used as secondary images on the co-simulations of the next iteration. We propose the use of Direct

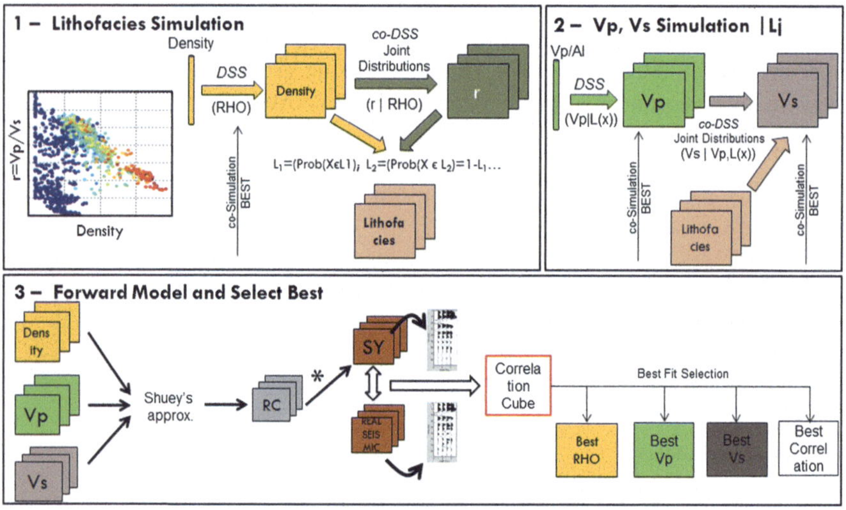

Fig. 1 Geostatistical AVO direct facies inversion workflow

Sequential Simulation (DSS) with Joint Probabilities (DSS with joint probability distribution) [2, 3] to generate the density, Vp, Vs and facies volumes, ensuring in this way the reproduction of their bivariate distributions. A facies classification needs to be performed a priori from, for example, the density versus Vp/Vs domain recurring to the available well-log. All the inverted models reproduce both the main spatial patterns, as revealed by the variograms, the probability distributions and bi-distributions estimated from the original well-log data and the well-log data at the well location. The develop algorithm may be briefly described as:

(1) Simulation of N_s density models recurring to DSS. Then, for each model the intermediate variable (e.g. r = Vp/Vs) is co-simulated (co-DSS with joint probability distribution) using the density volume as a secondary variable. For each duplet of images we classify it into a new facies volume (Fig. 1, step 1);
(2) On the second step, Vp is simulated with DSS conditioned to the previously derived facies cube. Vs models are then co-simulated with DSS with joint probability distributions (Fig. 1, step2);
(3) For all the simulated elastic models (*Ns* triplets of density, Vp and Vs) we compute synthetic seismograms, following Shuey's linear approximation [1], and compare their synthetic seismic responses with the observed one (Fig. 1, step 3);
(4) A "best" model is then built from the areas of models that produce the best local correlation coefficients between synthetic and real seismic;
(5) Return to (1) using the "best" density, Vp and Vs models as secondary variables in the co-simulation process.

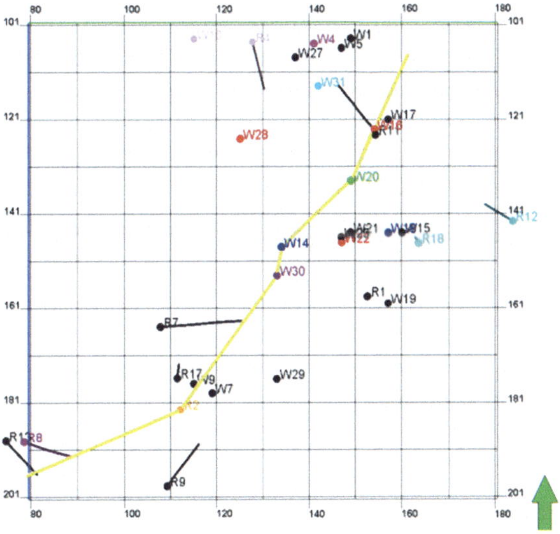

Fig. 2 Available pre-stack seismic and well-log data. Wells in *black* where not used to constraint the direct facies inversion but used exclusively used as blind tests to assess the convergence of the methodology. *Yellow line* shows the location of the vertical sections shown in Fig. 3

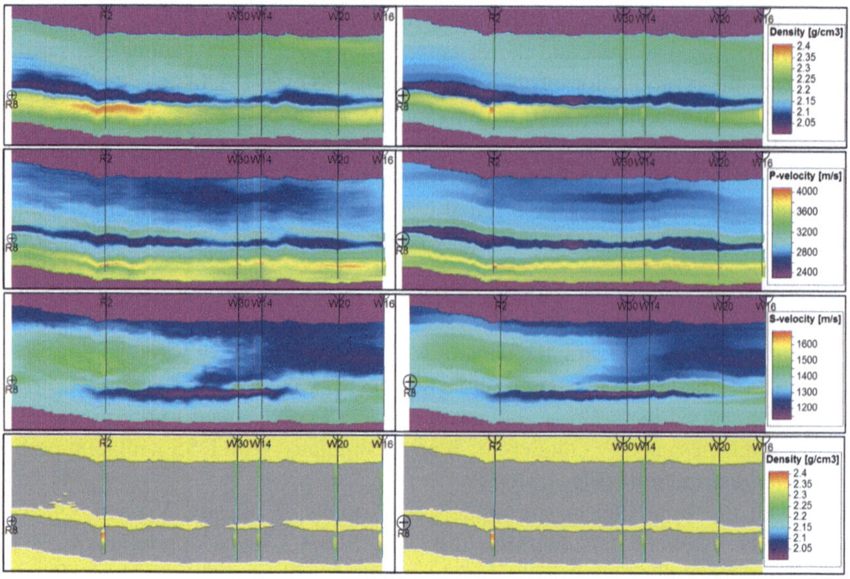

Fig. 3 Mean of all inverted model at the last iteration (on the *left*) and real elastic models (on the *right*), from *top* to *bottom*: density, p-wave velocity, s-wave velocity and facies model. For section location please see Fig. 2

3 Synthetic Seismic Example

The available dataset comprises a pre-stack 3D seismic volume (built from known density, Vp and Vs models) and a total of 32 wells from where only 15 were used to constraint the stochastic inversion (Fig. 2).

The best-fit inverse density, Vp, Vs and facies models produced a synthetic seismic with a global correlation coefficient of about 0.8 when compared against the real seismic. The mean model from all the simulated elastic models during the last iteration were compared with the original properties (Fig. 3), from where the real seismic was derived, and the reproduction of both large and small scale for all the properties is very accurate.

4 Conclusions

This inversion methodology is able to successfully invert pre-stack seismic data directly for density, Vp, Vs and facies models. Notice that the facies model is not derived from the inverted density, Vp and Vs models but is part of the inverse solution itself and therefore is constraint by both the well-log and the seismic data.

References

1. Shuey, R. T. (1985). A simplification of the Zoeppritz equations. *Geophysics*, *50*(4), 609–614.
2. Soares, A. (2001). Direct sequential simulation and cosimulation. *Mathematical Geology*, *33*(8), 911–926.
3. Horta, A., & Soares, A. (2010). Direct sequential co-simulation with joint probability distributions. *Mathematical Geosciences*, *42*(3), 269–292.

Optimized History Matching with Stochastic Image Transforming of a Deltaic Reservoir

Maria Helena Caeiro, Amilcar Soares, Vasily Demyanov and Mike Christie

1 Introduction

One of the major difficulties in modeling petroleum reservoirs is the characterization of the porosity and permeability of the reservoir, in spite of the impossibility to check results directly with the reality. The history matching is an inverse procedure that allows having a better knowledge of the reservoir and has a high importance for producer fields, because it enables the match of the reservoir properties (e.g. porosity and permeability) with the production from the wells. Many authors have been working in different history matching developments that contribute to the improvement of different methodologies [1–3]. Complex cases as a deltaic reservoir are interesting to explore and produce, but they require an exhaustive analysis and non-stationary modeling. Characterize the connected channelized structures is a compound task. The difficulty increases while perturbing at the same time the value of the variable and the shape of the channels. In this paper is presented a hybrid method encompassing the optimization firstly in the space of the anisotropy model parameters and secondly refines it in the space of the static models through global perturbation technique.

M. H. Caeiro (✉) · A. Soares
Instituto Superior Técnico, Technical University of Lisbon, Av. Rovisco Pais 1, 1049-001 Lisboa, Portugal
e-mail: helena.caeiro@ist.utl.pt

A. Soares
e-mail: asoares@ist.utl.pt

V. Demyanov · M. Christie
Institute of Petroleum Engineering, Energy Academy, Heriot Watt University, EH14 4AS Edinburgh, United Kingdom
e-mail: vasily.demyanov@pet.hw.ac.uk

M. Christie
e-mail: mike.christie@pet.hw.ac.uk

E. Pardo-Igúzquiza et al. (eds.), *Mathematics of Planet Earth*,
Lecture Notes in Earth System Sciences, DOI: 10.1007/978-3-642-32408-6_125,

2 Hybrid Optimization for Geostatistical History Matching. Deltaic Application

In this paper is presented an iterative methodology for optimized history matching, using the adaptive stochastic sampling in the multiparameter space, and direct sequential simulation with local anisotropy correction (DSS-LA) [4] as the engine for the image transforming of the internal properties of the reservoir. Therefore it can be considered has having a double optimization done in two distinguished phases, the former *optimization the multiparameter space* and the latter *optimization in the models space*. For case application was chosen a deltaic 2D slide of Stanford VI reservoir [5]. The optimization in the multiparameter space is based on the work presented by Ref. [6]. It consists in the creation of a model of anisotropy, defined by the main trends of the channels (direction of maximum anisotropy $-\theta$ and the ratio between the major and the minor amplitude distances −r), considering regions based on prior beliefs, integrated with the DSS-LA in a global iterative algorithm that also contains the dynamic simulation (Eclipse Black Oil Simulator) and the Neighborhood Approximation Algorithm (NA) [7]. This integrated iterative process is looped until the lowest misfit is achieved. The misfit value represents how well a model fits the data and the applied misfit function is the standard least square. The misfit function depends of well, dynamic match variables (well oil production rate—WOPR and the well water production rate—WWPR) and time step. The data standard deviation (σ) is 10 % of the average of history data values for each variable of each well and the history period corresponds to 5–387 days. Although the spatial patters parameters are optimized locally in the multiparameter optimization, the space of solutions of the local values of the internal properties can be still quite large. Therefore, was performed an optimization refinement based on the direct transformation of images approach with a method of global perturbation. The perturbation is performed by coDSS-LA based on the influence regions of each well. In the first iteration the coDSS-LA is conditioned to the porosity and permeability models obtained from the optimization in the multiparameter space. The dynamic responses of each of the realizations are obtained by Eclipse Black Oil Simulator and the respective mismatches evaluated through the standard least square objective function. Then, are composed the porosity and permeability models by merging the "best" parts from the respective realizations with lower local misfit. A correlation coefficient cube is also composed based on the local misfit values for each well. These models are going to constrain the next iteration and it loops until the stopping criterion is reached.

3 Results

Figures 1 and 2 show the comparison between the porosity and permeability models. Well P25 was chosen as reference to illustrate the production curves matches (Fig. 3). The improvement from multiparameter optimization to refinement

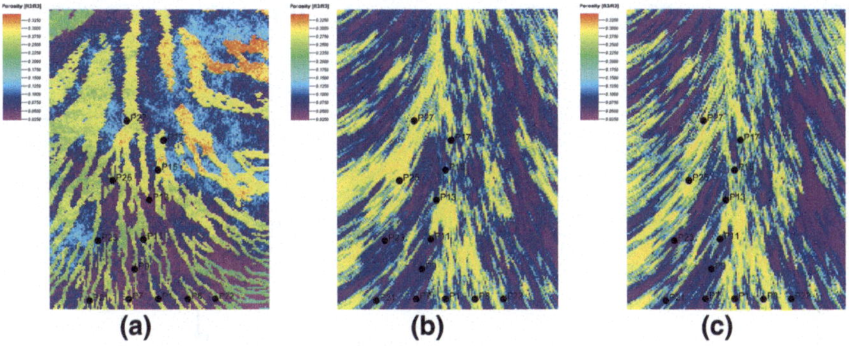

Fig. 1 Porosity Models: **a** True Case, **b** Multiparameter and **c** Refining optimizations

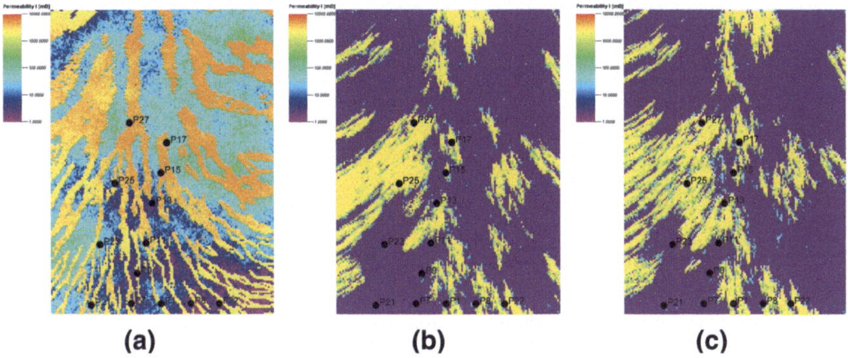

Fig. 2 Permeability Models: **a** True Case, **b** Multiparameter and **c** Refining optimizations

optimization is observed but there is still space for enhancement. The reproduction of the major geological features is achieved, however with less consistency for the permeability. The application was only done for a four models in the multiparameter optimization and three iterations with ten simulations for the refinement. This can be a drawback for this specific application due to time consuming problems of the methodology implementation.

4 Remarks

The goal of this paper is to demonstrate the impact of the optimization refinement, iteratively performed by conditioning the stochastic simulation with the porosity and permeability models. Globally, the results are promising and show the feasibility of the methodology to reproduce meander structures, like a delta front honoring the history production rates. Future work should consist in computation work,

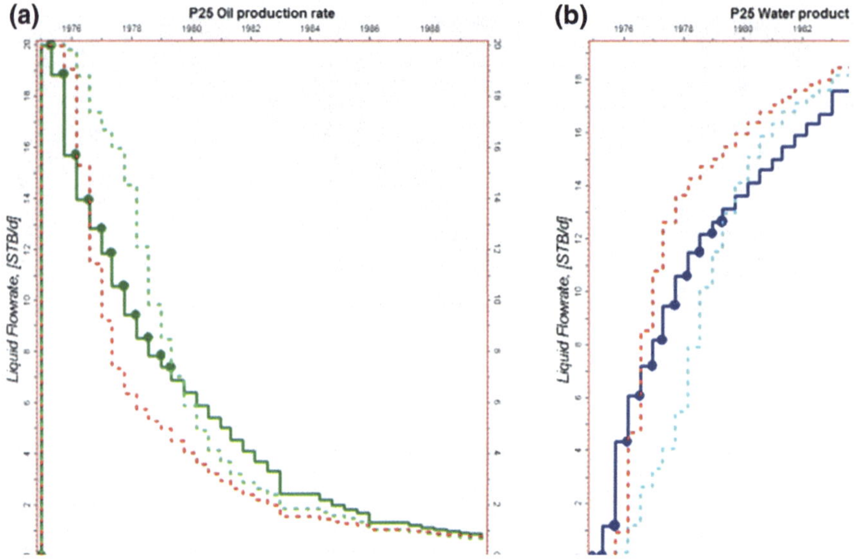

Fig. 3 Well P25 oil **a** and water **b** rate matches

namely increase the number of models produced so that optimization and convergence improves. Furthermore, a 3D application and the uncertainty evaluation of the obtained multiple models should be considered.

References

1. Mohamed, L., Christie, M., & Demyanov, V. (2010). Comparison Of stochastic sampling algorithms For uncertainty quantification. *SPE-119139-PA, SPE Journal*, *15*(1), 31–38.
2. Hajizadeh, Y., Christie, M., & Demyanov, V. (2011). Ant colony optimization for history matching and uncertainty quantification of reservoir models. *Journal of Petroleum Science and Engineering*, *77*(1), 78–92.
3. Christie, M., Eydinov, D. Demyanov, V., Talbot, J., Arnold, D., Shelkov, V. (2013). Use of multi-objective algorithms in history matching of a real field. SPE 163580, Reservoir Simulation Symposium, The Woodlands, Texas, USA.
4. Horta, A., Caeiro, M. H., Nunes, R., & Soares, A. (2010). Simulation of continuous variables at meander structures: application to contaminated sediments of a lagoon. In P. M. Atkinson & C. D. Lloyd (Eds.), *GeoENV VII—geostatistics for environmental applications, quantitative geology and geostatistics (vol. 16)*. New York: Springer.
5. Castro, S. A., Caers, J., & Mukerji, T. (2005). The Stanford VI reservoir, 18th Annual Report. Stanford Centre for Reservoir Forecasting, Stanford University.
6. Caeiro, M., Soares, A., Demyanov, V., & Christie, M. (2011). *Optimization of a geostatistical non-stationary model in history matching* (pp. 68–82). Salzburg: IAMG 2011 publication.
7. Sambridge, M. (1999). Geophysical inversion with a neighbourhood algorithm I—searching a parameter space. *Geophysical Journal International*, *138*, 479–494.

Conditioning 3D Object Based Models to a Large Number of Wells: A Channel Example

Jeff B. Boisvert and Michael J. Pyrcz

1 Introduction

Object-based methods (OBM) provide an important alternative to the cell-based geostatistical methods. With cell-based methods simulation proceeds by first assigning data to cells that coincide with hard data and then sequential conditional simulation along a random path over the remaining cells. Cell-based methods ensures perfect conditioning to hard data and good conditioning to trends given they do not contradict the data. Yet, such a method is limited in its ability to reproduce specific, crisp geometries and the resulting simulation does not preserve geometric information necessary for building internal trends within geometries.

Nevertheless, crisp, complicated geometries exist in some settings. Isolated sand filled channels encased in overbank shales are common to slope valley reservoirs in West Africa and Gulf of Mexico. Moreover, lateral accretion sets with consistent external geometries and complicated hierarchies of internal trends are common in the oil sands region of Canada. Comprehensive descriptions of these and other examples of hierarchies of reservoir geometry are available [1–3].

Conditioning remains a significant challenge in the application of OBM and methods have been proposed. References [4, 5] proposed two step processes. First objects are fit to the conditioning data and then objects are placed away from conditioning to honor trends. Reference [6] suggests rejection sampling followed with local corrections for conditioning. Care must be taken to prevent bias in the model, such as distinct statistics near and away from wells [7]. In this work, this is controlled using

J. B. Boisvert (✉)
Center for Computational Geostatics, Department of Civil and Environmental Engineering, University of Alberta, 3-133 Markin/CNRL, Edmonton, ABT6G 2w2, Canada
e-mail: jbb@ualberta.ca

M. J. Pyrcz
Chevron—Earth Science Research and Development, 1500 Louisiana Street, 27138, Houston, TX77002, USA
e-mail: mpyrcz@chevron.com

E. Pardo-Igúzquiza et al. (eds.), *Mathematics of Planet Earth*, Lecture Notes in Earth System Sciences, DOI: 10.1007/978-3-642-32408-6_126,

a multi-stage objective function. Also the well data interpretation should be checked to ensure it does not contradict the inferred object-based geometries. Checks include distributions of well element thickness and comparison of adjacent wells. Wells that fail to condition precisely may indicate contradictions.

At times object-based models are constructed without honoring local data. These unconditional models may be applied as TIs for MPS simulation. Nevertheless improved conditioning is needed for OBM. Further developments with parameterization and optimization provide a promising opportunity.

2 Optimization Methods

Four optimization techniques used are briefly reviewed. Note that the eventual goal of this optimization is to find a local minimum, which would be one channel, the global minimum is not required as there would only be one optimum channel.

(1) *Fminunc, Matlab*: A quasi-Newton method with an approximated hessian matrix. Specifically the Broyden–Fletcher–Goldfarb–Shanno method is used [8].
(2) *Hooke–Jeeves optimization*: The objective value is assessed in different directions. The direction with the minimum objective value is approximated [9].
(3) *Implicit filtering*: The only technique that includes constraints on the input variables. The algorithm uses coordinate searches as well as gradient approximation to generate parameters with a lower objective function value [10]
(4) *Simplex search method*: The Matlab function, *fminsearch*, implements [11].

The optimization techniques are referred to as Option 1, 2, 3, 4. None of the techniques require the specification of a gradient. Any object can be optimized. The methodology assumes wells have been coded into gross facies categories.

3 Methodology

A large number of objects, consistent with the data, are generated. Realizations are created by selecting an appropriate number of objects. First, objects are constructed from a set of parameters $OBJ(x_1, x_2, \ldots, x_p)$ where p $<$ 100 for computational reasons. The objective function is a weighted combination of: well mismatch $c_f \cdot f(o, v)$; conditioning $c_g \cdot g(o, v)$; and closeness bias penalty $c_h \cdot h(o, v)$.

$$obj\,(o, v) = c_f \cdot f\,(o, v) + c_g \cdot g\,(o, v) + c_h \cdot \boldsymbol{h}\,(o, v)$$

$$f\,(o, v) = \sum_{i=1}^{n} d\,(o, v_i); \quad g\,(o, v) = -\sum_{i=1}^{n} w\,(o, v_i); \, h\,(o, v) = r_w \sum_{i=1}^{n} \left(1 - \frac{b\,(o, v_i)}{c_b}\right)$$

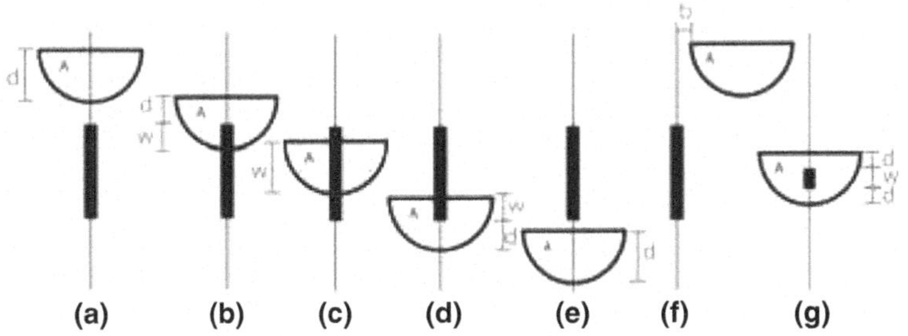

Fig. 1 Distances used for calculation of the objective function

where $d(o, v_i)$ is the distance between the object (o) and intersection i in the conditioning data (v_i). The objective function rewards objects that match conditioning data $w(o, v_i)$, the length of conditioning data matched. A penalty to objects, $b(o, v_i)$, not intersected by a well, but come close to the well, controls the bias for placing objects just outside well data (Fig. 1f). Object parameters are randomly initialized and an optimization algorithm is applied to condition .

4 Example

Channels are represented by two parabolas that meet at a thalweg. Channel center lines are defined by a cubic spline fit to a number of initial control points. The elevation of the channel is considered constant as modeling is conducted in depositional space; the extension to variable thicknesses is trivial. Crevasse spalys are added at locations where the derivative of the center line = 0 and a well indicates a crevasse splay. To demonstrate the methodology, a reasonable number of conditioning data (47) is used (Fig. 2) but similar results are seen for 100+ wells.

An issue in conditioning is that objects can be placed close to a well without a channel facies. The effect of introducing a penalty on this closeness is shown in Fig. 3; channel thickness is reduced to honor the well distance bias. The bias checks presented in Ref. [7] could be used to tune the penalty magnitude.

Figure 4 compares the generation of objects based on time and the ability of the algorithms to condition. There is some difference in the performance of the algorithms when weight is given to the conditioning data; option 1 seems to better match the constraints of the well data (Fig. 4) however, it is not able to obtain the same level of well conditioning as options 2 and 4. Option 4 performs nearly as well as option 2 for well conditioning, but takes considerably longer.

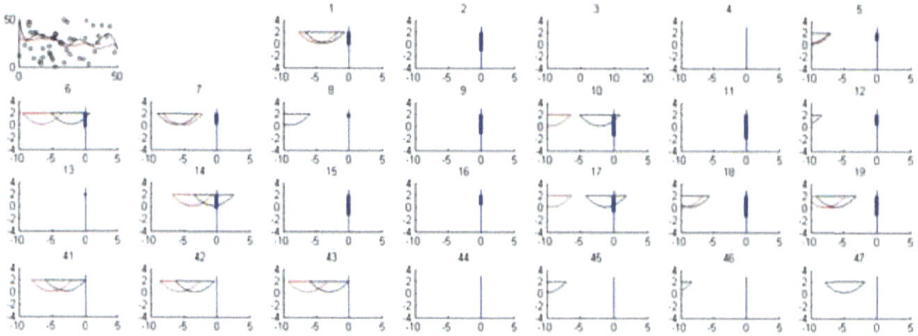

Fig. 2 Conditioned object (*black*) initial object (*red*). Add the optimized object with pen

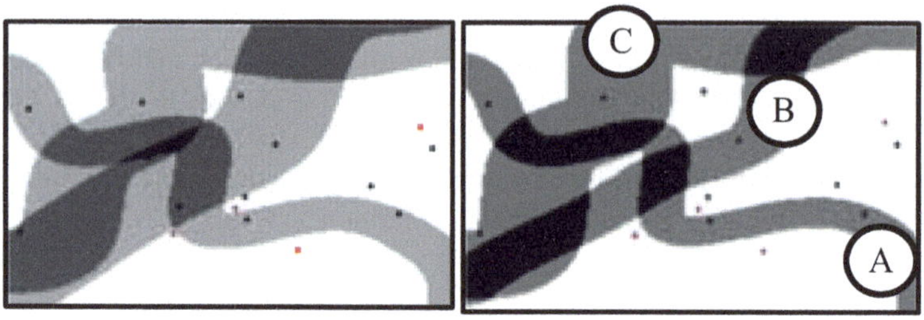

Fig. 3 Including a penalty (*right*) on the closeness of objects to non-channel wells

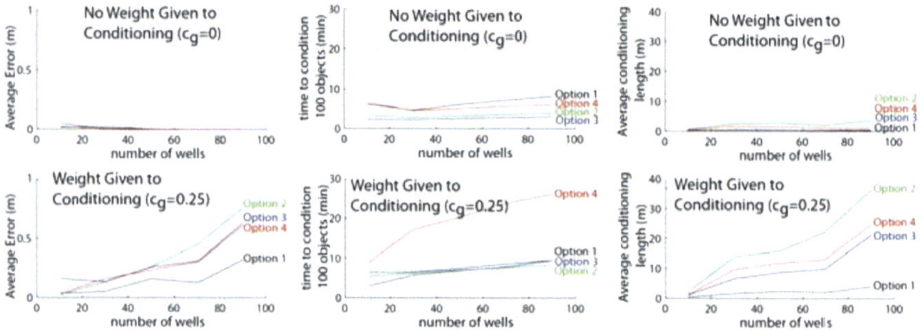

Fig. 4 CPU times and conditioning quality for the different optimization methods

5 Conclusion

The proposed objective function is simplistic but captures many aspects important for placing objects. Current optimization techniques can generate objects efficiently (order of seconds). Dense data slightly increases CPU time, but the proposed algorithms are effective for 100+ wells. Previous views have been that conditioning with dense data is not computationally efficient; this was not the case but further work is required for more complex objects and secondary data.

References

1. Walker, R. G., & James, N. P. (1992). *Facies Models: response to sea level changes*. St.John's: Geological Association of Canada.
2. Reading, H. G. (1996). *Sedimentary Environments: processes, facies and stratigraphy* (3rd ed.). Oxford: Black-well Science.
3. Galloway, W. E., & Hobday, D. K. (1996). *Terrigenouus clastic depositional systems: applications to fossil fuel and groundwater resources*. New York: Springer.
4. Shmaryan, L. E., & Deutsch, C. V. (1999). Object-based modeling of fluvial / deepwater reservoirs with fast data conditioning: methodology and case studies, in *SPE Annual Technical Conference and Exhibition*. Society of Petroleum Engineers.
5. Viseur, S., Shtuka, A., & Mallet, J. -L. (1998). New fast, stochastic, boolean simulation of fluvial deposits, in *SPE Annual Technical Conference and Exhibition*, New Orleans, LA, 1998. Society of Petroleum Engineers.
6. Pyrcz, M. J., Boisvert, J., & Deustsch, C. V. (2009). Alluvsim: a conditional event-based fluvial model. *Computers and Geosciences*. doi:10.1016/j.cageo.2008.09.02.
7. Hauge, R., & Syversveen, L. (2007). Well conditioning in object models. *Mathematical Geology*, *39*, 383–398.
8. Broyden, C. G. (1970). The convergence of a class of double-rank minimization algorithms. *Journal of the Institute of Mathematics and its Applications*, *6*, 76–90.
9. Hooke, R., & Jeeves, T. A. (1961). Direct search: solution of numerical and statistical problems. *Journal of the Association for Computing Machinery (ACM)*, *8*(2), 212–229.
10. Kelly, C. (2011). *Implicit filtering* (p. 184). Philadelphia: SIAM.
11. Lagarias, J. C., Reeds, J. A., Wright, M. H., & Wright, P. E. (1998). Convergence properties of the Nelder–Mead simplex method in low dimensions. *SIAM Journal of Optimization*, *9*(1), 112–147.

Modeling Channel Forms Using a Boundary Representation Based on Non-uniform Rational B-Splines

Jeremy Ruiu, Guillaume Caumon, Sophie Viseur and Christophe Antoine

1 Introduction

In most clastic reservoirs channels are prominent features and control the spatial layout of the main architectural elements which impact flow behavior [1, 2]. Modeling of these elements is typically achieved using object-based [3, 4] or multi-point geostatistical simulations [5, 6]. These methods often use simple objects descriptions which are not fully representative of the true 3D geometries encountered in nature. We propose a 3D channel object with a compact parametrization which allows for representing asymetric channel geometries and provides a curvilinear framework for modeling internal heterogeneities. This representation is based on Non-uniform Rational Basis Splines(NURBS) which consist in a smooth interpolation between a set of points [7]. NURBS are interesting for modeling abundant sedimentary structures because they can be edited with a small number of parameters and have a small storage footprint and interactive computation times. As an example, this method is applied to generate 3D training images, which can be used for MPS.

J. Ruiu (✉) · G. Caumon · C. Antoine
Georessources (UMR 7359)-ENSG, Université de Lorraine, Vandoeuvre-Lès-Nancy, France
e-mail: ruiu@gocad.org

G. Caumon
e-mail: caumon@gocad.org

S. Viseur
e-mail: viseur@cerege.fr

C. Antoine
e-mail: antoine@gocad.org

S. Viseur
CEREGE (UMR 7330), Aix Marseille University, 3 place Victor Hugo, 13331 Marseille cedex 03, France

E. Pardo-Igúzquiza et al. (eds.), *Mathematics of Planet Earth*,
Lecture Notes in Earth System Sciences, DOI: 10.1007/978-3-642-32408-6_127,

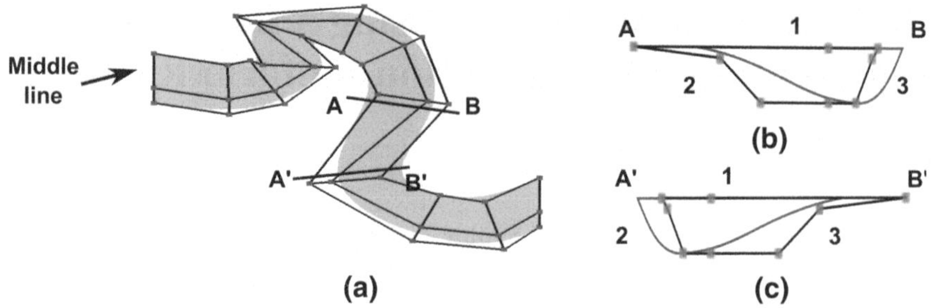

Fig. 1 Modeling of an asymmetric channel using 3 NURBS surfaces. *Black lines* are the control point net and the *grey points* are control points. **a** *Top view* of channel **b** *Side view* of the asymmetric channel **c** *Side view* of the channel with inversion of the asymmetry corresponding to the change in the channel curvature

2 Representing Channels with NURBS

To represent a channel form we propose to use three connected NURBS surfaces. Each surface is defined by a set of control points that represents a coarse approximation of the actual body geometry. The top surface control net is constructed around a backbone (Fig. 1a). The lateral surfaces are constructed starting from the top one and their points are placed along a quadratic shape (Fig. 1b). Common points between surfaces must remain colocated in order to maintain channel shape during deformations.

Using this representation, the seven following parameters define a channel shape:

- the length
- the width
- the height
- symmetric aspect ratio (formulated as a ratio of the width, i.e. if the value is 0.5 the channel is symmetric)
- the amplitude of the sinuosity orthogonally to the global channel direction
- the wavelength of the sinuosity orthogonally to the global channel direction
- a tortuosity factor used to obtain meandering aspects(i.e. if the value is 1, it authorizes meanders to be closed).

The first four parameters are used to construct the 3D geometry of a channel and the other ones enable a stochastic simulation of the channel back bone (Fig. 1a).

To stochastically generate several channels we propose to simulate the middle line (Fig. 1a) using the method developed by [3]. At first the amplitudes of displacement are calculated by performing an unconditional Gaussian simulation (Fig. 2a). Meandering aspects are then obtained using a growing factor (Fig. 2b) calculated by:

$$k\overrightarrow{G_i} = \alpha.\overrightarrow{T_i} + \beta.\overrightarrow{C_i} + \gamma.\overrightarrow{F_i} \tag{1}$$

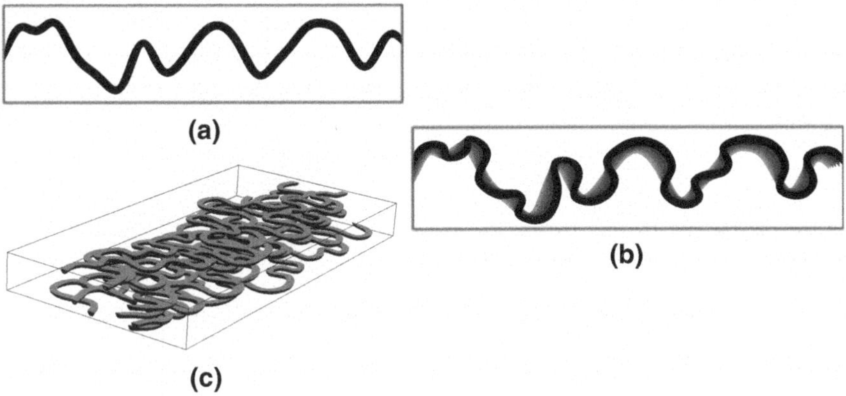

Fig. 2 Stochastic generation of channels. **a** The channel path is generated using an unconditional Gaussian simulation. **b** A growing factor is applied to obtain a meandering aspect. **c** Generation of 10 channels

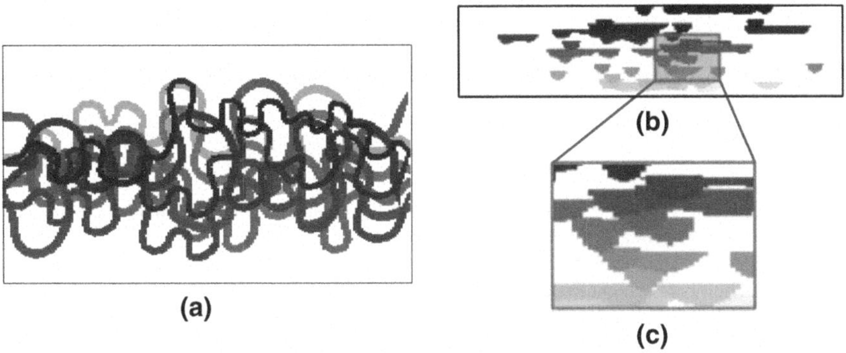

Fig. 3 3D MPS training image built from stochastically generated channels using NURBS. **a** *Map view* of the rasterized channel. **b** *Transversal section view*. **c** A zoom on the *transversal section view* shows that the rasterization is done according to the deposition order of channels

where $\overrightarrow{T_i}$ is the tangent and $\overrightarrow{C_i}$ the curvature at the projection of the control point i on the corresponding curve and $\overrightarrow{F_i}$ the vector representing the global flow direction. The constants α, β and γ are function of the tortuosity factor and the wavelength.

3 Application to Training Image Generation

Stochastically generated channels (Fig. 2c) are used to construct 3D training images by painting them on a regular Cartesian grid (Fig. 3). The bounding surfaces of each channel are transferred to the grid using the Brensenham rasterization method [8]. A scanning process fills the channel spaces in the grid (Fig. 3a, b). These two steps are performed following the depositional order to take into account the erosion processes (Fig. 3c).

4 Conclusion

A method is presented in this paper for effciently (1 s for 40 channels) generating geobodies from few parameters. It has potential for building training images in a very easy way. Future developments concern the reproduction of different geobodies geometries such as levees, point-bars, etc. as their spatial relationships (e.g. spatial organisation of these different bodies). They also have to deal with connection and hierarchic rules between these objects.

Acknowledgments We would like to thank, for their support, the industrial and academic members of the Gocad Consortium, ASGA, and Georessources, as well as Paradigm for providing the Gocad Software and API.

References

1. Haldorsen, H., & Lake, L. (1984). A new approach to shale management in field-scale models. *Old SPE Journal*, *24*(4), 447–457.
2. Jackson, M. D., Hampson, G. J., & Sech, R. P. (2009). Three-dimensional modeling of a shoreface-shelf parasequence reservoir analog, part 2: geologic controls on fluid flow and hydrocarbon recovery. *AAPG Bulletin*, *93*(9), 1183–1208. doi:10.1306/05110908145.
3. Viseur, S. (2001). Simulation stochastique basée-objet de dépôts fluviatiles. Ph.D. thesis, INPL, Nancy, France, p. 233.
4. Pyrcz, M., Boisvert, J., & Deutsch, C. (2009). Alluvsim: a program for event-based stochastic modeling of fluvial depositional systems. *Computers Geosciences*, *35*(8), 1671–1685. doi:10.1016/j.cageo.2008.09.012.
5. Strebelle, S. (2002). Conditional simulation of complex geological structures using multiple-point statistics. *Mathematical Geology*, *34*, 1–21.
6. Liu, Y., Harding, A., Abriel, W., & Strebelle, S. (2004). Multiple-point simulation integrating wells, three-dimensional seismic data, and geology. *AAPG Bulletin*, *88*(7), 905–921.
7. Piegl, L., & Tiller, W. (1995). *The NURBS book*. London: Springer.
8. Bresenham, J. E. (1965). Algorithm for computer control of a digital plotter. *IBM Systems Journal*, *4*(1), 25–30. doi:10.1147/sj.41.0025.

Moving Away from Distance Classifications as Measures of Resource Uncertainty

Ricardo A. Olea , James A. Luppens and Susan J. Tewalt

1 Introduction

For decades, the predominant practice in the characterization of uncertainty in coal mining assessments has been to use classifications based on distance from cells to the closest drill hole(s) as the criterion to report uncertainty in the estimation of resources [1, 2]. The main reason the practice has been accepted for such a long time is simplicity, despite not being able to provide accurate answers to critical issues such as: How good is my estimate of in-place tonnage? Theoretically and practically it has been illustrated multiple times that distance alone fails to adequately model uncertainty in coal and other mineral deposits because uncertainty depends only partly on distance to the closest data [3]. Other important factors are: (a) the geometry of the entire sampling, (b) the complexity of the geology, closely determined by the depositional environment and tectonic evolution, (c) internal discontinuities in the form of channels, erosional features and faults, and (d) complexity of the outer boundaries.

After tessellating the deposit into cells, distance classifications provide a breaking of total tonnage into four distance classes of cell tonnage: measured, indicated, inferred, and hypothetical. These four cumulative tonnages are used as the basis for personal conjectures about uncertainty in the assessments. This approach to characterizing uncertainty is overdue to be replaced by some mathematical formulation tested to provide a universal and realistic measure of uncertainty. Here, we propose superseding distance classifications as measure of uncertainty by two probability distributions based on stochastic simulation.

R. A. Olea (✉) · J. A. Luppens · S. J. Tewalt
U.S. Geological Survey, 12201 Sunrise Valley Drive, MS 956, Reston, VA 20192, USA
e-mail: olea@usgs.gov

E. Pardo-Igúzquiza et al. (eds.), *Mathematics of Planet Earth*,
Lecture Notes in Earth System Sciences, DOI: 10.1007/978-3-642-32408-6_128,

2 Methodology

Geostatistical stochastic simulation models uncertainty by generating multiple scenarios, called realizations in geostatistics, which honor the data and other important spatial styles of fluctuation that may be different for different methods [4, 5]. Despite a number of successful applications of geostatistics, its use in characterizing coal resources has moved slowly [6]. Stochastic simulation typically requires tessellation of the study area into small units, here called cells, which should not be sized less than the smallest detail of interest to model and not too large that, in too many instances, two or more data land in the same cell [7].

After reviewing the results of statistically experimenting with three deposits of completely different geologic characteristics [3, 8, 9], we recommend to even avoid mixed approaches [10], fully substituting distance classification methods with two probability distributions: one to characterize uncertainty of total resources and another to appraise uncertainty at the cell level. Both distributions are prepared from the same set of realizations; hence they are fully consistent and compatible.

The numerical distribution to model the uncertainty in total resources is straightforward to prepare. Each realization has an equal probability to be the true tonnage in the deposit. Thus, the addition of all cell tonnages in a realization provides one value for the numerical distribution of likely values of total tonnage. Collectively, a large number of realizations allows the modeler to numerically define the distribution.

The second distribution of interest is more laborious to prepare. This time, instead of working with the realizations individually, the first step requires choosing a location and collecting from all realizations the value at this selected site. Such values numerically define another distribution that now completely characterizes uncertainty in the tonnage at such cell. By repeating this process to all of the usually thousands of cells tessellating a study area, it is possible to have one distribution for every cell. However, it is not practical to present all these distributions as the final result. We simplified the modeling by retaining from each distribution only the length of the confidence interval defined by the 5th and the 95th percentiles, the 5–95 spread for short, which is a statistics characterizing dispersion in the values, and thus a measure of uncertainty. The collection of these thousands of cell spread values numerically defines the distribution of the 5–95 spread for the study area. One more step allows linking cell uncertainty to cell tonnage. For that purpose, the 5–95 spread values are ranked in increasing order and then each 5–95 spread value is paired with the tonnage for all cells with a less than or equal spread than the value being paired.

3 Example

Figure 1 shows results for the Canyon coalbed, Wyoming, taken from one of the test studies [9]. The extent of the study area is 945 km^2 (365 sq mi), with data at a fairly regular spacing of approximately 4.8 km (3 mi) in 7 columns and 11 rows. The cells

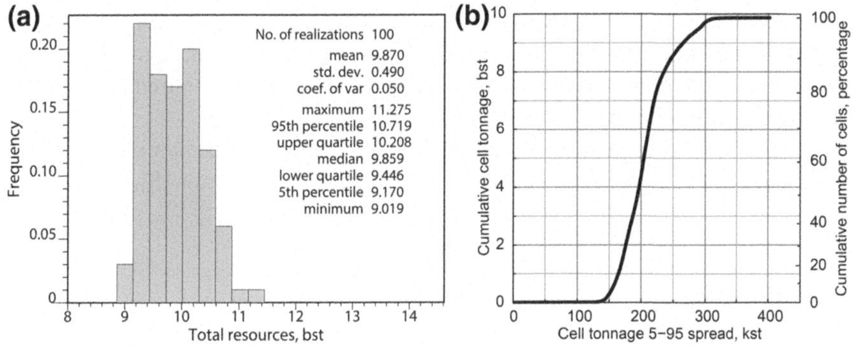

Fig. 1 Measures of uncertainty at two different scales: **a** histogram of likely values for total tonnage with bst standing for billion short tons; **b** distribution of the 5–95 confidence interval for all cells comprising the study area, with kst standing for thousand short tons

are squares with 122 m (400 ft) on the side. The realizations were generated using sequential indicator and Gaussian simulation. The range of the semivariogram for the indicators is 22.1 km (13.8 mi) and for the normal scores 14.2 km (8.9 mi). Certainly any other adequate simulation method could have been used.

Advantages offered by the two distributions are significant:

- Distance classifications leave it up to individual speculation to figure out the magnitude of the uncertainty in total resources according to the breaking of the total tonnage into four bins. A probability distribution such as the one in Fig. 1a allows the analysis of uncertainty in multiple ways, all according to standards universally accepted and understood across multiple disciplines. Distance classifications do not provide, for example, a direct answer to a question as simple and important as: What is the minimum tonnage to expect?
- Figure 1b links tonnage to a true measure of uncertainty: the length of a 90 % confidence interval. Under most practical circumstances, distance provides poor to no discrimination in terms of uncertainty [3].
- Figure 1b can be used to break the tonnage into classes, if desired. The cumulative distribution offers the advantage of allowing selection of the two main parameters that define a classification: number of bins and location of the dividers. Four bins can still be selected, but the choice is no longer compulsory. More or fewer bins can be chosen, if more meaningful in the analysis.
- Freedom to place the dividers implies that there is nothing equivalent to the inflexibility of placing the dividers at 0.16, 0.47 and 1.9 km (1/4, 3/4 and 3 mi) in the current USGS system for classifying coal resources [1]. For example, taking the 5-95 spread boundaries at 180 and 210 kst in Fig. 1b, the total tonnage is divided into a low uncertainty class with 2.4 bst, a medium class with 3.4 bst and a high uncertainty class with 4.1 bst. Mapping the 5–95 spread can help planning selective drilling to reduce uncertainty at high uncertainty areas in the deposit, which may be the result of poor control or more complex geology.

- Tonnage in bins based on distributions such as the one in Fig. 1b discriminates in terms of uncertainty without any overlapping in uncertainty from bin to bin. It has been illustrated more than once that errors in different distance-based bins may be hardly different one from each other [3].
- In our proposed approach, additional drilling has confirmed that new values indeed land about 90 % of the time inside the 5–95 confidence intervals [8].

4 Conclusions

Distance classification methods for appraising uncertainty are ad hoc approaches only acceptable in the mining of coal and a few other commodities. Distance methods fail to provide an effective, quantitative, or objective measure of uncertainty. Characterization of uncertainty in coal tonnages in terms of probability distributions is grounded in a standard body of knowledge long used in science and technology. Stochastic methods allow reaching numerous well tested and accepted conclusions, including many that are not possible to obtain from distance methods, such as: What is the smallest tonnage to expect for a given study area?

References

1. Wood Jr., G. H., Kehn, T. M., Carter, M. D., & Culbertson, W. C. (1983). Coal resources classification system of the U.S. geological survey. U.S. Geological Survey Circular 891, p. 65.
2. Coalfields Geology Council of NSW and Queensland Mining Council (2003). *Australia guidelines for estimating and reporting of inventory coal, coal resources and coal reserves* (p. 8). http://www.jorc.org/pdf/coalguidelines.p. Accessed Dec 2013.
3. Olea, R. A., Luppens, J. A., & Tewalt, S. J. (2011). Methodology for quantifying uncertainty in coal assessments with an application to a Texas Lignite deposit. *International Journal of Coal Geology*, *85*(1), 78–90.
4. Journel, A. G., & Kyriakidis, P. C. (2004). *Evaluation of mineral reserves—a simulation approach*. New York: Oxford University Press.
5. Caers, J. (2005). *Petroleum geostatistics* (p. 88). Richardson: Society of Petroleum Engineers.
6. Olea, R. A. (2013). Special issue on geostatistical and spatiotemporal modeling of coal resources. *International Journal of Coal Geology*, *112*, 179.
7. Hengl, T. (2006). Finding the right pixel size. *Computers and Geosciences*, *32*(9), 1283–1298.
8. Olea, R. A., & Luppens, J. A. (2012). Sequential simulation approach to modeling of multi-seam coal deposits with an application to the assessment of a Louisiana Lignite. *Natural Resources Research*, *21*(4), 443–459.
9. Olea, R. A., & Luppens, J. A. (2013). Modeling uncertainty in the resource assessment of a central area of the Gillette Coal Field, Wyoming. U.S. Geological Survey Scientific Investigations Report (in preparation).
10. de Souza, L. E., Costa, J. L. C. L., & Koppe, J. C. (2004). Uncertainty estimate in resource assessment: a geostatistical contribution. *Natural Resources Research*, *13*(1), 1–15.

Reliability Analysis of Least Squares Estimation and Prediction of a Non-parametric Discovery Process Model

Mi Shi-yun and Zhang Qian

1 Introduction

A nonparametric discovery process model (NDPM) which uses empirical distribution to characterize the probability distribution of field sizes was proposed by Kaufman [1]. Wang and Nair [2] proposed maximum likelihood techniques to estimate the parameters of NDPMs. Chen [3] developed a least squares approach to estimation of NPDM parameters. Unlike maximum likelihood techniques, the least squares approach has the advantage of allowing graphic display of goodness of fit between model predictions and real discovery sequences in a conventional way and provides graphic visualization to help the resource assessor to judge the accuracy of LSA parameter estimators. Chen's NDPM is indexed by two parameters (λ, β) defined above. We provide graphs of simulated estimators $(\hat{\lambda}, \hat{\beta})$ of (λ, β) as a functions of both sample size and population shape; i.e. in particular, we study their behavior when the in place size distribution is unimodal and when it is a mixture of two distinct in place size distributions. In both cases estimators appear to be robust.

2 Simulation

In addition to studying the behavior of $(\hat{\lambda}, \hat{\beta})$ as functions of population shape and sample size, our simulation design includes a study of the sensitivity and precision of LSA projections of undiscovered deposits sizes as functions of population shape and sample size.

M. Shi-yun (✉) · Z. Qian
Research Institute of Petroleum Exploration & Development, Beijing, China
e-mail: symi@petrochina.com.cn

Z. Qian
e-mail: zhangvqian@petrochina.com.cn

E. Pardo-Igúzquiza et al. (eds.), *Mathematics of Planet Earth*, 589
Lecture Notes in Earth System Sciences, DOI: 10.1007/978-3-642-32408-6_129,

2.1 Sample Sizes

Discovered field size data is simulated from a lognormal population with parameters $\mu = 0$ and $\sigma^2 = 1$. Datasets with sample sizes 20, 50, 200 are generated. Figure 1a, b and c display typical simulated empirical discovered field size histograms. For each simulated data set the two unknown parameters, (λ, β) are estimated by LSA of the NDPM. To understand the aggregate behavior of $(\hat{\lambda}, \hat{\beta})$ we investigate the presence of possible local minima in target function space: we employ a grid search of target function values as λ ranges from 0.5 to 5 in increments of 0.025 and β ranges from 0.55 to 2.2 in increments of 0.055. This yields the target function contour plot in Fig. 1b. One can see from Fig. 1b that when sample size is 20, the target function minimum is located in a neighborhood of $\lambda = 2.75, \beta = 1.1$. Similarly, Fig. 1d shows that when sample size is 50, the target function minimum is located in a neighborhood of $\lambda = 1.5, \beta = 0.55$. For sample size 500, Fig. 1f shows a minimum in the neighborhood of $\lambda = 3, \beta = 0.7$ These contour plots lead us to conclude that a LSA of NDPM yields a unique target function minimum close to the center of the dark blue area of each contour map.

2.2 Mixed Population

Discovered field size data is simulated in two ways: first, from in place deposit sizes generated by successive sampling of datasets generated by a Lognormal distribution and second, form in place deposit sizes generated from a mixture of Lognormal population densities:

$$f(y) = \frac{0.75}{\sqrt{2\pi}} e^{-(lny)^2/2} + \frac{0.25}{\sqrt{2\pi} \times 0.5} e^{-(lny+3)^2/(2\times 0.5^2)}. \qquad (1)$$

Figure 2a shows a histogram composed of log transformed deposit sizes generated by a sample of size 200 from mixed population dataset; Fig. 2b is the map of the calculated target function; Fig. 2c shows that projections of undiscovered deposit sizes for a mixed population dataset matches actual undiscovered deposit sizes quite well; Fig. 2d shows that this is true for population datasets of size 200 generated from a single Lognormal distribution as well. Figure 2c and d, lead us to conclude that the NDPM can handle data sets that obey mixed distribution as well as those from a single distribution.

2.3 Sensitivity Analysis

In order to study Chen's NDPM sensitivity to changes in parameter values and to changes in the structure of the model we conducted two simulation studies. The aim

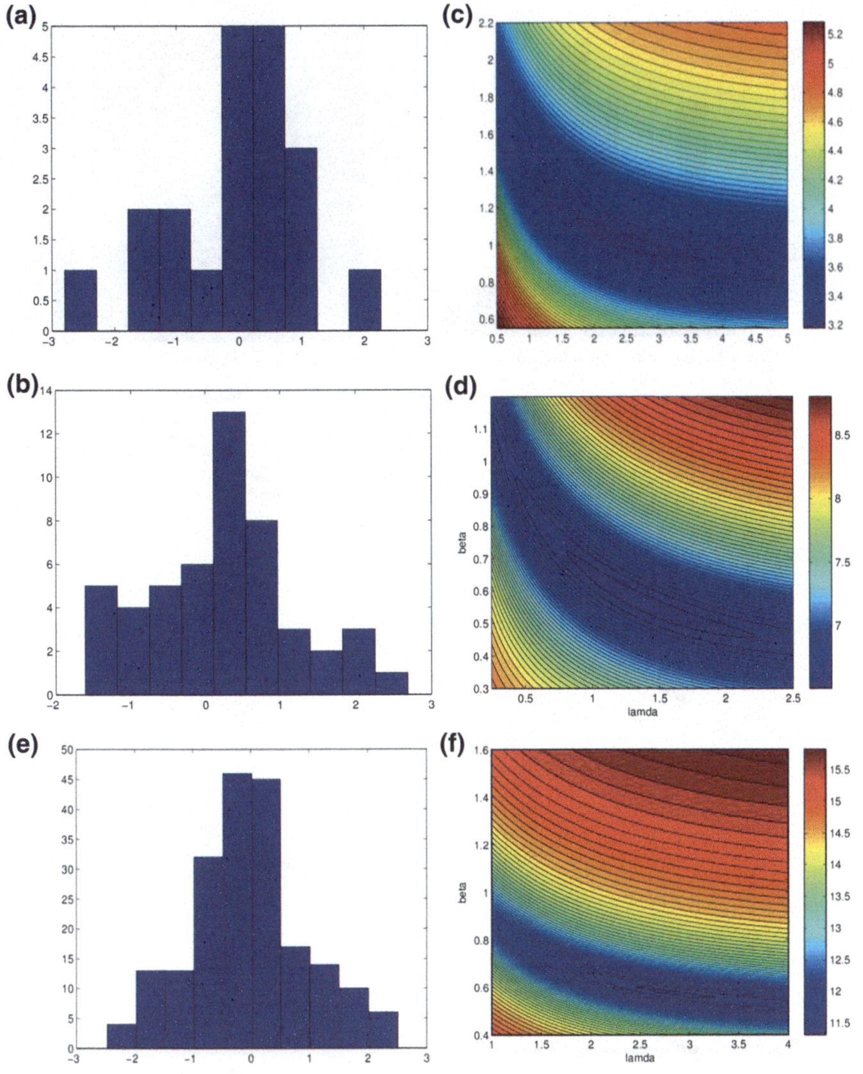

Fig. 1 Simulated logtransformed field size distribution and map of calculated target function

is to evaluate the sensitivity of (λ, β) to changes in model structure. The detailed procedure is given in the following.

Step 1 Simulate discovered field size data from a Lognormal population with parameters $\mu = 0$ and $\sigma^2 = 1$, and sample size = 200;

Step 2 Estimate parameters (λ, β) using Step 1 data;

Step 3 Fix λ, and vary β in a local area, then plot a prediction of resource potential yet to be discovered;

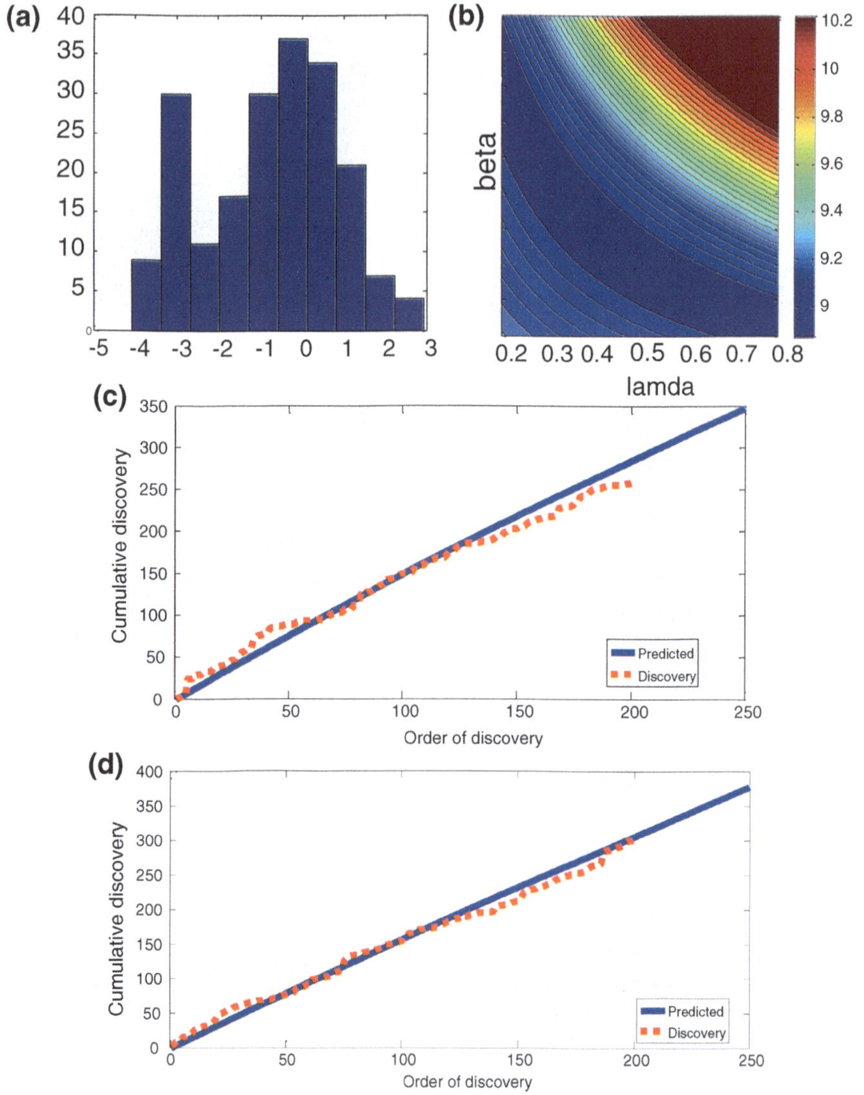

Fig. 2 Mixed population simulation results

Step 4 Fix β and plot the same prediction as in Step 3.

Results are shown in Fig. 3. Figure 3a shows results when λ is fixed at 0.1 and β ranges from 0.9 to 1.6 in steps of length 0.1. It can be seen from Fig. 3a that the results vary slightly. Similarly, Fig. 3b shows the results when β is fixed at 1.3 and λ is varied from 0.05 to 0.35 in steps of length 0.05. While Fig. 3b variability is modest, variability is larger than that shown in Fig. 3a.

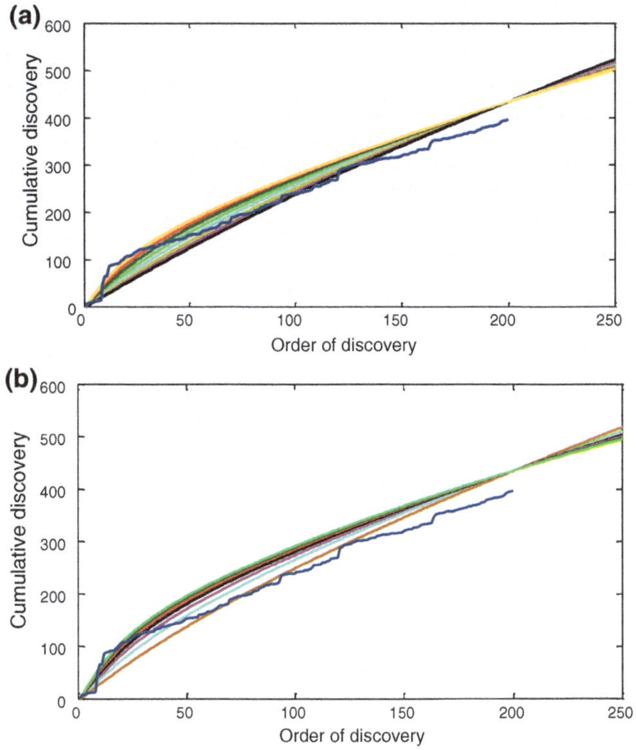

Fig. 3 Sensitivity analysis

3 Conclusions

Our sensitivity analysis of LSA of NDPM leads us to conclude that : (1) LSA of NDPM is effective over a rather large range of sample sizes. (2) It is robust in the sense that it yields rather accurate predictions of undiscovered potential for both univariate Lognormal in place size populations and for a mixture of Lognormal in place size populations; (3) figures for the changes of the parameters sensitivities.

Acknowledgments We gratefully acknowledge Professor Gordon Kaufman for useful comments. This research is supported by National Key Scientific and Technological Project "Research on Global Petroleum Resources Data System (2011ZX05028-004)".

References

1. Kaufman, G. M. (1965). Statistical analysis of the size distribution of oil and gas fields. In: *Symposium on Petroleum Economics and Evaluation*, (pp. 109–124).
2. Wang, P. C. C., & Nair, H. C. (1987). Resources Statistical analysis of oil and gas Discovery data. *Quantitative Analysis of Mineral and Energy 223*, 199–214.
3. Chen, G., Chen, Z., & Tzeng, P. (2010). *A non-parametric discovery process model - a least squares approach*. Open File 6253, Geological survey of Canada.

Application and Comparison of Discovery Model and Others in Petroleum Resource Assessment

Guo Qiu-Lin, Yan Wei and Chen Zhuo-Heng

1 Introduction

Three methods have been used in PetroChina's third National petroleum resource assessment in China: the Simple Pareto Model (SP), the Shifted and Truncated Pareto Model (STP) and the Lognormal Discovery Process Model (LPD). Results from the third national wide assessment were analyzed recently at PetroChina as preparation for its fourth coming resource assessment. The analysis presented here is based on a comparison of model outputs: the largest predicted field size (S), number of total fields (N) and total play resource (R) in a play. Some interesting findings provide useful insights for determination of methods to use in the forthcoming assessment exercise. In this paper, we briefly review of these methods and discuss merits and drawbacks of each via application to Zagros Basin plays.

Three Zagros Basin plays are selected to represent various exploration stages, field size distribution characteristics and exploration efficiency. The L play with 26 discovered fields and total discovered recoverable reserves of 8,102 MMBOE represents a relatively immature play. The largest field found is 933 MMBOE. The K play has a total of recoverable reserves of 47,973 MMBOE in 50 fields and the largest field size is 8,709 MMBOE, representing an intermediate level in exploration efficiency. The D play has 56 discoveries with largest discovered field of 29,165 MMBOE and total recoverable reserves of 13,349.9 MMBOE, representing the most mature play in this comparison.

G. Qiu-Lin · Y. Wei (✉)
Research Institute of Petroleum Exploration and Development, PetroChina,
100083 Beijing, China
e-mail: qlguo@petrochina.com.cn

Y. Wei
e-mail: yanwei123456@petrochina.com.cn

C. Zhuo-Heng
Geological Survey of Canada, 3033-33rd Street, Calgary, ABT2L 2A7, Canada
e-mail: Zhuoheng.Chen@NRCan-RNCan.gc.ca

E. Pardo-Igúzquiza et al. (eds.), *Mathematics of Planet Earth*,
Lecture Notes in Earth System Sciences, DOI: 10.1007/978-3-642-32408-6_130,

2 Methods

2.1 SP Distribution Model

The SP model is a simple transformation of a well-known distribution, the two-parameters Pareto distribution. Chen and Sinding-Larsen (1994) [1] define the representative size of the mth largest field in the finite population as

$$x_m = a\left(\frac{m}{N}\right)^{-\frac{1}{\theta}}, \tag{1}$$

where a is the lower economic cut off size and θ is the shape parameter. The ratio of the nth largest field to the mth largest field is

$$\frac{x_n}{x_m} = \left(\frac{m}{n}\right)^{\frac{1}{\theta}}, \tag{2}$$

By using (1) and (2), the number of fields in the play and total resources R are respectively

$$\begin{cases} N = \left(\frac{x_1}{x_N}\right)^{\theta} = \left(\frac{x_1}{a}\right)^{\theta} \\ R = x_1 \sum_{m=1}^{N} \left(\frac{1}{m}\right)^{\frac{1}{\theta}} \end{cases} \tag{3}$$

A deterministic method of Chen and Sinding-Larsen [1] works well for immature plays. Guo et al. [2] developed an iteration method based on an idea of Zhao [3] to capture prediction uncertainty.

2.2 STP Distribution Model

The density function for the STP is parameterized by Houghton [4] and Houghton et al. [5]. Based on a dataset of 2,600 oil & gas pools in the West Siberia Basin, Jin [6] derived the following version of STP density and cumulative functions:

$$\begin{cases} f(q) = \frac{\lambda(q_0+r)^{\lambda}}{(q+r)^{\lambda+1}} \\ F(q) = 1 - \left(\frac{q_0+r}{q+r}\right)^{\lambda} + \left(\frac{q_0+r}{q_{\max}+r}\right)^{\lambda} \end{cases} \tag{4}$$

Xu [7] developed a method for computing r and shape parameter . We can get as follows,

$$\lambda = \frac{1}{2}e^{-\frac{\gamma-4}{e\cdot\sigma}} + 0.5 \tag{5}$$

where σ is the standard deviation.

2.3 Discovery Process Model

The LDP can be written in terms of a joint density function in Eq. (6) with an underlying lognormal field size distribution.

$$\begin{cases} L(\theta) = \frac{N!}{(N-n)!} \prod_{j=1}^{n} f_\theta(x_j) \mathrm{E}_\theta \left[\prod_{j=1}^{n} \frac{x_j^\beta}{b_j + Y_{n+1}^\beta +, \cdots, + Y_N^\beta} \right] \\ f_\theta(x_j) = \frac{1}{x_j \sigma \sqrt{2\pi}} \exp\left[-\frac{1}{2} \left(\frac{\ln x_j - \mu}{\sigma} \right)^2 \right] \end{cases} \quad (6)$$

where L is the likelihood function and β is the exploration efficiency parameter. The mathematical details of this method can be found in Lee [8, 9].

3 Results and Discussion

All three methods are applied to three selected Zagros Basin plays. Output from the three methods are listed in Table 1. Assuming a lognormal parent distribution, Table 2 provides estimates of model parameters of the parent population in the three plays and to show general field size characteristics and exploration efficiency. Methods are compared in terms of total play resource(R), number of fields(N) and the largest field size (S) in the play. Results suggest that:

(a) The LDP model has the lowest estimates of total play resource and number of fields in general, but the largest predicted largest field size;
(b) The STP gives the largest play resource potential estimate in general and a moderate forecast of both number of fields and size of the largest field in a play;

Table 1 Comparison of the assessment results from the three different method in three petroleum plays of the Zagros Basin

Estimates	Method	Play L	Play K	Play D	Sum
Play resource	SP	14,623	80,560	225,034	320,217
	STP	16159	68,727	489,743	574,629
	LDP	15,430	78,047	208,900	302,377
Remaining resource	SP	6,611	32,589	91,284	130,484
	STP	8,057	20,754	355,993	384,804
	LDP	7,328	30,074	75,150	112,552
Number of fields	SP	86	330	683	1,099
	STP	63	79	256	398
	LDP	60	85	90	235
Largest field	SP	1,682	8,709	29,165	39,556
	STP	1,607	10,213	29,242	41,062
	LDP	2,051	18,890	30,730	51,671

Table 2 Estimated distribution parameters from the Lognormal Discovery Process Model (LDP) showing general characteristics of field distribution, discovery efficiency of the three plays in Zagros basin

Play L		Play K		Play D	
β	0.57	β	0.68	β	0.71
μ	4.75	μ	5.1	μ	6.5
σ^2	1.6	σ^2	3.9	σ^2	2.5
N	60	N	85	N	90

(c) The SP projects a large number of fields, particularly when both the number of discoveries and exploration efficiency are high, while the largest field size estimate is the smallest among the three models.

4 Conclusions

In general, the SP model appears to be suitable for immature exploration plays while the STP model performs better for mature plays. The LDP model is applicable to a wide scope of situations when number of discoveries is greater than 20. Here are some interesting findings that provide useful insights for determining which method to use in PetroChina's fourth national wide petroleum resource assessment:

(a) The SP model does not require a large amount of data and computation is straightforward. However, it typically leads to projection of a large number of fields.
(b) The STP model's probability of finding a field is proportional to a power of field size shifted by a parameter. This model forecasts a reasonably large number of fields. However, a large sample size is required to achieve a reasonable forecast.
(c) The LDP model appears to provide a reasonable projection of total resources R when the number of discoveries is greater than 20. However, it may project a conservative N along with an optimistic S, the largest field size in the play.

References

1. Chen, Z. H., Sinding-Larsen, & R., Sammatray, R. (1994). The interdependencies between volumetric variables and their impact on petroleum resource assessment. In *Proceedings of the 94 annual meeting of international association for mathematical geology*, 68–73, Quebec.
2. Guo, Q. L., Hu, S. Y., Zhang, Q. C., & Wang, X. H. (2003). The double serial method and its application in petroleum resource evaluation. *Petroleum Exploration And Development*, *30*, 50–52.
3. Zhao, X. D. (1988). *Quantitative evaluation of oil resources*. Beijing: Geological Press.
4. Houghton, J. C. (1988). Use of the truncated shifted pareto distribution in assessing size distribution of oil and gas fields. *Mathematical Geology*, *20*, 907–937.

5. Houghton, J. C., Dolton, G. L., Mast, R. F., Masters, C. D., & Root, D. H. (1993). U.S. Geological Survey estimation procedure for accumulation size distributions by play. *The American Association Petroleum Geologists Bulletin*, *77*, 454–466.
6. Jin, Z. J. (1995). A comparison study of five basic oil and gas pool size probability distribution models and its significance. *Acta Petrolei Sinica*, *16*, 6–13.
7. Xu, Z. M., Jin, Z. J., Sun, H. J., & Xu, J. X. (2010). The discussion about how to obtain the key parameters of generalized pareto distribution. *Petroleum Geology and Experiment*, *32*, 517–520.
8. Lee, P. J. (1998). Analyzing multivariate oil and gas discovery data. In A. Buccianti, G. Nardi & R. Polenza (Eds.), *Proceedings of the 4th annual conference of the international association of mathematical geology* (pp. 445–451). Italy.
9. Lee, P. J. (2008). *Statistical methods for estimating petroleum resources*. Oxford: Oxford University Press.

Handling Seismic Anomalies in Multiple Segment Prospects with Graphical Models

Gabriele Martinelli, Charles Stabell and Espen Langlie

1 Introduction

Bayesian Risk Modification (BRM) [4] is a standard statistical framework for de-risking exploration targets using seismic anomalies. When an anomaly is observed it translates exploration team assumptions into an increase or decrease in the Chance of Success (COS). A unique advantage of the Bayesian approach is that it provides a statistical framework for assessing COS along with resources generated by prospects with multiple targets with seismic anomalies (see [3]). This paper presents a new prospect-level BRM approach by introducing a parameter that captures the degree of dependence among seismic anomalies on targets in a prospect. We compare this new BRM approach with the classical risking framework. BRM fits well into any graphical model risking framework such as those presented in [1] and [2]. This allows straightforward Monte Carlo simulation of our risking procedure.

2 Methodology

In what follows, a discrete exploration target is a segment and a prospect is a collection of segments. Prospect COS is the chance that at least one segment contains hydrocarbons. A seismic anomaly indicator is defined to be a direct fluid indicator (DFI). Graphic models are designed to capture dependencies among seismic anomalies.

G. Martinelli (✉) · C. Stabell · E. Langlie
GeoKnowledge AS, A Schlumberger Company, Aslakveien 14, Oslo, Norway
e-mail: gmartinelli@slb.com

C. Stabell
e-mail: cstabell@slb.com

E. Langlie
e-mail: elanglie@slb.no

E. Pardo-Igúzquiza et al. (eds.), *Mathematics of Planet Earth*,
Lecture Notes in Earth System Sciences, DOI: 10.1007/978-3-642-32408-6_131,

A DFI dependence parameter is introduced to measure the degree of dependency among them.

2.1 A Graphical Model for Exploration Risk Factors

Consider first a single segment with *r* risk factors. If we assume mutual independence of risk factors, the segment's COS is the product of risk factor probabilities. Each risk factor is represented by a binary random variable. In turn, the segment's COS is a random variable whose multiple states represent success along with all possible failure scenarios (See [1]) whose conditional distribution (conditional probability table (CPT)) given the risk factors depends of course on the type of risk factors chosen.

In most geologic risk software (GeoX for example) up to 20 failure scenarios can be considered. Here we present an example with three risk factors—reservoir presence (RP), source presence (SP) and trap and seal (TS)—and two failure scenarios. Bottom nodes represent the segment's state. Top nodes in Fig. 1 below represent risk factors, i.e. the presence or absence of hydrocarbons (HC). The absence of either trap and seal or source, combined with presence of reservoir, leads to a water-filled reservoir, or brine case. Success of all three risk factors is essential for an oil (success) case. Figure 1, Left, shows the structure of CPT as described by Table 1. This is the risk dependency structure for the single risk factor reservoir presence (RP). The risk dependency group is composed of segments A, B and C.

Segments that belong to the same play or to the same petroleum system may share geological risk dependencies. It is useful to define a *prospect* as a group of segments that share risk and volumetric dependencies and can be developed as a single economic project. Risk dependencies among risk factors in different segments are encoded via an additional node with a shared probability that tunes the degree of dependency among segments. If the shared probability is equal to 1.0, risk elements are independent. If the shared probability is equal to the largest of risk factor prob-

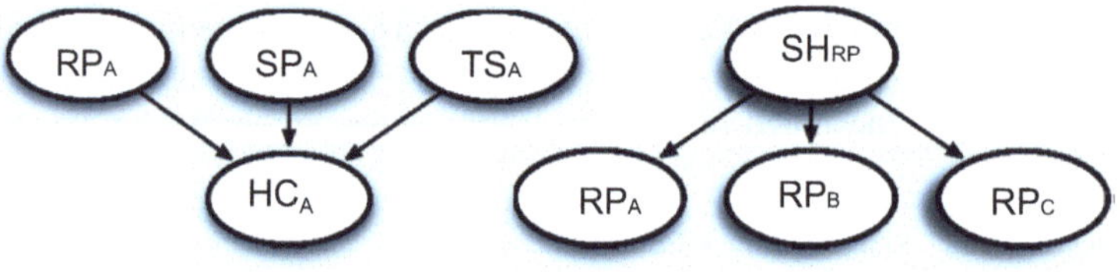

Fig. 1 *Left*: risk model for a single segment. *Top nodes* represent main risk factors reservoir presence (RP), source presence (SP) and trap and seal (TS). *Bottom nodes* represent a segment's state, i.e. the presence or absence of hydrocarbons (HC). On the *right*: risk dependency structure for a single risk factor reservoir presence (RP)

Table 1 Probabilities of success and failure cases

Risk factor			P(case)		
RP	SP	TS	Oil	Brine	Shale
1	1	1	1	0	0
0	1	1	0	0	1
1	0	1	0	1	0
0	0	1	0	0	1
1	1	0	0	1	0
0	1	0	0	0	1
1	0	0	0	1	0
0	0	0	0	0	1

abilities, dependencies are maximized. A simple model structure is shown in Fig. 1 Right.

2.2 A Graphical Model for Integrating Seismic Anomalies

Often identification of a seismic anomaly is what leads the exploration team to consider risking of a specific target. We treat identification of a new anomaly as information that modifies our original risk estimate. This new information appears as an added set of variables labeled Direct Fluid Indicators (DFIs). Given a segment, the conditional distribution of a DFI is assessed by the exploration team (Table 2). The probabilistic model is shown in Fig. 2.

Consider again multiple segments belonging to a prospect in which several anomalies are observed or not. A standard assumption is that DFI anomalies are independent given segment states. This assumption is often not valid: geological risk dependencies influence probabilities of outcome anomaly outcomes. For this reason we introduce probabilistic dependencies among DFI signals via a parameter k that captures the degree of dependency among seismic anomalies observed on segments within a defined DFI dependency group. The range of values of k is [0, 1]. $k = 0$

Table 2 CPT with the likelihood for all the success and failure scenarios

Case	$P(DFI\|case)$	$P(noDFI\|case)$
Oil	0.8	0.2
Brine	0.3	0.7
Shale	0.5	0.5

Fig. 2 Model that integrates the information provided from seismic anomalies. The related CPT is in Table 2

Table 3 CPT for the DFI nodes, with the likelihoods for success and failure scenarios integrated and the correlation parameter k

Case	DFIref	P(DFI ∣ case, DFIref)
Success	Present	0.8 + k*(1–0.8)
Failure	Present	0.3 + k*(1–0.3)
Success	Absent	0.8 – k*0.8
Failure	Absent	0.3 – k*0.3

corresponds to a group whose seismic anomalies are mutually independent. Seismic anomalies in a group with $k = 1$, are, on the other hand, functionally dependent, so observation of more than one of them provides no additional information. A group with $k = 0.5$ means seismic anomaly signals are probabilistically partially dependent. The larger the joint dependence among anomalies, the more these signals will resemble a reference anomaly indicator.

The parameter k is designed to capture the degree of dependency among seismic anomalies observed on segments via linear interpolation as shown in Table 3. Given likelihood values (for example 0.8 for the success case and 0.3 for the failure case) we weight each scenario (see Table 3).

This particular interpolation scheme does *not* imply that COS is necessarily a linear function of state probabilities.

The method can easily be extended to the case of multiple failure scenarios. The choice of the reference DFI indicator is automatic: the reference node is sampled uniformly among all the segments that belong to the prospect, in order to avoid

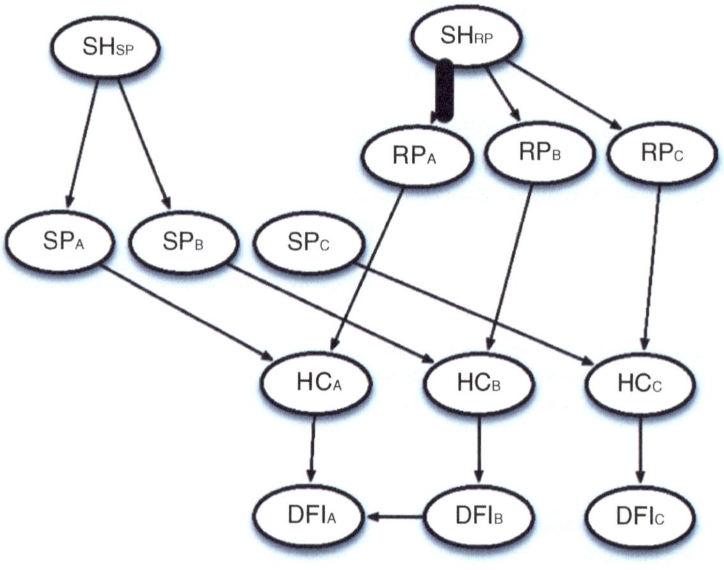

Fig. 3 Graphical representation of the whole risking model

biases. A representation of a typical entire risking model is shown in Fig. 3. In this example we have two risk factors with a different dependency group for each of them (source presence in C is independent), and three seismic anomalies. Two of them are correlated within a DFI dependency group that includes segments A and B.

3 From the Graphical Model to a Complete Monte Carlo Framework

A main benefit of a model that represents all risking components within the same graphical model, is that we can compute the probability of each possible scenario and, as a consequence, Monte Carlo sample efficiently to generate empirical probability distributions of volumetric parameters and risked and success volume distributions.

4 Summary and Conclusions

In conclusion, the new procedure extends BRM applicability to seismic anomalies in multiple segment prospects within any graphical model framework. Our procedure leverages benefits of Bayesian analysis. Computations are transparent and reproducible and easy to interpret and use.

References

1. Martinelli, G., Eidsvik, J., Hauge, R., & Drange-Forland, M. (2011). Bayesian networks for prospect analysis in the north sea. *AAPG Bulletin*, *95*(8), 1423–1442.
2. Martinelli, G., Eidsvik, J., Hauge, R., & Hokstad K. (2012). Strategies for petroleum exploration based on bayesian networks: A case study. SPE Paper 159722, SPE ATCE 2012.
3. Stabell, C.B., & Langlie, E. (2008). Handling seismic anomalies on multiple targets. In Back to Exploration, 2008 CSPG CSEG CWLS Convention.
4. Stabell, C.B., Lunn, S., & Breirem K. (2003). Making effective use of a dfi: A practical bayesian approach for risking prospects for seismic anomaly information. In SPE HC Economics and Evaluation Symposium, SPE 82020.

The North American Shale Resource: Characterization of Spatial and Temporal Variation in Productivity

Qudsia Ejaz and Francis O'Sullivan

Between 2005 and 2012, U.S. natural gas production from ultra-low permeability hydrocarbon-prone mud rock formations, often referred to as the "shale resource" [1], increased 20-fold to more than 570 Mm^3 per day, and now accounts for $\approx$33% of total U.S. gas output. These developments have had a profound impact on the U.S. energy sector. Despite it's rapid rise, the exploitation of the shale resource is still in it nascency, and knowledge of the precise production mechanisms remains limited [2]. A consequence of this is that the accurate economic characterization of the resource remains difficult. This paper examines spatial and temporal trends in the productivity of contemporary horizontal, hydraulically fractured wells within and between the major U.S. shale plays.

1 Shale Resource Performance: Intra- and Interplay Variability

The choice of shale well productivity metrics depends on available data. One widely used metric is the *peak production rate*.[1] It is a natural choice for assessing productivity since there is a strong correlation between a well's peak production rate and its economic performance. Such metrics report total gas production in some fixed period, hence we refer to them as *absolute* metrics. As gas production increases in proportion to a well's horizontal length, and these lengths vary significantly across the contemporary well ensemble, we define a corresponding *specific* metric. *Specific* productivity metrics reflect a well's performance normalized by the length of its horizontal or completed interval. By considering both absolute and specific metrics in

[1] Peak production rate refers to the highest recorded monthly output from a well. For the vast majority of horizontal shale wells, the peak production rate will be recorded during the first or second month of production.

Q. Ejaz · F. O'Sullivan (✉)
MIT Energy Initiative, Massachusetts Institute of Technology, 77 Massachusetts Avenue, Cambridge, MA 02139, USA
e-mail: frankie@mit.edu

E. Pardo-Igúzquiza et al. (eds.), *Mathematics of Planet Earth*,
Lecture Notes in Earth System Sciences, DOI: 10.1007/978-3-642-32408-6_132,

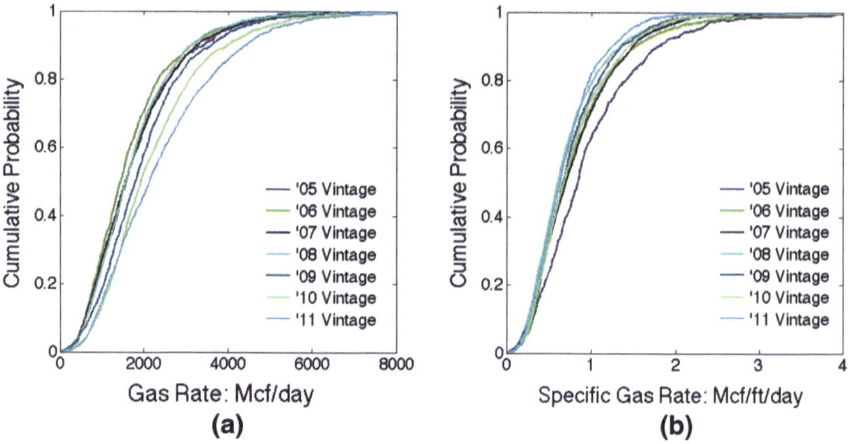

Fig. 1 Illustration of vintage-to-vintage and intra-vintage well-to-well variability in both the absolute and specific productivity of the Barnett shale's 2005–2011 horizontal well ensemble. **a** 2005–2011 vintage *absolute* peak production rate CDF's. **b** 2005–2011 vintage *specific* peak production rate CDF's

juxtaposition we are able to decouple the affect of changing horizontal lengths from underlying patterns in the shale resources' productivity.

In this paper we show that there is large intra-shale play variability in horizontal well absolute peak production rates, with the P80:P20 ratio lying between 2.5–3.5 in all major plays. Further, when disaggregated by vintage, the mean absolute peak production rate in each play generally increases year-on-year. However, intra-vintage well-to-well variability does not differ markedly from that of the overall well ensemble. These features, which are illustrated in Fig. 1a are important as they could be reasonably interpreted as evidence that there is a stochastic aspect to shale resource productivity, at least within the boundary of the current development paradigm.

The year-on-year increase in the mean absolute peak production rate shown in Fig. 1a is often presented as evidence of improving well productivity. Such interpretations are naive, primarily because an absolute metric fails to account for changes in horizontal well lengths over time. In this paper we argue that any assessment of shale resource productivity using horizontal well data should use *specific* metrics, as defined above. To support this, we show that in the contemporary U.S. shale plays, specific peak production rates are falling. Figure 1b shows the per-vintage specific peak production rate CDFs for the same ensemble. Interestingly, the figure shows that the mean specific peak production rate of the 2011 vintage is 29 % lower than in 2005. This result may support the notion that creaming also occurs over the lifetime of an unconventional resource play.

2 Spatial Patterns of Well Performance

We explore if there are portions of the active plays where clustering of higher performance wells is statistically significant. We use the Getis-Ord* tests [3] for this study, though the analysis can also be conducted using Anselin's index. As well productivity is reasonably represented by a lognormal distribution [4], we use the natural logarithm of the specific first 12 month production as the input parameter. The results of the Getis-Ord* test in the Barnett at the 10 km length scale show the "core" area where the higher Z-values are clustered. Wells are usually drilled at an 80 acre spacing.[2] The results of performing the analysis at a distance of 10 km, i.e. a circle of area 77630 acres, are shown in Fig. 2. These hot and cold spots persist when the data is partitioned by vintage, and when we use specific peak gas for a month or specific peak daily gas. Furthermore, as the neighborhood size of each well is increased beyond 10 km, e.g. to 100 km, the hot and cold spots persist and grow.

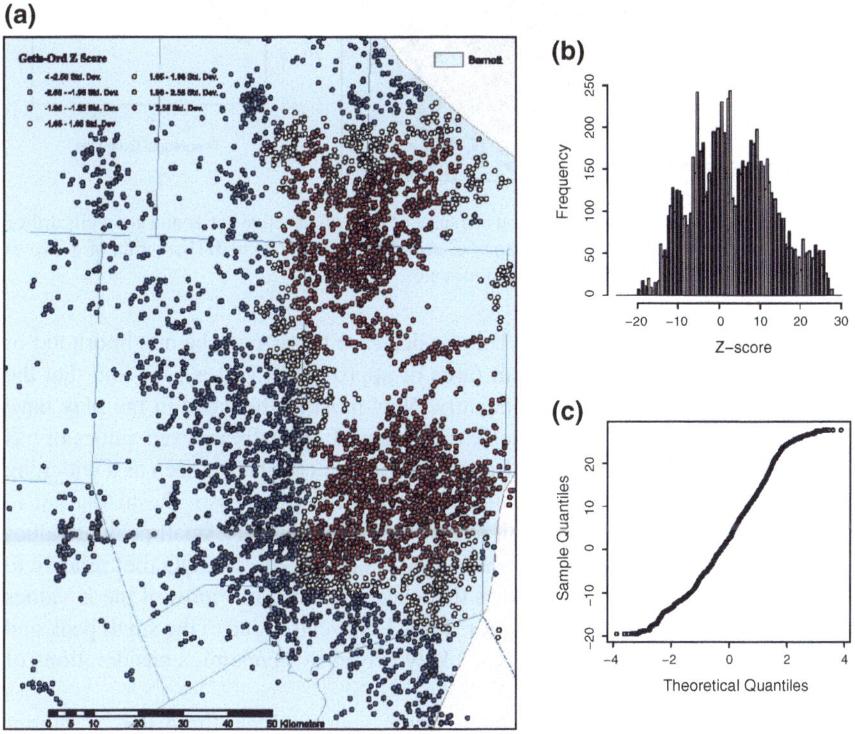

Fig. 2 Results of the local Getis-Ord* test at a scale of 1 km of all active horizontal gas wells drilled between 2005–2011 in the Barnett. **a** Map of Z-scores at 10 km scale. **b** Histogram of Z-scores at 10 km scale. **c** QQ-plot of Z-scores at 10 km scale

[2] 1 acre $= 4046.856\,m^2$; so A square of 80 acres will have a side of 570 m.

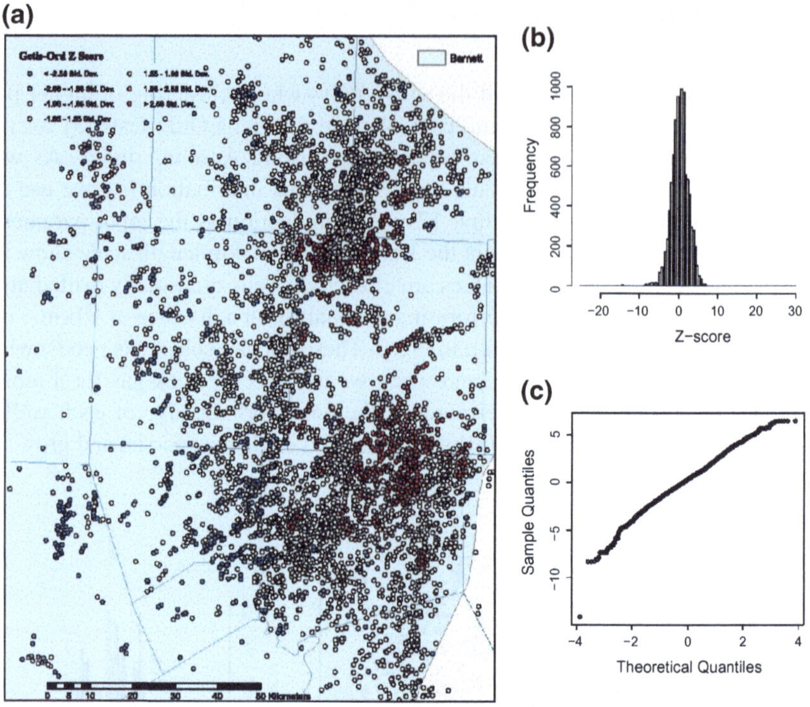

Fig. 3 Results of the local Getis-Ord* test at a scale of 1 km of all active horizontal gas wells drilled between 2005–2011 in the Barnett. **a** Map of Z-scores at a scale of 1 km. **b** Histogram of Z-scores at 1 km scale. **c** QQ-plot of Z-scores at 10 km scale

However, as shown in Fig. 3, when we decrease the size of the neighborhood of each well to a circle of radius 1 km (area of approx. 776.5 acres), we find that the hot and cold spots diminish significantly. Now most of the wells in the play have Z-scores that suggest that at this scale the observed, spatially indexed values of gas produced cannot be distinguished from a random pattern. This is seen as a widening band of yellow between the hot and cold spot in Fig. 3a. Also, the histogram in Fig. 3b shows all the Z-scores centered on zero with one very small peak at values below -5 (cold spot) and a small set of values outside 2σ. The QQ plot relative to a normal distribution in Fig. 3c shows that except for the tails, most of the Z values are normally distributed. The deviation at the lower end is due to the small peak and a truncation in observed values, most likely driven by economic considerations of the operators and producers.

This presence of clustering at large length scales, which diminishes at shorter scales leads us to the hypothesis that the large scale pattern can be described as a smooth mean process, $\mu(t, \mathbf{x})$. With this mean description, the productivity at a given location, $Y(t, \mathbf{x})$, is just a fluctuation, ε about the mean value, and is described by a lognormal distribution: $Y(t, \mathbf{x}) = \mu(t, \mathbf{x}) + \varepsilon(t, \mathbf{x})$. In this paper, we will explore the validity of such an assertion, with a characterization of the properties of $\varepsilon(t, \mathbf{x})$.

References

1. Jarvie, D. M., Hill, R. J., Ruble, T. E., & Pollastro, R. M. (2007). *AAPG bulletin, 91*(4), 475.
2. Anderson, D., Nobakht, M., Moghadam, S., & Mattar, L. (2010). In *SPE unconventional gas conference*.
3. Getis, A., & Ord, J. K. (2010). In L. Anselin & S. J. Rey (Eds.), *Perspectives on spatial data analysis* (pp. 127–145)., Advances in spatial science. Berlin: Springer. http://www.springerlink.com/content/t404h840q43m7r2v/abstract/.
4. Ejaz, Q., & O'Sullivan, F. Characterization of multi-scale trends in the productivity of the contemporary shale resource. Under preparation.

A Pyramid Scheme: Integrating Petroleum Systems Analysis into Probabilistic Petroleum Resource Assessments

Kirk G. Osadetz and Zhuoheng Chen

1 Introduction

The identification and appraisal of discovered (reserves and contingent resources) and undiscovered (prospective resources) petroleum resources are important for valuation and policy. The Petroleum Resource Management System (PRMS) provides reserve and resource valuation categories acceptable to the Security and Exchange Commission [7]. While the PRMS provides is useful for valuation it does not convey how technological improvement provides access to a larger resource-base. That is more effectively conveyed by the "resource pyramid" [6] that because of the power-law relationship between pyramidal height and volume, suggests that larger resource volumes occur in less commercial and more challenging reservoirs.

Petroleum supply has been transformed and revitalized by production from low permeability reservoirs using horizontal drilling and multiple-stage hydraulic fracturing [9, 11]. These technologies applied to low permeability reservoirs represent a major supply augmentation that challenges both reserve valuation methods and resource assessments. Current discussion focuses on developing an "appropriate technology" that classifies and values reserves and contingent resources hosted in low permeability plays. These accumulations are commonly described as "shale-hosted" with some proposing that value should be recognized even where the drilling has not occurred, because of the resource "continuity".

K. G. Osadetz (✉) · Z. Chen (✉)
Geological Survey of Canada: Calgary, 3303 – 33rd Street, NW, Calgary, ABT2L 1S2, Canada
e-mail: kosadetz@nrcan.gc.ca

Z. Chen
e-mail: zchen@nrcan.gc.ca

E. Pardo-Igúzquiza et al. (eds.), *Mathematics of Planet Earth*,
Lecture Notes in Earth System Sciences, DOI: 10.1007/978-3-642-32408-6_133,

We confirm that shale-hosted petroleum accumulations are previously discovered discrete contingent resources. We propose to model contingent resource characteristics of the unconventional accumulations considering petroleum system constraints in combination with a geographically based probabilistic resource assessment method [2] to define unconventional pools/fields that augment or modify the discovery sequence.

2 Shale-Hosted Petroleum Accumulations

Shale is an ambiguous term—it describes a fissile mudstone, but it is also used in place of group or formation to describe stratigraphic units, as in Colorado Shale or Bakken Shale. As a result shale gas and shale oil are ambiguous terms referring to low permeability reservoirs exploited using horizontal drilling and multiple stage hydraulic fracturing. Shale-hosted resources are often inferred "continuous". However, commercial reservoirs are generally not shale and the resources are not "continuous" [3, 11]. Shale-hosted resources have much well data available and unconventional fields may include conventional pools and production [4]. Assuming continuity, many assessors propose "unit-cell" assessment models. This approach does not consider: spatial reservoir variations resulting from geological processes; well performance variations that are large compared to the mean; and the many discrete small pools that cannot be assumed to be part of a continuous accumulation. By treating the accumulations as discrete we infer that pool class volumes follow a power-law relationship. Using petroleum system observations we constrain, the smallest pool size classes, unconventional accumulations and the total resource volume. If the discovered pool sizes are considered uncertain, but inferable, then probabilistic resource assessment methods. Including discovery sequence methods can be adapted to assess them.

3 Modeling Discovered Pool Sizes

Discovered conventional resources have, a discovery well, a geographic limit (petroleum/water contact), and an initial in-place volume estimate. These data are the key inputs to a discovery process analysis. The same data for previously non-commercial shale-hosted accumulations must be inferred to augment or modify the discovery sequence. Accumulation size "growth", "multiple imputation" and "modeling" provide three perspectives for characterizing these accumulations. Growth is an important feature of petroleum accumulations [5]. Reinson et al. [8] attempted to employ drill stem test data to augment the discovery sequence, but they could not associate pool volume estimates with test results. We also suggest that "growth" is inappropriate because there is no basis to infer that "growth" factors from the conventional resources apply to unconventional resources. Nor is multiple imputation [10] appropriate for estimating missing accumulation data, both because the previously non-commercial accumulation data are not missing randomly and because the available data sources are well-based, whereas the missing data is accumulation-based. Additionally, the boundaries of unconventional accumulations may include wells or accumulations thought part of the conventional discovery sequence, but which might be deleted once unconventional accumulations are recognized.

Modeling appears the best approach to estimating the contingent resource outside conventional accumulations. We propose the following inference method:

1. The general extent of accumulations is inferred using a geographical-based model [2], constrained by available data resulting in multiple equally probable petroleum distribution simulations. The method can be modified using wire-line log data, to include (1) discovered and developed conventional accumulations, together with wells indicating (2) presence for, or (3) lack of contingent resources.
2. Accumulation boundaries and size can be constrained: in the dip direction, by reservoir physical conditions and material properties such that the petroleum column pressure neither disrupts the top seal nor exceeds the trapping mechanism; and in the strike direction, from stratigraphic and lithologic constraints from conventional pools and stratigraphic analysis.
3. Additional constraints on accumulation boundaries and sizes are available from petroleum system data, such as extractable petroleum content. Where reliable pressure data indicates either a petroleum/water contact or a continuous petroleum column this can refine descriptions. Unfortunately data availability and quality is limited.
4. Potential accumulation characteristics are then estimated for each equally probable resource realization of the geographically based assessment. The realizations can be grouped by discovery well location and used to augment or modify the accumulation discovery history sequence.

4 Discussion

The advantage of the method proposed is that it uses petroleum system characteristics to infer discrete accumulations in unconventional reservoirs. Buoyancy from density differences between water and petroleum is a source of reservoir energy. By considering potential reservoir pressure the method indicates the best prospects for commercial development. This cannot be obtained from unit-cell "continuous" resource appraisal, as the inferred continuity does not extend to reservoir pressure if the seals are to remain un-fractured. The inferred accumulations augment or modify the discovery history, so that a discovery process analysis can vindicate the resource predictions. We hope to present an example application at the meeting.

5 Conclusions

Pressure data suggests that shale-hosted petroleum resources are not continuous and there are alternatives to "unit-cell" based resource assessment methods. Probabilistic methods can be employed to infer contingent resource volumes and accumulation boundaries that are additionally constrained by petroleum system information. If these inferred accumulations are used to update the discovery sequence the resource assessment can be vindicated using a discovery process analysis.

References

1. Burrus, J., Osadetz, K. G., Wolf, S., Doligez, B., Visser, K., & Dearborn, D. (1996). A two-dimensional regional basin model of Williston Basin hydrocarbon systems. *American Association of Petroleum Geologists Bulletin, 80*, 265–291.
2. Chen, Z., & Osadetz, K. G. (2006). Undiscovered petroleum accumulation mapping using model-based stochastic simulation. *Mathematical Geology, 38*, 1–16.
3. Hansen, W. B., & Long, G. I. W. (1991). Bakken production and potential in the U.S. and Canada: can the fairway be defined? In W. B. Hansen (Ed.), *1991 guidebook to geology and horizontal drilling of the Bakken Formation* (pp. 69–88). Montana: Montana Geological Society.
4. Hansen, W. B., & Long, G. I. W. (1991). Criteria for horizontal and vertical prospect generation in the Bakken Formation, Montana and North Dakota. In W. B. Hansen (Ed.), *1991 guidebook to geology and horizontal drilling of the Bakken Formation* (pp. 151–164), Montana: Montana Geological Society.
5. Klett, T. R. (2003). Graphic comparison of reserve-growth models for conventional oil and accumulation. USGS Bulletin, 2172-F.
6. Masters, J. A. (1984). Elmworth: case study of a deep basin gas field. *American Association of Petroleum Geologists Memoir, 38*, 316.
7. Purewal, S., Choquette, S., Gold, D., & Mallon, K. (2011). Guidelines for the application of the Petroleum Resource Management System. Society of Petroleum Engineers. http://www.spe.org/industry/docs/PRMS_Guidelines_Nov2011.pdf, p. 222.
8. Reinson, G. E., Lee, P. J., Warters, W., Osadetz, K. G., Bell, L. L., Price, P. R., et al. (1993). Part I: geological play analysis and resource assessment; in: Devonian gas resources of the Western Canada sedimentary basin. *Geological Survey of Canada Bulletin, 452*, 1–127.
9. Ross, D. J. K., & Bustin, R. M. (2008). Characterizing the shale gas resource potential of Devonian–Mississippian strata in the Western Canada sedimentary basin: application of an integrated formation evaluation. *AAPG Bulletin, 92*, 87–125.
10. Schafer, J. L. (1999). Multiple imputation: a primer. *Statistical methods in Medical Research, 8*, 1–15.
11. Shanley, K. W., Cluff, R. M., & Robinson, J. W. (2004). Factors controlling prolific gas production from low-permeability sandstone reservoirs: implications for resource assessment, prospect development, and risk analysis. *AAPG Bulletin, 88*, 1083–1121.

Application of a Least Square Non-parametric Discovery Process Model to Colorado Group Mixed Conventional and Unconventional Oil Plays, Western Canada Sedimentary Basin

Zhuoheng Chen, Kirk G.Osadetz and Gemai Chen

1 Introduction

"Halo" shale oils of the Western Canada Sedimentary Basin (WCSB) are unconventional "shale-hosted" oil resources [1] that differ from "sandwich" style shale oil found in the Middle Member of the Bakken Formation. "Halo oil" occurs either at the fringe of or between existing conventional discrete sandstone reservoirs and is commonly associated with conventional oil and gas pools in stratigraphic traps. These reservoirs are "tight" with porosity and permeability characteristics transitional between those of conventional pools in sandstones and true shales (i.e., mudrocks). Exploration and production in the WCSB have focused on high porosity and permeability conventional reservoirs during the last century. With rapid decline of conventional oil supply, halo oil plays have become a focus for new North American reserve growth due to application of horizontal drilling coupled with advanced multi-stage hydraulic fracturing. Halo oil reservoirs in the upper Cretaceous Colorado Group are among the most attractive exploration targets in the WCSB in recent years. A mixture of halo reservoirs with conventional reservoirs geographically and vertically makes resource assessment of the halo oil play a challenge.

In a recent methodological study, the Geological Survey of Canada developed a least squares non-parametric discovery process model, which is in principle suitable for assessing petroleum resource potential in a mixed play comprising petroleum accumulations with distinct size characteristics. The non-parametric model makes

Z. Chen (✉) · K. G. Osadetz
Geological Survey of Canada, 3303-33rd Street, NW, Calgary, AB,T2L 2A7, Canada
e-mail: zchen@nrcan.gc.ca

K. G. Osadetz
e-mail: kosadetz@nrcan.gc.ca

G. Chen
Department of Mathematics and statistics, University of Calgary, 2500 University Drive, NW, Calgary, AB,T2N 1N4, Canada
e-mail: gchen@math.ucalgary.ca

E. Pardo-Igúzquiza et al. (eds.), *Mathematics of Planet Earth*,
Lecture Notes in Earth System Sciences, DOI: 10.1007/978-3-642-32408-6_134,

no assumptions about the shape of a parent pool size distribution, and so permits flexibility in pool size distribution shape. This resolves the problem of how to handle multi-modal distributions arising from a set of petroleum accumulations composed of mixed populations that could come from both conventional and unconventional reservoirs within the same geographically restricted stratigraphic unit. It allows visualization of the *goodness-of-fit* between predicted and observed "creaming curves" and produces a parameter map that provides visual guidance for assessors enabling the selection among alternative parameter sets in order to generate assessments that are more consistent with available geological and geophysical constraints. Derivation of the methodology and its application to examples of single (uni-modal) population petroleum plays are described by Chen et al. [2]. The study presented here emphasizes application of the non-parametric model to a mixed petroleum play comprising conventional oil pools and halo oil accumulations in the Cardium Formation of the Colorado Group in WCSB.

2 Geological Setting and Data

The middle Cretaceous Colorado Group of the Western Canadian Sedimentary Basin (WCSB) consists predominantly of mudstone deposited during major marine inundations interspersed with relatively thin sandstone, siltstone and conglomerate beds [3]. The Colorado Group petroleum system contains a large portion of the light-medium oil and natural gas reserves within the Middle Jurassic to Cretaceous foreland basin succession in the WCSB. It is a supercharged, high efficiency, high impedance petroleum system. Where it is mature, source rocks of interspersed marine mudstone generate large volumes of crude oil. Crude oil and natural gas reserves are found in discrete conventional pools of high porosity-permeable sandstone reservoirs as well as in continuous extremely low porosity-permeability "tight" or "shale" reservoirs - so called "halo reservoirs". Since the first multi-stage fracturing horizontal well in the Cardium Formation of Colorado Group in 2007, a few hundred horizontal wells coupled with multi-stage fracturing were completed, some targeted at halo oil reservoirs at the end of 2011. To assess Cardium Formation oil resource potential, we define oil pools in conventional reservoirs and unconventional tight reservoirs in it as belonging to a single play with areal extent limited by stratigraphy and source rock maturity.

The discovery sequence comprises conventional pools recorded in ERCB's statistical data base [4] and a Cardium Formation ultimate halo oil reserve estimate (EURs). In this study, a conventional pool is defined as an oil accumulation with a common oil/water contact sharing the same reservoir pressure system, while a "halo pool" is defined by the drainage volume surrounding a horizontal well, whose boundary is determined by the envelope of the stimulated reservoir volume (SRV) created by hydraulic fracturing around a horizontal well. The "pool size" of a single halo oil pool is the EUR derived from extrapolation of production rates from the horizontal well. We have 378 conventional pools from ERCB's statistics [4] and 285 halo oil

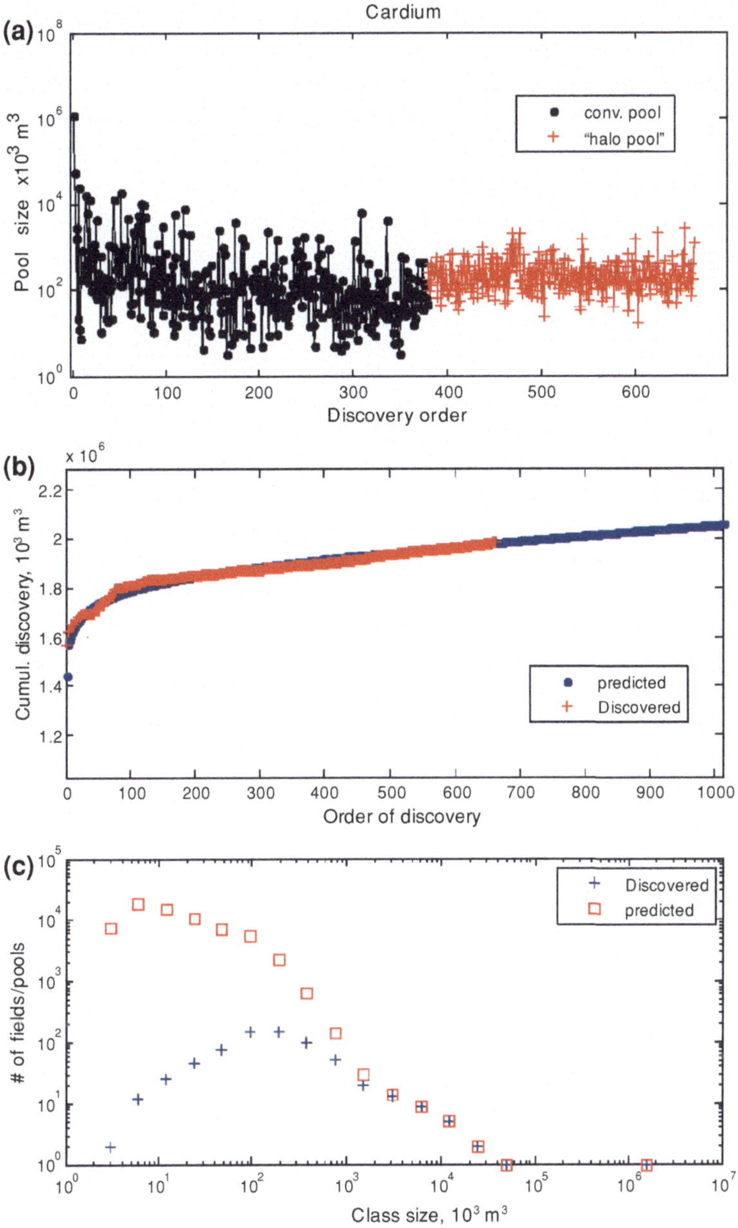

Fig. 1 **a** Discovery sequence of the Cardium Formation mixed oil play; **b** Creaming curve of cumulative discoveries comparing discovered and predicted reserves; **c** Comparison of discovered and predicted number of pools in size group

pools in tight reservoirs estimated from production data with production longer than 12 months. Total discovered oil in-place is $1,616.3 \times 10^6\,m^3$ in the 664 pools of the Cardium Formation. Assuming 10 % recovery, the total reserve in place of the 285 halo pools is $83.43 \times 10^6\,m^3$, contributing about 5 % to the total discovered oil in-place.

3 Results

A mixture of conventional and unconventional pools in ascending order of time of discovery defines a time series as input to the non-parametric model (Fig. 1a). Unconventional halo oil pools from tight reservoirs show distinct size characteristics as compared with the conventional pools in Fig. 1a. Application of the least square non-parametric discovery process model to the Cardium mixed oil play results in a point estimate of $4,111.9 \times 10^6\,m^3$ total of oil in-place. The creaming curve in Fig. 1b compares discovered cumulative reserve growth with the model prediction. The model predicts a large number of small pools (>67,000), most $<0.1 \times 10^6\,m^3$, presumably representing single well defined pools in tight reservoirs (Fig. 1c). The estimated total oil in-place resource is comparable to a recent estimate of $4,600 \times 10^6\,m^3$ of total oil in-place derived from a geological model-based simulation approach [5]. It is likely that the non-parametric discovery process method provides a conservative estimate of total oil resource in-place because it assigns weight only to observed (discovered) pool sizes. An undiscovered pool with size either greater than the largest or smaller than the smallest discovery is not predicted, a drawback of the non-parametric approach.

Acknowledgments The authors thank Prof. Gordon Kaufman of MIT and Peter Hannigan of GSC for their useful comments. This is ESS contribution # 20130042.

References

1. Clarkson, C. R., & Pedersen, P. K. (2011). Production analysis of Western Canadian unconventional light oil plays. CSUG/SPE Paper 149005.
2. Chen, G., Chen, Z., & Tzeng P. (2010). A non-parametric discovery process model – A least squares approach. Geological Survey of Canada Open File # 6253 (pp. 9, 10 Figures, 3 appendixes).
3. Creaney, S., Allan, J., Cole, K. S., Fowler, M. G., Brooks, P. W., Osadetz, K. G., et al. (1994). Petroleum Generation and Migration in the Western Canada Sedimentary Basin. In G. D. Mossop & I. Shetsen (Eds.), *Geological Atlas of the Western Canada Sedimentary Basin* (pp. 455–468). Calgary: Canadian Society of Petroleum Geologists and Alberta Research Council.
4. ERCB. ST98–2009. Alberta's energy reserves 2010 and supply/demand outlook 2010–2020(EB/OL). http://www.ercb.ca/docs/products/STs/st98_current.pdf
5. Chen, Z., & Osadetz, K.G. (2013) An assessment of tight oil resource potential in the Upper Cretaceous Cardium Formation, Western Canada Sedimentary Basin, Petroleum exploration and development, (in press)

A Further Investigation of Local Nonparametric Estimation Techniques in Shale Gas Resource Assessment

Emil D. Attanasi, Timothy C. Coburn and Philip A. Freeman

1 Introduction

As the industry's development strategies for shale gas resources continue to evolve, operators struggle to find ways to more precisely identify the drilling sites of highest productivity. So far, there is still considerable diversity and no universally recommended analytical procedures to remove the risks associated with drilling site selection. Operators may rely on ensembles of statistical procedures that can use the available data to predict potential productivity of undrilled sites.

This paper examines the application of one such set of tools to two mature shale gas plays in different geologic settings. The analysis uses location and expected ultimate recovery (EUR) of well and derived cell data from the Devonian Antrim Shale of the Michigan Basin, Michigan and the Mississippian Barnet Shale of the Fort Worth Basin, Texas to evaluate models that may be used to predict EURs at undrilled sites.

2 Local Nonparametric Predictive Models

Local nonparametric regression is a pragmatic approach to predictive modeling when the true distribution of some characteristic or phenomenon is not well understood. Such models can take many forms [1]. In this paper local nonparametric regression

E. D. Attanasi · P. A. Freeman
US Geological Survey, Reston, VA 22092, USA
e-mail: attanasi@usgs.gov

P. A. Freeman
e-mail: pfreeman@usgs.gov

T. C. Coburn (✉)
University of Tulsa, Tulsa, OK 74104, USA
e-mail: tim-coburn@utulsa.edu

E. Pardo-Igúzquiza et al. (eds.), *Mathematics of Planet Earth*,
Lecture Notes in Earth System Sciences, DOI: 10.1007/978-3-642-32408-6_135,

models are used to predict remaining shale gas resources in undrilled areas, where the predictions are weighted averages of the values of estimated ultimate recovery (EUR) associated with drilled "neighboring" sites. The components of the local nonparametric regression models consist of a weighting scheme (kernel) and the number of nearest neighboring points used in the calculation. The analytical form of the weighting schemes and the number of neighboring sites are generally chosen by cross-validation [1]. Analytic forms that have been tested include inverse distance, inverse distance squared and Stone regressions (see Table 1 in [2]); and sets of six, nine, and twelve neighboring points have been tried.

3 Shale Gas Study Areas

Two different shale gas study areas have been investigated: the Devonian Antrim Shale of the Michigan Basin and the Mississippian Barnett Shale of the Fort Worth Basin. The Antrim is relatively shallow and considered by some to have generally low productivity. The gas is considered to be biogenic in origin and the presence of natural fractures is crucial to its production. A recent review of some of the geological literature pertinent to the Antrim is given in [3].

The Barnett is deeper in comparison to the Antrim and the gas is thermogenic in origin. Natural fractures do not appear to be essential for production, and in some cases the presence of such fractures reduces well performance. On the other hand, geochemistry does appear to play a central role in reservoir performance.

4 Data

The Devonian Antrim study area encompassed Otsego and Montmorency Counties in Michigan. There were 5,890 producing and dry Devonian Antrim Shale wells (all vertical) in these counties. The wells were collected into 4,391 80-acre cells: 3,158 single-well cells, 1,013 two-well cells, 180 three-well cells and 40 cells having more than three wells. The 80-acre cell size was consistent with state well spacing regulations. The cell EURs were simply the sum of the EURs of the wells in each cell ([1] discusses the determination of well EURs). The maximum EUR per cell was 2.65 Bcf and the expected total recovery from all the drilled cells was 2.23 Tcf [3].

The Barnett data consisting of 1,385 wells (all vertical or deviated but not horizontal) were provided by an owner operator. The actual EURs (Bcf) were masked by a scaling factor and relative well locations were provided instead of actual locations. The wells were collected into 1,237 40-acre cells: 1,096 single-well cells and 134 two well-cells. The 40-acre cell size was consistent with recommended spacing, and maximum cell recovery was 6.8 EUR units.

5 Estimation Procedure and Performance

Selection of the best predictive model is based on a variation of the cross-validation procedures described in Ref. [2]. A random sample of 40 % of the observations was used as the training sample with the remaining 60 % used as the test sample. Predictions based on each of models were recorded along with statistics that indicate predictive performance. This procedure was repeated 101 times and the model chosen for further study was the one with the best predictive performance in terms of minimum root mean square predictive error (rmspe). For both the Antrim and Barnett data, the inverse distance model using twelve nearest neighbor points had the minimum rmspe.

The predictive accuracy of the procedure can be partially assessed from the cross-validation step. The average cell value in the Antrim was 0.508 Bcf and the average cell value in the Barnett was 1.674 (scaled Bcf). For the 101 test samples, the errors tended to average out to less than 3 % of the average cell value in both cases. The absolute values of the out-of-training-sample point prediction errors were also retained over the 101 iterations and amounted to 0.206 Bcf for Antrim cells and 0.658 (scaled Bcf) for Barnett cells. In summary, while the out-of-sample point prediction errors are not trivial, at many of the locations the bias in the total volumes is remarkably small.

6 Drill Site Selection and Extension to Horizontal Wells

Drill site selection is a logical application that follows from the estimation procedure described here. Three different decision strategies related to the siting of the next 500 wells were considered to illustrate the benefits of the approach. To start, 101 training sets of randomly selected cells were chosen to provide the ‘observed data’ to which these procedures could be applied. Each Antrim training set contained 439 observations and each Barnett training set contained 124 observations (in each case, 10 % of total cells).

Assuming a totally homogeneous and ubiquitous resource, one obvious strategy would be to avoid any model predictions altogether and drill in random order. This strategy (random drilling) can be operationalized by simply taking 10 random samples of 50 undrilled cells without replacement from the ‘observed data’ pool. Strategy 2 (static predictions) is to rank the set of undrilled sites on the basis of their model-predicted values. This strategy is invoked by selecting ten sets of 50 undrilled cells according to the size of their predicted values.

A third strategy (updated prediction) involves choosing a first set of 50 undrilled cells according to the size of their predicted values (Step 1, as in Strategy 2). For the remaining cells, revised predicted values are calculated (Step 2) by applying the algorithm to a combined file containing the first 50 values (Step 1 predictions) plus the original values in the training set, yielding predictions that are ‘updated.’ The next

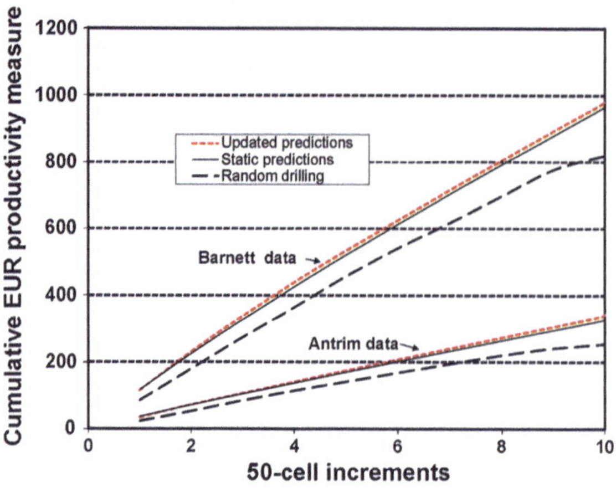

Fig. 1 Average cumulative EUR identified with three site selection strategies using increments of 50 cells: **a** random drilling, **b** static prediction, and **c** updated prediction. All results are based on the inverse-distance-squared / 12-nearest-neighbor-prediction algorithm, with a training set consisting of 10 % of total cells. For the Antrim (lower set), EUR is stated in Bcf. For the Barnett (higher set), EUR is a scaled value of Bcf

50 drilling locations are then chosen (Step 3) after rank ordering these recalculated predictions. This is repeated for 10 sets of 50 cells.

The three strategies were implemented on each of the 101 training samples. Figure 1 shows that more gas is found with static or updated prediction than with random drilling. The difference in results between the static and update strategies declines as the training set increases to 25 % of total observations. The difference grows when the training set represents only 5 % of the total.

The application of these tools may be extended to horizontal wells by judicious selection of cell sizes. EURs assigned to horizontal wells are shared by multiple contiguous cells, the allocation of which is accomplished in the gridding and data preparation stages. Model structures remain unchanged.

References

1. Hastie, T., Tibshirani, R., & Friedman, J. (2009). *Elements of Statistical learning—data mining, inference, and prediction* (2nd ed.). New York: Springer.
2. Coburn, T. C., Freeman, P. A., & Attanasi, E. D. (2012). Empirical methods for detecting regional trends and other spatial expressions in Antrim Shale gas productivity, with implications for improving resource projections using local nonparametric estimation techniques. *Natural Resources Research*, *21*, 1–21.
3. Montgomery, S. L., Jarvie, D. M., Bowker, K. A., & Pallastro, R. M. (2005). Mississippian Barnet Shale, Fort Worth basin, north-central Texas: gas-shale play with muti-trillion cubic foot potential. *AAPG Bulletin*, *89*, 155–175.

Fractal Analysis of AVO Seismic Attributes for Oil/Water Contact Identification

Sid-Ali Ouadfeul and Leila Aliouane

1 Introduction

The fractal analysis has been widely used for reservoir characterization, we cite for example the paper of Ouadfeul and Aliouane [1], it shows an application of the Wavelet-based multifractal analysis called the wavelet transform modulus maxima lines (WTMM) for lithofacies segmentation from well-logs data. Ouadfeul et al. [2] have published a chapter that contains many applications of the fractal analysis in exploration geophysics. In seismic field, the fractal analysis has been widely used. Ouadfeul and Aliouane [3] have analyzed the 3D Amplitude Versus Offset (AVO) data for heterogeneities analysis. Ouadfeul et al. [4] have used the fractal analysis for facies identification from 2D seismic data.

In this paper, we show that the fractal analysis can be used for fluid contact identification from seismic attributes. We start the paper by an introduction, after that we show the processing algorithm, the proposed idea is applied then to the synthetic AVO seismic data of a well located in the Algerian Sahara. We finalize the paper by results interpretation and conclusion.

2 Methodology

The proposed method is based on the estimation of the local Hölder exponent at maxima of the modulus of the continuous wavelet transform. Let us consider a signal S(t), the continuous wavelet transform CWT(a, b) is defined by:

S.-A. Ouadfeul (✉)
Algerian Petroleum Institute, IAP, Boumerdes, Algeria
e-mail: Souadfeul@ymail.com

L. Aliouane (✉)
LABOPHT, FHC, UMBB, Boumerdes, Algeria
e-mail: lil_aldz@yahoo.fr

E. Pardo-Igúzquiza et al. (eds.), *Mathematics of Planet Earth*,
Lecture Notes in Earth System Sciences, DOI: 10.1007/978-3-642-32408-6_136,

$$CWT(a,b) = \int_{-\infty}^{+\infty} S(t)\emptyset^{*}\left(\frac{t-b}{a}\right)dt$$

where a> 0 is the scale and b is a translation.

Ø is the analyzing wavelet, it must check the admissibility condition:

$$\int_{-\infty}^{+\infty} \emptyset(t)dt = 0$$

For some requirements, the analyzing wavelet should have a sufficient number of vanishing moments; it means that Ø is orthogonal at any polynomial of N − 1 degree.

The local Hölder exponent (h) is related by a power law to the modulus of the continuous wavelet transform for low scales.

$$|CWT(a,\ b)| = a^{h}$$

3 Application to Real Data

The proposed method is applied to the seismic synthetic AVO attributes of the Well OMP 80 located in the Algerian Sahara. The Oil/Water fluid contact is confirmed by the deep resistivity well-log. Synthetic seismic AVO attributes are firstly calculated using the Velocity of the Primary wave (Vp), Velocity of the Shear wave (Vs) and density logs (Rhob). These logs are presented in Fig. 1a, b and c respectively. The depth interval (2410 and 2500 m) is processed since it is the reservoir depth

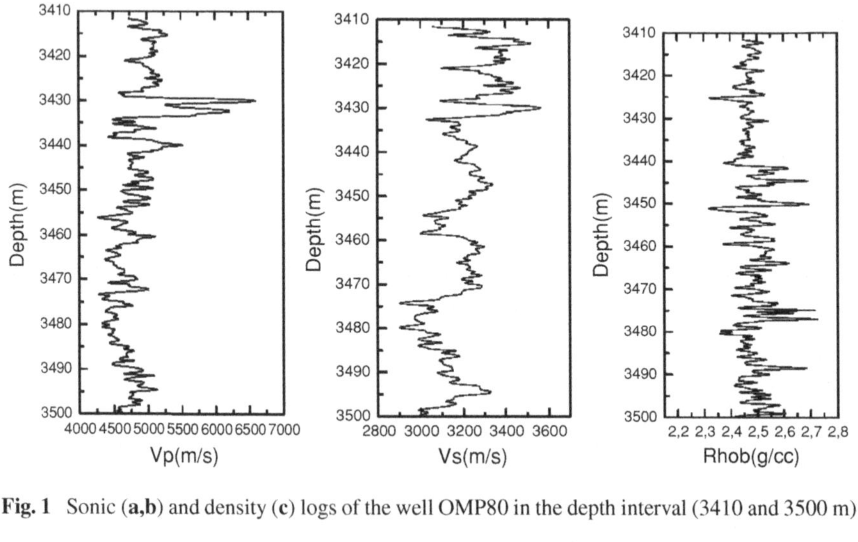

Fig. 1 Sonic (**a,b**) and density (**c**) logs of the well OMP80 in the depth interval (3410 and 3500 m)

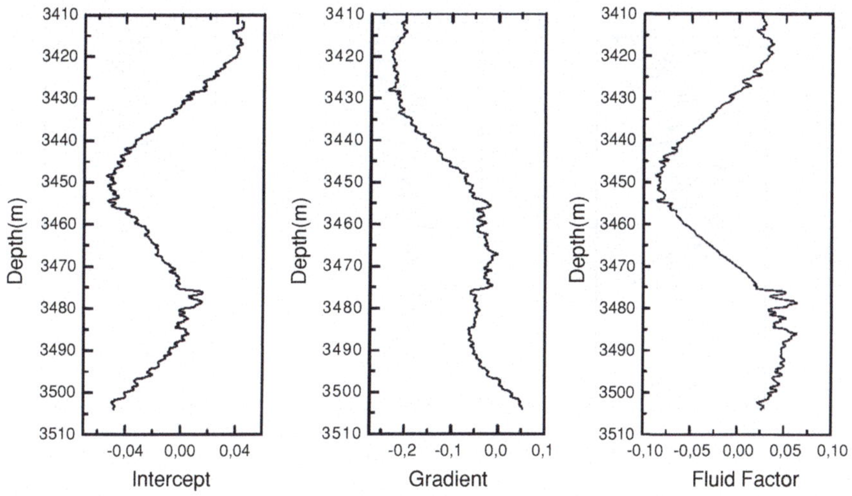

Fig. 2 Intercept, Gradient and Fluid Factor AVO synthetic seismic attributes

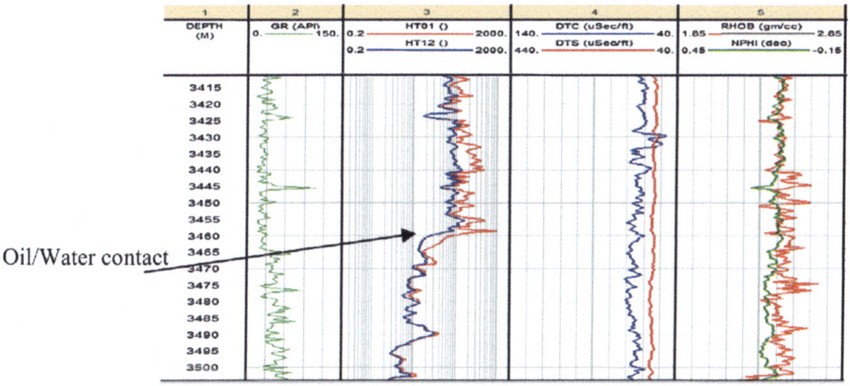

Fig. 3 Local Hölder exponents of the AVO attributes: **a** intercept, **b** gradient, **c** fluid factor

interval and it contains the fluid Oil/Water contact. The Ricker wavelet is used with a listening time of 12s. Figure 2 presents the Intercept, the Gradient and the Fluid Factor attributes. Local Hölder exponents estimated at maxima of the modulus of the continuous wavelet transform using linear fit of logarithm the modulus of the CWT versus logarithm the scales. Obtained results are presented in Fig. 3.

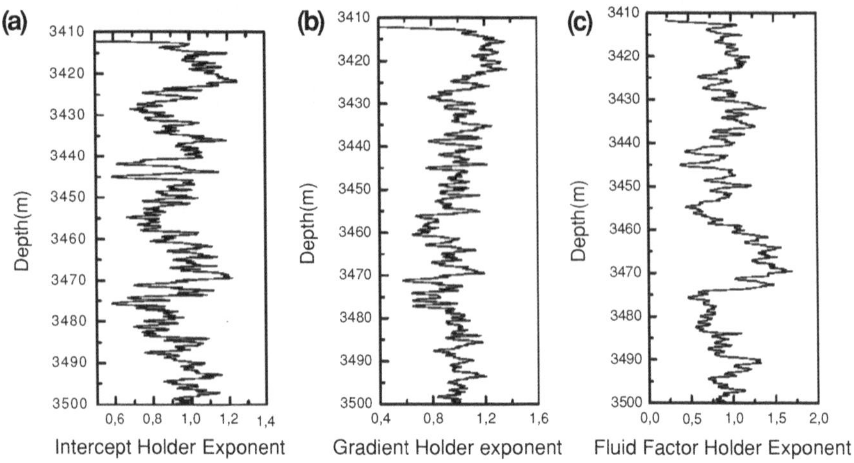

Fig. 4 Raw well-logs data of Well OMP80: *1*, Depth; *2*, Gamma ray; *3*, Deep resistivity; *4*, Compression and shear waves slowness; *5*, Density and neutron

4 Results Interpretation and Conclusion

Oil/Water contact is confirmed at the depth Z = 3460 m by the Deep resistivity (Rt) log (see Fig. 4), Gamma ray log clearly show that the main target is a sandy reservoir while Density and neutron porosity logs show the absence of Gaz. Rt log shows the presence of oil/water fluid contact. Estimated local Hölder exponents of the Intercept and the Fluid Factor AVO attribute show an increase of this exponent at the Depth Z = 3460 m, however no variation of the local Hölder exponent is observed for the Gradient AVO attribute. Obtained results show that the fractal analysis of seismic attributes can be greatly used to improve detection of fluid accumulations from surface seismic data.

References

1. Ouadfeul, S., & Aliouane, L. (2011). Multifractal analysis revisited by the continuous wavelet transform applied in lithofacies segmentation from well-logs data. *IJAPM*, *1*(1), 10–18.
2. Ouadfeul, S., Aliouane, L., & Boudella, A. (2012) Fractal and chaos in exploration geophysics, S. -A. Ouadfeul (Ed.), *Fractal analysis and chaos in geosciences* (pp. 155–174). Rijeka: InTech. ISBN 978-953-51-0729-3.
3. Ouadfeul, S., & Aliouane, L. (2012). Fractal analysis revisited by the continuous wavelet transform of AVO seismic data. *Arabian Journal of Geosciences*, *5*(6), 1469–1474.
4. Ouadfeul, S., Hamoudi, M., & Aliouane, L. (2012). A wavelet-based multifractal analysis of seismic data for facies identification. Application on the pilot KTB borehole. *Arabian Journal of Geosciences*, In Press.

Upscaling on Anelastic Vertically Heterogeneous Reservoirs

Alexey Stovas

1 Introduction

When considering numerical acoustic or elastic wave propagation in media containing small heterogeneities with respect to the minimum wavelength of the wavefield, being able to upscale physical properties (or homogenize them) is valuable for mainly two reasons: first, replacing the original discontinuous reservoir unit by a smooth and more simple one; second, it helps to understand what properties of a medium are really "seen" by the wavefield propagating through, which is an important aspect in an inverse problem approach. Upscaling means to replace of a heterogeneous volume with a homogeneous volume having effectively equivalent elastic or anelastic properties. Heterogeneity occurs at many scales, from pore scale to reservoir scale. Usually, the upscaling is related to the effective medium theory. To upscale the well log data measured at sonic frequencies is assumed that the medium is layered with layer thickness given by the logging step (commonly, 0.15 m). The standard method for upscaling was proposed by Backus [1] that gives the exact solution to compute the effective properties for a layered medium using assumptions that all constituents of the medium are linearly elastic and isotropic and there is no source of energy dissipation because of friction or viscosity. In this paper, I consider the Backus averaging method and the high frequency (HF) method to compute the effective properties from anelastic finely layered media. I propose the method to upscale the quality factor properties in zero-frequency and infinite-frequency limits. The proposed method is illustrated on the real well log data from the North Sea.

A. Stovas (✉)
Department of Petroleum Engineering and Applied Geophysics, Norwegian University of Science and Technology, S.P. Andersensvei 15A, N-7491 Trondheim, Norway
e-mail: alexey.stovas@ntnu.no

E. Pardo-Igúzquiza et al. (eds.), *Mathematics of Planet Earth*,
Lecture Notes in Earth System Sciences, DOI: 10.1007/978-3-642-32408-6_137,

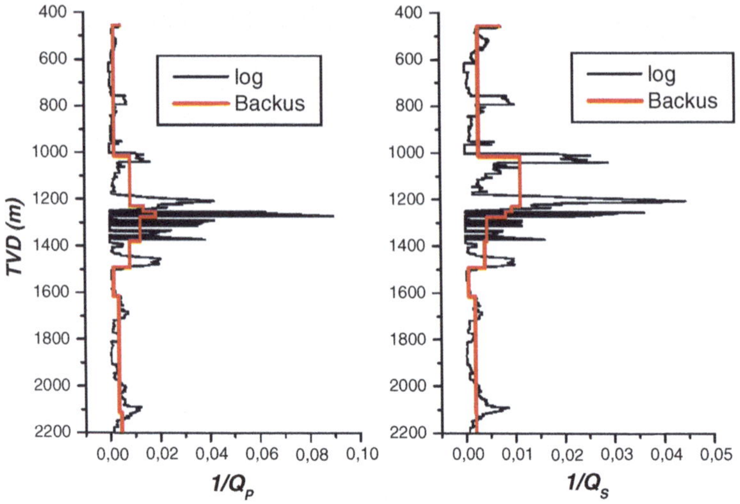

Fig. 1 Intrinsic and effective (Backus) quality factors inverse computed for P- (*left*) and S-wave (*right*) from anelastic well-log data

2 Low-Frequency Upscaling in a Finely Layered Medium

To perform Backus averaging for an anelastic isotropic medium with constant values for P- and S-wave quality factors, Q_P and Q_S, I need to link these factors to the stiffness coefficients c_{ij}. First, I reduce a VTI medium (transversely isotropic medium with a vertical symmetry axis) to an isotropic medium by setting

$$c_{11} = c_{33}, \quad c_{13} = c_{33} - 2c_{44}. \tag{1}$$

Therefore, the Backus averaging properties A_j, $j = 1, 4$, are reduced to

$$\begin{array}{l} A_1 = 4\left\langle \frac{c_{44}(c_{33}-c_{44})}{c_{33}} \right\rangle, \quad A_2 = \left\langle 1 - 2\frac{c_{44}}{c_{33}} \right\rangle \\ A_3 = \left\langle \frac{1}{c_{33}} \right\rangle, \quad A_4 = \left\langle \frac{1}{c_{44}} \right\rangle, \quad \rho_{ef} = \langle \rho \rangle \end{array}. \tag{2}$$

By using the definition of the quality factor from Carcione [2] for a standard linear solid model

$$\frac{1}{Q} = \frac{\operatorname{Im} V^2}{\operatorname{Re} V^2}, \tag{3}$$

the complex stiffness coefficients $\bar{c}_{33}$ and $\bar{c}_{44}$ can be defined as

$$\bar{c}_{33} = c_{33}\left(1 + \frac{i}{Q_P}\right), \quad \bar{c}_{44} = c_{44}\left(1 + \frac{i}{Q_S}\right), \tag{4}$$

where $i = \sqrt{-1}$. It might be more appropriate for frequency-independent treatment of Q to choose $1/2Q = \mathrm{Im}V^2/\mathrm{Re}V^2$ [14] instead of Eq. (3). In this case, we expect the smaller effect of attenuation.

Substituting (4) into (2) and computing the effective medium parameters results in complex vertical P- and S-wave velocities and anisotropy parameters. The effective quality factors can be defined as

$$\frac{1}{\tilde{Q}_P} = \frac{\left\langle \frac{Q_P}{c_{33}(1+Q_P^2)} \right\rangle}{\left\langle \frac{Q_P^2}{c_{33}(1+Q_P^2)} \right\rangle}, \quad \frac{1}{\tilde{Q}_S} = \frac{\left\langle \frac{Q_S}{c_{44}(1+Q_S^2)} \right\rangle}{\left\langle \frac{Q_S^2}{c_{44}(1+Q_S^2)} \right\rangle}. \tag{5}$$

3 Well-Log Example

To analyze the effect of intrinsic anelasticity, I choose the real well-log dataset from the North Sea. Figure 1 shows the intrinsic and effective Backus values for $1/Q_P$ and $1/Q_S$. The strong contrasts in the inverse quality factors nicely fit with the largest values for effective anelastic properties. The difference in Q inverse between these two limits is illustrated in Fig. 2. The larger difference is, the stronger contrast in elastic parameters is related to a given formation. The importance of using intrinsic anelasticity depends on the particular problem to be solved with a blocked effective model.

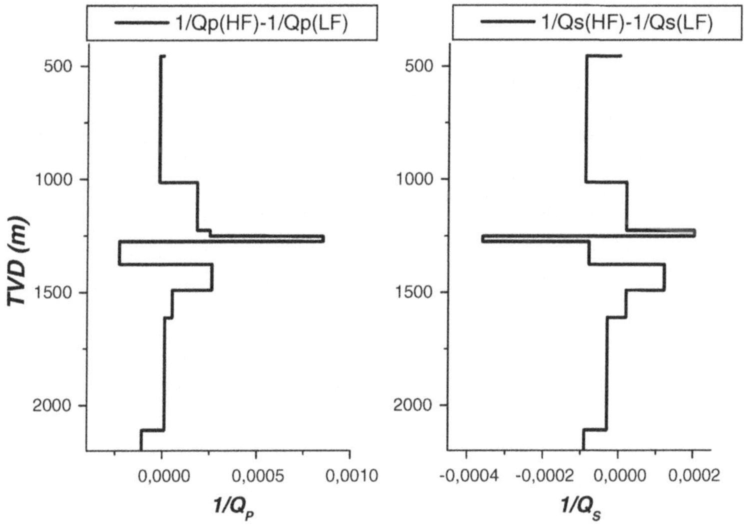

Fig. 2 Difference between the quality factors inverse from high- and low-frequency limits computed from anelastic version of well-log data

4 Conclusions

I developed the upscaling procedure for finely layered anelastic media and derived equations for computing of effective quality factor for P- and S-waves both in zero-frequency and infinite-frequency limits. The derived equations are applied for the real well-log data from the North Sea. The formations with strong vertical heterogeneity produce the largest effective attenuation. The proposed method can be used in computation of effective properties in heterogeneous anelastic reservoirs.

Acknowledgments I would like to acknowledge the ROSE project at NTNU.

References

1. Backus, G. E. (1962). Long-wave elastic anisotropy produced by horizontal layering. *Journal of Geophysical Research*, *67*, 4427–4440.
2. Carcione, J.M. (2007). Wave fields in real media: wave propagation in anisotropic, anelastic, porous and electromagnetic media. *Handbook of geophysical exploration*, vol 38. Amsterdam: Elsevier.
3. O'Connell, R. J., & Budiansky, B. (1978). Measures of dissipation in viscoelastic media. *Geophysical Research Letters*, *5*(1), 5–8.

Lithofacies Prediction from Well-Logs Data Using Hybrid Neural Network Model: A Case Study from Algerian Sahara

Sid-Ali Ouadfeul and Leila Aliouane

1 Introduction

The Artificial Neural Network (ANN) has been largely used for reservoir characterization from well-logs data. Ouadfeul and Aliouane [2] have combined between the Self-Organizing Map (SOM) and the Multilayer Perceptron (MLP) for better lithofacies classification from well-logs data. Obtained results show that the combined ANN machines are able to give more precise lithology. Ouadfeul et al. [3] have published a case study showing an application of the SOM for lithofacies classification from well-log data in the Trias Argilo-Gréseux. In this article, we used hybrid neural network model for lithofacies classification from well-logs data, the whole process will be applied to data of two wells located in Algerian Sahara.

2 The Self-organizing Map

The Self-organizing Map (SOM) is a kind of a neural network invented by the professor Teuvo Kohonen [1], it is constituted of two layers (Fig. 1). The SOM is based on an unsupervised learning based on the Euclidien distance and a neighborhood function, this function is generally given as a Gaussian. For better information about the training algorithm of the SOM machines, we invite readers to the paper of Kohonen [1] or the paper of Ouadfeul et al. [3].

S.-A. Ouadfeul (✉)
Algerian Petroleum Institute (IAP), Boumerdes, Algeria
e-mail: Souadfeul@ymail.com

L. Aliouane
LABOPHT, Faculté des Hydrocarbures et de la Chimie (FHC), Université M'Hamed Bougara, Boumerdés (UMBB), Boumerdes, Algeria
e-mail: lil_aldz@yahoo.fr

E. Pardo-Igúzquiza et al. (eds.), *Mathematics of Planet Earth*,
Lecture Notes in Earth System Sciences, DOI: 10.1007/978-3-642-32408-6_138,

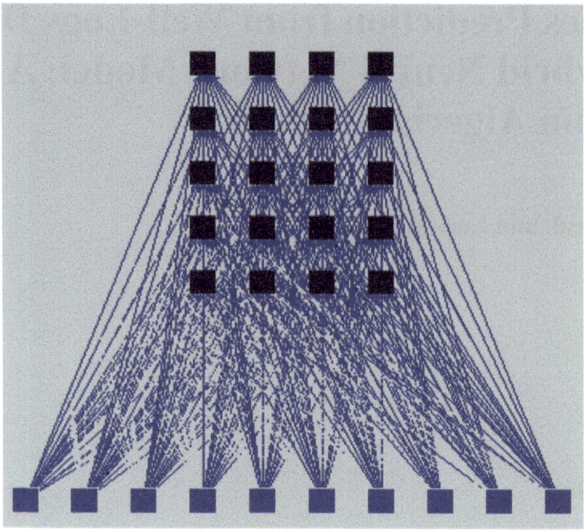

Fig. 1 Graphic schematization of the self-organizing map

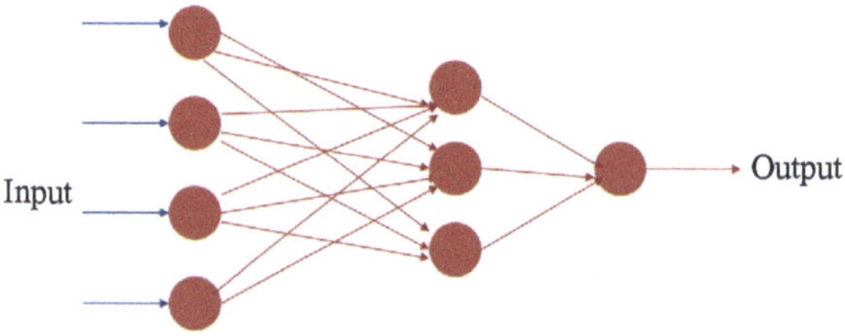

Fig. 2 An example of MLP with one hidden layer

3 The Multilayer Perceptron

The Multilayer Perceptron (MLP) is a kind of network based on a supervised learning, in this case we need a desired output for each input vector. The MLP is constituted of input and output layers and N hidden layers. It has been shown that one hidden layer is sufficient to resolve many scientific problems (see Fig. 2).

4 The Processing Algorithm

The proposed idea is based on the training of the MLP using the output of the SOM, the first step consists of training the Kohonen's map using the raw well-logs data as input. Well-logs data to be used are: The natural gamma ray (Gr), Slowness of the P wave (DT), neutron porosity (NPHI), Bulk density (RHOB) and the photoelectric

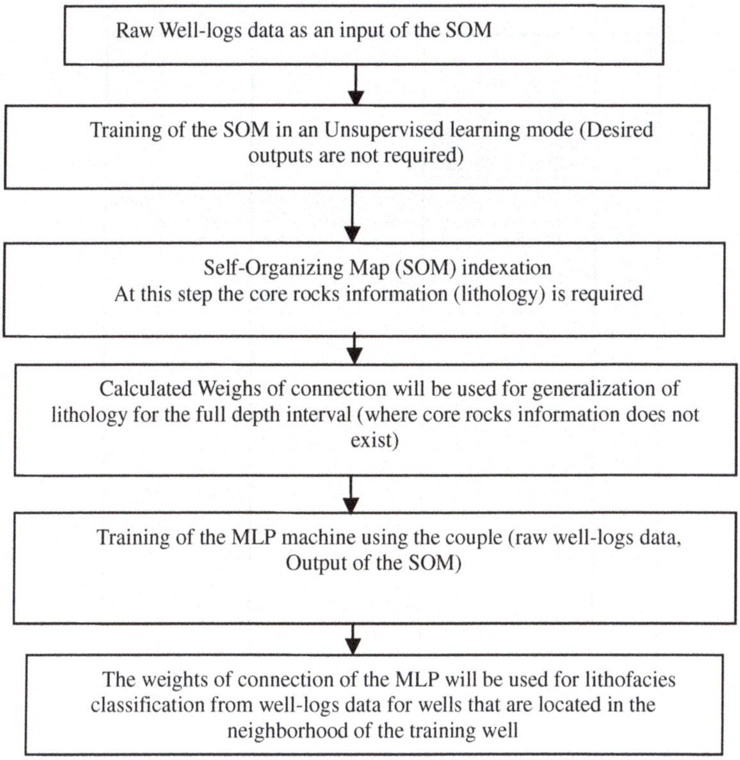

Fig. 3 Flow chart of the proposed processing algorithm of well-logs data

absorption coefficient (Pe). The obtained weights of connection by the training of the SOM will be used for propagation of the input and generalization of the lithology

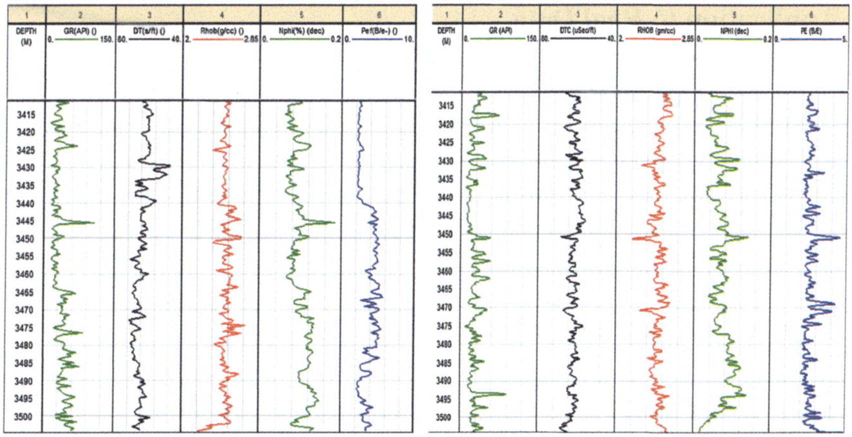

Fig. 4 Raw well-logs data of A1 and A2 wells: *2* Gamma ray (Gr) *3* Slowness of the P wave (DT) *4* Bulk Density (RHOB) *5* Neutron Porosity (Nphi) *6* Photoelectric absorption coefficient (Pe)

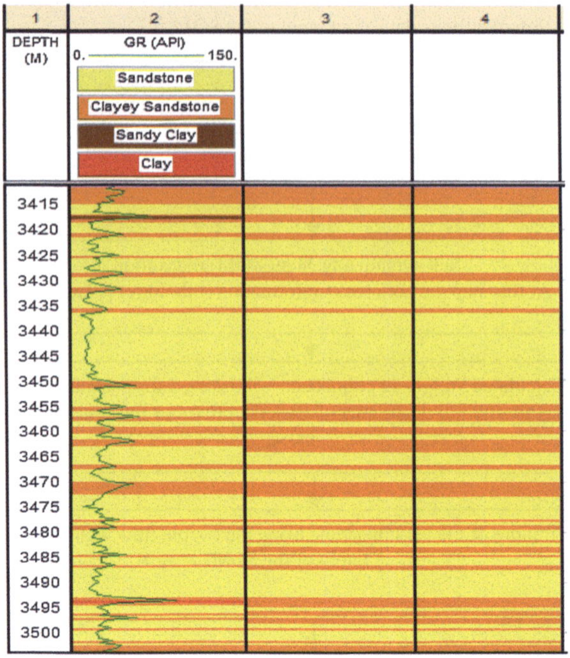

Fig. 5 Lithofacies classification in well A1 using: *1* Gama ray *2* SOM *3* SOM and MLP

for the full depth interval. The couple raw well-logs data, generalized lithology will be used for the training of the MLP. Figure 3 shows a detailed flow chart of the processing algorithm.

5 Application to Real Data

The proposed combination is applied to the raw well-logs data of two wells A1 and A2 located in Hassi Messaoud field (Algerian Sahara). Gamma ray and Slowness logs clearly shows that the main interval target is constituted of shale and sandstone intercalations. Four classes will be used for the SOM indexation which are: (1) Sandstone (2) Shaley Sandstone (3) Sandy Shale (4) Shale (Fig. 4).

6 Results, Interpretation and Conclusion

Application to wells A1 and A2 is down, well A1 is used as a pilot. At this step the weights of connection for the SOM and the hybrid MLP SOM are calculated. Figure 5 show four tracks, the first one is the depth, the second is the classification based only

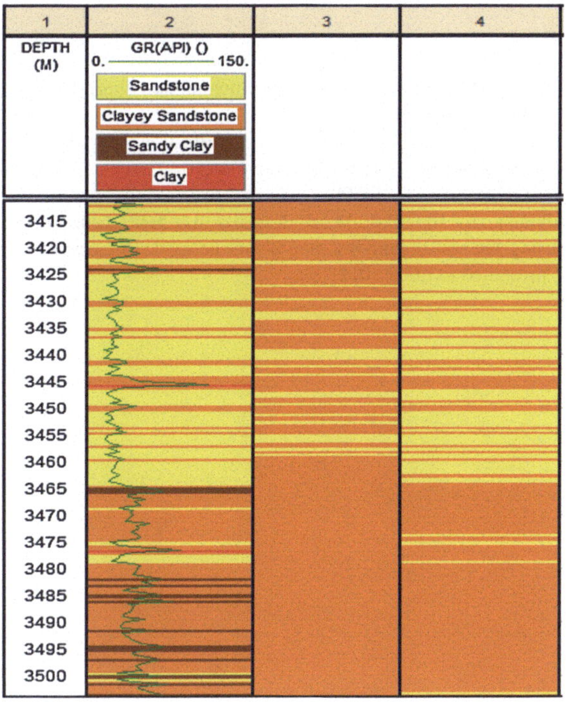

Fig. 6 Lithofacies classification in well A2 using: *1* Gama ray *2* SOM *3* SOM and MLP

on the Gamma ray log, the third track is the classification after the propagation of the input (raw well-logs data) in the SOM machine, while the last one is the lithofacies model after propagation in the MLP based on the output of the SOM as a desired output. To check the efficiency of each neural model we have generalized each of them for the second well A2, at this step we use the weights of connection calculated for each machine in the training step in A1 well to propagate the raw well-logs data of the second well. Obtained results clearly show that the MLP based on the SOM output as a desired output is able to give better results and a lithofacies model which is not very far from the lithofacies model based on the natural gamma ray log. By consequence, the proposed hybrid model can greatly enhance reservoir characterization and improve oil recovery (Fig. 6).

References

1. Kohonen, T. (1998). Self organization and associative memory springer series in information sciences. 2nd ed. Berlin: Springer.
2. Ouadfeul, S., & Aliouane, L. (2012). Lithofacies classification using the multilayer perceptron and the self-organizing neural networks. Lecture Notes in Computer Science, Vol. 7667, pp. 737–744. doi:10.1007/978-3-642-34500-5_87
3. Ouadfeul, S., Aliouane, L., Hamoudi, M., & Boudella, A. (2012). Lithofacies recognition from well-logs data using the self-organizing map neural network. A case study from the Berkine basin (Algeria), presented in EGU 2012.

Stochastic Simulation of the Morphology of Fluvial Sand Channel Reservoirs

Alexandra Kuznetsova, José A. Almeida and Paulo Legoinha

1 Introduction

The characterization of fluvial-type oil reservoirs using stochastic simulation algorithms [1] is a complex problem, because the morphologies of the channels are complex and are not fully characterized by the two-point statistics of geostatistical models [2, 3]. Among these morphological features, for example, are delta morphologies and meander-form floodplains where the channels form a complex 3D pattern of sediment accumulation representing the former waterway system.

To generate realistic 3D morphological models for fluvial sand channels reservoirs, the proposed methodology combines object and geostatistical simulation models to generate a set of equally probable scenarios. Those scenarios can be filled with petrophysical properties in a second step [4].

The workflow of the proposed methodology involves the following steps: (a) Construct histograms of the length, width, and height of sand channels and a 2D image of local orientations; (b) Simulate widths and heights of the hypothetical sand channels along their paths; (c) Generate random skeletons of the hypothetical channels and associate them with the width and height dimensions generated in the previous step (Boolean vector model); (d) Convert the Boolean vector model into a raster model and link regions of uncertainty according to distance to the skeletons; and (e) Using the method of Probability Field Simulation (PFS) [5], condition the Boolean model with a variogram and resolve the regions of uncertainty. The outputs

A. Kuznetsova · J. A. Almeida (✉) · P. Legoinha
CICEGe, FCT-Universidade Nova de Lisboa, 2829-516 Caparica, Portugal
e-mail: mestnaia@yandex.ru

J. A. Almeida
e-mail: ja@fct.unl.pt

P. Legoinha
e-mail: pal@fct.unl.pt

E. Pardo-Igúzquiza et al. (eds.), *Mathematics of Planet Earth*,
Lecture Notes in Earth System Sciences, DOI: 10.1007/978-3-642-32408-6_139,

are sets of binary images (channel—sand/not channel—shale). A synthetic case study for demonstrative purposes is presented in the next section.

2 Case Study

2.1 Initial Data and Stochastic Simulation of Width and Height

Reliable statistical starting data are a pre-requisite for the successful probabilistic modeling of fluvial reservoirs. Starting data for the proposed morphology consist of: (a) A histogram of lengths; (b) A histograms of widths (Fig. 1 Left) and heights and a ratio or a correlation coefficient of width–height; (c) An image of the local orientation angles (azimuths, 2D and dips, 3D) (Fig. 1 Right). In real case studies, this information should be provided by geological expertise, seismic or other sources.

The first processing step consists of the simulation of the widths and heights of the channels. This was performed here using conditional simulation (DSS) with the histograms of width and height, and a correlation coefficient of 0.7 between the two variables. Several realizations of 250 channels were simulated (Fig. 2 shows one example, in which each column of data represents a hypothetical channel).

2.2 Boolean Model of the Channels

The simulation of channels begins with a Boolean model. Again, several realizations of 250 skeletons/channels were simulated. After the simulation of the skeletons, width and height dimensions were applied and three regions (sand, transition, and

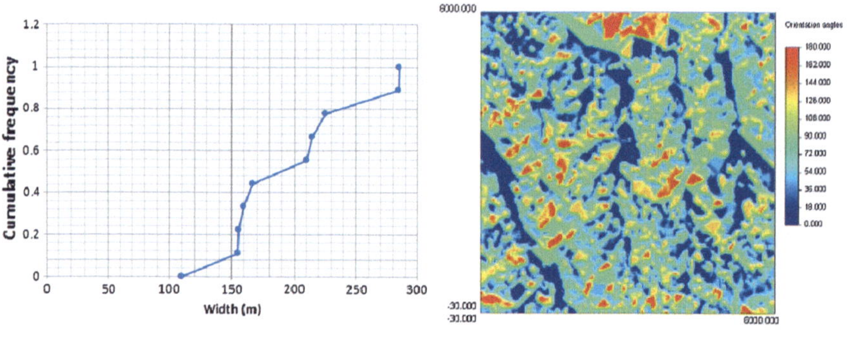

Fig. 1 *Left* cumulative histogram of channel widths. *Right* local orientation angles of the channels. Flow direction angles are calculated from a digital terrain model

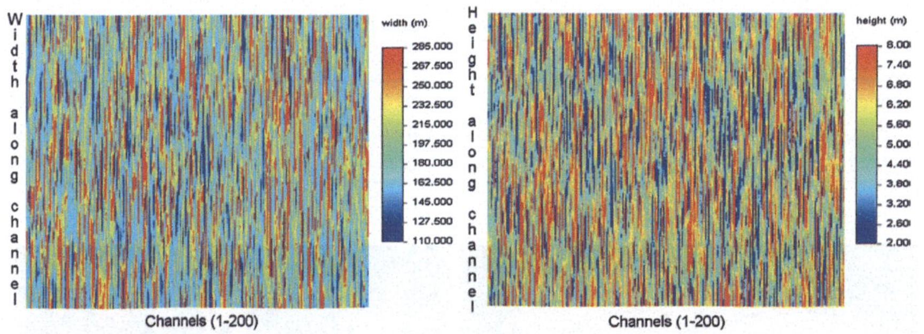

Fig. 2 *Left* simulated values of widths for 250 channels. *Right* simulated values of heights for 250 channels. A spherical variogram model with a range of 1200 m was used

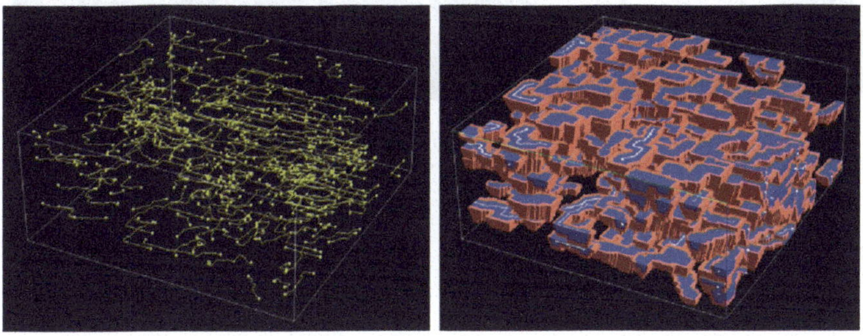

Fig. 3 *Left* One simulation of the skeletons for 250 channels. *Right* The corresponding three-region block model (*blue*—sand; *red*—transition sand/shale)

shale) were generated. Figure 3 illustrates one image of the skeletons and the corresponding block model.

2.3 Post-processing Using Probability Field Simulation

In a final step, PFS was applied for the simulation of shale/sand geometry conditioned to the previous object model regions (Fig. 3 Right) and a table of a priori probabilities of each region being sand/shale (Table 1) via the simulated probability maps with a uniform distribution, following the orientation angles of Fig. 1 Right.

The theoretical model of variogram used in PFS was defined in accordance with the proposed channel sizes: a spherical model with 3000 m in the direction of the channels, 200 m in the horizontal direction perpendicular to the channels, and 10 m in the vertical dimension. Figure 4 shows the final image of the channels.

Table 1 A priori probabilities of each region being sand/shale

Object model region	Sand	Shale
I-Shale	0.02	0.98
II-Transition	0.50	0.50
III-Sand	0.98	0.02

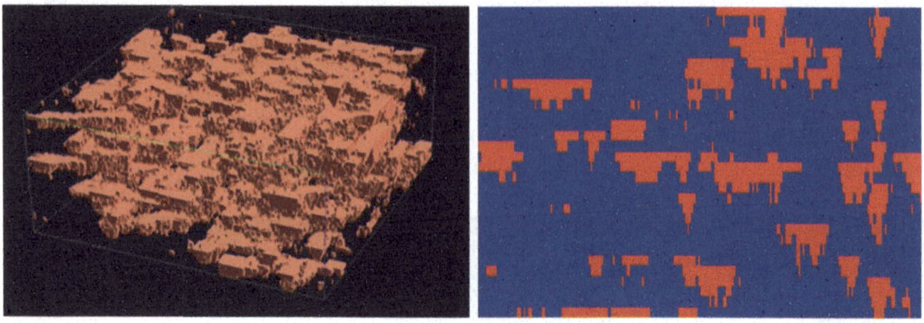

Fig. 4 *Left* one final simulation of the channels. *Right* intermediate cross-section view illustrating the distribution of channels and their half-ellipse shapes

3 Final Remarks

The obtained results for simulated scenarios of channels shows advantages over other methods: (a) The Boolean model imposes statistics of channel shape (half-ellipse section), size, and local orientation in a very efficient way; (b) Post-processing using PFS transforms the previous morphology obtained after the first stage of the method according to a variogram model, thereby adding realism. These morphological images, in a second step, can be filled with petrophysical property values, such as porosity and permeability.

References

1. Luis, J., & Almeida, J. A. (1997). *Stochastic characterization offluvial sand channels, Geostatistics Wollongong'96, Vol 1,* In Baafi, E. Y., & Schofieldpp, N. A. (pp. 477–488). The Netherlands: Kluwer Academic Publiaction.
2. Almeida, J.A. (1999). *Use of geostatistical models to improve reservoir description and flow simulation in heterogeneous oil fields*. PhD thesis IST, p. 161.
3. Almeida, J.A. (2010). Stochastic simulation methods for characterization of lithoclasses in carbonate reservoirs. *Earth Science Reviews, 101*,(3–4), 250–270.
4. Nunes, R., & Almeida, J. A. (2010). Parallelization of sequential Gaussian, indicator and direct simulation algorithms. *Computers Geosciences*, *36*(8), 1042–1052.
5. Froidevaux, R. (1993) Probability field simulation. In Soares, A. (ed.), *Geostatistics Troia'92*, vol 1 (73–84). Dordrecht: Kluwer Academic Publication.

Sorting Reservoir Models According to Flow Criteria: A Methodology, Using Fast Marching Methods and Multi-Dimensional Scaling

Gaétan Bardy and Pierre Biver

1 Introduction

With the development of powerful computers, it is become very easy to generate a large number of reservoir models (named realizations), using geostatistic algorithms [1]. The aim is to assess the uncertainty on the flow performance. But as the models become larger and complex, only a small number of realizations can be simulated dynamically. In practice, due to the flow simulation computing time, only a base case, sometime a low case and high case are simulated.

The principle of the methodology we propose is to use an alternative to full physics flow simulator: *the proxies.* They are faster because they use simplified physics and provide approached results. After that we compute the distance between all models, directly on the response curves or on a selected part of them. Using these distances, we propose to perform Multi-Dimensional Scaling and kernel clustering to select a small number of models which are representative of the entire set. At the end, we perform reference flow simulation on the selected realizations and some statistical analysis of the results to get quantiles (Q10, Q50 and Q90).

We illustrate our methodology on a real field case composed of a $130 \times 100 \times 61$ grid, 7 wells (3 injectors and 4 producers) and a set of 200 different models.

G. Bardy (✉)
Université de Lorraine, Rue du Doyen Marcel Roubault, 54018 Vandoeuvre-les-Nancy Cedex, France
e-mail: gaetan.bardy@total.com

G. Bardy · P. Biver
Total S.A., CSTJF, Avenue Larribau, 64018 Pau Cedex, France
e-mail: pierre.biver@total.com

E. Pardo-Igúzquiza et al. (eds.), *Mathematics of Planet Earth*,
Lecture Notes in Earth System Sciences, DOI: 10.1007/978-3-642-32408-6_140,

2 Approximation of the Flow Simulation

A lot of proxies are envisioned to approximate flow simulation. We will use two different ones: a proxy based on front propagation according to a connectivity cutoff and another proxy based on front propagation using Fast Marching Methods.

2.1 *Connectivity Proxy*

The connectivity proxy is very simple. A "front" is propagated from well perforations into all grid cells linked to the well with permeability above a given value [2]. The front starts with a numerical value equals to 0 and increments of 1 cell by cell. At the end, we are able to plot a curve of the hydrocarbon volume repartition as a function of the front index value. It is also possible to delay some wells by imposing their start value, which allows having different production phases.

2.2 *Fast Marching Methods Proxy*

The second proxy is based on the Fast Marching Methods (FMM) first described by J. A. Sethian [3]. FMM is used to solve the eikonal equation that describes the front propagation travel time.

In our approach, as in [4], we propagate a front from well perforation to the entire reservoir and we choose a pseudo-velocity equal to:

$$v = (k \times \rho \times g)/(\varnothing \times \nu) \tag{1}$$

where k is the permeability, ρ the fluid density, g the gravity acceleration, Ø the porosity and ν the fluid viscosity. After computation we got some curves of cell volume investigated by the front as function of a pseudo-time.

3 Model Selection

After getting all proxy simulation results, we propose to select a small number of models that are representatives of the initial set. For that we will use the workflow proposed by Scheidt [5].

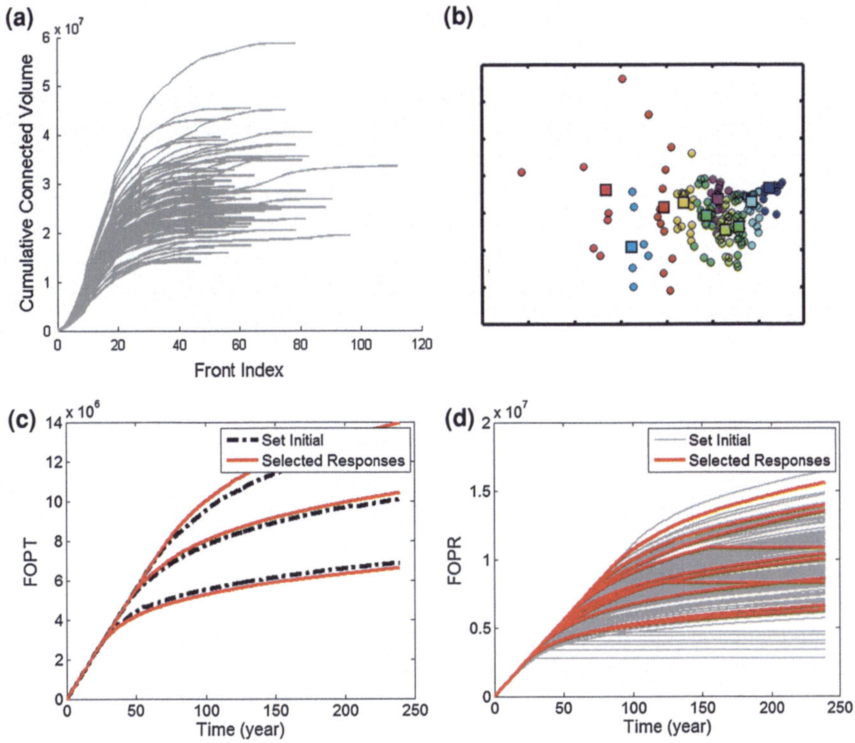

Fig. 1 **a** Cumulative curves get with the connectivity to well proxy for the 200 models. Ordinate represent the cumulative oil connected volume and abscise, the front value. **b** Result of the clustering: we choose 10 clusters and get their centers. **c** Eclipse® simulations for the 200 models (*gray curves*) and the 10 selected (*red curves*). **d** Comparison of the quantiles calculated with the original set of 200 Eclipse® models (*dot lines*) and quantiles computed on the 10 selected models (*red lines*)

3.1 Distance Computation and Multi-Dimensional Scaling

The first step of the workflow is to compute distances matrix of the models. For that, we use different distances such as Hausdorff distance, maximal square distance between each proxy response curve. Then the matrix is used by Multi-Dimensional Scaling to map each model into a Euclidian space [6]. As we use distances between curves, we got metric properties which allow us to use the classical Multi-Dimensional Scaling (MDS).

3.2 Kernel Transformation and Clustering

The main problem with MDS is that the result is not linear, in most cases, that is not convenient for analysis algorithms. To solve that, we use a basis kernel transformation to transform the Euclidian space in a feature space where the variations between models are more linear and then we use clustering method [7]. For the kernel transformation, we use a simple Gaussian function as Radial Basis Function (RBF) kernel. For the clustering we use a K-medoïd algorithm. The goal is to segregate the realizations in k groups or clusters and get the point close to the center. Finally we simulate dynamically the K centers and compute Q10, Q50, Q90.

4 Result Analysis and Further Work

The case presented in Fig. 1 has been realized with the connectivity proxy. To quantify the quality of our results, we compare the quantiles we get with our selection and those we get with the entire set of 200 Eclipse® profile. It shows that even with this very simple proxy we get results closes to the original quantiles.

Improving the velocity function of the proxy base on Fast Marching and integrating other production data (well fluid rate, pressure) will be the objects of a further work

References

1. Caers, J. (2011). *Modeling uncertainty in the earth sciences*. New York: Wiley.
2. Alabert F. G. & Monot V. (1992). Stochastic models of reservoir heterogeneity: Impact on connectivity and average permeability. In *SPE Annual Technical Conference and Exhibition*, pp. 355–370.
3. Sethian, J. A. (2008). *Level set methods and fast marching methods: Evolving interfaces in computational geometry, fluid mechanics, computer vision, and materials science*. New York: Cambridge University Press.
4. Hovadik, J. & Larue D. (1996). Predicting waterflood behavior by simulating earth models with no or limited dynamic data: From model ranking to simulating a billion-cell model. In Y. Z. Ma & P. R. La Pointe (Eds.), *Uncertainty analysis and reservoir modeling*. AAPG Memoir 96, pp. 29–55.
5. Scheidt C. & Caers J. (2007). A workflow for spatial uncertainty quantification using distances and kernels, Stanford Center for Reservoir Forecasting Annual Meeting.
6. Borg, I. & Groenen, P. (2005). *Modern multidimensional scaling: Theory and applications*. New-York: Springer.
7. Shawe-Taylor, J. & Cristianini N. (2004). *Kernel methods for pattern analysis*. Cambridge: Cambridge University Press.

Integrate Facies Clustering Feature Information in Reservoir Modeling

Yupeng Li and Lihui Geng

1 Introduction

As the facies spatial distribution is of major concern due to their influence on fluid flow, and thereby the production of oil and gas, a detailed facies model will be built first and then guide the modeling of the petrophysical properties [1].

Facies modeling can be done in a deterministic or a stochastic way. The indicator simulation, one of the geostatistical methods, is a common stochastic one [2]. For deterministic modeling, the geologist usually defines facies boundary lines over plan or cross sections based on the outcomes at well sites. In this view, facies modeling can be looked as a classification of the un-sampled locations based on the outcomes of surrounding locations. Hard and well known boundaries between different facies categories will be obtained after doing that kind of manually spatial classification. It would be a better approach in exploration stage and could integrate the geologists' conceptual understanding into the final facies distribution models. While, in mature reservoir, plenty of wells are drilled and it would be a challenge for geologists to integrate all the information together. Here, many clustering or classify algorithms in machine learning or artificial intelligence can be used in facies modeling [3]. The Ward's clustering approach based on the Mahalanobis distance is proposed for facies modeling in a mature reservoir in this study [4].

Y. Li (✉) · L. Geng
China University of Petroleum, No 18 Fuxue Road, Changping, Beijing 102200, China
e-mail: yupeng@cup.edu.cn

L. Geng
e-mail: Lihui@cup.edu.cn

E. Pardo-Ig£zquiza et al. (eds.), *Mathematics of Planet Earth*,
Lecture Notes in Earth System Sciences, DOI: 10.1007/978-3-642-32408-6_141,

2 The Clustering Method

In traditional facies modeling procedure, detailed sedimentological analysis is done before geostatistician step in. The facies types are defined for each single sand along the well path profile by sedimentologist. Then the geostatistical facies modeling approach such as cell based approach or object based approach will be used to construct the numerical facies model [5]. While, various information such as facies stacking pattern, thickness information, sedimentary background available for sedimentologist in the facies defining stage should also be used for geostatistician to build a geological realistic facies model which is more close to the true heterogeneity of underground reservoir. How to integrate all various information together as sedimentologist did in defining facies type along well path is a critical step. Thus, in this study, the facies modeling procedure starts from the facies classification in the single well vertical profile by using the Ward's clustering in facies modeling procedure.

It is assumed that each well has P features to support its facies type classification. For example, the possible classification features are: sand thickness, net-to-gross, buffer ratio, and/or well log profile shape which are noted as x_i, $i = 1, \ldots, P$. Here the Ward's method is used to do the classification based on the features. The Ward's clustering method is a hierarchical agglomerative method. If there are N wells in the domain, it begins with N types. Then searching the feature matrix for the most similar pair of types and reduce the number of types one by one through merging the most similar pairs. Perform those steps until appropriate types are obtained.

Assuming at certain clustering stage, the merged facies type total number is K types. And the groups are defined as G_k, $k = 1, \ldots, K$. Then, each class G_k would have a specific feature matrix which is written as:

	1	2	...	P
1	$x_{11}^{(k)}$	$x_{12}^{(k)}$	...	$x_{1p}^{(k)}$
2	$x_{21}^{(k)}$	$x_{22}^{(k)}$	...	$x_{2p}^{(k)}$
⋮	⋮	⋮	⋮	⋮
n_k	$x_{n1}^{(k)}$	$x_{n2}^{(k)}$	...	$x_{np}^{(k)}$
Mean	$\bar{x}_1^{(k)}$	$\bar{x}_2^{(k)}$	...	$\bar{x}_p^{(k)}$

The object in the Ward's clustering procedure is to find at each stage the two merged clusters will give the minimum increase in the total within group error sum of distance between the centroids of the merged clusters which is written as:

$$E(K) = \sum_{k=1}^{K} (\sum_{p=1}^{P} (\sum_{n=1}^{n_k} (x_{np}{}^{(k)} - \bar{x}_p^{(k)})^2)) \tag{1}$$

If a new unclassified well or sand along vertical profile which has a feature as $X = (x_1, x_2, \ldots, x_p)$ the Mahalanobis distance of this well to each of the facies type

Table 1 The predictor feature matrix from the study area

Facies	Mean thickness (m)	Net-to-gross (%)	Buffer ratio (%)	Porosity (%)	Permeability (mD)
Channel	5–12	0.6–0.8	0.3–0.5	0.12	200
Mouth bar	3–10	0.3–0.7	0.2–0.3	0.3	300
Over bank	1–4	0.2–0.5	$\leq$0.2	0.1–0.15	100
Plain mud	$\leq$1	$\leq$0.2	$\leq$0.1	0.1	$\leq$100

G_k center is calculated as:

$$D^2(X, G_k) = (X - \bar{X}^{(k)})'(\Sigma^{(k)})^{-1}(X - \bar{X}^{(k)}) \qquad (2)$$

where $\bar{X}^{(k)} = \bar{x}_1^{(k)}, \bar{x}_2^{(k)}, \ldots, \bar{x}_p^{(k)}$ and $\Sigma^{(k)}$ is the covariance of the feature matrix for facies type G_k. Then the facies type with the minimum distance to the type centroid will be picked as the facies outcome for this feature vector X.

3 Work Flow and Implementation Example

The proposed method is implemented in one region of a well-developed reservoir in west–north China. After nearly 70 years development, the wells in the reservoir are in a highly density. The average well departure is about 50–150 m away in this region. First, the Ward's cluster approach is applied in facies classification based on 12 wells with core samples. The input information for the classification are facies thickness, net-to-gross ratio, barriers thickness ratio along vertical direction, average porosity and permeability. Finally, four facies types are recognized from the input data. The feature of each facies is listed in Table 1.

It has a highly consistence with the detailed sedimentary research results which shows that the study zone has a braided river delta background. Type A is recognized as channel deposits. Type B is mouth bar inside of channel deposits. Type A and type B have a close thickness distribution but mouth bar shows a coarsening up well log profile. As there are too many wells to do facies analysis and also in order to avoid the objective bias in manually facies analysis, all the other nearly 500 wells are analyzed through calculating the Mahalanobis distance to each facies type centroid based on Eq. (2).

Using the feature distances of all the well locations as hard data, the cells of the research area are interpolated using the inverse distance interpolation approach. The variogram based kriging interpolation approach such as IK3D could also be used [6]. But, the inverse distance approach will take advantage of the highly sampled well data and doing spatial interpolation without variogram modeling. The interpolated distance maps for each facies are shown in Fig. 1. The final facies outcome for each location is the minimum distance to its cluster centroid which is shown in Fig. 1.

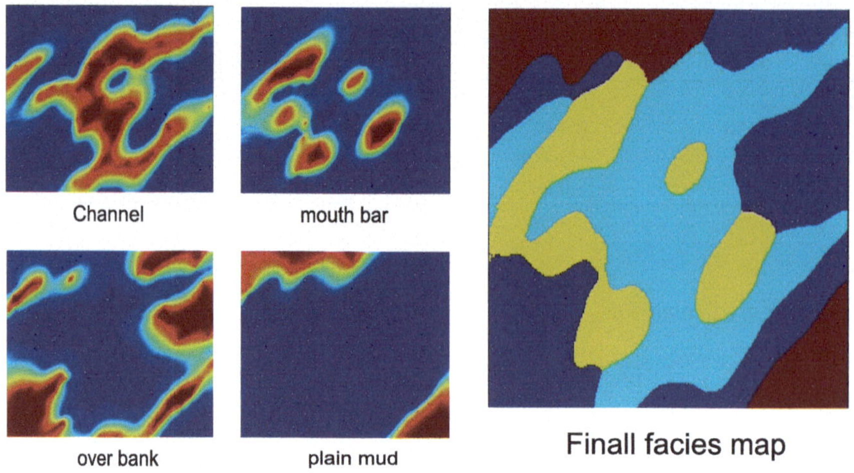

Fig. 1 The distance map of four facies and the final facies map

4 Conclusion and Discussion

It is proposed to do facies modeling before the facies assignment to the well location. The facies modeling approach should integrate the information provided to geologists when they do facies analysis along vertical well profile. Some spatial clustering artificial intelligence could be used to integrate those information in geostatistical facies modeling. The facies segmentation obtained by proposed approach will be used as constrain in the later petrophysical modeling for flow simulation. As for each facies spatial segment, the porosity and permeability macro heterogeneity characters are already characterized in the procedure of the spatial facies segmentation, it will ease the history matching works in flow simulation. Another advantage of this approach is using the inverse distance algorithm saves a large of variogram modeling work in facies modeling and petrophysical modeling.

References

1. Damsleth, E., Tjolsen, C. B., Omre, H., & Haldorsen, H. H. (1992). A two-stage stochastic model applied to a North Sea reservoir. *Journal of Petroleum Technology*, *44*(4), 402–408, 486.
2. Deutsch, C. V. (2006). A sequential indicator simulation program for categorical variables with point and block data: BlockSIS. *Computers and Geosciences*, *32*(10), 1669–1681.
3. Harff, J., & Davis, J. C. (1990). Regionalization in geology by multivariate classification. *Mathematical Geology*, *22*(5), 573–588.
4. Hervada-Sala, C., & Jarauta-Bragulat, E. (2004). A program to perform Ward's clustering method on several regionalized variables. *Computers and Geosciences*, *30*(8), 881–886.

5. Koltermann, C. E., & Gorelick, S. M. (1996). Heterogeneity in sedimentary deposits: A review of structure-imitating, process-imitating, and descriptive approaches. *Water Resources Research*, *32*(9), 2617–2658.
6. Deutsch, C. V., & Journel, A. G. (1998). *GSLIB: Geostatistical software library and user's guide* (2nd ed.). New York: Oxford University Press.

Permeability Prediction Using Artificial Neural Networks. A Comparative Study Between Back Propagation and Levenberg–Marquardt Learning Algorithms

Leila Aliouane, Sid-Ali Ouadfeul, Noureddine Djarfour and Amar Boudella

1 Introduction

Permeability governs the movement of fluids, gas and liquid, through pore spaces and networks in porous media. Its applications include enhancement of oil/gas recovery, management of water resources, CO_2 geological storage and geothermal energy extraction [6]. Most commonly, permeability is estimated from various well logs using either an empirical relationship or some forms of statistical regression (parametric or nonparametric). The empirical models may not be applicable in regions with different depositional environments without making adjustments to constants or exponents in the model. Also, significant uncertainty exists in the determination of irreducible water saturation and cementation factor in these models [3].

In recent years, nonparametric regression techniques such as Neural Network have been introduced to overcome the limitations of conventional methods [1] . In this context, a Multilayer Perceptron model has been used with two training algorithms for permeability estimation: back propagation and Levenberg Marquardt where petrophysical measurements and cores data of two wells from Algerian Sahara have been exploited. Well-logs data of the well well-A are used for the training of the two neural machines, at this step this well is used as a pilot and weights of connection are calculated and raw well-logs data of the depth reservoir interval [2195.8 m, 2223.5 m] are investigated. In this paper we present only data and results

L. Aliouane (✉) · N. Djarfour
Algerian Petroleum Institute, IAP, Boumerdes, Algeri
e-mail: lil_aldz@yahoo.fr

S. A. Ouadfeul (✉)
LABOPHT, FHC, UMBB, Avenue de l'indépendance, 35000 Boumerdes, Algeria
e-mail: Souadfeul@ymail.com

A. Boudella
Geophysics Department, FSTGAT, USTHB, Algeria

E. Pardo-Igúzquiza et al. (eds.), *Mathematics of Planet Earth*,
Lecture Notes in Earth System Sciences, DOI: 10.1007/978-3-642-32408-6_142,

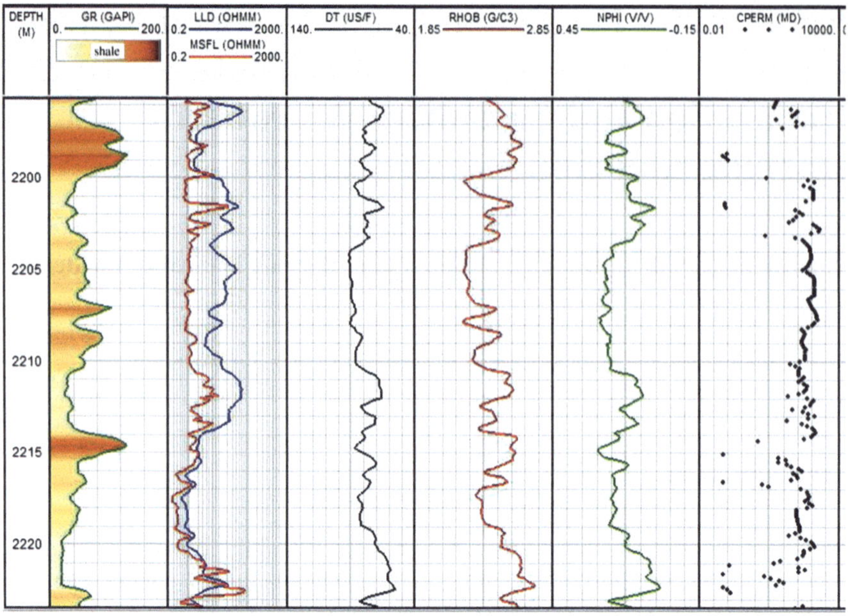

Fig. 1 Petrophysical parameters recordings of well-B

of the generalization well. Figure 1 is the petrophysical parameters recordings of this well (well-B).

2 Permeability Estimation Using Timur's Empirical Relationship

Different empirical approaches are used to describe the observed highly non-linear dependence of permeability to porosity by exponential or power-law relationships. [3] has presented the state of the art of permeability prediction by empirical models which are based on the correlation between permeability, porosity, and irreducible water saturation. In this paper, the Timur's relation is used [5, 7]. Figure 2c presents the core rocks permeability (CPERM) and the Timur's relation results (PERM_Timur). Comparison between PERM_Timur and CPERM clearly shows that this kind of empirical model is not able to provide good results. For this reason, we suggest the use of Artificial Neural Network (ANN) techniques to resolve this ambiguity; ANN does not require the knowledge of a permeability relationship.

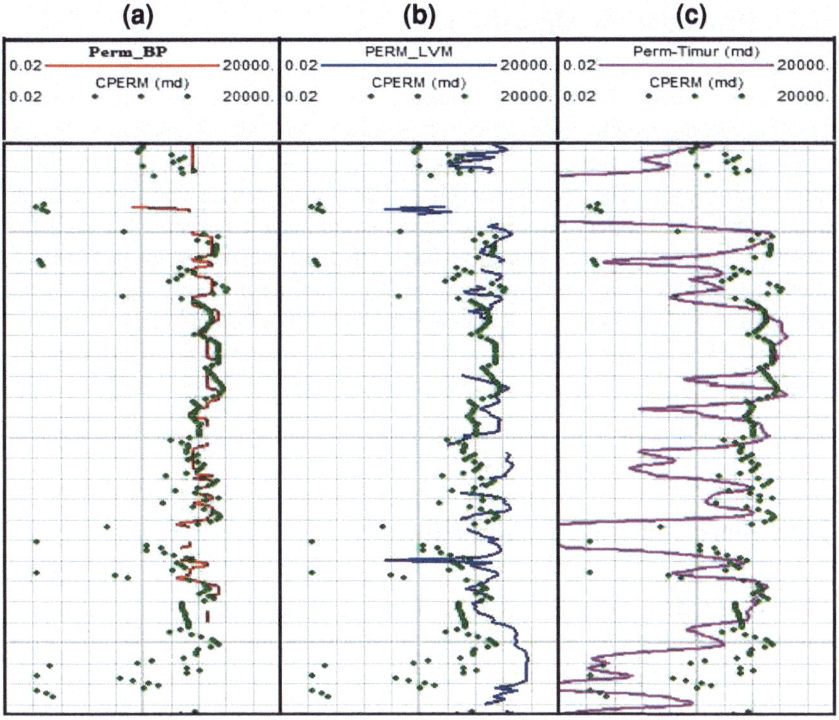

Fig. 2 Permeability prediction of a reservoir of well-B: **a** by Back propagation algorithm; **b** by Levenberg Marquardt algorithm; **c** by Levenberg-Timur's relation

3 The Multilayer Perceptron

Neural network is basically a parallel dynamic system of highly interconnected interacting parts based on neurobiological models. Neural network mimic somewhat the learning process of a human brain instead of using complex rules and mathematical routines. Here, the nervous system consists of individual but highly interconnected nerve cells called neurons. These neurons typically receive information or stimuli from the external environment. There are many types of network such as Multi Layer Perceptron (MLP) that commonly used in many practical applications. In this case, the structure of MLP is constituted with one input layer, one hidden layer and one output layer, inputs are the recordings of two wells well-A and well-B. The output is the calculated permeability using two learning algorithms: Back Propagation (Perm_BP) and Levenberg-Marquardt (Perm_LVM).

4 Back Propagation Algorithm

This training method uses Back Propagation to calculate derivatives of performance with respect to the weight and bias variables of the network. The network training function updates weight bias values according to gradient descent momentum and an adaptive learning rate. The adaptive learning starts with an initial value, then increases or decreases by multipliers in order to keep fast and stable learning. By adjusting its learning rate the network converge faster, thereby increasing the accuracy of predictions and shortening the training time. The learning process terminates when either the maximum and number of epochs is completed or the network sum-squared error drops below the min error goal set. Detailed explanation regarding this training method can be found in [2]. Obtained results (Perm_BP) using this kind of learning algorithm are shown in Fig. 2a.

5 Levenberg Marquardt Algorithm

The Levenberg Marquardt (LM) algorithm was designed to approach second-order training speed without having to compute the Hessian matrix [4]. When the performance function has the form of a sum of squares, the Hessian matrix can be approximated as follows [4]:

$$H = J^T.J$$

The gradient can be computed as: $g = J^T.e$

Where "J" is the Jacobian matrix, it contains the first derivatives of the network errors with respect to the weights and biases, and "e" is a vector of network errors.

Predicted values of permeability (Perm_LVM) using the LM learning algorithm are presented in Fig. 2b.

6 Results, Interpretation and Conclusion

Figure 2 shows the core rocks permeability (CPERM), predicted permeability using the Back propagation and Levenberg–Marquard algorithms and finally the Timur's relationship results. It is clear from this figure that the Timur's formula has not given good results. Calculated Root Mean Squares show a value of 0.17 for the Levenberg–Marquardt and 0.05 for the Back Propagation. By consequence the Back Propagation algorithm has proven its robustness to resolve this kind of petrophysical problem. We suggest the use of this kind of learning algorithm for permeability prediction from raw well-logs data rather than the Levenberg–Marquardt. By implementing our method, we have suggested an ANN scheme that can be used for permeability prediction of wells located in the neighborhood of the pilot well well-A.

References

1. Aliouane, L., Ouadfeul, S., Djarfour, N., & Boudella, A. (2012). Petrophysical parameters estimation from well-logs data using multilayer perceptron and radial basis function neural networks. In T. Huang et al. (Eds.), *ICONIP 2012, Part V. LNCS 7667* (pp. 730–736). Springer-Verlag.
2. Aqil, M., Kita, I., Yano, A., & Nishiyama, S. (2007). Neural networks for real time catchment flow modeling and prediction. *Water Resource Manage*, *21*, 1781–1796.
3. Balan, B., Mohaghegh, S., & Ameri, S. (1995). *State-of-the-art in permeability determination from well log data: Part 1—A comparative study, model development*. SPE. 30978. West Virginia: Eastern Regional Conference & Exhibition.
4. Demuth, H., & Beale, M. (1998). *Neural network toolbox for use with MATLAB, Users Guide, Version 3*. Massachusetts: The MathWorks, Inc.
5. Ellis, D. V., & Singer, J. M. (2007). *Well logging for earth scientists* (2nd ed.). Dordrecht: Springer.
6. Markov, M., Kazatchenko, E., Mousatov, A., & Pervago, E. (2010). Permeability of the Fluid-Filled Inclusions in Porous Media. *Transport in Porous Media*, *84*, 307–317.
7. Timur, A. (1968). An investigation of permeability, porosity, and residual water saturation relationship for sandstone reservoirs. *The Log Analyst*, *9*(4), 8.

A Method for Multi-Level Probabilistic History Matching and Production Forecasting: Application in a Major Middle East Carbonate Reservoir

Marko Maučec, Ajay Singh, Gustavo Carvajal, Seyed Mirzadeh, Steven Knabe, Richard Chambers, Genbao Shi, Ahmad Al-Jasmi, Harish Kumar Goel and Hossam El-Din

1 Introduction

Optimal improved oil recovery depends on the ability to use historical data to estimate volumes and locations of bypassed oil. In the description of geologically complex reservoirs, the geo-models' parameters are usually reconciled with pressure and multi-phase production data during history matching (HM) [1]. As an inverse problem, HM is highly non-linear and ill-posed, which means that, depending on the *prior*

M. Maučec (✉) · A. Singh · G. Carvajal · S. Mirzadeh · S. Knabe · R. Chambers · G. Shi
Halliburton Energy Services, 2107 City West Blvd., Houston TX 77042, USA
e-mail: marko.maucec@halliburton.com

A. Singh
e-mail: ajay.singh@halliburton.com

G. Carvajal
e-mail: gustavo.carvajal@halliburton.com

S. Mirzadeh
e-mail: mo.mirzadeh@halliburton.com

S. Knabe
e-mail: steven.knabe@halliburton.com

R. Chambers
e-mail: rchambers@lgc.com

G. Shi
e-mail: gshi@lgc.com

A. Al-Jasmi · H. K. Goel · H. El-Din
Kuwait Oil Company, 61008 Ahmadi, Kuwait
e-mail: AJASMI@kockw.com

H. K. Goel
e-mail: HKGoel@kockw.com

H. El-Din
e-mail: HDIbrahim@kockw.com

E. Pardo-Igúzquiza et al. (eds.), *Mathematics of Planet Earth*,
Lecture Notes in Earth System Sciences, DOI: 10.1007/978-3-642-32408-6_143,

information, we can obtain a set of non-unique solutions that honor both the prior constraints and the conditioning data. To quantify the uncertainty associated with the reservoir parameters, one must sample from the *posterior* distribution. The Bayesian methods [2] provide an efficient framework for performing this operation. Herewith, the posterior model $p_{m|d}(\mathbf{m}|\mathbf{d})$(i.e., the conditional probability of model parameter, **m**, given the observed data, **d**) is represented as proportional to the product of the prior $p_m(\mathbf{m})$ and the likelihood $p_{d|m}(\mathbf{d}|\mathbf{m})$models:

$$p_{m|d}\left(\mathbf{m}|\mathbf{d}\right) = \frac{p_{d|m}\left(\mathbf{d}|\mathbf{m}\right) p_m\left(\mathbf{m}\right)}{p_d\left(\mathbf{d}\right)} \tag{1}$$

In HM problems, the prior model corresponds to the static (geo-cellular) model and the likelihood model represents the dynamic response of reservoir simulator to the given static model. The normalization factor $p_d(\mathbf{d})$ is usually treated as a constant.

2 Methodology

The sequential Markov chain Monte Carlo (MCMC) algorithms provide the most statistically rigorous sampling of the posterior distribution, but when deployed in direct simulation, impose high computational cost. We implement a two-stage MCMC algorithm [3] that calculates an approximate likelihood by using streamline sensitivities [4]. These are obtained as derivatives of streamline time-of-flight with respect to reservoir parameter m, in our case the matrix permeability. Such an approach significantly increases the acceptance rate of traditional MCMC [5].

A pre-screening based on approximate likelihood calculations eliminates most of the rejected proxy models, and the exact MCMC is only performed on the accepted proposals, with improved acceptance rates. The approximate likelihood calculations using streamline sensitivities are fast and involve a linearized approximation around an already accepted state rather than an expensive flow simulation. Detailed mathematical derivations related to two-stage MCMC algorithms are given in [5, 6].

By design, probabilistic HM workflows can produce model realizations with non-geological features, and geological parameters may have an insignificant effect on the reservoir recovery. We identify optimal number of representative reservoir models and maintain the inherent uncertainty by dynamic model ranking [4] based on forecasted oil recovery factors (ORF) using the following steps:

- Forecast the ORF and quantify the (dis)similarity $D_{ij} = \sqrt{\sum (ORF_j - ORF_i)^2}$ between two model realizations i and j using multi-dimensional scaling and distance-based modeling [6].
- Group (cluster) the models with highly similar ORF using kernel k -means clustering [6] with an optimized number of data clusters.

- Identify lower, median, and higher oil-recovery models (i.e., P10, P50, P90) using the centroids of each cluster as representative models for production forecasting with the full physics simulator.

3 Case Study

The workflow is applied by HM in the pilot area of a major, structurally-complex, anticlinal Middle East carbonate reservoir, operated by Kuwait Oil Company (KOC). An ensemble of diverse high-resolution geo-model realizations with 1.4 million grid-cells was generated by honoring known geo-statistical semi-variograms and well-log constraints. The 3D, 3-phase, black-oil, single-porosity reservoir simulation model combines 49 real-time instrumented wells in five waterflood patterns to match 50 years of oil production and 12 years of water injection. In the first-level HM, the reservoir model is tuned to match reservoir pressure and oil production by adjusting rock-type properties and modifying pore volume and well transmissibility at well locations. The second level HM uses two-stage MCMC stochastic optimization to minimize the watercut misfit on a well-by-well basis. Our approach significantly accelerates the process by parameterizing permeability in the wave-number domain

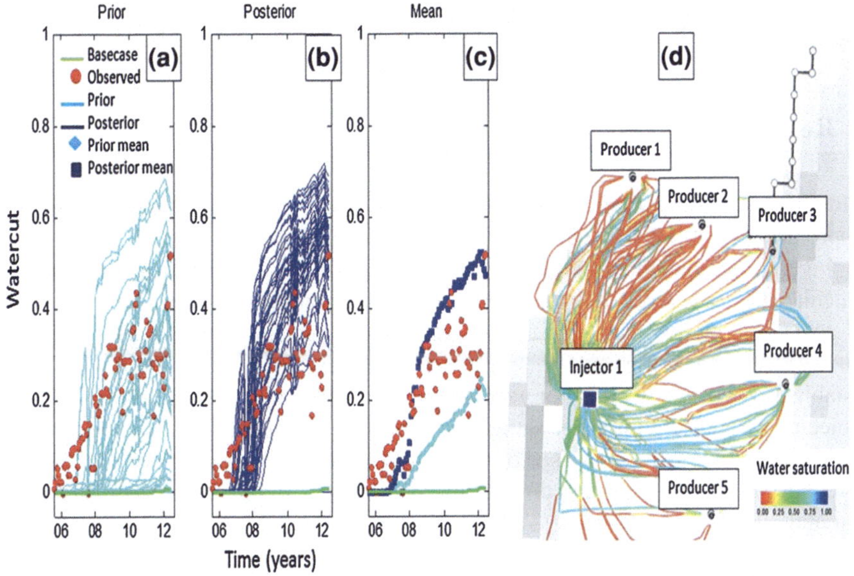

Fig. 1 Results of assisted HM: panels **a** and **b** show the watercut profiles from prior simulation (*light blue*) and posterior (*dark blue*) HM, panel **c** outlines the mean profiles averaged over 40 prior and posterior watercut curves. The *red dots* indicate the observed data. Panel **d** visualizes flow communication between the injector well and five associated producers

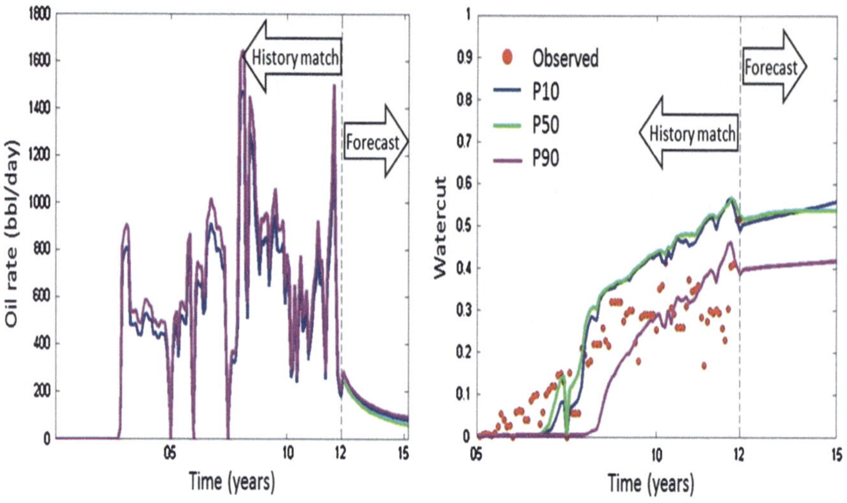

Fig. 2 Results of the production forecasting: oil rate (*left*) and watercut (*right*) forecasted profiles correspond to P10, P50 and P90 models identified by dynamic ranking of history-matched models based on 8-years of production forecasting

using Discrete Cosine Transform [6], implementing rigorous convergence diagnostics and deploying parallel and cluster computing.

4 Results

More than 2,000 probabilistic model inversion scenarios were executed, combining approximately 50 MCMC iterations with 40 geo-model realizations.

Figure 1 demonstrates a significant reduction in the watercut misfit between the calculated and observed data for a selected producing well. A dramatic improvement is achieved, relative to the base case static geo-model that initially did not produce water due to inconsistencies in reservoir connectivity. The flow communication established between the injector well and producer #4 is visualized using streamline trajectories labeled with water saturation in panel (d). The history matched models are dynamically ranked [6] based on forecasted oil recovery factors and models with P10, P50, and P90 oil recovery probability are identified. Figure 2 indicates the watercut and oil rate profiles, for selected models during the HM and 8-year production forecasting. Further results of this case study are presented in [7].

5 Conclusions

The workflow is implemented at the North Kuwait Integrated Digital Field (KwIDF) Project. It automatically interfaces the geo-modeling application with reservoir simulator and preserves the geological detail by updating high-resolution models with no upscaling. Traditional approaches do not incorporate such functionality. The workflow delivers a history matched reservoir model for waterflood optimization, automatically updates the model with production, completion and geological information quarterly, identifies the areas that require more data capture and selects the optimal number of models for production forecast.

References

1. Oliver, D., Chen, Y. & Nævdal, G. (2011). Updating Markov chain models using the ensemble Kalman filter. *Computers & Geosciences*, *15*(2), 325–344.
2. Lee, P. M. (1997). *Bayesian Statistics*. New York: Willey and Sons.
3. Efendiev, Y., Datta-Gupta, A., Ginting, V., Ma, X., & Mallick, B. (2005). An efficent two-stage Markov chain monte carlo method for dynamic data integration. *Water Resources Research*, *41*, W12423.
4. Datta-Gupta, A., & King, M. J. (2007). *Streamline Simulation: Theory and Practice*, SPE Textbook Series 11.
5. Maučec, M., Douma, S., Hohl, D., Leguijt, J., Jimenez, E. A., Datta-Gupta, A. (2007). Paper SPE-109943, Society of Petroleum Engineers, Richarson, Texas.
6. Maučec, M., Cullick, S., & Shi, G. (2011). Paper SPE-146748.
7. Maučec, M., Carvajal, G., Singh, A., Mirzadeh, S., Knabe, S., Chambers, R., et al. (2013). Paper SPE-13RCSC-P-314.

Causal Analysis and Data Mining of Well Stimulation Data Using Classification and Regression Tree with Enhancements

Srimoyee Bhattacharya, Marko Maučec, Jeffrey Yarus, Dwight Fulton, Jon Orth and Ajay Singh

1 Introduction

The well treatment program is an important part of the field development plan and certain variables like Job Pause Time (JPT) and fracture screen-out, can affect its efficiency. JPT is the time during which pumping is paused in-between subsequent treatments of a job and screen-out occurs when the fluid flow is restricted inside the fracture or throughout the perforation. We utilize data mining with Classification and Regression Tree (CART) [1] to find significant variables that affect the response variables and rank the predictor variables in the hierarchal order of their importance to control high JPT or screen-out. We validate with well-stimulation case studies and enhance the predictive capability by implementing normal score transform [2] and by dividing the large dataset into smaller groups, using *k*-means clustering [3].

S. Bhattacharya · M. Maučec (✉) · J. Yarus · D. Fulton · J. Orth · A. Singh
Halliburton Energy Services, 2107 City West Blvd, Houston, TX 77042, USA
e-mail: marko.maucec@halliburton.com

S. Bhattacharya
Shell International Exploration and Production, Houston, TX, USA
e-mail: srimoyee6@gmail.com

M. Maučec
e-mail: marko.maucec@halliburton.com

J. Yarus
e-mail: jeffrey.yarus@halliburton.com

D. Fulton
e-mail: dwight.fulton@halliburton.com

J. Orth
e-mail: jon.orth@halliburton.com

A. Singh
e-mail: ajay.singh@halliburton.com

E. Pardo-Igúzquiza et al. (eds.), *Mathematics of Planet Earth*,
Lecture Notes in Earth System Sciences, DOI: 10.1007/978-3-642-32408-6_144,

2 Methodology

The steps for applying CART on a particular dataset are outlined as follows:

(a) **Define** the response (output) variable y and input variables x_i and pre-process the data by de-noising and removing outliers. Assuming the normal distribution, outliers are identified as observations deviating by more than 3σ from the mean.
(b) **Perform CART analysis**. For the regression tree the splitting criteria is the minimized mean square error [1] and for the classification tree it can be for example, the maximized entropy [1]. The optimal level of tree-splitting is chosen such that cross-validation error is minimized at each node.
(c) **Quantify variable importance** as the measures for the contribution of a particular predictor variable in the tree formation. The data are divided into v sub samples of equal size $N_1, N_2, \ldots, N_v$ and the V-fold cross-validation error (ε_{CV}) for classification tree is defined as misclassification probability:

$$\varepsilon_{CV} = \frac{1}{N_v} \sum_{i=1}^{N_v} I(\widehat{y}_{i(N-N_v)} \neq y_i) \quad (1)$$

and for the regression tree as the mean square error:

$$\varepsilon_{CV} = \frac{1}{N_v} \sum_{i=1}^{N_v} (y_i - \widehat{y}_{i(N-N_v)})^2 \quad (2)$$

where $\widehat{y}_{i(N-N_v)}$ is the predicted value of sample i computed from the tree that is constructed using $N - N_v$ samples.

3 Application and Results

3.1 Classification Tree for the Fracture Screen-Out

The objective was to predict if a particular job would have screen-out or not, and to identify the critical causal factors. Both types of predictor variables are defined: categorical (engineer, customer, and crew) and continuous (depth, average rate and pressure, mass of proppant and clean volume). The corresponding tree is shown in Fig. 1. The tree growth continues until maximum purity is achieved or minimum samples constraint of 20 is reached in the terminal nodes. The analysis indicates that 3/10 terminal nodes have predicted screen-out occurrence. This is a very significant result since CART has separated cases of screen-out occurrence even though overall percentage of screen-out is only 7 %. Although the average pressure is the first splitting variable, it is below customer in ranking. This suggests high misclassification

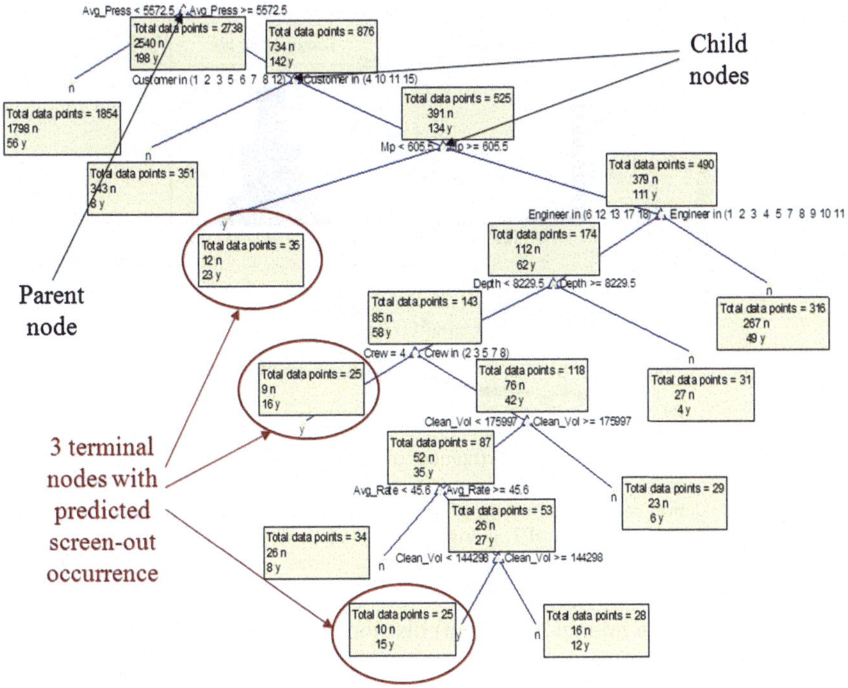

Fig. 1 Classification tree for screen-out with information for each node given in the *text box*. The generic elements of the tree structure and terminal nodes with identified fracture screen-out (occurrence labeled with "y") are also indicated

probability of customer over average pressure and indicates the possibility that the customer has a particular choice of proppant or is operating in a region which is not included in the analysis contributing towards its significance.

3.2 *Regression Tree for the Job Pause Time*

To predict Job Pause Time (JPT) and to find variables that influence or cause JPT, we define three categorical variables (customer, job configuration and first job year) and continuous variables as bottom-hole temperature and pressure, treatment pressure, slurry rate, acid rate and proppant concentration. This study results in a regression tree problem because response variable is continuous in nature. The tree consists of 113 terminal nodes and the optimum tree length of 7 nodes was found using V-fold cross-validation (see Fig. 2). One key observation is that the standard deviation is very high in most of the nodes making prediction unreliable. This is related to the distribution of the average JPT which is highly skewed (see insert a in Fig. 2). Normalizing the data prior to CART using a method like the Normal Score Transform

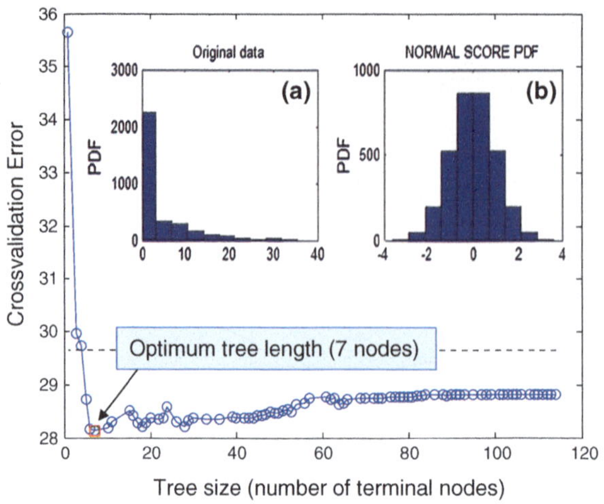

Fig. 2 Cross-validation error as a function of tree size (number of terminal nodes). The inserts represent the distribution of average JPT: **a** original dataset, **b** post-NST dataset

(NST) [2] results in a normal (Gaussian) distribution of average JPT (see insert b in Fig. 2). After performing *k*-means clustering [3] on the post-NST dataset, further improvements are observed: when using 5, 6 and 8 clusters the re-substitution error is reduced by 10, 14 and 18 %, respectively.

4 Conclusions

The application of CART analysis on well-stimulation data was successful in finding patterns in significant variables that affect the extreme values of JPT and in identifying the critical variables causing fracture screen-out. CART is a data driven, deterministic model and cannot calculate the confidence interval of the predicted response but the prediction capability can be enhanced by implementing normal score transform and by data clustering.

References

1. Breiman, L., Friedman, J. H., Olshen, R. A., & Stone, C. J. (1984). *Classification and regression trees*. Boca Raton: Chapman & Hall.
2. Everitt, B. S. (2002). *The Cambridge dictionary of statistics* (2nd ed.). Cambridge: Cambridge University Press.
3. Shawe-Taylor, J., & Cristianini, N. (2004). *Kernel methods for pattern analysis*. Cambridge: Cambridge University Press.

Seismic Data Interpretation Improvement by the Multiscale Analysis of Gravity Data Using the Wavelet Transform: Application to Algerian Sahara

Sid-Ali Ouadfeul and Leila Aliouane

1 Introduction

The wavelet transform has becoming a very useful tool for interpretation of gravity and magnetic potential field data. [1] have used the 2D continuous wavelet transform for structural boundaries delimitation from geomagnetic data. Another paper published by [2] shows an application of the 2D directional continuous wavelet transform (CWT) for mapping of geological contact from aeromagnetic data, obtained results are compared with the 3D Analytic signal solutions and show clearly robustness of the 2D CWT. [4] have analyzed the amplitude of the 3D analytic signal by the 2D continuous wavelet transform. The proposed method is versatile in case of high remanence of geomagnetic data. [3] have analyzed the amplitude of the horizontal gradient by the 2D CWT; the proposed combination is less sensitive to noise. In this paper, we show the role of multiscale analysis of a 2D gravity data of an area located in the Algerian Sahara for enhancement of petroleum exploration.

2 Wavelet Transform and Potential Field Data

The analogy between the continuous wavelet transform and upward continuation has been widely used for mapping of structural boundaries from potential gravity and magnetic data [2, 4]. By choosing an appropriate wavelet: for example the Poisson's Kernel [4], measurement of a potential field or its spatial derivatives can be processed

S.-A. Ouadfeul (✉)
Algerian Petroleum Institute, IAP, Boumerdes, Algeria
e-mail: Souadfeul@ymail.com

L. Aliouane
LABOPHT, FHC, UMBB, Boumerdes, Algeria
e-mail: lil_aldz@yahoo.fr

E. Pardo-Igúzquiza et al. (eds.), *Mathematics of Planet Earth*,
Lecture Notes in Earth System Sciences, DOI: 10.1007/978-3-642-32408-6_145,

as a wavelet transform. Indeed, this analysis unifies various classical techniques: it process gradients that have been upward to a range of altitudes. The expressions of various conventional operations on the potential field are well-designed in the wavelet domain. The most important is the equivalence between the concept of scaling and the upward. Indeed, the wavelet transform of a potential field F_0 (x, y) at a certain scale a = Z/Z_0 can be obtained from measurements made on the level Z_0 by:

1. Upward continue the measured field a level Z = a*Z_0
2. Calculation of the horizontal gradient in the plane (x,y)
3. Multiplication by a.

For a multiscale analysis of contacts, it is sufficient to look for local maxima of the modulus of the continuous wavelet transform (CWT) for different scales to get exact information about geological boundaries [1].

3 Application to Real Data

We have applied this idea to the gravity data of an area located in the Algerian Sahara. Figure 1a shows the Bouguer anomaly processed with a regular grid of 500 × 500 m. Firstly a 2D continuous wavelet transforms has been applied, the analyzing wavelet is the Poisson's Kernel [4]. Modulus of the 2D DCWT is presented for the smaller scale a = 705 m in Fig. 1b. The maxima of the modulus of the continuous wavelet transform are mapped for full range of scales. Figure 1c shows the map of these maxima.

4 Results Interpretation and Conclusion

Mapped maxima of the modulus of the 2D CWT are linked (see Fig. 2), obtained results are compared with the interpretation of the 2D seismic profile TGT06, Fig. 2 shows the location of this profile. It is clear that the 2D CWT is able to identify seven faults (F1, F2, F3, F4, F5, F5, F6 and F7) and the direction of F1 and F2, by consequence the multiscale analysis of the gravity data is a powerful tool that can greatly improve seismic data interpretation and enhance hydrocarbons exploration (Fig. 3).

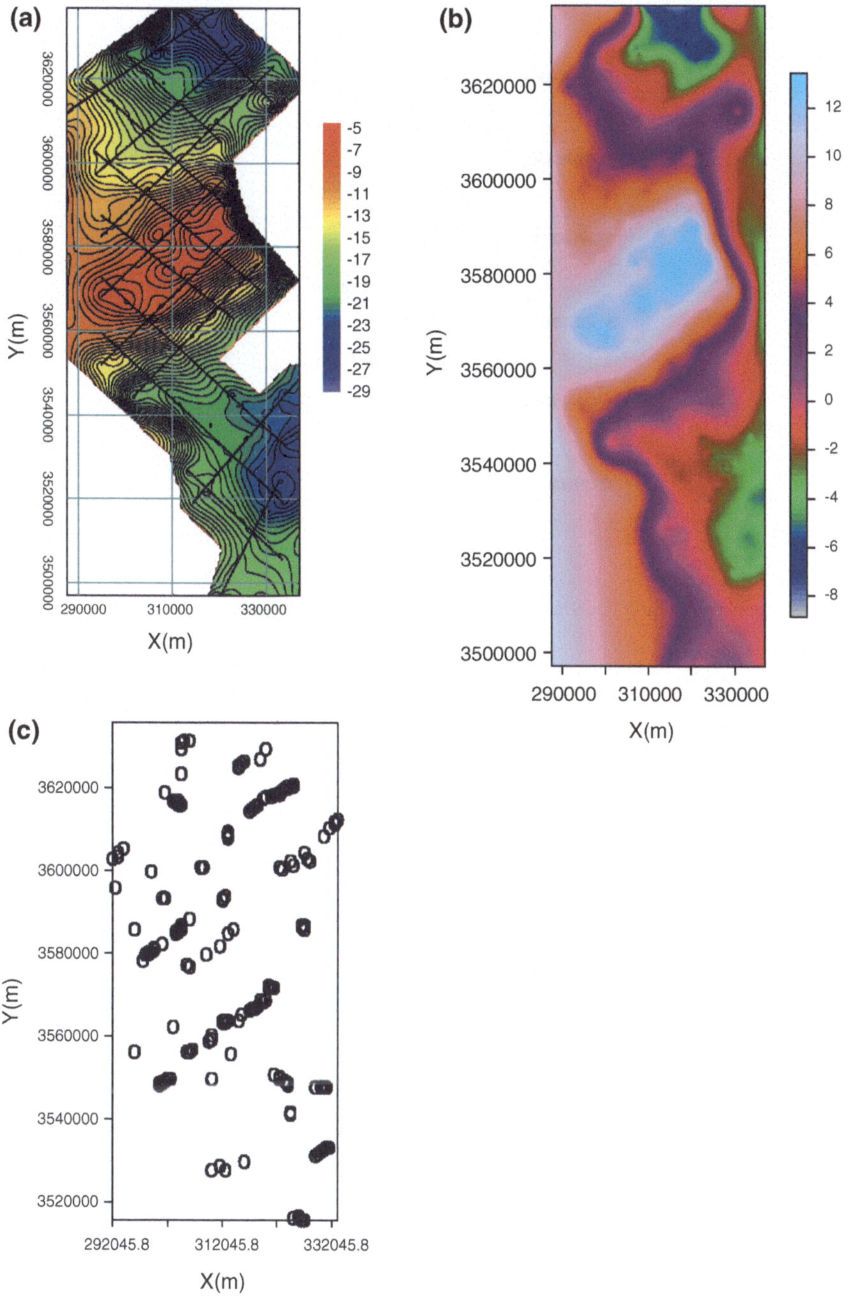

Fig. 1 Multiscale analysis of gravity data using the 2D DCWT. **a** Bouger anomaly in mGal, **b** modulus of the 2D DCWT for the lower scale a = 705 m, **c** mapped maxima of modulus of the 2D CWT

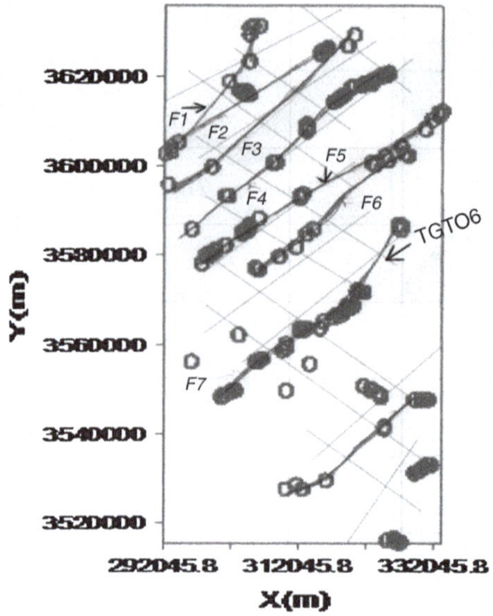

Fig. 2 Obtained structural boundaries by linking maxima of the modulus of the 2D CWT

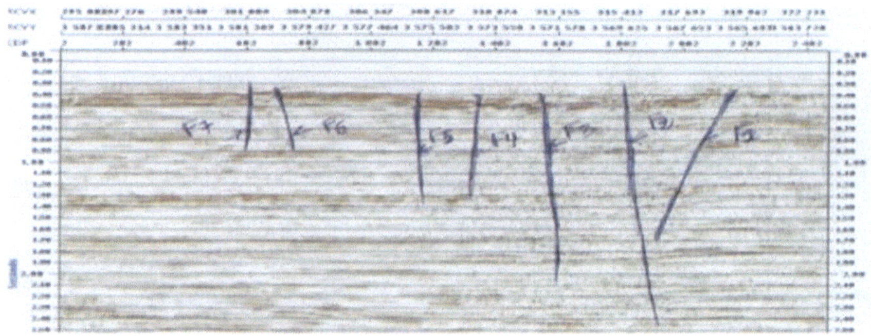

Fig. 3 Interpretation of the 2D seismic profile TGT06, faults are mapped from F1 to F7

References

1. Ouadfeul, S., Aliouane, L., & Eladj, S. (2010). Multiscale analysis of gomagnetic data using the continuous wavelet transform, SEG expanded abstract, Vol.29. doi:10.1190/1.3513065.
2. Ouadfeul, S., Hamoudi, M., Aliouane, L., & Eladj, S. (2012a). Aeoromagnetic data analysis using the directional continuous wavelet transform. *Arabian Journal of Geosciences*. doi:10.1007/s12517-011-0454-9.
3. Ouadfeul, S., & Aliouane, L. (2012b). Structural boundaries delimitation from aeromagnetic data using the horizontal gradient and the continuous wavelet transform, Istanbul 2012: *International Geophysical Conference and Oil & Gas Exhibition*, (pp. 1–4). doi:10.1190/IST092012-001.74.
4. Ouadfeul, S., & Aliouane, L. (2013). Pattern recognition of structural boundaries from aeromagnetic data using the 2D continuous wavelet transform and the 3D analytic signal. *Pattern Recognition in Physics*. doi:10.5194/prp-1-1-2013.

History Matching of Channelized Reservoir Using Ensemble Smoother with Clustered Covariance

Kyungbook Lee and Jonggeun Choe

1 Introduction

Ensemble Kalman filter (EnKF) has been researched for solving an inverse problem in history matching because it has lots of advantages over other optimization methods [1–4]. However, it has computational issues to apply to real fields because of many ensembles needed for reliable results [5]. Furthermore, it has shown overshooting and filter divergence problems when there are not sufficient ensembles, especially applied to non-Gaussian fields or improper ensemble designs [5–7].

Many researchers have emphasized the importance of Kalman gain to settle these problems and have researched to improve cross covariance in Kalman gain [8, 9]. Clustered covariance is also aimed at obtaining a more representative cross covariance.

Ensemble smoother (ES) has been focused after Skjervheim et al. [10] introduced it for history matching in reservoir parameters. It assimilates all available observation data at a time. Recent researches showed that ES reduces simulation time significantly with acceptable history matching results over EnKF [8, 11]. In this study, clustered covariance method is coupled with ES for channelized fields.

K. Lee (✉) · J. Choe
Department of Energy Systems Engineering, Seoul National University,
Seoul 151-742, South Korea
e-mail: book0108@snu.ac.kr

J. Choe
e-mail: johnchoe@snu.ac.kr

E. Pardo-Igúzquiza et al. (eds.), *Mathematics of Planet Earth*,
Lecture Notes in Earth System Sciences, DOI: 10.1007/978-3-642-32408-6_146,

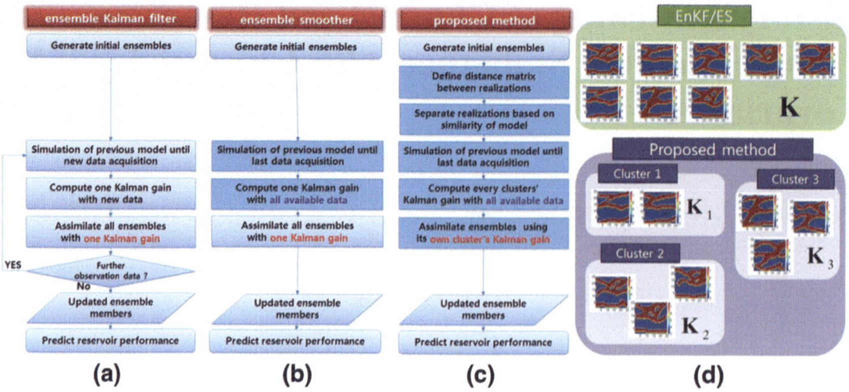

Fig. 1 The concept of EnKF, ES and the proposed method. Flow chart of (**a**) EnKF, (**b**) ES, (**c**) the proposed method, and (**d**) key idea of the proposed method for kalman gain

2 Methodology

The two standard ensemble methods, EnKF and ES, use the same principle equations and one Kalman gain for updating all ensembles. The difference is the way of assimilation. EnKF assimilates dynamic data whenever they are available (Fig. 1a), while ES does it once (Fig. 1b).

Many researchers have reported that the first iteration of Gauss—Newton method is similar to the equation of EnKF and ES [8, 9]. Covariance matrix has a role of sensitivity matrix in Gauss—Newton method. Therefore, it is difficult to set a representative covariance matrix for whole ensembles, especially for channelized fields. Small differences in initial ensembles make huge different responses in dynamic variables since channel streams have great effects on reservoir performances.

The proposed method is based on the ES algorithm with clustered covariance method. It separates initial ensembles using a distance-based method before assimilation (Fig. 1c). Hausdorff distance and k-means clustering method are used. Figure 1d shows the concept of the clustering method for initial ensembles. EnKF and ES use all initial ensembles to get Kalman gain even though they differ from each other. However, the proposed method groups analogous ensemble members in the same cluster, which results in more reliable Kalman gain for each cluster.

3 Results

The two standard ensemble methods and the proposed method are applied to a synthetic 2D channelized reservoir. Multipoint statistics is used to make a reference field and 200 ensembles. Training image has low sinuosity channel with sand facies ratio of 0.14 in 250 by 250 grid system. In the synthetic field, the channel streams have

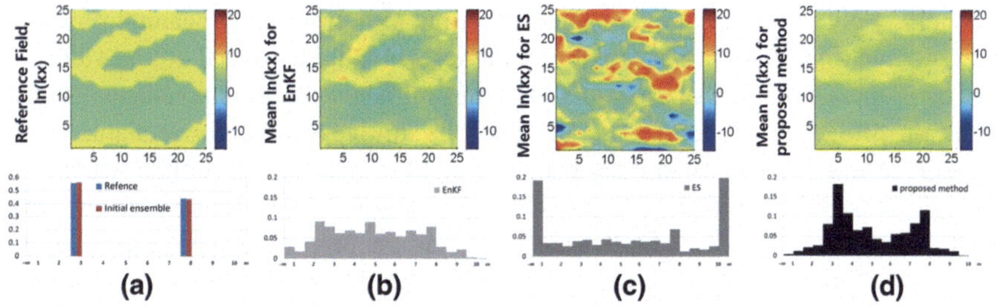

Fig. 2 The mean of all permeability models and the histogram of a single ensemble member at the end of assimilation. **a** The reference field, **b** EnKF, **c** ES, and **d** the proposed method

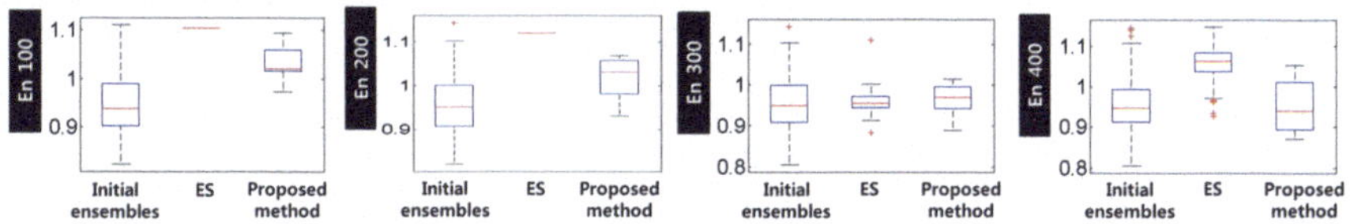

Fig. 3 The estimated oil production at 1,800 days. Ensembles size is changed from 100 to 400

2,000 md and background facies has 20 md. It is more realistic assumption comparing with previous research: 100, 1 md [7]. The three methods utilize 30 times available data until 600 days and predict production to 1,800 days.

The reference field has binomial distribution of values about 7.6 and 3 after log-transformation (Fig. 2a). The EnKF's result has overshooting values (Fig. 2b). The histogram of the model parameter loses binomial distribution and has a tendency to be a Gaussian distribution. In the case of ES method, there is obvious overshooting problem in the both of permeability map and histogram (Fig. 2c). However, the proposed method manages the overshooting problem (Fig. 2d). It keeps channel streams preserving prior information of geology. Moreover, the proposed method has the advantage of simulation efficiency over EnKF.

Predicted total oil productions are divided by true values to make box plots (Fig. 3). Default case (200 ensembles) shows that ES cannot predict reservoir performances properly. ES shows oscillating results upward and downward even though it contains the reference's results for 300 and 400 cases. The proposed method is less sensitive for ensembles size and it is still covering true performances at 100 ensembles case.

4 Conclusions

The standard EnKF, ES, and the proposed method are applied for history matching of a channelized reservoir. The standard ES loses channel distribution with severe overshooting problem and shows improper estimation of future production.

It cannot apply to channel field despite fast simulation time. EnKF cannot preserve the binomial distribution of permeability model with slightly overshooting values.

The proposed method successfully classifies initial ensembles into cluster groups based on similarity of geology model. It results in obtaining a more reliable cross covariance and updating model parameter without overshooting problem while preserving binomial distribution. It predicts oil production with reliable uncertainty for different ensemble sizes. Furthermore, ES and the proposed method spent about 4.5 % of the EnKF's computational time for 200 ensembles case in this study. The proposed method enables ES to apply to history matching of channel fields by enhancing accuracy of covariance matrix.

Acknowledgments The authors thank to research projects (2010201010092C, 2011201030001C, 10042556, 10038618).

References

1. Evensen, G. (1994). Sequential data assimilation with a nonlinear quasi-geostrophic model using Monte Carlo methods to forecast error statistics. *Journal of Geophysical Research, 99*(C5), 10143–10162.
2. Nævdal, G., & Vefring, E. H. (2002). Near-well reservoir monitoring through ensemble Kalman filter. *SPE/DOE Improved Oil Recovery Symposium*, Oklahoma, Tulsa, April 13–17, 2002.
3. Aanonsen, S. I., Nævdal, G., Oliver, D. S., Reynolds, A. C., & Valls, B. (2009). The ensemble Kalman filter in reservoir engineering: A review. *SPE Journal, 14*(3), 393–412.
4. Oliver, D. S., & Chen, Y. (2011). Recent progress on reservoir history matching: A review. *Computational Geosciences, 15*(1), 185–221.
5. Jafarpour, B., & McLaughlin, D. B. (2009). Estimating channelized-reservoir permeabilities with the ensemble Kalman filter: The importance of ensemble design. *SPE Journal, 14*(2), 374–388.
6. Lee, K., Jo, G., & Choe, J. (2011). Improvement of ensemble Kalman filter for improper initial ensembles. *Geosystem Engineering, 14*(2), 79–84.
7. Lee, K., Jeong, H., Jung, S., & Choe, J. (2013). Characterization of channelized reservoir using ensemble Kalman filter with clustering covariance. *Energy Exploration and Exploitation, 31*(1), 17–29.
8. Chen, Y., & Oliver, D. S. (2012). Ensemble randomized maximum likelihood method as an iterative ensemble smoother. *Mathematical Geosciences, 44*(1), 1–26.
9. Zafari, M., & Reynolds, A. C. (2007). Assessing the uncertainty in reservoir description and performance predictions with the ensemble Kalman filter. *SPE Journal, 12*(3), 382–391.
10. Skjervheim, J.-A., Evensen, G., Hove, J., & Vab, J. G. (2011). An ensemble smoother for assisted history matching. *SPE Reservoir Simulation Symposium*, The Woodlands, Texas, February 21–23, 2011.
11. Emerick, A. A., & Reynolds, A. C. (2012). Ensemble smoother with multiple data assimilation. *Computers & Geosciences*. doi:10.1016/j.bbr.2011.03.031.

Assessing the Probability of Training Image-Based Geological Scenarios Using Geophysical Data

Thomas Hermans, Jef Caers and Frédéric Nguyen

1 Introduction

A training image (TI) depicts the conceptual geological patterns and should be representative of the spatial heterogeneity. The construction of TIs is one of the most critical and important step of multiple-point statistics. If the geological context is generally known (e.g. alluvial aquifer), there remains in most cases considerable uncertainty on the characteristics of facies elements and on their relationships.

Geophysical methods may provide spatially distributed information on subsurface petrophysical properties, but due to their larger resolution, provide only indirect information on smaller scale features present in these TIs. In this work, we develop a methodology to verify the consistency of potential TIs with geophysical data. We demonstrate our proposed method in a case of electrical resistivity tomography (ERT) data within the context of the alluvial aquifer of the Meuse River in the area of Liege, Belgium.

T. Hermans (✉) · F. Nguyen
Applied Geophysics, University of Liege,
Chemin des Chevreuils 1, 4000 Liege, Belgium
e-mail: thomas.hermans@ulg.ac.be

F. Nguyen
e-mail: f.nguyen@ulg.ac.be

T. Hermans
F. R. S.-FNRS Research Fellow,

J. Caers
Department of Energy Resources Engineering, Stanford University,
367 Panama Street, Stanford CA 94305-2220, USA
e-mail: jcaers@stanford.edu

E. Pardo-Igúzquiza et al. (eds.), *Mathematics of Planet Earth*,
Lecture Notes in Earth System Sciences, DOI: 10.1007/978-3-642-32408-6_147,

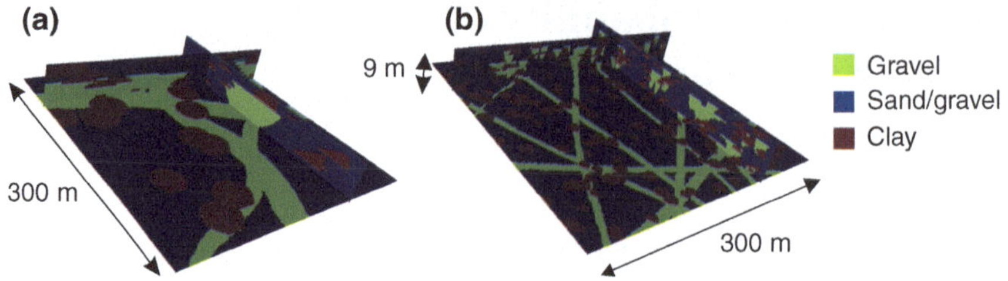

Fig. 1 Two examples of the six training image-based geological scenarios. **a** Alluvial aquifer with big gravel channels and big clay lobes. **b** Alluvial aquifer with small gravel channels and small clay lobes

2 Training Images as Geological Scenarios

Based on prior knowledge of the alluvial system, we describe geological heterogeneity using three facies (gravel, gravel in a sandy matrix, and clay/silt). We assume that gravel is present in the form of channels whereas clay is present in the form of lobes or lenses. Borehole analyses provide a prior proportion for each facies. From prior geological understanding on alluvial systems [e.g. 1], we generate training images with three different sizes for channels and two different sizes for lobes (six different geological scenarios or training images) (Fig. 1).

3 Evaluating Consistency with Geophysical Data

In each TI, we select randomly 12 sections (2D) and transform them into electrical resistivity models. Each facies is given a constant but uncertain electrical resistivity. Three different scenarios of resistivity distribution are offered to model this uncertainty. From these models, synthetic geophysical data sets (resistance data) are simulated. Next, these data sets are inverted with a least-square smoothness constraint algorithm to obtain TI-based inverted models. The inversion process causes loss of information inherent to geophysical methods. However, even if the models are smoothed, their specificity should still be apparent after inversion.

To verify the consistency of the TI-based inverted models, we compare them with two ERT inverted field datasets representative of the Meuse River alluvial aquifer. We calculate the Euclidean distance between any two inverted models (both TI-based and field-based) and visualize the results in a 2D space using multidimensional scaling (MDS) [2]. This represents about 60 % of the total variance. The field data models fall among the distribution of the synthetic models on the MDS-map (Fig. 2a), hence the presented TIs provide geophysical models close to observed field models. TIs are thus consistent with the available geophysical data.

In a second step, we perform a cluster analysis on the MDS-map to highlight which parameters are most impacting the geophysical response. In cluster 1 and 3,

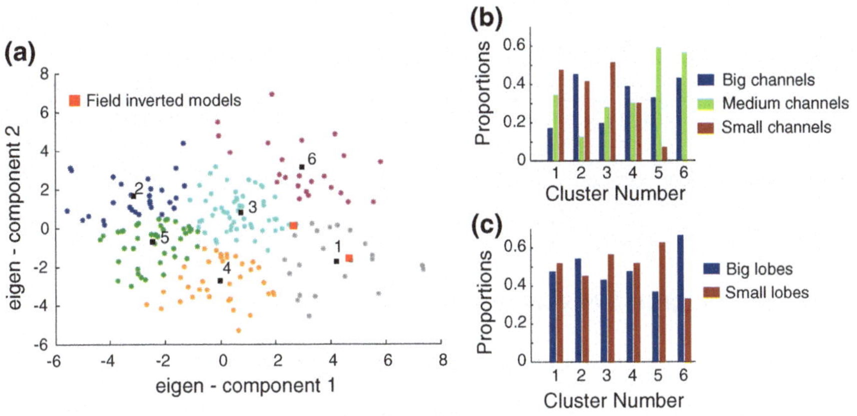

Fig. 2 **a** Multidimensional scaling map of TI-based and field models zoned in six clusters. **b** Proportion of models with big, medium and small gravel channels in each cluster. **c** Proportion of models with big and small clay lobes in each cluster

where field models are located, small channels are more abundant, suggesting that this kind of training image is more consistent with field data (Fig. 2b). In contrast, the size of clay lobes (Fig. 2c) is not a sensitive parameter since the proportions of models with big lobes and small lobes are close to 0.5 in clusters 1 and 3.

4 Assessing Probability of Scenarios

Based on this analysis, the probability of each geological scenario in the case of field data response is computed through kernel smoothing of the densities of models in the 2D MDS-map [3]. The aim is to calculate, through Bayes' rule, the probability $P(\Theta = \theta_i | \mathrm{D} = \mathrm{d})$ of a TI-based scenario θ_i given specific geophysical data d, i.e. a model of electrical resistivity in this example.

The likelihood of observing a specific geophysical data response d given a TI-based scenario θ_i, needed to apply Bayes' formula, is computed through kernel smoothing, where the kernel density estimation functions are approximated by a bivariate normal distribution using the coordinates of models of the 2D MDS-map.

The probability of each TI-based scenario is computed for the two selected ERT field models and shown in Table 1. The prior probability of each TI-based scenario is equal to 1/6, i.e. 0.1667. Scenarios with probabilities inferior to this value, e.g. scenarios with big channels, are less likely to occur. Geophysical data enable us to derive the most probable TI-based scenario for each particular field case: small channels and small lobes for the first; medium channels and big lobes for the second.

Table 1 Conditional probabilities for the 6 TI-based scenarios according to two different true field models based on the 2D-MDS map

Field data	BC/BL	BC/SL	MC/BL	MC/SL	SC/BL	SC/SL
Field model 1	0.0472	0.1061	0.1618	0.0613	0.1951	*0.4285*
Field model 2	0.0157	0.1497	*0.4259*	0.0252	0.1624	0.2211

(BC: big channels, MC: medium channels, SC: small channels/ BL: big lobes, SL: small lobes). The most probable TI-based scenarios are shown with italic values

5 Conclusion and Perspectives

We developed a methodology to verify the consistency of TI-based geological scenarios with geophysical models. This requires the specification of realistic values for the considered geophysical parameters and the choice of an adapted distance calculation to compare models using multidimensional scaling. In this case, we chose the Euclidean distance between models, but any measure, or combination of measures, or other geophysical attributes could be used.

For a single geological scenario, if a field model falls in the distribution of synthetic models, we state that this scenario is consistent with geophysical information. If it is not the case, one should try different scenarios to obtain a consistent training image. For multiple scenarios, a sensitivity analysis can be carried out through clustering to derive the most sensitive parameters. The probability of each geological scenario can be computed, giving some highlights for the selection of the best TI(s) to use for further simulations.

The methodology could be easily extended to 3D models, enabling to evaluate the consistency of more complex scenario. For example, it may be used to distinguish scenarios with channels from scenarios with elongated bars. However, it would require 3D field models which are more time consuming and more expensive to acquire.

If the consistency between TI and geophysics is confirmed, geophysical data may be further used as soft conditioning data in multiple-point statistics simulations or, the given training images and their posterior probabilities can be used on a Markov chain type inversion.

References

1. Gouw, M. J. P., & Erkens, G. (2007). Architecture of the Holocene Rhine-Meuse delta. (the Netherlands) - A result of changing external controls. *Geologie en Mijnbouw-Netherlands Journal of Geosciences*, *86*, 23–54.
2. Caers, J. (2011). *Modeling Uncertainty in the Earth Sciences*. Chichester: Wiley.
3. Park, H., Scheidt, C., Fenwick, D., Boucher, A., & Caers, J. (2013). History matching and uncertainty quantification of facies models with multiple geological interpretations. Computers and Geosciences. In Press. doi:10.1007/s10596-013-9343-5

Reservoir Modeling Combining Geostatistics with Markov Chain Monte Carlo Inversion

Andrea Zunino, Katrine Lange, Yulia Melnikova, Thomas Mejer Hansen and Klaus Mosegaard

1 Introduction

Reservoir modeling conditioned by recorded seismic reflection data is the most prominent geophysical technique to investigate the unknown properties of the subsurface. However, even if seismology produces good quality tomographic images, it still remains challenging to obtain a good picture of some particular properties such as porosity or permeability that are of most interest for oil and gas exploration. The link between elastic parameters and such properties lies in the complex relationships between, among others, intrinsic properties of rocks, mineralogy, and interaction with fluids which are usually described by a rock physics model [1]. Since these relationships are usually nonlinear and affected by uncertainty, it is difficult to invert seismic data directly for, e.g., porosity employing the standard optimization approaches because they generally rely on linearised models and simple scaling laws. Here we propose an approach based on a Markov chain Monte Carlo (McMC) technique which is able to combine rock physics modeling and reflection seismology to invert for porosity and facies of the subsurface. It takes into account the nonlinearities deriving from the rock physics model and moreover it provides an estimation of uncertainties on the unknown properties. Similar approaches have been studied before, see e.g., [2–4].

2 Overview of the Markov Chain Monte Carlo Inverse Method

We follow a probabilistic approach, in which all information is represented by probabilities, as described in [5], where the inverse problem consists in performing an

A. Zunino (✉) · K. Lange · Y. Melnikova · T. M. Hansen · K. Mosegaard
Department of Mathematical and Computational Geoscience (DTU Space) and CERE, Technical University of Denmark, Lyngby, Denmark
e-mail: anzu@dtu.dk

E. Pardo-Igúzquiza et al. (eds.), *Mathematics of Planet Earth*,
Lecture Notes in Earth System Sciences, DOI: 10.1007/978-3-642-32408-6_148,

indirect measurement of unobservable parameters of the subsurface given some measured quantities on the surface of the Earth. The solution to the inverse problem is the posterior distribution, a combination of the prior and likelihood functions describing all possible models and relative probabilities.

Our aim is then to explore the model space in order to obtain a collection of models which all fit the measured data and are consistent with the a priori information. Moreover we are interested in estimating the uncertainty on unknown model parameters. Markov chain Monte Carlo algorithms represent a natural choice to fulfill these requirements, so we construct a multi-step algorithm capable of sampling the posterior distribution. The ingredients necessary to sample solutions to this inverse problem are essentially two [6]: (I) an algorithm generating samples from a proposal distribution according to the available prior information and (II) a sampler of the likelihood function. The prior geological information is represented by one or multiple training images which supply the necessary information about geological patterns to the algorithm. The posterior distribution is finally sampled employing the extended Metropolis algorithm [6, 7] based on the degree of fit between measured and calculated seismograms. We consider Gaussian uncertainties and hence we utilize an L_2-norm for the misfit function.

Importance of Informed Priors: Geostatistics

One difficulty arising in high-dimensional space sampling is that a tremendous computational effort is needed to properly sample the posterior distribution. The huge size of model space, in fact, hampers the adoption of this kind of methodology in several cases. However, the use of proper informed priors can significantly improve the situation, reducing drastically the size of the model space to be sampled. This is obtained by employing an algorithm which generates models adhering to the prior knowledge so that only plausible models are taken into account in the sampling process. One recently introduced technique consists in generating realizations of a model exploiting the multiple-point statistics contained in prototype models. Specifically, the sequential Gibbs sampling method (see [8] and references therein) uses a sequential simulation approach where the algorithm learns the statistics from a training image which is scanned searching for recurring patterns. In principle, to increase the number of patterns, multiple training images may be used. A randomly selected hyper-rectangular volume of the model is then chosen to be re-simulated at each iteration of the Markov chain to propose a new model, where voxels are re-computed using sequential simulation conditioned on the rest of voxels [9].

3 Numerical Experiments

The target of our study is a synthetic reservoir model derived (but modified) from the Stanford VI-E model [10]. It consists of a 3D arrangement of $38 \times 50 \times 20$ voxels with size of 100, 100 and 4 m each respectively. Each voxel is parameterised with facies and porosity as the unknown parameters. Using the reservoir model derived from the Stanford VI-E model we constructed some "synthetic observations" by computing the seismograms to be inverted. In our case the forward model calculation consists of several steps. The first is the computation of the elastic properties from the facies and porosity of the subsoil. Then we compute the synthetic seismograms using a convolution approach.

The target zone of the reservoir is constituted by two facies, one representing sand (channel in a fluvial deposition system and oil-saturated) and the other representing shale (floodplain and brine-saturated). We assume the mineralogy to be known and describe it as consisting of four minerals (clay, quartz, feldspar, rock fragments) with known volume fraction in each facies but unknown porosity. The link between porosity and other petrophysical properties with the elastic moduli of the bulk rock for sand facies is modeled using the constant cement model [11] and the usual formula for isotropic V_P. An empirical law from [12] is used instead to compute V_P for shale facies.

Seismic modeling is carried out in the framework of the acoustic approximation, where the basic ingredients are the P-wave velocity and the density model. The seismic data are "recorded" at the surface on top of each pixel column as a zero-offset section. This in reality can correspond to data recorded at different source-receiver offset that have been processed such that they represent an equivalent zero-offset section which is easier to interpret. The wavelet is constructed from a Ricker function with 50 Hz peak frequency and is assumed to be known in the inversion process.

4 Results and Discussion

We ran $2 \cdot 10^6$ iterations, obtaining about $7 \cdot 10^5$ models, of which only one every 10^2 was retained to ensure independence of samples. Figure 1a shows one particular model from the solutions. We ended up with a collection of models representing samples of the posterior distribution which can be used to estimate subsurface properties and their relative probabilities/uncertainties. The solutions are used as a database that can be queried to obtain information on several different aspects since it represents the complete solution of the inverse problem. Here we show two examples of the kind of information which can be retrieved from the collection of models. The first is to compute the value of porosity at two different locations, obtaining histograms of possible values (Fig. 1b). The histogram tells us which range of values is most probable and, moreover, gives us an estimation of the uncertainty. The two histograms show a different behavior, one having a more pronounced peak, reflecting the different degree of resolving power. The second example is a map of the probability of having the sand facies on a slice of the 3D model at $z = 40$ m (Fig. 1c). The continuity of structures depicted in Fig. 1c is due to the prior information deriving from the geostatistical algorithm which takes into account the spatial continuity present in the training image. This example shows

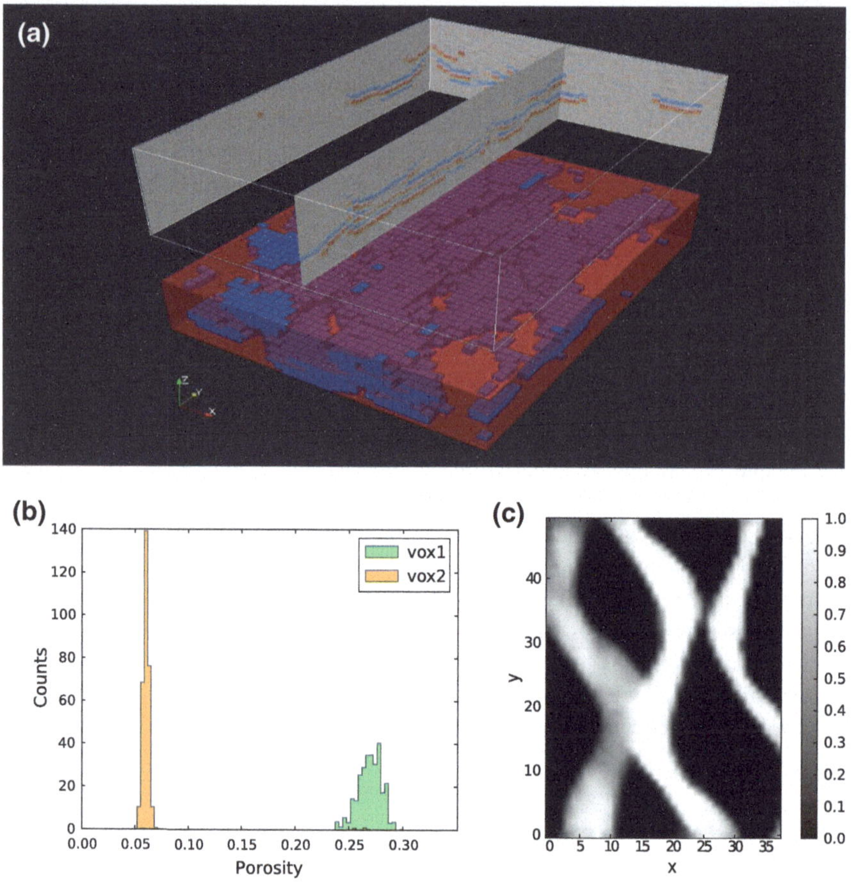

Fig. 1 **a** An example of a two-facies reservoir model from the collection of solutions with some slices through the volume of observed seismograms plotted on *top*. **b** Histogram of porosity for two voxels, one located at $(x, y, z) = (1500, 3500, 20)$ m and the other at $(1000, 1000, 48)$ m. **c** Probability of having sand (and hence a channel) on a 2D slice of the model at $z = 40$ m

how it is possible to retrieve more sophisticated information from the database of solutions that can result very useful for real problems applications. Again, the uncertainty, clearly imaged in this probability plot, is an integral part of the answer we were searching for.

References

1. Mavko, G., Mukerji, T., & Dvorkin, J. (2003). *The rock physics handbook*. Cambridge: Cambridge University Press.

2. González, E., Mukerji, T., & Mavko, G. (2008). Seismic inversion combining rock physics and multiple-point geostatistics. *Geophysics*, *73*(1), R11–R21.
3. Bosch, M., Mukerji, T., & Gonzalez, E. (2010). Seismic inversion for reservoir properties combining statistical rock physics and geostatistics: a review. *Geophysics*, *75*(5), 75A165–75A176.
4. Rimstad, K., & Omre, H. (2010). Impact of rock-physics depth trends and markov random fields on hierarchical bayesian lithology/fluid prediction. *Geophysics*, *75*(4), R93–R108.
5. Tarantola, A. (2005). *Inverse problem theory and model parameter estimation*. Philadelphia: SIAM.
6. Mosegaard, K., & Tarantola, A. (1995). Monte carlo sampling of solutions to inverse problems. *Journal of Geophysical Research*, *100*(B7), 12431–12447.
7. Metropolis, N., Rosenbluth, A., Rosenbluth, M., Teller, A., & Teller, E. (1953). Equations of state calculations by fast computing machines. *Journal of Chemical Physics*, *21*(6), 1087–1092.
8. Hansen, T., Cordua, K., & Mosegaard, K. (2012). Inverse problems with non-trivial priors: efficient solution through sequential gibbs sampling. *Computational Geosciences*, *16*, 593–611.
9. Strebelle, S. (2002). Conditional simulation of complex geological structures using multiple-point statistics. *Mathematical Geology*, *34*, 1–21.
10. Castro, S., Caers, J., & Mukerji, T. (2005). The Stanford VI reservoir: 118th annual report. Technical report. Stanford Center for Reservoir Forecasting, Stanford University, Palo Alto, CA.
11. Dvorkin, J., Nur, A., & Yin, H. (1994). Effective properties of cemented granular materials. *Mechanics of Materials*, *18*(4), 351–366.
12. Gardner, G., Gardner, L., & Gregory, A. (1974). Formation velocity and density: the diagnostic basics for stratigraphic traps. *Geophysics*, *39*(6), 770–780.

Geostatistical Inversion of 3D Post-stack Seismic and Well Data for the Characterization of Acoustic Impedance in Oil Fields

Fernando Alves, José A. Almeida and António Ferreira

1 Introduction

The 3D seismic inversion of petroleum reservoirs is a challenging problem, because the reservoirs are extensive in area, are often geologically complex, and in the early stages of characterization are sampled with a very limited number of wells where the seismic data are the only source of information [1].

Seismic inversion based on a model begins with a set of initial values of acoustic impedance, which are then modified until a good balance (correlation) between the actual seismic traces and the synthetic ones is found. This synthetic seismic trace is also known as a 1D convolutional model and is the basis of the proposed approach. Stochastic inversion based on geostatistics is recognised to be the best way to combine the seismic data and the data from the wells to obtain a cube of acoustic impedances that better represents the geology at depth.

In the present work a new geostatistical inversion approach, based on co-simulation from a cube of acoustic impedances obtained using a deterministic method, is formulated and tested. The initial deterministic solution is considered the optimal solution for the problem of the spectrum of acoustic impedance values available in the wells, seismic models, and real wavelets. In this context, we highlight the previous studies of [2, 3] and [4].

The approach involves the following steps: (1) Generate an initial deterministic acoustic impedance image (AI) and compute coefficients of reflection trace by trace (optimal solution); (2) A modified version of the sequential simulation and

F. Alves (✉) · J. A. Almeida · A. Ferreira
CICEGe, FCT-Universidade Nova de Lisboa, 2829-516 Caparica, Portugal
e-mail: fma18226@gmail.com

J. A. Almeida
e-mail: ja@fct.unl.pt

A. Ferreira
e-mail: amp-ferreira@fct.unl.pt

E. Pardo-Igúzquiza et al. (eds.), *Mathematics of Planet Earth*,
Lecture Notes in Earth System Sciences, DOI: 10.1007/978-3-642-32408-6_149,

co-simulation algorithm [5, 6] is applied to generate AI images conditioned on the reflection coefficients calculated via the deterministic image of AI and the AI observed in the wells; (3) The results are validated with particular regard to the correlation between the actual seismic data and synthetic seismic data obtained by convolution of the simulated AI images with the wavelet.

A modified version of the sequential simulation/co-simulation algorithm is the key feature of the proposed methodology namely the definition of the random path. As the seismic amplitudes are dependent on the entire trace of AI, simulation of AI within each trace begins in a random location at depth but the remaining nodes in the trace are selected up or down from the previous simulated nodes. If the node is the first of the trace a simulation of AI is performed, otherwise a co-simulation is conducted. The local value (soft data) is estimated via the homologous local reflection coefficient of the deterministic image taking into account the value at the node immediately up or down the previously simulated plus a local noise introduced as a local correlation value in co-simulation. This constraint in the path makes it possible to impose local reflection coefficients calculated from the deterministic solution maintaining the purpose of simulation in following a random path.

2 Case Study

2.1 Initial Data and Deterministic Inversion of AI

The initial dataset consisted of a 3D cube of seismic amplitudes (Fig. 1, Left), acquired through a campaign of seismic reflection and logs of 10 wells with AI, density, velocity, and porosity with a vertical resolution 1 ms, the same as the seismic data. The deterministic solution of the AI model based is illustrated in Fig. 1, Right.

Reflection coefficients were extracted from the AI deterministic solution (Fig. 2, Left). Synthetic seismic data obtained from AI well data were computed at well

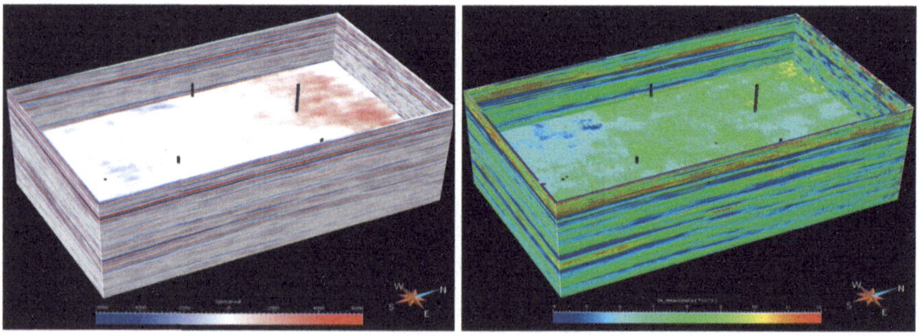

Fig. 1 *Left* 3D cube of seismic amplitudes. *Right* Deterministic model-based solution of AI

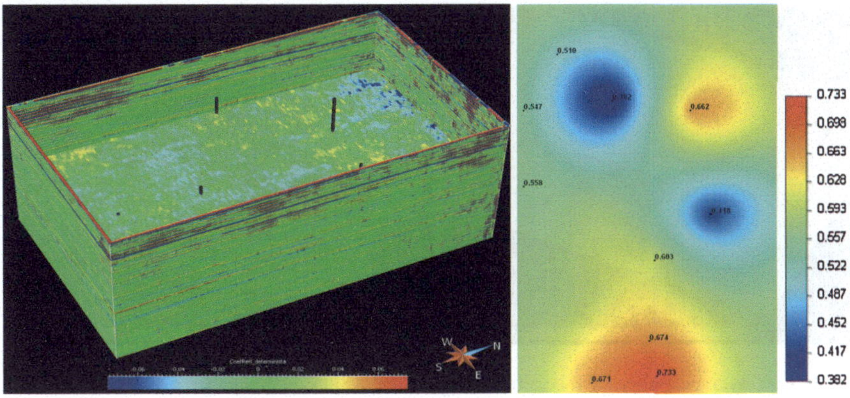

Fig. 2 *Left* Reflection coefficients of the deterministic AI. *Right* Aerial map of local correlation coefficients

locations and the local correlation between synthetic and actual data was calculated. Local correlation coefficients values were estimated by kriging over the entire area of the reservoir (Fig. 2, Right).

2.2 Stochastic Seismic Inversion of AI

A set of 200 AI realizations were simulated conditioned on AI well data and the reflection coefficients of Fig. 2 (Left). Figure 3 (Left) illustrates a 3D view of one realization of AI. For each image, local correlations between synthetic seismic data

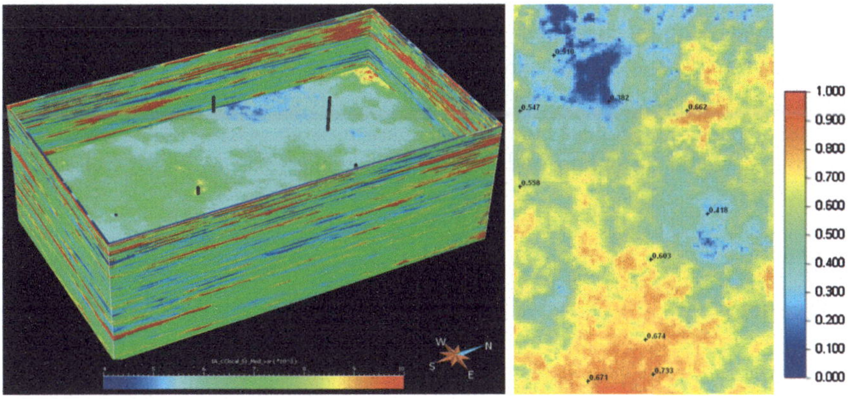

Fig. 3 *Left* One stochastic simulated image of AI. *Right* Average local correlation coefficients between synthetic seismic from AI simulated images and actual data

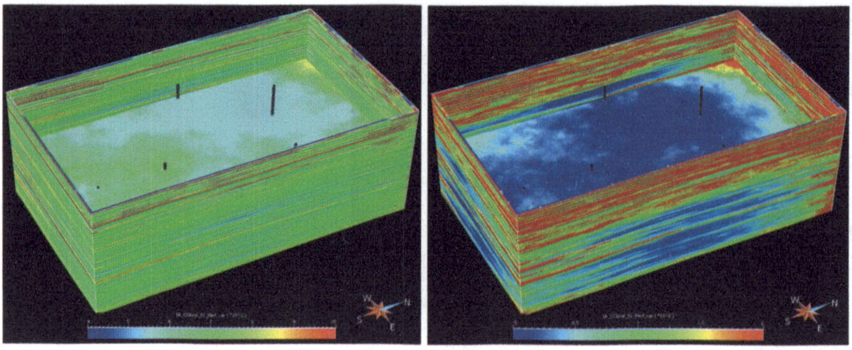

Fig. 4 *Left* Average image of the 200 realizations of AI. *Right* Image of variance of the 200 realizations of AI

and actual seismic data were also calculated and averaged (Fig. 3, Right). It is important to emphasize that the map of local correlations illustrated in Fig. 3, Right is similar to the objective map of correlations displayed in Fig. 2, Right.

Figure 4 shows an average image of AI from the 200 realizations (Left) and the corresponding image of variance (Right). The average image of AI is independent of the deterministic image but they coincide in the neighbours of the well data. It is important to remark that the AI solutions are very sensitive to reflection coefficients of the deterministic solution but it is not sensitive to the AI solution itself witch is a key issue in what concerns the efficient exploitation of the uncertainty space.

3 Final Remarks

The advantages of this proposed methodology include the following: (1) It is an efficient method. For a simulated image of AI it is necessary to make only a single simulation run. (2) The method explores the space of uncertainty. This is because the simulation is partly conditioned on the reflection coefficients obtained from the deterministic image. (3) It produces the general characteristics of a simulation, including histogram data, the variogram, and the coincidence of grid data with the well data at well locations.

References

1. Almeida, J. A. (2010). Stochastic simulation methods for characterization of lithoclasses in carbonate reservoirs. *Earth Science Reviews*, *101*(3–4), 250–270.
2. Soares, A., Diet, J. D., & Guerreiro, L. (2007). *Stochastic inversion with a global perturbation method. petroleum geostatistics*. Cascais, Portugal: EAGE.

3. Caetano, H. M. V. (2009). Integration of seismic information in reservoir models: Global Stochastic Inversion. PhD Thesis, IST-UTL.
4. Azevedo, L., Nunes, R. F., Almeida, J. A., Pinheiro, L. M., Caeiro, M. H., Correia, P. J., et al. (2012). Seismic attributes for constraining geostatistical seismic inversion. *Proceedings of the Ninth International Geostatistics Congress*. Oslo, Norway.
5. Alves, F. (2010) Inversão Estocástica de Sísmica 3D Post-Stack e Dados de Poço para a Caracterização de Reservatórios e Modelação de Propriedades. MSc thesis FCT-UNL.
6. Nunes, R., & Almeida, J. A. (2010). Parallelization of sequential Gaussian, indicator and direct simulation algorithms. *Computers & Geosciences*, *36*(8), 1042–1052.

The Effect of the Noise and the Regularization in Inverse Problems: Geophysical Implications

José Luis García Pallero, Juan Luís Fernández-Martínez, Zulima Fernández-Muñiz and Luis Mariano Pedruelo-González

1 Effect of the Noise in Nonlinear Inverse Problems

It has been shown in Fernández Martínez et al. [2] that in the case of nonlinear inverse problems the nonlinear region of equivalence has a curvilinear valley shape, and depending on the injectivity of the forward operator several basins of low misfit can coexist in the cost function landscape. To explore in detail how the noise affects the cost function landscape for this kind of problems, let us adopt a linearization of the forward operator $\mathbf{F}$ in a model $\mathbf{m}_0$ that belongs to the nonlinear equivalent region of value tol:

$$V_{tol} = \left\{\mathbf{m} : c_p(\mathbf{m}) = \|\mathbf{F}(\mathbf{m}) - \mathbf{d}^{obs}\|_2^2 \leq tol^2\right\}, \tag{1}$$

where $\mathbf{F} : \mathbb{R}^n \to \mathbb{R}^s$, $\mathbf{m} \in \mathbb{R}^n$, $\mathbf{d}^{obs} = \mathbf{d}^{true} + \delta\mathbf{d} \in \mathbb{R}^s$, and $\delta\mathbf{d}$ represents the noise. We have

$$\mathbf{F}(\mathbf{m}) = \mathbf{F}(\mathbf{m}_0) + \mathbf{JF}_{\mathbf{m}_0}(\mathbf{m} - \mathbf{m}_0) + o(\|\mathbf{m} - \mathbf{m}_0\|_2). \tag{2}$$

Substituting Eq. (2) into Eq. (1) the linearized hyper-quadric L_{tol} of value tol in $\mathbf{m}_0$ is

$$\Delta\mathbf{m}^T \mathbf{JF}_{\mathbf{m}_0}^T \mathbf{JF}_{\mathbf{m}_0} \Delta\mathbf{m} + 2(\Delta\mathbf{d} - \delta\mathbf{d})^T \mathbf{JF}_{\mathbf{m}_0} \Delta\mathbf{m} + \|\Delta\mathbf{d} - \delta\mathbf{d}\|_2^2 < tol^2, \tag{3}$$

J. L. G Pallero (✉)
ETSI en Topografía, Geodesia y Cartografía, Universidad Politécnica de Madrid, Madrid, Spain
e-mail: jlg.pallero@upm.es

J. L. Fernández-Martínez · Z.Fernández-Muñiz · L. M. Pedruelo-González
Departamento de Matemáticas, Universidad de Oviedo, Oviedo, Spain
e-mail: jlfm@uniovi.es

Z. Fernández-Muñiz
e-mail: zulima@uniovi.es

L. M. Pedruelo-González
e-mail: marianopg@educastur.princast.es

E. Pardo-Igúzquiza et al. (eds.), *Mathematics of Planet Earth*,
Lecture Notes in Earth System Sciences, DOI: 10.1007/978-3-642-32408-6_150,

where $\Delta\mathbf{m} = \mathbf{m} - \mathbf{m}_0$, $\mathbf{JF}_{\mathbf{m}_0}$ is the Jacobian of $\mathbf{F}$ in $\mathbf{m}_0$, and $\Delta\mathbf{d} = \mathbf{F}(\mathbf{m}_0) - \mathbf{d}^{true}$.

Taking into account [1] the singular value decomposition $\mathbf{JF}_{\mathbf{m}_0} = \mathbf{U}\Sigma_{\mathbf{m}_0}\mathbf{V}^T$, Eq. (3) can be written as

$$\Delta\mathbf{m}_V^T \Sigma_{\mathbf{m}_0}^T \Sigma_{\mathbf{m}_0} \Delta\mathbf{m}_V + 2\Delta\mathbf{b}_U^T \Sigma_{\mathbf{m}_0} \Delta\mathbf{m}_V + \|\Delta\mathbf{b}_U\|_2^2 < tol^2, \tag{4}$$

where $\Delta\mathbf{b}_U = \Delta\mathbf{d}_U - \delta\mathbf{d}_U = \mathbf{U}^T\left[\mathbf{F}(\mathbf{m}_0) - \mathbf{d}^{true} - \delta\mathbf{d}\right]$ and $\Delta\mathbf{m}_V = \mathbf{V}^T\Delta\mathbf{m}$.

Calling $r = \operatorname{rank}\left(\Sigma_{\mathbf{m}_0}\right)$ and μ_k the corresponding singular values of $\Sigma_{\mathbf{m}_0}$, the linearized equivalent region of value tol in model $\mathbf{m}_0$ is then

$$\sum_{k=1}^{r} (\mu_k \Delta m_{Vk} + \Delta b_{Uk})^2 = tol^2 - \sum_{k=r+1}^{s} \Delta b_{Uk}^2, \tag{5}$$

and for the free-noise case $\delta\mathbf{d} = \mathbf{0}$

$$\sum_{k=1}^{r} (\mu_k \Delta m_{Vk} + \Delta d_{Uk})^2 = tol^2 - \sum_{k=r+1}^{s} \Delta d_{Uk}^2. \tag{6}$$

The tol equivalent region in the noisy case, called now $tol^*(\mathbf{m}_0)$, will have the same size as the tol region for the free-noise case under the condition

$$tol^{*2}(\mathbf{m}_0) - \sum_{k=r+1}^{s} \Delta b_{Uk}^2 = tol^2 - \sum_{k=r+1}^{s} \Delta d_{Uk}^2, \tag{7}$$

then

$$tol^*(\mathbf{m}_0) = \sqrt{tol^2 + \sum_{k=r+1}^{s} \left(\delta d_{Uk}^2 - 2\Delta d_{Uk}\delta d_{Uk}\right)}. \tag{8}$$

This analysis depends on the level of the noise and on the model $\mathbf{m}_0$ that is considered. By continuity of the cost function there might exist a neutral line where $tol^*(\mathbf{m}_0; \delta\mathbf{d}) = tol(\mathbf{m}_0)$, so along it no deformation takes place. Inside the region limited by this line we will have $tol^*(\mathbf{m}_0; \delta\mathbf{d}) > tol(\mathbf{m}_0)$, and regions decrease in size; and outside $tol^*(\mathbf{m}_0; \delta\mathbf{d}) < tol(\mathbf{m}_0)$, that is, regions increase in size. In conclusion, noise deforms the topography of the cost function non-homogeneously, decreasing regions of low misfits and eventually increasing the regions of medium-high misfits. The higher order terms (particularly the Hessian) that have not been taken into account in Eq. (2) also induce a non-homogeneous deformation of the nonlinear equivalence region.

In the case of linear inverse problems the deformation due to noise is homogeneous, and tol^* does not depend on any model $\mathbf{m}_0$. Eq. (8) becomes

$$tol^* = \sqrt{tol^2 + \sum_{k=r+1}^{s} \left(\delta d_{Uk}^2 + 2d_{Uk}^{obs}\delta d_{Uk}\right)}, \tag{9}$$

2 Effect of the Regularization in Nonlinear Inverse Problems

The nonlinear equivalent region of value tol for the zero-order Tikhonov regularized inverse problem is

$$V_{tol} = \left\{\mathbf{m} : c_p(\mathbf{m}) = \|\mathbf{F}(\mathbf{m}) - \mathbf{d}^{obs}\|_2^2 + \varepsilon^2\|\mathbf{m} - \mathbf{m}^{ref}\|_2^2 \leq tol^2\right\}, \tag{10}$$

where ε^2 is the regularization parameter, and $\mathbf{m}^{ref}$ is a reference model. Performing a similar analysis as in Sect. 1, we arrive to the expression for the linearized hyper-quadric L_{tol}

$$\sum_{k=1}^{r}\left(\sqrt{\mu_k^2+\varepsilon^2}\Delta m_{Vk} - \frac{\varepsilon^2\Delta m_{Vk}^{ref} - \mu_k\Delta b_{Uk}}{\sqrt{\mu_k^2+\varepsilon^2}}\right)^2 + \varepsilon^2\sum_{k=r+1}^{s}\left(\Delta m_{Vk} - \Delta m_{Vk}^{ref}\right)^2 =$$
$$= tol^2 + \sum_{k=1}^{r}\frac{\left(\varepsilon^2\Delta m_{Vk}^{ref} - \mu_k\Delta b_{Uk}\right)^2}{\mu_k^2+\varepsilon^2} - \|\Delta\mathbf{b}_U\|_2^2 - \varepsilon^2\sum_{k=1}^{r}\Delta m_{Vk}^{ref2}. \tag{11}$$

Taking into account that the hyper-quadric for the non-regularized problem corresponds to Eq. (5), the tol equivalent region in the regularized case, called now $tol^*(\mathbf{m}_0)$, will have the same size than the tol region for the non-regularized case under the condition

$$\frac{\sqrt{tol^{*2}(\mathbf{m}_0) + \Delta}}{\sqrt{\mu_i^2+\varepsilon^2}} = \frac{\sqrt{tol^2 - \sum_{k=r+1}^{s}\Delta b_{Uk}^2}}{\mu_i}, \tag{12}$$

where

$$\Delta = \sum_{k=1}^{r}\frac{\left(\varepsilon^2\Delta m_{Vk}^{ref} - \mu_k\Delta b_{Uk}\right)^2}{\mu_k^2+\varepsilon^2} - \|\Delta\mathbf{b}_U\|_2^2 - \varepsilon^2\sum_{k=1}^{r}\Delta m_{Vk}^{ref2}. \tag{13}$$

Then we arrive to

$$tol^{*}(\mathbf{m}_0)_i = \sqrt{tol^2 + \frac{\varepsilon^2}{\mu_i^2}\left(tol^2 - \sum_{k=r+1}^{s} \Delta b_{Uk}^2\right) + \sum_{k=1}^{r} \frac{\varepsilon^2}{\mu_k^2 + \varepsilon^2}\left(\Delta b_{Uk} + \mu_k \Delta m_{Vk}^{ref}\right)^2}, \quad (14)$$

showing that the deformation of the linearized region of equivalence for the regularized problem is non-homogeneous and anisotropic, depending on the values of each principal axis stated by μ_i, and also on the selected model $\mathbf{m}_0$. Tikhonov's regularization might also be responsible for the introduction of local minima in the cost function landscape [2].

3 Implications in Applied Geophysics and Conclusions

The typical situation in geophysical inversion is to deal with data that has finite dimension, is insufficient in number, and always inaccurate due to the effect of noise. These features together with the modelling and numerical errors cause the inverse solution to be uncertain, and the geophysical inverse problem has to be solved in two different stages: solution optimization/search and appraisal. The appraisal stage (or uncertainty analysis of the solution) is motivated by the fact that the observed data does not contain enough information to determine a geophysical plausible unique solution. Practitioners tend to underestimate the uncertainty of inverse problems solutions relying on the fact that: (1) Uncertainty has in any case a random structure impossible to be known. (2) Regularization techniques are the panacea and cause the disappearance of the equivalent solutions. (3) Linearization techniques can provide a precise idea about the solution that has been adopted. All these three assumptions are not very precise, motivating in some cases the decisions that are taken based on these solutions are also incorrect, inducing big economic loses.

This paper provides new insights to understand uncertainty in the solution of inverse problem from a deterministic point of view. Further research is needed to improve the understanding about the uncertainty of the solution in inverse problems.

References

1. Aster, R. C., Borchers, B., & Thurber, C. H. (2012). *Parameter estimation and inverse problems*. Oxford: Academic Press.
2. Fernández Martínez, J. L., Fernández Muñiz, M. Z., & Tompkins, M. J. (2012). On the topography of the cost functional in linear and nonlinear inverse problems. *Geophysics*, *77*(1), W1–W15.

A Generalized Local Gradual Deformation Method for History Matching

Benjamin Marteau, DidierYu Ding and Laurent Dumas

1 Introduction

There is a large number of physical phenomena which lead to the presence of multiple scale heterogeneities in an oil field. Erosion and sedimentation for instance can be responsible for the creation of large geological structures such as large channels and meanders whereas lateral accretion and meter-scale cross-bedding can be observed more locally inside these structures. At the same time, the data available to characterize petrophysical properties of the reservoir are very sparse. In consequence, the complexity resulting from the numerous heterogeneities is very difficult to model. To improve the quality of a geological/geostatistical model, one can use dynamic data to reduce the uncertainty through history matching. To optimize the geostatistical model, parametrization techniques are usually used. In this work, we will give a very brief review of the gradual deformation method [3] and the domain deformation technique [2] and then propose a new approach which generalizes the gradual deformation by combining its advantages with those of the domain deformation.

B. Marteau
IFPEN UVSQ, Rueil Malmaison, France
e-mail: benjamin.marteau@ifpen.fr

D. Y. Ding
IFPEN, Rueil Malmaison, France
e-mail: didier-yu.ding@ifpen.fr

L. Dumas
UVSQ, Versailles, France
e-mail: Laurent.dumas@uvsq.fr

E. Pardo-Igúzquiza et al. (eds.), *Mathematics of Planet Earth*,
Lecture Notes in Earth System Sciences, DOI: 10.1007/978-3-642-32408-6_151,

2 A Generalized Gradual Deformation Technique

Geostatistical methods such as the Fast Fourrier Transform Moving Average [4] allow us to uncouple uncorrelated realizations from structured information (mean, variance, correlation length,...). A realization M of the geostatistical model is linked to a Gaussian white noise Z by an operator G: $M = G(Z)$.

2.1 Gradual Deformation Method

The local gradual deformation method [3] consists in combining locally several Gaussian white noises to obtain a new one. For example, when combining two realizations associated to the independant Gaussian white noises Z_0 and Z_1, we get:

$$Z = cos(t)Z_0 + sin(t)Z_1 \tag{1}$$

where Z is still a Gaussian white noise and can legitimately be used to generate a realization of the reservir model. This method can be extended to combine $(m + 1)$ Gaussian white noises by introducing m gradual deformation parameters.

2.2 Domain Deformation Method

Instead of combining two realizations in a given zone R_1, the domain decomposition method [2] uses one realization inside the zone, the other outside, and we parameterize the shape and size of the domain (Fig. 1) with a set of parameters t. The new Gaussian noise is therefore written by

$$Z(X) = \cos\left(\frac{\pi}{2}\alpha_{X,1}(t)\right) Z_0(X) + sin\left(\frac{\pi}{2}\alpha_{X,1}(t)\right) Z_1(X) \tag{2}$$

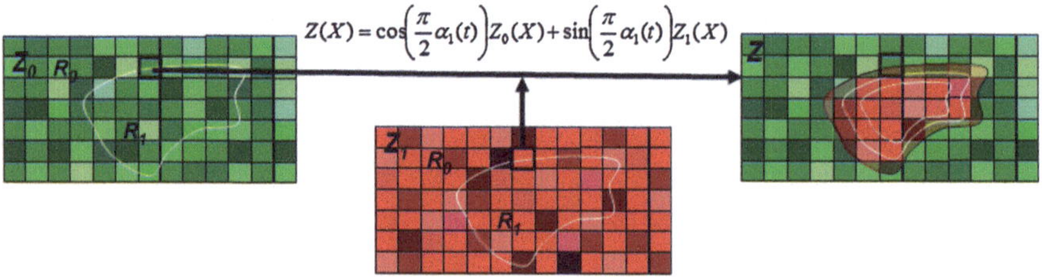

Fig. 1 Domain deformation

for each grid cell X with $\alpha_{X,1}(t) = \frac{V_1(t)}{V_0(t)+V_1(t)}$, $V_1(t)$ the volume of the grid cell X inside the zone R_1 and V_0 its complementary. $\alpha_{X,1}(t)$ is proportional to the volume of the grid cell X inside the domain. In particular, $\alpha_{X,1}(t) = 0$ if the grid cell is completely outside the domain, $\alpha_{X,1}(t) = 1$ if the grid cell is completely inside the domain. A new model realization is therefore created depending on the variation of the shape and size of the domain. This method is generalizable to n domains.

2.3 Generalization of the Local Gradual Deformation Method

The local gradual deformation method is only efficient if the local domains are well defined. However, there is currently no way to choose automatically good domains, the method is therefore dependant on an initial expert guess to define domains. On the other hand, the domain deformation method is a good approach to define suitable domains as shown in [2].

Our new method allows the optimization process to combine several realizations inside a domain while having the possibility to modify its shape and size. Let's start with a simple case where we want to combine 2 Gaussian white noises Z_0 and Z_1 in a single domain R_1 parameterized with t. $\alpha_{X,1}(t)$ is defined for each grid cell X as in the previous section. Similarly to the gradual deformation method, we define ρ_1^1 the parameter controlling the combination of the realizations inside R_1. We propose to optimize ρ and t for the following combination :

$$Z(X) = \cos(\pi\alpha_{X,1}(t)\rho_1^1)Z_0(X) + \sin(\pi\alpha_{X,1}(t)\rho_1^1)Z_1(X) \tag{3}$$

Here, the parameter $\alpha_{X,1}$ controls the shape and size of the domain while ρ_1^1 controls the combination of Z_0 and Z_1. Let's analyse some cases, consider a grid cell X :

- If X is completely outside R_1 then $\alpha_{X,1} = 0$ and $Z(X) = Z_0(X)$.
- If X is completely inside R_1, $\alpha_{X,1} = 1$ and $Z(X) = \cos(\pi\rho_1^1)Z_0(X) + \sin(\pi\rho_1^1)Z_1(X)$, which is exactly the gradual deformation method.
- If R_1 partially contains X, then a combination similar to the domain deformation method is applied.

In this simple case, we can see that this method actually allows a deformation of the domain simultaneously to the combination of 2 realizations inside R_1.

We can extend this one to deform $m+1$ model realizations inside n domains. Let's introduce $\alpha_{X,i}(t) = \frac{V_i}{\sum_{j=0}^{n} V_j}$ for each domain R_i and a gradual deformation parameter ρ_i^j for each couple domain/Gaussian white noise R_i/Z_j ($i = 1, \ldots, n$ and $j = 1, \ldots, m$). We then obntain a new model realization through the new Gaussian white noise Z :

$$Z(X) = \prod_{j=1}^{m} \cos(\pi \sum_{i=1}^{n} \alpha_{X,i}(t)\rho_i^j) Z_0(X) + \sum_{j=1}^{m} \sin(\pi \sum_{i=1}^{n} \alpha_{X,i}(t)\rho_i^j) \prod_{k=j+1}^{m} \cos(\pi \sum_{i=1}^{n} \alpha_{X,i}(t)\rho_i^k) Z_j(X) \quad (4)$$

We have here a new method which succesfully generalized the gradual deformation method and allowed a greater flexibility on the definition of the domains. By setting some of the gradual deformation parameters to constants, we can choose which domain to deform and which Gaussian white noise to combine in each zone. In addition, this approach is helpfull to initialize the model realization for optimizations in history matching.

This method was applied to a synthetic reservoir case. It was shown that in a given number of simulations, the greater flexibility of the method allowed us to decrease more the objective function despite having more parameters. Moreover, we can use optimization techniques adapted to partially separable functions to further decrease the computational cost [1].

References

1. Ding, D. Y., & McKee, F. (2013). Using partial separability of the objective function for gradient-based optimizations in history matching. *SPE Journal* (In press).
2. Ding, D. Y., & Roggero, F. (2010). History matching geostatistical model realizations using a geometrical domain based parameterization technique. *Mathematical Geosciences*, *42*(4), 413–432.
3. Hu, L. Y., Blanc, G., & Noetinger, B. (2001). Gradual deformation and iterative calibration of sequential stochastic simulations. *Mathematical Geology*, *33*(4), 475–489.
4. Le Ravalec, M., Noetinger, B., & Hu, L. Y. (2000). The FFT moving average (FFT-MA) generator: An efficient numerical method for generating and conditioning Gaussian simulations. *Mathematical Geology*, *32*(6), 701–723.

History Matching with Geostatistical Prior: A Smooth Formulation

Yulia Melnikova, Katrine Lange, Andrea Zunino, Knud Skou Cordua and Klaus Mosegaard

1 Introduction

History matching—inversion of reservoir production data for rock properties—is an ill-posed inverse problem with computationally expensive forward simulation. Highly non-linear relationship between data $\mathbf{d}^{\mathbf{obs}}$ and model parameters $\mathbf{m}$ result in non-uniqueness of solutions. With the aid of geostatistical prior information (usually in the form of training images **TI**) it has become possible to restrict the solution space drastically [1, 3]. The main challenge in history matching consists in minimizing the amount of forward simulations needed to achieve attractive solutions. In this work we present a new method for solving history matching problem using a probabilistic framework [7], searching for solutions deterministically.

2 Methodology

Our approach consists in integrating production data and prior information into a single *differentiable* objective function, minimizer of which has a high posterior value:

$$\mathbf{m}^{\text{HighPosterior}} = \underset{\mathbf{m}}{\operatorname{argmin}} \left\{ \frac{1}{2} ||\mathbf{d}^{\mathbf{obs}} - g(\mathbf{m})||^2_{C_D} + f^d(\mathbf{m}, \mathbf{TI}) \right\} \tag{1}$$

where the first term is a conventional data misfit term [7] and g represents forward simulation; the second term is the prior information misfit explained below (superscript d stands for differentiable). Solving Eq. 1 by a gradient-based technique for an

Y. Melnikova(✉) · E. Lange · A. Zunino · K. S. Cordua · K. Mosegaard
Department of Mathematical and Computational Geoscience (DTU Space) and CERE, Technical University of Denmark, Lyngby, Denmark
e-mail: yume@dtu.dk

E. Pardo-Igúzquiza et al. (eds.), *Mathematics of Planet Earth*,
Lecture Notes in Earth System Sciences, DOI: 10.1007/978-3-642-32408-6_152,

ensemble of starting models, we obtain a set of solutions with high-posterior value running a small number of forward simulations.

Lange et al. [4] define the prior misfit as the chi-square distance between multi-point statistics of the model and the training image, assuming both images to be discrete. Essentially, the statistics is the frequency distribution of multi-point patterns defined by a template. Contrary to Lange et al. [4], our formulation implies gradual change of the model parameters, therefore the prior misfit $f^d(\mathbf{m}, \mathbf{TI})$ should be defined for any continuous image. A continuous image can not be represented by the frequency distribution, however its differentiable approximation (a pseudo-histogram) can be computed.

Consider a set of size K of all unique discrete patterns observed in the training image. Then the pseudo-histogram of a continuous image is $H_j^{d,\mathbf{image}} = \sum_{i=1}^{N^{\mathbf{image}}} p_{ij}$, $j = 1, \cdots, K$, where $N^{\mathbf{image}}$ is the number of patterns in the continuous image and p_{ij} is a measure of similarity between continuous and discrete patterns:

$$p_{ij} = \frac{1}{(1 + Ad_{ij}^k)^s}. \tag{2}$$

Here $d_{ij} = ||pat_i^{\mathbf{image}} - pat_j^{\mathbf{TI},unique}||_2$ (Euclidean distance between pixel values of the corresponding patterns) and A, k, s are adjustable parameters. Per definition statistically similar images will have similar pseudo-histograms. Therefore we define the prior misfit as follows:

$$f^d(\mathbf{m}, \mathbf{TI}) = \sum_{j=1}^{K} \frac{(H_j^{d,\mathbf{TI}} - H_j^{d,\mathbf{m}})^2}{H_j^{d,\mathbf{TI}}}. \tag{3}$$

Use of the pseudo-histogram of the training image as a weight factor in Eq. 3 results in proper reproduction of the pattern statistics.

3 Numerical Example

We perform history matching on a 2D synthetic oil reservoir model. The goal is to obtain a set of permeability models having high posterior values. Others parameters, such as porosity, relative permeabilities and initial saturations, are assumed to be known. The reservoir model has 50 by 50 cells of size 10 by 10 m. Figure 1a shows the true permeability and locations of injectors (down-triangles) and producers (up-triangles). Production data are generated by running a forward simulation with the true model and applying 5 % Gaussian noise to the calculated water and oil rates. Physics of the flow (two-phase immiscible displacement) allows us to use few observations to perform history matching and spend less time computing sensitivities. We use only two measurements for each well (at 100 and 200 days), therefore 52

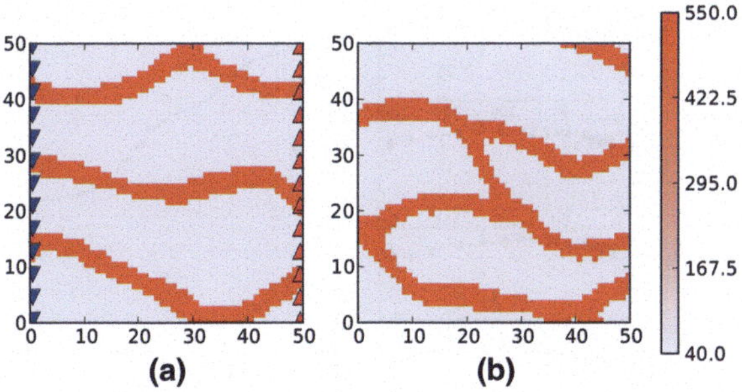

Fig. 1 **a** True permeability model (channels—500 mD, background—50 mD); **b** Training image (channels—500 mD, background—50 mD)

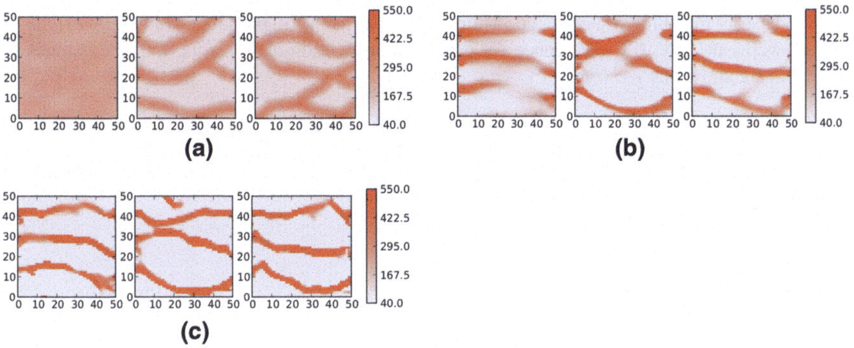

Fig. 2 **a** Starting permeability models; **b** Models after 30 iterations; **c** Solutions for each of the starting models above

measurements in total. However, we show the full history to assure the quality of history matching.

The prior information is given by a training image (Fig. 1b), which is an upscaled part of the Strebelle image [6]. A square template of 6×6 pixels is used for collecting pattern statistics. Parameters A, k and s (Eq. 2) are set to 100, 2 and 2 respectively (empirically optimal values). Starting models (Fig. 2b) are smoothed, upscaled parts of the Strebelle image [6]; after 30 iterations they are turned into models shown in Fig. 2b (LBFGS optimization algorithm was used [2]). Figure 2c shows final equally good solutions.

In all cases the pattern statistics of the training image is successfully reproduced, and the expected prior misfit value of 100 is achieved. Production data are resolved, since the data misfit value is everywhere $\approx N/2$, where N is the number of measurements [5]. Data matching and convergence plots (for the first solution) are shown in Fig. 3a, b respectively. Naturally, the convergence properties of the algorithm are dependent on the initial guess and quality of the production and statistical data. Multiple solutions found in this example is a natural consequence of the fact that the history matching problem is a strongly underdetermined problem. Thoroughly

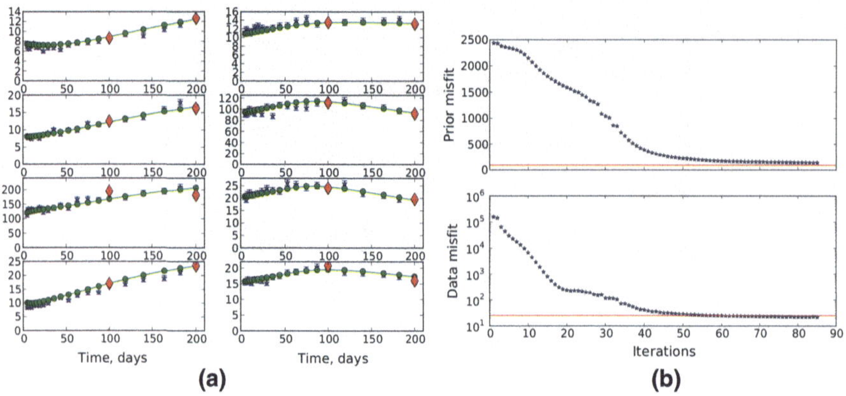

Fig. 3 **a** History matching: water injection rates (*left column*) and oil production rates (*right column*) of some wells, m^3/day; **b** Convergence for prior and production data misfits. Expected values of the misfits are shown by *red lines*

chosen initial guesses, obtained, for instance, from seismic data inversion, would be helpful in minimizing divergence of the solutions.

4 Discussion

We demonstrated how an ensemble of starting models can be gradually transformed into valuable solutions of the history matching problem. The suggested formulation has several advantages: (1) it guarantees prior-consistent solutions by including complex a priori information, (2) it allows using gradient-based optimization techniques, which save computational time, (3) it provides quantitative estimates of the data and prior information misfits and therefore allows us to distinguish between solutions as well as to choose the most reliable ones.

References

1. Caers, J. (2003). History matching under training-image-based geological model constraints. *SPE Journal*, *8*, 218–226.
2. Gao, G., & Reynolds, A. C. (2006). An improved implementation of the lbfgs algorithm for automatic history matching. *SPE Journal*, *11*(1), 5–17.
3. Jafarpour, B., & Khodabakhshi, M. (2011). A probability conditioning method (pcm) for nonlinear flow data integration into multipoint statistical facies simulation. *Mathematical Geosciences*, *43*, 133–164.
4. Lange, K., Frydendall, J., Cordua, K. S., Hansen, T. M., Melnikova, Y., & Mosegaard, K. (2012). A frequency matching method: solving inverse problems by use of geologically realistic prior information. *Mathematical Geosciences*, *44*, 783–803.

5. Oliver, D. S., Reynolds, A. C., & Liu, N. (2008). *Petroleum reservoir characterization and history matching*. New York: Cambridge University Press.
6. Strebelle, S. (2000). Sequential simulation drawing structures from training images. Ph.D Thesis, Stanford University.
7. Tarantola, A. (2005). *Inverse problem theory and methods for model parameter estimation*. Philadelphia: Society for Industrial and Applied Mathematics.

Factor Analysis for Metal Grade Exploration at Pallancata Vein in Peru

Jorge E. Gamarra-Urrunaga, Ricardo Castroviejo and Jesús Domínguez

1 Introduction

The epithermal Pallancata Vein is located ~520 km SE of Lima (Ayacucho department), at ~4,200 MASL. It crops out along 1.5 km, with N70° W strike and has a subvertical dip. Vein thicknesses are structurally controlled, both horizontally and vertically, varying from 1 m in areas of narrowing to 35 m in dilational areas [1]. The mine went into production in September 2007 and ranks sixth worldwide in silver production. It is the second largest silver mine in Peru and produced 263 t of Ag in 2011 (8.77 Moz [2]). The aim of the present study is to examine the geological information that can be gathered from the chemical data collected during the exploration of the deposit. The data are made of 35 elements which were analysed from cores of the 52 boreholes interesecting the vein. Only reliable analytical data from the vein have been used: 17 elements (*Ag*, As, *Au*, Ba, Cd, Co, Cr, *Cu*, Hg, Mo, Ni, P, *Pb*, S, *Sb*, Sr, *Zn*) have been selected, and logarithm values are used.

Two issues are relevant to this study. The first one is whether the analyses can provide a realistic picture of the evolving geochemistry of the ore solutions. The second one is the interpretation of this scenario from a metallogenic point of view and its use to support exploration. To answer the first question, a geological analysis of the vein, as well as a Factor Analysis study of the chemical data have been carried out [3]. Results from both approaches are consistent with a major mineralizing episode tightly relating Ag, Au, Pb, and Sb (*F1: Factor 1*), while Zn, Cd, and

J. E. Gamarra-Urrunaga (✉)
International Minerals Corporation, 230 South Rock Blvd, RenoNV 89502, USA
e-mail: jgamarraurrunaga@yahoo.com

R. Castroviejo (✉) · J. Domínguez
ETSI Minas, Universidad Politécnica de Madrid, c/ Ríos Rosas, 21, 28003 Madrid, Spain
e-mail: ricardo.castroviejo@upm.es

J. Domínguez (✉)
e-mail: jesus.dominguezR@upm.es

E. Pardo-Igúzquiza et al. (eds.), *Mathematics of Planet Earth*,
Lecture Notes in Earth System Sciences, DOI: 10.1007/978-3-642-32408-6_153,

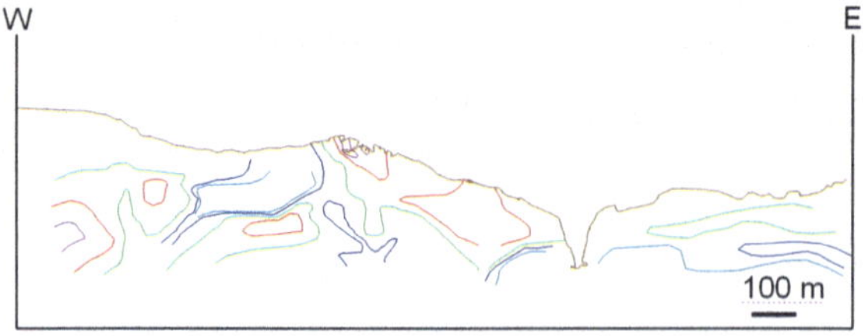

Fig. 1 Ag (logaritm of ppm) isocontent contours plotted on longitudinal section along the vein. Contour values: 2.2, 2.4, 2.6, 2.8, 3.0

Cu (*F2: Factor 2*) show no relationship to the Ag grades. A discussion of the spatial distribution of Factor 1 metals shall help answer the second question.

2 Element Distribution in the Vein

The spatial distribution in the vein of selected (log)metal contents shows that the Ag grade (Fig. 1) is consistent with Au and Pb values, but not with Zn or Cu values. To explain this, a statistical analysis of all data has been carried out. Factor analysis, computed with the programme R, defines two factors: *F1*, with loadings Ag: 0.875, Au: 0.896, Pb: 0.618, Sb: 0.855. *F2*, with loadings Cu: 0.603, Zn: 0.723, Cd: 0.808, and S: 0.734. The spatial distribution of F1 (Fig. 2) is similar to that of Ag (Au) but differs from F2, suggesting unrelated events for Ag (Au) ores and for Cu, Zn sulfides (chalcopyrite, sphalerite); even if the predictive value of these plots is limited, the conclusion is consistent with textural analysis of the ores (photomicrographs: [1, 3]).

3 3 Metal Ratios, Flow Paths and Exploration

Plotting ore metal distribution in a hydrothermal vein may show the flow paths of mineralisation, particularly if (log) metal ratios are used [4, 5]. These plots are suggested to assist exploration [5, 6], since subtle changes in metal ratios may point to unknown metal concentrations. Fluid chemistry variations of the evolving hydrothermal system can be related to fluid flow, under the assumptions that (i) metal precipitation will not begin until saturation is attained; (ii) saturation and therefore precipitation will be enhanced by cooling, and this is favoured by flow of the fluids to shallower levels or by lateral flow away from the source; (iii) pressure drops or other changes related to this process may act in the same way; (iv) the temperature, and therefore the time, of precipitation of a particular metal depends not only on its solubility, but also on its concentration: the precipitation of a very diluted metal will be delayed, as compared with the saturated metal. Under these assumptions, the resulting scenario

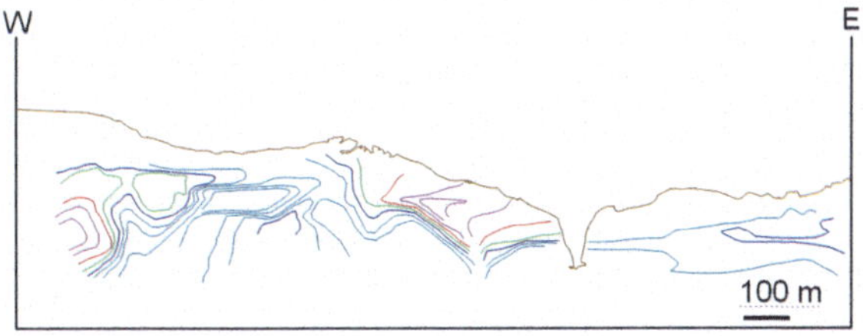

Fig. 2 Factor 1 (log) isovalue contours plotted on longitudinal section along the vein. Contour values: −1.0/0.0, 0.2, 0.4, 0.6, 0.8 − 1

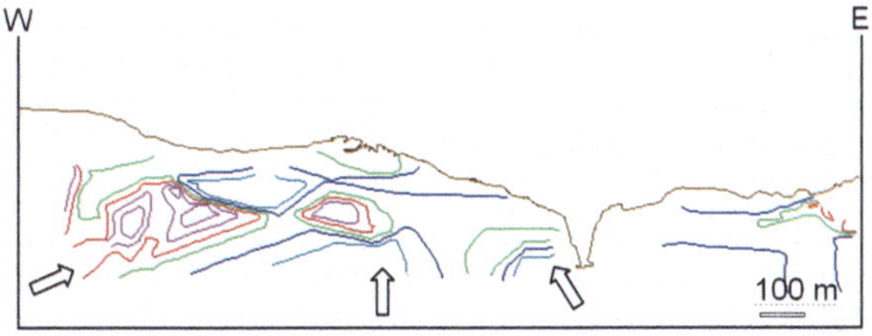

Fig. 3 Ratio logAg /logPb isovalue contours plotted on longitudinal section along the vein. Contour values: 0.9, 1.0, 1.1, 1.2, 1.4. (From chemical data in ppm)

(Fig. 3 for logAg/logPb) suggests a consistent upward flow of the hydrothermal fluids, progressing from the center and from both sides of the structure—arrows—and selectively precipitating Pb at lower levels, with a relative Ag enrichment at higher levels, consistent with known models based on experimental data [1, 7].

The spatial distribution of the ore is also reflected by the logAg/logPb ratio distribution (comp. Figs. 1 and 3). Au occurs as electrum or uytenbogaardtite, (Ag3AuS2), so it is tightly related to Ag [3], and log Au grade and logAu/logAg ratio show the same distribution as logAg grade and logAg/logPb ratio, respectively. The open loops in these log metal ratios suggests further ore to be expected in the lower center (possibly a Ag-impoverished, Pb-enriched core, consistent with current ore petrological models [1, 7]); in addition Ag is suggested westwards at deeper locations.

4 Conclusions

Factor analysis of all the data is consistent with a main event of Ag deposition (Factor 1, Fig. 2), and fits with the space distribution of metal ratios—compare Figs. 1, 2, 3,—as well as with current metallogenetic models [1, 7]. Overall geochemistry and fluid paths of the ore solutions are reflected by metal ratios (Fig. 3).The lower, open loop of the logAg/log Pb ratio distribution hints to a blind, deeper resource open for exploration.

Acknowledgments The research was funded, as part of a MSc Thesis of the first author, by EU grant ALFA II-0459FA. The authors state their thankful recollection of Prof. Enrique Chacón († 2012), UPM, for his advice and discussion of statistical data. Facilities and authorization to publish by the Co. Minera Oro Vega S.A.C. are gratefully acknowledged. Laboratory work was carried out in the *Lab*^{o}. *Microscopía Aplicada y Análisis de Imagen* (*rla* 207) of the Madrid School of Mines, UPM: funding through the Lab_Network of the Comunidad de Madrid (code UP0644) and through projects GR92-0135 and CGL2006-13688-C02-01 is also acknowledged.

References

1. Gamarra-Urrunaga, J., Castroviejo, R., & Bernhardt, H. J. (2013). Preliminary mineralogy and ore petrology of the intermediate-sulfidation pallancata deposit, Ayacucho. *Peru Canadian Mineralogist*, *51*, 67–91. doi:10.3749/canmin.51.1.67.
2. The Silver Institute, http://www.silverinstitute.org/production.php.
3. Gamarra Urrunaga, J. (2008). *Caracterización mineralógica y geoquímica de la Veta Pallancata.* Aplicaciones a la exploración minera, Parinacochas, Ayacucho, Peru. M.Sc. Thesis, Red DESIR Universidad Politécnica de Madrid.
4. Goodell, P. C., & Petersen, U. (1974). Julcani mining district, peru: A study of metal ratios. *Economic Geologists*, *69*, 347–361.
5. Castroviejo, R., Yparraguirre, J. A., & Chacón, E. (2007).Lithogeochemistry and fluid flow in the epithermal Veta Rublo base metal-silver deposit, Chonta Mine (Huancavelica, Perú). In J. Loredo (Ed.), *Exploring our environment, Proc. 23rd IAGS* (pp. 159–160). Oviedo (ISBN 978-84-690-6461-0).
6. Petersen, U., Noble, D. C., Arenas, M. J., & Goodell, P. C. (1977). Geology of the julcani mining district. *Peru Economic Geologists, 72*, 931–949.
7. Sack, R. O. (2005). Internally consistent database for sulfides and sulfosalts in the system Ag2S-Cu2S-ZnS-FeS-Sb2S3-As2S3: Update. *Geochimica et Cosmochimica Acta*, *69*, 1157–1164.

Compositional Block Cokriging

Raimon Tolosana-Delgado, Ute Mueller, K. Gerald van den Boogaart and Clint Ward

1 Illustration Data: The K Pit Deposit (Western Australia)

The Koolyanobbing greenstone belt (Southern Cross Province; Yilgarn craton) strikes NW and is approximately 35 × 8 km large. It is composed of a folded sequence of amphibolites, metakomatiites and intercalated banded iron formations (BIF) [1], and hosts at least 7 high-grade Fe deposits. The total pre-mining resources are estimated at more than 150 million metric tons at 58 % iron cutoff grade [2]. The K-deposit accounts for about 100 million tons thereof. It occurs where the main BIF horizon, striking 300° and dipping 70° NE, is offset by younger NNE striking faults [1]. The K deposit comprises high-grade (> 55 % Fe) magnetite, hematite and goethite hard ores and medium grade fault-controlled hematite-quartz breccia (45–58 %) and hematite-magnetite BIF (45–55 %)[2]. For ore reserve modeling the ore body was subdivided into seven domains. The definition of the domains took account of the north-northeast trending fault system. The domain considered here lies to the west of the through going faults, and is mainly formed by magnetite-hematite.

R. Tolosana-Delgado (✉) · K. G. van den Boogaart
Helmholtz Institute Freiberg for Resources Technology, Halsbrueckerstr. 34, 09599 Freiberg, Germany
e-mail: r.tolosana@hzdr.de

U. Mueller (✉)
Edith-Cowan University, 270 Joondalup Drive, Joondalup, WA 6027, Australia
e-mail: u.mueller@ecu.edu.au

K. G. van den Boogaart
Institute for Stochastics, Technical University "Bergakademie", Prueferstr. 9, 09599 Freiberg, Germany
e-mail: boogaart@hzdr.de

C. Ward
Cliffs Natural Resources, The Quadrant 1, William Street, Perth, WA 6000, Australia
e-mail: C.Ward@CliffsNR.com

E. Pardo-Igúzquiza et al. (eds.), *Mathematics of Planet Earth*,
Lecture Notes in Earth System Sciences, DOI: 10.1007/978-3-642-32408-6_154,

2 Compositional Data Analysis

A D-part composition $\mathbf{z} = [z_1, z_2, \ldots, z_D]$ can be identified with a vector of D positive components with total sum equal to or less than a fixed constant, usually $\kappa = 100\,\%$. The set of all D-part compositions is called the D part simplex, denoted by $\mathbb{S}^D$. Compositional data are known to be prone to the spurious correlation problem [3], which can be bypassed by analysing the data after a log-ratio transformation [4]. A simple option is to take one of the components as common reference, and compute the log-ratios of each variable against it. This transformation is known as the additive log-ratio transformation (alr) [4], and it corresponds to a representation of compositions in an oblique vector basis of the simplex $\mathbb{S}^D$ in the Aitchison geometry. The scores so obtained can be analysed with classical tools, as long as they are based on second-order moments.

In the case presented here, components Al_2O_3, Fe, SiO_2, P and loss on ignition (LOI) are considered. Although making up the major part of the total mass, they do not account for 100 %. A filler variable, fill $= 100 - (Al_2O_3 + \text{Fe} + SiO_2 + \text{P} + \text{LOI})$, is defined to complete the composition, thus $D = 6$. In the alr transformation the filler is chosen to be the common denominator. Each original variable is associated a log-ratio score $y_j = \ln(z_j/z_{\text{filler}})$. The transformation is invertible through the additive generalized logistic transformation (agl): with $j \in \{Al_2O_3, \text{Fe}, SiO_2, \text{P}, \text{LOI}\}$,

$$z_j = \frac{100 \cdot \exp(y_j)}{\alpha}, \quad z_{\text{filler}} = \frac{100}{\alpha}, \quad \alpha = 1 + \sum \exp(y_j). \tag{1}$$

3 Point Support-Geostatistics

We follow the log-ratio approach to geostatistics for point support compositional data proposed in [5]. Experimental direct- and cross-variograms of alr-transformed scores were calculated using 7 lag distance classes at a nominal spacing of 25 m and along several directions. The global rotation was determined to be a strike of 300°, a dip of 45° and a pitch of −10° (equivalent to rotations of 160° about Z, 10° about Y, and 45° about X, in mathematical convention). This zone dips slightly less than the deposit overall. Experimental variograms were recomputed for the rotated coordinates. A three-structure linear model of coregionalization was fitted to the experimental variograms, with a nugget effect and two spherical structures with ranges 30–150 m, and vertical anisotropy ratios of respectively 11/15 and 8/15. This is not shown for lack of space.

Following the formulation of cokriging in [6] isotopic cokriging estimates of the five alr scores were calculated. This approach provides, for each interpolation location x_0, a cokriging estimate $\mathbf{y}_0^*$ and a cokriging covariance matrix $\mathbf{S}_0$. Assuming multigaussianity, the true value $\mathbf{Y}_0$ follows a multivariate normal distribution with mean vector $\mathbf{y}_0^*$ and covariance matrix $\mathbf{S}_0$. This also fully specifies the distribution of $\mathbf{Z}_0 = \text{agl}(\mathbf{Y}_0)$ to be an additive logistic normal (ALN) [4]. According to [7] the conditional distribution of $\mathbf{Z}_0$ does not depend on the choice of log-ratio transformation , as long as the transformation is one-to-one, and a full cokriging is applied, with a complete set of direct and cross-variograms. The conditional distribution can be used to derive several point estimators, such as the direct inversion $\mathrm{agl}(\mathbf{y}_0^*)$ or the conditional expectation of $\mathbf{Z}_0$. The latter requires numerical integration [8].

4 Block-Support Cokriging Through Local Simulation

The preceding scheme can be applied to block-supports V too. The conventional multigaussianity assumptions underlying linear block cokriging provide an estimate of the average log-ratio within a block, denoted by $\mathbf{Y}_0(V)$. That would be obtained with any block cokriging algorithm. However, in mining applications one requires the conditional distribution of the *mass* $\mathbf{m}(V) = \int_{x \in V} \mathrm{agl}(\mathbf{Y}(x))dx$ of each component $\mathbf{m}(V)$ in a block V. In general, the mass is not given by the volumetric scaling of the average log-ratios: $\mathbf{m}(V) = \frac{|V|}{|V|}\int_{x\in V} \mathrm{agl}\left[\mathbf{Y}(x)dx\right] \neq |V|\cdot \mathrm{agl}\left[\frac{1}{|V|}\int_{x\in V}\mathbf{Y}(x)dx\right] = |V|\cdot \mathrm{agl}\left[\mathbf{Y}_0(V)\right]$.

However, the target quantity $\mathbf{m}(V)$ can be approximated by Monte Carlo simulation on a dense grid $\{x_1, \ldots, x_B\} \in V$ within the target block. For each simulation $\{\mathbf{Y}^{(s)}(x_1), \ldots, \mathbf{Y}^{(s)}(x_B)\}$ the mass $\mathbf{m}^{(s)}(V)$ can be approximated by summing the agl-transformed simulated values within the block. If a family of S simulations is available, then one obtains a set of values of $\{\mathbf{m}^{(s)}(V) : s = 1, \ldots, S\}$, which can be used to estimate its expectation, $\mathrm{E}[\mathbf{m}(V)] \approx \frac{1}{S}\sum_s \mathbf{m}^{(s)}(V) \approx \frac{|V|}{S\cdot B}\sum_s^S\sum_j^B \mathrm{agl}(\mathbf{Y}^{(s)}(x_j))$. Any geostatistical simulation procedure could serve, but LU decomposition is particularly suitable, since the number of simulation points is relatively small, as it must only be representative of the variability within the block, not of the regional variability; and with LU one can obtain a large number of simulations at once. As a result a very good approximation of $\mathrm{E}[\mathbf{m}(V)]$ and even of the whole distribution of $\mathbf{m}(V)$ can be generated.

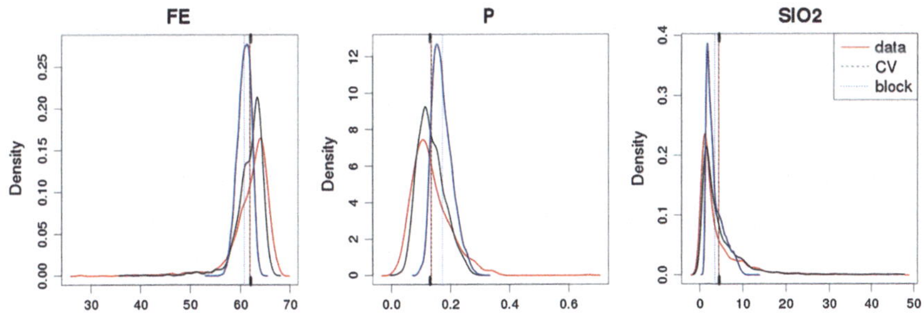

Fig. 1 Comparison of estimates of the conditional expectation of $\mathbf{Z}$ at point and block support with the original data for 3 out of the 6 variables considered. Point support estimates were obtained in a 10-fold cross-validation scheme. *Vertical lines* show each distribution mean. *Horizontal scale* in %

Figure 1 shows the results of this procedure for the study of selective mining unit blocks (SMU) of the data set, compared with the available data and point-support cokriging interpolations. Though this does not validate the goodness of the method in itself, it shows the expected averaging of block kriging, and the smoothing and information effects of kriging. Only block estimates for P show a certain bias with respect to the data.

References

1. Griffin, A. C. (1981). Structure and iron ore deposition in the Archaean Koolyanobbing Greenstone belt, Western Australia. *Geological Society of Australia, Special Publication 7*, 429–438.
2. Angerer, T., & Hagemann, S. (2010). The BIF-hosted high-grade iron ore deposits in the Archean Koolyanobbing Greenstone Belt, Western Australia: Structural control on synorogenic- and weathering-related magnetite-, hematite- and goethite-rich iron ore. *Economic Geology, 105*, 917–945.
3. Chayes, F. (1960). On correlation between variables of constant sum. *Jou. Geophys. Res., 65*, 4185–4193.
4. Aitchison, J. (1986). *The statistical analysis of compositional data*. London: Chapman & Hall Ltd.
5. Pawlowsky-Glahn, V., & Olea, R. A. (2004). *Geostatistical analysis of compositional data*. New York: Oxford University Press.
6. Myers, D. E. (1982). Matrix formulation of co-kriging. *Mathematical Geology14*, 249–257.
7. Tolosana-Delgado, R. (2006). *Geostatistics for constrained variables: Positive data, compositions and probabilities*. PhD dissertation, Girona, Spain.
8. Pawlowsky-Glahn, V. (1989). Cokriging of regionalized compositions. *Mathematical Geology, 21*, 513–521.

Milling Result Prediction

Stephan Matos Camacho, Thomas Leißner, Petya Atanasova, Andre Kamptner, Martin Rudolph, Urs Alexander Peuker and K. Gerald van den Boogaart

1 Introduction

The choice of the optimal milling parameters is a key problem in the first steps of mineral processing. If the milling is too coarse a large portion of the value mineral and waste minerals are still connected and effective separation is impossible. Finer milling will lead to particles consisting almost entirely of value or waste minerals, which can then be separated but costs to achieve finer grind sizes often rise exponentially and usually require additional milling capacity. This separation of valuable materials from gangue is called liberation and measured by the portion of valuable mineral that can effectively be separated from waste minerals in a given process. The precise definition of *liberation* varies from process to process, but it always depends on the 3D structure of the particle.

Usually the optimal milling parameters are determined experimentally. The experimental results usually include some measures of the ground product e.g. analysis using the MLA (Mineral Liberation Analyzer). One of the limitations of this approach is that the images only show a 2D section of the particles. Particles with a waste mineral attached above or below the 2D section will still be counted as liberated. Stereological corrections for this problem, relaying on simple grain shape, are e.g. discussed in [2] or [3].

S. M. Camacho (✉) · P. Atanasova · M. Rudolph · K. G. van den Boogaart
Helmholtz Institute Freiberg for Resource Technology, Halsbrücker Str. 34, 09599 Freiberg, Germany
e-mail: s.matos@hzdr.de

G. van den Boogaart
e-mail: boogaart@hzdr.de

T. Leißner · U. A. Peuker · K. G. van den Boogaart
TU Bergakademie Freiberg, Prüferstraße 9, 09599 Freiberg, Germany

A. Kamptner
UVR/FIA Gmbh, Chemnitzer Str. 40, 09599 Freiberg, Germany

E. Pardo-Igúzquiza et al. (eds.), *Mathematics of Planet Earth*,
Lecture Notes in Earth System Sciences, DOI: 10.1007/978-3-642-32408-6_155,

Here we propose an alternate paradigm. It allows prediction of the effect of milling to any size of liberation from a single MLA measurement on a coarsely milled sample of the material. This would make it possible to take small samples in a mining block, mill them with an analytical mill, take a single MLA sample and determine the optimal milling parameters for the local block within hours. This could help to introduce adaptive processing of the ore on a day to day basis.

2 The Two Mosaics

Our method is based on two random tessellations of the examined ore. One tessellation describes the partition of the ore into mineral grains. The other depicts the separation of the ore into particles during the milling process. In a first simplified model we could assume both tessellations being independent. However, in principle it is possible to model the second tessellation dependent on the first one.

The microstructure is modelled by a Bayesian statistical model i.e. as a parametric class of random mosaics in the sense of stochastic geometry, like e.g. in [7] or [6]. The prior distribution should represent the variability in the deposit. In our first synthetic example with three mineral phases we use Laguerre mosaics [5]. These mosaics are based on a marked point process representing "centers" of minerals. For simplicity we have chosen a Poisson process with an unknown density parameter λ_M. The prior for λ was modelled by a lognormal distribution. Another marked point process could be used to model the structure more precisely. A mineral phase is assigned to each of the points with a given probability, which forms a second parameter represented by the additive log ratio transform of the probabilities for the three phases, this allows the modelling of a different content of the different phase. To each of the centers we assign a normally distributed mark, modelling the different sizes of the different mineral phases. The expected value of the mark depends on the mineral phase. The contrasts between them form a third group of parameters that allow the modelling of the different mean sizes of the different mineral phases. A Laguerre mosaic is then formed by assigning each point in space to the center for which the squared distance minus the mark of the center is minimal.

The Laguerre mosaic model is extremely flexible (see e.g. [4], Theorem 2.4.3), and provides a basis for modelling far more complex microstructures In practice one should find appropriate mosaic models to model the microstructures in the deposit. Figure 1a shows a 2D slice of a modelled grain structure. This structure could be identified with EBSD imaging. Figure 1b shows the same structure with the three phases assigned. Such a figure could be seen in insitu MLA [1].

The milling or particle mosaic was modelled by a simple Poisson Voronoi mosaic, with the density parameter λ_P representing the coarseness of milling. The properties of the particle mosaic depend on the grinding parameters. Figure 1c shows the particle mosaic in the same area.

After milling these particles are free to move and can adopt any orientation. A sample is collected and they are embedded into epoxy resin in a random orienta-

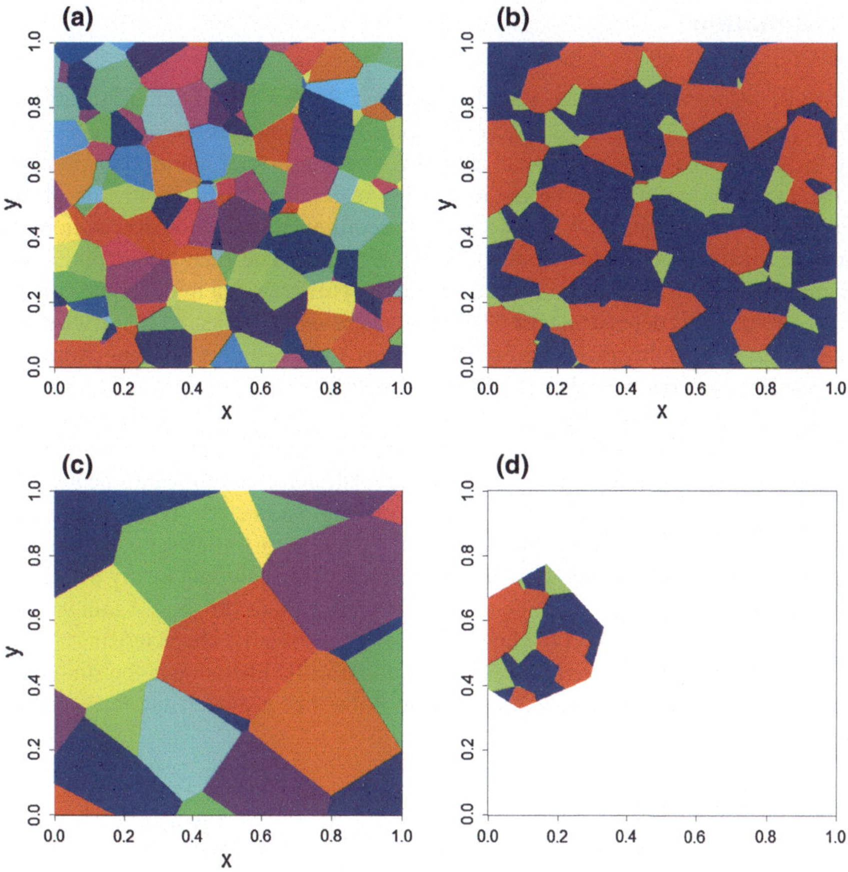

Fig. 1 Examples Mosaics: **a** grain structure, **b** microstructure with mineral phases, **c** the puzzle of particles after milling, **d** a resulting observable particle with internal mineral phase structure

tion and randomly oriented 2D sections of them can be observed in particle MLA images. Figure 1d shows such a particle as it could be seen. We assume that all further inference has to be made from a dataset of such 2D sections, typically several thousand.

It is suggested that coarsely grained material rather than insitu MLA images of the undistorted microstructure are used to get a representative sample of a larger ore volume, while still seeing enough local structure.

3 Estimation

To model the mosaic parameters must be estimated. For this spatial summary statistics observable on the 2D milled image are computed. Examples are: area portion of the phases, mean area of particle, mean area of continued mineral phases on a particle, apparent liberation in 2D. The key idea is to create a stochastic simulation of the parameters of the statistical mosaic model according to the prior distribution, in order to make a conditional simulation of the mosaic, and then to compute such characteristics of a 2D image of milled material. This creates a multivariate dataset with two stochastic dependent parts: The parameters and the observed characteristics. To this we fit a multivariate regression analysis for creating a mapping from the observed characteristics to the estimated parameters.

The prediction of 3D liberation is carried out in a similar way to the parameter estimation. For each simulated parameter a second simulation of the same random mosaic for the microstructure and a second simulation for a milling mosaic are created, which can however represent different milling parameters. Typically a much finer milling is used to ensure liberation. This gives rise to a dataset with two stochastic dependent classes of variables: the Liberation—according to any definition appropriate for further processing—and the observed characteristics. Again we use a regression model to predict the liberation from the observed characteristics.

Figure 2 shows the similarity of estimated and true 3D liberation in the simulated dataset, in (a) for the training dataset used to fit the linear model, and in (b) for an independently simulated dataset.

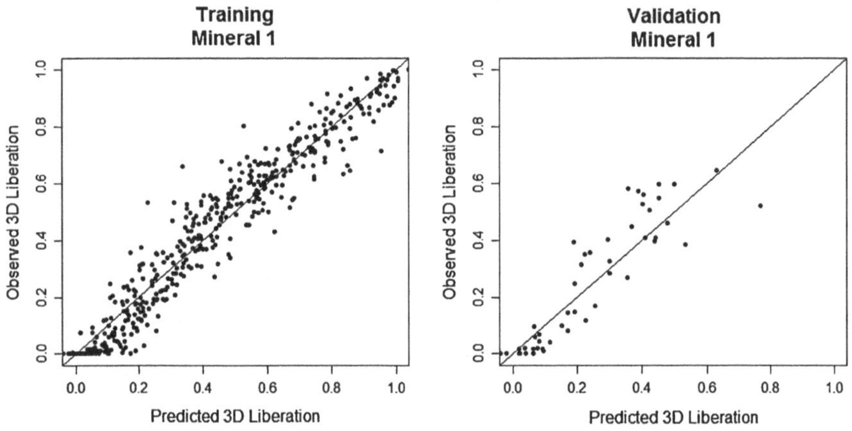

Fig. 2 Comparing the predicted with the actual liberation in the models. Note that this predicts the liberation for a different portion of the material and for a much finer milling

4 Discussion

This contribution offers a paradigm how liberation could be predicted from MLA measurements based on a stochastic geometric model of the microstructure and the milling process, and furthermore shows that this approach works well with a simulated example. For an application of the method a decent stochastic model of the microstructure and the milling is required. This might include more complex models, e.g. including dependence between neighbouring phases, complex inter growth, multiple domains, milling circuit operation and removing some particles during operation. It might also include the development of more sophisticated estimation procedures and tests.

References

1. Birtel, S., Tolosana Delgado, R., Matos Camacho, S., & van den Boogaart, K. G. (2013). *Statistics of MLA data: Proceedings of the IAMG 2013.*
2. Jones, M. P., & Shaw, J. L. (1973). Automated measurement and stereological assessment of mineral data for use in mineral technology. In: *Proceedings Xth International Mineral Proceedings Congress*, London, pp. 737–756.
3. King, R. P., & Schneider, C. L. (1993) An effective SEM-based image analysis system for quantitative mineralogy. *KONA Journal 11*, 165–177.
4. Lautensack, C. (2007). *Random Laguerre Tessellations.* Doctoral Thesis at the University of Kaiserslautern.
5. Okabe, A., Boots, B., Sugihara, K., & Chiu, S. N. (2000). *Spatial tessellations—concepts and applications of voronoi diagrams* (2nd ed.). Chichester: Wiley.
6. Schneider, R., & Weil, W. (2000). *Stochastic and integral geometry*. Berlin: Springer.
7. Stoyan, D., Kendall, W. S., & Mecke, J. (1995). *Stochastic geometry and its application* (2nd ed.). Chichester: Wiley.

A Geostatistical Study of Tertiary Coal Fields in Turkey

Fırat Atalay, A. Erhan Tercan, Bahtiyar Ünver, Mehmet Ali Hindistan and Güneş Ertunç

1 Introduction

Tertiary fields of Turkey cover an area of approximately 110,000 km^2 and are of great importance in terms of coal potential since the majority of lignite and subbituminous coal beds of Turkey was deposited during this period. Tertiary fields make up about 67.5 % of the total coal resources of Turkey which is 8,374 M tons. Tertiary fields are subject to a number of studies related to coal. Pasamehmetoğlu et.al. [1] carried out coal prospecting studies in various sites and determined target locations for coal. Görür et. al. [2] made an attempt to classify tertiary basins into seven categories according to their tectonic characteristics and discussed the geological evolution and coal potential of each basin. Tuncalı et.al. [3] studied the chemical and technological properties of tertiary coals. Temur et.al.

Tuncali et al. [4] studied Tertiary coals from the point of statistical view by using data generated by Tuncalı et. al. [3]. Toprak [5] studied petrographic properties of coal samples collected from a number of coal seams in Tertiary fields.

The purpose of this study is to describe the spatial distribution of Lower Calorific Value (LCV), Ash Content (AC) and Moisture Content (MC) for the Tertiary coals of Turkey at a global scale from a geostatistical perspective. Data subjected to this

F. Atalay (✉) · A. E. Tercan · B. Ünver · M. A. Hindistan · G. Ertunç
Department of Mining Engineering, Hacettepe University, 06800 Ankara, Beytepe, Turkey
e-mail: atalay@hacettepe.edu.tr

A. E. Tercan
e-mail: erhan@hacettepe.edu.tr

B. Ünver
e-mail: unver@hacettepe.edu.tr

M. A. Hindistan
e-mail: hmali@hacettepe.edu.tr

G. Ertunç
e-mail: gertunc@hacettepe.edu.tr

E. Pardo-Igúzquiza et al. (eds.), *Mathematics of Planet Earth*,
Lecture Notes in Earth System Sciences, DOI: 10.1007/978-3-642-32408-6_156,

study are taken from the study of Tuncalı et. al. [3] who collected the 187 samples primarily from the relatively big coal areas which belong to the government and private companies and then analyzed them for various quality variables. In this study, the tertiary field is divided into the 15,445 blocks each having 4000 × 4000 m dimensions and LCV, AC and MC of the each block are estimated by using geostatistical tools. In addition to the estimation, probability of LCV of each block being lower than the pre-determined cut-off value (1,600 kCal/kg) is calculated by using direct sequential simulation and areas with high uncertainty are determined.

2 Geostatistical Estimation and Simulation of the Tertiary Fields

In this study only terrestrial Tertiary fields are taken into consideration. For this purpose, geological map of Turkey (Fig. 1.) is digitized, only Tertiary fields are retained and then geostatistical estimation and simulation on these fields are performed by using SGeMS software.

Descriptive statistics for the LCV, AC and MC are given in Table 1. Experimental variograms for LCV, AC and MC are calculated and models are fitted. Parameters of the fitted variograms are given in Table 2.

Tertiary field is divided into the 15,445 block each having dimension 4000 × 4000 m and mean qualities of the blocks are estimated by ordinary kriging method. We did not consider co-kriging. To estimate one block minimum three, maximum 16 conditioning data are used. The conditioning data consist of coal quality analyses in Tertiary field. The results show that 99 % of the estimated values are in the range of 2,000 kCal/kg and 4,000 kCal/kg. Coals in Turkey have mid-calorific value. Ash con-

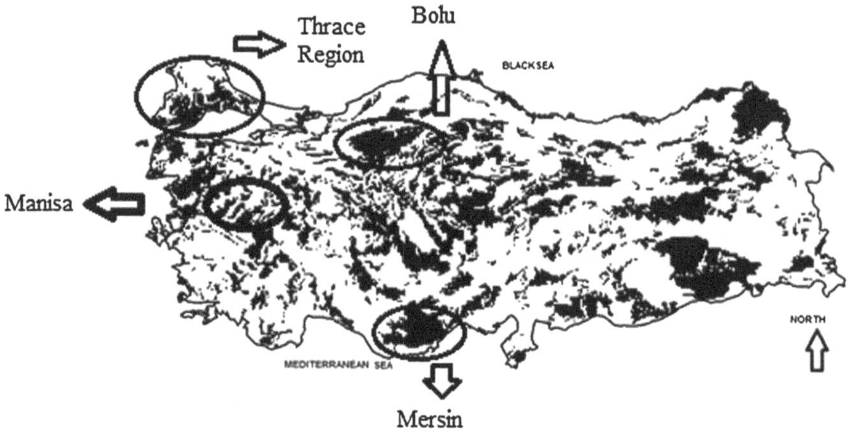

Fig. 1 Tertiary regions of Turkey (*in black*)

Table 1 Descriptive statistics of LCV, AC and MC

	LCV (kCal/kg)	Ash (%)	Moisture (%)
Number of data	187	187	187
Minimum	1185	5.21	1.20
Maximum	5574	56.09	57.66
Average	3186	20.39	25.45
Median	3103	18.42	24.49
Standard deviation	921	9.56	11.20
Skewness	0.30	0.29	0.19
Kurtosis	−0.21	0.79	−0.31

Table 2 Variogram model parameters of LCV, AC and MC

	LCV	Ash	Moisture
C0 (Nugget)	400,000	41	50
C1 (Sill)	450,000	48	75
Range(m)	140,000	33,000	120,000
Model	Spherical	Spherical	Spherical

tent generally ranges between 20 and 30 %. Ash content is relatively low in Thrace Region and Mersin region. Moisture content is the lowest in Manisa and Bolu regions.

Most of the coals produced in Turkey are sent to thermal power stations for the purpose of electrical energy production. In addition to estimation, LCV was simulated geostatistically to calculate the probability of Tertiary coals below mean LCV of coals fed to the thermal power stations. Mean LCV is equal to 1,600 kCal/kg and obtained from the study of Şenguler [6]. For this purpose Direct Sequential Simulation was used. For simulation of a single block a maximum of 16 conditioning data was used. LCV was simulated 100 times. For each block probability of LCV being lower than 1,600 kCal/kg is determined by using distribution function of 100 simulated values.

3 Results

Variogram of the LCV shows that spatial relation between the data disappears at 1,400 km distance. Estimation results shows that LCV is higher in western and central regions of Turkey than in the eastern and northern regions. The estimations show that LCV for 99 % of the coals in Turkey ranges from 2,000 kCal/kg to 4,000 kCal/kg. Therefore the probability of Tertiary coals being below than the LCV mean of coal fed to the thermal power plants is very low. When only LCV is considered, coals in Turkey are suitable as feed to thermal power stations. By using geostatistical simulation results, 95 % symmetric probability intervals for LCV were constructed for 15,445 blocks. Comparing the estimation results with the probability interval gives areas with high uncertainty. These are generally in central region of Turkey. Still there is no exhaustive research made in the areas of high potential coal occurrence in Tertiary areas. When considering this situation, there are yet potential undiscovered coal basins in Southeastern and Eastern regions of Turkey. Study in these regions will probably increase the number of coal occurrence in Turkey.

References

1. Paşamehmetoğlu A. G., Oskay M., Erler A., Lünel T., Güleç N., Dirik K., et al. (1983). Turkey coal master plan. *Turkey Coal Potentials and Researches, 4*, 285 p. TKE.
2. Görür N., Akkök R., Sakınç M., Ünalan G., & veYaltırak C. (1996). Türkiyenin Tersiyer Havzaları ve Kömür potansiyeli, Türkiye kömür arama hedeflerinin belirlenmesi ve arama yöntemlerinin saptanması, (Editorials : Önal G., Kurşun İ., Yazıcı H.), Yurt Madenciliğini Geliştirme Vakfı, İstanbul, 130 p.
3. Temur S., Orhan H., & ve Deli A. (2008). Statistical interpretation of some physical and chemical data of the tertiary coal deposits in turkey. *Geochemistry International, 46*(4),407–422. Pleiades Publishing.
4. Tuncalı E, Bilgin Ç., Yavuz N., Toprak S., Köker A., Gencer Z., et al. (2002). Türkiye Tersiyer Kömürlerinin Kimyasal ve Teknolojik Özellikleri, Maden Tekik ve Arama Genel Müdürlüğü, Ankara, 401 s.
5. Toprak, S. (2009). Petrographical properties of major coal seams in turkey and their formation. *International Journal of Coal Geolgy, 78*(4), 263–275.
6. Şengüler İ. (2010). Coal explorations in turkey: New projects and new reserves. In I. B. Morsi (Ed.), *Twenty-Seventh Annual International Pittsburgh Coal Conference* (55 p).

Placer Deposit: From Modeling to Evaluation

Nikolay Laverov, Irina Chizhova and Elena Matveeva

1 Introduction

The features of created multifactor multitag models (MMM) for search and estimation of ore placer deposits, first of all, are caused by specificity of the researched object—gold placer deposit grew out of durable evolution of ore deposit (occurrence), got to a field of exogenous processes activity. Therefore it was necessary to accept the originally division of all tags and criteria on the membership into two systems (Fig. 1):

- endosystem (ore hosted) represents a set of criteria and the tags, routed on revealing of potential ore objects—provinces, zones, ore regions, ore fields;
- exosystem (ore placer located) is put in correspondence to that level, where influence exogenous processes starts to play an appreciable role (field level).

2 Methodology of the Investigation

Methodology of the investigation included the following:

(1) compiling of a table containing all potentially informative tags;
(2) forming the database—coding the sites based on the principle of presence or absence of the analyzed tags;

N. Laverov · I. Chizhova (✉)
The Organization of Russian Academy of Sciences, Institute of Geology of Ore Deposits, Petrography, Mineralogy, and Geochemistry (IGEM RAS), Staromonetny lane, 35, 119017 Moscow, Russia
e-mail: tchijova@igem.ru

E. Matveeva
Federal state unitary enterprise all-russian scientific-research, institute of mineral resources named after N.M.Fedorovsky, (VIMS), Staromonetny lane, 31, 119017 Moscow, Russia

E. Pardo-Igúzquiza et al. (eds.), *Mathematics of Planet Earth*,
Lecture Notes in Earth System Sciences, DOI: 10.1007/978-3-642-32408-6_157,

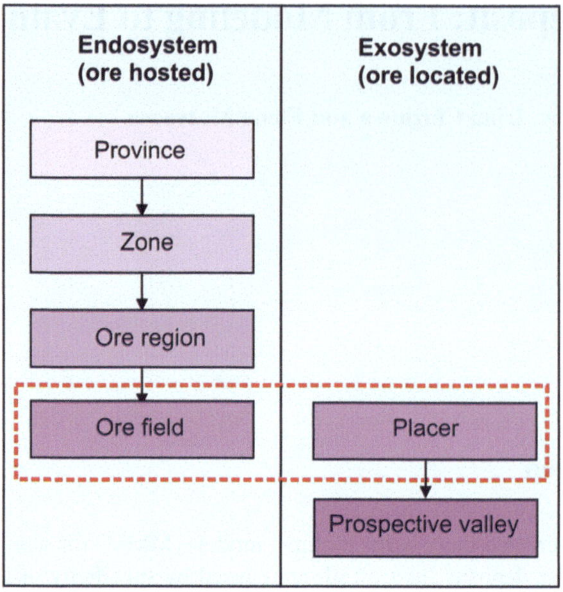

Fig. 1 Metallogenic taxa for placer objects

(3) choosing of the standard sites and compiling of the training populations;
(4) mathematical processing of data using various algorithms and pattern recognition programs, accounting the frequency and/or character of correlation of the tags;
(5) compiling of mathematical models of the determined classes from the most informative tags selected and calculation of their weights based on logic-informational analyses;
(6) checking the stability of the compiled models on the independent control populations;
(7) forming the knowledge base for the computer expert system. Previously this methodology was used while designing of computer expert system for epithermal gold-silver deposit prognostication (Okhotsk-Chuckchee Volcanic Belt, North-east Russia). All algorithms for mathematical processing of data are published in [1].

3 Logic-informational Analysis

We have applied logic-informational analysis [1, 2]: a special algorithm of the pattern recognition method, based on the study of variation object range and the similarity concept. The essence of this method is to outline the informative tags to distinguish different groups of data sets and estimate their significance for evaluation of this difference and similarity. The identification of the informative tags is based on the principle of the common features of sites, belonging to the same class. To distinguish the tags that are informative for description of class A_k and separate it from other classes, we have used the functions, based on: probability that tag j presents at the sites of class A_k and does not present at the sites of other classes; probability that tag j does not present at the sites of class A_k and presents at the sites of other classes.

They are used for construction of the decision rule for the site identification: the site belongs to that class which has the maximum value of the function. As a result, we have a relationship of coefficients for an examined site, which characterizes the similarity degree of this site with other classes.

4 The Region of Investigation

The area of researches covers a southeast part of the Aldan shield (the southeast of Siberian platform, Russia).

The gold mineralization manifestations, to which the gold placer linked on researched territory, are caused by a mesozoic stage of tectonic magmatic activation.

Purposely in the given investigation the attempt was made to rank the known ore placer objects and to define the complex of criteria and tags for productivity estimation.

5 Multifactor Multitag Models and Evaluation of Placer Deposit

While evaluation of area prospects the basic moment is the degree of manifestation of tectonic magmatic activation (TMA) and the presence of its direct and indirect tags.

Therefore the metatag "basis" was determined, which has included the various characteristic of a substratum: the age of host rocks, disjunctive violations, injectives, material constitution.

Further the object description was made taking into account that it was necessary to show specificity of TMA manifestation in the area under study. The following characteristics have been considered: "activation structures", "disjunctions", "injections" (the structure and the form of bodies), "ore", "material constitution", "size". The special sense is put in a metatag "morpho structure" as in offered model it occupies intermediate position on the junction of two systems according to the accepted definition of morphostructure.

Further the specific metatags are entered, which described the own features of exosystem and united with its formation as the placer located system: geomorphological characteristic of relief, specification of its parameters, structure of valleys, age and substance.

The evaluation of gold placers is carried out consistently by solution of following tasks:

1. An estimation of prospects of area;
2. In case of revealing of a prospecting area the estimation of productivity of perspective area is spent.

The mathematical processing of data has allowed us to compile logic-informational models (the set of tags, informative for indicated tasks, with weights of theirs importance for solving of the problem). For example, solving the first task, we have received the model "prospecting area" as the set of 14 tags from metatags: basis, ore, morphostrusture, substrat, structure of valleys—with weights in diapason from 0.046 to 0.58. The rule for making decision have allowed to determine correctly all placer deposits from database.

6 Conclusion

The technique has been developed. Using it the MMM for gold placer objects were developed. For receiving of models for placer fields of various productivity, the logic-informational modeling was carried out. Results are represented numerically, allowing its automated usage for creation of the knowledge base of the predicting expert system.

Acknowledgments The present research was executed under financial support of the Russian Foundation for Basic Research (project 12-05-90413Ukr_a).

References

1. Chizhova I. A. (1995). Logic-informational modeling while forecasting metallogenic analysis of prospective areas. In *Works of IGEM RAS. Modern problems of ore geology, petrology, mineralogy and geochemistry*. 4, pp. 59–84. Moscow: IGEM RAS. (in Russian).
2. Strujkov, S. F., Konstantinov, M. M., & Chizhova, I. A. (1999). Computer expert system for epithermal gold-silver deposit prognostication (Okhotsk-Chuckchee Volcanic Belt, Northeast Russia). *Natural Resources Research*, *8*(4), 315–343.

Uncertainty Assessment of the Orebodies Geometry by Using Block Indicator Simulation

Julia Carvalho, Pedro Correia, Sofia Menezes, Cláudia Peixoto and Amilcar Soares

1 Introduction

For the geological control of reserves estimation deterministic models are used to define the geometry and the boundaries of the oretypes within which geostatistical methodologies are applied for grades estimation. These oretypes morphology and boundaries are usually defined by expert geological knowledge or any deterministic polynomial interpolation that do not provide any measure of uncertainty which entails subjectivity to the uncertainty of the estimated grades inside them.

In this paper a new methodology is proposed—Block Indicator Simulation, BIS—to account for the uncertainty of the deterministic geological model, by incorporating the geometrical/geological model as a reference image, a block data with a given uncertainty attached, and the point hard data from drill holes and sampling.

2 Case Study Data

Neves-Corvo is an underground copper-tin mine, which is located in southern Portugal, a world-class massive sulphide deposit and one of the most important in Europe. The case study data comes from Zambujal, one of the five deposits of the mine, consisting of a $2 \times 2 \times 2$ m grided geological model of the deposit and point data from surface and underground drill holes and sampling. Zambujal consists of copper-poor, pyritic mineralization, which includes some zinc-rich massive sulfide lenses, with a

J. Carvalho (✉) · P. Correia · A. Soares
Center for Modeling Petroleum Reservoirs, Cerena/DECivil, Instituto Superior Técnico, Technical University of Lisbon, Lisbon 1049-001, Portugal
e-mail: jcarvalho@ist.utl.pt

S. Menezes · C. Peixoto
Somincor—Mining Company of Neves Corvo, Santa Bárbara Padrões 7780-409, Portugal

E. Pardo-Igúzquiza et al. (eds.), *Mathematics of Planet Earth*,
Lecture Notes in Earth System Sciences, DOI: 10.1007/978-3-642-32408-6_158,

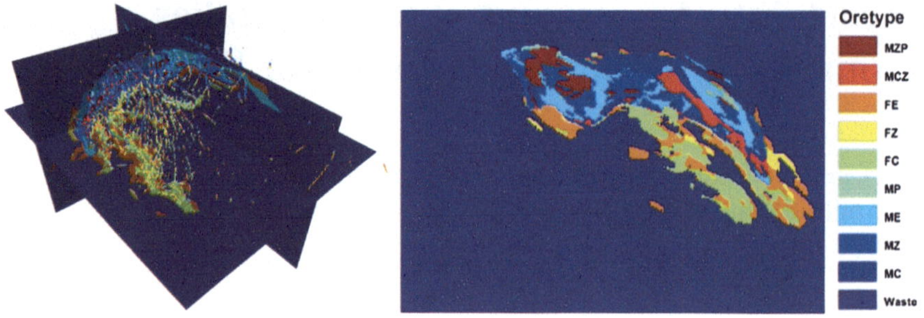

Fig. 1 Zambujal's 3D orebody with drill holes and sampling (*left*) and representative 2D section of the orebody—the reference image (*right*)

thickness of 53 m and plan dimensions of 550 m along strike and 600 m along dip [2]. Zambujal's orebody presents 9 different ore types, from massive (M) and stockwork (F) mineralisations of copper (C), zinc (Z), lead (P) and barren (E, waste with sulphides).

The oretypes geometry is defined with geological control and a polynomial interpolation (leapfrog3D®). From the 3D model of the orebody geometry, a representative 2D section was selected to test the proposed algorithm (Fig. 1).

3 Block Indicator Simulation with a Reference Image

A deterministic model of different oretypes is obtained by a polynomial interpolation (lipfrog) after the geological control. In this paper this 3D model is interpreted as a reference image (RI) to condition the uncertainty assessment of the oretypes geometry (Fig. 1). In a first step, this model, obtained with point "hard" data of the boreholes, is transformed into blocks, each with an attached uncertainty.

3.1 Block Indicator Simulation

A geostatistical simulation BIS is proposed to generate high resolution images of oretypes morphology, by integrating the previously defined "block" data (RI) and the "point" data from the experimental borehole samples. BIS follows the basic outline of Direct Block Sequential Simulation [1] for continuous variables.

The set of N oretypes are transformed in a vector of indicator random functions X_i, i = 1, N. The block indicator simulation follows the sequential path of regular grid of points. At each grid node location x_0, the following probabilities $p(x_0 \in X_i)$, i = 1, N, are estimated (block indicator kriging) with block and point data. The uncertainty attached to each block data is taken into account in the left hand side

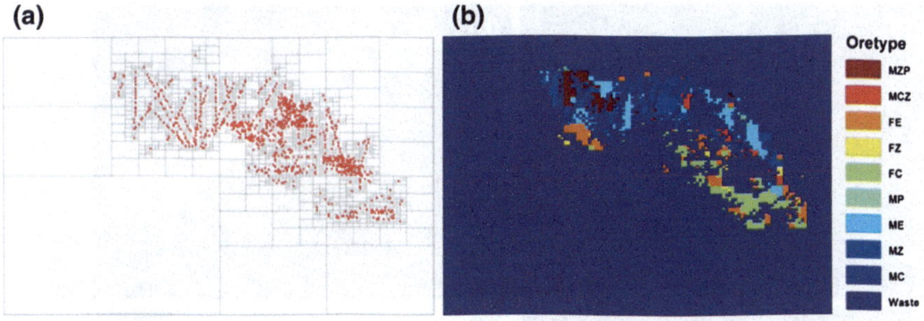

Fig. 2 Hard point data (*red points* in (**a**) and the representation of the block dataset (**b**)

indicator kriging matrix (point to point, point to block, block to block covariances in which the error from the blocks are introduced in the main diagonal of the block to block sub matrix). After the estimated probabilities, indicator values are drawn for the node x_0, and the method proceeds until all nodes are simulated. Uncertainty on the oretype boundaries are assessed at the "point" high resolution scale, based on the set of simulated 2D images.

3.2 Point and Block Data

BIS uses two datasets with different supports to produce simulations. The first is the "point" hard data (drill holes samples), the second the block data retrieved from the RI. The following methodology to characterize the blocks size and error was developed, based on the octrees concept, using a series of iterations (levels) as can be seen on Fig. 2. The idea is to produce smaller blocks with low uncertainty in the areas near the hard data and larger blocks with high uncertainty in the areas further from them.

Various methods can be used to attach uncertainty to a block. Since the larger blocks are farther from the "hard" data a linear function such as "larger the size, bigger the error" can be used. Also the (ordinary) indicator kriging variance (just with the "hard" data) can be used to attach uncertainty to each block.

4 Results and Conclusions

For the present study 100 block indicator simulations were performed and several results were retrieved, like the most likely oretype after 100 realizations (Fig. 3b). In order to map the spatial uncertainty and indentify what locations are more prone to uncertainty an error map was produced (1 minus the probability of the simulations to reproduce the same value as the RI). The error values range from 1 (all simulations

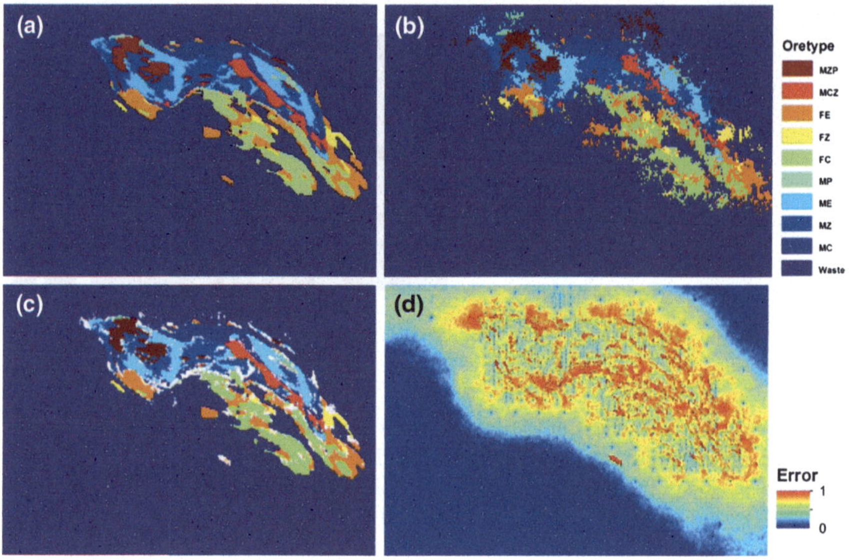

Fig. 3 **a** Reference Image (deterministic geological model); **b** the most likely simulated oretype; **c** RI with the identification of the most uncertain pixels in white; **b** error map

missed the value of the RI) to 0 (all simulations reproduce the value of the RI) (Fig. 3d). For a "hot spots" map an error of 0.8 and above was considered, i.e. where at least 80 % of the simulations produce a different oretype of the RI (Fig. 3c). An error (or contingency) table was also drawn between the RI (the deterministic geological model of the oretypes) and the most likely simulated map. The most likely map only delivers a different oretype than the one in the RI in 9.9 % of the pixels.

As expected this methodology has provided a map of the problematic areas (2.5 % of the study area in Fig. 3c) where there is greater difficulty to assert that a specific oretype should be present, such as in border zones or areas with many different small oretypes, or even a complex spatial pattern (highly variable local anisotropy).

References

1. Liu, Y., Journel, A. G. (2007). Geostatistical integration of coarse-scale and fine-scale data. In *Proceedings of the Annual Conference of IAMG* (P. 6). Beijing.
2. Relvas, J., Barriga, F., Ferreira, A., Noiva, P., Pacheco, N., Barriga, G. (2006). Hydrothermal alteration and mineralization in the neves-corvo volcanic-hosted massive sulfide deposit, Portugal. I. Geology, mineralogy, and geochemistry. *Economic Geology Society of Economic Geologists*, *101*, 753–790.

"Horse-Shoe" Cu-Au Porphyry Orebody Modeling Based on Blasthole Data Using Unfolding Technique

Mohamad Nur Heriawan, Loya Jirga and Anton Perdana

1 Introduction

Mineral exploration methods are intended to reveal the information regarding mineral resource, geological conditions and other associated information completely and accurately. This information enables the process of modeling and estimation, which in turn will help to improve the certainty level of mineral resource. Modeling and estimation processes vary greatly in terms of methods and problems. One of the problems is the folded ("horse-shoe") deposit. The folded form of deposit caused the estimation process to yield less accurate results. This is because of not stationarity distribution and also the significant differences of sample grade values will cause data misread. The solution to overcome this problem is a method called unfolding technique; see [1] and [2]. This process can be roughly described as an attempt to transform the coordinate system in which the folded deposit is relocated into a new one where the deposit is reformed into its original unfolded form. The unfolded deposit estimation result is known to yield more accurate result than the folded one, especially to generate the data to be more stationary.

M. N. Heriawan (✉)
Research Group of Earth Resources Exploration, Faculty of Mining and Petroleum Engineering, Bandung Institute of Technology, Jl. Ganesha 10, Bandung 40132, Indonesia
e-mail: heriawan@mining.itb.ac.id

L. Jirga
Department of Mining Engineering, Faculty of Mining and Petroleum Engineering, Bandung Institute of Technology, Jl. Ganesha 10, Bandung 40132, Indonesia
e-mail: eljirg@gmail.com

A. Perdana
Division of Geo Data and Modelling, PT. Freeport Indonesia, Tembagapura, Indonesia
e-mail: anton_perdana@fmi.com

E. Pardo-Igúzquiza et al. (eds.), *Mathematics of Planet Earth*,
Lecture Notes in Earth System Sciences, DOI: 10.1007/978-3-642-32408-6_159,

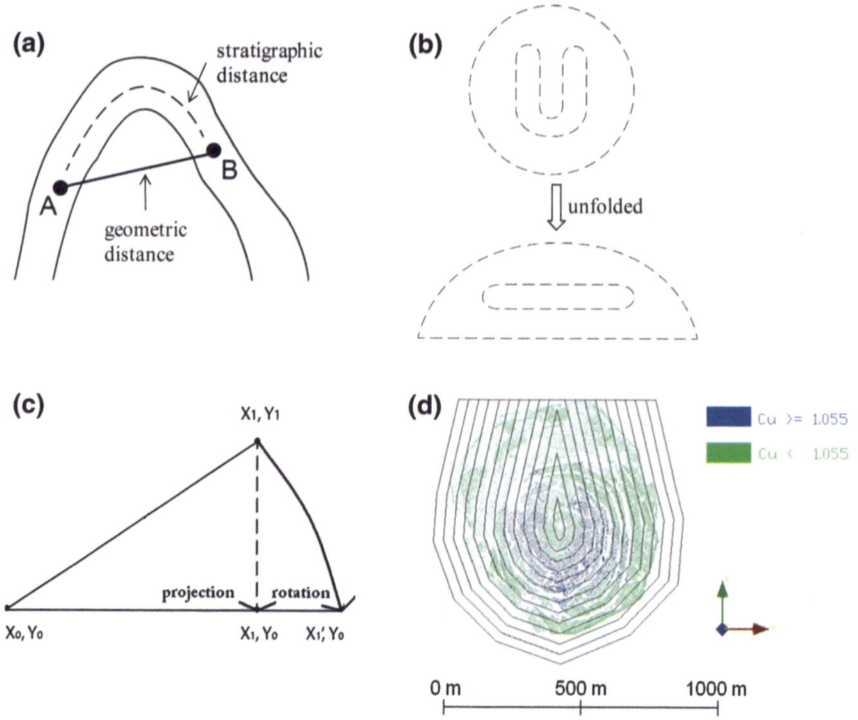

Fig. 1 **a** Geometric and stratigraphic distance between two points [2], **b** principle of unfolding technique, **c** projection and rotation coordinates transformation system, and **d** folding boundary on a "horse-shoe" likefolded deposit

2 Materials and Method

Fifteen meters data which originates from 3,585 m to 3,600 m elevation of a blasthole data in Grasberg Mine, Papua, Indonesia are examined. The sampled data are then prepared for the modeling and estimation process. Figure 1a illustrates two samples, i.e. A and B where each is located on one side of folded deposit. By using XYZ coordinate system, standard geometric distance of point A and B is gotten and is shown by the straight line. However from geological point of view, the real distance between these two points is shown by the broken line which follows the folded structure. In order to unfold the deposit, initial step is to know how the unfold process will be done. In this case, the unfolding process follows a simple guidance in form of boundaries (Fig. 1b).

How the folded boundaries transformed into unfolded lines are controlled by two methods of coordinate transformation. The first one is rotation and the second is projection (Fig. 1c). The boundaries are created firstly by following the rough form of folded deposit and then by using offset method, the boundary are duplicated on a

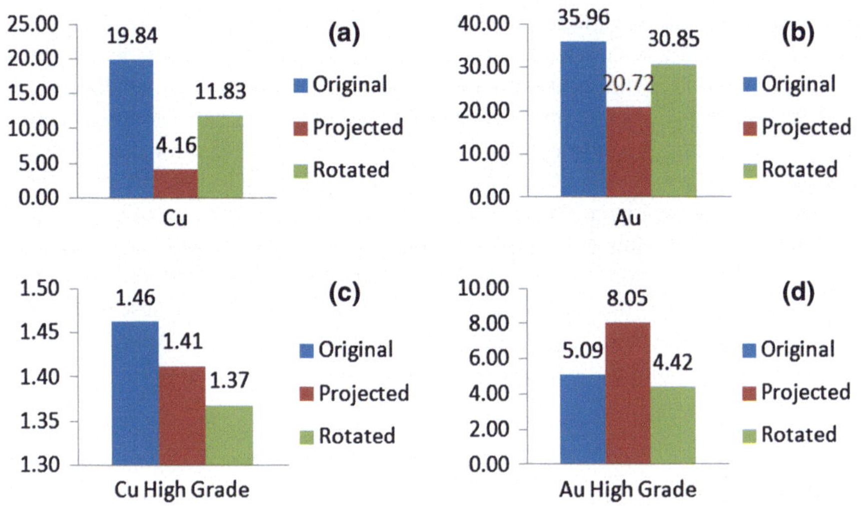

Fig. 2 **a-b** Mean of relative error for all Cu and Au estimates, **c-d** for high grade data only of Cu and Au estimates

set of distance until it covers the entirety of folded deposit such as shown in Fig. 1d. Each blasthole datum will then fall into one boundary. The blasthole data in each boundary will then undertake the unfolding process that transforms each folded form boundary into a straight line (in 2D plane). This process is then repeated for each 0.1 m drop in elevation until all 15 m is covered. Both the original blasthole data and the unfolded ones then undertake modeling and estimation processes.

3 Results and Discussion

The result of research shows that the original data set has larger variogram range than the projected unfolding data set and smaller than the rotated unfolding data set. This means that the rotated unfolding data set has a more homogenous data distribution compared to original data set and projected unfolding data set.

After variogram analysis was completed, kriging estimation can be performed to estimate the block. For original data set, grid was drawn following the extent of data distribution. For the two folded data, a different approach is required. For the unfolding rotated data, the grid were not drawn, instead the original grid data were unfolded and used for estimation. The purpose of this approach is so that the unfolding rotated grid can be refolded and then coincided to the original one. The estimation for unfolded projected data used grids that were drawn in the same way as the original data grids.

The statistical analysis shows that overall data with the smallest mean error (kriging variance) is unfolding projected data, while unfolding rotated data follows as second best. Variance kriging result from the unfolding processes and original data were analyses to know their relative error. Relative error can be used to define which modelling method is the most optimum. Figure 2a, b show that the smallest mean of relative error is owned by projected unfolding model. The mean of relative error for high grade data was also compared separately. Figure 2c, d show that the mean of relative error for high grade data were different compared to the all grade data. In high grade comparison, the lowest mean of relative error is owned by rotated unfolding model. Seeing data type aspect, high grade data were more representative to be used for determining the most optimum method. This is because the aim of this research is to estimate the "horse-shoe" deposit which is formed by these high grade data. By specifically observing mean of relative error of high grade data, it is known that the most optimum estimate derived from rotated unfolding model.

4 Conclusions

Some conclusions can be drawn from this research. Smallest mean kriging variance is owned by projected unfolding data with 11.55 ppm^2 for Au, 1.17 $\%^2$ for Cu, 11.50 ppm^2 for high grade Au and 1.14 $\%^2$ for high grade Cu. Statistical analysis from high grade data relative error showed that rotated unfolding method resulted the lowest relative error. The mean of relative error for that method was ± 4.42 % for Au and ± 1.30 % for Cu, while for the projected unfolding was ± 8.05 % for Au and ± 1.41 % for Cu, and for the original data was ± 5.09 % for Au and ± 1.46 % for Cu. From these results it is concluded that the optimum estimate of "horse-shoe" deposit derived from the rotated unfolding model.

Acknowledgments We are most grateful to PT. Freeport Indonesia for providing the dataset used in this research and also thank to Prof. Sutawanir Darwis and Mrs. Dila Puspita for their cordial discussion in the mathematical problems.

References

1. McArthur, G. J. (1988). Using geology to control geostatistics in the Hellyer deposit. *Mathematical Geology*, *20*(4), 343–366.
2. Newton, M. J. (1996). Structural analysis for folded deposit. *International Journal of Rock Mechanics and Mining Sciences and Geomechanics Abstracts*, *33*(6), 241 A.

A New Mathematical Approach to Model Trophic Dynamics of Mammalian Palaeocommunities. The Case of Atapuerca-TD6

Guillermo Rodríguez-Gómez, Jesús A. Martín-González, Idoia Goikoetxea, Ana Mateos and Jesús Rodríguez

1 Introduction

Trophic resource availability is one of the main constraints for the survival and dispersal of any species and it is generally accepted that animal resources were also essential to Pleistocene hominins in Europe [1, 2]. The ability of hominins to obtain animal resources from their environment is conditioned by the abundance of prey, their ecological characteristics, and the intensity of the competition with carnivores for these resources. Thus, several authors linked the survival opportunities of the first European hominins to their ability to compete with carnivores [3–5]. Two main factors should be taken into account to evaluate food resource availability for early Palaeolithic hunter-gatherers: (1) the amount of biomass that can be extracted from the populations of large herbivores and (2) the intensity of competition within the carnivore guild for those resources. Quantitative studies of the trophic dynamics of palaeocommunities have been used by palaeoecologists to understand trophic relationships in mammalian palaeocommunities from the Miocene [6] and the Pleistocene [7–9] and to evaluate the effects of extinctions and faunal turnover on ecosystem functioning during the mid-Pliocene [10]. Here we present a model that quantifies trophic resource availability for secondary consumers in a large mammal community, simulates resource distribution among those consumers and provides information about the intensity of competition inside the carnivore guild. We show the application of our model to the TD6 assemblage from the Atapuerca Gran Dolina, Burgos, Spain.

Departamento de Matemáticas y Computación, Universidad de Burgos, Burgos, Spain: Temporarily assigned to CENIEH.

G. Rodríguez-Gómez (✉) · I. Goikoetxea · A. Mateos · J. Rodríguez
Centro Nacional de Investigación sobre la Evolución Humana (CENIEH), Paseo Sierra de Atapuerca s/n, 09002 Burgos, Spain

J. A. Martín-González
Departamento de Matemáticas y Computación,Universidad de Burgos, Burgos, Spain

E. Pardo-Igúzquiza et al. (eds.), *Mathematics of Planet Earth*, Lecture Notes in Earth System Sciences, DOI: 10.1007/978-3-642-32408-6_160,

2 The Model

Our aim is to investigate the distribution of resources, primary consumer biomass, among secondary consumers in different scenarios and evaluate if the last may they reach viable densities. In order to achieve this aim, we propose a mathematical model. It provides the estimation of the biomass of primary consumers available to secondary consumers, i.e. the Total Biomass Output (TBO). Moreover, it is necessary to estimate the requirements of secondary consumers or Total Demanded Biomass (TDB). A detailed description of this model is provided in Rodríguez Gómez et al. [11] (Fig. 1).

2.1 Total Biomass Output (TBO)

2.1.1 Input

Physiological Variables

Input data are physical and physiological variables (adult body mass, body mass at birth, litter size, breeding interval, age at reproductive maturity, growth rate and lifespan) which are species specific. The values of these physiological variables are taken from the literature or estimated through allometric equations.

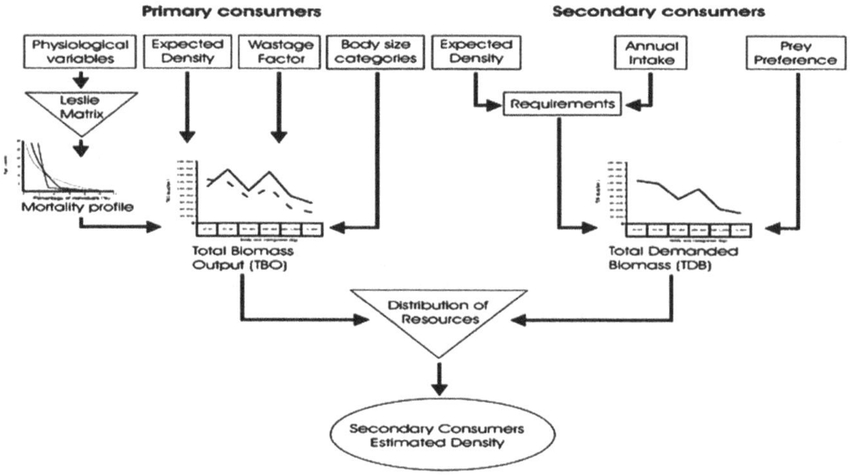

Fig. 1 Flow Diagram showing the components of the model used to evaluate trophic resource availability and intraguild competition

Expected Density

The model solutions are not dependent on population size, thus, an estimate of population density is needed to estimate the sustainable Biomass Output. Allometric equations are used to estimate primary consumers density e.g. [12–14].

Body Size Category

Each individual of a primary consumer species is classified into one of six body size categories according to its body mass at the age of death: 10–45, 45–90, 90–180, 180–360, 360–1,000, >1,000 kg see [15].

Wastage factor

We use the wastage factors from Viljoen [16] to account for the fact that a carcass includes a variable amount of non edible tissues (horns, bones, hide...).

2.1.2 Process: Leslie Matrix

Our model is developed under the assumption that all the variations in population size and composition may be taken as oscillations around a mean value that is constant through time, an assumption widely accepted in population dynamics studies [17]. We represent the average long term condition of every population using the Leslie Matrix [18, 19]:

$$X_{t+1} = \begin{bmatrix} a_1 & a_z & \cdots & a_n \\ b_1 & 0 & \cdots & 0 \\ \vdots & \ddots & \ddots & \vdots \\ 0 & \cdots & b_{n-1} & 0 \end{bmatrix} X_t \tag{1}$$

where a_i is fecundity, the per capita number of female offspring and b_i is the proportion of individuals that survive from age class i to age class i+1. The number of individuals of age class i died each year, whatever the cause is given by $1 - b_i$. The interval between age classes is one year and only female individuals are represented in a Leslie matrix. Leslie Matrix Inputs are physiological variables of section *Physiological Variables*. We assumed that sex ratio is equal to 1:1, that population profile is the same for males and females and that the survival rate is equal for both sexes. A further assumption is that sub-adult survival rate should be lower than adult survival rate [11].

Two additional conditions are introduced: (1) The population should be stable (i.e., the population size should be constant from year to year); (2) The population should be stationary (i.e., the age structure should be constant from year to year). These restrictions can be interpreted as the condition that makes the population structure stable and it occurs when the net reproduction rate (NRR) is equal to one.

The population profiles obtained from this model for every primary consumer population provide an estimation of the average sustainable biomass output by age classes (mortality profile), which are eventually translated into biomass per body size intervals. Total Biomass Output (TBO) is obtained as the sum of the biomass output in each size category from each primary consumer population.

2.1.3 Output

Combining the mortality profiles obtained in from the Leslie Matrix with the mean body size per age class, the estimated population density of the species and the wastage factor, the Total Biomass Output (TBO) is computed.

2.2 Total Demanded Biomass (TDB)

2.2.1 Input

Expected Density

Estimated using allometric equations e.g. [12–14].

Annual Intake

Carnivore intake rate was estimated using allometric equations e.g. [20].

Prey Preference

Six predation body size categories, equal to the primary consumer categories are defined and the preferences of each carnivore recorded: 10–45, 45–90, 90–180, 180–360, 360–1,000, >1,000 kg see [15].

2.2.2 Process: Requirements

The annual energetic requirements of a carnivore population by km^2 are obtained multiplying the individual annual intake by population density.

2.2.3 Output: Total Demanded Biomass (TDB)

Combining the Prey Preferences and the Requirements of each secondary consumer, Total Demanded Biomass (TDB) per body size categories is obtained.

2.3 Distribution of TBO Between Secondary Consumers

2.3.1 Input

Total Biomass Output (TBO) and Total Demanded Biomass (TDB) are the inputs for this process.

2.3.2 Process

Distribution of TBO between secondary consumers is based on the Proportional Predation Pressure (PPPij) of each species in each body size category. PPPij represents the relative amount of biomass demanded by the jth carnivore species from the ith prey body size category. It is calculated as the proportion of the total amount of biomass demanded from a prey body size category by all carnivores that corresponds to the requirements of a single carnivore species. See a detailed description of the computation of resource distribution in Rodríguez-Gómez et al. [11].

2.3.3 Output

This process provides estimated densities for the secondary consumers as output.

3 Application

We selected the TD6 assemblage from the Atapuerca Gran Dolina site (Burgos, Spain) (approximately 0.8 Ma.) because it has been considered as a key sample in resolving several palaeoeconomic issues related to the populations that inhabited Europe approximately one million years ago [21] and references therein. This is the single European site from this period where hominin remains have been found together with abundant faunal remains and a rich collection of lithic artifacts. Thus, the TD6 assemblage provides both biological information about the first European settlers and evidences of their nutritional and cultural activities.

We applied our model in TD6 to know if there was enough resources to maintain a stable human group in this ecosystem. Our results show that meat resources covered all secondary consumer requirements. Our results support the interpretation of a rich ecosystem at Atapuerca at the end of the Early Pleistocene that boasted a level of secondary production sufficient to maintain a well diversified guild of secondary consumers including a human population.

4 Conclusion

Mathematical modelling of palaeocommunity trophic dynamics is a useful tool for investigating food resource availability for Palaeolithic populations. Our model determines the age structures that make the populations of primary consumers stable, the average biomass that can be sustainably extracted in the long term and its distribution in body size categories. Thus, our model provides insights into competition for resources among secondary consumers. When the model is applied to the TD6 assemblage, the results suggest that resources were abundant enough to support the carnivore guild at maximum densities, i.e. it was a rich ecosystem for secondary consumers, including *Homo*.

Acknowledgments This research was funded by the MINECO project, CGL2012-38434-C03-02. G. Rodríguez-Gómez is the beneficiary of a pre-doctoral FPI Grant from the Spanish MICINN. Fieldwork at Atapuerca sites was funded by the Junta de Castilla y León.

References

1. Hublin, J. -J., & Richards, M. P. (2009). *The evolution of hominin diets subsistence. Vertebrate paleobiology and paleoanthropology series*. Leipzig: Springer.
2. Roebroeks, W. (2001). Hominid behaviour and the earliest occupation of Europe: An exploration. *Journal of Human Evolution*, *41*, 437–461.
3. Martínez-Navarro, B., & Palmqvist, P. (1996). Presence of the African Sabertoothed Felid Megantereon whitei (Broom, 1937) Mammalia, Carnivora, Machairodontinae in Apollonia-1 (Mygdonia Basin, Macedonia, Greece). *Journal of Archaeological Science*, *23*, 869–872.
4. Palombo, M. R. (2010). A scenario of human dispersal in the northwestern Mediterranean throughout the early to middle pleistocene. *Quaternary International*, *223–224*, 179–194.
5. Arribas, A., & Palmqvist, P. (1999). On the ecological connection between sabre-tooths and hominids: Faunal dispersal events in the lower Pleistocene and a review of the evidence for the first human arrival in Europe. *Journal of Archaeological Science*, *26*, 571–585.
6. Vizcaíno, S. F., Bargo, M. S., Kay, R. F., Fariña, R. A., Giacomo, M. D., Perry, J. M. G., et al. (2010). A baseline paleoecological study for the Santa Cruz Formation (late-early Miocene) at the Atlantic coast of Patagonia, Argentina. *Palaeogeography, Palaeoclimatology, Palaeoecology*, *2923–4*, 507–519.
7. Fariña, R. A. (1996). Trophic relationships among lujanian mammals. *Evolutionary Theory*, *11*(2), 125–134.
8. Palmqvist, P., Gröcke, D. R., Arribas, A., & Fariña, R. A. (2003). Paleoecological reconstruction of a lower Pleistocene large mammal community using biogeochemical ($\delta^{1}3C$, $\delta^{1}5N$, $\delta^{1}8O$, Sr: Zn) and ecomorphological approaches. *Paleobiology*, *29*(2), 205–229.
9. Bermúdez De Castro, J.M., Díez Fernández-Lomana, J.C., Mosquera Martínez, M., Nicolás Checa, M.E., Pérez Pérez, A., Rodríguez Méndez, J., et al. (1995). El nicho ecológico de los homínidos del Pleistoceno Medio de Atapuerca. *Complutum*, *6*, 9–56.

10. Vizcaíno, S. F., Fariña, R. A., Zárate, M. A., Bargo, M. S., & Schultzd, P. (2004). Palaeoecological implications of the mid-Pliocene faunal turnover in the Pampean region (Argentina). *Palaeogeography, Palaeoclimatology, Palaeoecology*, *213*(1–2), 101–113.
11. Rodríguez-Gómez, G., Rodríguez, J., Martín-González, J.Á., Goikoetxea, I., Mateos, A.: Modelling trophic resource availability for the first human settlers of Europe: The case of Atapuerca-TD6. *Journal of Human Evolution*. doi:10.1016/j.jhevol.2013.02.007 (In Press).
12. Damuth, J. (1981). Population density and body size in mammals. *Nature*, *290*, 699–700.
13. Silva, M., & Downing, J. A. (1995). The allometric scaling of density and body mass: a nonlinear relationship for terrestrial mammals. *The American Naturalist*, *145*(5), 704–727.
14. Silva, M., Brimacombe, M., & Downing, J. A. (2001). Effects of body mass, climate, geography, and census area on population density of terrestrial mammals. *Global Ecology and Biogeography*, *10*, 469–485.
15. Rodríguez, J., Rodríguez-Gómez, G., Martín-González, J. A., Goikoetxea, I., & Mateo, A. (2012). Predator-prey relationships and the role of homo in early pleistocene food webs in Southern Europe. *Palaeogeography, Palaeoclimatology, Palaeoecology*, *365–366*, 99–114.
16. Viljoen, P. C. (1993). The effects of changes in prey availability on lion predation in a natural ecosystem in northern. *Zoological Symposium*, *65*, 193–213.
17. Owen-Smith, N. (2010). The suite of population models. In: N. Owen-Smith (Ed.), *dynamics of large herbivore populations in changing environments* (pp. 34–62) Oxford: Wiley-Blackwell.
18. Leslie, P. H. (1945). On the use of matrices in certain population mathematics. *Biometrika*, *33*, 183–212.
19. Leslie, P. H. (1948). Some further notes on the use of matrices in population mathematics. *Biometrika*, *35*, 213–245.
20. Farlow, J. O. (1976). A consideration of the trophic dynamics of a late cretaceous large-dinosaur community (Oldman Formation). *Ecology*, *57*(5), 841–857.
21. Saladié, P., Huguet, R., Díez, C., Rodríguez-Hidalgo, A., Cáceres, I., Vallverdú, J., et al. (2011). Bermúdez De Castro, J.M., Carbonell, E.: Carcass transport decisions in Homo antecessor subsistence strategies. *Journal of Human Evolution*, *61*, 425–446.

Multifractals and Capacity Dimension as Measures of Disturbance Patch Dynamics in *Daedalus* Ichnofabrics

Carlos Neto de Carvalho and Andrea Baucon

1 Introduction

Behavior is the set of strategically and flexible responses of any phenotype for purposes of protection and transmission of its genetic legacy. This programmed responses allow the homeostatic development necessary to an organism exert some control over its ecosystem, which is intrinsically unpredictable. The fossilized remains of biological activity or ichnofossils are, in this context, the preserved solutions for the chaotic and aperiodic ecological constraints that define the evolution of behavior. The resulted programs were developed and modified by genetic pre-adaptations and by ecological parameters originally prevailing, that may have been preserved or even changed in the geological record by diagenesis imprint and tectonics. Ichnology, the discipline of paleontology focused on organism–substrate interaction dynamics (description, classification and interpretation), is a fundamental link between the evolutionary biological mechanisms and geological processes. In fact, ichnofossils represent the functional morphology and behavior of organism producers as well as the physical–chemical properties of the substrate where they were generated [1]. As ichnofossils are fossilized snapshots of behavioral and physiological functions, they translate organism producers condition to specific ecological situations. An ichnofabric includes all sediment structure and textural changes

C. Neto de Carvalho (✉) · A. Baucon
Geopark Naturtejo da Meseta Meridional—European and Global Geopark, Geology Survey of Idanha-a-Nova, Centro Cultural Raiano, Av. Joaquim Morão, 6060-101 Idanha-a-Nova, Portugal
e-mail: carlos.praedichnia@gmail.com

C. Neto de Carvalho
Geology Research Centre, University of Lisbon, C6, 3rd floor, office 6.3.57, 1749-016 Lisbon, Portugal

A. Baucon
Dipartimento di Scienze della Terra, Università di Milano, 20133 Milano, Italy
e-mail: andrea@tracemaker.com

E. Pardo-Igúzquiza et al. (eds.), *Mathematics of Planet Earth*, Lecture Notes in Earth System Sciences, DOI: 10.1007/978-3-642-32408-6_161,

resulting from bioturbation and bioerosion at all scales [2]. Methods have been developed to improve quantification of past and present biological sedimentary reworking. Quantifying bioturbation has been a very useful tool for supporting evolutionary theories of ecological radiation, among many other applications (see for example [3]). Descriptive or semi-quantitative methods were developed for analysis of ecospace used by benthic communities in shifting substrate environments based on bioturbation percentual indices, e.g. Refs. [3–6]. Our research is based in new methods for quantifying bioturbation and ichnofabrics, based on fractal geometry, which permits, at the same time, to recognize with precision the ways in which benthic communities organize and explore their ecological niches and the ecological succession with the temporal evolution of environmental parameters. As application, we used fossil domiciliary/feeding behavior patterns classified as *Daedalus halli*, occasionally very abundant in siliciclastics deposited in sandflat subtidal marine bottoms almost 480 million years ago that are commonly present in the Floian-age Armorican Quartzite Formation from Portugal. For this study we quantified dense ichnofabrics of *Daedalus* in the sections of Serra de Barreiras Brancas [7], and Angueira valley, near Serapicos. *Daedalus* ichnofossils are burrows of a still unknown worm-shaped producer that reworked sands in the three spatial directions by the coiled, protrusive or retrusive displacement of a vertical-to-oblique J-tube [8, 9] intersecting bedding planes normal or at high-angle, and reaching 50 cm deep.

2 Fractal and Multifractal Estimation of Bioturbation Spatial and Temporal Heterogeneities

One of the primary goals in the use of fractals is that statistically descriptive methods were developed in order to better describe natural forms. In this way, the measure of fractal dimension of certain ichnofossils is of particularly interest in more quantitative approaches to ichnotaxonomy [10–12]. Meanders, spirals or branches in feeding biogenic structures are ways of food processing and foraging that increase extraordinarily the explored area in a restricted volume, as well as collecting information from environmental stimuli. Fractal dimension in the ichnological context determines the probability of sediment to be covered by the producer' behavioral activity. Capacity fractal dimension estimation is particularly useful in the study of ichnofossils with complex patterns, which occur in sedimentary bedding planes. For fractal determination it is used the Box Counting theorem as described in Refs. [10, 11]. We used the free software FRACTALYSE 2.4.1 to estimate capacity dimension of monochrome sketches (.bmp) of *Daedalus halli*. Those images are covered with a grid of side a, the software counting the minimal number of squares $N(a)$ which includes bioturbation. Counting should be repeated M times for different a sizes (M $>$ 2 orders of magnitude). The absolute value of slope of the regression line adjusted to Richardson plot of $\log N(a)$ versus $\log a$ corresponds to Fractal Dimension by the relation

$$\log N(a) \propto \log(const) + D \log(1/a) \tag{1}$$

Standard deviation of error describes the adjustment quality of regression line. Approaching to 0, $\log N(a)$ and $\log(1/a)$ both reach a very high value when compared with $\log(const)$. In the limit when a tends to 0, it is obtained the exact definition of Hausdorff or Capacity Dimension

$$D_0 = \lim_{n \to 0} \frac{\log N(a)}{\log(1/a)} \tag{2}$$

Multifractal describes several processes with fractal signatures (fractals or pseudofractals) when characterized at multiple levels of information. Multifractal spectra can be used as a measure of spatial or sequential ichnofabric heterogeneities. It is based on the box counting method thus enabling quantitative analysis of ecospace occupation strategies translated for behavioral forms preserved on the bedding planes. The heterogeneity measure is given by a probability distribution which can be estimated, according to [13], as

$$P_\upsilon(a) = \frac{N_\upsilon(a)}{N} \tag{3}$$

where $N_\upsilon(a)$ is the number of identified bioturbations within the v-square grid and N is the total number of *Daedalus* in the digitalized image. This probability$v$ fits in the equation that quantifies density distribution in Ref. [14],

$$D(q) = \log_{n \to 0} \frac{1}{q-1} \frac{\log \sum_{\upsilon=1}^{N(a)} [p_\upsilon(a)]}{\log a} \tag{4}$$

where v identifies the square of side a and $p_\upsilon(a)$ registers the relative weight of v-square as expression of the total density. Generalized fractal dimensions spectra, $D(q)$, quantify non-uniformity in ichnofossil distribution density, which is dependent of q momentum [13].

3 Discussion: Evaluating Community Strategies in Space and Time for the *Daedalus* Producer

Multifractal D(q) anisotropies along the 40 m^2 bedding plane at Martim Preto shows spatial heterogeneity ($1,61 \pm 0.07 \leq D(q) \leq 1.89 \pm 0.14$). A non-uniform distribution of ichnofossils may be consequence of intraspecific low competition for resources, which actually does not promote the regular distribution of organisms. Otherwise, discrimination of a faint clustering by multifractal analysis could be a sign of stable and favorable environmental conditions during *Daedalus* fabrication by producers' population. This patchy distribution pattern opens the possibility for a model of meiofauna harvesting to explain *Daedalus halli* architecture based on the subtidal pump mechanism of interstitial water exchange by wave action.

Recurrence of tempestitic phenomena and its persistent effects in ecological succession can be evaluated from fractal data obtained in the Serapicos stratigraphic section. We measured bed-by-bed fractal frequence of *Daedalus* for 36 bedding planes. About 75.5 % of observed layers are bioturbated, with a Capacity Dimension mean value of $D_0 = 1.62 \pm 0.15$. Data shows that subtidal substrate settlement after each tempestitic event by *Daedalus* producers was intense, mostly multigenerational, with an exclusive and significant occupation of the emptied ecospace.

The presence of only one preserved behavioral strategy, the substrate depth affected by these structures, the high density of burrows and passive patchiness rates are characteristics of r-selected populations. Such large-scale and frequent disturbance events as storms, extreme on a gradient of disturbance intensities, were responsible for some of the earliest opportunistic behaviors in the fossil record. *Daedalus* was among the most resilient of them in the Early Paleozoic.

References

1. Pemberton, S. G., Frey, R. W., Saunders, T. D. A. (1990). Trace fossils. In D. E. G. Briggs, & P. R. Crowther (Eds.), *Palaeobiology: a synthesis* (pp. 355–361). Oxford : Blackwell Science.
2. Ekdale, A. A., & Bromley, R. G. (1983). Trace fossils and ichnofabrics in the Kjolby Gaard Marl, uppermost Cretaceous, Denmark. *Bulletin of the Geological Society of Denmark, 31*, 107–119.
3. Taylor, A., Goldring, R., & Gowland, S. (2003). Analysis and application of ichnofabrics. *Earth-Science Reviews, 60*, 227–259.
4. Reineck, H.-E. (1963). Sedimentgefuege im Bereich der suedlichen Nordsee. *Abhandl. der Senckenb. Naturforsch. Gesellsh., 505*, 1–138.
5. Droser, M. L., & Bottjer, D. J. (1986). A semiquantitative field classification of ichnofabric. *Journal of Sedimentary Petrology, 56*, 558–559.
6. Miller, M. F., & Smail, S. E. (1997). Multifractal characterization of microbially induced magnesian calcite formation in Recent tidal flat sediments. *Palaios, 12*, 391–396.
7. Sá, A. A., Meireles, C., & Coke, C. (2002). Concentração maciça de Daedalus labechei (Rouault) (Icnofóssil Ordovícico) no Alto do Martim Preto (Guadramil-Bragança): Património paleontológico a preservar e divulgar. In J. Civis , & J. A. González Delgado (Eds.), *XVIII Jornadas de la Sociedad Española de Paleontología/II Congreso Ibérico de Paleontología* (pp. 138–139). Salamanca: Libro de Resúmenes.
8. Durand, J. (1985). Le Grés Armoricain. Sédimentologie, Traces Fossiles, Milieux de dépôt. Mem. et Doc. du Centr. Armor. d'Etude Struct. des Socles 3, pp. 1–119.
9. Seilacher, A. (2000). Ordovician and Silurian arthrophycid ichnostratigraphy. In: M. A. Sola, & D. Worsley (Eds.), *Geological exploration in Murzuq Basin* (pp. 237–258). Amsterdam: Elseiver.
10. Gibert, J. M., Jeong, K., & Martinell, J. (1999). Ethologic and ontogenetic significance of the Pliocene trace fossil Sinusichnus sinuosus from the northwestern Mediterranean. *Lethaia, 32*, 31–40.
11. Neto de Carvalho, C. (2006). Roller coaster behaviour in the Cruziana rugosa group from Penha Garcia (Portugal): implications for the feeding program of Trilobites. *Ichnos, 13*(4), 255–265.
12. Baucon, A. (2010). Da Vinci's Paleodictyon: the fractal beauty of traces. *Acta Geologica Polonica, 60*(1), 3–17.

13. Kropp, J., Block, A., Bloh, W., Klenke, T., & Schellnhuber, H. J. (1997). Multifractal characterization of microbially induced magnesian calcite formation in Recent tidal flat sediments. *Sedimentary Geology*, *109*, 37–51.
14. Hentschel, H. G. E., & Procaccia, I. (1983). The infinite number of generalized dimensions of fractals and strange atractors. *Physica D*, *8*, 4435–4444.

Non-Linear Thermo-Mechanics of Folding in Geomaterials

Martin K. Paesold, Ali Karrech, Tim Dodwell, Klaus Regenauer-Lieb, Andrew P. Bassom, Alison Ord and Bruce E. Hobbs

1 Introduction

Geologists and engineers rely on the anomalies within the Earth's crust to predict its history and content in terms of profitable resources. The mechanisms that produce anomalies such as folds, boudinage, fractures and/or faults are the subject of considerable research efforts. In particular, the mechanics of folding has been studied extensively [1–4]. However, most of the existing approaches ignore the variations of temperature, large transformations (deformations and rotations) and/or the coupling between the different processes that are involved.

In this paper, a non-linear thermodynamics description of folding in geomaterials under external loading is presented. The coupled equations of motion and heat transfer are integrated numerically using the finite element method. To this end, a novel frame indifferent material description suitable for finite strain and large rotations is employed [5, 6]. Hence, the temporal evolution of folds under various conditions is studied.

M. K. Paesold (✉)
School of Mathematics and Statistics, University of Western Australia, Crawley, Australia
e-mail: 21102578@student.uwa.edu.au

A. Karrech
School of Civil and Resource Engineering, University of Western Australia, Crawley, Australia

T. Dodwell
Centre for Nonlinear Mechanics, University of Bath, Bath BA2 7AY, UK

K. R. -Lieb
Earth Science and Resource Engineering, CSIRO, Kensington, Australia

A. P. Basson
School of Mathematics and Statistics, University of Western Australia, Crawely, Australia

A. Ord · B. E. Hobbs
School of Earth and Envirnoment, University of Western Australia, Crawely, Australia

E. Pardo-Igúzquiza et al. (eds.), *Mathematics of Planet Earth*,
Lecture Notes in Earth System Sciences, DOI: 10.1007/978-3-642-32408-6_162,

2 Problem Definition

The problem at hand consists of solving the fully coupled thermo-mechanical behaviour of geomaterials subjected to external loading. At high temperatures, these materials exhibit permanent elasto-visco-plastic deformations due to plate tectonics. In this paragraph, we describe the governing equations as well as the constitutive material behaviour within the context of frame indifferent finite strain and large rotations (see also [5, 6]). The differential equation describing the stress distribution reads:

$$\nabla \cdot \boldsymbol{\sigma} + \rho \mathbf{f} = 0 \quad \text{with} \quad \boldsymbol{\sigma} \cdot \mathbf{n} = \mathbf{T}_d \text{ at } \Gamma_\sigma, \quad \mathbf{u} = \mathbf{u}_d \text{ at } \Gamma_u \tag{1}$$

where ρ is the material density, $\mathbf{f}$ is the body force, Γ_u and Γ_σ are the surfaces where the Dirichlet and Neumann boundary conditions are applied: $\mathbf{u}_d$, prescribed displacement, and $\mathbf{T}_d$, prescribed surface traction.

The equation of heat transfer can be derived from the conservation of energy [5, 6], which results in:

$$\rho C_p \frac{dT}{dt} = \nabla \kappa \nabla T + \chi \boldsymbol{\sigma} : \mathring{\mathbf{h}}^p \quad \text{with} \quad \nabla T = \mathbf{q}_d \text{ at } \Gamma_q, \quad T = T_d \text{ at } \Gamma_T \tag{2}$$

where C_p is the heat capacity, κ is the thermal conductivity, χ is a coefficient which describes the fraction of dissipation generating heat, $\mathbf{h}^p$ is the plastic deformation and Γ_q and Γ_T are the frontiers with prescribed heat flux $\mathbf{q}_d$ and temperature T_d, respectively. The notation $\mathring{(\cdot)}$ refers to the corotational rates which will be explained in detail in the next paragraph.

The motion of a deforming body is described by the deformation mapping $\varphi(\mathbf{X}, t) = \mathbf{x}$ that relates points in a reference configuration, $\mathbf{X}$, with points in the current configuration $\mathbf{x}$ at time t. Based on the deformation mapping, the Hencky strain tensor is defined:

$$\mathbf{h} = \left(\ln \mathbf{F}\mathbf{F}^t\right)/2 \qquad \text{with} \qquad \mathbf{F} = \partial\varphi/\partial\mathbf{X}. \tag{3}$$

An independent observer of the deforming materials would measure a corotational rate of the strain tensor $\mathring{\mathbf{h}} = \dot{\mathbf{h}} + \mathbf{h}\Omega - \Omega\mathbf{h}$ with Ω a spin. A spin that gives objective corotational rates is:

$$\Omega = \mathbf{w} + \sum_{A,B=1,A\neq B} \left(\frac{\lambda_A + \lambda_B}{\lambda_A - \lambda_B} + \frac{2}{\ln\lambda_A - \ln\lambda_B}\right) \mathbf{p}_A \mathbf{D} \mathbf{p}_B, \tag{4}$$

with $\mathbf{p}_A = \mathbf{n}_A \otimes \mathbf{n}_A$ the eigenprojection of $\mathbf{F}\mathbf{F}^t$, λ_A is the corresponding eigenvalue, and $\mathbf{w}$ and $\mathbf{D}$ are the classic spin and rate of deformation, respectively.

The materials studied here undergo elasto-visco-plastic deformations. The elastic behaviour is given by $\sigma = \mathbf{C} : (\mathbf{h} - \mathbf{h}^p)$, where $\mathbf{C}$ denotes the fourth-order elasticity tensor which is a function of Young's modulus, E, and Poisson's ratio, ν. The onset of visco-plastic behaviour is defined by the elasticity envelope:

$$f(\sigma, \dot{\varepsilon}^v) = \sigma_{eq} - \sigma_0 - \psi^{-1}(\dot{\varepsilon}^v) = 0. \tag{5}$$

Here, ψ^{-1} is the inverse of the considered power-law rheology $\dot{\varepsilon}^v = \psi(\sigma) = A_p \sigma^n \exp(-Q/(RT))$ where the factor A_p, the exponent n and the activation energy Q are material constants. The temperature is denoted with T and the gas constant with R. The equivalent stress and yield stress are denoted by σ_{eq} and σ_0, respectively.

3 Numerical Application

In order to study the temporal evolution of folds, we simulated a feldspathic vein embedded in quartzite using the finite element method as implemented in ABAQUS. The frame indifferent material description outlined above was implemented in the Fortran user material subroutine UMAT. The samples studied are 5 km long, 2 km thick and subjected to a strain rate of 10^{-12} Hz. The material properties used in our study are listed in Table 1. The heat capacity depends polynomially on temperature.To initialize the simulations, elements were chosen randomly and the assigned Young's modulus, E, and yield stress, σ_0, perturbed. The prescribed boundary condition are a 30 K/m geotherm, geostatic pressure, gravity and the load due to the weight of the overlying material. The element type CPE8T with approximate size 30 m was used. The simulation time was set to 9×10^{11} s ($\sim$ 28′000 a).

In our simulations, the depth of the sample is varied and therefore pressure and temperature. The thickness of the feldspathic layer is chosen to be 2 m, 10 m, 20 m or 40 m. As an example, Fig. 1 shows a folded feldspathic layer buried at a depth of 10 km. The resulting fold is aperiodic and irregular.

Table 1 Typical material properties as documented in literature

Parameter	A_p	n	Q	ρ	α	E	ν	κ
unit	(MPa^{-n}/s)	-	(kJ/mol)	(kg/m^3)	(10^{-5}/K)	(GPa)	-	(J/(m s K))
Feldspar	3.3×10^{-4}	3.2	238	2620	1.53	103	0.3	1.7
Quartz	6.32×10^{-12}	4	135	2648	2.43	75	0.125	1.6

The thermal expansion coefficient is α

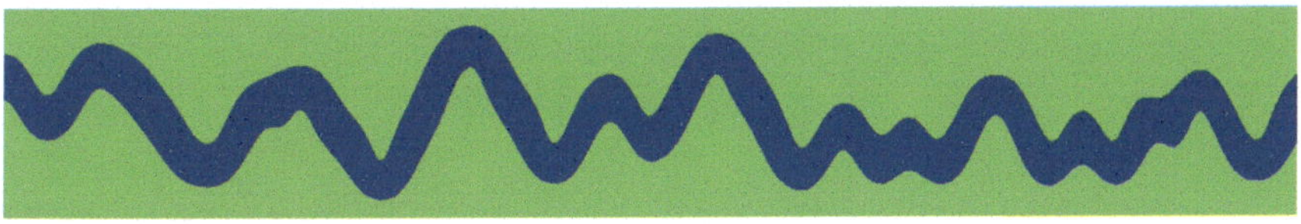

Fig. 1 Folded feldspathic vein (*blue*) embedded in quartzite (*green*). The initial thickness of the feldspathic vein is 2 m. The sample was buried at a depth of 10 km. The lateral dimension of the deformed sample is 500 m

4 Conclusions

A material description was presented that is suitable for finite strain and large rotations, and therefore for geological scenarios. Implementing this material description in the finite element method enabled us to simulate the temporal evolution of folds in geomaterials. Based on these simulations, the bifurcation behaviour of folding in dependence of temperature and pressure was investigated. Aperiodic folds were observed.

Acknowledgments This work is supported by the Australian Postgraduate Award and the International Postgraduate Research Scholarship.

References

1. Biot, M. A. (1965). *Mechanics of incremental deformations*. New York: Wiley.
2. Johnson, A. M., & Fletcher, R. C. (1994). *Folding of viscous layers*. New York: Columbia University Press.
3. Hobbs, B., Regenauer-Lieb, K., & Ord, A. (2008). Folding with thermal-mechanical feedback. *Journal of Structural Geology*, *30*(12), 1572.
4. Hudleston, P. J., & Treagus, S. H. (2010). Information from folds: A review. *Journal of Structural Geology*, *32*(12), 2042.
5. Karrech, A., Regenauer-Lieb, K., & Poulet, T. (2011). Frame indifferent elastoplasticity of frictional materials at finite strain. *International Journal of Solids and Structures*, *48*(34), 397.
6. Karrech, A., Regenauer-Lieb, K., & Poulet, T. (2011). Continuum damage mechanics for the lithosphere. *Journal of Geophysical Research*, *116*, B04205.

Analytical and Numerical Investigation of 3D Multilayer Detachment Folding

Naiara Fernández and Boris Kaus

1 Introduction

Folding is a common mode of deformation in geology when mechanically layered rocks are subjected to compression. They occur on a wide range of scales from mm-scale to 10's of kilometers in places where upper crustal rocks are compressed as a result of tectonic plate convergence (such as in the Zagros mountains, Iran). The physical instability that results in the formation of folds has been studied for different rheologies: e.g. elastic, viscous and visco-elastic [1], power law [2] and more recently, also for visco-elasto-plastic [3]. However, most of the studies focus on the problem of a competent layer embedded in a matrix [1] or the case of a multilayer system embedded in a matrix [4], and therefore they consider only two different material properties. This is appropriate for small-scale folds, but on a crustal scale it was recently demonstrated that the viscosity contrast between various sedimentary layers also plays an important role [3]. Here, we therefore focus on the case of a multilayer system overlying a matrix or lower detachment layer [4, 5] affected by gravity with three different material viscosities: lower detachment or salt layer (η_s), overlying weak layers (η_w) and competent layers (η_c).

N. Fernández (✉) · B. Kaus (✉)
Institute of Geosciences, Johannes Gutenberg University Mainz, J.-J.-Becher Weg 21, 55128 Mainz, Germany
e-mail: fernande@uni-mainz.de

B. Kaus
e-mail: kaus@uni-mainz.de

E. Pardo-Igúzquiza et al. (eds.), *Mathematics of Planet Earth*, Lecture Notes in Earth System Sciences, DOI: 10.1007/978-3-642-32408-6_163,

2 Methods

The equations that describe slow moving geological processes are the incompressible Stokes equations for layers with strongly varying (Newtonian) viscosity. These equations can be solved analytically for a multilayer set up (assuming small deformations), and numerically using a (parallel) finite element method.

The analytical solution we use is a thick plate stability analysis [6]. While thin plate is a valid approximation for one layer problem in which the shear stresses between the competent layer and matrix can be omitted in certain cases, the thick plate is a more suitable approach for the multilayer case where the shear stresses occurring at different interfaces cannot be ignored.

The numerical models are performed using Lithosphere and Mantle Evolution Model (LaMEM), which is a 3D finite element code that solves the governing equations in a velocity-pressure formulation on massively parallel supercomputers using iterative multigrid preconditioners. In the simulations shown here, we use higher-order elements in a Lagrangian manner and do not remesh which assures high accuracy.

3 Multilayer Folding Phase Diagrams

The 2D semi-analytical solution is used to create mechanical phase diagrams in the parameter space defined by two viscosity ratios ($R1 = \eta_c / \eta_s$ and $R2 = \eta_c / \eta_w$) that are larger than one in nature (Fig. 1). Such diagrams show that different multilayer

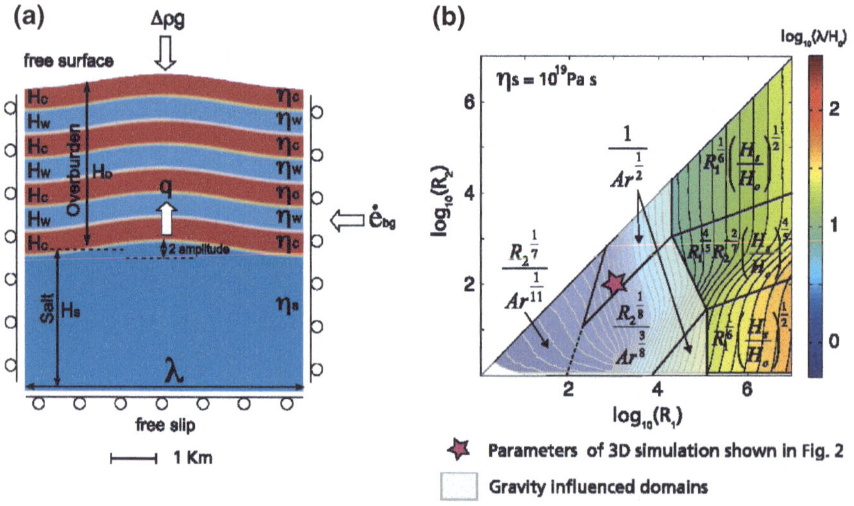

Fig. 1 **a** Model setup. **b** Phase diagram of normalized wavelength in the $R_1 - R_2$ space for $\eta_s = 10^{19}$ Pas, based on over 6000 results of the analytical solution. Parameters are defined in text

folding modes occur, each characterized with a scaling law for dominant wavelength (λ) and growth rate (q) that accounts for all model parameters (salt viscosity η_s, competent layer viscosity η_c, weak layer viscosity η_w, background strain rate $\dot{e}_{bg}$, density difference between rocks and air $\Delta\rho$, gravitational acceleration g, salt layer thickness H_s, and overburden thickness H_o). Four non-dimensional parameters occur: $R_1 = \eta_c/\eta_s$, $R_2 = \eta_c/\eta_w$, H_s/H_o and the Argand number [5] $Ar = (\Delta\rho g H_o)/(2\eta_c \dot{e}_{bg})$.

Seven folding domains have been defined (Fig. 1). The three domains with the highest R_1 values are independent of the Argand number, and thus independent on gravity, but require unrealistically large viscosity contrasts (which are most likely limited to 10^4-10^5). The other four folding domains within the low R_1 range all depend on the Argand number, and only two of them depend on R_2. In the Argand number dependent domains the lower detachment layer must be below a critical value for the real detachment-folding mode to occur (defined as the folding mode controlled mainly by the H_s/H_o ratio). These results are in agreement with the folding modes defined for a two-layer system in [5]. For Earth-like viscosity contrasts if $R_1 < 10^5$, gravity is thus likely to play a role and as a consequence of the dependence in the Argand number, the deformation rate with which the mountain belt was deformed has an effect on the dominant wavelength that formed.

4 3D Numerical Simulations

We tested the validity of the phase diagrams beyond the initial folding stages by performing several 2D and 3D numerical forward simulations using LaMEM. The topography from the numerical simulations was extracted and analyzed during strain evolution, using both curvature analysis and 1D Fourier analysis both parallel and orthogonal to the compression orientation. The dominant wavelength predicted with the analytical methods was compared to the one obtained in the simulations, which shows that the calculated dominant wavelength is well preserved and clearly observable after a shortening of 30 % (Fig. 2). Somewhat more pronounced than in 2D simulations, there is a range of wavelengths around the dominant wavelength that occurs. Moreover, the folds that develop have a range of aspect ratios that vary from elongated and sinuously perturbed to nearly egg-shaped. In addition, several high-resolution 3D numerical simulations were performed to study the evolution of 3D multilayer detachment folds under conditions that might be relevant to natural examples of fold-belts and study the interaction between different individual folds.

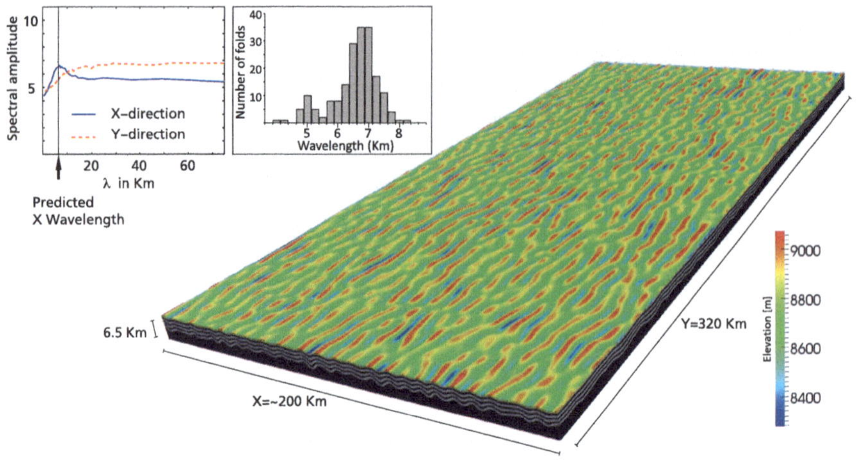

Fig. 2 Example of a 3D detachment folding simulation performed with LaMEM on 1024 cores. Resolution is $513 \times 513 \times 27$ nodes (high order Q2P-1 element). Inset shows the spectral analysis, as well as a histogram of fold spacing of the obtained topography

5 Conclusions

Several folding modes, that are applicable in a multilayer system overlying a lower detachment layer, have been defined. The equations of the dominant wavelength and growth rate for each of those domains and their boundaries were extracted using a 2D analytical solution. Analytically derived results were tested beyond the initial stages of folding by using numerical methods.

Numerical simulations show that the fold wavelength selected during strain in 3D multilayer folding is in agreement with the analytically calculated values. Furthermore, statistics of the numerical simulations exhibit a normally distributed fold wavelength around a dominant one in the direction parallel to the main compression and a large range of fold aspect ratios due to lack of wavelength selectivity in the direction orthogonal to the main compression. The different ways of interaction between folds as they laterally propagate, further influences the wide variety and range of final fold geometries.

Acknowledgments Funding was provided by the ERC under the European Community's Seventh Framework program (FP7/2007-2013) Grant agreement #258830. Numerical simulations were performed in JUQUEEN supercomputer of the Forschungszentrum Jülich, Germany

References

1. Biot, M. A. (1961). Theory of folding of stratified viscoelastic media and its implications in tectonics and orogenesis. *GSA Bulletin*, *72*(11), 1595–1620.

2. Fletcher, R. C. (1974). Wavelength selection in the folding of a single layer with power-law rheology. *American Journal of Science*, *274*(9), 1029–1043.
3. Yamato, P., et al. (2011). Dynamic constraints on the crustal-scale rheology of the Zagros fold belt. *Iranian Journal of Geology*, *39*(9), 815–818.
4. Ramberg, H. (1964). Selective buckling of composite layers with contrasted rheological properties, a theory for simultaneous formation of several orders of folds. *Tectonophysics*, *1*(4), 307–341.
5. Schmalholz, S. M., Podladchikov, Y. Y., & Burg, J. P. (2002). Control of folding by gravity and matrix thickness: Implications for large-scale folding. *Journal of Geophysical Research*, *107*(B1), 2005.
6. Johnson, A. M., & Fletcher, R. C. (1994). *Folding of viscous layers*. New York: Columbia University Press.

Mechanics of Fold Development in Pure- and Simple Shear

Maria-Gema Llorens, Paul D Bons, Albert Griera and Enrique Gomez-Rivas

1 Introduction

Folds are structures commonly used to unravel rock deformation history. They typically indicate orientation and amount of shortening (e.g., [1–4]). However, much more information can potentially be extracted from fold analysis, like the viscosity contrast between folded layers and matrix, kinematic vorticity and non-linearity of the viscosity [5]. Additionally, one can find indications on whether folds have been shortened and stretched again [6, 7]. In this contribution we investigate the relationships between the different parameters that determine fold geometries and their mechanical implications, in order to be able to improve the use of folds to unravel rock mechanics, kinematics and rheology.

The viscosity or competence contrast between folding layers and matrix (m) is assumed to be the main factor determining the initial wavelength and amplification rate of the developing folds [8–10]. Additionally, non-linear viscous behaviour when the strain rate is proportional to the stress to the power n, also influences fold geometry [10–12].

M.-G. Llorens (✉) · P. D. Bons · E. Gomez-Rivas
Department of Geosciences, Eberhard Karls University of Tübingen, Tübingen, Germany
e-mail: maria-gema.llorens-verde@uni-tuebingen.de

P. D. Bons
e-mail: paul.bons@uni-tuebingen.de

E. Gomez-Rivas
e-mail: enrique.gomez-rivas@uni-tuebingen.de

A. Griera
Departament de Geologia, Universitat Autònoma de Barcelona, Barcelona, Spain
e-mail: albert.griera@uab.cat

E. Pardo-Igúzquiza et al. (eds.), *Mathematics of Planet Earth*,
Lecture Notes in Earth System Sciences, DOI: 10.1007/978-3-642-32408-6_164,

2 Numerical Simulation Method and Experimental Setup

We use the 2D numerical modeling platform ELLE, including the finite-element module BASIL [13], to simulate the development of single-layer folds. Layer and matrix are defined by a contiguous set of polygons, to which mechanical properties are assigned. All the boundaries of the model are periodic, so that it is a unit cell that can be repeated infinitely in all directions. In simple shear experiments, rigid-plate velocity conditions are applied to the horizontal boundaries with a constant incremental shear strain of $\Delta\gamma = 0.025$. Contrary to simple-shear simulations, the pure-shear models are initially not square but rectangular. Velocity boundary conditions (~1% incremental shortening) are applied to all boundaries. All the simulations presented in this paper have an initial random noise of layer thickness to noise amplitude ratio of 40, as in [5, 7], a constant viscosity contrast of $m = 50$ and a strain-independent, power-law viscous rheology with a stress exponent (n) of either 1 or 3.

In simple shear simulations, the layer is originally inclined 14° with respect to the shear plane, and sheared to $\gamma = 4$, which results in 75 % shortening of the layer. In pure shear it is parallel to the maximum shortening direction. Simulations with a linear viscosity ($n = 1$) reached 75 % shortening and those with $n = 3$ reached 55 % shortening.

3 Results

Stress distribution for $n = 1$ simulations are shown in Fig. 1a, b (simple shear) and Fig. 1c, d (pure shear), and the maximum deviatoric stress in the layer as a function of layer shortening is shown in Fig. 2a. Maximum deviatoric stresses are higher in power-law materials than in linear-viscosity materials. Maximum deviatoric stress in pure shear is initially double than that of simple-shear folding, but decreases markedly towards the simple-shear value after about 20 % strain. In case of $n = 3$ (Fig. 1e, f, simple shear, and Fig. 1g, h, pure shear) the difference in maximum deviatoric stress between pure shear and simple shear is less pronounced and both decrease steadily during the first 30 % of shortening (Fig. 2b).

Stress localizations are not regularly distributed at the initial stages of folding. After a certain amount of shortening stress concentrations determine the future fold hinges.

4 Discussion and Conclusions

Although similar final fold geometries are developed in simple and pure shear conditions (Fig. 1), differences in stress evolution can be observed (Fig. 2). Pure shear is characterized by an initial stage of high stress followed by a strong

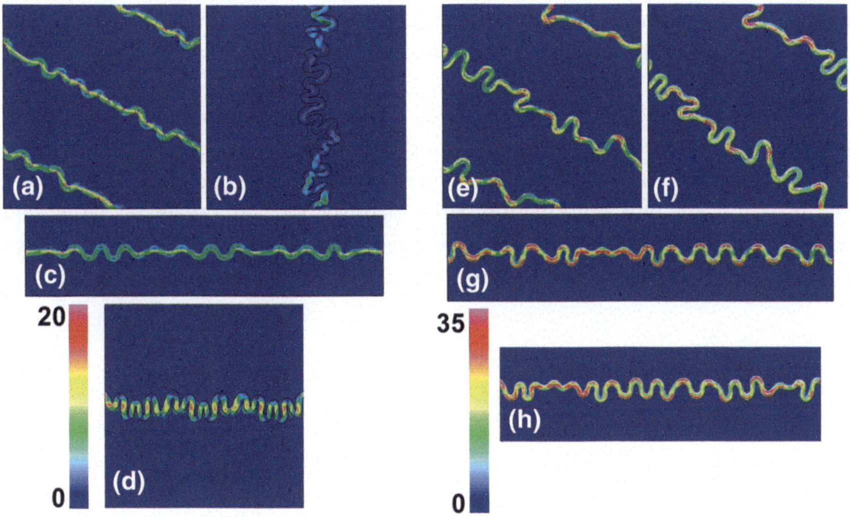

Fig. 1 Plots of maximum deviatoric stress of the simple shear experiment with linear viscosity at **a** 35 % and **b** 75 % of shortening. Pure-shear experiment with linear viscosity at **c** 35 % and **d** 75 % of shortening. Simple shear with non-linear viscosity at **e** 35% and **f** 55 % of shortening and the same for pure shear at **g** 35 % and **h** 55 % of shortening

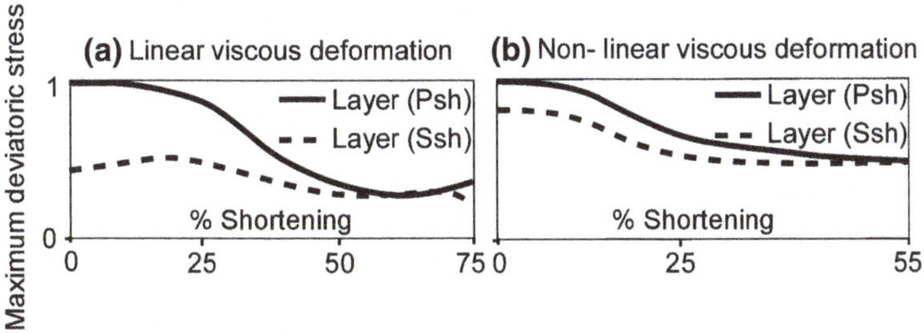

Fig. 2 Maximum deviatoric stress in the folded layer as a function of layer shortening in Psh and Ssh, and for **a** linear ($n = 1$) and **b** power-law ($n = 3$) viscosity models. Stress values are normalized with respect to the maximum layer-averaged deviatoric stress in the linear and non-linear simulations, respectively

weakening after fold perturbations develop. An increase of stress at large strain indicates that folds become locked and shortening is now accommodated by fold limb stretching. In case of pure shear, the folded layer is always parallel to the maximum shortening direction, whereas the fold train rotates relative to the maximum shortening direction in simple shear. Folding in simple shear shows relative lower stresses and the geometrical strain weakening is less pronounced. For the linear case, a strain hardening-softening behavior in the stress evolution is observed. Peak stresses are not observed at the beginning, but when the layer envelope is at ~45°, the maximum shortening strain rate is reached. The non-linear case shows an evolution similar to the pure shear cases, without a stress maximum at ~45°, maybe due to the strain-rate softening behavior of power law materials with $n > 1$. Further studies are still required.

Although similar in geometry, the folds are less regular in simple shear than in pure shear. This appears to be related to the organization of stress perturbations that seed the folds. In pure shear, the folded layer maintains a constant orientation with respect to the bulk deformation. Stress perturbations are organized earlier and more regularly than in simple shear, where the folding layer continuously changes its orientation relative to the bulk deformation.

References

1. Hudleston, P. J. (1986). Extracting information from folds in rocks. *Journal of Geological Education, 34*, 237–245.
2. Hudleston, P. J., & Lan, L. (1993). Information from fold shapes. *Journal of Structural Geology, 15*, 253–264.
3. Bobillo-Ares, N. C., Toimil, N. C., Aller, J., & Bastida, F. (2004). FoldModeler: A tool for the geometrical and kinematical analysis of folds. *Computer and Geosciences, 30*, 147–159.
4. Bastida, F., Aller, J., Bobillo-Ares, N. C., & Toimil, N. C. (2005). Fold geometry: A basis for their kinematical analysis. *Earth-Science Reviews, 70*, 129–164.
5. Llorens, M.-G., Bons, P. D., Griera, A., Gomez-Rivas, E., & Evans, L. A. (2013). Single layer folding in simple shear. *Journal of Structural Geology, 50*, 209–220.
6. Means, W. D. (1999). Reversed structures and bounce structures: Are they recognizable? Are they real? *Journal of Structural Geology, 21*, 917–921.
7. Llorens, M.-G., Bons, P. D., Griera, A., & Gomez-Rivas, E. (2013). When do folds unfold during progressive shear? *Geology, 41*, 563–566.
8. Biot, M. A. (1961). Theory of folding of stratified viscoelastic media and its implication in tectonics and orogenesis. *Geological Society of America Bullettin, 72*, 1595–1632.
9. Ghosh, S. K. (1966). Experimental tests of buckling folds in relation to strain ellipsoid in simple shear deformations. *Tectonophysics, 3*, 169–185.
10. Fletcher, R. C. (1974). Wavelength selection in the folding of a single layer with power-law rheology. *American Journal of Science, 274*, 1029–1043.
11. Abassi, M. R., & Mancktelow, N. S. (1992). Single layer buckle folding in non-linear materials - I. Experimental study of fold development from an isolated initial perturbation. *Journal of Structural Geology, 14*, 85–104.
12. Mühlhaus, H. B., Hobbs, B. E., & Ord, A. (1994). The role of axial constraints on the evolution of folds in single layers. In H. J. Siriwardane & M. M. Zaman (Eds.), *Computer methods and advances in geomechanics 1* (pp. 223–231). Rotterdam: Balkema.
13. Bons, P. D., Koehn, D. & Jessell, M.W. (Eds.). (2008). *Microdynamics simulation*. Lecture notes in earth sciences (Vol. 106). Springer-Verlag: Berlin.

Stochastic Modelling of the 3D Geometry of a Faulted and Folded Deep Carbonate Aquifer: Loma de Úbeda (Southern Spain)

Javier Heredia, Eulogio Pardo-Igúzquiza and Antonio González-Ramón

1 Introduction

The olive plantations in Úbeda (Spain) stand out in the world. Because of it, the irrigation has grown strongly, increasing the water demand. At the beginning the water was taken from the unconfined part of the Jurassic carbonate aquifer of the Loma de Úbeda, then it started the exploitation of the deep confined sector and later the underlying Triassic aquifer was exploited. The impact of pumping was notable due to the lack of a management plan. The Geological Survey of Spain (IGME) has carried out hydrogeological studies to define a sustainable exploitation of the aquifer system. Flow modelling has allowed the sustainable management of the resources [1]. Such flow modelling required a geometrical model of the carbonate aquifer.

The carbonate aquifer, together with the underlying Triassic formations (Keuper and Buntsandstein), constitute the Deep Aquifer System of the Loma de Úbeda. The extension of the carbonates is 880 km^2 and one third, located in the north, is the unconfined sector, which is strongly eroded and is crossed by the Guadalimar river. The carbonates have a slight dip, from 5 to 15°, towards the S–SE and 1–2 km to the south of its outcrop they are at 100 m depth. 40 % of the confined sector lies between 400 and 800 m. The carbonates are very tectonized by large fractures, regional faults, folds and flakes that compartmentalize and connect it with the Triassic formations (Fig. 1). The mean thickness of the confined sector is 96 m.

J. Heredia (✉) · E. Pardo-Igúzquiza · A. González-Ramón
Spanish Geological Survey, C/ Ríos Rosas,23, 28003 Madrid, Spain
e-mail: j.heredia@igme.es

E. Pardo-Igúzquiza
e-mail: e.pardo@igme.es

A. González-Ramón
e-mail: a.gonzalez@igme.es

E. Pardo-Igúzquiza et al. (eds.), *Mathematics of Planet Earth*,
Lecture Notes in Earth System Sciences, DOI: 10.1007/978-3-642-32408-6_165,

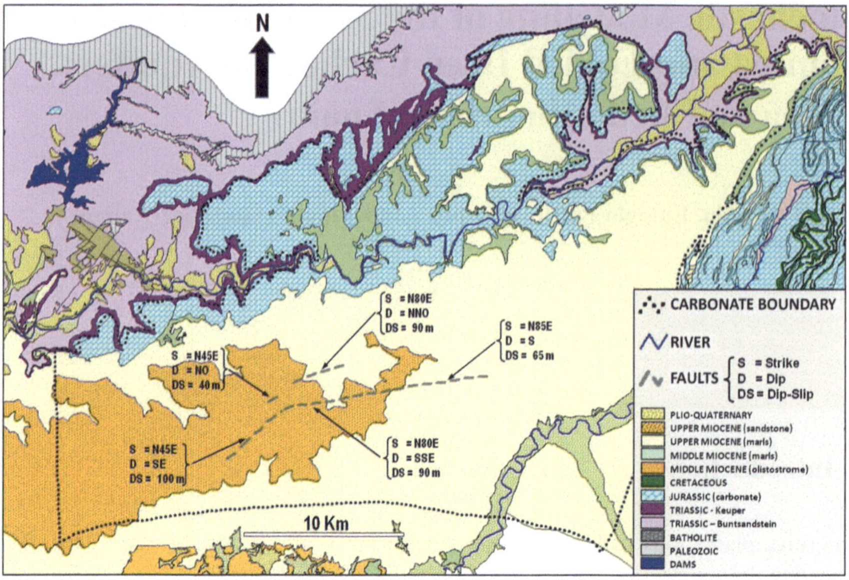

Fig. 1 Geology of the Úbeda Deep Aquifer System

2 Methodology and Data

The geometric modelling of the carbonates deals with two complex aspects: (1) beds are not flat-lying but dip due to folds and (2) the lack of spatial continuity because of large regional faults whose displacements are of the order o larger than the layer thickness. The three-dimensional geometrical model constructed tries to simulate the "geological evolution" of the carbonates: (a) geometrical reconstruction of folded beds without discontinuities; (b) introduction of discontinuities because of the regional faults and (c) erosion process.

First of all, the top and base surfaces of the studied stratigraphic units were reconstructed using the kriging interpolation algorithm without taking into account the faults. Next, the following steps were performed: (1) study of the changes in dip of the surface employing surface variograms, (2) obtaining a directional experimental variogram according to the dip variations, (3) fitting of a theoretical variogram model, (4) checking the fitted model by cross-validation and (5) interpolation using the variogram fitted model. The thickness of the carbonates was obtained by the subtracting the height of the top and the base surfaces. Later on the reconstructed top was redefined considering the regional faults. Using the beds' dip and the displacement along the faults, new preliminary coordinates of the cut off points of the top surface in the hangingwall and footwall of the faults were estimated. The final position of the top surface was calculated using steps 1 to 5 above. The coordinates of the points within the base surface affected by the faults was calculated subtracting the calcu-

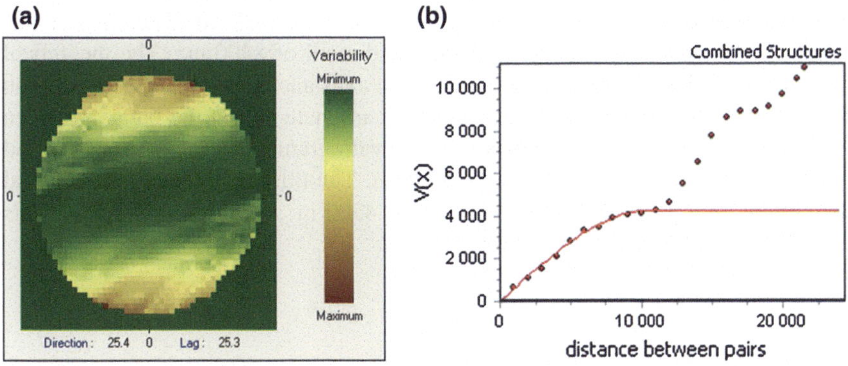

Fig. 2 **a** Surface variogram and **b** directional variograms along direction N86.5E

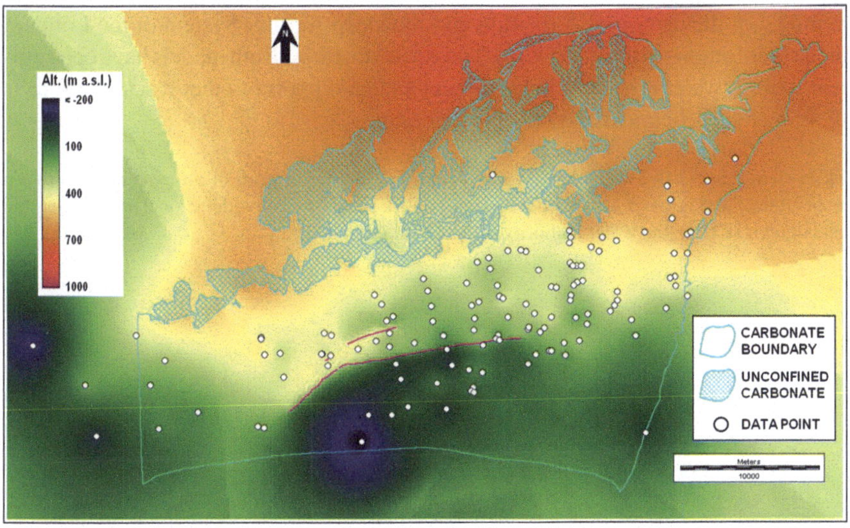

Fig. 3 Reconstruction of the carbonates top surface considering the regional faults

lated thickness to the top surface by the faults. Finally, digital elevation model was superimposed onto the limestone outcrops depicted on a topographic map.

The data available were 143 drill holes and the positions of the top and base of the carbonates contact at the outcrops of the carbonates. The geostatistics software Gstat [2] and Idrisi ANDES [3] were used for the algebraic operations.

3 Results

Only a brief synthesis is presented here regarding the application of the methodology above to the top of the carbonates. In the reconstruction without taking into account the faults, the experimental variogram indicates a range of 20 km with a direction of minimum variability along the direction N86E. The fitted model was an spherical model with a range of 35 km and a sill of 8400 m^2. The theoretical variogram describes the experimental one up to a distance of 12 km. This restriction is imposed when interpolating the top of the carbonate. When the faults are taken into account (Figs. 2 and 3) the experimental variogram has a range of 22 km and minimum variability along the direction N86E. The fitted model was an spherical model with a range of 10.55 km and a sill of 4300 m^2. The theoretical model fits very well the experimental model up to 11 km.

4 Conclusion

The geometrical model of the carbonates has allowed the development of a numerical model of the carbonate formation and the Deep Aquifer System of the Loma of Úbeda that comprises the underlying Triassic formations. Both models have been efficient tools in water resources management and as support of other research projects involving hydrochemical and isotopic techniques.

Acknowledgments This work has been developed in the framework of an agreement between the Instituto Geológico y Minero de España (IGME) and the Confederación Hidrográfica del Guadalquivir (CHG), the water authority of the Guadalquivir Basin.

References

1. Heredia, J., Manzano, M., Ortega, L., González-Ramón, A., Rodríguez Arévalo, J., & Muñoz de la Varga, D. (2012). Numeric check of the preliminary inputs of ^{222}Rn of the conceptual model of the deep aquifer system of Ubeda (Jaén, Spain). *7th Portuguese–Spanish Assembly of Geodesy and Geophysics*. San Sebastián, Spain.
2. Pedesma, E. J., & Wesseling, C. G. (1998). Gstat: a program for geostatistical modeling, prediction and simulation-. *Computers and Geosciences*, *24*(1), 17–31.
3. Eastman, R. (2006). *IDRISI Andes guide to GIS and image processing*. Worcester: Clark University.

3D Geometry of an Active Shale-Cored Anticline in the Western South Caspian Basin

Idaira Santos-Betancor, Juan Ignacio Soto, Lidia Lonergan, Ismael Sánchez-Borrego and Carlos Macellari

1 Introduction

The South Caspian Basin (SCB) is a back-arc Neogene basin dominated by the Arabia-Eurasia convergence and the Tethys closure during Cenozoic (e.g., [1]). Several active fold-thrust belts of arcuate form encircle the basin, which is floored by a Jurassic oceanic crust and contains a thick sedimentary section (>20 km). Remarkable oil and gas deposits are generated in the Maykop Unit (Early Oligocene to Early Miocene) and stored in the Productive Series (PS; Late Messininan to Early Pliocene, ~5.9–3.1 Ma). In the western SCB the PS is up to 10 km thick and is buried by the post-Productive Series: Akchagyl, Apsheron and Gelasian units (post-PS; Late Pliocene to Present; <3.1 Ma) [2].

Buckle folds of Pliocene-to-Present age deform the sedimentary cover of the SCB. Rapid sedimentation, folding and subsidence generated overpressure in the shale-rich Maykop Unit and triggered mud diapirism and volcanism (e.g., [1, 3]).

I. Santos-Betancor (✉) · J. I. Soto
Instituto Andaluz de Ciencias de la Tierra and Departamento de Geodinámica, Universidad de Granada, Granada, Spain
e-mail: idaira@ugr.es

J. I. Soto
e-mail: jsoto@ugr.es

L. Lonergan
Department of Earth Science and Engineering, Imperial College London, London, UK
e-mail: l.lonergan@imperial.ac.uk

I. Sánchez-Borrego
Departamento de Estadística e Investigación Operativa, Universidad de Granada, Granada, Spain
e-mail: ismasb@ugr.es

C. Macellari
Repsol Exploración S.A., Madrid, Spain

E. Pardo-Igúzquiza et al. (eds.), *Mathematics of Planet Earth*,
Lecture Notes in Earth System Sciences, DOI: 10.1007/978-3-642-32408-6_166,

Our aim is to reconstruct the 3D shape of an anticline through seismic interpretation and with nonparametric regression methods estimate the folding history. One of the singularities of this structure is that the fold core is cut by complex mud-diapir structures.

2 3D Geometry

We reconstruct the 3D geometry of the fold and of the mud-diapir structures through seismic interpretation of a post-stack, depth-migrated seismic volume settled in the western margin of the SCB, in offshore Azerbaijan. Interpretation has been tied with well and logging data. Repsol has provided this dataset.

Structural depth-contour maps of the different seismic horizons allow us to depict the internal geometry of the fold, alike the shape of the mud diapirs and the position and nature of the associated faults, as double-vergence reverse surfaces in the core and syn-sedimentary collapse normal faults along the crest (cf. Fig. 1).

The studied fold is a non-cylindrical, NNW-SSE anticline with curved axial surfaces and deforms up to the seafloor (Fig. 1). Fold vergence and shape change along strike from symmetrical box-like profiles to asymmetric. It has two culminations pierced by teardrop mud diapirs connected with sub-vertical welds. The internal geometry shows two contrasting sequences, separated by a regional and erosive unconformity: (1) the PS units, with an approximately constant thickness between fold flanks, which corresponds mostly to the pre-growth sequence; and (2) the post-PS units with progressive thinning towards the fold crest that correspond to the syn-growth epoch (Fig. 1).

3 Shortening Estimates

To evaluate the deformation history, we have used a key seismic section where the fold profile is non-pierced by mud (Fig. 1). In this section the geometry of the different sedimentary layers is fitted by nonparametric methods. With these results we compute subsequently two shortening magnitudes: curvimetric shortening (S_C) and planimetric shortening (S_A) (e.g., [4]). S_C compares the actual length of the folded surface with its original length in the undeformed situation, whereas S_A uses the area of structural relief of the deformed horizon and the horizon height above a reference surface. The results are expressed in Fig. 2 that compares S_C with S_A for the overall set of seismic horizons.

Figure 2 contains most of our findings regarding the folding history. (1) It is confirmed that the PS and post-PS units correspond to the pre- and syn-folding sequences, respectively, although shortening seems to initiate towards the end of PS (H9 to PS-top) as is evidenced by thinning of these sequences towards the fold crest and the slight change in slope obtained by S_C and S_A trends. (2) There are some

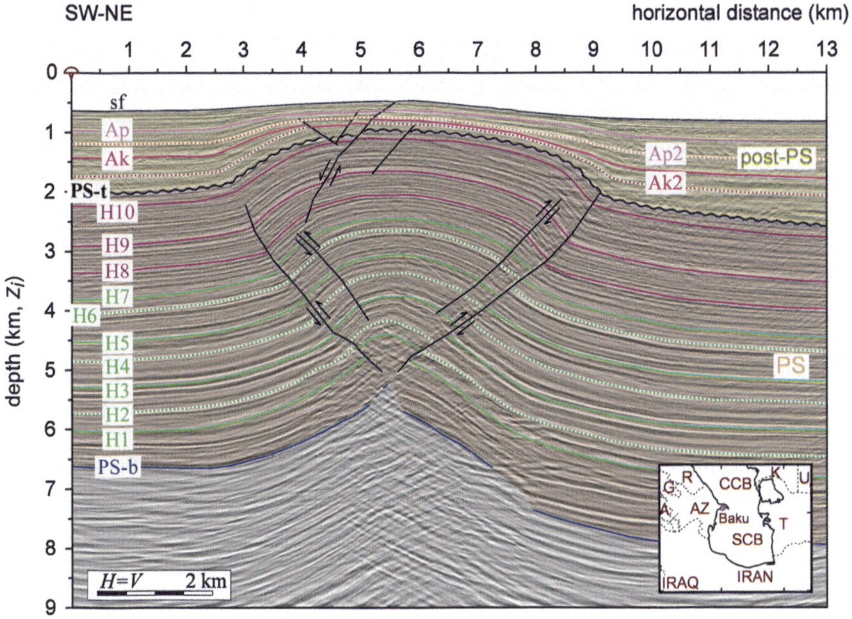

Fig. 1 Seismic inline in depth to show the profile across one of the fold culminations. The fold geometry here illustrates a box-like profile with associated, double-vergence reverse faults. *Inset* shows the geographic situation of the South Caspian basin (SCB). PS: Productive Series, PS-b and PS-t: *bottom* and *top* of the PS sequences, respectively; Ak: Akchagyl; Ap: Apsheron; sf: seafloor

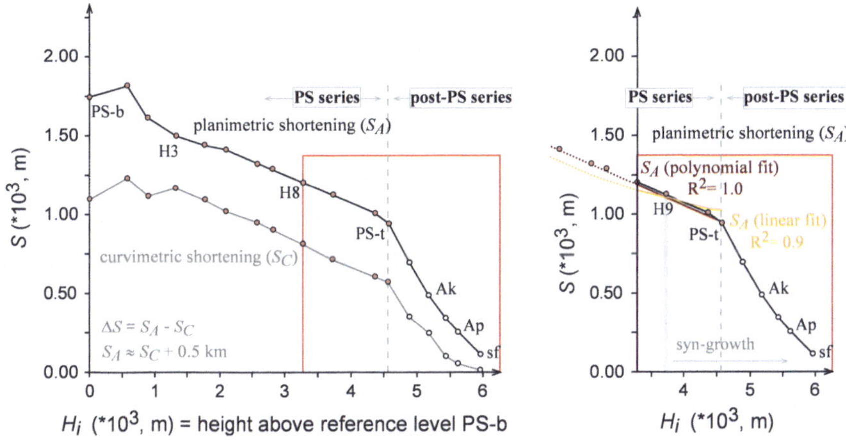

Fig. 2 *Left* Comparison between S_C and S_A estimates against height of the horizons above the PS-bottom (H_i). *Right* Best polynomial and linear fits for the PS units. Results correspond to section shown in Fig. 1. Abbreviations like in Fig. 1

discrete folding pulses during the syn-sedimentary or post-PS period (<3.1 Ma). (3) Curvimetric shortening minimizes the deformation magnitude computed through the planimetric shortening ($S_A \approx S_C + 0.5$ km), thus suggesting a significant material influx during folding. These results need to be extended to other fold sections, but the abundance of mud-escape structures in the fold culminations reinforces that volume does not maintain constant during deformation.

4 Conclusions

The structure resembles a detachment fold-type with a low-dipping decoupling layer. We have implemented the standard length and area-balanced algorithms for detachment folds to estimate the shortening magnitudes and rates, also inferring that the décollement surface is at 9.4 km depth, presumably located within the thick and weak Maykop Formation. The 3D geometry configuration departs from the classic detachment-fold model mostly because: (1) folding occurred simultaneously to mud diapirism, whilst tilting and differential subsidence toward the basin centre, (2) deformation has a probable strike-slip component partitioned along the axial surface; and (3) volume does not remain constant during deformation.

Maximum shortening estimates are obtained in regions perforated by the overpressured mud. We document two pulses of folding: (1) an early, minor event towards the end of the deposition of the PS (~3.4–3.2 Ma) and (2) fold-growth occurred during post-PS with a significant pulse near the end of the deposition of Akchagyl (~1.7 Ma).

References

1. Devlin, W. J., Cogswell, J. M., Gaskins, G. M., Isaksen, G. H., Pitcher, D. M., Puls, D. P., et al. (1999). South Caspian basin: young, cool, and full of promise. *GSA Today*, *9*(7), 1–9.
2. Abreu, V., & Nummedal, D. (2007). Miocene to quaternary sequence stratigraphy of the south and central Caspian basin. In P. O. Yilmaz & G. H. Isaksen (Eds.), *Oil and gas of the greater Caspian Sea* (Vol. 55, pp. 65–86). Tulsa: AAPG.
3. Allen, M. B., Vincent, S. J., Alsop, G. I., Ismail-zadeh, A., & Flecker, R. (2003). Late cenozoic deformation in the South Caspian region: effects of a rigid basement block within a collision zone. *Tectonophysics*, *366*(3–4), 223–239.
4. Gonzalez-Mieres, R., & Suppe, J. (2006). Relief and shortening in detachment folds. *Journal of Structural Geology*, *28*, 1785–1807.

Gravity-Instabilities Processes and Mass-Transport Complexes During Folding: The Case of a Shale-Cored Anticline in the Western South Caspian Basin

Idaira Santos-Betancor, Juan Ignacio Soto, Lidia Lonergan and Carlos Macellari

1 Introduction

Mass-transport complexes (MTCs) are failed mass of sediments mobilized from unstable slopes towards deepwater areas (e.g.,[1]). Seismic reflection data evidences that seafloor on the western South Caspian Basin (SCB) has been unstable since the Pliocene, subjected to different gravity instability processes induced by active tectonics, high and rapid sedimentary loading and mud diapirism (e.g.,[2]).

We analyze the uppermost sedimentary section (<2 km) in a single anticline where widespread MTCs are registered. The sedimentary package is composed by the Akchagyl (Late Pliocene to Pleistocene; <3.1 Ma), Apsheron (Early Pleistocene to Late Pleistocene; <1.7 Ma) and Gelasian (Late Pleistocene to Holocene; <0.8 Ma) units [3]. Basinal deposition within Akchagyl and Apsheron units comes from turbidity currents and deepwater lacustrine deposits. Sedimentation during deposition of the Gelasian Unit corresponds to prograding clinoforms and deepwater fans in large shelf-edge deltas. These units lie above a regional unconformity that corresponds to the top of the Productive Series (PS; Late Messinian to Early Pliocene, ~6–3.1 Ma). Here we referred collectively Akchagyl to Gelasian units as the post-PS sequence.

I. Santos-Betancor (✉) · J. I. Soto
Instituto Andaluz de Ciencias de la Tierra (CSIC-Univ. Granada) and Departamento de Geodinámica,Universidad de Granada, Granada, Spain
e-mail: idaira@ugr.es

J. I. Soto
e-mail: jsoto@ugr.es

L. Lonergan
Department of Earth Science and Engineering, Imperial College London, London, UK
e-mail: l.lonergan@imperial.ac.uk

C. Macellari
Repsol Exploración S.A., Madrid, Spain

E. Pardo-Igúzquiza et al. (eds.), *Mathematics of Planet Earth*,
Lecture Notes in Earth System Sciences, DOI: 10.1007/978-3-642-32408-6_167,

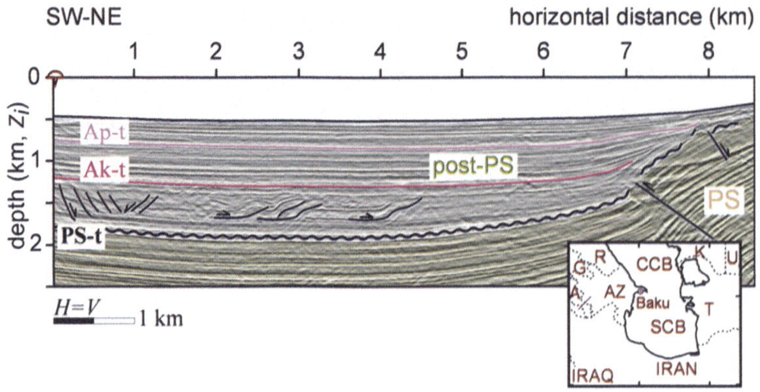

Fig. 1 Detailed shallow structure of the studied fold in its western flank showing a MTC within the Akchagyl Unit. Inset shows the geographic situation of the South Caspian Basin (SCB). *Ak-t* top of the Akchagyl Unit, *Ap-t* top of the Apsheron Unit, *PS* Productive series, *PS-t* top of the PS sequences

This study is conducted through the detailed interpretation of a depth-migrated seismic cube. Our research aims to unravel the most-recent evolution of a shale-cored anticline in the western margin of the SCB. We look for establishing the relationships between sedimentation, folding, and gravity-instability processes on the seafloor.

2 Mass-Transport Deposits and Processes

We have inspected the detailed geometry of the most-recent sedimentary sequences of a single anticline located in offshore Azerbaijan, in the west of the SCB. Seismic interpretation has been conducted in a 3D seismic cube through horizontal slices, vertical sections, and reflection attributes maps. The cube is a post-stacked, depth migrated dataset that has been provided by Repsol.

The folded structure deforms congruently most of the PS sequences, whereas the post-PS units thin progressively towards the fold crest (Fig. 1). This observation demonstrates that deposition of the post-PS units (i.e., during the last 3 my) accompanied fold growth, i.e. they correspond to syn-growth sequences. It is a NNW-SSE sigmoidal anticline that varies along strike from symmetric to asymmetric shapes. This fold presents two culminations, both cored by mud diapirs.

Within the post-PS sequences there are packages with an overall sigmoidal geometry and an internal laminated seismic fabric with imbricate elements (Fig. 1). These bodies correspond mostly to mass-transport complexes (MTCs), in where the down-slope migration of unconsolidated sediments develop internally normal faulting in proximal areas and thrusting in the down-slope, frontal domains. We have identified different MTCs, associated with the high-slope region of SE-prograding deltas (sub-perpendicular to the section shown in Fig. 1). These complexes are developed

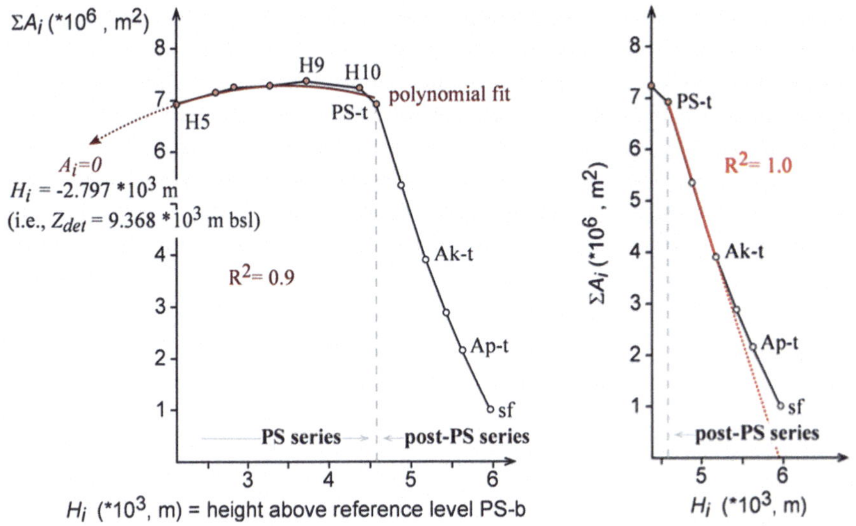

Fig. 2 *Left* plot with the A_i–H_i relationship and trend of the planimetric shortening. *Dots* correspond to seismic horizons within the PS and post-PS sequences. *Right* detailed plot for the post-PS sequences showing a change after deposition of the Akchagyl Unit. *sf* seafloor

particularly in three major intervals: near the top of the Akchagyl Unit and within the Apsheron and Gelasian units.

Accordingly to their geometry and internal fabric, we have identified three major types of sediment instability processes and MTCs: (1) large slides, with concave erosive scars in the uphill domains and limited imbrications in the convex downslope front; (2) debris flows, that document movement of disaggregated and probably unconsolidated sediments from curved headwall scars; and (3) slides, seen as isolated and small bodies with incoherent seismic facies. Numerous MTCs of type 1 are developed within the Akchagyl Unit, with maximum thickness of 350 m, whereas debris flows are more dominant in the Apsheron and Gelasian units, especially along the eastern fold limb.

3 Folding History

Shortening has been calculated to unravel the fold growth history and to precise the most-recent deformation history, with particular attention to the fold-growth during the last 3.1 Ma, epoch that corresponds to the sedimentation of the post-PS sequences.

We have calculated the planimetric shortening magnitude in a representative fold section, by means of the known relationship in folds between structural relief of the deformed horizons (A_i) and the horizon height above a reference surface (H_i) [4]. The shape of the different seismic horizons has been fitted through nonparametric statistics to obtain a complete distribution across the fold profile. The results are shown in Fig. 2. These plots summarize most of our findings regarding the folding history. (1) The pre- and syn-growth sequences correspond mostly to PS and post-PS, respectively. Folding started probably at about 3.4 Ma (corresponding to the interval H9 to PS-top). (2) The best fit (polynomial) of the A_i–H_i relationship for PS suggests a deep detachment level at about 9.4 km depth that occurs when $A_i = 0$. (3) Detailed inspection of the post-PS series evidences a change in the slope of the planimetric shortening, indicating a decrease in the shortening magnitude and associated fold growth after deposition of the Akchagyl Unit (Late Pliocene to Pleistocene; <1.7 Ma).

4 Conclusions

Following these results, we infer that the abundance and maturity of the MTCs developed within the Akchagyl Unit are triggered by an important episode of fold uplift (3.1–1.7 Ma), which deformed the seafloor. In contrast, other MTCs, like isolated debris flows and slides, registered during Apsheron and Gelasian units (<1.7 Ma), were developed in periods of relatively folding quiescence and may be related to instability processes linked to delta progradation and shelf edge migration.

References

1. Posamentier, H. W., & Martinsen, O. J. (2011). The character and genesis of submarine mass-transport deposits: insights from outcrop and 3D seismic data. In R. C. Shipp &Weimer, P.Posamentier, H.W.(Eds.), *Mass-transport deposits in deepwater settings* (Vol. 96, pp. 7–38). Tulsa: SEPM Special Publication.
2. Richardson, S. E. J., Davies, R. J., Allen, M. B., & Grant, S. F. (2011). Structure and evolution of mass transport deposits in the South Caspian Basin. *Azerbaijan, Basin Research, 23*(6), 702–719.
3. Abreu, V., & Nummedal, D. (2007). Miocene to quaternary sequence stratigraphy of the South and Central Caspian Basin. In P. O. Yilmaz & G. H. Isaksen (Eds.), *Oil and gas of the Greater Caspian Sea* (Vol. 55, pp. 65–86). Tulsa: AAPG.
4. Gonzalez-Mieres, R., & Suppe, J. (2006). Relief and shortening in detachment folds. *JSG, 28*, 1785–1807.

Domaining Bi-modal Data Sets Geostatistically Using a Directional Neighborhood Search

Steffen Brammer

1 Introduction

Given is an area that hosts a broad N-S orientated system of relatively low values (domain 1), and a narrow, obliquely cross-cutting system of relatively high values (domain 2)—see Fig. 1a. When sampled on a regular grid (here 5×5 units), the resulting data set is bi-modal with the populations of the two domains mixed together (Fig. 2a). Domaining is required prior to any estimation or simulation run. However, domaining with conventional methods (e.g. manually digitized regions or by selected cut-off values) can be sometimes difficult or misleading. For instance, in this example, the N-S trend of the low graded domain 1 might not be recognized (see Fig. 1b). Geostatistical methods exist that offer applicable solutions. References [1, 2], for example, give overviews over possible approaches with cluster data analysis. This paper proposes to split the mixed population with statistical methods, based on the assumption that, if the individual underlying populations are normally distributed, the resulting domains have to be normally distributed, too. Since purely statistical methods do not consider the spatial properties of the individual samples and misallocations are consequently inevitable, a geostatistical component such as a directional neighborhood search would need to be factored into the algorithm.

2 Split of Mixed Populations by Statistical Methods

The statistics disciplines offer several methods to identify and split mixed populations (see e.g. [3]). This study follows in principle the method described by [4, 5]. It requires the underlying populations to be normally distributed—which, in this example, is

S. Brammer (✉)
Brazca Ltd, Montréal 32250, France
e-mail: sb@brazca.com

E. Pardo-Igúzquiza et al. (eds.), *Mathematics of Planet Earth*,
Lecture Notes in Earth System Sciences, DOI: 10.1007/978-3-642-32408-6_168,

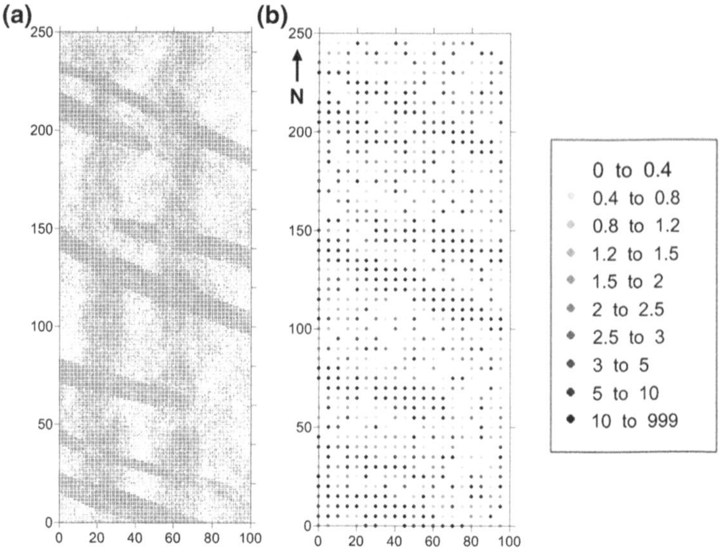

Fig. 1 **a** (*left*) Bi-modal system of a N–S orientated low grade domain and an obliquely cross-cutting high grade domain; **b** (*right*) Samples taken on a regular grid of 5 × 5 units

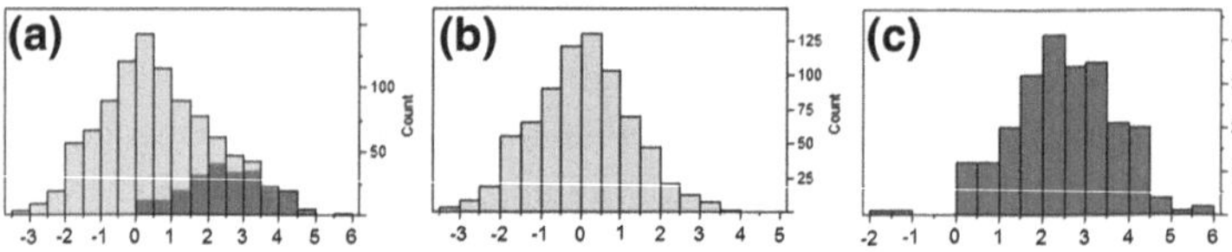

Fig. 2 Histograms of log-transformed sample data: **a** (*left*) Single data set with population 1 (*pale green*) and population 2 (*dark green*); **b** (*middle*) Population 1; **c** (*right*) Population 2

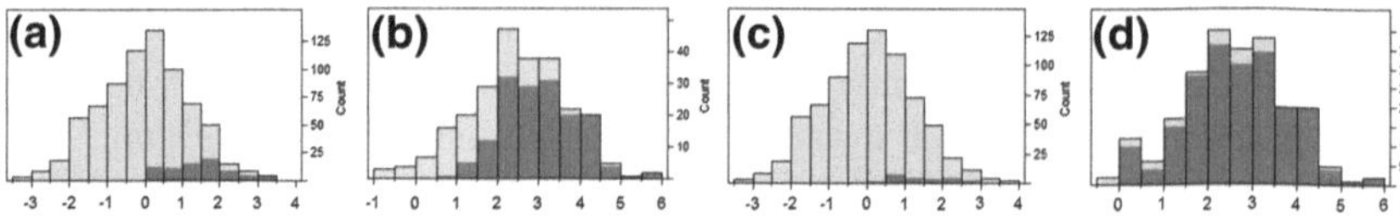

Fig. 3 Histograms of log-transformed sample data after split showing original domain 1 (*pale green*) and original domain 2 (*dark green*): **a-b** Population 1 (*left*) & 2 (*middle left*) without neighborhood search; **c-d** Population 1 (*middle right*) and Population 2 (*right*) with directional neighborhood search

achieved by log-transformation of the original data—and is based on an initial visual estimate of means and spreads of the two populations. Normal fits are forced over the data, and the respective density probabilities and the probability ratio calculated. The probability ratio of each sample is compared with a randomly generated dummy and the sample assigned accordingly to one of the two domains. This process is repeated with iterative optimization of the input parameters until the two resulting domains are both normally distributed. This can be tested with a goodness-of-fit analysis. Figure 3a, b show the data set after the split. Although the statistical moments mean, standard deviation and number of samples of the resulting domains are very similar to those of the original domains, 175 or 17.5 % of the samples have been assigned to the wrong domain.

3 Directional Neighborhood Search

To improve the allocation of samples during the split, a directional neighborhood search has been introduced into the process. Appropriate search parameters were determined by calculating the mean of the variances of the neighborhoods for various search ranges, directions and anisotropy ratios. In this case, best results were obtained with a search direction of 112.5° which is approximately parallel to the orientation of the high grade domain, an anisotropy ratio of 3:1 and a range of 15 units. For each sample, the mean grade of the (log-transformed) samples within this search ellipse was then calculated and compared with the means of the two domains to establish the coefficient C_m as follows:

$$C_m = ((X_2 - X_1)/2) - X_N \tag{1}$$

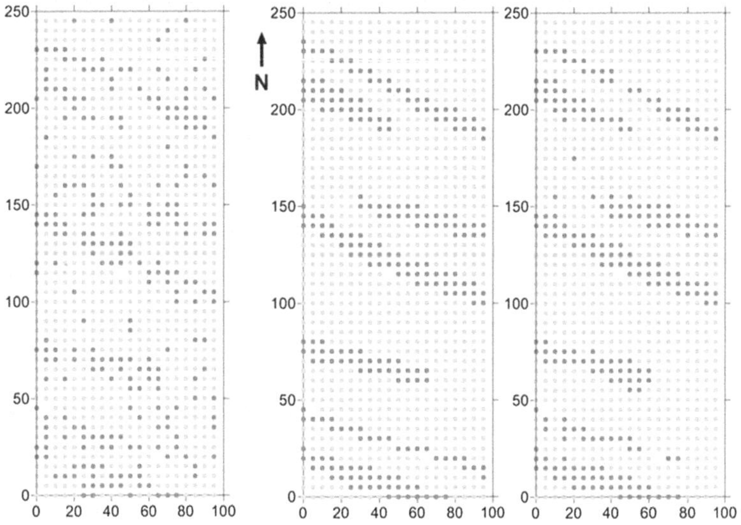

Fig. 4 Bi-modal data set showing domain 1 (*light dots*) and domain 2 (*dark dots*): **a** (*left*) after split without neighborhood search; **b** (*middle*) original data; **c** (*right*) after split with neighborhood search

where X_1 is the mean of population 1, X_2 the mean of population 2 and X_N the mean of the neighborhood. C_m decreases with increasing means of the neighborhood and is factored into the splitting algorithm. This way, samples of relative high grade within a neighborhood of low values are penalized and vice versa. Figure 3c, d show the data after the split using a directional neighborhood search.

Comparison with the original data (Fig. 4) suggests that the introduction of a directional neighborhood search improves the domaining process significantly. Applications for the proposed method are 2-dimensional, bi-modal data sets such as geochemical soil samples (e.g. for environmental studies or mineral exploration), particularly when the data is only quantitative without any additional qualitative information that could otherwise assist in conventional domaining.

References

1. Ambroise, C., Dang, M, & Govaert, G. (1995). Clustering of spatial data by the EM algorithm. In Soares, A., et al. (Eds.) *Quantitative Geology and Geostatistics*, Vol. 9, (pp. 493–504). Kluwer, Dordrecht.
2. Romary, T., et al. (2012). Domaining by clustering multivariate geostatistical data. In Abrahamsen, et al. (Eds.) *Geostatistics Oslo 2012*, (pp. 455–466). Dordrecht: Springer.
3. McLaughlan, G. J., & Peel, D. (2000). *Finite mixture models*. New York: Wiley.
4. Clark, I., & Garnett, R. H. T. (1974). Identification of multiple mineralization phases by statistical methods. *Transactions of The Institution of Mining and Metallurgy*, *83*(809), A43–A52.
5. Glynn, E. F. (2007). Mixtures of Gaussian. Stowers Institute for Medical Research. http://research.stowers-institute.org/efg/R/Statistics/MixturesOfDistributions/index.htm.

FRISER-IRMIX Database: A Web-Based Support System with Implications in Planetary Mineralogical Studies, Ground Temperature Measurements and Astrobiology

Jesús Martínez-Frias, María Serrano Rubio, F. Javier Martín-Torres, María Paz Zorzano, José A. Rodríguez-Manfredi, Javier Gómez-Elvira and REMS team

1 Introduction

Web-based databases are used as "real" support systems to help to decide how to proceed with respect to a specific (or a set of) scientific subject at different levels. This is the case of the FRISER-IRMIX database: a web-based support system with implications in planetary mineralogical studies, ground temperature measurements and astrobiology, which is developed in the framework of the scientific activities of the NASA-MSL-Rover Environmental Monitoring Station (REMS).

The Mars Science Laboratory (MSL) payload is composed of a set of instruments including REMS, which is a contribution of Spain. The Centro de Astrobiología (CAB) led the development of the instrument. REMS is a suite of different sensors to record pressure, humidity, air and ground temperature, UV radiation and wind speed and direction [1] REMS is an instrument composed of four modules. Two modules, located on the Remote Sensor Mast (RSM) of the MSL (Curiosity) rover,

J. Martínez-Frias (✉) · M. S. Rubio · F. J. Martín-Torres · M. P. Zorzano
J. A. Rodríguez-Manfredi · J. Gómez-Elvira · REMS Team
Centro de Astrobiología, CSIC-INTA, NASA Astrobiology Institute, Ctra. Ajalvir, km. 4, 28850 Torrejón de Ardoz, Madrid, Spain
e-mail: jmfrias@cab.inta-csic.es

M. S. Rubio
e-mail: serranorm@cab.inta-csic.es

F. J. Martín-Torres
e-mail: javiermt@cab.inta-csic.es

M. P. Zorzano
e-mail: zorzanomm@cab.inta-csic.es

J. A. Rodríguez-Manfredi
e-mail: manfredi@cab.inta-csic.es

J. Gómez-Elvira
e-mail: gomezej@cab.inta-csic.es

E. Pardo-Igúzquiza et al. (eds.), *Mathematics of Planet Earth*,
Lecture Notes in Earth System Sciences, DOI: 10.1007/978-3-642-32408-6_169,

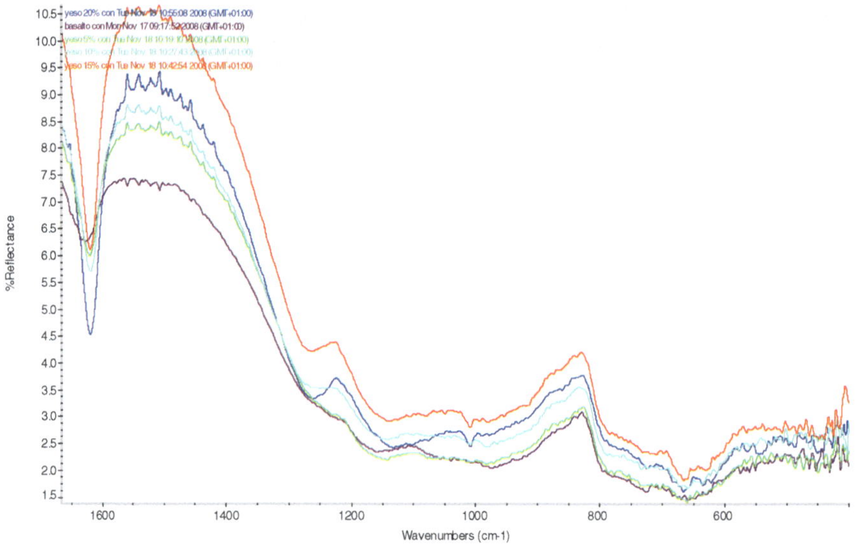

Fig. 1 FTIR reflectance spectra of basaltic powder and gypsum (mixed different amounts). *Purple* basalt. *Green* Gypsum (5 %). *Light Blue* Gypsum (10 %). *Red* Gypsum: 20 %. *Dark Blue* Gypsum: 20 %

are devoted to measuring the wind vector, the ground and air temperatures and the relative humidity of the air. The third module, located on the rover deck, is dedicated to measuring the flux of solar ultraviolet radiation at various bands. The fourth module, located in the rover body, is measuring the atmospheric pressure [2]. REMS started collecting data on sol 1 (a Mars solar day has a mean period of 24 h 39 min 35.244 s, and is customarily referred to as a "sol" in order to distinguish this from the roughly 3 % shorter solar day on Earth), and has been operating continuously since sol 10, with the exception of sol 28 which focused on engineering activities [3]. Specifically, the FRISER-IRMIX database was build in connection with the science related to the REMS-Ground Temperature Sensor (GTS). Nevertheless, the spectroscopic results (Fig. 1) and their systematic arrangement in the database can also be useful for other more general mineralogical and astrobiological studies [4, 5].

2 FTIR Spectra. Interpretation of Mars' Surface Temperature Measurements

The currently used IR spectral data of Martian dust, rocks and sediments allow the interpretation of Martian spectra accounting for the spectra of different minerals and lithologies, and those of their alteration and weathering products, but these are terrestrial spectra. These spectra were measured, considering different mineralogical

Table 1 Example describing one of the samples: CAB-GY-20CS a mixture of gypsum and basalt

Features	Description
Label	Official label of the powder sample in the database (CAB-GY-20CS)
Basalt/Mineral mixture	Gypsum (GY)
% Basalt	80 (mixing amount)
% Mineral	20 (mixing amount)
Wt - Basalt	1.2007 (weight of the powder)
Wt - Mineral	0.3016 (weight of the powder)
Origin/Locality	Sorbas cave (interior), Almería
FTR spectrum	Hyperlink to the spectrum (in jpg format)
Other information	e.g. texture, sampling, etc.

mixtures (Fig. 1), and covering the specific working wavelength range of the REMS-GTS in the NASA's MSL (Curiosity) rover, which is currently operating on Mars. This research was performed taking into account the size of the Martian regolith ($<45\,\mu m$) and the wavelength range of the REMS' GTS, from 3.34 to 25 microns, with a resolution of 2500000 nm. Wider percentages of each mineral (0–100 %) were considered in a previous study [5]. In a second stage, all measurements were restricted to small amounts ($<20\,\%$) of the water-related minerals.

The IR emission of the ground depends on the kinetic surface temperature of the ground and the emissivity of the materials in the area of observation of the detector. The basalt reflectance percentage increases or decreases, even up to 100 %, depending on the mixing ratios of the different minerals [5].

Thus, an accurate knowledge of the emissivity is essential to retrieve the ground temperature correctly. This unequivocally confirms the need for considering the chemical–mineralogical assemblages (and their textures) for any investigation and interpretation of Mars surface environment, and hence the scientific significance of the database as a web-based support system with implications in planetary mineralogical studies, ground temperature measurements and astrobiology.

3 FRISER-IRMIX Database

The FRISER-IRMIX (http://auditore.cab.inta-csic.es/friser-irmix) contains FTIR spectra of a set of selected minerals (oxides, oxi/hydroxides, sulfates, chlorides, carbonates, sulfides, opal, clays, etc) and basalt (as the main and widespread volcanic Martian rock), that are more representative of Martian conditions.

At present, FRISER-IRMIX comprises more than 400 FT-IR reflectance spectra and circa 3,000 cells of data. All minerals are included into two general groups (by alphabetical order): 1) minerals from A to H, and 2) from J to Z, facilitating the rapid search for the specific spectrum or set of spectra in the database.

The FRISER-IRMIX database is in continuous development, and it is planned to incorporate new minerals and mineral mixtures. It is fully available for scientific collaboration. The following features are described for each mineral (or mixture) sample (see Table 1).

References

1. Gómez-Elvira, J., Haberle, B., Harri, A., Martinez-Frias, J., Renno, N., Ramos, M., et al. (2011). In: 4th *International Workshop on the Mars Atmosphere: Modeling and Observations*. Paris, France.
2. Gomez-Elvira, J., Armiens, C., Castañer, L., Domínguez, M., Genzer, M., Gómez, F., et al. (2012). The Environmental Sensor Suite for the Mars Science Laboratory Rover. *Space Science Reviews, 170*(1–4), 583–640.
3. Gómez-Elvira, J. Armiens, C., Carrasco, I., Genzer, M., Gómez, F. Harri, A-M., et al. (2013) (The REMS Team and the MSL Science Team). Rover environmental monitoring station. Overview of First 100 sols on Mars. In: *44th LPSC, the Woodland*, Texas, USA.
4. Sebastián, E., Armiens, C., Gómez-Elvira, J., Zorzano, M. P., Martinez-Frias, J., Esteban, B., et al. (2010). The ground temperature sensor of the rover environmental monitoring station: a pyrometer for Mars ground temperature measurement. *Sensors, 10*, 9211–9231.
5. Martín-Redondo, M. P. (2009). Sebastian Martínez, E., Fernández Sampedro, M.T. Armiens, C., Gómez-Elvira, J. and Martinez-Frias, J. FTIR reflectance of selected minerals and their mixtures: implications for the ground temperature-sensor monitoring on Mars surface environment (NASA/MSL-Rover Environmental Monitoring Station). *Journal of Environmental Monitoring, 11*, 1428–1432.

Spectral Analysis for Anomaly Detection in the Central Volcanic Range, Costa Rica. Implications for Planetary Geology

Juan Gregorio Rejas, Ruben Martínez, Miguel Marchamalo, Javier Bonatti and Jesús Martínez-Frías

1 Introduction

Image spectrometry has proven to be efficient in the characterization of minerals based on statistical methods using specific reflection and absorption bands. Hyperspectral configurations in SWIR and VNIR have been successfully used for mapping hydrothermal alteration materials in different geological scenarios [1]. Previous works have shown the applications of ranges of the spectrum in MIR (Medium Infrared) and TIR, combined with other sources [2].

On the other hand, detection of spectral anomalies aims at extracting automatically pixels that show significant responses in relation to their surroundings. Several methods have been developed over the last decades, allowing a better understanding of the relationships between hyperspectral data dimensionality and the optimization of search procedures [3, 4] as well as the sub-pixel differentiation of the spectral mixture and its implications in anomalous responses.

J. G. Rejas (✉)
NationalInstitute for Aerospace Technology, INTA, Ctra. de Ajalvir km 4 s/n,
28850 Torrejón de Ardoz, Spain
e-mail: rejasaj@inta.es

J. G. Rejas · R. Martínez · M. Marchamalo
Technical University of Madrid, UPM, Ramiro de Maeztu, 7, 28040 Madrid, Spain
e-mail: ruben.martinez@upm.es

M. Marchamalo
e-mail: miguel.marchamalo@upm.es

J. Bonatti
Costa Rica University, UCR, CampusUCR-San Pedro de Montes de Ocas s/n,
0125 SanJosé, Costa Rica
e-mail: jbonatti2011@gmail.com

J.Martínez-Frías
Astrobiology Centre, CSIC-INTA, Ctra. de Ajalvir km 4 s/n, 28850 Torrejón de Ardoz, Spain
e-mail: jmfrias@cab.inta-csic.es

E. Pardo-Igúzquiza et al. (eds.), *Mathematics of Planet Earth*,
Lecture Notes in Earth System Sciences, DOI: 10.1007/978-3-642-32408-6_170,

In the current work, we present a new approach to the study of absorption and reflection spectra bands, as well as spectral outliers significance in anomaly detection, as a tool for automated spectral analysis in Planetary Geology, based on a case of study in the Central Volcanic Mountain Range (Costa Rica).

2 Hyperspectral Data and Test Site

Costa Rica is emplaced on a triple union point of tectonic plates in which the Nazca, Cocos and Caribean plates converge. As a result, this zone is a very active volcanic region defined by a set of faults, which generate different tectonic blocks. The Central Volcanic Mountain Range runs NW–SE and is formed by a row of volcanic cones, comprising by pyroclastic rocks, andesitic, pyroxenic and basaltic lavas [5].

We used reflectance images of the HyMAP sensor [6] and the MASTER sensor [7], acquired on March 7, 2005, on the CVR and surroundings. The scene HyMAP (125 channels between 0.4589 and 2.491 μm) has a dimension of 710 $\times$ 2415 pixels, with a spatial resolution of 15 m. The MASTER scene is 1650 $\times$ 4466 pixels, with spatial resolution of 9 m and radiometric resolution of 16-bit. The 50 MASTER image channels are grouped in a port of 25 channels in the VNIR-SWIR (0.463–2.427 μm), and 25 channels between 3.075 and 13 μm in the MIR and TIR.

2.1 Pre-Processing

An algorithm MNF (Minimum Noise Fraction) has been applied to reduce noise of the MASTER and HyMAP images. Channels sensors presenting a high signal/noise ratio were rejected, channels ranging from 62 to 65 and 125 in the case of HyMAP, and from 16 to 19, 25 to 41 and 50 in the MASTER images.

Up to 49 mineral samples were collected during two sampling campaigns carried out in August 2010 and February 2012. Spectral measurements were performed in situ and at laboratory by an USB400 and an ASD FieldSpec Hi-Res spectroradiometers. The spectra allowed the minerals in the samples to be identified, and were utilized to check the HyMAP (Fig. 1) and MASTER reflectance images by an empirical linear regression.

3 Spectral Analysis and Anomaly Detection

Several minerals were detected remotely by their spectral responses (absorbance and reflectance) in the Turrialba Volcano area [1]. The channels of the HyMAP enabled minerals such as illite (6, 25, 105, 108), alunite (6, 25, 105, 116), kaolinite (6, 81, 108, 116) and kaolinite + smectite + monmorollonite (6, 81, 108, 119) to be

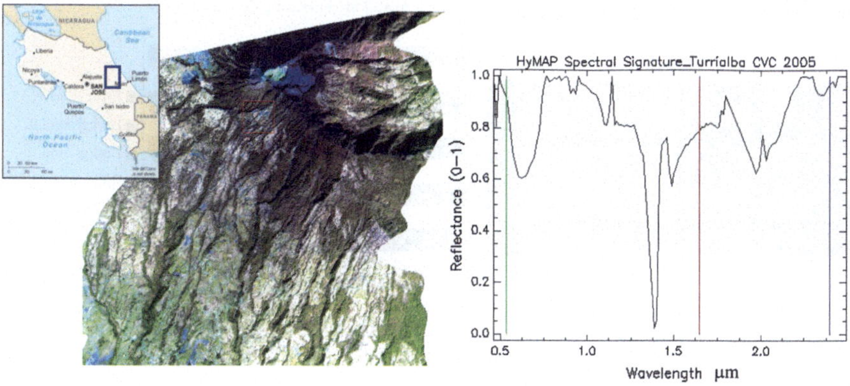

Fig. 1 Location of the study area (*map*) and HyMAP rgb 81, 6, 119 combination of the Turrialba volcano detail in the CVR (*left*) with a spectral measurement point example (*red frame*). Spectral signature extracted from the hyperspectral image (*right*)

detected. In the case of the MASTER images, alunite (4, 8, 21, 24), illite (4, 13, 22, 24) and kaolinite (4, 13, 22, 24) were identified.

Anomalies obtained for RX algorithm have been verified by using a method based on *projection pursuit* [4] and a subspace RX method [3]. The computation in all methods has been carried out separately for spectral ranges of reflective and emissive channels. For MASTER images we calculated a thermal index [1] profiting from the capability of the sensor to discriminate between different covers in the emissive spectrum.

The relationship between the spectral anomalies and the diagnostic bands has been studied. In order to do that, it has been linearly adjusted a sample space of 35 pair of tie points, which had spectral anomaly and diagnostics bands of minerals. All regressions have been calculated at a confidence level of 95 %, removing in each adjustment the sample values that showed unusual residues and which correspond mainly with clouds and shadows. Among all linear models obtained, the RX shows the highest R-squared value (73.12 %), for the absortion bands of illite from MASTER. The P value obtained was less than 0.05 in all cases, except for RX and absortion bands of the kaolinite and montmorollonite from HyMAP, suggesting great statistical significance of the method. The highest correlation coefficients from the regressions (−0.83 and 0.85) indicate a moderately strong relationship.

The difference between spectral anomalies with and without VNIR diagnostic bands for the HyMAP case is 0.75 %, and for the MASTER case (Fig. 2) is 0.84 %, while the difference with respect to the background was −2.52 and −2.32 %, respectively.

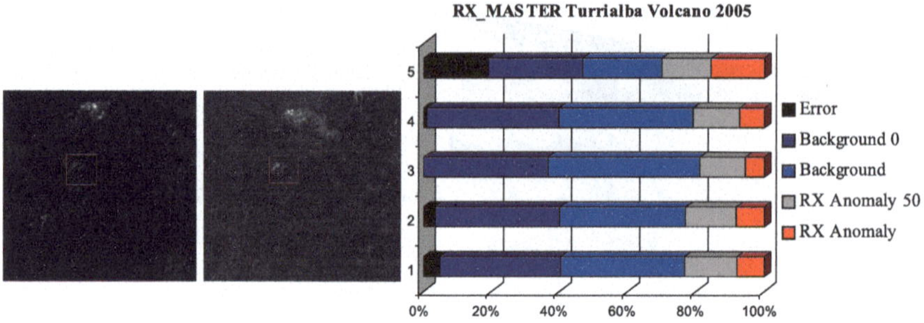

Fig. 2 RX anomalies from VNIR (*left*) and TIR MASTER chs. (*middle*). Comparison of RX anomaly detection methods for MASTER: *1*, All channels; *2*, VNIR; *3*, Only VNIR diagnostic bands; *4*, VNIR without diagnostic bands; *5*, TIR

4 Conclusions

Results of the standard RX, projection pursuit and subspace anomaly detectors were compared in spectral ranges of different image technologies into a terrestrial analog for Planetary Exploration. Higher concentrations of clay minerals, in scenarios where the sources of error are minimized, are correlated with the anomalies calculated in the reflective spectral range. The spectral mixing associated directly with the spatial resolution, has an impact in a significantly way on the characterization of the natural and semi-urban backgrounds, and thus in the spectral anomalies.

This paper improves the prospection processes searching areas and automatically detecting minerals of interest by virtue of their surface material characteristics, which is highly relevant for Planetary Geology.

References

1. Rejas, J. G., Martínez-Frías, J., Bonatti, J., Martínez, R., & Marchamalo, M. (2012). Anomaly detection and comparative analisys of hydrothermal alteration materials trough hyperspectral multisensor data in the Turrialba volcano. *XXII ISPRS International Conference*, Melbourne, Australia, eISSN 2194–9034.
2. Nasrabadi, N. M. (2008). Multisensor joint fusion and detection of mines using SAR and hyperspectral. *Proceedings of IEEE Sensors 2008 Conference*.
3. Borghys, D., Achard, V., Kasen, I., & Perneel, C. (2012). Comparative evaluation of hyperspectral anomaly detection methods in scenes with diverse complexity. *Proceedings of OPTRO2012 Symposium on Optronics in Defence and Security*, Paris, France.
4. Malpica, J. A., Rejas, J. G., & Alonso, M.-C. (2008). A projection pursuit algorithm for anomaly detection in hyperspectral imagery. *Pattern Recognition*, *41*(11), 3313–3327.
5. Alvarado, G. E. (1991). Resumen cronoestratigráfico de las rocas ígneas de Costa Rica basado en dataciones radiométricas. *Journal of South American Earth Sciences*, *6*(3), 151–168.

6. Cocks T., Jenssen, R., Stewart, A., Wilson, I., & Shields, T. (1998). The HyMAP airborne hyperspectral sensor: the system, calibration and performance. *Proceedings of 1st EARSeL Workshop on Imaging Spectroscopy*, EARSeL, Paris, pp. 37–43.
7. Hook, S. J., Myers, J. J., Thome, K. J., Fitzgerald, M., & Kahle, A. B. (2001) The MODIS, ASTER airborne simulator (MASTER)—a new instrument for earth science studies. *Remote Sensing of Environment*, *76*(1), 93–102.

A Mathematical Algorithm to Simulate the Growth and Transformation of Framboidal Pyrite: Characterization of the Biogenic Influence in Their Size Distributions

Raúl Merinero Palomares, Rosario Lunar Hernández, Francisco-Javier González-Sanz, Luis Somoza Losada and Jesús Martínez-Frías

1 Introduction

Framboidal morphology represents the most dominant pyrite texture observed in sedimentary environments [1]. Framboids are microscopic spherical clusters of equidimensional and equimorphic microcrystals. The formation of framboidal pyrite is explained by extremely high supersaturation levels [2]. The limited microcrystal size (0.1–2 μm) and the multiple (up to 10^6) pyrite nuclei evidence that a high nucleation rate/crystal growth rate ratio is necessary for framboid formation [3, 4]. The formation of micro-environments with high pyrite supersaturation could be related with the close relationship observed between organic matter and sedimentary framboidal pyrite [3]. Many experimental works has been achieved to explain the formation of framboidal pyrite being the models of aggregation of iron monosulfide microcrystals and their subsequent replacement into pyrite the most accepted [2]. The main problem of these models is the explanation for the spherical morphology and the role of microorganisms in their development. Although some authors explain the spherical

R. M. Palomares (✉) · R. L. Hernández
Crystallography and Mineralogy Department, Geology Faculty of Complutense University of Madrid, C/ Jose Antonio Novais, 2, 28040 Madrid, Spain
e-mail: rmeriner@ucm.es

R. L. Hernández
e-mail: lunar@ucm.es

F-J. González-Sanz · L. S. Losada
Spanish Geological Survey, C/ Ríos Rosas, 23, 28003 Madrid, Spain
e-mail: fj.gonzalez@igme.es

L. S. Losada
e-mail: l.somoza@igme.es

J. Martínez-Frías
Astrobiology Center, CSIC-INTA, associated to the NASA Astrobiology Institute, Ctra de Ajalvir, km. 4, 28850 Torrejón de Ardoz (Madrid), Spain
e-mail: jmfrias@cab.inta-csic.es

E. Pardo-Igúzquiza et al. (eds.), *Mathematics of Planet Earth*,
Lecture Notes in Earth System Sciences, DOI: 10.1007/978-3-642-32408-6_171,

shape by interaction forces, including gravitational, electrostatic, van der Waals and magnetic forces [3], other authors consider that the presence of biogenic spherical surfaces supports the development of spherical framboids [5].

The crystal-size distributions (CSD) or frequency-size distributions have distinctive shapes which can convey information about growth and subsequent transformations of framboids [6]. In particular, biogenic framboidal pyrite presents narrower CSD with lower size variability than abiogenic framboids [5].

To resolve the subject about the spherical shape and the role of microorganisms and organic matter in the formation of framboidal pyrite we implemented a mathematical algorithm to model the growth of framboidal pyrite by microcrystal aggregation. The algorithm was executed under diverse conditions simulating natural mechanisms with the aim of explain the role of spherical organic surfaces as the origin of the framboidal morphology and the base of the microcrystal aggregation. We also stressed the role of organic matter in the formation of narrow CSD.

2 Algorithm Details and Results

The mathematical algorithm subject of this research was implemented in C programming language. The thousand of calculated data was subsequently managed with Microsoft Excel and the statistical R software. The growth of a large number of framboids was simulated varying the following basic parameters: an initial diameter (D) that simulate the microcrystal aggregation on a pre-existent spherical surface, the packed factor (f), the diameter (d), the habit (h) and the number (n) of microcrystals. The volume and the diameter of each framboid were calculated from these basic parameters. An initial diameter of $D = 0$ implies aggregation without control of pre-existent surfaces, whereas values of D similar to microorganisms involved in framboidal pyrite formation simulate their role in the microcrystal aggregation inducting the spherical shape of framboids. The f and d parameters was introduced as variables to simulate some natural conditions. In sedimentary conditions it is expected that f was lower than cubic close packing because of the high amount of organic matter and other small substances trapped inside framboidal structure [7]. The high pyrite supersaturation level required to form framboids implies the development of microcrystals with small diameter. However, in sedimentary conditions, the formation of thousands of microcrystals decrease drastically the pyrite saturation and therefore it is expected that d decrease as the growth of the framboids advances [4]. Additional typical growth processes observed in natural framboidal pyrite are implemented: dissolution and subsequent overgrowth by aggregation of microcrystals, the formation of radiating elongated outer crystals around framboids and homogenization processes by interstitial spaces infilling [7].

The algorithm consists of two main stages: growth in open system followed by growth in closed system (constant volume) simulating dissolution and the subsequent processes described above. Growth in open system was implemented according to the Law of Proportional Effects that consists of a size dependent growth [8]. Growth in

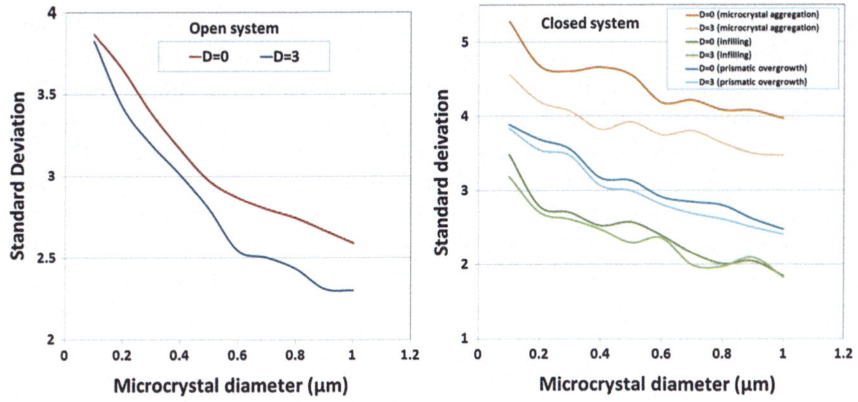

Fig. 1 Standard deviation variation in function of the initial diameter (D), microcrystal diameter and closed system growth mechanisms. In open system (*left*) initial values of D near microorganism sizes generated populations with lower variability. In closed system (*right*) infilling of interstitial voids generated populations with the lowest variability and microcrystal aggregation generated populations with the highest variability

open system generated CSD with lognormal shape. Simulations with $D>0$ generated CSD with lower variability but the same shape than that obtained with $D = 0$ (Fig. 1). Similar CSD was obtained decreasing the f and d parameters during growth, but with lower variability for the same average size.

Growth in closed system implies that the volume was constant, and therefore the growth of framboids is controlled by the dissolution of previously formed microcrystals. In each step of growth a critical radius was estimated according with Ostwald ripening [8] and framboids with lower size than critical radius were dissolved. A supply of reactants equivalent to the microcrystals dissolved was incorporated into framboids by one of these processes: infilling of interstitial voids, overgrowth by microcrystal aggregation or formation of outer elongated crystals. Growth in closed system transformed the shape of CSD into symmetric or negatively skewed and decreases the variability (Fig. 1). The infilling of interstitial voids generated the CSD with the lowest variability for the same average framboidal size.

3 Concluding Remarks

The microcrystal aggregation on a pre-existent spherical surface generates theoretical populations of framboids with narrow CSD similar to that observed in sedimentary conditions. Organic matter-rich environments (simulated decreasing the packed factor) produce narrower CSD than microcrystal aggregation on a pre-existent spherical surface. The same effect is obtained by the decay of pyrite supersaturation (simulated decreasing the diameter of microcrystals). Growth under constant volume conditions (closed system) also reduces the variability of sizes, although in this case the shape of CSD changes from lognormal to symmetrical or negatively skewed. Under these conditions the infilling of interstitial voids is more effective generating narrow CSD than other simulated mechanism in closed system. Mathematical computer simulation of the growth of framboidal pyrite is a useful and easy tool to generate large populations simulating natural sedimentary conditions where framboidal is the most dominant pyrite texture. Our results show that the presence of pre-existent spherical surfaces with sizes similar to that of microorganisms involved in the formation of sedimentary pyrite and the growth in organic matter-rich conditions can control the development of the narrow lognormal CSD observed in nature and the spherical shape of framboids.

References

1. Wilkin, R. T., Barnes, H. L., & Brantley, S. L. (1996). The size distribution of framboidal pyrite in modern sediments: An indicator of redox conditions. *Geochimica et Cosmochimica Acta, 60*, 3897–3912.
2. Ohfuji, H., & Rickard, D. (2005). Experimental syntheses of framboids—A review. *Earth-Science Reviews, 71*, 147–170.
3. Wilkin, R. T., & Barnes, H. L. (1996). Pyrite formation by reactions of iron monosulfides with dissolved inorganic and organic sulfur species. *Geochimica et Cosmochimica Acta, 60*, 4167–4179.
4. Butler, I. B., & Rickard, D. (2000). Framboidal pyrite formation via the oxidation of iron(II) monosulphide by hydrogen sulphide. *Geochimica et Cosmochimica Acta, 64*, 2665–2672.
5. Popa, R., Kinkle, B. K., & Badescu, A. (2004). Pyrite framboids as biomarkers for iron-sulfur systems. *Geomicrobiology Journal, 21*, 193–206.
6. Merinero, R., Lunar, R., Somoza, L., Díaz-del-Río, V., & Martínez-Frías, J. (2009). Nucleation, growth and oxidation of framboidal pyrite associated with hydrocarbon-derived submarine chimneys: Lessons learned from the Gulf of Cadiz. *European Journal of Mineralogy, 21*, 947–961.
7. Sawlowicz, Z. (2000). Framboids: From their origin to application. *Prace Mineralogiczne, 88*, 1–80.
8. Eberl, D. D., Drits, V. A., & Srodon, J. (1998). Deducing growth mechanisms for minerals from the shapes of crystal size distributions. *American Journal of Science, 298*, 499–533.

Computational Models and Simulations of Meteor Impacts as Tools for Analysing and Evaluating Management of Crisis Scenarios

Jesús Martínez-Frías and Alain Leppinette Malvite

1 Introduction

The Earth is in constant interaction with outer space, and we must do our best to understand the scientific and social implications of this fact. The geological record evidences that cosmic impacts have accompanied the geobiological evolution of our planet (Fig. 1). Astronomic and astrogeological studies indicate that around 70 % of potential impactors are of asteroidal type (the rest is cometary). Numerous studies [1–8] allow to estimate that the amount of the potentially dangerous NEO (Near Earth Objects) which could close up to our planet are: (a) circa 150 million of 10 m; (b) around 300,000 of 100 m; (c) 10,000 of 500 m, and (d) approximately 2,000 massive objects of 1 km of diameter. Fortunately, 75 % of the large asteroids (of more than 1 km) are already localized and none of them has a dangerous orbit.

There is a general agreement that the largest risk of impact (and the highest difficulty of observation and monitoring) is focussed on "small objects" with sizes ranging from 10 to 300 m [3–6]. The energy released is huge: around 10 Mt TNT for an object of 50 m (in its atmospheric entry). The transition from 100 to 200 m practically multiplies by 10 the final value (from 75 to 600 Mt). Computer simulation and modelling offer new insight into the formation of impact craters, not only helping to understand ancient impact events (Fig. 1), but also providing a way for analyzing and evaluating management issues related to future crisis scenarios.

J. Martínez-Frías (✉) · A. L. Malvite
Centro de Astrobiología, CSIC-INTA, associated to the NASA Astrobiology Institute, Ctra. Ajalvir, km. 4,Torrejón de Ardoz, 28850 Madrid, Spain
e-mail: jmfrias@cab.intacsic.es

L. Malvite
e-mail: lepinettema@cab.inta-csic.es

E. Pardo-Igúzquiza et al. (eds.), *Mathematics of Planet Earth*,
Lecture Notes in Earth System Sciences, DOI: 10.1007/978-3-642-32408-6_172,

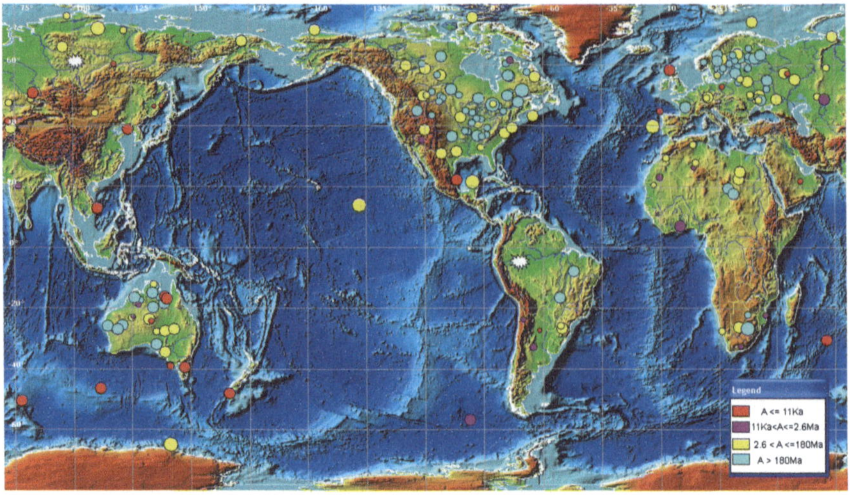

Fig. 1 Map of confirmed impact structures on the Earth surface (180 structures) and on the oceanic *bottom* (24 structures). Size of circles is proportional to the crater diameter. *White stars* depict the epicenters of two largest air blasts of the twentieth century - Tunguska (1908) and Brasil (1930). Courtesy of Expert Database on the Earth Impact Structures (EDEIS), Tsunami Laboratory, ICMMG SD RAS, Novosibirsk, 2006 [9]

2 Computerized Simulations and Evaluation of Crisis Scenarios

A set of 20 possible crisis scenarios, related to different types of impactors, have been simulated and evaluated. In our model, two compositional types of impactors have been considered (silicate-rich -stony and metallic - irons). We have considered: (a) different sizes of the impactors, ranging from 10 to 300 m (10, 50, 100, 200 and 300 m as their terminal size (striking earth's surface); (b) vertical impacts, and (c) a velocity of 12 km/s. Regarding the substrates, they were simulated as (dry) rocky substrate and rocky substrate (with water). In order to estimate the energy (Mt) and seismicity (Richter scale), which correspond to each impact scenario, the results were combined with the simulation data from the model of Collins et al. [10].

As a reference for assessing the management of the social crisis scenario [11] originated by the collision, it is important to note that the atomic bomb of Hiroshima had an explosive energy equal to 13 kt TNT [12]. It wiped out 90 % of the city and approximately 80,000 people were killed as a direct result of the blast. Figures 2 and 3 displays two examples of computerized simulated impact craters.

In the same way that it is extremely important to monitor the Near Earth Objects, it is also crucial to study the effects of these collisions by the characterization of the terrestrial impact craters and by their computerized simulation. Very recent events, such as the Chelyabinsk meteor event (Russia, 15 february, 2013) [13] confirm the

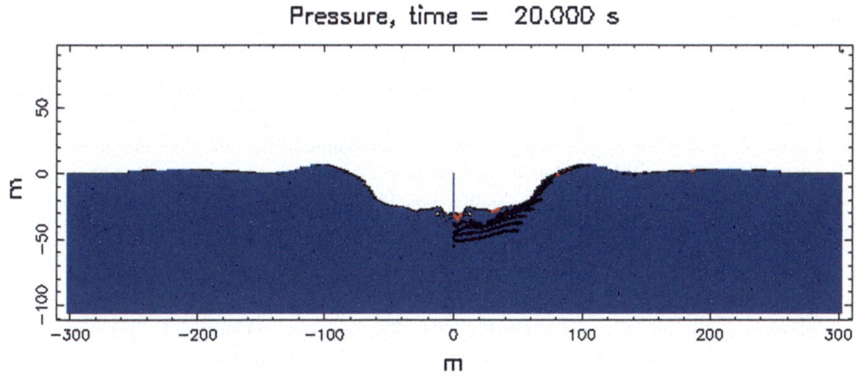

Fig. 2 Impact crater generated by a silicate-type asteroidal object. Ø = 10 m. v = 12 km/s. Rocky substrate (dry). Energy of impact = 0.32 MegaTons. Ejecta: mostly blocked by the atmosphere. Associated seismicity: 4,3 Richter scale. No thermal radiation. Scarce vaporization. Simple crater

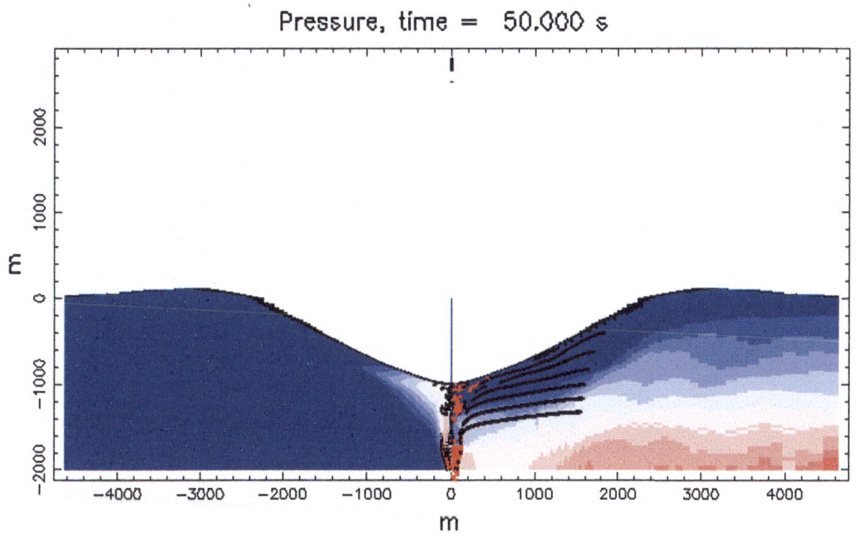

Fig. 3 Impact crater generated by a metallic-type asteroidal object. Ø = 300 m. v = 12 km/s. Rocky substrate (with water). Energy of impact= 9.83×10^2 MegaTons. Ejecta: they would create a layer of 32,8 m thick., and their mean average size would be 98.8 m. Associated seismicity: 6.8 Richter scale. No thermal radiation. Scarce vaporization. Crater in transition from simple to complex

significance of being aware about these astrogeological hazards, as well as of having appropriate tools and protocols which help us minimize their effects.

References

1. Napier, W. M., & Clube, S. V. M. (1979). A theory of terrestrial catastrophism. *Nature*, *282*, 455–459.
2. Hughes, D. W. (2000). A new approach the the calculation of the cratering rate of the Earth over the last 125 +/- 20 Myr. *Monthly Notices Royal Astronomical Society*, *317*, 429–437.
3. Near Earth Object Program. NASA. Sentry Risk Table. http://neo.jpl.nasa.gov/risk/.
4. Morrison, D., (Ed.). (1992). The Spaceguard Survey: Report of the NASA International Near-Earth-Object Detection Workshop. NASA Publication http://impact.arc.nasa.gov.
5. Chapman, C. R., & Morrison, D. (1994). Impacts on the earth by asteroids and comets: Assessing the hazard. *Nature*, *367*, 33–39.
6. Chapman, C., Durda, D.D., & Gold, R.E. (2001). The comet/asteroid impact hazard: A systems approach. SwRI White Paper: www.boulder.swri.edu/clark/neowp.html.
7. Kring, D. A. (2003). Environmental consequences of impact cratering events as a function of ambient conditions on Earth. *Astrobiology*, *3*(1), 133–152.
8. Chapman, C. R. (2004). The hazard of near-Earth asteroid impacts on Earth. *Earth and Planetary Science Letters*, *222*, 1–15.
9. Impact Holoce Working Group (IHWG) (2013). http://tsun.sscc.ru/hiwg/.
10. Collins, G., Melosh, H. J., & Marcus, R. Impact Earth. http://www.purdue.edu/impactearth (2010).
11. Harry, S., & Truman Library & Museum. (2009). *U. S. strategic bombing survey: The effects of the atomic bombings of Hiroshima and Nagasaki, June 19, 1946*. President's secretary's file, truman Papers. 2. Hiroshima, page 11 of 51. Retrieved March 15, 2009.
12. Blair, B.D. (2003). Decision model for potential asteroid impacts. Research paper EB560 - decision analysis, division of economics and business, colorado school of mines.
13. Yeomans, D., Chodas, P. (2013 1 March). *Additional details on the large fireball event over Russia on February 15, 2013*. NASA/JPL near-earth object program office. Retrieved 2 March 2013.

In-situ Raman Analysis of the Precipitation Sequence of Sulphate Minerals Using Small Droplets: Application to Rio Tinto (Spain)

Fernando Rull, Francisco Sobrón, Julia Guerrero, Jesús Medina, Gloria Venegas, Fernando Gázquez and Jesús Martínez-Frías

1 Introduction

Rio Tinto (Huelva, SW Spain) is considered a modern model of formation of sulphates, linked to significant acidophilic biogenic activity [1]. The river's basin lies on the Iberian Pyrite Belt (IPB) that is an arcuate belt, 250 km long and 25–70 km wide, in the southwest of the Iberian Peninsula. The IPB is one of the largest volcanogenic massive sulphide provinces over the world. Sulphates come from aqueous alteration of iron-rich sulphide minerals of the IPB. To understand the mineral precipitation sequence and the role of the bacterial activity on this process a detailed study of the Rio Tinto physicochemical system has been undertaken. This study comprises chemical composition, chemical species distribution at

F. Rull (✉) · F. Sobrón · J. Guerrero · J. Medina · G. Venegas · F. Gázquez · J. Martínez-Frías
Unidad Asociada Universidad de Valladolid—CSIC—Centro de Astrobiología, Valladolid, Spain
e-mail: rull@fmc.uva.es

F. Sobrón
e-mail: sobron@iq.uva.es

J. Guerrero
e-mail: juliagf18@hotmail.com

J. Medina
e-mail: medina@fmc.uva.es

G. Venegas
e-mail: glovedeva@hotmail.com

F. Gázquez
e-mail: f.gazquez@ual.es

J. Martínez-Frías
e-mail: jmfrias@cab.inta-csic.es

F. Rull · J. Medina · G. Venegas · J. Martínez-Frías
Centro de Astrobiología, CSIC-INTA, Associated to the NASA Astrobiology Institute,
Ctra de Ajalvir, km. 4 , 28850 Torrejón de Ardoz, (Madrid), Spain

E. Pardo-Igúzquiza et al. (eds.), *Mathematics of Planet Earth*,
Lecture Notes in Earth System Sciences, DOI: 10.1007/978-3-642-32408-6_173,

equilibrium and mineral precipitation sequences using mainly spectroscopic techniques. Experiments at the field using portable instruments have been compared with those performed at the laboratory using a simulator of the riverbank evaporation conditions and also using small droplets of acidic water deposited on different substrates. In addition, computer simulation methods allowing chemical speciation, chemical equilibrium and sequence precipitation calculations have been developed to compare with the experimental results.

2 Spectroscopic Results at the River

In-situ Raman spectroscopy performed at the riverbanks (Fig. 1) allowed up to 10 minerals, mainly hydrated sulphates, to be identified in the Rio Tinto efflorescences (Table 1). In particular, Copiapite and Coquimbite group minerals were found, in addition to other low hydrated sulphates, such as rhomboclase, rozenite and szomolnokite.

3 Spectroscopic Results at the Evaporation-Precipitation Simulator

In Fig. 2a it is depicted the simulator used for evaporating natural waters from the river.

Fig. 1 **a** and **b** In-situ Raman spectroscopy. **c** and **d** Examples of sulphate efflorescences on the riverbanks

Table 1 Minerals found by Raman spectroscopy in the Rio Tinto efflorescences and experiments carried out from natural water

Time (days)	Minerals
1	Ferricopiapite
3	Coquimbite
5	Copiapite
7	Magnesiocopiapite
13	Hematite
16	Rozenite
16	Szomolnokite
18	Rhomboclase
30	Metavoltine

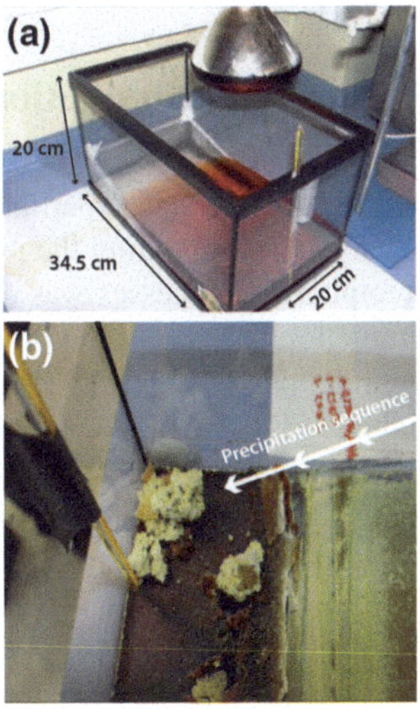

Fig. 2 **a** Evaporation-precipitation simulator used in the laboratory. **b** Efflorescences generated after evaporation of the Rio Tinto water

The mineralogy of the efflorescences formed in the evaporation simulator (Fig. 2b) was similar to that of the natural samples taken from the riverbanks in the Rio Tinto setting (Table 1). Additionally, magnesiocopiapite, Al-copiapite and szomolnokite were found in the simulator experiments. Nevertheless, the exact precipitation sequence of these minerals was difficult to be described because of some of them precipitated at the same time.

4 Spectroscopic Results on Small Droplets

We have designed a new series of experiments in which small droplets of natural waters from the river evaporated at room temperature on different substrates (Fig. 3), giving rise to a mineralogical sequence as a function of time. The results obtained at the micro-scale have enabled to simulate the precipitation sequence occurred in the riverbanks (Table 2) in a more detail.

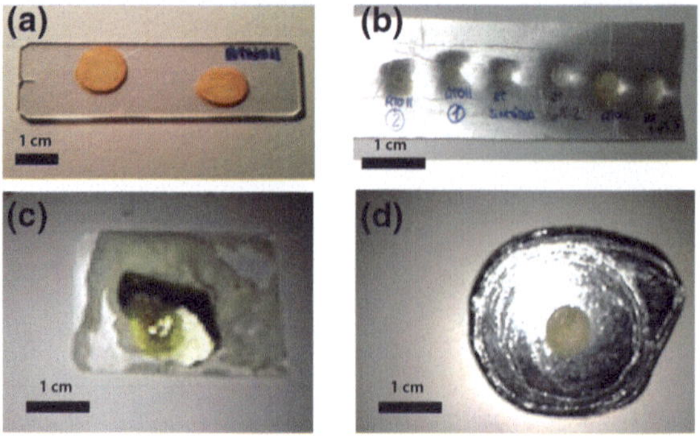

Fig. 3 Small droplets on glass **a**, aluminum **b**, pyrite **c** and zinc **d** substrates

5 Computer Simulations of Evaporation Processes

In order to check the experimental results obtained and to produce new and powerful tools for the simulating the precipitation, computer simulations of the evaporation-precipitation processes was undertaken. Fig. 4 illustrates an example the computer simulation process of the precipitation of a double salt of Mg and Fe sulphate in a small droplet of water. The initial volume is 5 mm^3 at 23 °C and the drop geometry is shown in Fig. 4a. Figure 4b shows the evolution of the molar fraction of the different chemical species in solution as function of time. Remarkably, the behavior of Fe and Mg is different, therefore illustrating the differences in the precipitation sequence.

Table 2 Precipitation sequence obtained from micro-scale evaporation experiments of a Rio Tinto water sample on glass substrate along 30 days

Minerals	Formula
Baryte	$BaSO_4$
Copiapite[a, b]	$Fe^{2+}Fe_4^{3+}(SO_4)_6(OH)_2 \cdot 20(H_2O)$
Coquimbite[a, b]	$Fe_2^{3+}(SO_4)_3 \cdot 9H_2O$
Epsomite	$MgSO_4 \cdot 7H_2O$
Ferricopiapite[b]	$(Fe_{2/3}^{3+}Fe_4^{3+}(SO_4)_6(OH)_2 \cdot 20H_2O)$
Gypsum	$CaSO_4 \cdot 2H_2O$
Halotrichite	$Fe^{2+}Al_2(SO_4)_4 \cdot 22H_2O$
Jarosite[a]	$(KFe_3^{3+}(OH)_6(SO_4)_2)$
Melanterite	$(Fe^{2+}SO_4 7H_2O)$
Rhomboclase	$H_5Fe^{3+}O_2(SO_4)_2 \cdot 2H_2O$
Rozenite[a, b]	$(Fe^{2+}SO_4 \cdot 4H_2O)$

[a] Minerals detected by in-situ Raman
[b] Minerals found in the simulator experiments

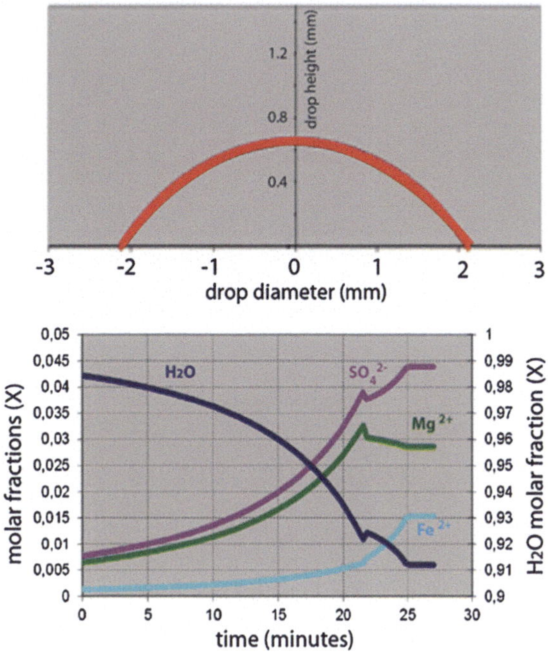

Fig. 4 **a** Initial geometry of the small drop of 5 mm^3 aqueous solution at 23 °C. **b** Molar fraction evolution as a function of time of the different chemical species during the evaporation process (*dark blue* water; *purple* SO_4^2; *green* Mg^{2+}; *light blue* Fe^{2+})

6 Conclusions

The application of Raman spectroscopy to the analysis of precipitation process is a powerful tool for the in-situ study at the molecular scale of the different processes leading to the mineral phase sequences. The results obtained with the simulators at the laboratory using acidic waters from Rio Tinto allow identifying all the minerals previously found in natural conditions at the riverbanks. Nevertheless, the small droplets system is more reliable, providing very precise time-information on the mineral sequence precipitation. The micro-Raman analysis of these small systems in combination with computer simulation tools represents a very promising way to analyze the mineral evaporation-precipitation processes in aqueous solutions.

Acknowledgments Financial support was made available through Project 'UMBRELLA' (7th Framework Program of the European Union).

Reference

1. Fernández-Remolar, D., Morris, R., Gruener, J. E., Amils, R., & Knoll, A. H. (2005). The Rio Tinto Basin, Spain: Mineralogy, sedimentary geobiology, and implications for interpretation of outcrop rocks at Meridiani Planum, Mars. *Earth Planetary Science Letters Journal, 240*, 149–167.

Signal Analysis by Means of Multi-Scale Methods

Christian Blick

1 Introduction

The multi-scale method to decorrelate seismic data sets such as 3-D velocity and density models as well as the corresponding migration results of a possible site for a geothermal power plant is based on the *inhomogeneous wave equation* [2]

$$(\Delta + k^2)V(x) = F(x). \tag{1}$$

The variable $k = \frac{\omega}{c}$ is called *wavenumber* and is defined as *angular velocity* ω divided by the *velocity of the medium* c. Further on, the functions F and V are linked by the integral equation

$$V(x) = \int_\Gamma F(y)G(\Delta + k^2, \|x-y\|)dy, \tag{2}$$

where the *fundamental solution* $G(\Delta + k^2, \|x - y\|)$ of the inhomogeneous Helmholtz equation is given by

$$G(\Delta + k^2, \|x-y\|) = -\frac{\exp(ik\|x-y\|)}{4\pi\|x-y\|}. \tag{3}$$

Christian Blick (✉)
Geomathematics Group, University of Kaiserslautern, Kaiserslautern, Germany
e-mail: blick@mathematik.uni-kl.de

E. Pardo-Igúzquiza et al. (eds.), *Mathematics of Planet Earth*,
Lecture Notes in Earth System Sciences, DOI: 10.1007/978-3-642-32408-6_174,

2 Decomposition of Seismic Data Sets

In general, we have two options to decompose a seismic data set given as a sample of a function $F(x)$, $x \in \mathbb{R}$. We can either calculate and decompose the function $V(x)$ given via Eq. (2) or we can decompose the function $F(x)$ directly.
Further on, the choice of the wave number k plays an important role and depends on the type of the data set, which should be decorrelated. If we are faced with velocity or density models, k should be chosen as equal to zero. On the other hand, if we want to decorrelate a migration result, k should be chosen as a positive real number. Here, the most important part is that by choosing different wave numbers, we are able to highlight different rock formations in the ground as can be seen in Fig. 1c. In the following, we concentrate on the case $k > 0$.

2.1 Decorrelation of V

In order to decorrelate the function V, we use a regularization technique [3] on the fundamental solution of the Helmholtz equation (Eq. 3) by applying Taylor Expansion of order $n+2$, $n \in \mathbb{N}$ in a ball of radius ρ around x leading to the scaling function

$$S_\rho^n(\Delta + k^2, \|x-y\|) = \begin{cases} -\frac{\exp(ik\|x-y\|)}{4\pi\|x-y\|} & , \quad \|x-y\| \geq \rho \\ -\frac{\exp(ik\|x-y\|)}{4\pi}\left(\frac{1}{\|x-y\|} + \frac{(\rho-\|x-y\|)^{n+2}}{\|x-y\|\rho^{n+2}}\right) & , \quad 0 < \|x-y\| < \rho \\ -\frac{n+2}{4\pi\rho} & , \quad \|x-y\| = 0 \end{cases} . \tag{4}$$

Hence,

$$V(x) = \lim_{\rho \to 0} \int_\Gamma F(y) S_\rho^n(\Delta + k^2, \|x-y\|)dy \tag{5}$$

and we can decorrelate the function V by

$$V(x) = \int_\Gamma F(y) S_{\rho_j}^n(\Delta + k^2, \|x-y\|)dy + \sum_{m=j}^{\infty} \int_\Gamma F(y) \Psi_m^n(\Delta + k^2, \|x-y\|)dy, \tag{6}$$

where ρ_j is a null sequence and the *wavelet function* $\Psi_m^n(\Delta + k^2, \|x-y\|)$ regarding the *scalling parameter* m is given by

$$\Psi_j^n(\Delta + k^2, \|x-y\|) = S_{\rho_{j+1}}^n(\Delta + k^2, \|x-y\|) - S_{\rho_j}^n(\Delta + k^2, \|x-y\|). \tag{7}$$

2.2 Decorrelation of F

In order to decorrelate the function F directly, we calculate the corresponding scaling function

$$B_\rho^n(\Delta + k^2, \|x-y\|) = \frac{(\Delta + k^2)S_\rho^n(\Delta + k^2, \|x-y\|)}{\|(\Delta + k^2)S_\rho^n(\Delta + k^2, \|x-y\|)\|_{L^1(\mathbb{R}^3)}} \tag{8}$$

given by

$$B_\rho^n(\Delta + k^2, r) = \begin{cases} \frac{\exp(ikr)(\rho-r)^n(n+2)(n+1+2ki(r-\rho))}{4\pi r \rho^{n+2}\|(\Delta+k^2)S_\rho^n(\Delta+k^2,\|x-y\|)\|_{L^1(\mathbb{R}^3)}} &, |r| \geq \rho \\ 0 &, \text{else} \end{cases} . \tag{9}$$

The normation of the function $B_\rho^n(\Delta + k^2, \|x-y\|)$ is important in order to be able to compare different scaling functions. The decomposition can now be obtained by substituting the functions V by F and S by B in Sect. 2.1.

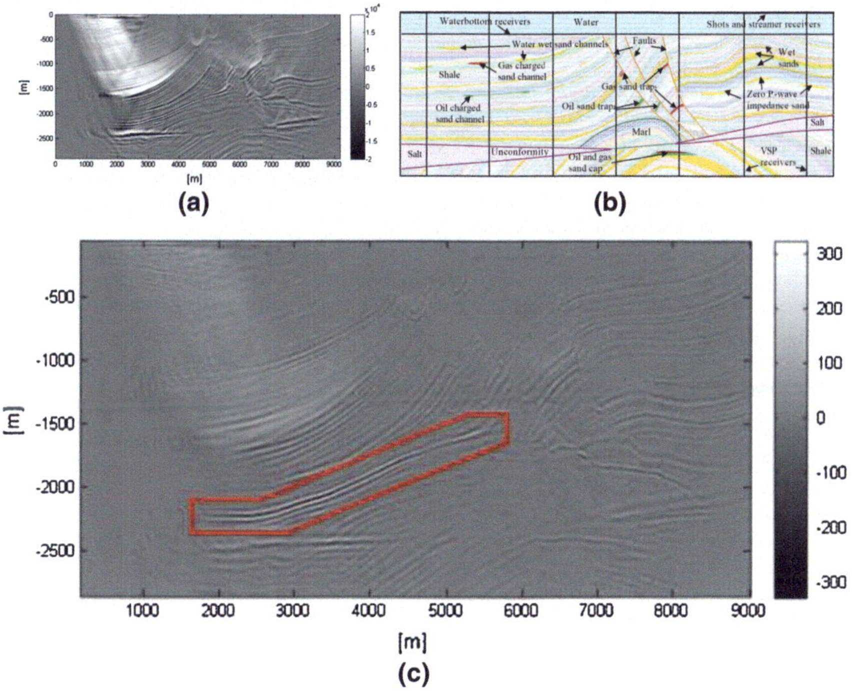

Fig. 1 Migration result (**a**), interpretation of the Marmousi data set (**b**) [4, 5] and exemplary detail function (**c**) of the decomposition of V for parameters $k = 0.099,\ j = 4$ and $\rho = 1500 \cdot 2^{-j} m$ [7]

3 Numerical Results

In order to test the 3-D multi-scale method, we used the 2-D Marmousi model [4, 5] and created our 3-D data set by putting the 2-D data behind each other. A 2-D method was already discussed in [4, 6]. Figure 1a, b shows the migration result of the Marmousi data set and it's interpretation, Fig. 1c an exemplary detail function of the decomposition of V corresponding to the parameters $k = 0.099$ and $j = 4$ given our 3-D data set as input. Our results show, that by choosing different wave numbers, we are able to repress different rock formations while highlighting a specific one. Further on, the method is able to filter out the noise in the upper left corner of the initial data set. Specifically, for the chosen parameters, we are able to highlight the formation above the salt structure in the lower left corner, which extends above the marl dome in the center of the figure (red box).

References

1. Yilmaz, Ö. (2001). *Seismic data analysis*. Tulsa: Society of Exploration Geophysicists.
2. Mller, C. (1957). *Grundprobleme der mathematischen theorie elektromagnetischer schwingungen*. Berlin: Springer.
3. Freeden, W., & Gerhards, C. (2012). *Geomathematically oriented potential theory*. London: Chapman & Hall/CRC.
4. Ilyasov., M. (2011). *A tree algorithm for helmholtz potential wavelets on non-smooth surfaces: Theoretical background and application to seismic data processing*. Ph.D.thesis: Geomathematics Group, University of Kaiserslautern, Kaiserslautern.
5. Computation of MULTI-VALUED TRAVELTIMES in the Marmousi model, Department of Computational and Applied Mathematics, RICE University, Houson. http://www.caam.rice.edu/benamou/testproblem.html. Accessed 27 Mar 2013.
6. Augustin, M., Freeden, W., Gerhards, C., Mhringer, S., & Ostermann, I. (2012). Mathematische methoden in der geothermie. *Mathematics Semesterberichte*, *59*, 1–28.
7. Freeden, W., & Blick, C. (2013). *Signal decorrelation by means of multi-scale methods. World of mining*, (in print).

Modeling and Simulation of Forest Fire Spreading

Sarah Eberle

1 Mathematical Modeling

We consider a similar model of forest fire spreading as Asensio and Ferragut [1].

$$\frac{\partial T}{\partial t} = -\mathbf{v}\nabla T + D\Delta T + A\left(Y \exp\left(-\frac{B}{T - T_\infty}\right) - h(T - T_\infty)\right), \tag{1}$$

$$\frac{\partial Y}{\partial t} = -bY \exp\left(-\frac{B}{T - T_\infty}\right), \tag{2}$$

with temperature of fuel T, time t, wind velocity $\mathbf{v}$, diffusion coefficient D, pre-exponential factor of reaction A, mass fraction of fuel Y, coefficient due to modified Arrhenius law B, natural convection coefficient h, disappearance rate of fuel b, and ambient temperature T_∞.

2 Numerical Solution

2.1 Space and Time Discretization

We use for the space discretization a collocation method based on the sums:

$$u = \sum_{i=1}^{I} \phi(x, z_i)u_i, \quad y = \sum_{i=1}^{I} \phi(x, z_i)y_i, \quad x \in X, \; z \in Z, \; I \text{ number of points}, \tag{3}$$

S. Eberle (✉)
Geomathematics Group, University of Kaiserslautern, Kaiserslautern, Germany
e-mail: eberle@mathematik.uni-kl.de

E. Pardo-Igúzquiza et al. (eds.), *Mathematics of Planet Earth*,
Lecture Notes in Earth System Sciences, DOI: 10.1007/978-3-642-32408-6_175,

where ϕ is the trial function, X a grid representing the collocation points and Z a grid consisting of the centers of the trial functions (for more details the reader is referred to Eberle et al. [2]). Then, we apply the ansatz (3) for the temperature T and mass fraction Y and plug it in Eqs. (1) and (2):

$$\frac{\partial u}{\partial t} = \sum_{i=1}^{I} (-\mathbf{v}\,\nabla\phi_i + D\Delta\phi_i - Ah\phi_i)\,u_i + AhT_\infty + A\sum_{i=1}^{I}\phi_i\,y_i \exp\left(-\frac{B}{T-T_\infty}\right), \tag{4}$$

$$\frac{\partial y}{\partial t} = -b\sum_{i=1}^{I}\phi_i\,y_i \exp\left(-\frac{B}{T-T_\infty}\right). \tag{5}$$

The time discretization is done by a Crank-Nicolson-scheme.

2.2 *Stabilization*

The above introduced solution scheme yields strongly oscillating results in the convection dominated case (Gibbs phenomenon). Thus, the method needs to be stabilized. Here, we follow the procedure of flux corrected transport of Kuzmin, Löhner, Turek [3]. In doing so, we apply the stabilization exemplary for the temperature T.

Step (1) We start with the approximation of the initial conditions and determine the according coefficients u_0 by solving the system

$$Mu_0 = T_0, \tag{6}$$

where $M = m_{ij}$ is the mass matrix given by $m_{ij} = \phi(x_i, z_j)$.

The coefficients are needed for the space discretization within the time-stepping scheme.

Step (2) Next, we consider the so-called "low-order" problem and define the lumped mass matrix M_L by

$$m_{ii} = \sum_j m_{ij} \quad \text{for} \quad i = j. \tag{7}$$

Step (3) After that we have a look at the "high-order" problem, which means we construct the operator K^H given by

$$k_{ij}^H(\phi) = -\mathbf{v}\nabla\phi(x_i, z_j) + D\Delta\phi(x_i, z_j), \tag{8}$$

which describes the convection and diffusion.

Step (4) Artificial diffusion is added now and we define the diffusion operator in the same way as by Möller [4]

$$d_{ii} = -\sum_{j \neq i} d_{ij}, \quad d_{ij} = d_{ji} = \max\{0, -k_{ij}^H, -k_{ji}^H\} \text{ for } i < j \tag{9}$$

and the low-order operator $K^L = K^H + D$.
Step (5) The right-hand side of our convection-diffusion-reaction-problem (1) is represented by the reaction term q and we call its coefficients q_{n-1}.

$$q = A\left(Y \exp\left(-\frac{B}{T - T_\infty}\right) - h(T - T_\infty)\right). \tag{10}$$

Step (6) Following the procedure in [4] we make an approximation of the coefficients of the collocation method by

$$\overline{u} = u_{n-1} - \frac{\Delta t_n}{2} M_L^{-1}(K^L u_{n-1} - q_{n-1}). \tag{11}$$

Step (7) Next, we modify the right-hand side of problem (1) by applying Zalesak's algorithm [5] for which we need to calculate the residuum r and the weights α to get q_{n-1}^*. The algorithm considers only the next neighbors i of every collocation point

$$P_j^+ = \sum_{i \neq j} \max_{i=1,\dots,N} \{0, r_{ij}\}, P_j^- = \sum_{i \neq j} \min_{i=1,\dots,N} \{0, r_{ij}\}, \tag{12}$$

$$Q_j^+ = \max\{0, \max_{i=1,\dots,N} (\overline{u}_j - \overline{u}_i)\}, Q_j^- = \min\{0, \min_{i=1,\dots,N} (\overline{u}_j - \overline{u}_i)\}, \tag{13}$$

$$R_j^+ = \min_{i=1,\dots,N} \left\{1, \frac{m_i Q_i^+}{P_i^+}\right\}, R_j^- = \min_{i=1,\dots,N} \left\{1, \frac{m_i Q_i^-}{P_i^-}\right\}, \tag{14}$$

$$\alpha_{ij} = \min_{i=1,\dots,N} \left\{1, \frac{R_i^+}{R_j^-}\right\} \text{ for } r_{ij} > 0 \text{ and } \alpha_{ij} = \min_{i=1,\dots,N} \left\{1, \frac{R_i^-}{R_j^+}\right\} \text{ else.} \tag{15}$$

Step (8) Now we are able to determine the coefficients

$$u_n = u_{n-1} - \frac{\Delta t_n}{2} M_L^{-1}(K^L u_{n-1} - q_{n-1} - q_{n-1}^*) \tag{16}$$

Step (9) Finally, we use these coefficients to get solutions for the temperature T and the mass fraction of the fuel Y with the stabilized method.

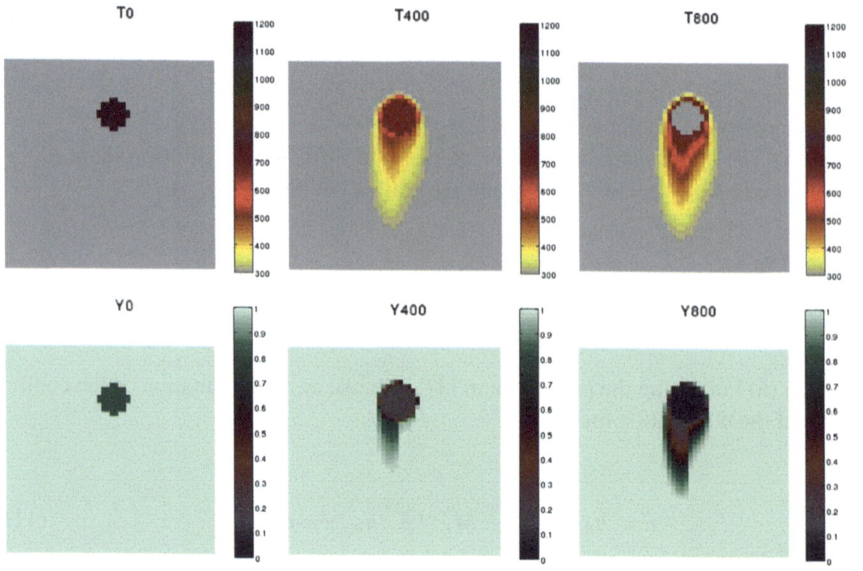

Fig. 1 In the first row the temperature of fuel T in $[K]$ is plotted for the initial temperature, after 400 and 800 time steps and accordingly in the second row we see the mass fraction of fuel Y

3 Numerical Simulation

Figure 1 shows first simulations for two different fuel types (type 1 on the left-hand side and type 2 on the right-hand side) and wind directed to the south. We can see the fire spreads faster for the fuel type 1 and due to the wind its shape is elliptic.

References

1. Asensio, I., & Ferragut, L. (2002). On a wildland fire model with radiation. *International Journal for Numerical Methods in Engeneering*, *54*, 137–157.
2. Eberle, S., Freeden, W., Matthes, U. *Forest fire spreading*. Handbook of geomathematics (2nd ed.). Springer (in preperation).
3. Kuzmin, D., Löhner, R., Turek, S. (Eds.) (2012). *Flux-corrected transport. scientific computation* . New York: Springer.
4. Möller, M. (2008). *Adaptive High-Resolution Finite Element Schemes*. Ph.D. Thesis, Technische Universität Dortmund. Germany.
5. Zalesak, S.T. (1979). *Fully multidimensional flux-corrected transport algorithms for fluids*. London: Academic Press, Inc.

The Finite Pointset Method (FPM) and an Application in Soil Mechanics

Jörg Kuhnert and Isabel Ostermann

1 Introduction

The numerical idea of the meshfree Finite Pointset Method (FPM) is the use of a generalized finite difference algorithm solving PDEs in fluid dynamics and continuum mechanics based on a cloud of numerical points with simple neighborhood relationship using balls of interaction (see, e.g., [1, 5–7]). These points carry the required physical information (such as temperature, velocity, or pressure) and move with flow velocity in a Lagrangian approach. The partial time derivatives are represented by first or second order time integration. The spatial derivatives are modeled by weighted moving least squares (WMLS) approximations. There are major advantages of the method. The Lagrangian approach on the point-cloud-basis gives rise to efficiently solve free surface, multiphase, and moving domain applications. No additional computation effort is required for mesh generation and adaptation. For each point, only neighborhood lists, given by the balls of interaction, are required. Moreover, the independence of the character of the PDEs allows an easy handling of a large variety of material laws (viscous, elastic, plastic, mixture of the previous, hypoplastic/barodetic).

Laboratory tests—such as the triaxial test—are typically used in soil mechanics to test new material laws, such as the nonlinear barodesy model developed by the Division of Geotechnical and Tunnel Engineering at the University of Innsbruck, Austria (see [2, 3]). The nonlinear barodesy model for the stress tensor necessitates a specific local linearization in order to be integrated within the FPM framework. Furthermore, the constant confining pressure acting on the soil sample during the test has to be properly incorporated as a numerical boundary condition.

J. Kuhnert (✉) · I. Ostermann
Fraunhofer ITWM, Fraunhofer-Platz 1, 67663 Kaiserslautern, Germany
e-mail: joerg.kuhnert@itwm.fraunhofer.de

I. Ostermann
e-mail: isabel.ostermann@itwm.fraunhofer.de

E. Pardo-Igúzquiza et al. (eds.), *Mathematics of Planet Earth*,
Lecture Notes in Earth System Sciences, DOI: 10.1007/978-3-642-32408-6_176,

2 Technical Idea of FPM

In classical finite element methods, meshes consist of connected node points in order to provide elements of various types, which represent a topological basis for the simulation. Compared to this, meshfree FPM is based on unconnected points filling the computation area. The neighborhood relationship is defined by balls of influence of radius h, also called smoothing length. h may be space and time dependent. The neighborhood graph is recomputed in each time cycle and used for the WMLS-approximations. The maintenance of the point cloud consists in adding and removing points (hole filling, point clustering) in order to provide a convenient local point pattern. This is crucial for achieving best approximation quality.

The meshfree approximation of partial derivatives is given by $\partial^i f/(\partial x_j)^i \approx \sum_{k=0}^{N} c_k^{ij} \cdot f_k$. The stencil entries $c_k^{ij}, k = 0...N$ are determined by the WMLS procedure with the help of Lagrange multipliers. The f_k are the discrete neighbor function values and N is the number of neighbors of the point under concern. Together with the time integration scheme, they enable the discretization of the considered PDEs.

3 Integration of Barodesy in FPM

The integration of the nonlinear barodesy model (cf. [2, 3]) in FPM for a 3D triaxial test is one of the latest developments (see [4] for the details). The triaxial test is a common method to extract and classify mechanical parameters of granular materials. These are useful to predict the soil behavior in engineering applications. A typical triaxial test apparatus (see Fig. 1) is characterized by a cylindrical soil sample enclosed in a thin rubber membrane which is placed inside a pressure chamber between two plates. The bottom plate is fixed; the upper plate can move vertically and apply a load to the sample.

The starting point of the numerical model in FPM is the discretized equation of motion for the velocity $\mathbf{v}$, the current time step size Δt, the density ρ, the stress tensor $\mathbf{T}$, and the vector of body forces $\mathbf{g}$: $\frac{\mathbf{v}_i^{n+1}-\mathbf{v}_i^n}{\Delta t} = \frac{1}{\rho_i}(\nabla^{\mathrm{T}}\mathbf{T}_i^{n+1})^{\mathrm{T}} + \mathbf{g}_i$ with i denoting the index of the numerical point and n, $n+1$ denoting the considered time steps. As the stress tensor at time level $n+1$ is unknown, an adequate estimation of this quantity is required. We use the local linearization of the nonlinear law of barodesy based on estimated future compression and shear moduli (C and μ, repsectively) given by $\mathbf{T}_i^{n+1} = \mathbf{T}_i^n + \Delta t \cdot \frac{1}{3}C_i^{n+1}(\nabla^{\mathrm{T}}\mathbf{v}_i^{n+1})\mathbf{I} + \Delta t \cdot 2\mu_i^{n+1}\mathbf{D}_i^{n+1}$, where $\mathbf{I}$ is the identity matrix and $\mathbf{D}$ is the symmetric part of the velocity gradient. Combining the above equations with the penalty formulation we employ in FPM, with the discretized barodesy model, and with the discretized formulation for the void ratio e, we obtain the following system:

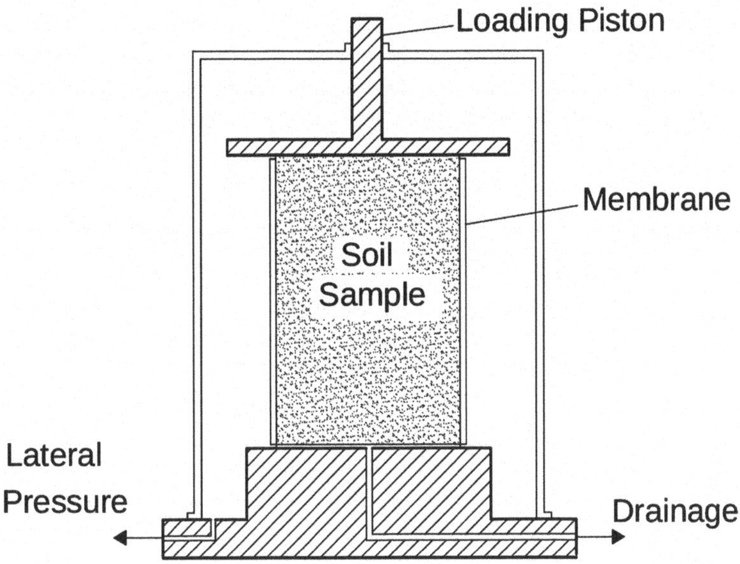

Fig. 1 Triaxial test apparatus (see [4])

$$\frac{\mathbf{v}_i^{n+1}-\mathbf{v}_i^n}{\Delta t}+\frac{1}{\rho_i}\nabla\varepsilon_i^{n+1}-\frac{1}{\rho_i}\left(\nabla^{\mathrm{T}}(2\hat{\eta}_i^{n+1}\mathbf{D}_i^{n+1})\right)^{\mathrm{T}}=\hat{\mathbf{g}}_i, \tag{1}$$

$$\nabla^{\mathrm{T}}\mathbf{v}_i^{n+1}+\nabla^{\mathrm{T}}\left(\frac{\Delta t_{\mathrm{virt}}}{\rho_i}\nabla\varepsilon_i^{n+1}\right)-\frac{1}{\rho_i}\left(\frac{\partial\rho}{\partial p}\right)_i^{n+1}\frac{\varepsilon_i^{n+1}}{\Delta t}=0,$$

$$\mathbf{T}_i^{n+1}=\mathbf{T}_i^n+\Delta t\cdot\mathbf{H}(\mathbf{T}_i^n,\mathbf{D}_i^{n+1},e_i^{n+1}),$$

$$e_i^{n+1}=\frac{\Delta t\cdot\mathrm{tr}(\mathbf{D}_i^{n+1})+e_i^n}{1-\Delta t\cdot\mathrm{tr}(\mathbf{D}_i^{n+1})}.$$

We have the identity $C=3\rho(\partial p/\partial\rho)$. The numerical viscosity is $\hat{\eta}=\eta+\Delta t\cdot\mu$ (η being the dynamic viscosity) and the resulting body forces are $\hat{\mathbf{g}}=\frac{1}{\rho}(\nabla^{\mathrm{T}}\mathbf{T}^{\mathbf{n}})^{\mathrm{T}}+\mathbf{g}$. $\Delta t_{\mathrm{virt}}=\mathcal{O}(\rho h^2/\hat{\eta})$ is a penalty term (choosing $\Delta t_{\mathrm{virt}}=0$ results in a direct scheme of weak compressibility, which however is more difficult to solve). $\mathbf{H}$ represents the nonlinear model of barodesy. The first two equations are solved simultaneously for the unknowns $\mathbf{v}^{n+1}$ and ε^{n+1}. In the following two steps, they are used to determine the new (future) stress tensor and void ratio at time level $n+1$. Further details on estimations of the compressibility and shear moduli can be found in [4].

Apart from the local linearization of the barodesy model the incorporation of the physical boundary condition at the free surface of the sample is crucial. During the simulation, the free surface is subject to a constant confining pressure. We represent this by the determination of the required symmetric part of the velocity gradient at the free surface such that the pressure condition is satisfied and, subsequently, the determination of the required velocity at the free surface (for details see [4]).

We compare the results of our simulations with laboratory tests. Furthermore, the comparison with the stress-strain curves obtained by the simulated element test in 1D as well as the comparison of the 3D simulated normal component of the stress tensor with the given confining pressure are of special interest.

4 Conclusion

The basic numerical and technical ideas of the meshfree simulation tool FPM for applications in computational fluid dynamics and continuum mechanics are presented. One of the latest applications of FPM is the simulation of a standard laboratory test, namely the triaxial test, used in soil mechanics to classify granular materials and test new material laws such as the barodesy model. The integration of the nonlinear material law of barodesy in FPM necessitates the local linearization of this material law. Furthermore, the reflection of the setup of a triaxial test leads to a specific formulation for the velocity boundary condition at the free surface of the soil sample.

Acknowledgments The authors are supported by the "Deutsche Forschungsgemeinschaft (DFG)" (project number KU 1430/7-1) in the scope of the joint research project "Untersuchungen der Anwendbarkeit von netzfreien numerischen Simulationsmethoden auf Probleme der Geotechnik und Geomechanik" in cooperation with Prof. C. Vrettos, University of Kaiserslautern, and Prof. D. Kolymbas, University of Innsbruck.

References

1. Hietel, D., Junk, M., Kuhnert, J., & Tiwari, S. (2005). Meshless methods for conservation laws. In G. Warnecke (Ed.), *Analysis and numerics for conservation laws* (pp. 339–362). Berlin: Springer.
2. Kolymbas, D. (2011). Barodesy: A new hypoplastic approach. *International Journal for Numerical and Analytical Methods in Geomechanics, 36*, 1220–1240.
3. Kolymbas, D. (2012). Barodesy: A new constitutive frame for soils. *Géotechnique Letters, 2*, 17–23.
4. Ostermann, I., Kuhnert, J., Kolymbas, D., Chen, C-H., Polymerou, I., Šmilauer, V., et al. (2013). *Meshfree generalized finite difference methods in soil mechanics - part I: Theory*. Berlin: Springer. (Submitted to *International Journal on Geomathematics*)
5. Tiwari, S., Antonov, S., Hietel, D., Kuhnert, J., & Wegener, R. (2007). A meshfree method for simulations of interactions between fluids and flexible structures. In M. Griebel & M. Schweitzer (Eds.), *Meshfree methods for partial differential equations III*. Berlin: Springer.
6. Tiwari, S., & Kuhnert, J. (2005). A numerical solution scheme for solving incompressible and low mach number flows by finite pointset method. In: M. Griebel & M. Schweitzer (Eds.), *Lecture notes in computational science and engineering* (Vol. 43, pp. 191–206). Berlin: Springer.
7. Tiwari, S., & Kuhnert, J. (2007). Modeling of two-phase flows with surface tension by finite pointset method (FPM). *Journal of Computational and Applied Mathematics, 203*(2), 376–386.

Geostatistical Estimation of the Crustal Thickness Inferred from the Geometry of Monogenetic Volcanoes (Central Mexico Case)

Raúl Pérez-López, Carolina Guardiola-Albert and José Luis Macias

1 Introduction

Monogenetic volcanic field is a dense population of volcanoes that occur in a single eruption. In general, these volcanic fields are related with forearc basins in tectonic convergence boundaries and mostly it encompasses hundreds to thousands of square kilometers. The Michoacán-Guanajuato Volcanic Field (MGVF) is located at the central part of the Trans-Mexican Volcanic Belt (TMVB) (Fig. 1). This volcanic arc is related to the subduction of the Rivera and Cocos tectonic plates underneath the North America Plate. The middle geometrical axis of the TMVB does not trend with a direction parallel to the Middle American Trench (MAT, NW-trending), but trends with a NW-SE direction from the west Pacific margin to the central part of the TMVB (Morelia city), turning to E-W from Morelia to the East Atlantic coast of Mexico (close to Veracruz city), and showing a deviation of 15° W from the MAT (Fig. 1). This orientation is related to the shape of the 100 km depth of the subducted plate [1, 2].

The TMVB represents the largest monogenetic field around the world. More than one thousand of monogenetic edifices were mapped and studied (petrological and geochemical analysis) by [3]. The TMVB shows a complex spatial pattern of migrating mafic pulses from Middle Miocene (11 Ma) to Pliocene (3.5 Ma) [1]. The main eruptive style of the TMVB is bimodal, active polygenetic volcanoes building stratovolcanoes as Popocatépetl and Quaternary monogenetic fields mainly composed by cinder cones.

R. Pérez-López (✉) · C. Guardiola-Albert
IGME—Geological Survey of Spain, Ríos Rosas 23, 28003 Madrid, Spain
e-mail: r.perez@igme.es

C. Guardiola-Albert
e-mail: c.guardiola@igme.es

J. L. Macias
UNAM Campus Morelia, Morelia, Mexico
e-mail: macias@geofisica.unam.mx

E. Pardo-Igúzquiza et al. (eds.), *Mathematics of Planet Earth*,
Lecture Notes in Earth System Sciences, DOI: 10.1007/978-3-642-32408-6_177,

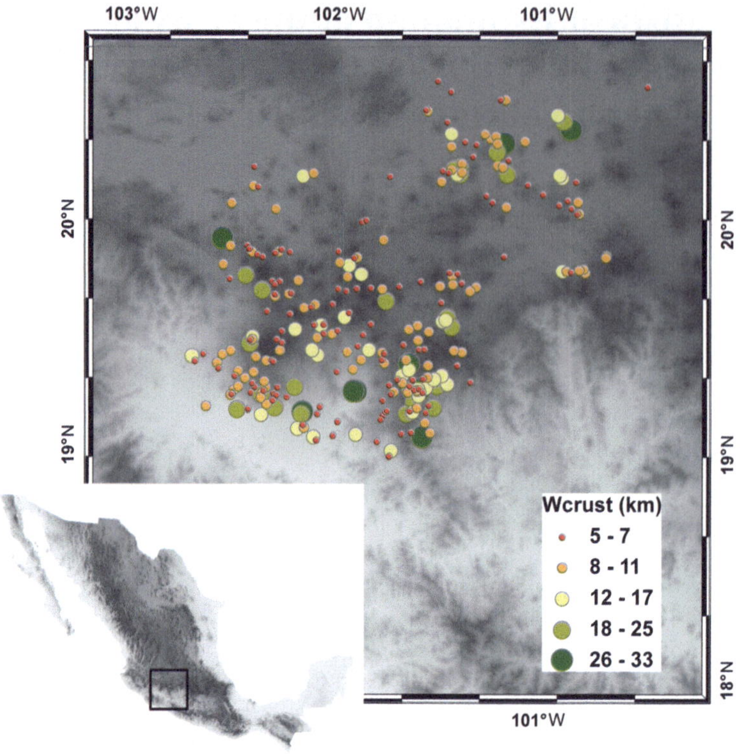

Fig. 1 Spatial distribution of monogenetic volcanoes within the Michoacán Guanajuato volcanic field (Mexico)

1.1 Geometry of Monogenetic Volcanoes

Most of the MGVF cinder cones are spatially distributed across a flat area [3] in a database of 1,042 cinder cones within the central part of Mexico (mainly Morelia, Guanajuato Estado de Mexico and Queretaro). The estimated morphometric parameters which define the volcanic cone are: (a) the basal diameter (Wco), (b) the crater diameter (Wcr) and the height of the edifice from the bottom to the top (Hco). The [3] database comprises these values from: (1) 1:50,000 scale topographic maps from the INEGI (Instituto Nacional de Estadística y Geografía Mexicano), series E13B, E14A and F14C. The contour interval for these maps is 20 m. The resolution of theses maps is 50 m, lesser than the minimum Wco value (200 m). (2) Field work to take measures of petrology and morphometric parameters for minor cones [3]. For further explanation of the morphometry of cinder and scoria cones (monogenetic) see [3].

1.2 Relationship Between m-Volcanoes and Crustal Thickness

We have obtained a power law empirical relationship between the basal diameter of a monogenetic volcano and the continental crustal thickness from the data of [5] and [6]. The Eq. (1) shows this empirical relationship:

$$\text{Wcrust} = 5.81^{*}\text{Wco}^{2.59}. \quad (1)$$

where Wcrust represents the thickness of the continental crust and Wco represents the basal diameter of the monogenetic vent. Other authors have related the special vent distribution of monogenetic volcanoes with the crustal thickness [7]. In this sense, we highlight that the spatial distribution of Wco obey a power law [8], suggesting a natural phenomena.

2 Geostastitical Analysis

Wcrust values are a function of spatial position. In the realm of geostatistics, these Wcrust values can therefore considered as values of a "regionalized variable" [9].The variogram, which originates from the theory of regionalized variables developed by Matheron [9], is used to measure the spatial dependence of neighboring observations for any continuously varying phenomenon. In this study experimental variograms are used to model the underlying spatial structure for Wcrust data. The variogram analysis has revealed one main geometric anisotropy structure within the NW trending. This direction is parallel to the MAT.

With the modeled variograms the geostatistical techniques of kriging [9] and direct sequential simulation [10] are applied to discrete the crustal thickness over all the studied area. For this last approximation Wcrust values are taken to be the median of 300 equiprobable realizations. Resulting Wcrust maps are shown in Fig. 2.

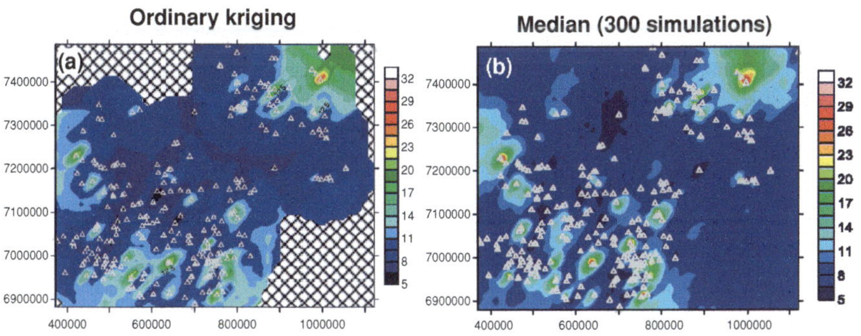

Fig. 2 Wcrust maps obtained with ordinary kriging (**a**) and direct sequential simulation (**b**) Locations of Wcrust data are indicated with *grey triangles*

3 Conclusions

Geoestatiscal analysis is a power tool for searching spatial anisotropies related with geodynamic processes. The empirical relationship between the basal diameter and the continental crustal thickness allows mapping a discrete population of single Wcrust values in a wide area. In the particular case of the Michoacán Guanajuato Volcanic field, we have obtained a population data set of Wcrust bounded between 5 and 33 km width. These values area lesser though congruent which those obtained by [11].

References

1. Ferrari, L. (2004). Slab detachment control on mafic volcanic pulse and mantle heterogeneity in central Mexico. *Geology, 32*, 77–80.
2. Pardo, M. & Suárez, G. (1995). Shape of the subducted Rivera and Cocos plates in southern Mexico: seismic and tectonic implications. *Journal of Geophysical Research, 100* 12,357–12,373.
3. Hasenaka, T., & Carmichael, I. S. E. (1985). The cinder cones of Michoacán- Guanajuato. Central México: Their age, volume and distribution, and magma discharge rate. *Journal of Volcanology and Geothermal Research, 25*, 105–124.
4. Dóniz, J., Romero, C., Coello, E., Guillén, C., Sánchez, N., García-Cacho, L., et al. (2008). Morphological and statistical characterisation of recent mafic volcanism on Tenerife (Canary Islands, Spain). *Journal of Volcanology and Geothermal Research, 173*, 185–195.
5. Smith, P. J. (1973). *Topics in geophysics* (p. 246). Cambridge, MA: MIT Press.
6. Fedotov, S. A. (1976). Ascent of basic magmas in the crust and the mechanism of basaltic fissure eruption. *International Geology Review, 20*, 33–48.
7. Mazzarini, F. (2007). Vent distribution and crustal thickness in stretched continental crust: The case of the Afar Depression (Ethiopia). *Geosphere, 3*, 152–162.
8. Pérez-López, R., Legrand, D., Garduño-Monroy, V. H., Rodríguez-Pascua, M. A., & Giner-Robles, J. L. (2011). Scaling laws of the size-distribution of monogenetic volcanoes within the Michoacán-Guanajuato Volcanic Field (Mexico). *Journal of Volcanology and Geothermal Research, 201*, 65–72.
9. Matheron, G. (1963). Principles of geostatistics. *Economic Geology, 58*(8), 1246–1266.
10. Caers, J. (2000). Adding local accuracy to direct sequential simulation. *Mathematical Geology, 32*(7), 815–850.
11. Molina-Garza, R., & Urrutia-Fucugauchi, J. (1993). Deep crustal structure of central Mexico derived from interpretation of Bouguer gravity anomaly data. *Journal of Geodynamics, 17*(4), 181–201.

Mathematical Modelling of Geochemical Processes Applied to Cenozoic Iberian Volcanics: A Review

Raúl Benito and Jesús Martínez-Frías

1 Introduction

Cenozoic volcanism of the Iberian Peninsula is mainly of two types: alkaline basaltic volcanism in four provinces (Central Spain, NE Spain, SE Spain and Gulf of Valencia) and calc-alkaline to ultrapotassic volcanism (SE Spain) [1].

Volcanic rocks formed by the crystallization of magmas generated by partial melting of source materials, which have mostly been affected by previous contamination processes. In some cases, later magma mixing (and/or assimilation of crustal material) occurred. The mathemathical modelling of such processes, by trace elements and isotope geochemistry, has allowed to establish genetic hypotheses to each province.

2 Volcanic Rocks: Origin and Main Geochemical Signatures

Melilitites, nephelinites and alkaline basalts occur in Central Spain. Their origin is explained as the result of variable degrees of partial melting of a rather homogeneous mantle source, which is enriched in highly incompatible trace elements [2]. The geochemical signatures of this mantle resemble the European Asthenopsheric reservoir (EAR) [3].

Basanites and alkaline basalts crop out in NE Spain. They are considered the product of partial melting from a lithospheric enriched mantle source. It is assumed the

R. Benito (✉)
Museo Nacional de Ciencias Naturales, CSIC, CJose Gutierrez Abascal, 2,
28006 Madrid, Spain
e-mail: raul.benito@csic.es

J. Martínez-Frías
Centro de Astrobiología, CSIC-INTA, Associated to the NASA Astrobiology Institute,
Ctra de Ajalvir, km. 4, Torrejón de Ardoz, 28850 Madrid, Spain
e-mail: jmfrias@cab.inta-csic.es

E. Pardo-Igúzquiza et al. (eds.), *Mathematics of Planet Earth*,
Lecture Notes in Earth System Sciences, DOI: 10.1007/978-3-642-32408-6_178,

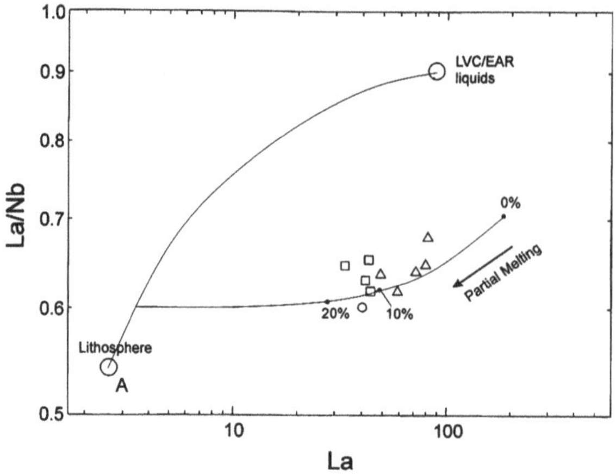

Fig. 1 La/Nb-La diagram for the primary magmas of NE Spain (*triangles*: leucite basanites; *squares*: nepheline basanites; *circles*: olivine basalts). The composition of the LVC/EAR-derived liquids is the average of calatrava samples [2]. The composition of the mantle lithosphere A (La = 2,6 ppm; La/Nb = 0,54) is from [5]. *Lines* represent the calculated partial melting model (percentages indicate degrees of melting) from a hybrid source made up of the A starting lithospheric composition and ≈5 % of LVC/EAR-derived liquids (after [4])

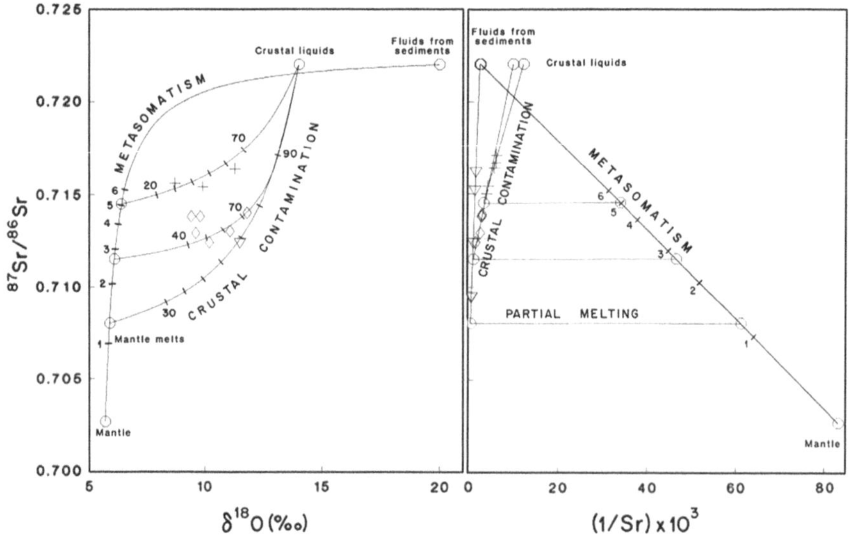

Fig. 2 $^{87}Sr/^{86}Sr$-$\delta^{18}O$ and $^{87}Sr/^{86}Sr$-1/Sr x 10^3 diagrams of the metasomatism of the mantle and crustal contamination model for the high-K calc-alkaline lavas of El Hoyazo (*diamonds*) and Mar Menor (*crosses*), and shoshonites of Vera (*inverted triangles*) (SE Spain). *Numbers* indicate the percentages of participation of the sediment-derived fluids and the crustal liquids, respectively (after [7])

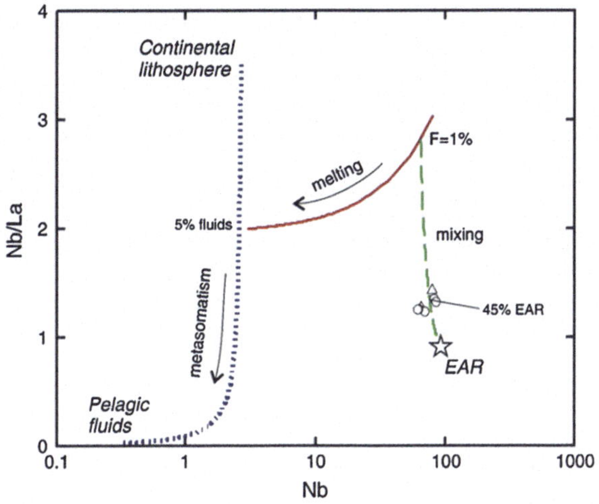

Fig. 3 Nb/La-Nb diagram for the petrogenetic model by the alkali basalts from SE Spain. Participation of pelagic sediment fluids is 5 %. Degree of melting of mantle lithosphere source is 1 % (after [9])

existence of previous metasomatism related to liquids derived from a sublithospheric mantle source similar to EAR (Fig. 1) [4].

Calc-alkaline to ultrapotassic magmas occur in SE Spain. They formed by partial melting of a lithospheric mantle source after metasomatic processes by fluids coming from subducted pelagic sediments [6].

It is assumed an interaction of melts with Betic basement-bearing crustal liquids (Fig. 2) to generate the high-K calc-alkaline and shoshonitic magmas [7, 8].

Alkali basalts (the latest volcanics in SE Spain) result from the interaction between primitive melts produced from a sublithospheric mantle source, similar to EAR and liquids derived from the overlying lithospheric mantle, heterogeneous with metasomatized portions by fluids derived from sediments (Fig. 3) [9].

Nevertheless, this model does not consider other processes which have probably intervened, such as mixing with shoshonitic or ultrapotassic magmas and assimilation of metamorphic rocks by the basaltic magma [10–12].

References

1. López Ruiz, J., Cebriá, J.M., & Doblas, M. (2002). Cenozoic volcanism I: The Iberian peninsula E. In: W. Gibbons & M. T. Moreno (Eds.), *The Geology of Spain* (pp. 417–438). London: Geological Society.
2. Cebriá, J. M. (1995). López Ruiz, J.: Alkali basalts and leucitites in an extensional intracontinental plate setting: The late cenozoic calatrava volcanic province (central Spain). *Lithos*, *35*, 27–46.

3. Cebriá, J. M., & Wilson, M. (1995). Cenozoic mafic magmatism in Western/Central Europe: A common European asthenospheric reservoir? *Terra Nova Abstract Supplement, 7*, 162.
4. Cebriá, J. M., López Ruiz, J., Doblas, M., Oyarzun, J., Hertogen, J., & Benito, R. (2000) Geochemistry of the quaternary alkali basalts of garrotxa (NE volcanic province, Spain): A case of double enrichment of the mantle lithosphere. *Journal of Volcanology and Geothermal Research, 102*, 217–235.
5. McDonough, W. F. (1990). Constraints on the composition of the continental lithosphere mantle. *Earth Planetary Science Letters, 101*, 1–18.
6. López Ruiz, J., & Wasserman, M. D. (1991) Relación entre la hidratación/desvitrificación y el $\delta^{18}O$ en las rocas volcánicas neógenas del sureste de España. *Estudios Geologicos*, 47, 3–11.
7. Benito, R., López Ruiz, J., Cebriá, J.M., Hertogen, J., Doblas, M., & Oyarzun, J. (1999) Sr and O isotope constraints on source and crustal contamination in the high-K calc-alkaline and shoshonitic neogene volcanic rocks of SE Spain. *Lithos, 46*, 773–802.
8. Benito, R., Martínez-Frías, J., Lunar, R.,& Wolf, D. (1998) El Hoyazo: a unique garnet-rich volcanic complex in southeast Spain. *Transactions of the Institution of Mining and Metallurgy : (Sect. B: Applied Earth Science), 107*, 150–164.
9. Cebriá, J. M. (2009). López Ruiz, J., Carmona, J., Doblas, M.: Quantitative petrogenetic constraints on the Pliocene alkali basaltic volcanism of the SE Spain Volcanic Province. *Journal of Volcanology and Geothermal Research, 185*, 172–180.
10. Turner, S. P., Platt, J. P., George, R. M. M., Kelley, S. P., Pearson, D. G., & Nowell, G. M. (1999). Magmatism associated with orogenic collapse of the Betic-Alboran domain, SE Spain. *Journal of Petrology, 40*, 1011–1036.
11. Duggen, S., Hoernle, K., Van Den Bogaard, P., & Garbe-Schönberg, D. (2005). Post-collisional transition from subduction- to intraplate-type magmatism in the westernmost mediterranean: evidence for continental-edge delamination of subcontinental lithosphere. *Journal of Petrology, 46*, 1155–1201.
12. Aparicio, A., & García, R. (2009). Basanites and crustal contamination in the Cartagena volcano field (South-East of Spain). *Icfai University Journal of Earth Science, 3*, 23–40.

Management of Geological Field Data: The Open Solution MO2GEO FieldModule

Lars Schimpf and Wolfgang Gossel

1 Introduction

Today's high-performance mobile devices make it possible to use them for geological data-acquisition in the field. A few companies already developed different software solutions for this field of application, often combined with special hardware. There are three main reasons for the development of a new and separate application:

1. All of the commercial software is proprietary. Therefore it is impossible to customize them for the special demands of users or working groups.
2. Another problem of commercial software are the expenses.
3. The data acquisition itself causes different issues. Geological data are managed heterogeneously based on diverse national standards. These standards differ technically and in their recording structure of geological features.

A platform independent Open Source software which allows the use of different common standards leads to the improvement according to local specifications and even the setup of own symbolizations. Of course it follows the philosophy of Open Source and so it also enables the possibility of a science oriented usage for everyone who is interested. These reasons led to the development of the MO2GEO FieldModule which is part of the modular MO2GEO (move to geological modeling [1]) framework.

L. Schimpf · W. Gossel (✉)
Department of Hydrogeology and Environmental Geology, Institute of Geosciences and Geography, Martin-Luther University Halle, Von-Seckendorff-Platz 3, 06120 Halle, Germany
e-mail: lars.schimpf@outlook.com

L. Schimpf
e-mail: wolfgang.gossel@geo.uni-halle.de

E. Pardo-Igúzquiza et al. (eds.), *Mathematics of Planet Earth*,
Lecture Notes in Earth System Sciences, DOI: 10.1007/978-3-642-32408-6_179,

Table 1 Standards implemented in the MO2GEO FieldModule

Standard	Application
Symbolschlüssel geologie [2]	Stratigraphy, different attributes
BoreholeML [4]	Borehole description
Hierarchical classification of rocks [5]	Consolidated and unconsolidated rocks
Guidelines for soil description [6]	Soil description
International stratigraphic chart [7]	Stratigraphy
BGS vocabularies [3]	Structural geology, different attributes
Digital cartographic standard for geologic map symbolization [8]	2D/3D drawing, signatures
GeoNames [9]	Geography

2 Standards

The advent of electronic data management in the earth sciences and herewith related digital storage and processing of geological data made it necessary to develop appropriate data models. This includes a standard glossary of definitions and issues. Usually geological data are saved as understandable plain text in the field. However the individual phrases lead to difficulties during the digital storage and recognition. To avoid these problems different national vocabularies were developed (e.g. *Symbolschlüssel Geologie* [2] in Germany or *British Geological Survey vocabularies* [3] in Great Britain). But in the scope of increasing international collaboration it becomes necessary to develop international standards to facilitate data exchange and collaboration (*OneGeology-Europe* [10], *GeoSciML* [11]). However there is no generally accepted international standard. For this reason it was decided for this project to let it up to the user which standards should be used. So the preinstalled standards (Table 1) can be substituted by the user preferred ones if necessary.

3 FieldModule

The MO2GEO FieldModule is a complex Open Source tool for management (storage, handling and visualization) of geological data. The basic concept of the software is the project related management of different data (e.g. boreholes, outcrops). To accomplish this, all of the data are stored and managed in a particular objectrelational PostgreSQL database. The MO2GEO FieldModule is not only a database management tool but also capable to visualize the collected field data. Reworking geological data for a two- or three-dimensional representation is also the basis for the creation of cross sections, which may constitute a significant improvement of he field work and following static geological modelling processes.

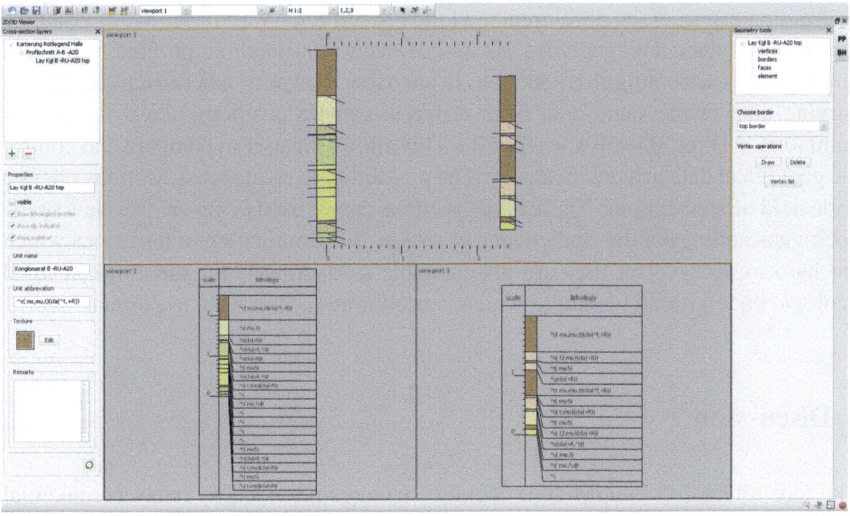

Fig. 1 A *cross section* created from two lithological profiles in the shown 2D

3.1 Database

As described before, the MO2Geo FieldModule database (m2gfm-DB) is the backbone of the application. The project oriented design of the program and the typical dependencies of geological data require an adapted data model. An optimal solution seemed to be the usage of an object-relational database combined with hierarchical structures. At the moment the database consists of 69 different tables which can be divided into two main categories: attribute and object tables. Attribute tables contain standardized single information and are used to describe user-defined objects to accomplish comparable base data.

3.2 User Interface and Visualization

The usage of standards for a controlled input in the context of avoiding formal and technical mistakes requires sophisticated database- and program structures and often results in complex user interfaces. Another problem is also the common discrepancy in applications to be either user friendly or convenient. Splitting the input mask into several forms and sub-forms offers a satisfying solution. This results in having a single input form for every corresponding database table. The outcome of this is a clear data frame showing only the most necessary information but with the possibility to show the relevant sub-forms.

Another aspect of the development of the FieldModule was the ability to visualize the recorded data. This includes 2D and 3D views of borehole profiles, spots and outcrops and connecting cross-sections in up to four viewports where each can display varying data or the same data from different camera positions like it is used by common CAD- or 3D-software (Fig. 1). This allows the user to compare and control the geological data in a practical way. The possibility to create cross-sections opens a wide field of applications. These cross-sections can be used to get an overview of the geological settings of the studied area or to ease the examination of territories which are hard to access. But they are also suitable as data basis for three-dimensional geologic underground modeling which is becoming increasingly important.

4 Discussion

The tool MO2GEO FieldModule includes features provided by many commercial solutions. Furthermore it offers the ability of customizing standards and visualization of the collected data. The further development of FieldModule will be an embedding into GIS and the static and dynamic modelling of geological structures.

References

1. Gossel, W. (2011). MO2GEO: an OpenSource software approach for geological modelling. In *Proceedings IAMG Konferenz*, Salzburg, vol. 6.
2. Preuss, H. (2010). *Symbolschlüssel geologie: symbole für die dokumentation und verarbeitung geologischer feld und aufschlussdaten* (6th ed.). Hannover: Landesamt für Bergbau, Energie und Geologie Hannover.
3. British Geological Survey (2012). Vocabularies: further information. http://www.bgs.ac.uk/data/vocabularies/details.html. Accessed 10 Jun 2012.
4. Staatliche Geologische Dienste Deutschlands (2012). InfoGEO.de: BoreholeML. http://www.infogeo.de/home/boreholeML?lang=1. Accessed 27 Mar 2013.
5. Isaak, S. (2011). Erfassung geologischer Sammlungsobjekte in aktuellen Datenbankmanagementsystemen (unter besonderer Berücksichtigung von Specify 6), Diplomarbeit Martin-Luther-Universität Halle-Wittenberg.
6. Food and Agriculture Organization of the United Nations (2006). Guidelines for soil description, 4th edn. Rom.
7. International Commission on Stratigraphy (2010). International stratigraphic chart. http://www.stratigraphy.org. Accessed 27 Mar 2013.
8. U.S. Geological Survey (2006). FGDC digital cartographic standard for geologic map symbolization (PostScript implementation). http://pubs.usgs.gov/tm/2006/11A02/. Accesssed 27 Mar 2013.
9. GeoNames Team (2012). GeoNames geographical database. http://www.geonames.org/. Accessed 27 Mar 2013.
10. OneGeology (2010). OneEurope OneGeology: applying geoscience for society. http://www.onegeology.org/docs/Oneeurope_Onegeology_eBook/flash.html. Accesssed 27 Mar 2013.
11. Commission for the Management and Application of Geoscience Information GeoSciML Cookbook (2012). How to Map Data to GeoSciML Version 2. http://www.geosciml.org/geosciml/2.0/documentation/cookbook/GeoSciML_Data_CookBook_V2.pdf. Accessed 27 Mar 2013.

Swiss Data Models for Geology: from 2D Data Towards Geological 3D Models

Cristina Salomè Michael, Nils Oesterling, Stefan Strasky and Roland Baumberger

1 Introduction

After more than 80 years of traditional geological paper map production, geological vector datasets became increasingly important. The growing interest in geological GIS applications in the field of earth sciences, such as natural hazards prevention, resource management, land use planning, has heightened the need for well-structured vector data and their proper documentation. Moreover, according to the Swiss Legislation [1] geological surveying data belong to the official basic geodata. Therefore it has to be supplied to the broad public and minimal data models have to be developed.

In order to do so, the following two-step approach was applied. In a first step geological paper maps were scanned and georeferenced for use in GI-Systems. For such raster graphics meta-information (type, perimeter etc.) is organised by a data model and will be provided in XML-format along with the map data. In a second step the raster graphics were vectorised using a simple data model, which is optimised for data acquisition. However, this data model does not efficiently support data analyses. Moreover, due to the following reasons, semantic homogeneity of the existing geological vector data is poor: (a) geological mapping was carried out by a large number of geologists with different styles of interpretations. (b) geological theories have changed over time. (c) guidelines for well-structured geological data were missing.

To overcome these restrictions, the Swiss Data Model Geology (SDMG-2D) was developed. This data model not only defines the structure (geometries, attributes, data types etc.) of geological vector datasets but also defines a standard vocabulary.

C. S. Michael (✉) · N. Oesterling (✉) · S. Strasky · R. Baumberger
Federal Office of Topography swisstopo,Swiss Geological Survey, Seftigenstrasse 264, Ostermundigen CH-3084 Wabern, Switzerland
e-mail: salome.michael@swisstopo.ch

N. Oesterling
e-mail: nils.oesterling@swisstopo.ch

E. Pardo-Igúzquiza et al. (eds.), *Mathematics of Planet Earth*,
Lecture Notes in Earth System Sciences, DOI: 10.1007/978-3-642-32408-6_180,

Since subsurface potentials like geothermal energy, carbon dioxide storage and other underground activities shift into the focus of public interest, also 3D models are getting an indispensable tool for subsurface planning. The transition from 2D to 3D has a direct influence not only on technical requirements but also on existing data, their structure and semantics. Therefore, for geological 3D models the SDMG-2D has been extended to 3D (SDMG-3D).

2 Swiss Data Model Geology: 2D Data

The development of a conceptual data model for geological vector data in 2D has started in 2008. It is mainly based on the objects of the existing paper maps of the Geological Atlas of Switzerland 1:25,000 and its symbol description.

The primary focus of geological data, which are structured by the SDMG-2D, is to allow users to analyse seamless vector data for entire Switzerland in GIS applications. Therefore, the SDMG-2D provides a description of all relevant object types as well as their characteristics and relationships. In order to meet such customer requirements as best as possible, swisstopo has involved external experts from the main target groups of geological vector data (e.g. cantonal and federal administration, private offices, universities) in the modelling process.

The multilingual (German, French, Italian, English) documentation of the SDMG-2D [2] consists of an object catalogue, which describes geological object types together with their attributes and attribute values. The object types are structured in eight themes and 50 classes (Fig. 1). Additionally, lookup tables with a standardised vocabulary for rock types, tectonic units and the geological time scale is included. Standardised lithostratigraphic terms are under preparation.

In addition to a "prosa" description, the SDMG-2D is documented in UML format and INTERLIS. The latter is prescribed by Swiss Legislation [1], so that basic geodata and their data models have to be provided in INTERLIS-format.

In order to facilitate data acquisition according to the SDMG-2D and to guarantee data consistency swisstopo has developed in collaboration with Centre de Recherche sur l'Environnement Alpin (CREALP) the ToolMap2-software [3].

3 Swiss Data Model 3D Geology: 3D Models

Also for geological 3D models a description of their geometric and semantic structure are required. For this reason the SDMG-3D [4] was developed. In contrast to 2D, where geological formations are represented as polygons in planar view, in 3D they appear as volumetric bodies in space. Therefore, the "shell", describing the envelope of a discrete "geo-body", and the "volume", describing the sum of a finite number of subdivisions (e.g. Voxel) of a "geo-body", were introduced as new geometry types. Figure 2 shows the applied geometry changes.

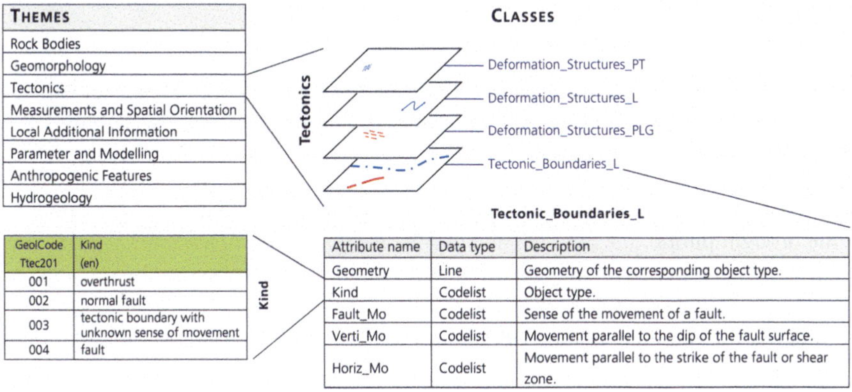

Fig. 1 The SDMG-2D is organised in eight themes. Each theme contains several classes, as illustrated for the theme "Tectonics". Further, each class comprises various attributes, whose valid values are listed in lookup tables (e.g. "Kind")

Apart from geometry, modifications of specific thematic classes and attributes were performed. Basically, the number of existing classes of the SDMG-2D was reduced to 25. Additionally a new theme for subsurface bodies, comprising two classes, was created. The newly introduced classes describe natural (e.g. caves) and artificial (e.g. tunnels) subsurface objects, respectively.

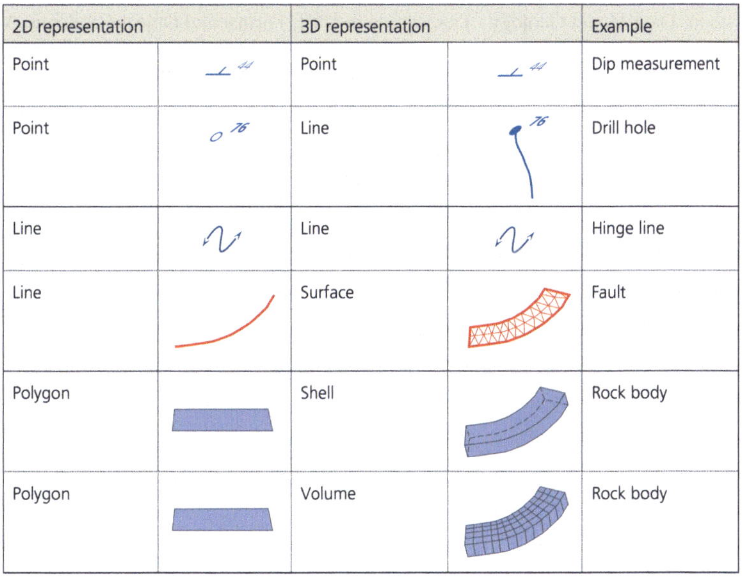

Fig. 2 Geometry changes from 2D to 3D

4 Outlook

The development of the above described data models is an important basis for various topics and activities of swisstopo. The following mid- to long-term activities are planned:

- Storing seamless geological 2D vector data in a spatial database and provide it to the brought public.
- Developing a portrayal model for system neutral description and exchange of geological symbology.
- Adjust the Swiss data models with data models on the European and international level.
- Storing geological 3D models in a spatial database (GST framework [5]) and provide it to the brought public.
- Developing a geological information system in order to face future challenges in relationship to subsurface planning.
- Extending the existing data models to the fourth dimension, in order to model kinematic evolutions over time.

References

1. The Federal Assembly of the Swiss Confederation, Federal Act of 5 October 2007 on Geoinformation (Geoinformation Act, GeoIA), SR 510.62 (2007).
2. Swisstopo: Data Model Geology - Description in UML Format and Object Catalogue, Version 2.1 (2012).
3. Strasky, S., Vandelli, A., Schreiber, L., Sartori, M., Ornstein, P., & Möri, A. (2011). TOOLMAP2 - a powerful tool for digital mapping. *Swiss Bulletin für angewandte Geologie, 16/2*, 87–95.
4. Swisstopo: Data Model 3D Geology - Extension to the Data Model Geology, Version 1.0 (2010).
5. GiGa infosystems: http://www.giga-infosystems.com/

Virtual Collaboration of Geoscientists of Russian Academy of Sciences in the Far East of Russia

Vera V. Naumova, Mike I. Patuk and Valentin V. Nichepkov

1 Introduction

Scientific organizations in the Russian Academy of Sciences in the Russian Far East are located in Vladivostok, Khabarovsk, Blagoveshchensk, Magadan, Petropavlovsk-Kamchatsky and Yuzhno-Sakhalinsk. The large-area dispersion requires that information that obtained from widely-separated research groups and scientists must be integrated for joint research, discussions, conferencing, and sharing of analytical equipment. Modern information technologies provide means for integration of data from widely-disperesed scientists [1]. Following are proposed solutions.

1. Virtual science laboratories: virtual toolkits for research and experiments; topical virtual labs; and remote access to analytical equipment.
2. Virtual science environments: Internet, including E-libraries, data infrastructure and data centers; data-processing infrastructures; video conferencing; scientific data sharing and processing; scientific management.

This article describes a system to design and implement a virtual science environment to allow widely-separated scientists to communicate for joint research. The system will enable research by utilizing various information technology methods to integrate telecommunication, data sharing, data-processing, media environment, geographic informational systems, data centers, applied applications, and analytical equipment.

V. V. Naumova (✉) · M. I. Patuk · V. V. Nichepkov
Far East Geological Institute, Far Eastern Branch, Russian Academy of Sciences (FEGI FEB RAS), Vladivostok, Russia
e-mail: naumova@fegi.ru

E. Pardo-Igúzquiza et al. (eds.), *Mathematics of Planet Earth*,
Lecture Notes in Earth System Sciences, DOI: 10.1007/978-3-642-32408-6_181,

2 System Goals

1. Support widely-separated scientists with modern communication, such as instant messaging or group conferencing;
2. Provide basic solutions for synchronous processing of files;
3. Enable remote access to scientific equipment in the Far Eastern Branch of Russian Academy of Sciences (FEB RAS);
4. Enable remote access to computer processing resource, library catalogs, E-libraries, databases, satellite images, GPS databases, and remote-monitoring systems.

The System will be based on: Microsoft unified communications solutions; Polycom video conferencing; and Internet science services.

With the use of modern video conferencing, remote access to databases, and data processing, we will be able to build a virtual environment for collaboration between widely-dispersed scientists (Fig. 1).

3 Hardware- and Software-Solutions for the System

For creating the System, Polycom-Microsoft solutions video conferencing will be employed at varous levels: desktop, laboratories, and conferences This type of integration combines various Microsoft solutions, such as Lync, Exchange Server, Sharepoint Server, and Polycom solutions for use of HD voice and video communication,and flexible image sharing. The quality of communication will be similar to face-to-face contact for both within-institution interactions and for a B2B environment.

Base hardware software equipment includes: Polycom RMX-200 server for video conferencing; Polycom RSS 2000 server for video conference recording and streaming; and Polycom CMA 4000 solution for video conferencing.

A preliminary version of the System is stre gthened with the Microsoft server components, such as Exchange Server for E-Mail, Lync Server for unified communication, and Sharepoint for Intranet portals. These software systems utilize Active Directory service for all users. The Microsoft Lync Server 2013 enables efficient communication for interaction, which considerably simplifies dialog and improves quality of collaborations. It combines standard communications for users.

The primary function of Microsoft Sharepoint 2013 is rapid creation of internal websites for virtual laboratories, collaborative projects, programs, and events, and personal cabinets for scientists. A test version of the System is based on a server with VMWARE ESXI virtualization system supporting five virtual machines.

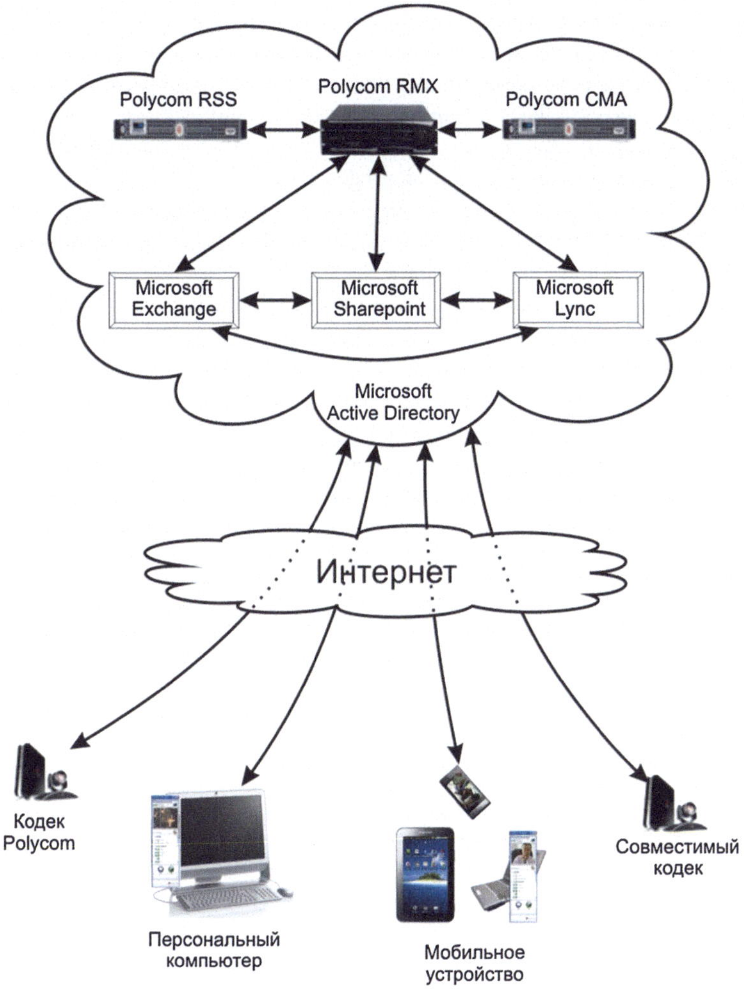

Fig. 1 Schematic virtual science environment for collaboration between widely-dispersed earth scientists in the FEB RAS

4 Remote Access Services to Science Resources

A virtual scientific environment with remote access to computational, analytical and information resources (using servers of the Far Eastern Branch of RAS) will allow widely-separated scientist to share scientific resources for investigations. Research scientists of the FEB RAS institutes have accumulated enormous amounts of scientific information. It is critical that institutes systematize and archive available data into databases, GIS files, information retrieval systems, and E-libraries. Because new methods of data acquisition are readily accessible and because efficiency increases, the bulk of data steadily grows. Therefore transition to new digital technologies of data collecting, processing, and distribution is needed. This will enable reception of original data from ground based remote sensing, from the field sites, from satellites, from surface or airborne laser scanning , from digital and electronic geodetic monitors, and from aerial cameras. New digital environments for scientific data storage will lead to state-of-the-art information processing.

Significant informational resources are owned by the Far Eastern Branch of RAS: E-libraries and repositories; archives of satellite images; and GIS spatial databases for geology, volcanology and geography.

Organization of a common entry point to widely-distributed and heterogeneous science information resources will offer new opportunities for scientific research between widely-dispersed earth scientists. For a preliminary version of the System, a Subsystem is proposed for remote access to the analytical equipment of the FEB RAS Centers for collective use. Proposed technological solutions imply transition from remote monitoring devices to conducting remote analyses with analytical equipment. This type of access will be enabled through a Common entry point to FEB RAS analytical equipment. Authorized users will be able to: be trained with analytical equipment; conduct experiments with their own samples with remote access to equipment, or by orders without personal participation; monitor experiment performance as a distant observer; and choose a remote access mode to perform research.

Reference

1. Naumova, V. V. (2013). Virtual science environments for joint work of the geographically distributed scientists. *Informational Technologies*, *4*, 46–51.

Geological 3D Modeling (Processes) and Future Needs for 3D Data and Model Storage at Geological Survey of Finland

Eevaliisa Laine

1 Introduction

At the Geological Survey of Finland (GTK), geological 3D modeling has traditionally been related with ore modeling. Presently, regional 3D bedrock models are being built from different parts of the Precambrian bedrock of Finland. Similarly, the 3D modeling of Quaternary deposits overlying Precambrian bedrock is going on. In future, it will be important to be able to to build and present combined 3D models of the Precambrian crystalline rock overlaid by the Quaternary deposits in many engineering and environmental geological investigations.

Different approaches and software are needed because of differences in data and modeled 3D geometries. 3D modeling process have been divided into data processing and import to 3D modelling tools, actual 3D modeling and visualization and, finally, to the 3D model storage associated with a description of a modeling process and evaluation of the reliability of the model.

2D data, such as geological observations and geological maps, and 3D data, such as drill hole data, geological sections and elevation models, are already available and stored directly in digital format. At GTK, researchers are searching and extracting these data using ArcMap (ESRI product) for their target areas. In future, GTK also needs a 3D database for storing 3D objects from geological, geophysical and geochemical modelling. It should be possible to extract geometrical objects in simple ASCII data with the meta-data, links to reports describing the 3D modeling process and properties of the used data and applications. In addition, the grids should be transformable into hexaedric grids for physical modeling such as 4D modeling in thermal and geomechanical applications. In general, there are needs for interchanging 3D information between GTK and mining companies, state organizations and consulting firms.

E. Laine (✉)
Geological Survey of Finland, Betonimiehenkuja 4, FI-02150 Espoo, Finland
e-mail: eevaliisa.laine@gtk.fi

E. Pardo-Igúzquiza et al. (eds.), *Mathematics of Planet Earth*,
Lecture Notes in Earth System Sciences, DOI: 10.1007/978-3-642-32408-6_182,

There has been thoughts for building geotechnical 3D database in Finland (e.g. [5]). Gustavsson [1] made a report of 3D storage possibilities at GTK.

2 Present 3D Modeling at GTK

Computerized 3D ore modeling has been done since 1980 at GTK. One of the recent 3D models of Koistinen is presented in the Fig. 1.— a solid model of Syväjärvi lithium deposit. The used software was GEMS.

Nowadays, important 3D modeling applications are done within many different applications such as groundwater modeling, 3D modeling of the subsurface in urban areas, 3D modeling in evaluation of environmental risks and damages, 3D modeling of dimension stones and 3D modeling of stone aggregate resources.

In addition to the geometrical geological models, 3D grids presenting distribution of physical and chemical properties are important for example in computation of groundwater flow, ore resource estimates and thermal capacity of the bedrock. Figure 2 illustrates Ni distribution in the Vuonos ore in Eastern Finland.

The important task is also to popularize geological research results using illustrative 3D models.

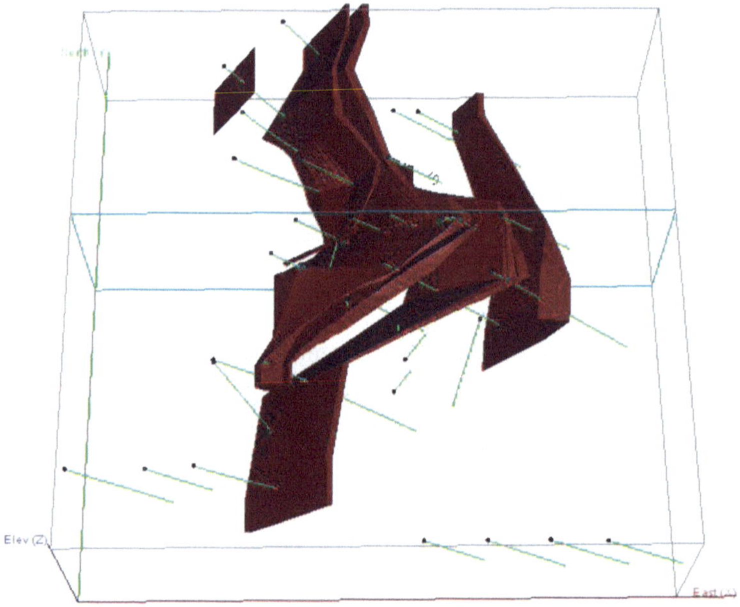

Fig. 1 A solid model of the Syväjärvi lithium deposit seen downwards to north Length of the deposit is 500 m. [4]

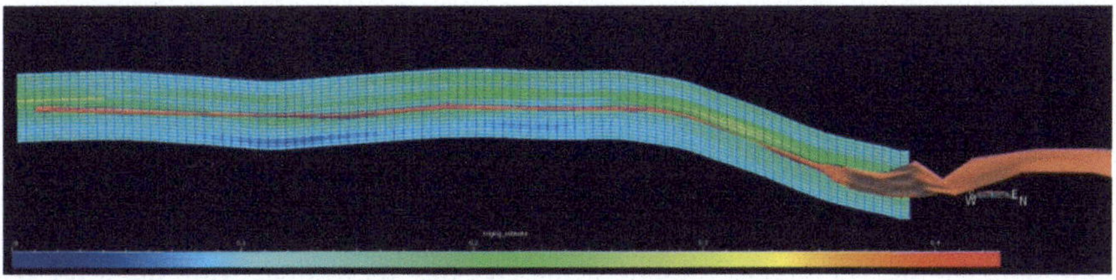

Fig. 2 Vuonos ore and gridded Ni content [2]

3 3D Modeling Processes and Future Need for the 3D Data and Model Storage

At GTK, the planning of the 3D model storage is based on the analysis of the 3D modeling processes that begin by data import and organization. The main other components in 3D modeling are actual 3D model building, analyzing and presenting uncertainties and finally data storage, reporting and possibly publication of the models in web.

The data import process is perhaps the most time consuming phase in 3D modelling. This is not because it is not well documented but because it is always difficult to combine data from different sources having different resolutions and aims than the ongoing 3D modelling work. A good example is the simplification of detailed drill core information for regional scale geological 3D models. 2D data (and 3D data~drill hole data, elevation models) are already available at GTK and stored directly in digital format and it is managed in different data systems from which information is mainly shared as different digital information products.

3D bedrock modeling process depend on the scale and purpose of the study. In 3D regional modeling the geological ideas and theories are essential because the data are sparse and in greater depths the only data sources are deep geophysical soundings that may be interpreted in many different ways. In mining sites, dense drilling gives a possibility to build more reliable models. Even then, the structures between drill holes can be drawn in many different ways according to different geological interpretations. In many practical applications, such as in nuclear waste site investigations, groundwater modeling or rock engineering, it is important to estimate rock fracturing in 3D. Connected rock fractures act as water conduits and, in general, fracturing affects the rock mechanical properties. Each specific geologic 3D modeling process will need its own workflow.

3D modeling of Quaternary deposits differs from the Precambrian formations with the fact that, in most cases, it is possible to define a soil layer stratigraphy as nearly horizontal layers young upwards. However, this is not always the case, for example, along the moraine ridges soil structures are very complicated. 3D models for several practical applications require the combination of bedrock models with surficial formations. Examples of such applications are groundwater and environmental modeling.

In Finland there are already organized data storage for 3D models in mines and within nuclear waste site studies. For example, within nuclear waste studies at Olkluoto there exists already 3D geological data storage (e.g. [3]). In general, there will be needs for interchanging 3D information between GTK and mining companies, state organizations and consulting firms.

Based on the preliminary studies of GTK 3D processes it seems reasonable to begin with saving 3D models directly using specific 3D formats from the used 3D modeling software together with metadata and 3D modeling report. In the future, more general 3D database should be built but it is possible that software based and general 3D database should both exist so that the general 3D database could represent more scientific 3D model storage with carefully analysed uncertainties and publications.

References

1. Aaltonen, I. (ed.), Lahti, M., Engström, J., Mattila, J., Paananen, M., Paulamäki, S., Gehör, S., Kärki, A., Ahokas, T., Torvela, T. & Front, K., 2010. Geological model of the Olkiluoto Site, Version 2.0. Posiva Oy, Working Report 2010–70, 580 p.
2. Nils Gustavsson (ed.), 2009. 3D-tiedonhallinnan kehittäminen GTK:ssa. Arcive report. (In Finnish). Koistinen, Esko 2011a. Malmiesiintymän 3D-mallinnus Gemcom GEMSohjelmistolla - Projektin perustaminen. 86 s. Geological Survey of Finland, Archive report, 18/2011. (in Finnish).
3. Koistinen, Esko; Seppänen, Hannu; Ahtola, Timo; Kuusela, Janne 2010. Mineral resource assessment and 3D modelling of the Syväjärvi lithium pegmatite deposit in Kaustinen, Western Finland. 20 s. + 21 liites. Geological Survey of Finland, Archive report, M19/2323/2010/45. (in Finnish).
4. Laine, E. (ed.) 2012. 3D modeling of polydeformed and metamorphosed rocks: the old Outokumpu Cu-Co-Zn mine area as a case study. Geological Survey of Finland. Report of Investigation 195, 77 pages, 66 figures and 1 table.
5. Vähäaho, Ilkka, 1998. From geotechnical maps to three-dimensional models. Tunnelling and Underground Space Technology, Volume 13, Issue 1, January-March 1998, Pages 51–56.

Estimation of Parameters in Random Dynamical Systems

Silke Konsulke, K. Gerard van den Boogaart, Fellix Ballani, Markus Franke and Martin Sauke

1 Introduction

In ecology and other geosciences we are often confronted with random dynamical systems such as forest systems [4], limnic systems [7], ground water reactive transport systems, ...whose dynamic is created from a complex interaction of different parts and is only partially deterministic. Typically these situations are modelled through a computer simulation. These models are, however, typically still dependent on some parameters such as the survival probability of seedlings under certain given conditions, coefficients describing the dependence of a bacteria population growth on nitrite concentration, variogram parameters of the hydraulic conductivity,....

Our aim is to propose a general technique for the estimation of parameters of such system models where, on the one hand, least squares fitting is not feasible because of the random character of the systems and, on the other hand, likelihoods are not computable due to the algorithmic character of the best available model.

S. Konsulke (✉) · K. G. van den Boogaart (✉)
Helmholtz Institute Freiberg for Resource Technology, Halsbrücker Straße 34, 09599 Freiberg, Germany
e-mail: s.konsulke@hzdr.de

K. G. van den Boogaart
e-mail: boogaart@hzdr.de; boogaart@math.tu-freiberg.de

K. G. van den Boogaart · F. Ballani · M. Franke · M. Sauke
Institute for Stochastics, TU Bergakademie Freiberg, Freiberg, Germany

F. Ballani
e-mail: ballani@math.tu-freiberg.de

M. Franke
e-mail: markus.franke@math.tu-freiberg.de

M. Sauke
e-mail: sauke@math.tu-freiberg.de

E. Pardo-Igúzquiza et al. (eds.), *Mathematics of Planet Earth*,
Lecture Notes in Earth System Sciences, DOI: 10.1007/978-3-642-32408-6_183,

Our introductory example is a simple stochastic predator-prey model as the archetype of an ecological model. In this model the different development of the numbers of predators ($Y(t)$) or prey ($X(t)$) depend on the number of individuals already present and on the interaction of the different individuals. Like in [1] and [3] the system is modelled by a stochastic differential equation adding some disturbance (dW) to the usual ordinary differential equations predator-prey model:

$$\begin{aligned} dX &= (a \cdot X - b \cdot X \cdot Y)dt + \sigma dW_X \\ dY &= (c \cdot X \cdot Y - d \cdot Y)dt + \sigma dW_Y \end{aligned} \tag{1}$$

For a general introduction to stochastic differential equations as one of the possible examples of random dynamical systems see e.g. [5]. Here the parameter $\theta = (a, b, c, d, \sigma)$ describes birth and death rates, the interaction of the two species and the disturbance. The numbers of predators and prey can be observed with some observation error, the parameters cannot. Therefore it is necessary to estimate the parameters.

2 The SystemsStat-Package

We have developed an R-package (SystemsStat) which is intended to work flexibly with various kinds of system models. The central idea is to provide a standard interface allowing different system models as input. It provides different estimation algorithms only depending on simulations. One example of such an approach is: Create a Bayesian statistical model by adding a simulatable prior distribution for the parameters. Define a linear space of candidate functions for estimators and get the optimal estimator in the subset by some sort of stabilized linear model estimation [6]. This approach has been motivated by research on nonlinear geostatistics and is related to an approach of approximated Bayesian computation [2].

For applying the package we need to define a simulation function `sim` which simulates the system in question Then we need to define the parameter space object (`ParSpace`) through a simulation function of the prior distribution of the parameters. The prior used for the simulations shown in Fig. 1 is a lognormal distribution.

In order to model incomplete observations and measurement errors we define an observation simulation function `Obs` transforming a simulated situation to corresponding observations, in our case by adding a normally distributed "measurement" error to the simulated population sizes.

In order to integrate our predator-prey model into the package the three ingredients described above are put together into a model object, with a simple instruction:

```
SysMod  <- SysModel(X=sim, theta=ParSpace, obs=Obs)
```

For the before mentioned simple estimation scheme we need to define a set of informative statistics `mystats` as a function of the observation. In our example we selected the estimated coefficients of a multivariate linear model predicting each

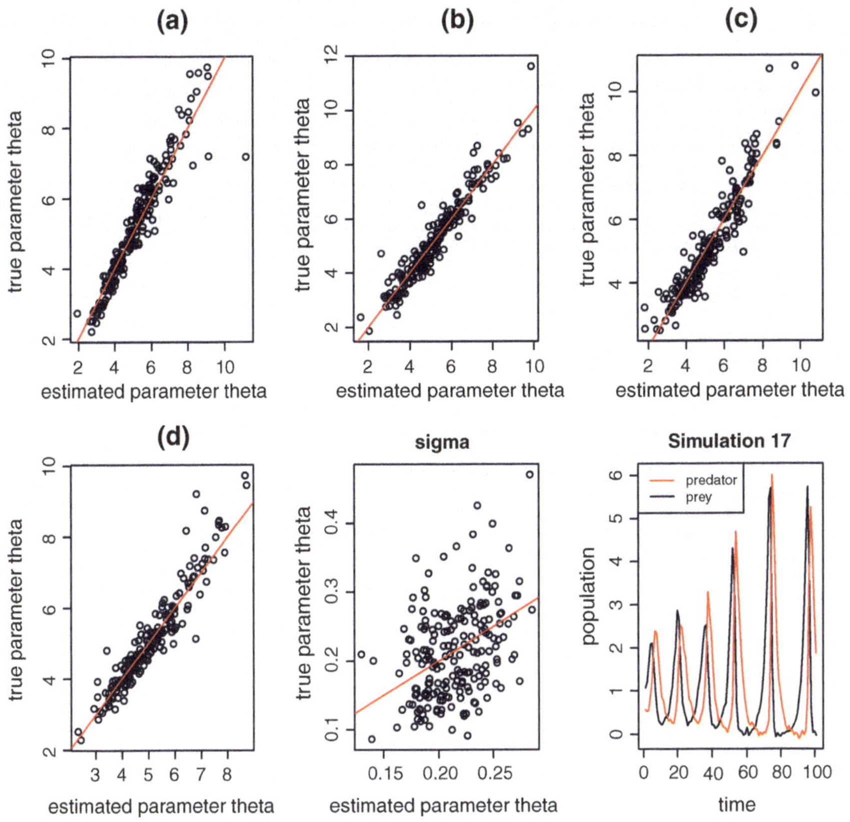

Fig. 1 The *lower right figure* shows an example simulation of the model. The other panels show the performance of the estimator for the different model parameters. The estimation of σ from the small data set of population observations with random errors at 101 time points is difficult. The prior is thus used more heavily in the approximation of a Bayesian estimation leading to the usual bias of Bayesian estimators weighting bias against estimation error

observation from its preceding observation and the entries of the estimated variance matrix of the residuals of these model. Defining a good set of statistics typically requires some experience. An estimator can than be generated by a simple command

```
est <- SysLMEstimator(SysMod,stat=mystat)
```

Here all the necessary simulations and the model fitting are done automatically based on the model information provided through the system model `SysMod`. The computed estimator can than be applied to a set of observations by a simple command:

```
thetaHat <- SysEstimate(est,TheRealWorldObservations)
```

Other more elaborate estimation techniques based on the (generalized) method of moments and quasi-likelihood inference will also be available as part of the package.

For this paper we drew 200 realizations from the model, applied the estimator to each of the realizations and compared the estimated parameters $\hat{\theta}$ with the parameters θ actually used. The result can be seen in Fig. 1. An example simulation is shown in the bottom right panel. There can be seen that the parameters $a, \ldots, d$ are estimated quite well. The parameter σ has a bias which can be explained with the Bayesian character of the approach.

The actual R-code used, as well as directions where to find the SystemsStat-package, can be requested from the authors.

3 Outlook

The new SystemsStat package shall provide a uniform way to analyse data from random systems without closed form likelihoods. The idea is to make the statistical analysis of random systems as easy as linear model analysis was in the 1960s by providing a general toolbox including descriptive analysis, estimation, parameter tests, goodness of fit tests, and diagnostics.

References

1. Abundo, M. (1991). A stochastic model for predator-prey systems: basic properties, stability and computer simulation. *Journal of Mathematical Biology*, *29*(6), 495–511.
2. Blum, M. G. B. & François, O. (2010). Non-linear regression models for approximate bayesian computation. *Statistics and Computing*, *20*(1), pp 63–73. doi:10.1007/s11222-009-9116-0
3. Gard, T. C., & Kannan, D. (1976). On a stochastic differential equation modeling of prey-predator evolution. *Journal of Applied Probability*, *13*(3), 429–443.
4. Grimm, V., & Railsback, S. F. (2004). *Individual-based modeling and ecology*. Princeton: Princton University Press.
5. Øksendal, B. (2003). Stochastic differential equations, An introduction with applications (6th ed.). Springer: New York. *XXIV*, 360 p. ISBN 978-3-642-14394-6.
6. Rösch, K. (2009). Parameterschätzung in individuenbasierten stochastischen Waldmodellen, Diploma thesis, TU Bergakademie Freiberg.
7. Trolle, D. et. al. (2012). A community-based framework for aquatic ecosystem models, Opinion Paper, *Hydrobiologia, 683*, 25–34. doi:10.1007/s10750-011-0957-0

Nonlinear Estimation with Gaussian Kriging and Riemann Sums

K. Daniel Khan

1 Introduction

The objective of this paper is to introduce a practical way of constructing nonlinear spatial estimates of geological random variables (RVs) from kriging estimates. Indicator Kriging (e.g., [1]) and Disjunctive Kriging (e.g., [2]) are existing approaches, but both methods are practically demanding [1, 3]. This paper presents a more direct way to compute nonlinear conditional expected values, including all (Gaussian) higher-order moments. Furthermore, a new general solution to nonlinear change of support is developed through the proposed approach.

2 Gaussian Kriging and Riemann Sums

The unbiased expected value $E[\cdot]$ of a nonlinear function $\varphi(y)$ with respect to the probability density function (pdf) $f(y)$ is given by

$$E[\varphi(y)] = \int_{-\infty}^{\infty} \varphi(y) f(y) dy. \tag{1}$$

Spatial estimations of $E[\varphi(\mathbf{u})]$ are computable by applying Riemann sums to Eq. 1 using Gaussian kriging estimates, $y(\mathbf{u})$. This approach was proposed for powers of Gaussian variables in [4] as

K. D. Khan (✉)
XXX, 5-709 Luscombe Pl, Victoria, BC, Canada
e-mail: k.danielkhan@gmail.com

K. D. Khan
Saudi Aramco,31311 Dhahran, Saudi Arabia

E. Pardo-Igúzquiza et al. (eds.), *Mathematics of Planet Earth*,
Lecture Notes in Earth System Sciences, DOI: 10.1007/978-3-642-32408-6_184,

$$E\left[z(\mathbf{u})^m\right] = E\left[\varphi(y(\mathbf{u}))^m\right] = \int_{-\infty}^{\infty} \varphi(y)^m f(y(\mathbf{u}))dy. \tag{2}$$

Equation 2 is not useful if the function $\varphi(y)$ is unknown. A practical method is given here for general application. The normal scores transform, or Gaussian anamorphosis, is an empirical tabulation of the function $z = \tilde{\varphi}(y)$ (e.g., [1]), which can be embedded into the discretized Eq. 2 as

$$E\left[z(\mathbf{u})^m \approx \widehat{z^m}(\mathbf{u})\right] = \sum_i \left[\tilde{\varphi}(y)\right]_i^m f(y(\mathbf{u}))_i \Delta y_i. \tag{3}$$

This process is simply Kriging followed by Riemann Sums (KRS). KRS yields the conditional moments $E\left[z(\mathbf{u})^m | y(u_\alpha); \alpha = 1, \ldots, n\right]$ of any order, m, of the conditional nonlinear function $f(z(\mathbf{u})|\cdot)$ of the Gaussian pdf $f(y(\mathbf{u}))$.

2.1 A Generalized Nonlinear Solution for Change of Support

The nonlinear expectation theory expressed in Eq. 2 applied to the problem of estimating blocks from point samples can be expressed as [5],

$$E\left[\varphi(y(\mathbf{u}))^m\right] = \int_{-\infty}^{\infty} \left[\int_{-\infty}^{\infty} \varphi(y) f_{BKV}(y(\mathbf{u}))dy\right] c\, f_v(y(\mathbf{u}))dy. \tag{4}$$

Equation 4 suggests two Gaussian pdfs are needed: $f_{BKV}(y(\mathbf{u}))$ is the pdf of the blocks with the block kriging conditional mean $\hat{y}(\mathbf{u})$ and the block kriging variance $\hat{\sigma}^2_{BKV}$, and $f_v(y(\mathbf{u}))$ is the pdf accounting for the additional ensemble variance component σ_v^2 of points within the blocks. To express Eq. 4 as a single integral, a convolution of the pdfs in Eq. 4 is proposed and tested for the first time here.

Suppose $f(y(\mathbf{u})) * g(y(\mathbf{u}))$ is a convolution of the two pdfs in Eq. 4 such that $f(y(\mathbf{u})) = f_{BKV}(y(\mathbf{u}))$ and $g(y(\mathbf{u})) = f_v(y(\mathbf{u}))$, with mean $\mu_g = 0$. The convolution $f(y(\mathbf{u})) * g(y(\mathbf{u}))$ is done following the convolution theorem

$$F^{-1}\left[F(f(y(\mathbf{u})))F(g(y(\mathbf{u})))\right] = f(y(\mathbf{u})) * g(y(\mathbf{u})), \tag{5}$$

where F and F^{-1} here represent the Fourier and inverse Fourier transforms, respectively, and $F\left[f(y(\mathbf{u}))\right]$ is the characteristic function of $f(y(\mathbf{u}))$. Equation 5 is solved for the following product of Gaussian characteristic functions:

$$F^{-1}\left[\left(\frac{1}{\sqrt{2\pi}}e^{i\pi_f\omega-\frac{\omega^2\sigma_f^2}{2}}\right)\left(\frac{1}{\sqrt{2\pi}}e^{-\frac{\omega^2\sigma_g^2}{2}}\right)\right]=\frac{1}{2\pi\sqrt{\sigma_f^2+\sigma_g^2}}e^{-\frac{(y-\mu_f)}{2\left(\sigma_f^2+\sigma_g^2\right)}}. \quad (6)$$

Equation 6 yields a convolved Gaussian function which requires normalization by a factor $\sqrt{2\pi}$ to provide a pdf that integrates to unity, as

$$\sqrt{2\pi}(f(y(\mathbf{u}))*g(y(\mathbf{u})))=\frac{1}{2\pi\sqrt{\sigma_f^2+\sigma_g^2}}e^{-\frac{(y-\mu_f)}{2(\sigma_f^2+\sigma_g^2)}}. \quad (7)$$

The convolved Gaussian pdf of Eq. 7 has a mean equal to the Gaussian block kriging mean $\hat{y}(\mathbf{u})=\mu_f$ and total variance, $(\sigma_{\cdot v}^2+\sigma_{BKV}^2(\mathbf{u}))=\sigma_f^2+\sigma_g^2$. Eq. 2 is then directly applicable as KRS from Gaussian block kriging as

$$E\left[z_v(\mathbf{u})^m\right]=\sqrt{2\pi}\int_{-}^{\infty}\varphi(y)^m f(y(\mathbf{u}))*g(y(\mathbf{u}))dy. \quad (8)$$

Equation 8 yields unbiased volume-averaged moments of the nonlinear averaging variable $z(\mathbf{u})$.

2.2 *Verification*

A 1-D lognormal random field profile was used for the KRS tests (Fig. 1). The analytical benchmark is obtained by kriging the Gaussian data and applying the well-known lognormal corrections for points (see e.g., [6]) and blocks [5]. The nonlinear estimates obtained from KRS are identical to the analytical solutions (Fig. 2). An important result is that heteroscedasticity is reproduced as a general outcome of the KRS algorithm.

3 Discussion and Conclusion

The idea of KRS is to "directly" obtain unbiased expected values of nonlinear transformations of Gaussian RVs. The practical value of the approach is that nonlinear estimates of stochastic parameters with any shape of conditional pdfs are available from a single Gaussian kriging. The power of the KRS approach is its generality, which is shown to extend to nonlinear upscaling of Gaussian block estimates via a new solution for non-linear change of support. The main advantage of KRS over direct kriging (e.g., [7]) is that KRS returns estimates that have the correct heteroscedasticity. The new solution for non-linear change of support in Sect. 2.1 is

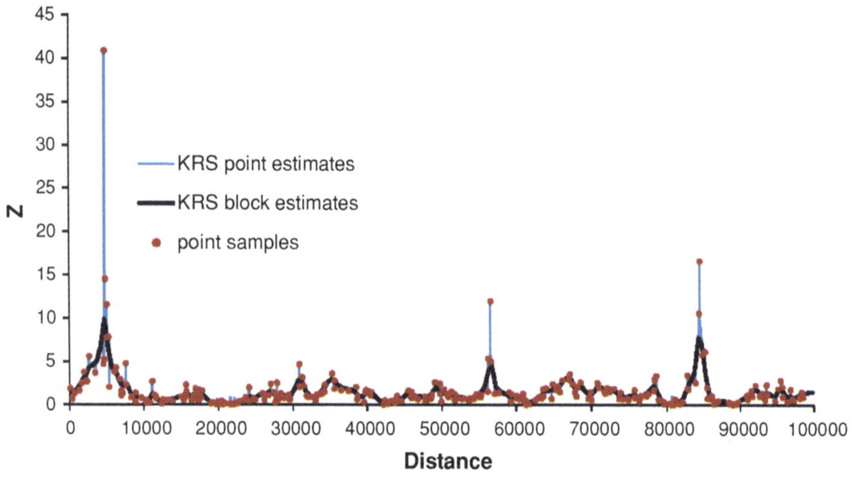

Fig. 1 Profile for kriging examples with data from a synthetic lognormal reference field

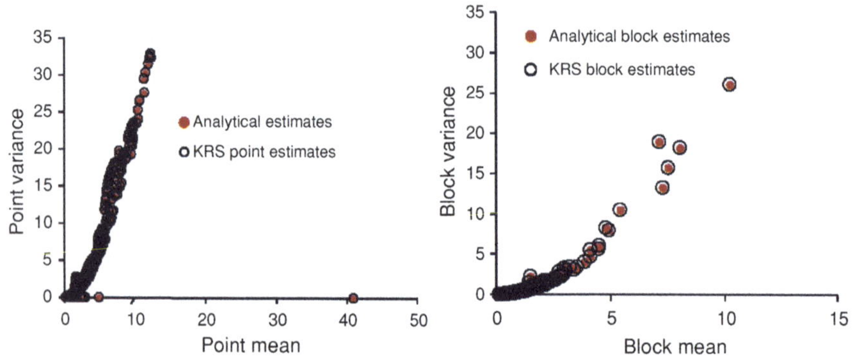

Fig. 2 Plots of kriging mean versus kriging variance to verify point KRS (*left*) and block KRS (*right*) against analytical lognormal corrections

an ensemble upscaling solution applicable to nonlinearly-averaging heterogeneous fields used in stochastic geological systems models, such as permeability fields in groundwater or petroleum flow models.

References

1. Deutsch, C. V., & Journel, A. G. (1998). *GSLIB geostatistical software library and user's guide* (p. 369). Oxford: Oxford University Press.
2. Armstrong, M., & Matheron, G. (1986). Disjunctive kriging revisited—Part I. *Mathematical Geology*, *18*(8), 711–728.

3. Ortiz, J., Oz, B., & Deutsch, C. V. (2005). A step by step guide to bi-Gaussian disjunctive kriging. In O. L. Leuangthong & C. V. Deutsch (Eds.), *Geostatistics Banff 2004* (pp. 1097–1102). Netherlands: Springer.
4. Vargas-Guzmán, J. A. (2004). Geostatistics for power models of Gaussian fields. *Mathematical Geology*, *36*(3), 307–322.
5. Vargas-Guzmán, J. A. (2005). Change of support of transformations: Conservation of lognormality revisted. *Mathematical Geology*, *37*(6), 551–567.
6. Wackernagel, H. (2003). *Multivariate geostatistics* (p. 387). Berlin: Springer.
7. Journel, A. G. (1999). Conditioning geostatistical operations to nonlinear volume averages. *Mathematical Geology*, *31*(8), 931–953.

Modeling Trace Element Concentration in Vertical Regolith Profile Over Mineral Deposits

Qiuming Cheng

1 Introduction

There have been substantial efforts in the literature to study formation mechanisms of geochemical anomalies in deeply weathered and transported terrains with regolith or other land cover [1–5]. Several mechanisms that assume the upward migration of ions or fine particles to the surface have been suggested, including electrochemistry, diffusion-and-convection, evaporation-transpiration-convection, cyclical dilatancy pumping, vegetation and biochemistry, and nanoparticles in geogasses. Studies have also reported experimental and applied research on the migration of elements from mineral deposits and through fractures [6, 7]. There are, however, relatively few quantitative studies of concentration distribution of elements in regolith caused by vertical element migration.

2 Differential Equations for Modeling of Element Concentration

There are many types of dynamic processes observed in earth system such as ocean waves, clouds, and cascade subduction of plate tectonics that can be considered as multiplicative cascade processes (MCM). MCM can cause irregular energy and density distribution of multiscale singularities that can be characterized by multifractal models [8]. The geochemical patterns observed in the surface media or long vertical profile of regolith over mineral deposits, can be viewed as the end products of "cascade geo-processes" of various types of geochemical mechanisms operated in

Q. Cheng (✉)
State Key Lab of Geological Processes and Mineral Resources,
China University of Geosciences, Beijing 100083, China
e-mail: qiuming@yorku.ca

Q. Cheng
Department of Earth and Space Science and Engineering, Department of Geography, York University, Toronto M3J1P3, China

E. Pardo-Igúzquiza et al. (eds.), *Mathematics of Planet Earth*,
Lecture Notes in Earth System Sciences, DOI: 10.1007/978-3-642-32408-6_185,

a region with upward migration of ore materials from depth to the surface [9]. From a nonlinear dynamic point of view, the element concentration profile, C(x), in regolith over mineral deposit can be characterized by the following nonlinear differential equation [9]

$$\frac{dC(x)}{dx} = -\frac{1}{\lambda(x)} C^{\gamma}(x) \quad (1)$$

where the coefficient $\lambda(x)$ is a function of scale (vertical distance) x and γ is an index of nonlinearity. The coefficient function $1/\lambda(x)$ can be approximated by means of a Taylor series expansion of $1/[\lambda(1/x)]$ which gives a combination of four different functions: Gaussian exponential, normal exponential, power-law and exponential with inverse time, as the solution of model (1):

$$C(x) = ce^{-\lambda_{-1}x^2} e^{-\lambda_0 x} x^{-\lambda_1} e^{\lambda_2/x} \quad (2)$$

where $C(x)$ is a function with five parameters that can be can be readily estimated by multivariate linear regression.

3 Data from a W-Mo Mineral Deposit in Inner Mongolia, China

Two areas with the cover being deserts and grasslands were chosen for this study. Here we will only show an example from the Dalaimiao district, northeastern Inner Mongolia, China. In this mineral district, about 70 % of the area is covered by up to 100 m Tertiary to Quaternary sediments. Mesozoic intrusions are associated with the hydrothermal mineralization of W and Mo in the area. There are extensive zones of hydrothermal alteration in the sedimentary rocks around the buried intrusions of finegrained granite. The topography of the area is relatively flat. Several drill cores were selected to study the migration of elements in the regolith. Figure 1a shows the main lithologic units observed in drill hole #WRZK520-11 including approximately 10 m soil at the surface section (yellow colored section in Fig. 1a). The main ore bodies are found at a depth of approximately 400 m below the surface. The drill core, including soils in the top section, was systematically analyzed using pXRF on-site and in the lab. Approximately 33 samples from the 10-m section of regolith were analyzed using the pXRF device for multiple elements S, K, Fe, Zn, Pb, Ca, Ti, Mn, Cu, V, Zr, Sr, Rb, Th, As, Ni, Ba, Cs, Te, Sb, Sn, Mo, V, Au, and Ag. Some samples have missing data for Mo, W, Au and Ag which were excluded in the following data analysis.

The plot in Fig. 1b shows the data of Cu with horizontal axis representing the vertical distance from the surface of weathered rocks beneath the regolith. Dots represent the actual values of Cu and curves are fitted by Least Squares (LS) with different functions: red curve is from a combined exponential and power-law function, $Cu = 32.16x^{-1.09}e^{0.08x}$, $R^2 = 0.504$; black solid line is from a power-law

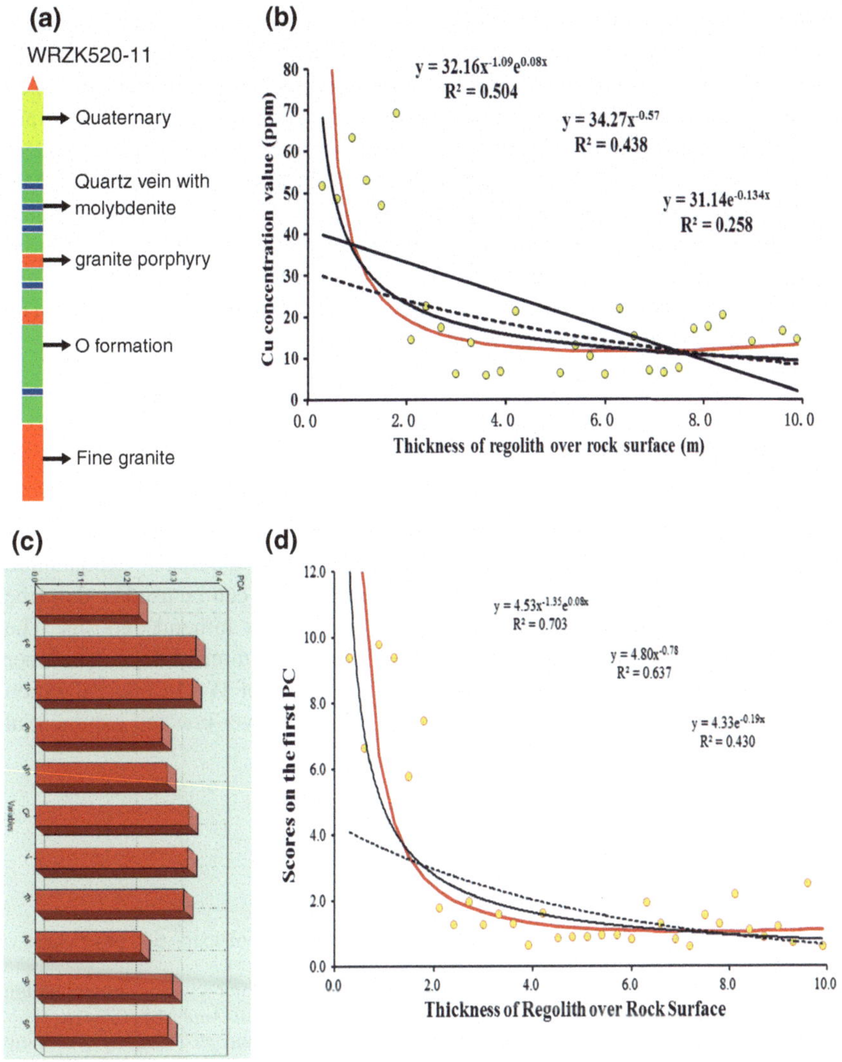

Fig. 1 **a** Lithologic units in drill hole #WRZK520-11 in the Wurinitu mineral deposit in Dalaimioa district; **b** trend of Cu values. The *horizontal axis* represents the vertical distance from the surface of weathered rocks beneath the regolith; **c** loadings of elements on the first principal component (PC); and **d** trend of scores on the first PC with a constant 2.5 added to scores in order to avoid negative values. *Dots* represent actual values and *red curves* are fitted by LS with a combined function, *black solid lines* with power-law functions, and *dashed lines* with exponential functions, respectively

function, $Cu = 34.27x^{-0.57}$, $R^2 = 0.438$; and dashed line is from an exponential function, $Cu = 31.14e^{-0.134x}$, $R^2 = 0.258$, where R is the correlation coefficient of LS. It can be seen that the combined function gives a slightly better fit in comparison with the results obtained from a power-law function. Most of the selected elements associated with W and Mo mineralization including Cu, Pb, Zn, Fe, Th, Mn, V, As, and Sb depict similar trends which can be fitted by LS with either combined powerlaw and exponential function or power-law function along. Ca and S do not show systematic changes in the profile. Furthermore, applying principal component analysis to the dataset gives three main components (PCs) {Cu, Pb, Zn, Fe, Th, Mn, V, As, Sb}, {Mn, As, -K, -Sn}, and {K, Mn, -Cu, -Pb, -V, -Th}. The first PC is related to Mo and W mineralization in the area. The scores on the first PC show a similar trend as observed from Cu as shown in Fig. 1b, d. In order to avoid negative values for logarithmic transformation, a constant 2.5 was added to the value of scores. The scores on other two PCs do not show clear trend.

4 Concluding Remarks

The combined form of solution derived from nonlinear differential equation (1) not only shows the association of the four types of functions: Gaussian, exponential, power-law, and exponential with inverse distance, but also can be used to describe element concentration distribution in vertical regolith over mineral deposits. The dataset obtained from the regolith over the Wurinitu W-Mo mineral deposit in Inner Mongolia has demonstrated that the vertical distribution of W-Mo mineralization associated elements can be fitted either by a combined power-law and exponential functions or by power-law functions along.

References

1. Mann, A. W. (2010). Strong versus weak digestions: ligand-based soil extraction geochemistry. *Geochemistry: Exploration, Environment, Analysis, 10*, 17–26.
2. Cohen, D. R., Kelley, D. L., Anand, R., & Coker, W. B. (2010). Major advances in exploration geochemistry, 1998–2007. *Geochemistry: Exploration, Environment, Analysis, 10*, 3–16.
3. Anand, R. R., Robertson, I. D. M. (2012). The role of mineralogy and geochemistry in forming anomalies on interfaces and in areas of deep basin cover: implications for exploration. *Geochemistry: Exploration, Environment, Analysis, 12*, 45–66.
4. Wang, X., Zhang, B., & Liu, X. (2012). Nanogeochemistry: Deep-penetrating geochemical exploration through cover. *Earth Science Frontiers, 19*, 101–112.
5. Dunn, C. E. (2007). *Biogeochemistry in Mineral Exploration: Handbook of Exploration and Environmental Geochemistry* (Vol. 9). Amsterdam: Elsevier.
6. Goldberg, I. S. (1998). Vertical migration of elements from mineral deposits. *Journal of Geochemical Exploration, 61*, 191–202.
7. Carrigan, C. R., Heinle, R. A., Hudson, G. B., Nitao, J. J., & Zucca, J. J. (1996). Trace gas emissions on geological faults as indicators of underground nuclear testing. *Nature, 382*, 528–531.
8. Cheng, Q., & Agterberg, F. P. (2009). Singularity analysis of ore-mineral and toxic trace elements in stream sediments. *Computers & Geosciences, 35*, 234–244.
9. Cheng, Q. (2012). Singularity theory and methods for mapping geochemical anomalies caused by buried sources and for predicting undiscovered mineral deposits in covered areas. *Journal of Geochemical Exploration, 122*, 55–70.

Unified Principles for Nonlinear Nonstationary Random Fields in Stochastic Geosciences

José A. Vargas-Guzmán

1 Introduction

A motivation for solving stochastic partial differential equations (spde's) is the modeling of clastic sedimentary basins, which are relevant to hydrocarbon resources. Sedimentology and stratigraphy are related to diffusion processes starting from formulations with Markov chains [4]. Furthermore, sedimentary clastic deposition can be modeled with flow and transport spde, containing diffusion terms. The results of spde's are usually non-stationary models.

An issue is that spde Monte Carlo solutions for large models are slow. For example, spde's are routinely used for hydrocarbon and water flow modeling. Monte Carlo is still the most popular method because it is simple. Several realizations of a geostatistically predicted parameter field are input to a numerical nonlinear algorithm, repeatedly. Although spde's are essential to handle uncertainty and risk, Monte Carlo solutions may become unmanageable in high-resolution models containing billions of cells, and hundreds of realizations. Furthermore, the statistical information is extracted by computing ensemble averages. Unfortunately, the convergence of moments to ensemble parameters in non-stationary models is slow.

Another complex issue for flow and transport spde models in sedimentary basins is that matching observed rocks at wells is non-trivial. A similar difficulty has also been observed in reservoir flow modeling, where the multiple Monte Carlo realizations conditional to the input parameter data do not match simultaneously the response data (e.g., pressure and flux) observed at wells. Although "iterative" inverse problem solutions are available from hydrology and geophysics (e.g., [1, 7]), large number of cells and multiple realizations, may preclude such computer intensive iterative approaches.

J. A. Vargas-Guzmán (✉)
Saudi Aramco, P.O. Box 13734, Dhahran 31311, Saudi Arabia
e-mail: ja.var.guz@gmail.com

E. Pardo-Igúzquiza et al. (eds.), *Mathematics of Planet Earth*,
Lecture Notes in Earth System Sciences, DOI: 10.1007/978-3-642-32408-6_186,

Alternatives to Monte Carlo methods need to be examined, in order to speed up spde solutions for gigantic high resolution models. As a consequence, using Itô's integration [2, 3] appears very attractive; however, Itô did not develop corrections for multiple integrations. Therefore, a generalization of higher-order integration is needed, for handling geostatistics. In addition, unification with Riemann–Stieltjes averaging is necessary for direct estimation of conditional probability density function (PDF) parameters, at each spatial location. Proper stochastic integration might enable faster and unbiased stochastic solutions of PDFs, integrating data for the input and the response, allowing for large models.

2 Itô's Stochastic Integration Generalized

It is well known that $y_1(x) = \int_{-\infty}^{x} \partial W(x)$ is a Wiener process [10], the increments $\partial W(x)$ represent Gaussian white noise. The Itô's stochastic integral is $y_2(x) = \int_{-\infty}^{x} y_1(x) \cdot \partial y_1(x)$. If the (domain) increments were constant, classic calculus would hold as $y_2(x) = \frac{1}{2}\left(y_1^2(x)\right)$, see also [5] for generalized covariances and kriging for integrated spatial processes. It is also known that direct classic integration is inapplicable to heterogeneous differentials. Itô [2] solved the problem adding a linear correction term $\frac{1}{2}\sigma^2 \cdot x$ to the traditional (one dimensional) result $\frac{1}{2}\left(y_1^2(x)\right)$, at location it x, where $\sigma^2 \sim \operatorname{var}(\Delta y_1(x))$, and $\Delta y_1(x) \approx \partial y_1(x)$. Itô's integration should not be applied recursively with successive $\frac{1}{2}\sigma^2 \cdot x$ linear corrections. Itô's general solution is his famous lemma or bivariate truncated Taylor series [3], where one domain variable is time, and the second is the random component. It is hard to apply such lemma to spde (e.g., flow/transport) and unknown non-stationary models.

Following another venue, the response differentials for a spde can be predicted from input data, using case specific relations with stationary parameter models. Furthermore, heuristic successive integral extensions beyond $y_2(x)$ are as follows:

$$y_3(x) = \int_{-\infty} \left(\int_{-\infty} y_1(x).\partial y_1(x)\right) \cdot \partial y_1(x) = \int_{-\infty} y_2(x).\partial y_1(x)$$

yields

$$y_3(x) = \frac{1}{2 \cdot 3}\left(y_1^3(x) + 3\langle y_1(x)\rangle\sigma^2 \cdot x\right) \tag{1}$$

This result is integrated again,

$$y_4(x) = \frac{1}{2 \cdot 3 \cdot 4}\left(y_1^4(x) + 6\langle y_1(x)\rangle^2\sigma^2 \cdot x + 3\sigma^2\sigma^2 \cdot x^2\right) \tag{2}$$

and again

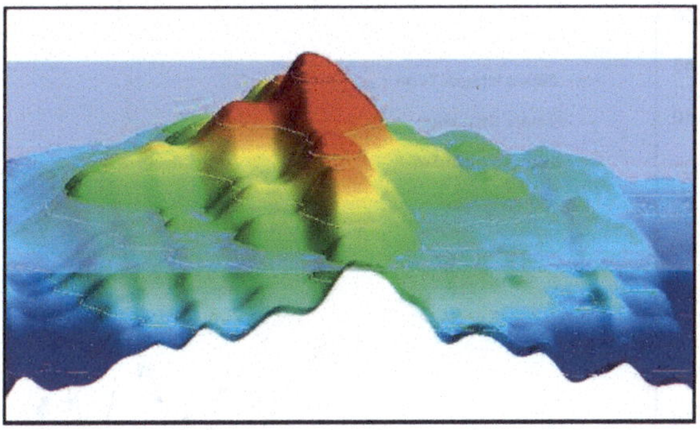

Fig. 1 Stochastic integration of a geostatistically simulated stationary difference model

$$y_5(x) = \frac{1}{2 \cdot 3 \cdot 4 \cdot 5}\left(y_1^5(x) + 10\langle y_1(x)\rangle^3(x)\sigma^2 \cdot x + 15\langle y_1(x)\rangle\sigma^2\sigma^2 \cdot x^2\right) \quad (3)$$

The angular brackets represent the conditional average, (over the realization). The correction terms behave as drift functions of the spatial coordinates, with parametric coefficients. A generalization represents

$$y_n = \int_{-\infty}^{\infty} \cdots \int_{-\infty}^{\infty}\left(\int_{-\infty}^{\infty}\left(\int_{-\infty}^{\infty} y_1(x) \cdot \partial y_1(x)\right) \cdot \partial y_1(x)\right) .\partial y_1(x) \cdots \partial y_1(x)$$

Successive derivatives of $y_n(x)$ with respect to the domain variable reboot to $y_1(x)$.

Flow and transport in 3D space can be integrated along 1D streamtubes, e.g., [6]. Streamtubes are enabled by considering a collection of streamlines, allowing fast integration using Eqs. 1, 2, 3 and so forth.

Furthermore, Itô's integration has been generalized as a Principle of Stochastic Integration (PSI), which is fundamental for computing expected values and higher-order moments of non-stationary models following spde.

3 Numerical Verification

The integration of geostatistically simulated second order difference random variables enables the construction of a non-stationary folded structural model with elevation boundary conditions, Fig. 1.

An example for a fifth order integral of proportions is provided in Fig. 2. One Monte Carlo numerical trapezoidal summation (yellow) can be compared to the numerically direct, and therefore faster, almost overlapping result from Eq. 3 (red). Just for illustration, the other curves are the classic calculus integral (black), plus the small effect of the linear correction (brown). Note the quadratic function of space (or

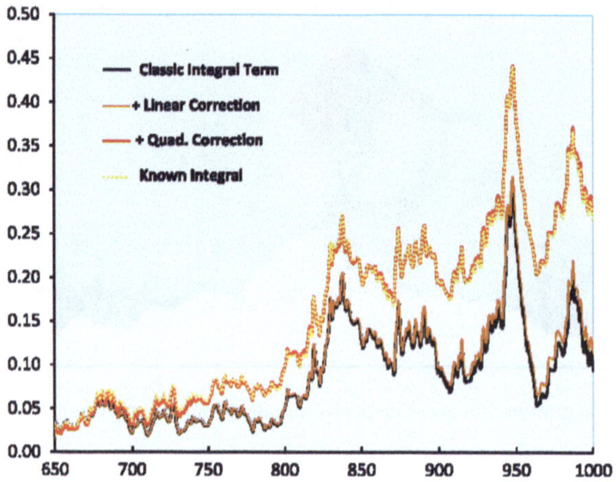

Fig. 2 Fifth order stochastic integration for a random variable

time) carries on a significant correction for this order of integration. Since Eq. 3 is a function of mean and variance, this model is based on second order statistics, but can be applied to integrate non-Gaussian noise by including cumulant terms using polynomials [9].

4 Polynomials from Riemann Stieltjes Integration

Equations 1, 2 and 3 contain monomial nonlinear terms $\varphi(y(x))$. The nonlinear averaging of a "conditional" input random variable is computed by integration $E\left[\varphi\left(y(x)\right)\right] = \int_{-\infty}^{\infty} \varphi\left(y(x)\right) \partial F\left(y(x)\right)$, where $\partial F\left(y(x)\right) / \partial y(x)$ is the conditional probability density function (PDF) with input kriging parameters. For example, the exponential is decomposed as Taylor series yielding a combination of kriged moments

$$E\left[\exp\left(y_1(x)\right)\right] = 1 + E\left[y_1(x)\right] + \frac{1}{2}E\left[y_1^2(x)\right] + \frac{1}{2!\cdot 3!}E\left[y_1^3(x)\right] + \frac{1}{2!\cdot 3!\cdot 4!}E\left[y_1^4(x)\right]\cdots \tag{4}$$

Polynomial solutions for the Gaussian kriging moments $E\left[y_1^n(x)\right]$ are in [8]. Expectations of nonlinear monomial terms in Eqs. 1, 2, and 3 are consistent with each term in Eq. 4 from Riemann–Stieltjes integration. The expressions for non-Gaussian moments are in [9]. This result couples nonlinearity to non-stationarity principles.

5 Conclusion

This study has unified stochastic integration and nonlinear averaging principles enabling an approach for stochastic geoscience models. Heterogeneous differences are calibrated to match the response data, at boundary conditions. The solutions contain nonlinear and drift terms. The drift components are functions of space (or time), and probability distribution parameters, of the differentials (i.e., noise). The nonlinear expected values follow Riemann–Stieltjes averages. If the drift terms are eliminated, the models resort to classic nonlinear collocated transformation models.

References

1. Hansen, T. H., Journel, A., & Mosegaard, K. (2006). Linear inverse gaussian theory of geostatistics. *Geophysics*, *71*(6), 101–111.
2. Itô, K. (1944). Stochastic integral. *Proceedings of the Imperial Academy of Tokyo*, *20*(8), 519–524.
3. Itô, K. (1951). Multiple wiener integral. *Journal of the Mathematical Society of Japan*, *3*, 157–169.
4. Krumbein, W. C., & Dacey, M. F. (1969). Markov chains and embedded markov chains in geology. *Mathematical Geology*, *1*(1), 79–96.
5. Matheron, J. (1973). The intrinsic random functions and their applications. *Advances in Applied Probability*, *5*, 439–468.
6. Thiele, M. R. (1994). *Modeling Multiphase Flow in Heterogeneous Media Using Streamtubes*. Ph.D. Dissertation,: Stanford University.
7. Vargas-Guzmán, J. A., & Yeh, T. C. J. (2002). The successive linear estimator: A revisit. *Advances in Water Resources*, *25*, 773–781.
8. Vargas-Guzmán, J. A. (2004). Geostatistics for power models of gaussian fields. *Mathematical Geology*, *36*(3), 307–322.
9. Vargas-Guzmán, J. A. (2012). Heavy tailed probability distributions for non-gaussian simulations with higher-order cumulant parameters predicted from sample data. *Stochastic Environmental Research and Risk Assessment*, *26*(6), 765–776.
10. Wiener, N. (1938). The homogeneous chaos. *American Journal of Mathematics*, *60*(4), 897–936.

The manufacturer's authorised representative in the EU is Springer Nature Customer Service Centre GmbH, Europaplatz 3, 69115 Heidelberg, Germany. If you have any concerns regarding our products, please contact ProductSafety@springernature.com

Printed and bound by CPI Group (UK) Ltd, Croydon, CR0 4YY
01/10/2026
02225852-0020